Cognitive Technologies

Managing Editors: D.M. Gabbay J. Siekmann

Springer-Verlag Berlin Heidelberg GmbH

Sankar K. Pal   Lech Polkowski
Andrzej Skowron (Eds.)

Rough-Neural Computing

Techniques for Computing with Words

With 128 Figures and 79 Tables

Springer

Editors

Prof. Dr. Sankar K. Pal
Indian Statistical Institute
Machine Intelligence Unit
203 Barrackpore Trunk Road
Calcutta 700 035, India

Prof. Dr. Lech Polkowski
University of Warmia and Mazury
Department of Mathematics
and Computer Science
Zolnierska 14
Olsztyn, Poland

Prof. Dr. Andrzej Skowron
Warsaw University
Institute of Mathematics
Banacha 2
02-097 Warsaw, Poland

Managing Editors

Prof. Dov M. Gabbay
Augustus De Morgan Professor of Logic
Department of Computer Science
King's College London
Strand, London WC2R 2LS, UK

Prof. Dr. Jörg Siekmann
Forschungsbereich Deduktions- und
Multiagentensysteme, DFKI
Stuhlsatzenweg 3, Geb. 43
66123 Saarbrücken, Germany

Library of Congress Cataloging-in-Publication Data

Rough-neural computing: techniques for computing with words / Sankar K. Pal, Lech Polkowski, Andrzej Skowron (eds.).
p. cm. – (Cognitive Technologies)
Includes bibliographical references and index.
ISBN 978-3-642-62328-8 ISBN 978-3-642-18859-6 (eBook)
DOI 10.1007/978-3-642-18859-6

1. Soft computing. 2. Rough sets. 3. Neural computers. 4. Natural language processing (Computer science). I. Pal. Sankar K. II. Polkowski, Lech. III. Skowron, Andrzej. IV. Cognitive Technologies (Berlin, Germany)
QA76.9.S63.R68 2003 006.3–dc21 2002030320

ACM Computing Classification (1998): F.1, I.2, F.4

ISSN 1611-2482

http://www.springer.de

Originally published by Springer-Verlag Berlin Heidelberg New York in 2004
Softcover reprint of the hardcover 1st edition 2004

Cover design: KünkelLopka, Heidelberg
Typesetting: Camera-ready by authors
Printed on acid-free paper 45/3142 SR- 543210

To my Jamaibabu
Prof. R.C. Bhattacharyya

Sankar K. Pal

To

Maria and Martin

Lech Polkowski

To

Anna and Aleksandra

Andrzej Skowron

Foreword

It is my great pleasure to welcome a new book on rough sets edited by the prominent scientists Sankar K. Pal, Lech Polkowski, and Andrzej Skowron. The objective of the book is to explore a new paradigm in soft computing: granulation of knowledge. This paradigm has a long history that is related not only to computing and information but also to its sources in the long-lasting controversy between the continuous versus the discrete — of utmost importance to the foundation of mathematics and physics. The idea has been revived recently by Prof. Zadeh in the context of computation, under the name "computing with words."

Granulation of information is inherently associated with the concept of rough sets and is derived from the indiscernibility relation generated by data. Thus, the paradigm has its natural mathematical model in rough set theory that can be used theoretically and practically to support this approach.

The book stresses a newly emerging approach to the problem of knowledge granulation, called rough-neural computing. The approach combines rough and fuzzy set theory, neural networks, and evolutionary computing and is meant to be used to study and explore the granular structure of complex information. The book covers a very wide spectrum of topics related to its main paradigm and gives fair state-of-the-art information about the subject. It can be recommended to all those working in this fascinating area of research.

The authors and the editors deserve the highest appreciation for their outstanding work.

Warsaw, November 25, 2002 *Zdzisław Pawlak**

---

*Member of the Polish Academy of Sciences, Professor in the Institute of Theoretical and Applied Informatics, Polish Academy of Sciences, Gliwice, Poland and in the University of Information Technology and Management, Warsaw, Poland.

Foreword

Edited by Profs. S.K. Pal, L. Polkowski and A. Skowron, "Rough-Neural Computing: Techniques for Computing with Words," or RNC for short, is a work whose importance is hard to exaggerate. A collection of contributions by prominent experts in their fields, RNC addresses some of the most basic issues centering on the concept of information granularity.

In essence, information granularity is a concomitant of the bounded ability of sensory organs, and ultimately the brain, to resolve detail and store information. As a consequence, human perceptions are, for the most part, intrinsically imprecise. More specifically, perceptions are f-granular in the sense that: (a) the boundaries of perceived classes are fuzzy; and (b) the values of perceived attributes are granular, with a granule being a clump of values drawn together by indistinguishability, similarity, proximity or functionality. In effect, information granulation may be viewed as a human way of achieving data compression. Furthermore, it plays a key role in implementation of the strategy of divide-and-conquer in human problem-solving.

Information granules may be crisp or fuzzy. For example, crisp granules of age are the intervals $[0,1]$, $[1,2]$, $[2,3]$, $\ldots$, while the fuzzy intervals labeled young, middle-aged and old are fuzzy granules. Conventionally, information granules are assumed to be crisp. In fuzzy set theory and fuzzy logic, information granules are assumed to be fuzzy. Clearly, crisp granularity may be viewed as a special case of fuzzy granularity, just as a crisp set may be viewed as a special case of a fuzzy set.

Information granularity plays a key role in Prof. Pawlak's pioneering theory of rough sets (1982). At first glance, it may appear that the concepts of rough set and fuzzy set are closely related. In fact, this is not the case. A fuzzy set is a class with unsharp boundaries, whereas a rough set is a crisp set with imprecisely described boundaries. In recent years, however, the theory of rough sets has been extended in many directions, leading to the concepts of fuzzy rough set and rough fuzzy set. I should like to suggest that rough sets be called Pawlak sets in honor of Prof. Pawlak.

An extension of rough set theory which is developed in RNC relates to what is referred to as rough-neural computing – an important extension which generalizes neural computing, opening the door to many new applications that are described in

Parts III and IV of the book, which deal, respectively, with application areas and case studies.

More broadly, RNC extends the theory of Pawlak sets to granular computing and, more particularly, to computing with words and perceptions. In what follows, I will take the liberty of focusing my comments on the extension to computing with words and perceptions (CWP).

In contrast to computing with numbers (CN), the objects of computation in CWP are words, propositions and perceptions described in a natural language. In science, there is a deep-seated tradition of striving for the ultimate in rigor and precision. Words are less precise than numbers. Why and where, then, should words be used in preference to numbers?

There are three principal rationales. First, when the available information is not precise enough to justify the use of numbers. Second, when there is a tolerance for imprecision which can be exploited to achieve tractability, robustness and low solution cost. And third, when the expressive power of words is higher than the expressive power of numbers. In CWP, what is employed in this instance is Precisiated Natural Language (PNL).

In large measure, the importance of CWP derives from the fact that it opens the door to computation and reasoning with information which is perception- rather than measurement-based. Perceptions play a key role in human cognition, and underlie the remarkable human capability to perform a wide variety of physical and mental tasks without any measurements and any computations. Everyday examples of such tasks are driving a car in city traffic, playing tennis and summarizing a story. There is an enormous literature on perceptions, spanning psychology, linguistics, philosophy and other fields. But what cannot be found in this literature is a theory in which perceptions are objects of computation.

A key idea in CWP is that of dealing with perceptions not directly but through their descriptions in a natural language. In this way, computation with perceptions is reduced to computation with words and propositions drawn from a natural language. This is where Precisiated Natural Language (PNL) enters into the picture.

More specifically, as was noted earlier, perceptions are intrinsically imprecise. More concretely, perceptions are f-granular. F-granularity of perceptions is passed on to their descriptions in a natural language. The implication is that to be able to compute with perceptions it is necessary to be able to compute with f-granular propositions. This is what conventional predicate-logic-based meaning-representation methods cannot do.

The ability of PNL to deal with f-granularity derives from a key idea. Specifically in PNL, the meaning of a proposition, *p*, drawn from a natural language, NL, is represented as a generalized constraint on a variable which is implicit in *p*. More concretely, the meaning of *p* is precisiated through translation of *p* into a precisiation language. In the case of PNL, the precisiation language is the Generalized Constraint Language (GCL). Thus, the elements of GCL are generalized constraints. One of the generalized constraints is the Pawlak set constraint. This is the link between PNL and Prof. Pawlak's theory of rough sets.

In RNC, PNL is not an object of explicit exposition or discussion. But the concepts and techniques described in RNC have a direct bearing on computing with words. In this perspective, RNC lays the groundwork for PNL and PNL-based methods in computing with words and perceptions.

The concepts and techniques described in RNC point to a new direction in information processing and decision analysis—a direction in which the concept of information granularity plays a pivotal role. One cannot but be greatly impressed by the variety of issues and problems which are addressed in RNC. Clearly, rough-neural computing in combination with computing with words provides new and powerful tools for the conception, design and utilization of intelligent systems. The editors, the contributors and the publisher, Springer-Verlag, have produced a book that excels in all respects. They deserve our thanks and congratulations.

Berkeley, November 25, 2002 *Lotfi A. Zadeh**

---

*Professor in the Graduate School and Director, Berkeley Initiative in Soft Computing (BISC), Computer Science Division and the Electronics Research Laboratory, Department of EECS, University of California, Berkeley, CA 94720-1776, USA.

Preface

Lotfi Zadeh has recently pioneered a research area known as "computing with words" (CW) and explained the computational theory of perception (CTP). The objective of this new research is to build foundations for future intelligent machines and information systems that perform computations on words (names of concepts in natural language) rather than on numbers. The main notion of this approach is related to information granulation. Information granules are understood as clumps of objects (points) that are drawn together by indistinguishability, similarity, or functionality. Information granulation and methods for constructing relevant information granules are crucial for exploiting tolerance of imprecision and uncertainty to achieve tractability, robustness, and low production costs for future intelligent systems. Several approaches to the formulation of information granule concepts have been proposed so far. Granular computing (GC) is one such computing paradigm based on information granule calculi.

This book is dedicated to a newly emerging approach to CW and CTP, called rough-neural computing (RNC). In RNC, computations are usually performed on information granules. The foundations of RNC are built in soft computing frameworks comprising synergistic hybridization of rough set theory, rough mereology, fuzzy set theory, neural networks, and evolutionary computing.

Any approach to information granulation should make it possible to define complex information granules (e.g., granules relevant to spatial and temporal reasoning in autonomous systems). The following facts are especially important in the process of complex information granule construction:

(i) Any concept of an information granule considered should reflect and describe its inherent vagueness in formal terms.
(ii) Target granules cannot be constructed (induced) directly from input granules but rather are constructed in a many-stage process.
(iii) The schemes of new granule construction also interpreted as approximate reasoning schemes (AR schemes) should be robust with respect to input granule deviations.
(iv) For real-life applications, adaptive AR schemes become a necessity.

To deal with vagueness, one can adopt soft computing methods developed by fuzzy or rough set theoretical approaches and their different integrations. Information granules are represented in this book by parameterized formulas over signatures of relational systems, called information granule systems. In such systems, we emphasize the role of inclusion and closeness of information granules, to a degree, on the basis of a rough mereological approach. Information granule systems are relational systems over which information granules can be interpreted. Target information granule systems represent approximation spaces. These spaces are generalizations of approximation spaces used in rough set theory. The second aspect in the above list is related to several issues, such as reasoning from measurements to perception, multilayered learning of concept approximations, and fusion of information coming from different sources. Methods of searching for AR schemes are investigated using rough mereological tools. In general, they return hierarchical schemes for constructing information granules.

Among important topics discussed in this book are different methods for specifying operations on information granules. Such operations are used for constructing relevant information granules from experimental data and background knowledge. These are the basic components of methods aimed at constructing hierarchical schemes of information granules. In the more general case, we deal with network-like structures, transforming and exchanging information granules or information about them. Such networks are called rough-neural networks (RNN), and they are a generalization of AR schemes represented by trees. One of the important aspects of the approach to information granule calculi, as described in this book, is its strong connection with multiagent systems. For example, constructing AR schemes is closely related to ideas of cooperation and conflict resolution in multiagent systems. Moreover, agents exchanging information granules and information about them perform operations on information granules represented in languages they "understand." Hence, granules received in argument ports by a given agent as arguments of his/her operation should be approximated by properly tuning the approximation spaces. These spaces create interfaces between agents. Rough-neural networks are analogous to neural networks. They perform computations on information granules. The process of tuning parameters in rough-neural networks corresponds to adjusting weights in neural networks. Parameters of rough-neural networks are related, in particular, to approximation spaces used in their construction. By tuning these parameters, one can expect to induce relevant target information granules. The relevance of target information granules can be measured by carefully selected criteria. For example, one can use measures based on inclusion (closeness) degrees of granules and/or some other measures related to their sizes.

The methods that induce rough-neural networks using rough sets in combination with other soft computing tools create a core for RNC. These methods show that connectionist and symbolic approaches can work complementarily and not competitively. For example, derived AR schemes are aimed at representing patterns sufficiently included in target complex concepts. The structures of AR schemes are derived using both approaches. Symbolic reasoning is used to get the structure of schemes from data and background knowledge. A connectionist approach is used for tuning the parameters of such structures.

One of the most important research directions of RNC concerns relationships between information granules and words or phrases in a natural language. Investigating such relationships leads to methods for inducing (from data and background knowledge) rough-neural networks approximating reasoning schemes in natural language. This creates a strong link between the approach presented and the approaches directed toward operating and reasoning with perception-based information, such as CW and CTP. In particular, one can interpret rough-neural networks that approximate reasoning schemes in natural language as schemes of approximate reasoning from measurement to perception.

Formally, the robustness of a rough-neural network means that such a network produces a higher-order information granule, which is a clump (e.g., a set) of information granules rather than a single granule. The inputs for such networks are clumps of deviations (up to an acceptable degree) from some standard input information granules to networks. In general, rough-neural networks should be robust with respect to different sources of noise. One of the basic research directions in RNNs that we have in mind in this book is developing strategies for adaptive rough-neural network construction.

This book presents recent advances made in RNC by researchers from different countries. In the first part, the foundations of information granule calculus are discussed. Such a calculus based on a rough mereological approach creates a basis for synthesizing and analyzing rough-neural networks. Recent results on the foundations of RNC are included. The reader can find an introduction to a rough set theoretical approach together with an explanation of why a generalization of approximation spaces, used so far in rough set theory, has been introduced. Close relationships of rough set approaches with multivalued (especially with three valued) logic are also presented. The second part shows how different integrations of soft computing tools can help to induce information granules. Special emphasis is given to methods based on hybridization of rough sets with neural techniques. Such techniques are crucial for developing RNC methods in synthesizing complex information granules. The reader can find how different approaches to constructing information granules based on fuzzy sets, rough sets, rough fuzzy sets, and other soft computing paradigms can work in synergy. Moreover, the way different approaches can be combined with symbolic approaches like nonmonotonic reasoning, deductive databases, and logic programming is presented. Methods for constructing interfaces between experimental knowledge and symbolic knowledge are discussed. The necessity to use statistical tools in information granule construction is underlined. Selected application areas for RNC and CW are discussed in the third part. Modeling methods for complex sociological situations or sociological games and semantic models for biomedical reasoning are included. Finally, the last part of the book consists of several case studies illustrating recent developments based on RNC. This includes problems in signal analysis, medical data analysis, and pattern recognition.

It is worthwhile mentioning that from a logical point of view, research in RNC is closely related to the pragmatic aspects of natural language. As an example of such a pragmatic aspect investigated in RNC, one can consider the attempts made to understand concepts by means of experimental data and reasoning schemes in natural language. Another example would be communication between agents using different languages for information granule representation.

We do hope that this self-contained book will encourage students and researchers to join a fascinating journey toward building intelligent systems.

The book has been very much enriched thanks to forewords written by Prof. Zdzisław Pawlak, the founder of rough set theory, and Prof. Lotfi A. Zadeh, the founder of fuzzy set theory and more recently of the new paradigms CW and CTP. We are honored to have their contributions and we would like to express our special gratitude to both of them.

We are grateful to our colleagues who prepared excellent chapters for this volume. Our special thanks go to: Prof. Leonard Bolc, a member of the advisory board of the series; Ms. Ingeborg Mayer and Ms. Joylene Vette-Guillaume of Springer for their help with the production of this volume; and Dr. Sinh Hoa Nguyen, Dr. Hung Son Nguyen, and Ms. Grażyna Domańska for preparing the LaTeX version of the book. We are grateful to the copy editors of Springer for their devoted help in producing this volume, and especially to Mr. Ronan Nugent.

Our work was helped by grants from the State Committee for Scientific Research (KBN) of the Republic of Poland, Nos. 8T11C 02519 (A. Skowron) and 8T11C 02417 (L. Polkowski). A. Skowron was, moreover, supported by a grant from the Wallenberg Foundation.

Warsaw, June 2003

Sankar K. Pal
Lech Polkowski
Andrzej Skowron

Contents

List of Contributors

Mohua Banerjee
Department of Mathematics
Indian Institute of Technology
Kanpur 208 016
INDIA
mohua@iitk.ac.in

Maciej Borkowski
University of Manitoba
Department of Electrical and Computer Engineering
Winnipeg, Manitoba R3T 5V6
CANADA
maciey@ee.umanitoba.ca

Tom R. Burns
Uppsala Theory Circle
Department of Sociology
University of Uppsala
Box 821
75108 Uppsala
SWEDEN
tom.burns@soc.uu.se

Mihir K. Chakraborty
Department of Pure Mathematics
University of Kolkata
35, Ballygunge Circular Road
Kolkata 700019
INDIA
mihirc99@cal3.vsnl.net.in

Andrzej Czyżewski
Technical University of Gdańsk
Sound and Vision Engineering Department
Narutowicza 11/12, 80-952 Gdańsk
POLAND
andrzej@sound.eti.pg.gda.pl

Biswarup Dasgupta
Machine Intelligence Unit
Indian Statistical Institute
203 B. T. Road, Kolkata 700108
INDIA
biswarupdg@yahoo.com

Patrick Doherty
Department of Computer and Information Science
Linköping University
58183 Linköping
SWEDEN
patdo@ida.liu.se

Ju-Zhen Dong
Knowledge Information Systems Laboratory
Department of Information Engineering
Maebashi Institute of Technology
460-1, Kamisadori-Cho
Maebashi City 371
JAPAN
ning@mc.kcom.ne.jp

Jan Doroszewski
Department of Biophysics and Biomathematics
Medical Center of Postgraduate Education

Marymoncka 99, 01-813 Warszawa
POLAND
jandoro@cmkp.edu.pl

Ivo Düntsch
Department of Computer Science
Brock University
St. Catherines, Ontario, L2S 3AI
CANADA
duentsch@cosc.brocku.ca

Günther Gediga
Institut für Evaluation und
Marktanalysen, Brinkstr. 19
49143 Jeggen
GERMANY
gediga@eval-institut.de

Anna Gomolińska
University of Białystok
Department of Mathematics
Akademicka 2, 15-267 Białystok
POLAND
anna.gom@math.uwb.edu.pl

Linda K. Goodwin
Department of Information Services
and the School of Nursing
Duke University
Durham, NC 27710
USA
goodw010@mc.duke.edu

Jerzy W. Grzymala-Busse
Department of Electrical Engineering
and Computer Science
University of Kansas
Lawrence, KS 66045
USA
jerzy@lightning.eecs.ukans.edu

Witold J. Grzymala-Busse
RS Systems, Inc.
Lawrence, KS 66047
USA
witek@argus.rs-systems.com

Masahiro Inuiguchi
Department of Electronics and
Information Systems
Graduate School of Engineering
Osaka University
2-1 Yamadaoka, Suita, Osaka 565-0871
JAPAN
inuiguti@eie.eng.osaka-u.ac.jp

Jarosław Kachniarz
Soft Computer Consultants
34350 US19N
Palm Harbor, FL 34684
USA
jk@softcomputer.com

Bożena Kostek
Technical University of Gdańsk
Sound and Vision Engineering
Department
Narutowicza 11/12, 80-952 Gdańsk
POLAND
bozenka@sound.eti.pg.gda.pl

Chunnian Liu
School of Computer Science
Beijing Polytechnic University
Beijing 100022
P.R. CHINA
bpvliu@public.bta.net.cn

Witold Łukaszewicz
The College of Economics and
Computer Science
Wyzwolenia 30, 10-106 Olsztyn
POLAND
witlu@ida.liu.se

Jan Małuszyński
Department of Computer
and Information Science
Linköping University
58183 Linköping
SWEDEN
janma@ida.liu.se

Pabitra Mitra
Machine Intelligence Unit
Indian Statistical Institute
203 B. T. Road, Kolkata 700108
INDIA
pabitra_r@isical.ac.in

Hung Son Nguyen
Institute of Mathematics
Warsaw University
Banacha 2, 02-097 Warsaw
POLAND
son@mimuw.edu.pl

Tuan Trung Nguyen
Institute of Mathematics
Warsaw University
Banacha 2, 02-097 Warsaw
POLAND
nttrung@mimuw.edu.pl

Setsuo Ohsuga
Department of Information and Computer Science
School of Science and Engineering
Waseda University
3-4-1 Okubo Shinjuku-Ku, Tokyo 169
JAPAN
ohsuga@fd.catv.ne.jp

Sankar K. Pal
Machine Intelligence Unit
Indian Statistical Institute
203 B. T. Road, Kolkata 700108
INDIA
sankar@isical.ac.in

Zdzisław Pawlak
Institute of Theoretical and Applied Informatics
Polish Academy of Sciences
Bałtycka 5, 44-100 Gliwice
POLAND
and
University of Information Technology and Management
Newelska 6, 01-477 Warsaw
POLAND
zpw@ii.pw.edu.pl

Witold Pedrycz
Department of Electrical & Computer Engineering
University of Alberta
Edmonton, Alberta T6G 2M7
CANADA
and
Systems Research Institute
Polish Academy of Sciences
Newelska 6, 01-447 Warsaw
POLAND
pedrycz@ee.ualberta.ca

James F. Peters
Department of Electrical and Computer Engineering
University of Manitoba
Winnipeg, Manitoba R3T 5V6
CANADA
jfpeters@ee.umanitoba.ca

Lech Polkowski
Polish–Japanese Institute of Information Technology
Research Center
Koszykowa 86, 02-008 Warsaw
POLAND
and
Department of Mathematics and Computer Science
University of Warmia and Mazury
Żołnierska 14a, 10-561 Olsztyn
POLAND
polkow@pjwstk.edu.pl

Sheela Ramanna
University of Manitoba
Department of Electrical and Computer Engineering

Winnipeg, Manitoba R3T 5V6
CANADA
sramanna@io.uwinnipeg.ca

Ewa Roszkowska
University in Białystok
Faculty of Economics
Warszawska 63, 15-062 Białystok
POLAND
and
Białystok School of Economics
Choroszczańska 31, 15-732 Białystok
POLAND
erosz@w3cache.uwb.edu.pl

Władysław Skarbek
Department of Electronics and Information Technology
Warsaw University of Technology
Nowowiejska 15/19, 00-665 Warsaw
POLAND
and
Multimedia Group, Altkom Akademia S.A.
Stawki 2, 00-193 Warsaw
POLAND
w.skarbek@ire.pw.edu.pl

Andrzej Skowron
Institute of Mathematics
Warsaw University
Banacha 2, 02-097 Warsaw
POLAND
skowron@mimuw.edu.pl

Jarosław Stepaniuk
Department of Computer Science
Białystok University of Technology
Wiejska 45A, 15-351 Białystok
POLAND
jstepan@ii.pb.bialystok.pl

Zbigniew Suraj
University of Information Technology and Management
H. Sucharskiego 2, 35-225 Rzeszów
POLAND
zsuraj@wenus.wsiz.rzeszow.pl

Roman Swiniarski
San Diego State University
Department of Mathematical and Computer Sciences
5500 Campanile Drive
San Diego, CA 92182
USA
rswiniar@sciences.sdsu.edu

Andrzej Szałas
The College of Economics and Computer Science
Wyzwolenia 30, 10-106 Olsztyn
POLAND
andsz@ida.liu.se

Tetsuzo Tanino
Department of Electronics and Information Systems
Graduate School of Engineering
Osaka University
2-1 Yamadaoka, Suita, Osaka 565-0871
JAPAN
tanino@eie.eng.osaka-u.ac.jp

Shusaku Tsumoto
Department of Medicine Informatics
Shimane Medical University
School of Medicine
89-1 Enya-cho Izumo City
Shimane 693-8501
JAPAN
tsumoto@computer.org

Aida Vitória
Department of Science and Technology
Linköping University
60174 Norrköping
SWEDEN
aidvi@itn.liu.se

Piotr Wojdyłło
Department of Mechanical and Aeronautical Engineering
University of California
One Shields Avenue
Davis, CA 95616
USA
pwoj@mimuw.edu.pl

Jakub Wróblewski
Polish-Japanese Institute of Information Technology
Koszykowa 86, 02-008 Warsaw
POLAND
jakubw@mimuw.edu.pl

Yiyu Yao
Department of Computer Science
University of Regina
Regina, Saskatchewan S4S 0A2
CANADA
yyao@cs.uregina.ca

Xinqun Zheng
Department of Electrical Engineering and Computer Science
University of Kansas
Lawrence, KS 66045
USA
jerzy@lightning.eecs.ukans.edu

Ning Zhong
Department of Information Engineering
Maebashi Institute of Technology
460-1, Kamisadori-Cho
Maebashi City 371
JAPAN
zhong@maebashi-it.ac.jp

Part I

Rough Sets, Granular Computing, and Rough-Neural Computing: Foundations

Introduction to Part I

Part I opens the volume with chapters dealing with the foundations of the concept of rough-neural computing (RNC) perceived throughout this book. RNC combines the idea of a neural network as a network of intelligent computing units with the ideas derived from rough set theory of the approximation of exact and rough concepts and with Zadeh's ideas of knowledge granulation and computing with words. In other words, computations in RNC are performed in a neural net-like manner with objects described in terms of rough set theory, fuzzy set theory, etc.

Part I begins with the chapter entitled *Elementary Rough Set Granules: Toward a Rough Set Processor* by Pawlak. It provides the basic notions of rough set theory, discusses their probabilistic counterparts, and concludes with an account of rough set processor architecture.

The chapter, *Rough-Neural Computing: An Introduction*, by Pal, Peters, Polkowski, and Skowron introduces the basic ideas of RNC. Readers will find a general discussion of knowledge granulation in a distributed environment followed by a specific case of approximation spaces. Some models of rough neurons, viz., interval-based rough neurons, approximation neurons, decider neurons, and rough-fuzzy approximation neural nets, are then surveyed.

The concept of knowledge granulation and RNC is further examined by Skowron and Stepaniuk in their chapter, *Information Granulation and Rough-Neural Computing*. Mechanisms for granule formation are discussed first and then operations on information granules are studied in detail from the point of view of general approximation spaces. Rough-neural networks for performing computations on information granules via methods of approximation spaces are introduced, their mechanisms are analyzed thoroughly, and possible applications are pointed out.

The basic ideas and schemes pertaining to granulation and granular computing as well as to partial containment as a basis for object classification have been developed in the framework of rough mereology. The chapter, *A Rough-Neural Computation Model Based on Rough Mereology*, by Polkowski brings forth an outline of rough mereology with an account of Leśniewski mereology and ontology. A model for rough-neural computing based directly on rough mereology is introduced and analyzed.

Theoretical ideas derived in the above chapters are further developed by Pedrycz in his chapter, *Knowledge-Based Networking in Granular Worlds*. Granules (clusters) of knowledge are formed here via the fuzzy-c-means method, and the granulation process is discussed from the point of view of a distributed system of collaborative databases, each serving the role of an autonomous granular world. Emphasis is put, from the point of view of applications, on security and confidentiality of transactions.

Information granulation in constructing adaptive classification systems is the subject of the chapter, *Adaptive Aspects of Combining Approximation Spaces*, by Wróblewski. In the framework of parameterized approximation spaces, the process of constructing an adaptive classification algorithm over a distributed system (network) of approximation spaces is described and analyzed from the theoretical and application points of view.

The chapter by Banerjee and Chakraborty, titled *Algebras from Rough Sets*, deals with a formal study of approximation spaces from the standpoint of predominantly algebraic structures. Here, the authors survey such structures as Heyting algebras, Stone algebras, and Łukasiewicz algebras and point to their logical interpretations, showing the richness of a variety of formal structures encoded in approximation spaces.

Chapter 1
Elementary Rough Set Granules: Toward a Rough Set Processor

Zdzisław Pawlak

Institute of Theoretical and Applied Informatics, Polish Academy of Sciences,
Bałtycka 5, 44-100 Gliwice, Poland
and
University of Information Technology and Management,
Newelska 6, 01-477 Warsaw, Poland
zpw@ii.pw.edu.pl

Summary. In this chapter, the basics of the rough set approach are presented, and an outline of an exemplary processor structure is given. The organization of a simple processor is based on elementary rough set granules and dependencies between them. The rough set processor (RSP) is meant to be used as an additional fast classification unit in ordinary computers or as an autonomous learning machine. In the latter case, the RSP can be regarded as an alternative to neural networks.

1 Introduction

Rough set theory [4] has proved its effectiveness in drawing conclusions from data [6]. However, to take full advantage of the theory in data analysis, adequate processor organization is necessary. The architecture of such processors was proposed first in [4]. In this chapter, another proposal for rough set processor organization is presented.

Rough-set-based data analysis starts from a decision table, which is a data table. The columns of a decision table are labeled with attributes; the rows are labeled with objects of interest; and attribute values are entered in the data cells of the table. Attributes of the decision table are divided into two disjoint groups called condition and decision attributes, respectively. Each row of a decision table induces a decision rule, which specifies the decision (action, results, outcome, etc.) if some conditions are satisfied. If a decision rule uniquely determines a decision in terms of conditions, the decision rule is certain. Otherwise the decision rule is uncertain. Decision rules are closely connected with approximations, which are basic concepts of rough set theory. Roughly speaking, certain decision rules describe the lower approximation of decisions in terms of conditions, whereas uncertain decision rules refer to the upper approximation of decisions.

Two conditional probabilities, called the certainty and the coverage coefficient, are associated with every decision rule. The certainty coefficient expresses the conditional probability that an object belongs to the decision class specified by the decision rule, given that it satisfies the conditions of the rule. The coverage coefficient gives the conditional probability of the reasons for a given decision.

It turns out that the certainty and coverage coefficients satisfy Bayes' theorem. This gives new insight into the interpretation of Bayes' theorem, showing that Bayes' theorem can be used differently for drawing conclusions from data than the use offered by classical Bayesian inference philosophy [5].

This idea is at the foundation of rough set processor organization. In this chapter, the basics of rough set theory are presented, and an outline of an exemplary processor structure is given. The rough set processor is meant to be used as a "rough" classifier, or as a learning machine, and can be regarded as an alternative to neural networks.

2 Information Systems and Decision Tables

In this section we define the basic concept of rough set theory: information systems. The rudiments of rough set theory can be found in [4, 6]. An information system is a data table whose columns are labeled with attributes, rows are labeled with objects of interest, and attribute values are entered in the data cells of the table.

Formally, the *information system* is a pair $S = (U,A)$, where U and A are nonempty finite sets called the *universe* of objects and the set of *attributes*, respectively, such that $a : U \rightarrow V_a$, where V_a is the set of all *values* of a, called the *domain* of a, for each $a \in A$. Any subset B of A determines a binary relation $I(B)$ on U, which will be called an *indiscernibility relation*, and is defined as follows:

$$(x,y) \in I(B) \quad \text{if and only if} \quad a(x) = a(y) \quad \text{for every} \quad a \in A,$$

where $a(x)$ denotes the value of the attribute a for the element x. Obviously $I(B)$ is an equivalence relation. The family of all equivalence classes of $I(B)$, i.e., a partition determined by B, will be denoted by $U/I(B)$, or simply by U/B; an equivalence class of $I(B)$, i.e., the block of the partition U/B containing x will be denoted by $B(x)$.

If (x,y) belongs to $I(B)$, we will say that x and y are *B-indiscernible objects (indiscernible with respect to B)*. Equivalence classes of the relation $I(B)$ (or blocks of the partition U/B) are referred to as *B-elementary sets* or *B-elementary granules*.

If we distinguish in the information system two disjoint classes of attributes, called *condition* and *attribute decision*, respectively, then the system will be called a *decision table* and will be denoted by $S = (U,C,D)$, where C and D are disjoint sets of condition and decision attributes, respectively, and $C \cup D = A$. $C(x)$ and $D(x)$ will be referred to as the condition class and the decision class induced by x, respectively.

Thus the decision table describes decisions (actions, results etc.) taken when some conditions are satisfied. In other words, each row of the decision table specifies a decision rule that determines decisions in terms of conditions.

An example of a simple decision table is shown in Table 1. In the table, *age*, *sex*, and *profession* are condition attributes, whereas *disease* is the decision attribute.

Table 1. Decision table

Decision rule	*Age*	*Sex*	*Profession*	*Disease*
1	*Old*	*Male*	*Yes*	*No*
2	*Med.*	*Female*	*No*	*Yes*
3	*Med.*	*Male*	*Yes*	*No*
4	*Old*	*Male*	*Yes*	*Yes*
5	*Young*	*Male*	*No*	*No*
6	*Med.*	*Female*	*No*	*No*

The table contains data on the relationship among age, sex, and profession and a certain vocational disease. Decision tables can be simplified by removing superfluous attributes and attribute values, but we will not consider this issue in this chapter.

3 Decision Rules

In what follows, we will describe decision rules more exactly. Let $S = (U,C,D)$ be a decision table. Every $x \in U$ determines a sequence

$$c_1(x),\ldots, c_n(x), d_1(x),\ldots, d_m(x),$$

where $\{c_1,\ldots,c_n\} = C$ and $\{d_1,\ldots,d_m\} = D$. The sequence will be called a *decision rule induced by* x (in S) and denoted by

$$c_1(x),\ldots,c_n(x) \rightarrow d_1(x),\ldots,d_m(x),$$

or in short, $C \underset{x}{\rightarrow} D$. The number $supp_x(C,D) = |A(x)| = |C(x) \cap D(x)|$ will be called the *support* of the decision rule $C \underset{x}{\rightarrow} D$, and the number,

$$\sigma_x(C,D) = \frac{supp_x(C,D)}{|U|},$$

will be referred to as the *strength* of the decision rule $C \underset{x}{\rightarrow} D$, where $|X|$ denotes the cardinality of X. Another decision table is shown in Table 2. This decision table can be understood as an abbreviation of a bigger decision table containing 1100 rows. Support of the decision rule means the number of identical decision rules in the original decision table.

Table 2. Support and strength

Decision rule	*Age*	*Sex*	*Profession*	*Disease*	*Support*	*Strength*
1	*Old*	*Male*	*Yes*	*No*	200	0.18
2	*Med.*	*Female*	*No*	*Yes*	70	0.06
3	*Med.*	*Male*	*Yes*	*No*	250	0.23
4	*Old*	*Male*	*Yes*	*Yes*	450	0.41
5	*Young*	*Male*	*No*	*No*	30	0.03
6	*Med.*	*Female*	*No*	*No*	100	0.09

With every decision rule $C \underset{x}{\rightarrow} D$, we associate the *certainty factor* of the decision rule, denoted $cer_x(C,D)$ and defined as follows:

$$cer_x(C,D) = \frac{|C(x) \cap D(x)|}{|C(x)|} = \frac{supp_x(C,D)}{|C(x)|} = \frac{\sigma_x(C,D)}{\pi[C(x)]},$$

where $\pi[C(x)] = \frac{|C(x)|}{|U|}$.

The certainty factor may be interpreted as a conditional probability that y belongs to $D(x)$, given y belongs to $C(x)$, symbolically $\pi_x(D|C)$. If $cer_x(C,D) = 1$, then $C \underset{x}{\rightarrow} D$ will be called a *certain decision rule*; if $0 < cer_x(C,D) < 1$, the decision rule will be referred to as an *uncertain decision rule*. We will also use a *coverage factor* of the decision rule, denoted $cov_x(C,D)$ [7] defined as

$$cov_x(C,D) = \frac{|C(x) \cap D(x)|}{|D(x)|} = \frac{supp_x(C,D)}{|D(x)|} = \frac{\sigma_x(C,D)}{\pi[D(x)]},$$

where $D(x) \neq \emptyset$ and $\pi[D(x)] = \frac{|D(x)|}{|U|}$. Similarly,

$$cov_x(C,D) = \pi_x(C|D).$$

If $C \underset{x}{\rightarrow} D$ is a decision rule, then $D \underset{x}{\rightarrow} C$ will be called an *inverse decision rule*. Inverse decision rules can be used to give *explanations* (*reasons*) for a decision.

4 Approximation of Sets

Suppose we are given an information system $S = (U,A)$, $X \subseteq U$, and $B \subseteq A$. Our task is to describe the set X in terms of attribute values from B. To this end, we define two operations assigning to every $X \subseteq U$ two sets $B_*(X)$ and $B^*(X)$ called the *B-lower* and the *B-upper approximation* of X, respectively, and defined as

$$B_*(X) = \bigcup_{x \in U} \{B(x) : B(x) \subseteq X\} \text{ and}$$

$$B^*(X) = \bigcup_{x \in U} \{B(x) : B(x) \cap X \neq \emptyset\}.$$

Hence, the B-lower approximation of a set is the union of all B-granules that are included in the set, whereas the *B-upper* approximation of a set is the union of all B-granules that have a nonempty intersection with the set. The set

$$BN_B(X) = B^*(X) - B_*(X),$$

will be referred to as the *B-boundary region* of X.

If the boundary region of X is the empty set, i.e., $BN_B(X) = \emptyset$, then X is *crisp* (*exact*) with respect to B; in the opposite case, i.e., if $BN_B(X) \neq \emptyset$, X is referred to as *rough* (*inexact*) with respect to B.

There is an interesting relationship between approximations and decision rules. Let $C \underset{x}{\rightarrow} D$ be a decision rule. The set,

$$\bigcup_{y \in D(x)} \{C(y) : C(y) \subseteq D(x)\},$$

is equal to the lower approximation of the decision class $D(x)$, by condition classes $C(y)$, whereas the set,

$$\bigcup_{y \in D(x)} \{C(y) : C(y) \cap D(x) \neq \emptyset\},$$

is equal to the upper approximation of the decision class by condition classes $C(y)$.

That means that approximations and decision rules are two different methods for expressing imprecision. Approximations are better suited to expressing topological properties of decision tables, whereas rules describe hidden patterns in data in a simple way.

5 Probabilistic Properties of Decision Tables

Decision tables have important probabilistic properties that are discussed next.

Let $C \underset{x}{\rightarrow} D$ be a decision rule, and let $\Gamma = C(x)$ and $\Delta = D(x)$. Then the following properties are valid:

$$\sum_{y \in \Gamma} cer_y(C,D) = 1, \tag{1}$$

$$\sum_{y \in \Delta} cov_y(C,D) = 1, \tag{2}$$

$$\pi[D(x)] = \sum_{y \in \Gamma} cer_y(C,D) \cdot \pi[C(y)] = \sum_{y \in \Gamma} \sigma_y(C,D), \tag{3}$$

$$\pi[C(x)] = \sum_{y \in \Delta} cov_y(C,D) \cdot \pi[D(y)] = \sum_{y \in \Delta} \sigma_y(C,D), \tag{4}$$

$$cer_x(C,D) = \frac{cov_x(C,D) \cdot \pi[D(x)]}{\sum_{y \in \Delta} cov_y(C,D) \cdot \pi[D(y)]} = \frac{\sigma_x(C,D)}{\pi[C(x)]}, \tag{5}$$

$$cov_x(C,D) = \frac{cer_x(C,D) \cdot \pi[C(x)]}{\sum_{y \in \Gamma} cer_y(C,D) \cdot \pi[C(y)]} = \frac{\sigma_x(C,D)}{\pi[D(x)]}, \tag{6}$$

that is, any decision table satisfies (1) – (6). Observe that (3) and (4) refer to the well-known *total probability theorem*, whereas (5) and (6) refer to *Bayes' theorem*. Thus, to compute the certainty and coverage factors of decision rules according to formula (5) and (6), it is enough to know only the strength (support) of all decision rules. The strength of decision rules can be computed from data or can be a subjective assessment.

These properties will be used as a basis for the rough set processor organization. The certainty and coverage factors for the decision table presented in Table 2 are shown in Table 3.

Table 3. Certainty and coverage factors

Decision rule	*Strength*	*Certainty*	*Coverage*
1	0.18	0.31	0.34
2	0.06	0.40	0.13
3	0.23	1.00	0.43
4	0.41	0.69	0.87
5	0.03	1.00	0.06
6	0.09	0.60	0.17

Let us observe that, according to formulas (5) and (6), the certainty and coverage factors can be computed employing only the strength of decision rules. In Table 2, decision rules 3 and 5 are certain, whereas the remaining decision rules are uncertain. This means that middle-aged males having a profession and young males not having a profession are certainly healthy. Old males having a profession are most probably ill (probability = .69), and middle-aged females not having a profession are most probably healthy (probability = .60).

The inverse decision rules say that healthy persons are most probably middle-aged males having a profession (probability = .43) and ill persons are most probably old males having a profession (probability = .87).

6 Decision Tables and Flow Graphs

With every decision table, we associate a *flow graph*, i.e., a directed acyclic graph defined as follows: to every decision rule $C \underset{x}{\rightarrow} D$, we assign a *directed branch* x connecting the *input node* $C(x)$ and the *output node* $D(x)$. The strength of the decision rule represents a *throughflow* of the corresponding branch. The throughflow of the graph is governed by formulas (1) – (6).

Formulas (1) and (2) say that the outflow of an input node or an output node is equal to their respective inflows. Formula (3) states that the outflow of the output node amounts to the sum of its inflows; whereas formula (4) says that the sum of the outflows of the input node equals its inflow. Finally, formulas (5) and (6) reveal how throughflow in the flow graph is distributed between its inputs and outputs. The flow graph associated with the decision table presented in Table 2 is shown in Fig. 1.

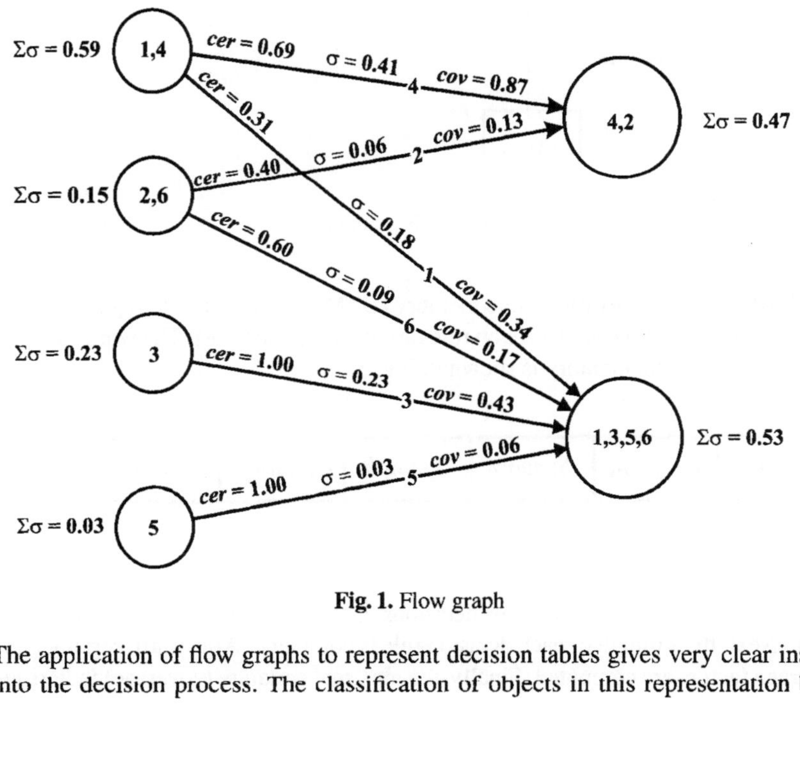

Fig. 1. Flow graph

The application of flow graphs to represent decision tables gives very clear insight into the decision process. The classification of objects in this representation boils down to finding the maximal output flow in the flow graph; whereas, the explanation of the decisions is connected to the maximal input flow associated with the given decision (see also [1] and [6]).

7 Rough Set Processor

To make the most of rough set theory in data analysis, a special microprocessor, the RSP, is necessary, to speed up the classification process. The RSP should perform operations pointed out by the flow graph of a decision table, that is, first it should compute its strengths from the supports of decision rules and afterward compute the certainty and coverage factors of all decision rules. Finally, the maximal certainty (coverage) factor should be computed, pointing out the most probable decision class (reason) for the classified object.

Many hardware implementations of this idea are possible. An example of a simplified RSP structure is depicted in Fig. 2.

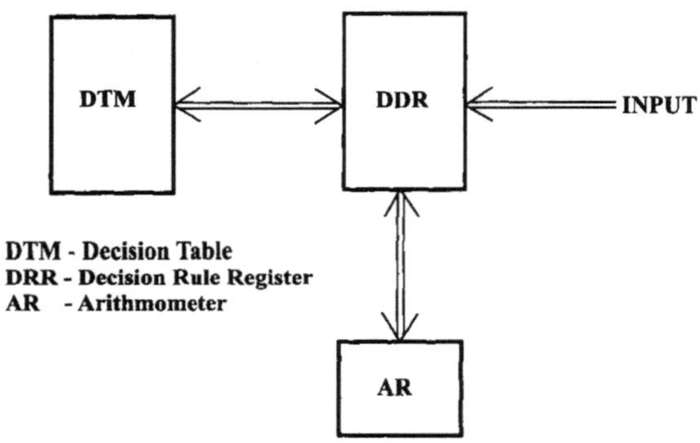

Fig. 2. RSP structure

The RSP consists of decision table memory (DTM), a decision rule register (DRR) and an arithmometer (AR). Decision rules are stored in the DTM. The word structure of the decision table memory is shown in Fig. 3.

Condition	*Decision*	*Support*	*Strength*	*Certainty*	*Coverage*

Fig. 3. Word structure

At the initial state, only conditions, decisions, and support of each decision rule are given. Next, the strength of each decision rule is computed. Afterward, certainty and coverage factors are computed. Finally, the maximal certainty (coverage) factors are ascertained. They will be used to define the most probable decision rules (inverse decision rules) induced by data. Let us also observe that the flow graph can easily be implemented as an analogue electrical circuit.

8 Conclusion

Rough-set-based data analysis consists of discovering hidden patterns in decision tables. It is shown that decision tables display basic probabilistic features; particularly, they satisfy the total probability theorem and Bayes' theorem. Moreover, rough set theory allows us to represent the above theorems in a very simple way using only the strengths of the decision rules. This property allows us to represent decision tables in the form of a flow graph and interpret the decision structure of the decision table as throughflow in the graph. The flow graph interpretation of decision tables can be employed as a basis for rough set processor organization.

References

1. M.Berthold, D.J. Hand. *Intelligent Data Analysis: An Introduction*. Springer, Heidelberg, 1999.
2. J. Łukasiewicz. Die Logischen Grundlagen der Wahrscheinlichkeitsrechnung. Kraków, 1913. In L. Borkowski, editor, *Jan Łukasiewicz: Selected Works*, North–Holland, Amsterdam and Polish Scientific Publishers, Warsaw, 16–63, 1970.
3. A. Kanasugi. A design of architecture for rough set processor. In *Proceedings of the International Workshop on Rough Set Theory and Granular Computing (RSTGC 2001)*, volume 5(1/2) of the *Bulletin of International Rough Set Society*, 201–204, 2001.
4. Z. Pawlak. *Rough Sets: Theoretical Aspects of Reasoning about Data*. Kluwer, Dordrecht, 1991.
5. Z. Pawlak. Rough sets and decision algorithms. In *Proceedings of the 2nd International Conference on Rough Sets and Current Trends in Computing (RSCTC 2000)*, LNAI 2005, 30–45, Springer, Berlin, 2001.
6. A. Skowron. Rough sets in KDD (plenary talk at IFIP 2000). In *Proceedings of the Conference on Intelligent Information Processing (IIP 2000)*, 1–17, Publishing House of Electronic Industry, Beijing, 2001.
7. D. Ślęzak. Data model based on approximate bayesian network. In B. Bouchon-Meunier, J. Gutierrez-Rios, L. Magdalena, R.R. Yager, editors, *Technologies for Constructing Intelligent Systems 2: Tools*, 313–326, Physica, Heidelberg, 2002.
8. S. Tsumoto, H. Tanaka. Discovery of functional components of proteins based on PRIMEROSE and domain kowledge hierarchy. In *Proceedings of the 3rd Workshop on Rough Sets and Soft Computing (RSSC'94)*, 280–285, San Jose, CA, 1995.

References

Chapter 2
Rough-Neural Computing: An Introduction

Sankar K. Pal,[1] James F. Peters,[2] Lech Polkowski,[3] Andrzej Skowron[4]

[1] Machine Intelligence Unit, Indian Statistical Institute, 203 B.T. Road, Kolkata 700108, India
sankar@isical.ac.in

[2] Department of Electrical and Computer Engineering, University of Manitoba, Winnipeg, Manitoba R3T 5V6, Canada
jfpeters@ee.umanitoba.ca

[3] Polish-Japanese Institute of Information Technology Research Center, Koszykowa 86, 02-008 Warsaw, Poland
and
Department of Mathematics and Computer Science, University of Warmia and Mazury, Żołnierska 14a, 10-561 Olsztyn, Poland
polkow@pjwstk.edu.pl

[4] Institute of Mathematics, Warsaw University, Banacha 2, 02-097 Warsaw, Poland
skowron@mimuw.edu.pl

Summary. This chapter presents a new paradigm for neural computing that has its roots in rough set theory. Historically, this paradigm has three main threads: production of a training set description, calculus of granules, and interval analysis. This paradigm gains its inspiration from the work of Pawlak on rough set philosophy as a basis for machine learning and from work on data mining and pattern recognition by Swiniarski and others in the early 1990s. The focus of this work is on the production of a training set description and inductive learning using knowledge reduction algorithms. This first thread in rough-neural computing has a strong presence in current neural computing research. The second thread in rough-neural computing has two main components: information granule construction in distributed systems of agents and local parameterized approximation spaces (see Sect. 2.2 and Chap. 3). A formal treatment of the hierarchy of relations of being a part to a degree (also known as approximate *rough mereology*) was introduced by Polkowski and Skowron in the mid- and late-1990s. Approximate rough mereology provides a basis for an agent-based, adaptive calculus of granules. This calculus serves as a guide in designing rough-neural computing systems. A number of touchstones of rough-neural computing have emerged from efforts to establish the foundations for granular computing: cooperating agent, granule, granule measures (e.g., inclusion, closeness), and approximation space parameter calibration. The notion of a cooperating agent in a distributed system of agents provides a model for a neuron. Information granulation and granule approximation define two principal activities of a neuron. Included in the toolbox of an agent (neuron) are measures of granule inclusion and closeness of granules. Agents (neurons) acquire knowledge by granulating (fusing) and approximating sensor inputs and input (granules) from other agents. The second component of the granular form of rough-neural computing is a new approach to training agents (neurons). In this new paradigm, training a network of agents (neurons) is defined by algorithms for adjusting parameters in the parameter space of each agent. Parameters accessible to rough neurons replace the usual scalar weights on (strengths-of-) connections between neurons. Hence, learning in a rough neural network is defined relative to local parameter adjustments. In sum, the granule construction paradigm provides a model for approximate reasoning by systems of communicating agents. The third thread in rough-neural computing stems from the introduction of a rough set approach to interval analysis by Banerjee, Lingras, Mitra, and Pal in the later part of the 1990s. This work has led to a variety of new rough-neural computing computational models. This chapter gives a brief presentation of an agent (neuron)-based calculus of granules. The design of different kinds of rough neurons is considered. Architectures of a number of different rough-neural computing schemes are also considered.

1 Introduction

The hint that rough set theory provides a good basis for neuro-computing can be found in a discussion about machine learning by Zdzislaw Pawlak in 1991 [26]. Inductive learning is divided into two phases that are reminiscent of training in classical neural computing: closed-world training and open-world training. Closed-world training focuses on constructing a pair (R_0, U_0), where R_0 is an initial set of classification rules and U_0 is an initial set of classified objects (initial universe). For each object $x \in U_0$, an agent is able to classify x based on identified features of the object (e.g., color, shape, weight, velocity). The aim of open-world training is to achieve complete knowledge of the universe by constructing (R_c, U_c), where R_c is created either as an initial set of classification rules or by modifying old rules and U_c is a complete set of classified objects (complete universe). A particular condition vector of feature values provides the basis for a decision in classifying an object in the set U_c. To some extent, this form of training is analogous to selecting a training set used to calibrate a neural network. During open-world training, an agent attempts to use R_0 to classify further (possibly new) objects by finding the condition vector in R_0 that most closely matches the experimental condition vector for a new object. In effect, the condition vectors in R_0 provide a "codebook" to define the space of input patterns. The trick is to use the codebook to identify the feature pattern of each new object. When an object x cannot be classified using R_0, a new classification rule $\chi \Rightarrow d_\chi$ is formulated, R_c is augmented to reflect the knowledge about the changing universe (i.e., $R_c = R_0 \cup \{\chi \Rightarrow d_\chi\}$), and U_0 is augmented with newly classified objects ($U_c = U_0 \cup \{x\}$). The inductive learning method resembles learning vector quantization and self-organizing feature maps described in [1, 11]. The approach outlined above is simplified. For more advanced discussions on the rough set approach to inductive learning, refer to Chaps. 3 and 25.

The studies of neural networks in the context of rough sets [2, 5, 10, 14, 19, 21, 22, 24, 35–37, 41, 57, 60–63, 69, 74] and granular computing [12, 31, 40, 44–46, 48, 52, 53, 55, 74] are extensive. An intuitive formulation of information granulation was introduced by Zadeh [70, 71]. Practical applications of rough-neural computing have recently been found in predicting urban highway traffic volume [14], speech analysis [19, 24], classifying the waveforms of power system faults [10], signal analysis [36], assessing software quality [31], and control of autonomous vehicles [20]. In its most general form, rough-neural computing provides a basis for granular computing. A rough mereological approach to rough neural networks springs from an interest in knowledge synthesized (induced) from successive granule approximations performed by neurons (cooperating agents) [44]. The distributed agent model for a neural network leads naturally to nonlayered neural network architectures, that is, it is possible for an agent (neuron) to communicate granules of knowledge to other agents (neurons) in its neighborhood rather than following the usual restricted model of a movement of granules "upward" from neurons in one layer to neurons in a higher layer. For this reason, the distributed agent model for rough-neural computing is reminiscent of the Wiener internuncial pool model for message-passing between neurons in the human nervous system [68] and, more recently, the swarm intelligence model [4].

This chapter is organized as follows. An overview of a granular approach to rough-neural computing is presented in Sect. 2. A number of different forms of neurons are briefly described in Sect. 3. The architectures of hybrid forms of neural networks are described in Sect. 4.

2 Granular Approach to Rough-Neural Computing

A brief introduction to a rough-neural computing model based on an adaptive calculus of granules is given in this section. Information granule construction and parameterized approximation spaces provide the foundation for the model of rough-neural computing [44]. A fundamental feature of this model is the design of neurons that engage in knowledge discovery. Mechanically, such neurons return granules (synthesized knowledge) derived from input granules.

2.1 Adaptive Calculus of Granules

To facilitate reasoning about rough neural networks, an *adaptive calculus of granules* has been introduced [40, 44, 48, 49]. The calculus of granules is a system for approximating, combining, describing, measuring, reasoning about, and performing operations on granules by intelligent computing units called agents. In the calculus of granules, the term *information granule* (or *granule*, for short) denotes an assemblage of objects aggregated together by virtue of their indistinguishability, similarity, or functionality. Intuitively, a granule is also called a clump [70]. The term *calculus* comes from G.W. v. Leibniz, who thought of a calculus as an instrument of discovery inasmuch as it provides a system for combining, describing, measuring, reasoning about, and performing operations on objects of interest such as terms in a logical formula in a logical calculus or infinitesimally small quantities in differential calculus [3, 13]. The calculus of classes described by Alfred Tarski [67] shares some of the features found in a calculus of granules. The term *class* is synonymous with *set*, an assemblage of distinct entities, either individually specified or satisfying certain specified conditions [6] (e.g., equivalence class of y consisting of all objects equivalent to y). In measure theory, a class is a set of sets [8]. The element of a class is a subset. It is Georg Cantor's description of how one constructs a set that comes closest to what we have in mind when we speak of a granulation, that is, a set is the result of collecting together certain well-determined objects of our perception or our thinking into a single whole (the objects are called elements of a set) [7]. In a calculus of classes, the kinds of classes (e.g., empty class, universal class), relations between classes (e.g., inclusion, overlap, identify), and operations on classes (∪, ∩, -) are specified. Similarly, a calculus of granules distinguishes among kinds of granules (e.g., elementary granules, set-, concept-, and granule-approximations), relations among granules (e.g., inclusion, overlap, closeness), and operations on granules (e.g., granule approximation, decomposition). It should be observed that in the case of information granules, we cannot use crisp equality in comparing granules. Instead, we are forced to deal with similarity, closeness, and being a part of a whole to a degree, concepts in considering relations between granules.

Calculus of granules includes a number of features not found in the calculus of classes, namely, a system of agents, communication of granules of knowledge between agents, and the construction of granules by agents. To some extent, the new calculus of granules is similar to the agent-based, value-passing calculus of communicating systems proposed by Robin Milner [17, 18]. In Milner's system, an agent is an independent process possessing input and output ports. Agents communicate via channels connecting the output (input) port of one agent with the input (output) port of another agent. Milner's calculus is defined by a tuple $(A, L, Act, X, V, K, J, \varepsilon)$ where A is a set of names; L, a set of labels; Act, a set of actions; X, a set of agent variables; V, a set of values; K, a set of agent constants; J, an indexing set; and ε is a set of agent expressions. This calculus includes a grammar for formulating expressions. Even though adaptivity, granules of knowledge, information granulation, parameterized approximations, and hierarchy of relations of being a part, to a degree (fundamental features of the calculus of granules), are not found in Milner's calculus, it is possible to enrich Milner's system to obtain a variant of the calculus of granules.

The fundamental feature of a granulation system is the exchange of information granules of knowledge between agents by transfer functions induced by rough mereological connectives extracted from information systems. A calculus of granules has been introduced to provide a foundation for the design of information granulation systems. The keystone in such systems is the granularity of knowledge for approximate reasoning by agents [42]. Approximate reasoning on information granules is not only caused by inexactness of information that we have but also by the fact that we can gain efficiency in reasoning if it is enough to deliver approximate solutions, sufficiently close to ideal solutions. An agent is modeled as a computing unit that receives input from its sensors and from other agents, acquires knowledge by discovering (constructing) information granules and by granule approximation, learns (improves its skill in acquiring knowledge), and adapts (adjusts in granulation parameters of predicates in response to changing, for example, sensor measurements and feedback from other agents). For two finite sets $X, Y \subseteq U$ (universe of an information system), we define standard rough inclusion using

$$\mu(X,Y) = \frac{card\,(X \cap Y)}{card(X)} \text{ if } X \text{ is nonempty, and } \mu(X,Y) = 1, \text{ otherwise.} \quad (1)$$

A simple granule of knowledge of type (μ, B, C, tr, tr') has the form (α, α') where μ is the standard rough inclusion, B and C are subsets of A (attributes, that is, sensors, of an information system), and $tr, tr' \in [0,1]$ are thresholds on functions defined with respect to μ such that $\mu([\alpha]_B, [\alpha']_C) \geq tr$ and $\mu([\alpha']_C, [\alpha]_B) \geq tr'$. For example, we assert that $Gr(\mu, B, C, tr, tr', \alpha, \alpha')$ is true when (α, α') is a (μ, B, C, tr, tr') granule of knowledge. There are several sources of adaptivity in the scheme defined by a calculus of granules. First, there is the possibility that changes can be made in parameters μ, B, C, tr, tr' in the granulation predicate $Gr(\mu, B, C, tr, tr', \alpha, \alpha')$ for any agent $ag \in Ag$ (set of agents). Second, new granules can be constructed by any agent in response to a changing environment. Third, new rough inclusion measures can be instituted by an agent by changing, for example, the parameters in a t-norm and an s-norm used in defining μ. The possibility that any agent can make one or more of these changes paves the way toward an adaptive calculus of granules [42]. A recently formulated rough-fuzzy neural network has partially realized this idea with an adaptive threshold relative to a set of real-value attributes without employing rough inclusion [19].

Each agent (neuron) distills its knowledge from granulated (fused) sensor measurements, from granulated signals from other agents, and from approximate reasoning in classifying its acquired granules. An agent communicates its knowledge through channels connected to other agents. An agent (neuron) learns by adjusting accessible parameters in response to feedback from other agents. Let Ag be a nonempty set of agents. In describing the elements of a calculus of granules, we sometimes write U instead of $U(ag)$, for example, where U [and $U(ag)$] denotes a nonempty set of granules (universe) known to agent $ag \in Ag$. Similarly, when it is clear from the context, we sometimes write Inv, St, A, M, L, $link$, O, AP_O, Unc_rel, Unc_rel, H, Dec_rule, lab as a shorthand for $Inv(ag)$, $St(ag)$, $A(ag)$, $M(ag)$, $L(ag)$, $Link(ag)$, $O(ag)$, $AP\_O(ag)$, $Unc\_rel(ag)$, $Unc\_rule(ag)$, $H(ag)$, $Dec\_rule(ag)$, respectively. The calculus of granules establishes a scheme for a distributed system of agents that is characterized by the following tuple:

$$\begin{aligned} Scheme = (&U, Inv, St, Ag, L_{rm}, A, M, L, link, \\ &O, AP\_O, Unc\_rel, Unc\_rel, H, Dec\_rule, lab), \end{aligned} \quad (2)$$

where U denotes a nonempty set of granules (universe) known to agent $ag \in Ag$, Inv denotes an inventory of elementary objects available to ag, St a set of standard objects for ag, Ag a set of agents, L_{rm} a rough mereological logic [12], A an information system of ag, M a pre-model of L_{rm} for ag, L a set of unary predicates of ag, $link$ a string denoting a team of agents communicating objects (input) to an agent for granulation, O a set of operations of an agent, Unc_rel a set of uncertainty relations, H a strategy for producing uncertainty rules from uncertainty relations, Dec_rule a set of granule decomposition rules, and lab a set of labels (one for each agent $ag \in Ag$). The calculus of granules provides a computational framework for designing neural networks in the context of a rough set approach to approximate reasoning and knowledge discovery. The original idea of an open-world model for inductive learning by agents [26] has been enriched by considering a distributed system of agents that stimulate each other by communicating granules of knowledge gleaned from granules received from other agents.

An approximate rough mereology with its own logic L_{rm} (syntax, grammar for its formulas, axioms, and semantics of its models) provides a formal treatment of being a part in a degree. This paves the way toward a study of granule inclusion degree testing and measures of the closeness of granules implemented by cooperating agents [44]. The calculus of granules is considered adaptive to the extent that the construction of information granules by a distributed system of interacting agents will vary in response to variations in the approximate reasoning by agents about their input signals (input granules). Agents usually live and learn inductively in an open system like that described by Pawlak [26]. Let (Inv, Ag) denote a distributed system of agents where Inv denotes an inventory of elementary objects and Ag is a set of intelligent computing units (agents). Let $ag \in Ag$ be an agent endowed with tools for reasoning and communicating with other agents about objects within its scope. These tools are defined by components of the agent label (denoted lab) such that

$$\begin{aligned} lab(ag) = [&\mathbf{A}(ag), \mathrm{M}(ag), L(ag), Link(ag), St(ag), O(ag), \\ &AP\_O(ag), Unc\_rel(ag), Unc\_rule(ag), H(ag), Dec\_rule(ag)], \end{aligned} \tag{3}$$

where

- $\mathbf{A}(ag) = [U(ag), A(ag)]$ is an information system relative to agent ag, where the universe $U(ag)$ is a finite, nonempty set of granules containing elements of the form $(\alpha, [\alpha])$ such that α is a conjunction of descriptors and $[\alpha]$ denotes its meaning in $\mathbf{A}(ag)$ [26]. It is also possible that the objects of $U(ag)$ are complex granules.
- $M(ag) = [U(ag), [0,1], \mu_0(ag)]$ is a premodel of L_{rm} with a rough inclusion $\mu_0(ag)$ on the universe $U(ag)$. The notation L_{rm} denotes a rough mereological logic [42].
- $L(ag)$ is a set of unary predicates (properties of objects) in a predicate calculus interpreted in the set $U(ag)$. Further, formulas of $L(ag)$ are constructed as conditional formulas of logics L_B where $B \subset U(ag)$.
- $Link(ag)$ is a collection of strings of the form $ag_1 ag_2 \ldots ag_k ag$ denoting a team of agents such that $ag_1 ag_2 \ldots ag_k$ are the children of agent ag in the sense that ag can assemble complex objects (constructs) from simpler objects sent by agents $ag_1, ag_2, \ldots, ag_k$ [19].
- $St(ag) = \{st(ag)_1, \ldots, st(ag)_n\} \subset U(ag)$ is the set of standard objects at ag.
- $O(ag) \subseteq \{o \mid o : U(ag_1) \times (ag_2) \times \cdots \times U(ag_k) \to U(ag) \textit{ is operation at } ag\}$.

- $AP\_O(ag)$ is a collection of pairs of the form

$$\langle o(ag,t), \{AS_1[o(ag),in], \ldots, AS_k[o(ag),in], AS[o(ag),out]\}\rangle,$$

where $o(ag,t) \in O(ag)$, k is the arity of $o(ag)$, $t = ag_1, ag_2, \ldots, ag_k \in Link(ag)$, $AS_i[o(ag,t),in]$ is a parameterized approximation space corresponding to the i-th argument of $o(ag,t)$ and $AS[o(ag,t),out]$ is a parameterized approximation space for the output of $o(ag,t)$. The meaning of $o(ag,t)$ is that an agent performs an operation enabling the agent to assemble from objects $x_1 \in U(ag_1)$, $x_2 \in U(ag_2)$, ..., $x_k \in U(ag_k)$ the object $z \in U(ag)$ that is an approximation defined by $AS[o(ag,t),out]$ of $o(ag,t)(y_1,y_2,\ldots,y_k) \in U(ag)$ where y_i is the approximation of x_i defined by $AS_i[o(ag,t),in]$. One may choose here either a lower or an upper approximation. For more details, refer to Chap. 3.
- $Unc\_rel(ag)$ is a set of uncertainty relations unc_rel_i of type

$$\begin{aligned}[o_i(ag,t), \rho_i(ag), ag_1, \ldots, ag_k, ag,\\ \mu_o(ag_1), \ldots, \mu_o(ag_k), \mu_o(ag),\\ st(ag_1)_i, \ldots, st(ag_k)_i, st(ag)_i]\end{aligned} \quad (4)$$

of agent ag where $ag_1, ag_2, \ldots, ag_k \in Link(ag)$, $o_i(ag,t) \in O(ag)$ and ρ_i is such that $\rho_i[(x_1,\varepsilon_1), \ldots, (x_i,\varepsilon_k), (x,\varepsilon)]$ holds for $x \in U(ag)$, $x_1 \in U(ag_1)$, ..., $x_k \in U(ag_k)$, $\varepsilon, \varepsilon_1, \ldots, \varepsilon_k \in [0,1]$ iff $\mu_o[x_j, st(ag_j)_i] \geqslant \varepsilon_j$, $j = 1, \ldots, k$ for the collection of standards $st(ag_1)_i, \ldots, st(ag_k)_i, st(ag)_i$ such that $o_i(ag,t)[st(ag_1)_i, \ldots, st(ag_k)_i] = st(ag)_i$. Values of the operation o are computed in three stages. First, approximations of input objects are constructed. Next, an operation is performed. Finally, the approximation of the result is constructed. A relation unc_rel_i provides a global description of this process. In practice, unc_rel_i is composed of analogous relations corresponding to the three stages. The relation unc_rel_i depends on parameters of approximation spaces. Hence, to obtain satisfactory decomposition (similarly, uncertainty, and so on) rules, it is necessary to search for satisfactory parameters of approximation spaces. This search is analogous to weight tuning in traditional neural computations.
- $Unc\_rule(ag)$ is a set of uncertainty rules unc_rule_i of type,

$$\begin{aligned}&\textbf{if } o_i(ag,t)[st(ag_1)_i, \ldots, st(ag_k)_i] = st(ag)_i \text{ and}\\ &\qquad x_1 \in U(ag_1), \ldots, x_k \in U(ag_k) \text{ satisfy the conditions}\\ &\qquad\quad \mu_o[x_j, st(ag_j)_i] \geqslant \varepsilon(ag_i) \text{ for } i = 1, \ldots, k,\\ &\textbf{then } \mu_o[o_i(ag,t)(x_1, \ldots, x_k), st(ag)_i] \geqslant f_i[\varepsilon(ag_1), \ldots, \varepsilon(ag_k)],\end{aligned} \quad (5)$$

where $ag_1, ag_2, \ldots, ag_k \in Link(ag)$ and $f_i : [0,1]^k \rightarrow [0,1]$ is a so-called rough mereological connective. Uncertainty rules provide functional operators (approximate mereological connectives) for propagating uncertainty measure values from the children of an agent to the agent. The application of uncertainty rules is in negotiation processes where they inform agents about plausible uncertainty bounds.

- $H(ag)$ is a strategy that produces uncertainty rules from uncertainty relations.
- $Dec\_rule(ag)$ is a set of decomposition rules,

$$[\Phi(ag_1),...,\Phi(ag_k),\Phi(ag)], \tag{6}$$

of type $[o_i(ag,t),ag_1,...,ag_k,ag]$ of agent ag, where

$$\Phi(ag_1) \in L(ag_1),...,\Phi(ag_k) \in L(ag_k),\Phi(ag) \in L(ag), \tag{7}$$

$ag_1, ag_2, \ldots, ag_k \in Link(ag)$, and there exists a collection of standards $st(ag_1)_i$, ..., $st(ag_k)_i, st(ag)_i$ such that $o_i(ag,t)[st(ag_1)_i, \ldots, st(ag_k)_i] = st(ag)_i$ and these standards satisfy $\Phi(ag_1), \ldots, \Phi(ag_k), \Phi(ag)$, respectively. Decomposition rules are decomposition schemes, that is, such rules describe the standard $st(ag)_i$ and standards $st(ag_1)_i, \ldots, st(ag_k)_i$ from which the standard $st(ag)_i$ is assembled under o_i relative to predicates that these standards satisfy.

It has been pointed out that there is an analogy between calculi of granules in distributed systems and rough-neural computing [44]:

1. An agent *ag* with input and output ports creating communication links with other agents provides a model for a neuron η (analogously, agent *ag*) with inputs supplied by neurons $\eta_1, \ldots, \eta_k$ (analogously, agents $ag_1, \ldots, ag_k$), responds with output by η, and η is designed with a parameterized family of activation functions represented as rough connectives. In effect, a neuron resembles the model of an agent proposed by Milner [17].
2. Values of rough inclusions are analogous to weights in traditional neural networks.
3. Learning in a system governed by an adaptive calculus of granules is in the form of back propagation where incoming signals are assigned a proper scheme (granule construction) and a proper set of weights in negotiation and cooperation with other neurons.

2.2 Granules in Distributed Systems

In this section, the fulfillment of an ontology of approximate reasoning stems from the consideration of granular computing in the context of parameterized approximation spaces as a realization of an adaptive granule calculus. This realization is made possible by introducing a parameterized approximation space in designing a reasoning system for an agent. A step toward the realization of an adaptive granule calculus in a rough-neural computing scheme is described in this section and is based on [44]. In a scheme for information granule construction in a distributed system of cooperating agents, weights are defined by approximation spaces. In effect, each agent (neuron) in such a scheme controls a local parameterized approximation space.

Let us now consider a definition of a parameterized approximation space. A parameterized approximation space is a system

$$AS_{\#,*,\$} = (U, I_{\#}, R_{*}, \nu_{\$}), \tag{8}$$

where #,*, \$ denote vectors of parameters, U is a nonempty set of objects, and

- $I_{\#} : U \rightarrow \wp(U)$ is an *uncertainty function* where $\wp(U)$ denotes the power set of U; $I_{\#}(x)$ is called the *neighborhood* of $x \in U$;
- $R_{*} \subseteq \wp(U)$ is a family of *parameterized patterns*;
- $\nu_{\$} : \wp(U) \times \wp(U) \rightarrow [0,1]$ denotes *rough inclusion.*

The uncertainty function defines for every object x in U a set of similarly described objects. A constructive definition of an uncertainty function can be based on the assumption that some metrics (distances) are given on attribute values. The family R_{*} describes a set of (parameterized) patterns (e.g., representing, for fixed values of parameters, the sets described by the left-hand sides of decision rules). A set $X \subseteq U$ is definable on $AS_{\#,*,\$}$ if it is a union of some patterns. The rough inclusion function $\nu_{\$}$ defines the value of inclusion between two subsets of U. In particular, for any neighborhood, its inclusion degree in a given pattern can be computed. Moreover, for classifiers, the degree of inclusion of patterns in decision classes can be estimated. The neighborhood $I_{\#}(x)$ can usually be defined as a collection of objects close to x. Also note that for some problems, it is convenient to define an uncertainty set function of the form $I_{\#} : \wp(U) \rightarrow \wp(U)$. This form of uncertainty function works well in signal analysis, where we want to consider a domain over sets of sample signal values.

For a parameterized approximation space $AS_{\#,*,\$}$ and any subset $X \subseteq U$, the lower and upper approximations of X in U based only on an uncertainty function and rough inclusion are defined as follows:

$$LOW\left(AS_{\#,*,\$}, X\right) = \{x \in U \mid \nu_{\$}\left(I_{\#}(x), X\right) = 1\} \text{ [\textit{lower approximation}]}, \tag{9}$$

$$UPP\left(AS_{\#,*,\$}, X\right) = \{x \in U \mid \nu_{\$}\left(I_{\#}(x), X\right) > 0\} \text{ [\textit{upper approximation}]}. \tag{10}$$

However, if one would like to consider the approximation of concepts in an extension U' of U by taking patterns and their inclusion degrees in the concepts, the definition of concept approximation should be changed. The reader can find more details on concept approximations in Chaps. 3, 6, and 25.

Sets of objects that are collections of objects defined by an uncertainty function or patterns from a data table are examples of information granules. A parameterized approximation space can be treated as an analogy to a neural network weight (see Fig. 1). In Fig. 1, $w_1, \ldots, w_n$, Σ, f denote the weights, aggregation operator, and activation function of a classical neuron, respectively, whereas $AS_1(P), \ldots, AS_k(P)$ denote parameterized approximations spaces where agents process input granules $G_1, \ldots, G_k$ and O denotes a (parameterized) operation from a given set of operations that produce the output of a granular network. The parameters in P of an approximation space should be learned to induce the relevant information granules.

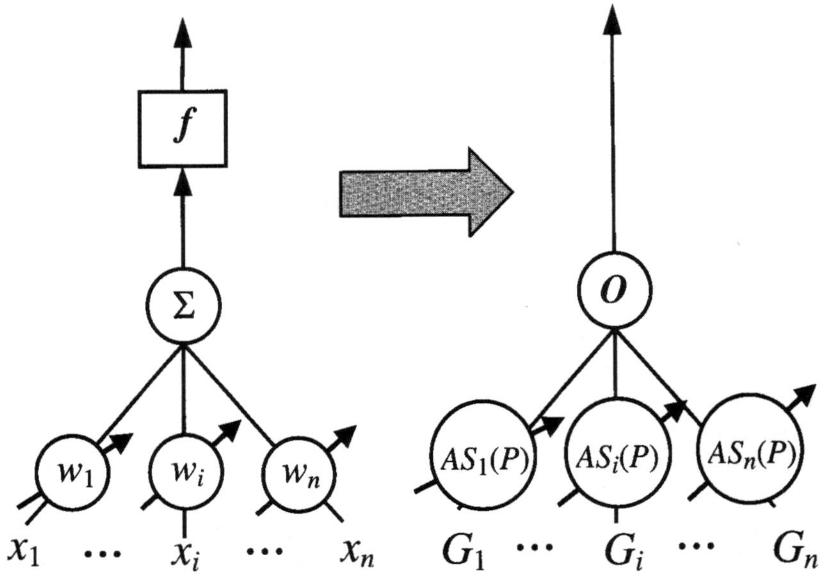

Fig. 1. Comparison of classical and granular network architectures

3 Rough Neurons

The term *rough neuron* was introduced in 1996 by Lingras [15]. In its original form, a rough neuron was defined relative to upper and lower bounds, and inputs were assessed relative to boundary values. Hence, this form of neuron might also be called a boundary value neuron. This form of rough neuron has been used in predicting urban high-traffic volumes [4]. More recent work considers rough-neural networks (RNNs) with neurons that construct rough sets and output the degree of accuracy of an approximation [35, 36]. This has led to the introduction of approximation neurons [36] and their application in classifying electrical power system faults [10], signal analysis [37], and in assessing software quality [29]. An information granulation model of a rough neuron was introduced by Skowron and Stepaniuk in the late 1990s. This model of a rough neuron is inspired by the notion of a cooperating agent (neuron) that constructs granules; perceives by measuring values of available attributes, granule inclusion, granule closeness, and by granule approximation; learns by adjusting parameters in its local parameter space; and shares its knowledge with other agents (neurons). A rough-fuzzy multilayer perceptron (MLP) useful in knowledge encoding and classification was introduced in 1998 by Banerjee, Mitra, and Pal [2]. The study of various forms of rough neurons is part of a growing number of papers on neural networks based on rough sets. Transducers discussed in Chap. 8 transforming rough set arguments into rough sets can also be considered as rough neurons.

3.1 Set Approximation

Rough set theory offers a systematic approach to set approximation [26]. To begin, let $S = (U, A)$ be an information system where U is a nonempty, finite set of objects and A is a nonempty, finite set of attributes, where $a : U \rightarrow V_a$ for every $a \in A$. For each $B \subseteq A$, there is associated an equivalence relation $Ind_A(B)$ such that

$$Ind_A(B) = \{(x, x') \in U^2 \,|\, \forall a \in B.a(x) = a(x')\}. \tag{11}$$

If $(x, x') \in \mathrm{Ind}_A(B)$, we say that objects x and x' are indiscernible from each other relative to attributes from B. The symbol $[x]_B$ denotes the equivalence class of $\mathrm{Ind}_A(B)$ defined by x. Further, partition symbol $\mathrm{U/Ind}_A(B)$ denotes the family of all equivalence classes of relation $\mathrm{Ind}_A(B)$ on U. For $X \subseteq U$, the set X can be approximated only from information contained in B by constructing a B-lower and B-upper approximation denoted by $\underline{B}X$ and $\bar{B}X$, respectively, where

$$\underline{B}X = \{x | [x]_B \subseteq X\} \text{ and } \bar{B}X = \{x | [x]_B \cap X \neq \emptyset\}. \tag{12}$$

3.2 Rough Membership Set Function

In this section, a set function form of the traditional rough membership function introduced in [54] is applied. Let $S = (U, A)$ be an information system, $B \subseteq A$, and let $[u]_B$ be an equivalence class of an object $u \in U$ of $\mathrm{Ind}_A(B)$. A set function $\mu_u^B : \wp(U) \rightarrow [0, 1]$ defined by (13)

$$\mu_u^B(X) = \frac{card(X \cap [u]_B)}{card([u]_B)} \tag{13}$$

for any $X \in \wp(Y)$, $u \in U$, is called a *rough membership function.*

A rough membership function provides a classification measure inasmuch as it tests the degree of overlap between the set X and the equivalence class $[u]_B$. The form of rough membership function presented above is slightly different from the classical definition [27], where the argument of the rough membership function is an object u and the set X is fixed. For example, let $X_{B_{\mathrm{approx}}} \in \{\overline{B}X, \underline{B}X\}$ denote a set approximation. Then, we compute the degree of overlap between $X_{B_{\mathrm{approx}}}$ and $[u]_B$ by

$$\mu_u^B(X_{B_{\mathrm{approx}}}) = \frac{card([u]_B \cap X_{B_{\mathrm{approx}}})}{card([u]_B)}. \tag{14}$$

In the sequel, we also write $\mu_{u,B}(X_{B_{\mathrm{approx}}})$ instead of $\mu_u^B(X_{B_{\mathrm{approx}}})$.

3.3 Decision Rules

In deriving decision system rules, the discernibility matrix and discernibility function are essential. Given an information system $S = (U, A)$ with n objects, the $n \times n$ matrix (c_{ij}), called the discernibility matrix of S [denoted $M(S)$], is defined as

$$c_{ij} = \{a \in A \mid a(x_i) \neq a(x_j)\}, \text{ for } i, j = 1, \ldots, n. \tag{15}$$

A discernibility function $f_{M(S)}$ for the system S is a Boolean function of m Boolean variables $a_1^*, \ldots, a_m^*$ corresponding to attributes $a_1, \ldots, a_m$, respectively, and defined by

$$f_{M(S)}(a_1^*, \ldots, a_m^*) = \wedge\{c_{ij}^* \mid 1 \leq j < i \leq n, c_{ij} \neq \emptyset\} \text{ where } c_{ij}^* = \{a^* \mid a \in c_{ij}\}. \tag{16}$$

Precise conditions for decision rules can be extracted from a discernibility matrix as in [43, 47]. For the information system $S = (U, A)$, let $B \subseteq A$ and let $\wp(V_a)$ denote the power set of V_a, where V_a is the value set of a. For every $\delta \in A - B$, a decision function $d_\delta^B : U \to \wp(V_\delta)$ is defined in (17) as in [56]:

$$d_\delta^B(u) = \{v \in V_\delta \mid \exists u' \in U, (u', u) \in Ind_B \text{ and } \delta(u') = v\}. \tag{17}$$

In other words, $d_\delta^B(u)$ is the set of all elements of the decision column δ of S such that the corresponding object is a member of the same equivalence class as argument u. The next step is to determine a decision rule with a minimal number of descriptors on the left-hand side. Pairs (a, v), where $a \in A, v \in V$, are called *descriptors*. A decision rule over the set of attributes A and values V is an expression of the following form:

$$a_{i_1}(u_i) = v_{i_1} \wedge \ldots \wedge a_{i_j}(u_i) = v_{i_j} \wedge \ldots \wedge a_{i_r}(u_i) = v_{i_r} \underset{S}{\Rightarrow} d(u_i) = v, \tag{18}$$

where $u_i \in U$, $v_{i_j} \in V_{a_{i_j}}$, $v \in V_d$, $j = 1, \ldots, r$ and $r \leq card(A)$. The fact that a rule is true is indicated by writing it in the following form:

$$(a_{i_1} = v_{i_1}) \wedge \ldots \wedge (a_{i_r} = v_{i_r}) \underset{S}{\Rightarrow} (a_p = v_p). \tag{19}$$

In practice also are important rules that are true in S to the degree in which the set defined in S by the left-hand side of the rule is included in the set defined in S by the right-hand side of the rule. The left- and right-hand sides of rules are information granules in S. Then the degree mentioned above can be interpreted as the degree of inclusion of such information granules. The decision rules can also be treated as information granules (see Chap. 3).

Let $RED(S)$ be a reduct set generated from a decision system S, e.g., a set of local reducts with respect to objects [12]. [1] For decision system S, the set of decision

[1] Note that there are many different kinds of reducts and methods of selection of relevant reducts used in constructing data description models (see Chap. 25).

rules constructed with respect to a reduct $R \in RED(S)$ is denoted by $OPT(S,R)$. Then the set $OPT(S)$ of all decision rules derivable from reducts in $RED(S)$ is the following set:

$$OPT(S) = \cup\{OPT(S,R) \mid R \in RED(S)\}. \tag{20}$$

3.4 Interval-Based Rough Neuron

An *interval-based rough neuron* was introduced in 1996 [15]. A brief introduction to this form of rough neuron is given in this section. Rough neurons are defined in the context of rough patterns. Objects such as a fault signal or daily weather can be described by a finite set of features (e.g., amplitude, type of waveform, high-frequency component, rainfall, temperature) characterizing each object. The description of an object is an n-dimensional vector, where n is the number of features used to characterize an object. A pattern is a class of objects based on the values of some features of objects belonging to the class.

Let x be a feature variable in the description of an object. Further, let $\underline{x}, \overline{x}$ represent upper and lower bounds of x. In a rough pattern, the value of each feature variable x is specified by $\underline{x}, \overline{x}$ (called rough values). Rough values are useful in representing an interval or set of values for a feature, where only the upper and lower bounds are considered relevant in a computation. This form of rough neuron can be used to process intervals in a neural network.

Let $r, \underline{r}, \overline{r}$ denote a rough neuron, lower neuron, and upper neuron, respectively. A rough neuron is a pair $(\underline{r}, \overline{r})$ with three types of connections: i/o connections to $\underline{r}$, i/o connections to $\overline{r}$, and connections between $\underline{r}$ and $\overline{r}$. In effect, a rough neuron stores the upper and lower bounds of input values for a feature and uses these bounds in its computations. Let in_i, out_j, w_{ij} denote the input to neuron i, the output from neuron j, and the strength of the connection between neurons i and j, respectively. The input to an upper, lower, or conventional neuron i is calculated as a weighted sum as

$$in_i = \sum_{j=1}^{n} w_{ij} out_j \quad \text{(neuron} j \text{ is connected to neuron } i). \tag{21}$$

Assuming the subscript $i = \underline{r}$, we obtain the input to a lower neuron, and for $i = \overline{r}$, we obtain the input to an upper neuron. Let t be a transfer function used to evaluate the input to an upper (lower) neuron. Then the output of an upper (lower) neuron is computed as in (22) and (23), respectively:

$$out_{\overline{r}} = \max\left[t\left(in_{\overline{r}}\right), t\left(in_{\underline{r}}\right)\right]; \tag{22}$$

$$out_{\underline{r}} = \min\left[t\left(in_{\overline{r}}\right), t\left(in_{\underline{r}}\right)\right]. \tag{23}$$

The output of a rough neuron will be computed from

$$rough\_neuron\_output = \frac{out_{\overline{r}} - out_{\underline{r}}}{average(out_{\overline{r}}, out_{\underline{r}})}. \tag{24}$$

The inputs to rough neurons considered in [15] are related to deviations in measurements of some attribute value. One can consider another case when deviations of a real function defined, e.g., on the lower approximation of a given set X, are used to define inputs to neurons. Another possibility to consider is deviations of rough membership function values on elements of a tolerance class (see Sect. 3.5).

3.5 Approximation Neurons

This section considers the design of rough neural networks based on set approximations and rough membership functions, and hence, this form of network is called an *approximation neuron* (AN). The approximation neuron was introduced in [9], and elaborated in [35, 36]. Preliminary computations in an AN are carried out with a layer of approximation neurons, which construct rough sets and where the output of each approximation neuron is computed with a rough membership function. This section considers ANs constructed with one type of rough neuron: the approximation neuron. Let $B, F, F_{B_{\text{approx}}}$, $\lfloor f \rfloor_B$ denote a set of attributes, a finite set of neuron inputs (this is an archival set representing past stimuli, a form of memory accessible to a neuron), a set approximation, and an equivalence class containing measurements derived from known objects, respectively. The basic computation steps performed by an approximation neuron are illustrated in Fig. 2.

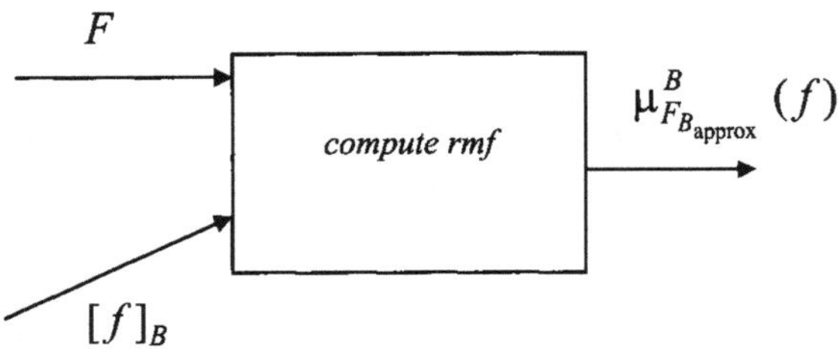

Fig. 2. Approximation neuron

The approximation neuron measures the degree of overlap of a set $\lfloor f \rfloor_B$ and $F_{B_{\text{approx}}}$. Let us consider a more general case when instead of an indiscernibility class $\lfloor f \rfloor_B$, an input is defined by a more general information granule (see Chap. 3), i.e., τ-tolerance class of $\lfloor f \rfloor_B$, (a family $\{\lfloor f' \rfloor_B : f \tau f'\}$). The output of a neuron is defined by two numbers representing the deviation of the rough membership function on elements of the tolerance class. Other forms of rough neurons are described in [37].

3.6 Decider Neuron

The notion of a *decider neuron* was introduced in [35, 36] and applied in [37]. A decider neuron implements a collection of decision rules by (i) constructing a condition vector $c_{\exp}$ from its inputs, which are rough membership function values, (ii) discovering the rule $c_i \Longrightarrow d_i$ with a condition vector c_i that most closely matches an input condition vector $c_{\exp}$, and (iii) outputs $AND(1-e_i, d_i)$ where $d_i \in \{0,1\}$, and relative error $e_i = \|c_{\exp} - c_i\| / \|c_i\| \in [0,1]$ where $\|\cdot\|$ denotes the vector length function. When $e_i = 0$, then $y_{\text{rule}} = AND(1-e_i, d_i) = d_i$, and the classification is successful. If $e_i = 1$, then $y_{\text{rule}} = AND(1-e_i, d_i) = 0$ indicates the relative error in an unsuccessful classification. A flow graph showing the basic computations performed by a decider neuron is given in Fig. 3.

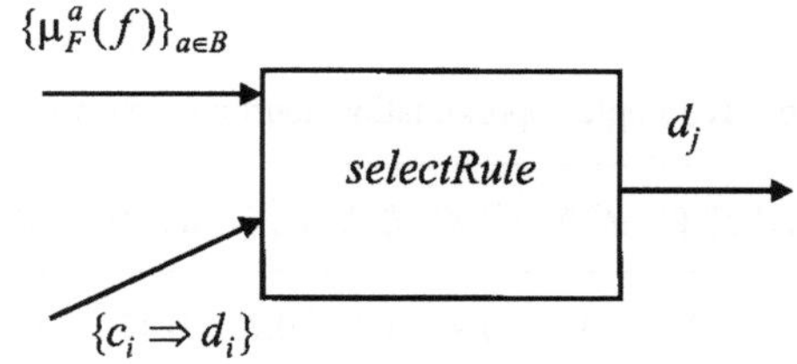

Fig. 3. Flow graph for decider neuron

The set $rmf = \{\mu_F^a(f)\}_{a\in B}$ consists of approximation neuron measurements in response to the stimulus provided a new object f requiring classification. The elements of the set rmf are used by a decider neuron to construct an experimental condition vector $c_{\exp}$. A second input to a decider neuron is the set $R = \{c_i \Longrightarrow d_i\}$. The elements of the set R are rules that have been derived from a decision table using rough set theory. Let *selectRule* denote a process that implements an algorithm to identify a condition vector in one of the rules of R, that most closely matches $c_{\exp}$.

Let us observe that one of the input of decider neuron is an information granule represented by a set of decision rules. The rough neuron considered is used to measure the degree of closeness of object f to an information granule represented by the set of decision rules.

It is worthwhile mentioning that the case considered is very simple. Instead of a rough membership function, computed relative to attributes from B, one should consider a relevant family $C_1,\ldots,C_k$ of subsets of B. For classifiers based on decision rules, such subsets are defined by reducts local with respect to objects [12]. [2]

3.7 Architecture of Approximation Neural Networks

Approximation neural networks (ANN) are well suited to solving classification problems where nuances in a feature space over time can be gleaned from rough approximations. This form of rough neural computation has been successfully applied

[2] In the more general case, one can consider, instead of *rmf* inclusion, degrees of input information granules into the information granules representing relevant patterns for decision classes (see Chaps. 3 and 25).

in two applications: classifying the waveforms of electrical power system faults ([9], [10, 37] and in determining the number of changes required in a software system based on quality measurements [31]. In this section, a brief summary of the results of the software quality study are given.

Sample computations with a set of 11 approximation neurons are given in Table 1. The first 11 columns of Table 1 are rough membership function outputs, and the last column is a target value for the aggregate of the rmf values. The target column in Table 1 serves as a decision column.

Table 1. Sample approximation neuron output values

$\mu_u^{a_1}(X)$	$\mu_u^{a_2}(X)$	$\mu_u^{a_3}(X)$	$\mu_u^{a_4}(X)$	$\mu_u^{a_5}(X)$	$\mu_u^{a_6}(X)$	$\mu_u^{a_7}(X)$	$\mu_u^{a_8}(X)$	$\mu_u^{a_9}(X)$	$\mu_u^{a_{10}}(X)$	$\mu_u^{a_{11}}(X)$	**Target**
1	1	1	1	1	0.9	0.9	0.72	0.47	0.66	1	**1**
1	1	1	1	1	0.9	0.9	0.72	0.4	0.66	1	**1**
						...					
1	1	1	1	1	0.9	0.927	0.72	0.47	0.66	1	**1**
0.68	0.89	0.74	0.78	0	0	0	0	0.57	0	0.23	**0**

During calibration of the ANN, adjustments are made to the weights (strengths of connections) associated with approximation neurons relative to a probabilistic sum of the rmf values and target value. For each test input, the output of an ANN indicates the probability that the test input belongs to the set of measurements associated with the ith life cycle product attribute. Let Σ denote an output neuron (see Fig. 4) that computes the sum of all weighted outputs from the approximation neurons in the first layer. A software metric (e.g., reusability, complexity, maintainability [6]) is used to define each life cycle product quality attribute.

The application of the rough mereological approach in software quality measurement is described in [38]. The basic idea is to characterize the global quality assessment of a given software product using a vector of rough mereological distances (cf. rows of Table 1) of quality measurements relative to a chosen set of industry standards. Sample calibrations of a rough membership function neural network are shown in Figs. 5 and 6. In the case considered, constructed information granules are degrees of inclusion of patterns in decision classes. They are computed as linear combinations of rough membership degrees of a given input u in the set X.

3.8 Architecture of Approximation-Decider Neural Network

The output of the approximation neural network in Fig. 4 serves as a module in testing the quality measurements of a particular life cycle product relative to a particu-

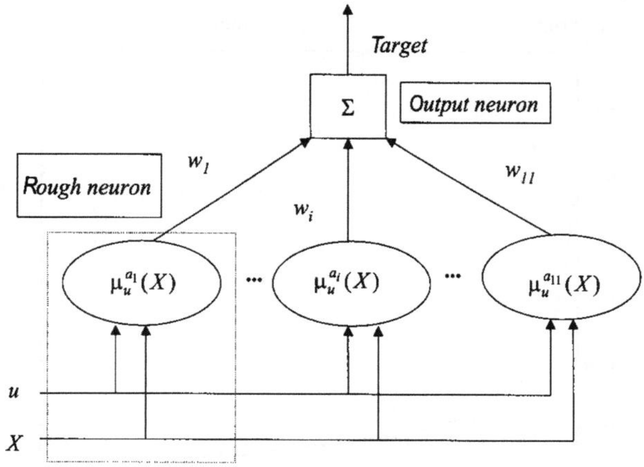

Fig. 4. Approximation neural network

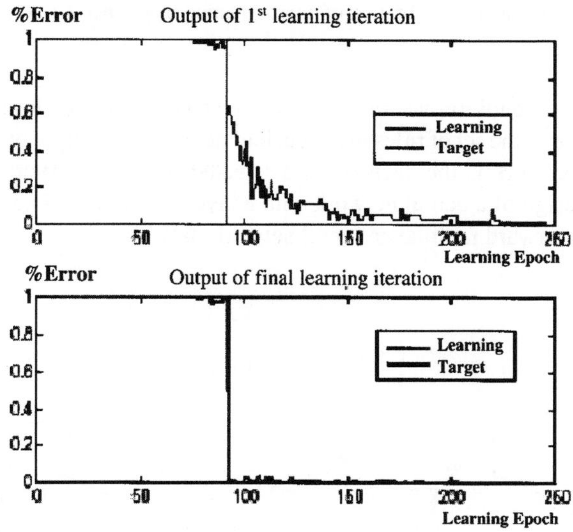

Fig. 5. Sample calibration of approximation neural network

lar number of changes required to correct product quality deficiencies. For the study of the quality of life cycle products, 17 such approximation neural networks were constructed. The outputs of the k ($k = 17$) ANNs form a condition vector of the form $[e_1\ e_2 \ldots e_k]$, where e_i denotes an experimental approximation neuron output (rough membership function) value. During training, a test set of condition vectors and corresponding decisions (specified number of changes decided on for each condition vector) was constructed. A set of decision rules is then derived and incorporated in a neural network output neuron that is called a decider neuron (see Fig. 7). During

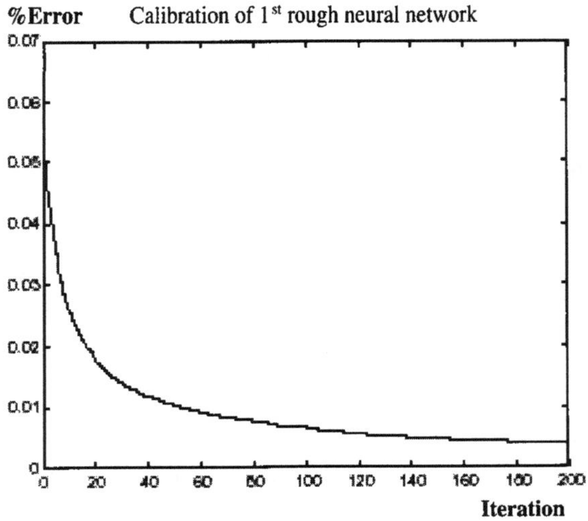

Fig. 6. Sample calibration of approximation neural network (cd.)

testing, each test condition vector $[t_1 \ldots t_k]$ is matched with the closest training set condition vector in the set of rules in the decider neuron. The output of the composite rough-neural network is the decision of a selected rule in the decider neuron (see Fig. 8). The design of a neural network with a layer of rule-based neurons has been used in a feed-forward multilayer neural network [64].

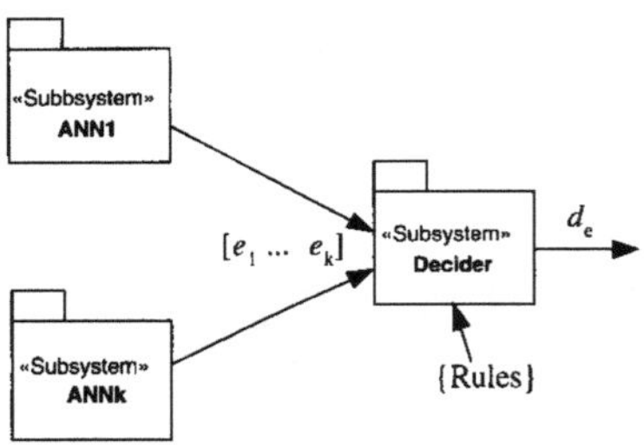

Fig. 7. Training network

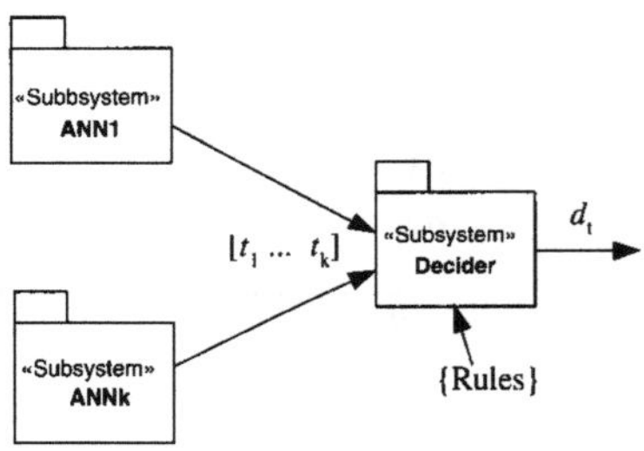

Fig. 8. Testing network

For the experiment described in this section, 17 approximation neural network modules (a total of 187 approximation neurons with a structure like that in Fig. 4) were incorporated into the design of a composite neural network with a single neuron in the output layer, namely, a decider neuron. After calibration of the subnetworks ($ANN_1, \ldots, ANN_k$), formation of a decision table, and derivation of decision rules, the result is the training network in Fig. 7. During testing, the network in Fig. 8 is used to find a rule with a condition vector with the best match to a rule residing in the decider neuron. As a result, we obtain a basis for making an approximate decision about the number of changes required for the particular life cycle product being evaluated. The results of sample training sessions are shown in Fig. 9.

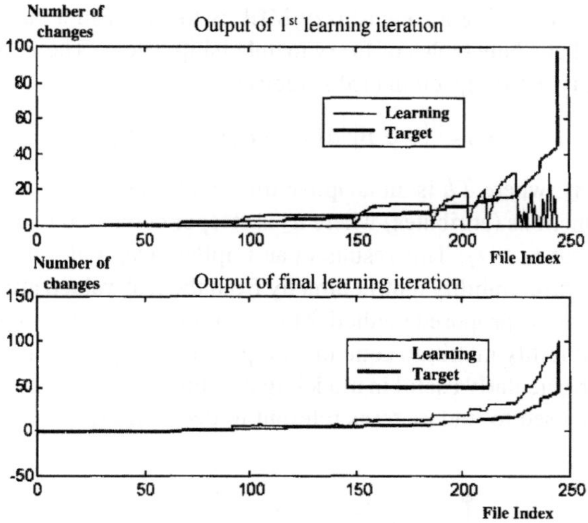

Fig. 9. RNN performance

4 Hybrid Neural Networks

A number of hybrid neural networks with architectural designs based on rough set theory and more traditional neural structures have been proposed: *rough-fuzzy MLP* [19], *evolutionary rough-fuzzy MLP* [24], *interval-based rough-fuzzy networks* [15], and *approximation rough-fuzzy networks* [37]. It should also be mentioned that it is common to use rough set theory as the basis for preprocessing inputs to a neural network (see Chap. 25). In this section, two recent forms of hybrid networks are briefly described: a rough-fuzzy multilayer perceptron neural network [19] and a rough-fuzzy approximation neural network [29]. In rough-fuzzy neural networks, some patterns are extracted using rough set methods. Next, such patterns can be fused using fuzzy logic rather than classical propositional connectives.

4.1 Rough-Fuzzy MLP

The rough-fuzzy multilayer perceptron neural network introduced in [19], was developed for pattern classification. This form of MLP combines both rough sets and fuzzy sets with neural networks for building an efficient connectionist system. In this hybridization, fuzzy sets help in handling linguistic input information and ambiguity in output decision, whereas rough sets extract domain knowledge for determining network parameters.

4.2 Architecture of Rough-Fuzzy MLP Network

The first step in designing a rough-fuzzy MLP is to establish a basis for working with real-valued attribute tables of fuzzy membership values. The traditional model of a discernibility matrix given in (15) is replaced by

$$c_{ij} = \{a \in B | \, |a(x_i) - a(x_j)| > Th\} \tag{25}$$

for $i, j = 1, ..., n_k$, where Th is an adaptive threshold. Let a_1, a_2 correspond to two membership functions (attributes) where a_2 is steeper compared to a_1 (see Fig. 10). It is observed that $r_1 > r_2$. This results in an implicit adaptivity of Th while computing c_{ij} in the discernibility matrix directly from the real-valued attributes. Herein lies the novelty of the proposed method. Moreover, this type of thresholding also enables the discernibility matrix to contain all representative points/clusters present in a class. This is particularly useful in modeling multimodal class distributions. Rough set methods are used to find patterns relevant to decision classes. While designing

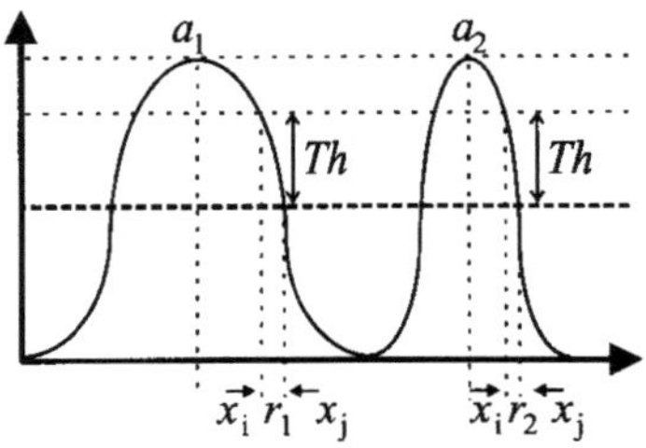

Fig. 10. Illustration of adaptive thresholding of membership functions

the initial structure of the rough-fuzzy MLP, the union of the rules of l classes is considered. The input layer consists of $3n$ attribute values (it is assumed that each attribute can have three fuzzy values: *low, medium, high*) whereas the output layer is represented by l classes. The hidden layer nodes model the first level (innermost) operator in the antecedent part of a rule, which can be either a conjunct or a disjunct. The output layer nodes model the outer level operands, which can again be either a conjunct or a disjunct. For each inner level operator, corresponding to one output class (one dependency rule), one hidden node is dedicated. Only those input attributes that appear in this conjunct/disjunct are connected to the appropriate hidden node, which in turn is connected to the corresponding output node. Each outer level operator is modeled at the output layer by joining the corresponding hidden nodes. Note that a single attribute (involving no inner level operators) is directly connected to the appropriate output node via a hidden node, to maintain uniformity in rule mapping.

Modular Training A method of learning the parameters of rough-fuzzy MLP has recently been described [24] using the modular concept that is based on the divide and conquer strategy. This provides accelerated training and a compact network suitable for generating a minimum number of classification rules with high certainty values. A new concept of a variable genetic mutation operator is introduced for preserving the localized structure of the constitutive knowledge-based subnetworks while they are integrated and evolved.

4.3 Rough-Fuzzy Approximation Neural Network

A rough-fuzzy approximation neural network gains its name from the fact that it contains neurons designed using rough set theory connected to various forms of neurons designed using fuzzy set theory (see, e.g., Fig. 11).

4.4 Architecture of a Sample Rough-Fuzzy Neural Network

The sample rough-fuzzy neural network described in this section consists of four layers (see Fig. 11). The first layer of the network in Fig. 11 contains approximation neurons connected to very basic fuzzy neurons (layer 2) that compute the degree of membership of rough neuron outputs in various distributions. Layer 2 neurons are connected to AND neurons (also called logic neurons [29]). Let x be an approxima-

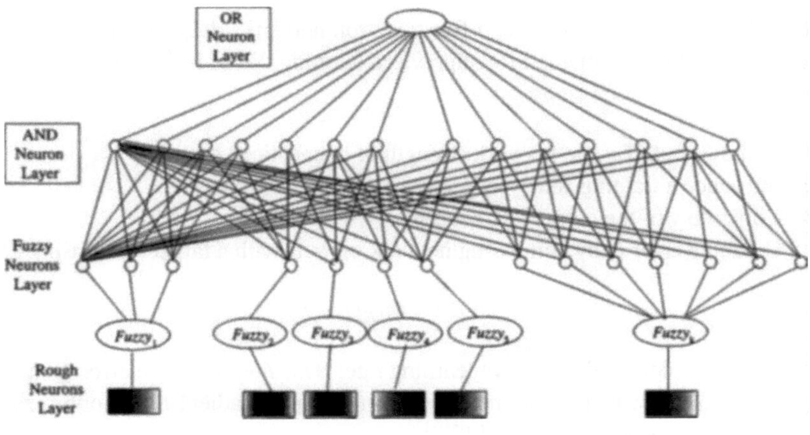

Fig. 11. Rough-fuzzy neural network

tion neuron output, and let $f(x)$ be the degree of membership of x in a particular distribution. Then defined parameters r and w denote a cutoff (reference point) and strength of connection (weight), respectively. Then, we define a fuzzy implication $r \to x$ by

$$(r \to x) = \begin{cases} \frac{x}{r} & \text{if } x < r \\ \min\left(1, \frac{x}{r}\right) & \text{otherwise.} \end{cases}$$

Let t ("and" usually interpreted as *min*) and s ("or" interpreted as a probabilistic *sum* here) denote t-norm and s-norm operators from fuzzy set theory [29]. An AND neuron has output z defined as

$$z = \mathop{T}_{i=1}^{n} [x_i s w_i] = \min[x_1 s w_1, \ldots, x_n s w_n].$$

The model for an AND neuron is specialized relative to fuzzy implication as

$$z = \min_{i=1}^{n} \{[r_i \to f(x)_i] + w_i - [r_i \to f(x)_i] w_i\}.$$

The output layer of the network in Fig. 11 consists of OR neurons that aggregate the information gleaned from connections to AND neurons. The model for an OR neuron is

$$y = \mathop{S}_{i=1}^{n} [z_i t u_i] = (z_1 t u_1) s \ldots s (z_n t u_n).$$

In the model for an OR neuron, u_i denotes the strength of connection between the OR neuron and the ith AND neuron.

4.5 Supervised Learning Approach to Calibration of Rough-Fuzzy Neural Network

The calibration scheme for rough-fuzzy neuron networks described in this section employs the standard method for supervised learning. What follows is a brief summary of the calibration steps:

1. Initialize cutoff r and network strength of connections w and u.
2. Introduce a training set.
3. Compute y of the output OR neuron.
4. Compute error Q by comparing network outputs with a target value using (26).

$$Q = target - y. \tag{26}$$

5. Let $\alpha > 0$ denote the positive learning rate. Based on the value error Q in (26), adjust the r, w, and u parameters using the usual gradient-based optimization method suggested in (27) and (28):

$$param(new) = param - \alpha \frac{\partial Q}{\partial param}, \tag{27}$$

$$\frac{\partial Q}{\partial param} = \frac{\partial Q}{\partial y} \frac{\partial y}{\partial param}. \tag{28}$$

A more detailed explanation of how one trains a network containing combinations of AND and OR nodes is given in [31].

5 Concluding Remarks

A scheme for designing rough-neural networks based on an adaptive calculus of granules for distributed systems of cooperating agents has been presented. This scheme is defined in the context of an approximate rough mereology, granule construction and granule approximation algorithms, measures of granule inclusion and closeness, and local parameterized approximation spaces. Adaptivity is also a feature of this scheme, where agents can change local parameters in response to changing signals from other agents and from the environment. A number of models of rough neurons have been proposed. Four such models have been briefly described in this chapter: interval-based neurons, approximation neurons, decider neurons, and rough-fuzzy MLPs. Four rough-neural computing architectures have also been briefly considered: approximation neural network, approximation-decider neural network, rough-fuzzy neural network, and rough-fuzzy MLP network.

Acknowledgments

The research of James Peters has been supported by the Natural Sciences and Engineering Research Council of Canada (NSERC) research grant 185986. Lech Polkowski was supported by a grant from the State Committee for Scientific Research of the Republic of Poland (KBN), No. 8T11C02417. The research of Andrzej Skowron has been supported by the State Committee for Scientific Research of the Republic of Poland (KBN), research grant No. 8T11C02519 and by a Wallenberg Foundation grant.

References

1. M.A. Arbib. The artificial neuron. In E. Fiesler, R. Beale, editors, *Handbook of Neural Computation*, B1.1–B1.7, Institute of Physics Publishing, Bristol, 1997.
2. M. Banerjee, S. Mitra, S.K. Pal. Rough fuzzy MLP: Knowledge encoding and classification. *IEEE Transactions on Neural Networks*, 9(6): 1203–1216, 1998.
3. I.M. Bocheński. *A History of Formal Logic*. Chelsea, New York, 1956.
4. E. Bonabeau, M. Dongo, G. Theraulaz. *Swarm Intelligence: From Natural to Artificial Systems*. Oxford University Press, Oxford, 2000.
5. B. Chakraborty. Feature subset selection by neuro-rough hybridization. In *[74]*, 519–572, 2001.
6. N.E. Fenton, S.L. Pfleeger. *Software Metrics: A Rigorous & Practical Approach*. PWS, Boston, 1997.
7. W. Gellert, H. Kustner, M. Hellwich, H. Kastner. *The VNR Concise Encyclopedia of Mathematics*. Van Nostrand, London, 1975.
8. P.R. Halmos. *Measure Theory*. Van Nostrand, London, 1950.
9. L. Han, R. Menzies, J.F. Peters, L. Crowe. High voltage power fault–detection and analysis system: Design and implementation. In *Proceedings of the Canadian Conference on Electrical & Computer Engineering (CCECE'99)*, 1253–1258, Edmonton, 1999.
10. L. Han, J.F. Peters, S. Ramanna, R. Zhai. Classifying faults in high voltage power systems: A rough-fuzzy neural computational approach. In *[73]*, 47–54, 1999.

11. T. Kohonen. The self-organizing map. In: *Proceedings IEEE*, 78: 1464–1480, 1990.
12. J. Komorowski, Z. Pawlak, L. Polkowski, A. Skowron. Rough sets: A tutorial. In *[25]*, 3–98, 1999.
13. G.W. Leibniz. In L. Couturat, editor, *Opuscles et Fragments Inedits de Leibniz*, 256, Félix Alcan, Paris, 1903.
14. P.J. Lingras. Fuzzy-rough and rough-fuzzy serial combinations in neurocomputing. *Neurocomputing*, 36: 29-44, 2001.
15. P.J. Lingras. Rough neural networks. In *Proceedings of the 6th International Conference on Information Processing and Management of Uncertainty (IPMU'96)*, 1445–1450, Universidad da Granada, Granada, 1996.
16. P.J. Lingras. Comparison of neofuzzy and rough neural networks. *Information Sciences. An International Journal*, 110: 207–215, 1998.
17. R. Milner. *Communication and Concurrency*. Prentice-Hall, Upper Saddle River, NJ, 1989.
18. R. Milner. *Calculus of Communicating Systems*. Report number ECS-LFCS-86-7 of Computer Science Department, University of Edinburgh, 1986.
19. S. Mitra, P. Mitra, S.K. Pal. Evolutionary modular design of rough knowledge-based network with fuzzy attributes. *Neurocomputing: An International Journal*, 36: 45–66, 2001.
20. H.S. Nguyen, A. Skowron, M.S. Szczuka. Situation identification by unmanned aerial vehicle. In *[74]*, 49–56, 2001.
21. H.S. Nguyen, M. Szczuka, D. Ślęzak. Neural networks design: Rough set approach to real-valued data. In *Proceedings of the 1st European Conference on Principles and Practice of Knowledge Discovery in Databases (PKDD'97)*, LNAI 1263, 359–366, Springer, Berlin, 1997.
22. T. Nguyen, R.W. Swiniarski, A. Skowron, J. Bazan, K. Thagarajan. Applications of rough sets, neural networks and maximum likelihood for texture classification based on singular decomposition. In *Proceedings of the 3rd International Workshop on Rough Sets and Soft Computing (RSSC'94)*, 332–339, San Jose, CA, 1994.
23. S.K. Pal, S. Mitra. Multi-layer perceptron, fuzzy sets and classification. *IEEE Transactions on Neural Networks*, 3: 683–697, 1992.
24. S.K. Pal, P. Mitra. Rough Fuzzy MLP: Modular evolution, rule generation and evaluation. *IEEE Transactions on Knowledge and Data Engineering* (in press).
25. S.K. Pal, A. Skowron, editors. *Rough-Fuzzy Hybridization: A New Trend in Decision Making*. Springer, Singapore, 1999.
26. Z. Pawlak. *Rough Sets: Theoretical Aspects of Reasoning about Data*. Kluwer, Dordrecht, 1991.
27. Z. Pawlak, A. Skowron. Rough membership functions. In R. Yager, M. Fedrizzi, J. Kacprzyk, editors, *Advances in the Dempster–Shafer Theory of Evidence*, 251–271, Wiley, New York, 1994.
28. W. Pedrycz, F. Gomide. *An Introduction to Fuzzy Sets: Analysis and Design*. MIT Press, Cambridge, MA, 1998.
29. W. Pedrycz, L. Han, J.F. Peters, S. Ramanna, R. Zhai. Calibration of software quality: Fuzzy neural and rough neural computing approaches. *Neurocomputing: An International Journal*, 36: 149–170, 2001.
30. Z. Pawlak, J.F. Peters, A. Skowron, Z. Suraj, S. Ramanna, M. Borkowski. Rough measures and integrals: A brief introduction. In *[72]*, 375–379, 2001.
31. W. Pedrycz, J.F. Peters. Learning in fuzzy Petri nets. In J. Cardoso, H. Scarpelli, editors. *Fuzziness in Petri Nets*, 858–886, Physica, Heidelberg, 1998.

32. J.F. Peters, A. Skowron, J. Stepaniuk. Rough granules in spatial reasoning. In *Proceedings of the Joint 9th International Fuzzy Systems Association (IFSA) World Congress and 20th North American Fuzzy Information Processing Society (NAFIPS) International Conference*, 1355–1361, Vancouver, BC, 2001.
33. J.F. Peters, S. Ramanna. A rough set approach to assessing software quality: Concepts and rough Petri net model. In *[25]*, 349–380, 1999.
34. J.F. Peters, W. Pedrycz. *Software Engineering: An Engineering Approach.* Wiley, New York, 2000.
35. J.F. Peters, A. Skowron, Z. Suraj, L. Han, S. Ramanna. Design of rough neurons: Rough set foundation and Petri net model. In *Proceedings of the International Symposium on Methodologies for Intelligent Systems (ISMIS 2000)*, LNAI 1932, 283–291, Springer, Berlin, 2000.
36. J.F. Peters, A. Skowron, L. Han, S. Ramanna. Towards rough neural computing based on rough membership functions: Theory and application. In *[74]*, 604–611, 2001.
37. J.F. Peters, L. Han, S. Ramanna. Rough neural computing in signal analysis. *Computational Intelligence*, 1(3): 493–513, 2001.
38. L. Polkowski, A. Skowron. Approximate reasoning about complex objects in distributed systems: Rough mereological formalization. In W. Pedrycz, J.F. Peters, editors, *Computational Intelligence in Software Engineering. Advances in Fuzzy Systems-Applications and Theory 16*, 237–267, World Scientific, Singapore, 1998.
39. L. Polkowski, A. Skowron. Rough mereology: A new paradigm for approximate reasoning. *International Journal Approximate Reasoning*, 15(4): 333–365, 1996.
40. L. Polkowski, A. Skowron. Calculi of granules based on rough set theory: Approximate distributed synthesis and granular semantics for computing with words. In *[73]*, 20–28, 1999.
41. L. Polkowski, A. Skowron. Rough-neuro computing. In *[74]*, 57–64, 2001.
42. L. Polkowski, A. Skowron. Towards adaptive calculus of granules. In *Proceedings of the 6th International Conference on Fuzzy Systems (FUZZ-IEEE'98)*, 111–116, Anchorage AK, 1998.
43. A. Skowron, C. Rauszer. The discernibility matrices and functions in information systems. In R. Słowiński, editor, *Intelligent Decision Support: Handbook of Applications and Advances of the Rough Sets Theory*, 331-362, Kluwer, Dordrecht, 1992.
44. A. Skowron. Toward intelligent systems: Calculi of information granules. *Bulletin of the International Rough Set Society*, 5(1/2):9–30, 2001.
45. A. Skowron. Approximate reasoning by agents in distributed environments. In N. Zhong, J. Liu, S. Ohsuga, J. Bradshaw, editors, *Intelligent Agent Technology: Research and Development. Proceedings of the 2nd Asia-Pacific Conference on IAT (APCIAT 2001)*, 28–39, World Scientific, Singapore, 2001.
46. A. Skowron. Approximate reasoning by agents. In *Proceedings of the 2nd International Workshop of Central and Eastern Europe on Multi-Agent Systems (CEEMAS 2001)*, LNAI 2296, 3–14, Springer, Berlin, 2002.
47. A. Skowron, J. Stepaniuk. Decision rules based on discernibility matrices and decision matrices. In *Proceedings of the 3rd International Workshop on Rough Sets and Soft Computing (RSSC'94)*, 602–609, San Jose, CA, 1994.
48. A. Skowron, J. Stepaniuk. Information granules in distributed environment. In *[73]*, 357–365, 2001.
49. A. Skowron, J. Stepaniuk, S. Tsumoto. Towards discovery of information granules. In *Proceedings of the 3rd European Conference on Principles and Practice of Knowledge Discovery in Databases (PKDD'99)*, LNAI 1704, 542–547, Springer, Berlin, 1999.

50. A. Skowron, J. Stepaniuk. Tolerance approximation spaces. *Fundamenta Informaticae*, 27: 245–253, 1996.
51. A. Skowron, J. Stepaniuk. Information granules and approximation spaces. In *Proceedings of the 7th International Conference on Information Processing and Management of Uncertainty in Knowledge-Based Systems (IPMU'98)*, 1354–1361, Paris, 1998.
52. A. Skowron, J. Stepaniuk. Information granules: Towards foundations of granular computing. *International Journal of Intelligent Systems*, 16(1): 57–104, 2001.
53. A. Skowron, J. Stepaniuk. Information granule decomposition. *Fundamenta Informaticae*, 47(3/4): 337–350, 2001.
54. A. Skowron, J. Stepaniuk, J.F. Peters. Approximation of information granule sets. In *[74]*, 65–72, 2001.
55. A. Skowron, J. Stepaniuk, J.F. Peters. Hierarchy of information granules. In H.D. Burkhard, L. Czaja, H.S. Nguyen, P. Starke, editors, *Proceedings of the Workshop on Concurrency, Specification and Programming (CSP 2001)*, 254–268, Warsaw, 2001.
56. A. Skowron, Z. Suraj. A parallel algorithm for real-time decision making: A rough set approach. *Journal of Intelligent Information Systems*, 7: 5–28, 1996.
57. R.W. Swiniarski. *RoughNeuralLab, software package*. Developed at San Diego State University, San Diego, CA, 1995.
58. R. Swiniarski, F. Hunt, D. Chalvet, D. Pearson. Prediction system based on neural networks and rough sets in a highly automated production process. In *Proceedings of the 12th System Science Conference*, Wrocław, Poland, 1995.
59. R. Swiniarski, F. Hunt, D. Chalvet, D. Pearson. Intelligent data processing and dynamic process discovery using rough sets, statistical reasoning and neural networks in a highly automated production systems. In *Proceedings of the 1st European Conference on Application of Neural Networks in Industry*, Helsinki, 1995.
60. R.W. Swiniarski. Rough sets and neural networks application to handwritten character recognition by complex Zernike moments. In *Proceedings of the 1st Internatonal Conference on Rough Sets and Current Trends in Computing (RSCTC'98)*, LNAI 1424, 617–624, Springer, Berlin, 1998.
61. R.W. Swiniarski, L. Hargis. Rough sets as a front end of neural networks texture classifiers. *Neurocomputing: An International Journal*, 36: 85–103, 2001.
62. M. S. Szczuka. Refining classifiers with neural networks. *International Journal of Intelligent Systems*, 16(1): 39-55, 2001.
63. M. S. Szczuka. Rough sets and artificial neural networks. In L. Polkowski, A. Skowron, editors, *Rough Sets in Knowledge Discovery 2: Applications, Cases Studies and Software Systems*, 449–470, Physica, Heidelberg, 1998.
64. M.S. Szczuka. Function approximation by neural networks with application of rough set methods. Master's thesis, Faculty of Mathematics, Informatics and Mechanics, Warsaw University, 1995 (in Polish).
65. M.S. Szczuka. Symbolic methods and artificial neural networks in classifier construction, Ph.D. dissertation, Faculty of Mathematics, Informatics and Mechanics, Warsaw University, 2000 (in Polish).
66. M.S. Szczuka. Rough set methods for constructing artificial neural networks. In B.D. Czejdo, I.I. Est, B. Shirazi, B. Trousse, editors, *Proceedings of the 3rd Biennial Joint Conference on Engineering Systems Design and Analysis (ESDA'96)*, 9–14, Montpellier, France, 1996.
67. A. Tarski. In *Introduction to Logic and to the Methodology of Deductive Sciences*, IV, 68–78. Oxford University Press, New York, 1965.
68. N. Wiener. *Cybernetics or Control and Communication in the Animal and the Machine*, 2nd ed. MIT Press, Cambridge, MA, 1961.

69. P. Wojdyłło. Wavelets, rough sets and artificial neural networks in EEG analysis. In *Proceedings of the 1st International Conference on Rough Sets and Current Trends in Computing (RSCTC'98)*, LNAI 1424, 444–449, Springer, Berlin, 1998.
70. L.A. Zadeh. Fuzzy logic = computing with words. *IEEE Transactions on Fuzzy Systems*, 4: 103–111, 1996.
71. L.A. Zadeh. A new direction in AI: Toward a computational theory of perceptions. *AI Magazine*, 22(1): 73–84, 2001.
72. T. Terano, T. Nishida, A. Namatame. S. Tsumoto, Y. Ohsawa, T. Washio, editors. *New Frontiers in Artificial Intelligence. Joint JSAI 2001 Workshop Post Proceedings*, LNAI 2253, Springer, Berlin, 2001.
73. N. Zhong, A. Skowron, S. Ohsuga, editors. *New Directions in Rough Sets, Data Mining, and Granular–Soft Computing*, LNAI 1711, Springer, Berlin, 1999.
74. W. Ziarko, Y.Y. Yao, editors. *Proceedings of the 2nd International Conference on Rough Sets and Current Trends in Computing (RSCTC 2000)*, LNAI 2005, Springer, Berlin, 2001.

Chapter 3
Information Granules and Rough-Neural Computing

Andrzej Skowron[1] and Jarosław Stepaniuk[2]

[1] Institute of Mathematics, Warsaw University, Banacha 2, 02-097 Warsaw, Poland
skowron@mimuw.edu.pl

[2] Department of Computer Science, Białystok University of Technology, Wiejska 45A, 15-351 Białystok, Poland
jstepan@ii.pb.bialystok.pl

Summary. In this chapter we discuss the foundations of rough-neural computing (RNC). We introduce information granule systems and information granules in such systems. Information granule networks, called approximate reasoning schemes (AR schemes), are used to represent information granule constructions. We discuss the foundations of RNC using an analogy of information granule networks with neural networks. RNC is a basic paradigm of granular computing (GC). This paradigm makes it possible to tune AR schemes to construct relevant information granules, e.g., satisfying a given specification to a satisfactory degree. One of the goals of our project is to develop methods based on rough-neural computing for computing with words (CW).

1 Introduction

Information granules are intuitively described in the literature as collections of entities that are arranged together due to their similarity, functional adjacency, or indiscernibility relation. The process of forming information granules is referred to as information granulation. Information granulation belongs to intensively studied topics in soft computing (see, e.g., [42–44]). One of the recently emerging approaches to deal with information granulation, called granular computing, is based on information granule calculi (see, e.g., [27,35]). The development of such calculi is important for making progress in many areas such as object identification by autonomous systems (see, e.g., [4,40]), web mining (see, e.g., [10]), spatial reasoning (see, e.g., [6]), and sensor fusion (see, e.g., [3,22]). One of the main goals of GC is to achieve computing with words (see, e.g., [42–44]). The granular computing paradigm, as opposed to numeric-computing, is knowledge oriented [14,39]. Computations in granular computing are performed on information granules. Developing methods of GC is also crucial for making progress in knowledge discovery and data mining. The main reason for this is that knowledge-based processing is a cornerstone of knowledge discovery and data mining.

There is a need to develop information granulation tools for constructing complex information granules. For example, in spatial and temporal reasoning, one should be able to determine if a road is safe on the basis of sensor measurements [40] or to classify situations in complex games such as soccer [37]. These complex information granules constitute a form of information fusion. Any calculus of complex information granules should make it possible to

- Deal with the vagueness of information granules.
- Develop strategies for inducing multilayered schemes of complex granule construction.
- Derive robust (stable) information granule construction schemes with respect to deviations of the granules from which they are constructed.
- Develop adaptive strategies for reconstructing induced schemes for complex information granule synthesis.

To deal with vagueness, one can adopt fuzzy set theory [41] or rough set theory [19] either separately or in combination [20]. The second requirement is related to the problem of understanding reasoning from measurements relative to perception [43], to concept approximation learning in layered learning [37], and to fusion of information from different sources [42–44]. Methods of searching for approximate reasoning schemes as schemes of new information granule construction have been investigated using rough mereological tools [24,25,27,29,31]. In general, those methods return hierarchical schemes for new information granule construction. The process of AR schemes construction is related to ideas of cooperation, negotiation, and conflict resolution in multiagent systems [2,9]. Among important topics studied in relation to AR schemes are methods for specifying operations on information granules. In particular, AR schemes are useful in constructing information granules from data and background knowledge and in supplying methods for inducing these hierarchical schemes of information granule construction. One of the possible approaches is to learn such schemes using evolutionary strategies [13]. The robustness of the scheme means that any scheme produces a higher order information granule that is a clump (e.g., a set) of close information granules rather than a single information granule. Such a clump is constructed by means of the scheme from input clumps defined by deviations (up to acceptable degrees) of input information granules from standard (prototype) granules.

It is worthwhile mentioning that modeling complex phenomena requires us to use complex information granules representing local models (perceived by local agents) that are fused. This process involves negotiations between agents [9] to resolve contradictions and conflicts in local modeling. This kind of modeling will become more and more important in solving complex real-life problems that we cannot model using traditional analytical approaches. If the latter approaches can be applied to modeling of such problems, they lead to exact models. However, the necessary assumptions used to build them for complex real-life problems often make the resulting solutions *too far* from reality to be accepted as solutions of such problems.

Using multiagent terminology, let us also observe that local agents perform operations on information granules they *understand.* Hence, granules received as operation arguments from other agents should be approximated by properly tuned approximation spaces creating interfaces between agents. The process of tuning the approximation space [29,33] parameters in AR schemes corresponds to the tuning of weights in neural networks. The methods for inducing AR schemes to transform information granules into developed information granules using rough set [12,19] and rough mereological methods [24,29] in hybridization with other soft computing approaches (i.e., neural networks [30], fuzzy sets [20,41,44], and evolutionary programming [13,17]) create a core for rough-neural computing. In RNC, computations are performed on information granules by schemes analogous to neural networks. The aim of such schemes is, for example, to construct information granules satisfying, at least to a satisfactory degree, a given specification or to preserve some invariants during computation.

One of the basic research directions in RNC concerns relationships between information granules and words (linguistic terms) in a natural language. Another direction of the research concerns the possibility of using induced AR schemes that match, to a satisfactory degree, reasoning schemes in natural language. Further research in this direction will create strong links between RNC and CW.

In this chapter, we discuss the foundations for RNC. We introduce information granule systems and information granules in such systems. Any granule system consists of a parameterized formula set and a finite parameterized relational system in which the semantics of such formulas is defined. Information granules in a given granule system are elements of a parameterized formula set.

We present several examples of information granule systems and information granules in such systems. Different kinds of information granules and inclusion (closeness) measures between information granules are discussed in the following sections of this chapter. Networks of information granule construction are represented by approximate reasoning schemes. We discuss the foundations of the RNC paradigm using an analogy of information granule networks to neural networks.

The rough-neural computing paradigm is a basic paradigm of GC. One of the goals of our project is to develop methods based on RNC for computing with words.

2 Motivation: Illustrative Example

In this section, we consider an example that shows the need for information granulation, construction of AR schemes, and RNC. The example is related to estimating a situation on a road on the basis of sensor measurements made by an unmanned helicopter [40]. Let us assume that we would like to estimate whether the situation on the road is dangerous (see Fig. 1).

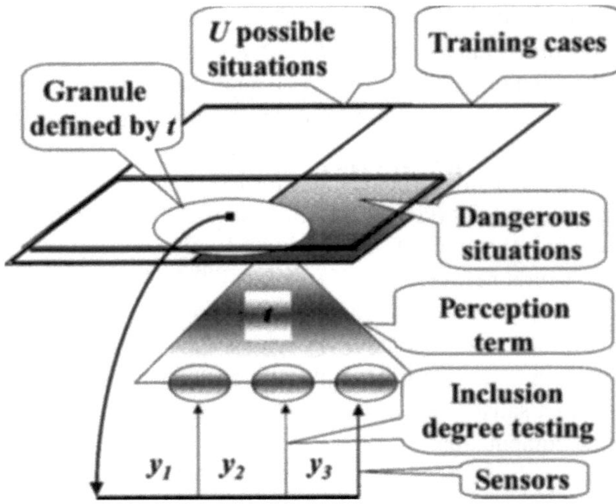

The problem is how to induce large patterns, sufficiently included in the concept *dangerous* in terms of sensor measurements. Such patterns can be used to approximate the concept. This can be done on the basis of *perception trees* representing reasoning schemes in natural language, based, for example, on behavior of cars on the road. We expect that AR schemes (or their clusters) can be constructed along such perception trees. They will make it possible to estimate whether sensor measurements corresponding to a given situation are sufficiently close to input information granules of such schemes. If it is true, one can conclude that it is highly probable that the analyzed situation is dangerous.

3 Information Granule Systems

In this section, we present a basic notion of our approach, i.e., the information granule system. Any information granule system is a tuple,

$$S = (G, R, Sem), \tag{1}$$

where

1. G is a set of parameterized formulas, called *information granules*;
2. R is a (parameterized) relational structure; and
3. *Sem* is the semantics of G in R.

We assume that with any information granule system the following are associated:

1. H, a set of *granule inclusion degrees* with a partial order relation $\leq$ that defines on H a structure used to compare the inclusion degrees; we assume that H contains the lowest degree 0 and the largest degree 1.

2. $\nu_p \subseteq G \times G$, a binary relation that is *a part to a degree at least p* between information granules from G, called *rough inclusion*. (Instead of $\nu_p(g, g')$, we also write $\nu(g, g') \geq p$).

The components of an information granule system are parameterized. This means that we deal with parameterized formulas and a parameterized relational system. The parameters are tuned to make it possible to construct finally relevant information granules, i.e., granules satisfying specification or/and some optimization criteria. Parameterized formulas can consist of parameterized subformulas. The value set of parameters labeling a subformula defines a set of formulas. By tuning parameters in an optimization process and/or information granule construction, a relevant subset of parameters is extracted and used for constructing the target information granule.

There are two kinds of computations on information granules: computations on information granule systems and computations on information granules in such systems. The first aim at constructing of relevant information granule systems defining parameterized approximation spaces for concept approximations used on different levels of target information granule construction. The goal of the second is to construct information granules across such information granule systems to obtain target information granules, e.g., satisfying a given specification (at least to a satisfactory degree).

Examples of complex granules are tolerance granules created by similarity (tolerance) relation between elementary granules, decision rules, sets of decision rules, sets of decision rules with guards, information systems, or decision tables [27,31,35]. The most interesting class of information granules consists of information granules that approximate concepts specified in natural language by experimental data tables and background knowledge.

One can consider as an example of the set H of granule inclusion degrees the set of binary sequences of fixed length with the relation ν defined by the lexicographical order. This degree structure can be used to measure the inclusion degree between granule sequences or to measure the matching degree between granules representing classified objects and granules describing the left-hand sides of decision rules in simple classifiers [29]. However, one can consider more complex degree granules assuming the degree of inclusion of granule g_1 in granule g_2 as the granule representing a collection of common parts of these two granules g_1 and g_2.

New information granules can be defined by operations performed on already constructed information granules. Examples of such operations are set theoretical operations (defined by propositional connectives). However, there are other operations widely used in machine learning or pattern recognition ([15]) for constructing classifiers. These are the $Match$ and $Conflict\_res$ operations [29]. The $Match$ operation is used to construct a granule describing the matching result of elementary granules describing classified objects by granules representing the left-hand sides of decision rules. The $Conflict\_res$ is an operation producing from this matching granule the resulting granule, e.g., identifying a relevant decision class for any classified object. It is worthwhile mentioning yet another important class of operations, namely, those defined by data tables called decision tables [35]. From these decision tables, decision rules specifying operations can be induced. More complex operations on information granules are so called transducers [4]. They have been introduced to use background knowledge (not necessarily in the form of data tables) in constructing new granules. One can consider theories or their clusters as information granules. Reasoning schemes in natural language define the most important class of operations on information granules to be investigated. One of the basic problems of such operations and schemes of reasoning is how to approximate them by available information granules, e.g., constructed from sensor measurements.

In an information granule system, the relation ν_p as a part to a degree at least p has a special role. It satisfies some additional natural axioms and, additionally, some axioms of rough mereology [25]. It can be shown that the rough mereological approach, built on the basis of the relation to be part, to a degree, generalizes the rough set and fuzzy set approaches. Moreover, such relations can be used to define other basic concepts, such as closeness of information granules, their semantics, the indiscernibiliy and discernibility of objects, information granule approximation and approximation spaces, the perception structure of information granules as well as the notion of ontology approximation. One can observe that the relation to be part to a degree can be used to define operations on information granules corresponding to a generalization of already defined information granules.

Let us finally note that new information granule systems can be defined using already constructed information granule systems. This leads to a hierarchy of information granule systems.

4 Basic Examples of Information Granule Systems

In the following sections, we present examples of information granule systems. Each of them will be specified by a set of information granules, relational structure, and semantics of information granules. We will also discuss some operations on information granule system that make it possible to construct new relevant information granule systems in searching for the target information granule systems representing approximation spaces.

4.1 Elementary Information Granule Systems

Let us consider an example from [19]. We define a language $L_{\mathcal{A}}$ used for elementary granule description, where $\mathcal{A} = (U, A)$ is an information system that is a relational structure for the example discussed. The syntax of $L_{\mathcal{A}}$ is defined recursively by

1. $(a \in V) \in L_{\mathcal{A}}$, for any $a \in A$ and $V \subseteq V_a$.
2. If $\alpha \in L_{\mathcal{A}}$, then $\neg\alpha \in L_{\mathcal{A}}$.
3. If $\alpha, \beta \in L_{\mathcal{A}}$, then $\alpha \wedge \beta \in L_{\mathcal{A}}$.
4. If $\alpha, \beta \in L_{\mathcal{A}}$, then $\alpha \vee \beta \in L_{\mathcal{A}}$.

The semantics of formulas from $L_{\mathcal{A}}$ with respect to an information system $\mathcal{A}$ is defined recursively by

1. $Sem_{\mathcal{A}}(a \in V) = \{x \in U : a(x) \in V\}$.
2. $Sem_{\mathcal{A}}(\neg\alpha) = U - Sem_{\mathcal{A}}(\alpha)$.
3. $Sem_{\mathcal{A}}(\alpha \wedge \beta) = Sem_{\mathcal{A}}(\alpha) \cap Sem_{\mathcal{A}}(\beta)$.
4. $Sem_{\mathcal{A}}(\alpha \vee \beta) = Sem_{\mathcal{A}}(\alpha) \cup Sem_{\mathcal{A}}(\beta)$.

Elementary granules. In an information system $\mathcal{A} = (U, A)$, elementary granules are defined by $EF_B(x)$, where EF_B is a conjunction of selectors (descriptors) of the form $a = a(x)$, $B \subseteq A$ and $x \in U$. For example, the meaning of an elementary granule $a = 1 \wedge b = 1$ is defined by

$$Sem_{\mathcal{A}}(a = 1 \wedge b = 1) = \{x \in U : a(x) = 1 \;\&\; b(x) = 1\}.$$

In the elementary information granule system discussed, the set of elementary granules consists of a set of conjunctions of selectors, the set H of inclusion degrees is a subset of $[0,1]$ and $\nu_p(EF_B, EF'_B)$ if and only if

$$\frac{card\left[Sem_{\mathcal{A}}(EF_B) \cap Sem_{\mathcal{A}}(EF'_B)\right]}{card\left[Sem_{\mathcal{A}}(EF_B)\right]} \geq p.$$

The number of conjuncts in the granule can be taken as one of parameters to be tuned; this is well known as the drooping condition technique in machine learning [15].

One can extend the set of elementary granules assuming that if α is any Boolean combination of descriptors over A, then $(\overline{B}\alpha)$ and $(\underline{B}\alpha)$ define the syntax of elementary granules, too, for any $B \subseteq A$. The reader can find more details on granules defined by rough set approximations in [36].

One can consider extension of elementary granules defined by a tolerance relation. Let $\mathcal{A} = (U, A)$ be an information system, and let τ be a tolerance relation on elementary granules of $\mathcal{A}$. Any pair (α, τ) is called a τ-*elementary granule*. The semantics $Sem_{\mathcal{A}}[(\alpha, \tau)]$ of (α, τ) is the family $\{Sem_{\mathcal{A}}(\beta) : (\beta, \alpha) \in \tau\}$. Parameters to be tuned in searching for a relevant tolerance granule can be its support (represented by the number of supporting objects) and its degree of inclusion (or closeness) in some other granules as well as parameters specifying the tolerance relation.

Certainly, one can consider other forms of elementary granules. For example, one can define an elementary information granule as any tuple consisting of a Boolean combination of descriptors or its semantics in a given information system. The choice depends on applications.

4.2 Operations on Information Granule Systems

The simplest operations on information granule systems are set theoretical operations. For, example if $S = (G, R, Sem)$ and $S' = (G', R', Sem')$ are two information granule systems, then $pair(S, S')$ is an information granule system in which information granules are pairs (α, α') where $\alpha \in G$, $\alpha' \in G'$ and the semantics of (α, α') is defined by $(Sem(\alpha), Sem'(\alpha'))$. Analogously, other set theoretical operations, such as $tuple(S_1, \ldots, S_k)$ or $set(S_1, \ldots, S_k)$ can be defined. One can also define projection and extension operations, helping to solve feature extraction and selection problems.

First, let us discuss some examples of possible applications of such constructions of new information granule systems. Granules defined by rules in information systems are examples of information granules in $pair(S, S')$. Let $\mathcal{A} = \mathcal{A}'$ be an information system and let (α, β) be a new information granule in $pair(S, S')$ received from the rule *if* α *then* β, where α, β are elementary granules in S, S', respectively. If the right-hand sides of rules represent decision classes, then the among parameters to be tuned in classification is the number of conjuncts on the left-hand sides of rules. A typical goal is to search for the minimal number of such conjuncts (corresponding to the largest generalization), which still guarantees a satisfactory degree of inclusion in a corresponding decision class [12,15]. One can now consider the new information granule system $set\,[pair(S, S'), \ldots, pair(S, S')]$ in which information granules represent sets of rules. An important problem in machine learning is the problem of searching for a granule of the smallest cardinality sufficiently close to one given in such a system, i.e., a searching problem for representating of a given rule collection by another set of rules of sufficiently small cardinality and sufficiently close to the collection.

In examples presented, we have discussed parameterized information granules. We have pointed out that the process of parameter tuning is used to induce relevant (for a given task) information granules. In particular, the process of parameter tuning is performed to obtain a satisfactory degree of inclusion (closeness) of information granules.

We have not yet discussed how the rough inclusion relations in the resulting information granule systems are defined. In the following section, we discuss inclusion and closeness relations for information granules.

4.3 Examples of Granule Inclusion and Closeness

In this section, we will discuss inclusion and closeness of different information granules. Inclusion and closeness are basic concepts related to information granules [27,35]. Using them, one can measure the closeness of a constructed granule to a target granule and the robustness of the construction scheme with respect to deviations of information granules that are components of the construction. For details and examples of closeness relations, refer to [27,35].

The choice of inclusion or closeness definition depends very much on the area of application and data analyzed. This is the reason that we have decided to introduce a separate section with this more subjective (or task oriented) part of granule semantics.

The inclusion relation between granules g, g' to the degree at least p (i.e., $\nu(g,g') \geq p$), will be denoted by $\nu_p(g,g')$. By $\underline{\nu}_p(g,g')$, we denote the inclusion of g in g' to the degree at most p, i.e., that $\nu(g,g') \leq p$ holds. Similarly, the closeness relation between granules g, g', to the degree at least p, will be denoted by $cl_p(g,g')$. By p, we denote a vector of parameters (e.g., from the interval $[0,1]$ of real numbers).

A general scheme for constructing of hierarchical granules and their closeness can be described by the following recursive metarule: if granules of order $\leq k$ and their closeness have been defined, then the closeness $cl_p(g,g')$ (at least to the degree p) between granules g, g' of order $k+1$ can be defined by applying an appropriate operator F to closeness values of components of g, g', respectively.

Elementary granules. We have introduced the simplest case of granules in information system $\mathcal{A} = (U, A)$. They are defined by $EF_B(x)$, where EF_B is a conjunction of selectors of the form $a = a(x)$, where $a \in B \subseteq A$ and $x \in U$. Let

$$G_{\mathcal{A}} = \{EF_B(x) : \emptyset \neq B \subseteq A \ \& \ x \in U\}. \tag{2}$$

In the standard rough set model [19], elementary granules describe indiscernibility classes with respect to some subsets of attributes. In a more general setting [33], tolerance (similarity) classes are described. The crisp inclusion of α in β is defined by

$$Sem_{\mathcal{A}}(\alpha) \subseteq Sem_{\mathcal{A}}(\beta), \tag{3}$$

where $\alpha, \beta \in \{EF_B(x) : B \subseteq A \ \& \ x \in U\}$ and $Sem_{\mathcal{A}}(\alpha)$ and $Sem_{\mathcal{A}}(\beta)$ are sets of objects from $\mathcal{A}$ satisfying α and β, respectively. The noncrisp inclusion, known in KDD [1] for association rules, is defined by two thresholds t and t':

$$support_{\mathcal{A}}(\alpha,\beta) = card\left[Sem_{\mathcal{A}}(\alpha \wedge \beta)\right] \geq t, \tag{4}$$

$$accuracy_{\mathcal{A}}(\alpha,\beta) = \frac{support_{\mathcal{A}}(\alpha,\beta)}{card\left[Sem_{\mathcal{A}}(\alpha)\right]} \geq t'. \tag{5}$$

Elementary granule inclusion in a given information system $\mathcal{A}$ can be defined using different schemes, e.g., by

$$\nu_{t,t'}^{\mathcal{A}}(\alpha,\beta) \text{ if and only if } support_{\mathcal{A}}(\alpha,\beta) \geq t \text{ and } accuracy_{\mathcal{A}}(\alpha,\beta) \geq t' \tag{6}$$

or

$$\nu_t^{\mathcal{A}}(\alpha,\beta) \text{ if and only if } accuracy_{\mathcal{A}}(\alpha,\beta) \geq t. \tag{7}$$

The closeness of granules can be defined by

$$cl_{t,t'}^{\mathcal{A}}(\alpha,\beta) \text{ if and only if } \nu_{t,t'}^{\mathcal{A}}(\alpha,\beta) \text{ and } \nu_{t,t'}^{\mathcal{A}}(\beta,\alpha) \text{ hold.} \tag{8}$$

Decision rules as granules. One can define inclusion and closeness of granules corresponding to rules of the form **if** α **then** β by using accuracy coefficients. Having such granules $g = (\alpha,\beta)$, $g' = (\alpha',\beta')$, one can define the inclusion and closeness of g and g' by

$$\nu_{t,t'}^{\mathcal{A}}(g,g') \text{ if and only if } \nu_{t,t'}^{\mathcal{A}}(\alpha,\alpha') \text{ and } \nu_{t,t'}^{\mathcal{A}}(\beta,\beta'). \tag{9}$$

Closeness can be defined by

$$cl_{t,t'}^{\mathcal{A}}(g,g') \text{ if and only if } \nu_{t,t'}^{\mathcal{A}}(g,g') \text{ and } \nu_{t,t'}^{\mathcal{A}}(g',g). \tag{10}$$

Another way of defining the inclusion of granules corresponding to decision rules is as

$$\begin{aligned}\nu_{t}^{\mathcal{A}}((\alpha,\beta),(\alpha',\beta')) &\text{ if and only if} \\ &\nu_{t_1,t_2}^{\mathcal{A}}(\alpha,\alpha') \text{ and } \nu_{t_1,t_2}^{\mathcal{A}}(\beta,\beta') \text{ and } t = w_1 \cdot t_1 + w_2 \cdot t_2,\end{aligned} \tag{11}$$

where w_1, w_2 are some given weights satisfying $w_1 + w_2 = 1$ and $w_1, w_2 \geq 0$.

Extensions of elementary granules by tolerance relation. For extensions of elementary granules defined by similarity (tolerance) relation, i.e., granules of the form (α,τ), (β,τ), one can consider the following inclusion measure:

$$\begin{aligned}\nu_{t,t'}^{\mathcal{A}}((\alpha,\tau),(\beta,\tau)) &\text{ if and only if} \\ &\nu_{t,t'}^{\mathcal{A}}(\alpha',\beta') \text{ for any } \alpha',\beta' \text{ such that } (\alpha,\alpha') \in \tau \text{ and } (\beta,\beta') \in \tau\end{aligned} \tag{12}$$

and the following closeness measure:

$$cl_{t,t'}^{\mathcal{A}}((\alpha,\tau),(\beta,\tau)) \text{ if and only if } \nu_{t,t'}^{\mathcal{A}}((\alpha,\tau),(\beta,\tau)) \text{ and } \nu_{t,t'}^{\mathcal{A}}((\beta,\tau),(\alpha,\tau)).$$

It can be important for some applications to define the closeness of an elementary granule α and the granule (α,τ). The definition reflecting an intuition that α should be a representation of (α,τ) sufficiently close to this granule is the following:

$$cl_{t,t'}^{\mathcal{A}}(\alpha,(\alpha,\tau)) \text{ if and only if } cl_{t,t'}^{\mathcal{A}}(\alpha,\beta) \text{ for any } (\alpha,\beta) \in \tau. \tag{13}$$

Sets of rules. An important problem related to association rules is that the number of such rules generated even from a simple data table can be large. Hence, one should search for methods of aggregating close association rules [5,38]. This can be defined as searching for some close information granules. Let us consider two finite sets *Rule_Set* and *Rule_Set'* of association rules defined by

$$Rule\_Set = \{(\alpha_i, \beta_i) : i = 1, \ldots, k\} \text{ and } Rule\_Set' = \{(\alpha'_i, \beta'_i) : i = 1, \ldots, k'\}. \tag{14}$$

One can treat them as higher order information granules. These new granules,

$$Rule\_Set, Rule\_Set', \tag{15}$$

can be treated as close to the degree at least t (in $\mathcal{A}$) if and only if there exists a relation *rel* between sets of rules *Rule_Set* and *Rule_Set'* such that

1. For any $Rule \in Rule\_Set$ there is $Rule' \in Rule\_Set'$ such that $(Rule, Rule') \in rel$ and *Rule* is close to *Rule'* (in $\mathcal{A}$) to the degree at least t.
2. For any $Rule' \in Rule\_Set'$ there is $Rule \in Rule\_Set$ such that $(Rule, Rule') \in rel$ and *Rule* is close to *Rule'* (in $\mathcal{A}$) to the degree at least t.

Another way of defining the closeness of two granules G_1, G_2 represented by sets of rules can be described as follows: Let us consider again two granules *Rule_Set* and *Rule_Set'* corresponding to two decision algorithms. We denote by $I(\beta'_i)$ the set $\{j : cl_p^{\mathcal{A}}\left(\beta'_j, \beta'_i\right)\}$ for any $i = 1, \ldots, k'$.

Now, we assume $\nu_p^{\mathcal{A}}(Rule\_Set, Rule\_Set')$ if and only if for any $i \in \{1, \ldots, k'\}$ there exists a set $J \subseteq \{1, \ldots, k\}$ such that

$$cl_p^{\mathcal{A}}\left(\bigvee_{j \in I(\beta'_i)} \beta'_j, \bigvee_{j \in J} \beta_j\right) \text{ and } cl_p^{\mathcal{A}}\left(\bigvee_{j \in I(\beta'_i)} \alpha'_j, \bigvee_{j \in J} \alpha_j\right), \tag{16}$$

and for closeness, we assume

$$cl_p^{\mathcal{A}}\left(Rule\_Set, Rule\_Set'\right) \text{ if and only if } \tag{17}$$
$$\nu_p^{\mathcal{A}}\left(Rule\_Set, Rule\_Set'\right) \text{ and } \nu_p^{\mathcal{A}}\left(Rule\_Set', Rule\_Set\right).$$

For example, if the granule G_1 consists of rules: **if** α_1 **then** $d = 1$, **if** α_2 **then** $d = 1$, **if** α_3 **then** $d = 1$, **if** β_1 **then** $d = 0$, **if** β_2 **then** $d = 0$ and the granule G_2 consists of rules: **if** γ_1 **then** $d = 1$, **if** γ_2 **then** $d = 0$, then $cl_p(G_1, G_2)$ if and only if

$$cl_p(\alpha_1 \vee \alpha_2 \vee \alpha_3, \gamma_1) \quad \text{and} \quad cl_p(\beta_1 \vee \beta_2, \gamma_2).$$

One can consider a searching problem for a granule *Rule_Set'* of minimal size such that *Rule_Set* and *Rule_Set'* are close.

Certainly, one would like to provide closeness of rules on the extension on the universe U of objects used for inducing rules. Hence, rule grouping is not so simple as in the above example. Rule grouping should be tuned to preserve such constraints.

Granules defined by sets of granules. The previously discussed methods of inclusion and closeness definition can be easily adopted for granules defined by sets of already defined granules. Let G,H be sets of granules.
The inclusion of G in H can be defined by

$$\nu_{t,t'}^{\mathcal{A}}(G,H) \text{ if and only if for any } g \in G \text{ there is } h \in H \text{ for which } \nu_{t,t'}^{\mathcal{A}}(g,h) \quad (18)$$

and the closeness by

$$cl_{t,t'}^{\mathcal{A}}(G,H) \text{ if and only if } \nu_{t,t'}^{\mathcal{A}}(G,H) \text{ and } \nu_{t,t'}^{\mathcal{A}}(H,G). \quad (19)$$

Let G be a set of granules, and let φ be a property of sets of granules from G [e.g., $\varphi(X)$ if and only if X is a tolerance class of a given tolerance $\tau \subseteq G \times G$.] Then, $P_{\varphi}(G) = \{X \subseteq G : \varphi(X) \text{ holds}\}$. Closeness of granules $X,Y \in P_{\varphi}(G)$ can be defined by

$$cl_t(X,Y) \text{ if and only if } cl_t(g,g') \text{ for any } g \in G \text{ and } g' \in H. \quad (20)$$

We have the following examples of inclusion and closeness propagation rules:

$$\frac{\text{for any } \alpha \in G \text{ there is } \alpha' \in H \text{ such that } \nu_p(\alpha,\alpha')}{\nu_p(G,H)}, \quad (21)$$

$$\frac{cl_p(\alpha,\alpha'), cl_p(\beta,\beta')}{cl_p[(\alpha,\beta),(\alpha',\beta')]}, \quad (22)$$

$$\frac{\text{for any } \alpha' \in \tau(\alpha) \text{ there is } \beta' \in \tau(\beta) \text{ such that } \nu_p(\alpha',\beta')}{\nu_p[(\alpha,\tau),(\beta,\tau)]}, \quad (23)$$

$$\frac{cl_p(g,g'), cl_p(h,h')}{cl_p[(g,h),(g',h')]}, \quad (24)$$

where $\alpha,\alpha',\beta,\beta'$ are elementary granules and g,h,g',h' are finite sets of elementary granules.

One can also present other cases for measuring the inclusion and closeness of granules in the form of inference rules. The exemplary rules have a general form, i.e., they are true in any $\mathcal{A}$ (under the chosen definition of inclusion and closeness). Some of them are derivable from others. We will see in the next part of the chapter that there are also some operations of new granule construction specific for a given information granule system. In this case, one should extract the specific inference rules from existing data.

Information granules defined by inclusion and closeness measures. Let us observe that inclusion (closeness) measures can be used to define new granules that are approximations or generalizations of existing ones. Assume that g,h are given information granules and ν_p is the inclusion measure (where $p \in [0,1]$). A (h,p)-approximation of g is an information granule $Apr(\nu,h,p)$ represented by a set $\{h' : \nu_1(h',h) \wedge \nu_p(h',g)\}$. Now, the lower and upper approximations of given information granules can be easily defined [33] (see Sect. 11.3).

4.4 Target Information Granule Systems: Approximation Spaces

One of an interesting class of information granules consists of *classifiers* (see also Sect. 6.1 in Chap. 25). One can observe that sets of decision rules generated from a given decision table $DT = (U,A,d)$ can be interpreted as information granules. First, one can construct granules G_j corresponding to each particular decision $j = 1,\ldots,r$ by taking a collection $\{g_{ij} : i = 1,\ldots,k_j\}$ of left-hand sides of decision rules for a given decision, where k_j is the number of decision rules for decision j. Next, one can construct from them a collection $G = \{G_1,\ldots,G_r\}$ represented by one granule. Let E be a set of elementary granules over $\mathcal{A} = (U,A)$. We can now consider a granule denoted by $H(e,G)$ for any $e \in E$ that is a collection of coefficients ε_{ij}, where $\varepsilon_{ij} = 1$ if $Sem_{\mathcal{A}}(e) \subseteq Sem_{\mathcal{A}}(g_{ij})$ and 0, otherwise. Hence, the coefficient ε_{ij} is equal to 1 if and only if granule e matches granule g_{ij} in $\mathcal{A}$. Denote now by $Conflict\_res$ a function (resolving conflict between decision rules recognizing elementary granules) defined on granules of the form $H(e,G)$ with values in the set of possible decisions $1,\ldots,r$. Then, $Conflict\_res\,[H(e,G)]$ is equal to the decision predicted by the classifier $Conflict\_res\,[H(\bullet,G)]$ on the input granule e.

Hence, one can see that classifiers can be treated as special cases of granules. The parameters to be tuned are voting strategies, matching strategies of objects against rules, as well as other parameters discussed above, such as the closeness of classifier granule in a target granule.

Note that classifiers are information granules in relevant information granule systems. These information granule systems can be treated as approximation spaces. They are targets in computations on information granule systems. The relevant approximation spaces for a given task are information granule systems in which relevant information granules for concept approximation can be selected.

Let us look more deeply into the structure of approximation spaces in the framework of information granule systems. Such information granule systems satisfy the following conditions related to their information granules, relational structure, and semantics:

1. Semantics consists of two parts, namely, relational structure R and its extension R^*.
2. Different types of information granules can be identified: (a) object granules (denoted by x), (b) neighborhood granules (denoted by n with subscripts), (c) pattern granules (denoted by *pat*), and (d) decision class granules (denoted by c).
3. There are decision class granules $c_1,\ldots,c_r$ with semantics in R^* defined by a partition of object granules into r decision classes. However, only the restrictions of these collections on the object granules from R are given.
4. For any object granule x, there is a uniquely defined neighborhood granule n_x.

5. For any class granule c, there is constructed a collection granule $\{(pat,p) : \nu_p^R(pat,c)\}$ of pattern granules labeled by maximal degrees to which *pat* is included in c (in R).
6. For any neighborhood granule n_x, there is distinguished a collection granule $\{(pat,p) : \nu_p^R(n_x,pat)\}$ of pattern granules labeled by maximal degrees to which n_x is at least included in *pat* (in R).
7. There is a class of $Classifier$ functions transforming collection granules (corresponding to a given object x) described in the two previous steps into the power set of $\{1,\ldots,r\}$. One can assume that object granules are the only arguments of $Classifier$ functions if other arguments are fixed.

The classification problem is to find a $Classifier$ function defining a partition of object granules in R^* as close as possible to the partition defined by decision classes. Any such $Classifier$ defines the lower and the upper approximations of any family of decision classes $\{c_i\}_{i\in I}$, where I is a nonempty subset of $\{1,\ldots,r\}$, by

$$\underline{Classifier}(\{c_i\}_{i\in I}) = \{x \in \bigcup_{i\in I} c_i : \emptyset \neq Classifier(x) \subseteq I\}, \tag{25}$$

$$\overline{Classifier}(\{c_i\}_{i\in I}) = \{x \in U^* : Classifier(x) \cap I \neq \emptyset\}. \tag{26}$$

The positive region of the $Classifier$ is defined by

$$POS(Classifier) = \underline{Classifier}(\{c_1\}) \cup \ldots \cup \underline{Classifier}(\{c_r\}). \tag{27}$$

The closeness of the partition defined by the constructed $Classifier$ and the partition in R^* defined by decision classes can be measured, e.g., by using the ratio of the positive region size of the $Classifier$ to the size of the object universe. The *quality of* the $Classifier$ can be defined by taking into account, as usual, only objects from $U^* - U$:

$$Quality(Classifier) = \frac{|POS(Classifier) \cap (U^* - U)|}{|(U^* - U)|}. \tag{28}$$

One can observe that approximation spaces have many parameters to be tuned to construct the approximation of high-quality class granules .

5 Examples of Operations on Information Granules

New granules can be generated by operations. One can distinguish several classes of operations on information granules [35]. In this section, we discuss some important examples of such operations.

5.1 Generalization

Generalization operations are very important in inducing concept description in machine learning, pattern recognition, knowledge discovery, and data mining applications [15]. A general form of such operations can be defined using similarity, inclusion, or closeness relations. One can define the generalization of an information granule g relative to a given family of information granules G and measure ν by

$$Gen(g,G,\nu) = Make\_granule(\{h \in G : \nu(g,h)\}) \tag{29}$$

Some special cases can be obtained assuming $\nu = \nu_p$ or $\nu = cl_p$, i.e., assuming that ν is an inclusion degree relation to the degree at least p or a closeness relation to the degree at least p. In the former case, one can obtain a cluster of information granules sufficiently covering a given information granule g. In the latter case, we obtain a cluster of information granules close to a given information granule g as a generalization of g. The operation *Make_granule* constructs a new granule from a collection of granules. In the simplest case, when granules in the collections are object sets, this operation can be defined as the set theoretical union.

Let us observe that this general definition can be applied, e.g., to an information granule representing a classifier over a set of elementary granules E and to its generalization to a superset of E. One can treat the classifier $Classifier$ across a set E of elementary granules as a collection of pairs $[e, Classifier(e)]$. Such a classifier can be extended to a new classifier, $Classifier^*$, over a superset E^* of E. The inclusion degree of $Classifier^*$ in the concept being approximated can be expressed as a ratio of the number of the same pairs in both classifiers for elementary granules from $E^* - E$ to the number of all elementary granules in $E^* - E$. The operation *Make_granule* can be interpreted as a fusion operation of different extensions of the classifier, $Classifier$, to some supersets of the initial set of elementary granules E.

5.2 Operations Defined by Decision Tables

Operations on information granules are often (partially) specified by means of data tables. Let us discuss such a specification in more detail. We assume that any (partial) operation $f : G_1 \times \ldots \times G_k \to H$ with arguments from the sets $G_1, \ldots, G_k$ of information granules and values in the set H of information granules is partially specified by a data table (information system [19]). Any row of the data table corresponds to an object that is a tuple $[g_1, \ldots, g_k, f(g_1, \ldots, g_k)]$, where $(g_1, \ldots, g_k)$ belongs to the domain of f. The attribute values for a given object consist of

1. Values of attributes from sets $A_{G_1}, \ldots, A_{G_k}$ on information granules $g_1, \ldots, g_k$ (attributes are extracted from some preassumed feature languages $L_1, \ldots, L_k$).
2. Values of attributes characterizing relations among information granules $g_1, \ldots, g_k$ specifying constraints under which the tuple $(g_1, \ldots, g_k)$ belongs to (a relevant part of) the domain of f.
3. Values of attributes selected for the information granule $f(g_1, \ldots, g_k)$ description.

In this way, partial information about the function f is given. In our considerations, we assume that objects indiscernible by condition attributes are indiscernible by decision attribute, i.e., the decision table $DT = (U, A, d)$ considered is consistent [19]. We also assume, that the representation is consistent with a given function

on information granules, i.e., any image obtained by f of the Cartesian product of indiscernibility classes defined by condition attributes is included in a decision indiscernibility class.
Now, we explain in what sense the decision table $DT = (U, A, d)$ can be treated as partial information about the function $f : G_1 \times \ldots \times G_k \rightarrow H$. For $i = 1, \ldots, k$, let

$$G_i^{DT} = \{g_i \in G_i : \text{ there exists in } DT \text{ an object } (g_1, \ldots, g_i, \ldots, g_k, h)\}. \tag{30}$$

One can define H^{DT} in an analogous way. The decision table DT defines a function

$$f_{DT} : G_1/IND(A_{G_1}) \times \ldots \times G_k/IND(A_{G_k}) \rightarrow H^{DT}/IND(d)$$

by

$$f_{DT}([g_1]_{IND(A_{G_1})}, \ldots, [g_k]_{IND(A_{G_k})}) = [h]_{IND(d)} \text{ if and only if } \tag{31}$$
$$(g_1, \ldots, g_k, h) \text{ is an object of } DT,$$

where $[g_i]_{IND(A_{G_i})}$ denotes the A_{G_i}-indiscernibility class defined by g_i for $i = 1, \ldots, k$. We assume that a consistency modeling condition for f is satisfied, namely,

$$f([g_1]_{IND(A_{G_1})} \times \ldots \times [g_k]_{IND(A_{G_k})}) = f_{DT}([g_1]_{IND(A_{G_1})}, \ldots, [g_k]_{IND(A_{G_k})}) \tag{32}$$

for any $(g_1, \ldots, g_k) \in G_1^{DT} \times \ldots \times G_k^{DT}$ where the left-hand side in (32) denotes the image obtained by f of the set $[g_1]_{IND(A_{G_1})} \times \ldots \times [g_k]_{IND(A_{G_k})}$. The function description can be induced from such a data table by interpreting it as a decision table with the decision corresponding to the attributes specifying the values of the function f.

Certainly, the induced description should be experimentally verified and can be achieved by searching for relevant parameters of data tables, such as relevant features (attributes) and decomposition methods for information granules.

6 Granulation of Relational Structures

In this section, we discuss the information granulation process in a logical framework. Such a process returns information granules of different kinds.

Let us fix some basic notation [7]. Among the basic concepts that will be used are relational structure M of a given signature Sig with a domain Dom and a language L of signature Sig.

There is one more very important concept for information granulation, namely, *neighborhood function*, i.e., any function

$$\mathcal{N} : Dom \longrightarrow Pow^{\omega}(Dom), \tag{33}$$

where

- $Pow^{\omega}(Dom) = \bigcup_{k \in \omega} Pow^k(Dom)$.
- $Pow^1(Dom) = Pow(Dom)$ and $Pow^{k+1}(Dom) = Pow\left[Pow^k(Dom)\right]$ for any non-negative integer k.[1]

To explain this concept, let us consider an information system $\mathcal{A} = (U,A)$ as an example of a relational structure. A neighborhood function $\mathcal{N}_{\mathcal{A}}$ of $\mathcal{A} = (U,A)$ is defined by $\mathcal{N}_{\mathcal{A}}(x) = [x]_A$ for $x \in U = Dom$ where $[x]_A$ denotes the A-indiscernibility class of x (see Chaps. 1 and 8). Hence, the neighborhood function forms basic granules of knowledge about the universe corresponding to objects.

One can approximate sets by means of such neighborhoods (see Chap. 1). It is possible to consider a more general case by taking as the indiscernibility relation a similarity relation between objects instead of an equivalence relation. Such a similarity relation can be defined on objects by means of a similarity of vectors of attribute values on the objects. In this case, the neighborhood function value for a given object x is equal to the similarity class of x consisting of all objects similar to x. We show an example of a neighborhood function returning values more complex than elements of $Pow(Dom)$. Let us consider a case when the neighborhood function values are from $Pow^2(Dom)$. Assume that together with an information system, $\mathcal{A} = (U,A)$ is also given a similarity relation τ defined on vectors of attribute values. This relation can be extended to objects. An object $y \in U$ is similar to a given object $x \in U$ if the attribute value vector on x is τ-similar to the attribute value vector on y. Now consider a neighborhood function defined by $\mathcal{N}_{\mathcal{A},\tau}(x) = \{[y]_A : x\tau y\}$.

One might ask why we need such detailed information about the neighborhood function value as in the last case, i.e., a family of indiscernibility classes. One choice will be to take the union of these indiscernibility classes. However, more detailed information is often needed to define some basic relations on information granules, namely, inclusion and closeness of (generated and target) information granules, or, using such a family of elementary information granules, to define some relevant patterns. It can be important for some applications to know about information granules created by a family of indiscernibility classes not only a degree to which its union is included in the target concept but also degrees into which fusions of some subfamilies[2] of such classes are included in a given target concept.

Let us consider more examples. They are related to the granulation of relational structure M by neighborhood functions. We would like to show that, due to the relational structure granulation, we obtain new information granules of more complex structure and in consequence, more general neighborhood functions than those discussed above. Hence, basic granules of knowledge about the universe corresponding to objects can have more complex structures.

---

[1] The power set of X is denoted by $Pow(X)$.

[2] For example, a family of neighborhoods included in the target information granules (concept) to a satisfactory degree.

Assume that a relational structure M and a neighborhood function $\mathcal{N}$ are given. Let us assume at the beginning that the domain of $f_{\mathcal{N}}$ is equal to $Pow(Dom)$. The aim is to define a new relational structure $M_{\mathcal{N}}$ called the *$\mathcal{N}$-granulation* of M. This is done by granulating all components of M by means of $\mathcal{N}$.

We restrict our considerations to two examples. Let us consider first a binary relation $r \subseteq Dom \times Dom$. There are numerous possibilities to define a relation $r_{\mathcal{N}}$ from r. The choice depends on applications. Let us list some possible definitions of $r_{\mathcal{N}}$:

$$\begin{aligned}
&r_{\mathcal{N}}[\mathcal{N}(x), \mathcal{N}(y)] \text{ iff } \mathcal{N}(x) \times \mathcal{N}(y) \subseteq r, \\
&r_{\mathcal{N}}[\mathcal{N}(x), \mathcal{N}(y)] \text{ iff } [\mathcal{N}(x) \times \mathcal{N}(y)] \cap r \neq \emptyset, \\
&r_{\mathcal{N}}[\mathcal{N}(x), \mathcal{N}(y)] \text{ iff } card\{[\mathcal{N}(x) \times \mathcal{N}(y)] \cap r\} \geq s \cdot card[\mathcal{N}(x)], \\
&\text{where } s \in [0,1] \text{ is a threshold.}
\end{aligned} \tag{34}$$

In this way some patterns for pairs of objects are created. Such patterns can be used to approximate a target concept (or concept on an intermediate level) over objects composed of pairs (x, y). Certainly, to induce high-quality approximations it is necessary to search for relevant patterns for concept approximation. This problem is discussed in Sect. 7.

Now, let us consider a function f from M and some possible $\mathcal{N}$-granulations $f_{\mathcal{N}}$ of $\mathcal{N}$. [3]

$$\begin{aligned}
f_{\mathcal{N}}[\mathcal{N}(x), \mathcal{N}(y)] &= \{\mathcal{N}(z) : z = f(x', y') \text{ for some } x' \in \mathcal{N}(x), y' \in \mathcal{N}(y)\}, \\
f_{\mathcal{N}}[\mathcal{N}(x), \mathcal{N}(y)] &= \bigcup_{x' \in \mathcal{N}(x), y' \in \mathcal{N}(y)} \{\mathcal{N}(z) : z = f(x', y')\}, \\
f_{\mathcal{N}}[\mathcal{N}(x), \mathcal{N}(y)] &= \bigcup_{x' \in \mathcal{N}(x), y' \in \mathcal{N}(y)} \{\mathcal{N}(z) : card(\mathcal{N}(z)) \geq s \text{ and } z = f(x', y')\},
\end{aligned} \tag{35}$$

where $s \in [0,1]$ is a threshold.

One can consider the values of $f_{\mathcal{N}}$ as generators of patterns used for the target concept approximation. An example of pattern language can be obtained by considering the results of set theoretical operations on neighborhoods.

Observe that in the first example the values of function $f_{\mathcal{N}}$ are in $Pow^2(Dom)$. Hence, one could extend the neighborhood function and the relation granulation on this more complex domain. Certainly, this process can be continued, and more complex patterns can be generated. On the other hand, it is also necessary to bound the depth of exploring $Pow^{\omega}(Dom)$. This can be done by using the rough set approach. For example, after generating patterns from $Pow^2(Dom)$, one should, in a sense, reduce them to $Pow(Dom)$ by considering some operations from $Pow^2(Dom)$

[3] We assume that f has two arguments.

into $Pow(Dom)$ returning the relevant patterns for the target concept approximation. Such a reduction is necessary, especially if the target concepts are elements of the family $Pow(Dom)$.

We have discussed relations of inclusion and closeness in this chapter. They can also be used for granulation of relational structures. Here, it is worth mentioning that such relations between information granules must be considered not only due to the incomplete information about objects. Many real-life problems can often be expressed as searching for construction of information granules (from some elementary ones) satisfying a given specification (which can also be treated as information granule) to a satisfactory degree. This allows making the construction (reasoning) process more efficient and robust with respect to deviation of parameters, e.g., related to input information granules.

7 Rough Set Approach to Inductive Reasoning

In this section, we would like to explain in more detail computations on information granule systems aiming at constructing relevant approximation spaces (see Sect. 4.4).

One can consider the rough set approximations of decision classes in decision systems (see Chap. 1). Such approximations can be used to obtain models of decision classes. However, in inductive reasoning, we would like to approximate concepts over universe of objects, say U^∞, wider than the universe U of objects in a given decision system. In other words, assuming $U \subset U^\infty$, we would like to approximate concepts over U^∞ that are extensions of decision classes in a given decision system. In this section, we present a general searching scheme for approximation spaces relevant to approximation of such concepts and show how to induce in them *classifiers* approximating those concepts. This is a basic scheme for machine learning, pattern recognition, data mining, and knowledge discovery [11,15].

The main observation is that, in the case considered, it is also necessary to induce a relevant approximation space. Such a space is usually different from the partition defined by the conditional attributes of a given decision system. It consists of some subsets of U^∞, called neighborhoods of objects. It should be emphasized that neighborhoods usually create a covering of U^∞, not necessarily a partition. They are defined by *patterns* chosen from some relevant *pattern languages*. In practical applications, it is often necessary to specify a given model using its particular description in a pattern language. Different descriptions may have substantially different properties outside of U, e.g., patterns for disjoint decision classes can have different nonempty intersections in $U^* - U$. To indicate that a given model is specified by a particular description, we use the term *description model*.

The structure of the pattern languages and the patterns themselves should be discovered. The whole process is quite complex and is illustrated in Fig. 2, where

- $\mathcal{A} = (U, A, d)$ denotes a decision system.[4]
- $\mathcal{A}_{\mathrm{train}}$ and $\mathcal{A}_{\mathrm{test}}$ are training and testing subsystems of $\mathcal{A}$, respectively.
- $\mathcal{L} = \{L_i\}_{i \in I}$ is a family of pattern languages.
- $\mathcal{Q} = \{Q_j\}_{j \in J}$ is a family of quality measures for description models.
- M is a description model covering objects in U.
- C is a classifier obtained from M covering (almost) the whole universe U^{∞}.

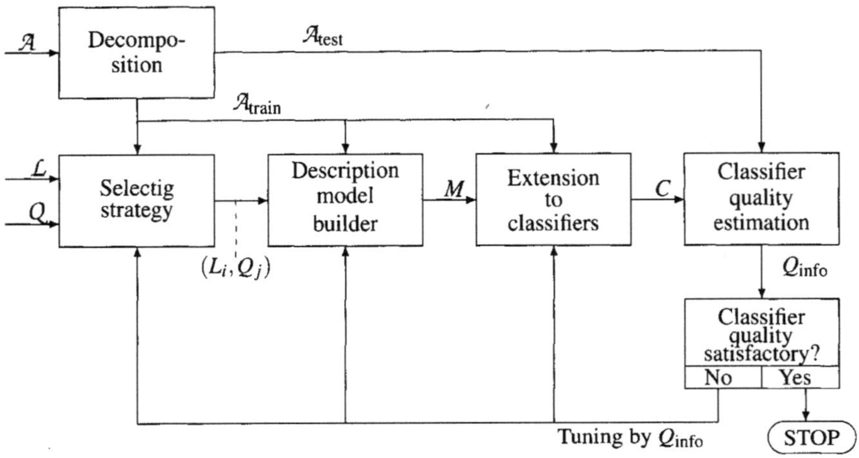

Fig. 2. Approximation space and classifier construction using rough sets.

Elements of L_i are formulas called *patterns*. Patterns, in a given decision system, define sets of objects in which they are satisfied. Description models describe the decision classes of $\mathcal{A}$ by using patterns from L_i and some inclusion measures of those patterns in decision classes. Description models can be built by means of, e.g., decision rules over descriptors from L_i. As a typical example, one can consider the language of patterns consisting of conjunctions of descriptors over a selected set of attributes. More complex pattern language can include conjunctions of formulas that are disjunctions of descriptor conjunctions.

Quality measures can be used as criteria for tuning the model. For given L_i and Q_j, one can search for a description model using patterns from L_i which is (sub-) optimal w.r.t. the measure Q_j. However, the goal is to induce the relevant description model for the induced classifier, covering the whole universe of objects.

[4] For simplicity in reasoning, we assume that $\mathcal{A}$ does not change in time.

This, in particular, requires tuning parameters of the description quality measure. There are many ways to specify quality measures. For example, a measure Q_j, can be specified using the *minimum description length principle* (see Chap. 25), where one estimates the quality of approximation as well as the size of the description model defined. The minimum description length principle requires choosing a description of the smallest size from those of the same approximation quality. In this case, the quality measure depends on two arguments. The first, represents the quality of approximation (e.g., using the positive region of decision classes or entropy measure). The second represents the values of some measures based on the model size. A proper balance between these two arguments is generally obtained by using training data. Tuning may involve thresholds for degrees of inclusion of patterns from L_i in decision classes or for the positive region size. The use of the notion of inclusion to a satisfactory degree allows one to reduce the size of the positive region description, compared to descriptions based on crisp inclusion.

The whole process presented in Fig. 2 can be viewed as a searching process for a relevant approximation space. As we have mentioned before, such an approximation space consists of neighborhoods of objects from U as well as inclusion relations making it possible to measure degrees of inclusion (or closeness) of such neighborhoods in other information granules.

The induced description model should be extended to a classifier of all objects from the whole universe of objects U^{∞}, not only from U. [5] Recall that, for any object to be classified, it is necessary to compute its degree of inclusion in any pattern from the description model. For new objects (outside of U), these degrees can suggest conflicting decisions and, together with the degrees of pattern inclusion in decision classes, create input for conflict resolution strategy necessary to compute the classifier output.

Next, the induced classifier is tested on objects from $\mathcal{A}_{test}$. Information Q_{info} about the classifier behavior quality is returned from the classifier quality estimation module. If Q_{info} shows that the classifier quality is unsatisfactory, it is used to tune parameters in different modules presented in Fig. 2 and to reconstruct the classifier to a new one with better quality. In addition, matching strategies for objects and patterns, as well as parameters for conflict resolution strategy, can also be tuned. The parameters involved in the tuning process can, for instance, be inclusion degree thresholds, parameters characterizing approximation quality, or parameters measuring the description model size.

The approximation spaces discussed are examples of information granules. They can be obtained as results of complex computations on information granule systems.

[5] In Sect. 4.4 and Chap. 25, we discuss a classifier structure.

8 AR schemes and Rough-Neural Networks

AR schemes are the basic constructs used in RNC. Such schemes can be derived from parameterized productions representing robust dependencies on data. Algorithmic methods for extracting such productions from data are discussed in [24,31], [34]. The left-hand side of each *production* (see Fig. 3) is (in the simplest case) of the form

$$\left(st_1(ag),(\varepsilon_1^{(1)},\ldots,\varepsilon_r^{(1)})\right),\ldots,\left(st_k(ag),(\varepsilon_1^{(k)},\ldots,\varepsilon_r^{(k)})\right) \tag{36}$$

and the right-hand side is of the form $[st(ag),(\varepsilon_1,\ldots,\varepsilon_r)]$ for some positive integers k,r.

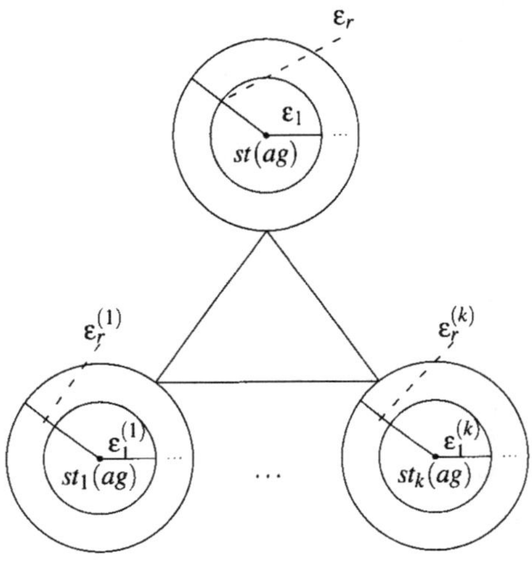

Fig. 3. Parameterized production

Such a production represents information about an operation o that can be performed by the agent ag. In a production, k denotes the arity of operation. The operation o represented by the production transforms standard (prototype) input information granules $st_1(ag),\ldots,st_k(ag)$ into the standard (prototype) information granule $st(ag)$. Moreover, if input information granules $g_1,\ldots,g_k$ are close to

$$st_1(ag),\ldots,st_k(ag)$$

to degrees at least $\varepsilon_j^{(1)},\ldots,\varepsilon_j^{(k)}$, then the result of operation o on information granules $g_1,\ldots,g_k$ is close to the standard $st(ag)$ to a degree at least ε_j where $1\le j\le k$. Standard (prototype) granules can be interpreted in different ways. In particular, they can correspond to concept names in natural language.

The productions described above are basic components of a reasoning system over an agent set Ag. An important property of such productions is that they are expected to be discovered from available experimental data and background knowledge. Let us also observe that the degree structure is not necessarily restricted to reals from the interval $[0,1]$. The inclusion degrees can have a structure of complex information granules used to represent the degree of inclusion. It is worthwhile mentioning that the productions can also be interpreted as constructive descriptions of some operations on fuzzy sets. The methods for such constructive description are based on rough sets and Boolean reasoning [12,19].

AR schemes can be treated as derivations obtained by using productions from different agents. The relevant derivations generating AR schemes satisfy a so-called *robustness* (or *stability*) condition. This means that at any node of derivation the inclusion (or closeness) degree of a constructed granule to the prototype (standard) granule is higher than that required by the production to which the result should be sent (see Fig. 4). This makes it possible to obtain a sufficient robustness condition for all derivations. For details, refer to [26,27].

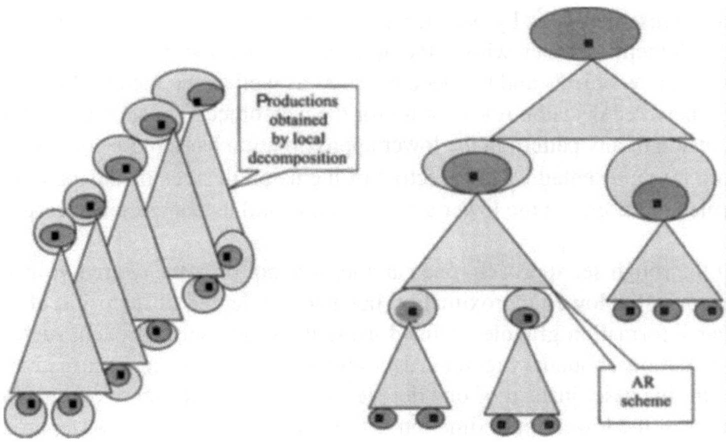

Fig. 4. Productions and AR schemes

AR schemes are discovered from data and background knowledge by using symbolic reasoning. After that they can be used as in connectionist approach. If sensor measurements are sufficiently close to inputs of AR schemes, then with high probability, one can predict that an analyzed object (situation) belongs to a target concept (the concept of a *dangerous* situation on a road considered earlier). Hence, symbolic and connectionist approaches can work as complementary not competitive approaches.

When standards are interpreted as concept names in natural language and a reasoning scheme in natural language over the standard concepts is given, the corresponding AR scheme represents a cluster of reasoning (constructions) approximately following (by means of other information granule systems) the reasoning in natural language. In the following section, we discuss different approaches to standard information granule definition.

8.1 Standards Represented by Rough Sets

In the simplest case, standards can be represented by lower approximations of concepts. The degree of inclusion of a pattern supported by objects from the set $X \subseteq U$ in the lower approximation supported by the objects from the set $Y \subseteq U$ can be measured by the ratio $|X \cap Y|/|X|$ where U is the set of objects in a given decision table (representing the training sample).

However, if the lower approximation is intended to describe the concept in an extension of the training sample U, then inductive reasoning should be used to find an approximation of this lower approximation of the concept. Such approximations can be represented, e.g., by decision rules describing the lower approximation and its complement together with a method making it possible to measure matching degrees of new objects and the decision rules as well as the method for conflict resolution between decision rules voting for the new objects. In such cases, the degree of inclusion of any pattern in the lower approximation has a more complex structure and can be represented by two vectors of inclusion degrees of this pattern in decision rules representing the lower approximation and its complement, respectively.

Using the rough set approach, one can measure not only the degree of inclusion of a concept in the lower approximation but also the degree of inclusion of a concept in other information granules defined using the rough set approach, such as upper approximations, boundary regions, or complements of upper approximations of concepts. In this case, instead of one degree, one should consider a vector of degrees. However, if the lower approximation is too small, then such a lower approximation cannot be treated as a standard of good quality and it may be necessary to consider other kinds of standards that can be constructed using, e.g., a rough-fuzzy approach or classifier construction methods.

8.2 Standards Corresponding to Rough-Fuzzy Sets

The approach presented in the previous section can be extended to concepts defined by fuzzy sets. We will show that the dependencies between linguistic variables can be modeled by productions. Using the rough-fuzzy approach, one can search for dependencies between lower approximations of differences between relevant cuts of fuzzy sets modeling linguistic variables. The productions built along such dependencies make it possible to model dependencies between linguistic variables.

Moreover, the approximate reasoning on linguistic variables can be modeled by approximate reasoning schemes derived from productions.

We are now going to describe rough-fuzzy granules. We assume that if X is an information granule, e.g., a set of objects, then its upper and lower approximations with respect to any subset of attributes in a given information system or decision table are information granules, too. Let us see now how such information granules can be used to define fuzzy concept [41] approximations in a constructive way.

Let $DT = (U, A, d)$ be a decision table where the decision d is the fuzzy membership function ν restriction to the objects from U. Consider reals $0 < c_1 < \ldots < c_k$ where $c_i \in (0,1]$ for $i = 1,\ldots,k$. Any c_i defines c_i-cut by $X_i = \{x \in U : \nu(x) \geq c_i\}$. Assume that $X_0 = U$ and $X_{k+1} = X_{k+2} = \emptyset$. A *rough-fuzzy granule* (*rf-granule*, for short) corresponding to $(DT, c_1, \ldots, c_k)$ is any granule $g = (g_0, \ldots, g_k)$ such that for some $B \subseteq A$,

$$\begin{aligned} &Sem_B(g_i) = \left[\underline{B}(X_i - X_{i+1}), \overline{B}(X_i - X_{i+1})\right], \text{ for } i = 0,\ldots,k, \text{ and} \\ &\overline{B}(X_i - X_{i+1}) \subseteq (X_{i-1} - X_{i+2}), \text{ for } i = 1,\ldots,k, \end{aligned} \tag{37}$$

where $\underline{B}$ and $\overline{B}$ denote the B-lower and B-upper approximation operators, respectively [19], and $Sem_B(g_i)$ denotes the semantics of g_i.

Any function $\nu^* : U \rightarrow [0,1]$ satisfying the conditions

$$\begin{aligned} &\nu^*(x) = 0, \text{ for } x \in U - \overline{B}X_1, \\ &\nu^*(x) = 1, \text{ for } x \in \underline{B}X_k, \\ &\nu^*(x) = c_{i-1}, \text{ for } x \in \underline{B}(X_{i-1} - X_i), \text{ and } i = 2,\ldots,k-1, \\ &c_{i-1} < \nu^*(x) < c_i, \text{ for } x \in (\overline{B}X_i - \underline{B}X_i), \text{ where } i = 1,\ldots,k, \text{ and } c_0 = 0, \end{aligned} \tag{38}$$

is called a B-approximation of ν.

Assume that a rule *if* α *and* β *then* γ is given, where α, β, γ are linguistic variables. The aim is to develop a searching method for rough-fuzzy granules g^1, g^2, g^3 approximating, to satisfactory degrees, α, β, γ, respectively, and at the same time, making it possible to discover association rules of the form *if* α' *and* β' *then* γ' with sufficiently large support and confidence coefficients, where α', β', γ' are some components (e.g., the lower approximations of differences between cuts of fuzzy concepts corresponding to linguistic variables) of granules g^1, g^2, g^3 (modeling linguistic variables), respectively. Searching for such patterns and rules is a complex process with many parameters to be tuned. For given linguistic rules, the relevant cuts for fuzzy concepts corresponding to them should be discovered. Next, the relevant features (attributes) should be chosen. They are used to construct approximations of differences between cuts. Moreover, relevant measures should be chosen to measure the degree of inclusion of object patterns in the lower approximations constructed. One can expect that these measures are parameterized and that the relevant parameters should be discovered in the process of searching for productions.

Certainly, in searching for relevant parameters in this complex optimization process, evolutionary techniques can be used. The quality of discovered rules can be measured as a degree to which discovered rule *if* α' *and* β' *then* γ' approximates the linguistic rule *if* α *and* β *then* γ. This can be expressed by such parameters as degrees of inclusion of patterns α', β', γ' in α, β, γ, their supports, etc.

Let us observe that for a given linguistic rule, it will be necessary to find a family of rules represented by discovered patterns which together create an information granule sufficiently close to a modeled linguistic rule. One can also search for more general information granules representing clusters of discovered rules

$$if\ \alpha'\ and\ \beta'\ then\ \gamma'$$

approximating the linguistic rule

$$if\ \alpha\ and\ \beta\ then\ \gamma.$$

These clustered rules can be of higher quality. Certainly, this makes it necessary to discover and tune many parameters relevant to measuring the similarity or closeness of rules.

The problem discussed is of great importance in classifying situations by autonomous systems on the basis of sensor measurements [40]. Moreover, this is one of the basic problems to be investigated for hybridization of rough and fuzzy approaches.

8.3 Standards Corresponding to Classifiers

For classifiers, we obtain another possibility. Let us consider information granules corresponding to values of terms $Match(e, \{G_1, \ldots, G_k\})$ for $e \in E$ (see also Sect. 6.1 in Chap. 25 and [31]), where E is a set of elementary granules and $G_1, \ldots,$ G_k are granules corresponding to patterns described by the left-hand sides of decision rules for k decision classes. Any such granule defines a probability distribution on a set of possible decisions (extended by the value corresponding to *no decision predicted*). The probability for each such value is obtained simply as a ratio of all votes for the decision value determined by this information granule and the number of objects. Some probability distributions can be chosen as standards. this means that instead of the lower approximations, one can use such probability distributions. Certainly, it can sometimes be useful to choose not one such standard but a collection of them. Now, one should decide how to measure the distances between probability distributions. Using a chosen distance measure, e.g., Euclidean or a more advanced one developed in statistics, it is possible to measure the degree of closeness of classified objects e, e' using the probability distributions corresponding to them. The next steps in constructing an approximate reasoning rule based on classifiers is analogous to the discussed before.

One of the most interesting cases occurs when standards are interpreted as concepts from natural language. In this case, measures of inclusion and closeness can be based on semantic similarity and closeness relations rather than on statistical properties. Constructing such measures is a challenge. This case is strongly related to the CW paradigm. The productions discovered can, to a satisfactory degree, be consistent with reasoning steps performed in natural language.

9 Rough-Neural Networks

Rough-neural networks are schemes for information granule construction in a distributed environment. They perform computations on information granules representing concepts rather than on numbers. These information granules can be constructed, e.g., using a rough set approach, rough-fuzzy approach, or an approach based on classifiers.

Rough-neural networks have several types of parameters to be tuned. Let us list some of them assuming that we use a rough set approach for concept approximations:

- parameters of approximation spaces used to approximate concepts transformed by operations performed by agents.
- parameters of approximation spaces located in interfaces between agents used to approximate, by one agent, concepts communicated by some other agents.
- degrees of inclusion and closeness of patterns constructed along the network to concept approximations in the network (see Fig. 5).
- parameters of approximation spaces for pattern approximations.

Let us observe that for uncertainty rules illustrated in Fig. 5, the approximation spaces related to agents ag, ag_1, ag_2 are also involved in discovery from data of function f, called a mereological connective [24]. Rough mereological connectives make it possible to propagate uncertainty coefficients. If objects (or patterns) x_1, x_2 delivered by agents ag_1, ag_2 are close to standards $st(ag_1), st(ag_2)$ to degrees at least $\varepsilon_1, \varepsilon_2$, then the object (pattern) x constructed by operation O from x_1, x_2 is close to $st(ag)$ to degree at least $f(\varepsilon_1, \varepsilon_2)$. Hence, if $f(\varepsilon_1, \varepsilon_2) \geq \varepsilon$, where ε is a given admissible uncertainty threshold for ag, then an estimate given by a rough mereological connective is sufficient.

From the above considerations, it follows that a rough-neural network is a complex structure with different sorts of parameters to tune. The goal of tuning these parameters is to learn structures that can compute relevant information granules with acceptable accuracy.

We would like to comment on approximation spaces creating interfaces between communicating agents. These parameterized approximation spaces can be treated

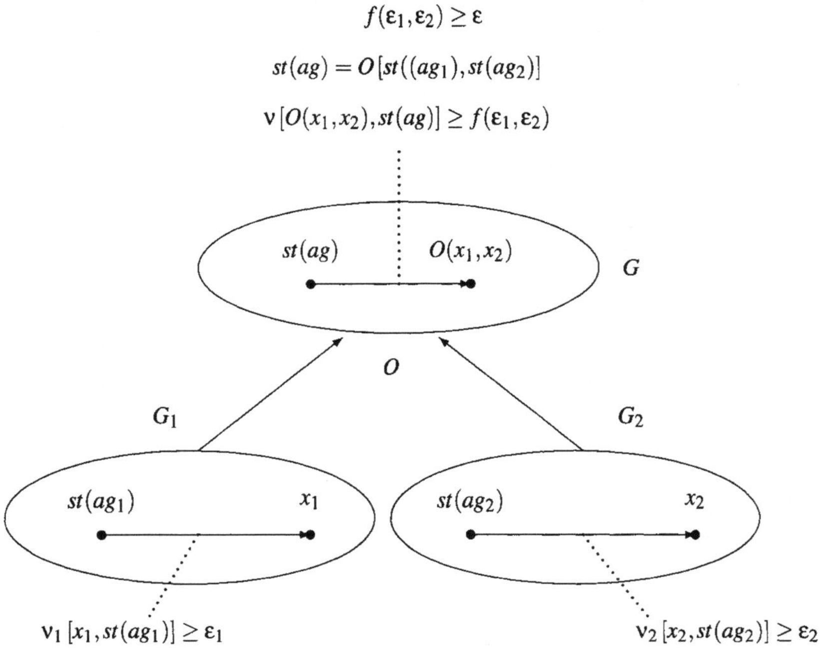

Fig. 5. Uncertainty rules

as analogous to neural network weights. These parameters should be learned to induce the relevant information granules.

Extending AR schemes for synthesizing information granules by adding interfaces represented by approximation spaces and used for approximating information granules exchanged between agents, we obtain granule construction schemes that can be treated as generalizations of neural network models. The main idea is that granules sent by one agent to another are not, in general, exactly understandable by the receiving agent because these agents are using different languages and usually there is no translation (from the sender language to the receiver language) preserving the exact semantic meaning of formulas. Hence, it is necessary to construct interfaces that will make it possible to approximately understand received granules. These interfaces can be, in the simplest case, constructed on the basis of information exchanged by agents and stored in the form of decision data tables. From such tables, the approximations of concepts can be constructed using a rough set approach. In general, it is a complex process because a high-quality approximation of concepts can often be obtained only in dialogue (involving negotiations, conflict resolutions, and cooperation) among agents. In this process, the approximation can be constructed gradually when the dialogue is progressing. In our model, we assume that for any n-ary operation $o(ag)$ of agent ag, there are approximation spaces $AS_1[o(ag),in],\ldots,AS_n[o(ag),in]$ that will filter (approximate) the granules received by the agent for performing operation $o(ag)$. In turn, the granule sent by an agent after performing the operation is filtered (approximated) by the approximation space $AS[o(ag),out]$. These approximation spaces are parameterized. The parameters are used to optimize the size of neighborhoods in these spaces as well as the inclusion relation. A granule approximation quality is taken as the optimization criterion. Approximation spaces attached to any operation of ag correspond to neuron weights in neural networks, whereas operation performed by an agent ag on information granules corresponds to an operation realized on vectors of real numbers by a neuron (see Fig. 1 in Chap. 2). A parameterized approximation space can be treated as an analogy to a neural network weight. In this figure, $w_1,\ldots,w_n,f$ denote weights, aggregation operator, and activation function of a classical neuron, respectively, whereas

$$AS_1(P),\ldots,AS_k(P)$$

denote parameterized approximations spaces where agents process input granules $G_1,\ldots,G_k$ and O denotes an operation (usually parameterized) that produces the output of a granular network. The parameters P of approximation spaces should be learned to induce the relevant information granules.

We call extended schemes for complex object construction *rough neural networks*. The problem of deriving such schemes is closely related to perception [43]. The stability of such networks corresponds to the resistance to noise of classical neural networks.

Let us observe that in our approach deductive systems are substituted by production systems of agents linked by approximation spaces, communication strategies, and mechanism for deriving AR schemes. This revision of classical logical notions seems to be important for solving complex problems in distributed environments.

10 Extracting AR Schemes from Data and Background Knowledge

In this section, we present some methods of information granule decomposition aimed at extracting decomposition rules from data. We restrict our considerations to methods based only on experimental data. This approach can be extended to information granule decomposition methods using background knowledge [36].

The search methods discussed in this section return local granule decomposition schemes. These local schemes can be composed by using techniques discussed in the previous section. The schemes of granule construction received (which can also be treated as approximate reasoning schemes) also have the following property: if input granules are sufficiently close to input concepts, then the output granule is sufficiently included in the target concept, provided this property is preserved locally [27].

The above may be formulated in terms of a synthesis grammar [28] with productions corresponding to local decomposition rules. The relevant derivations over a given synthesis grammar represent approximate reasoning schemes. Note that synthesis grammars reflect processes in multiagent systems in which agents are involved in cooperation, negotiation, and conflict-resolving actions when attempting to provide a solution to the specification of a problem. Complexities of membership problems for languages generated by synthesis grammars may be taken *ex definitione* as complexities of the underlying synthesis processes.

We show that in some cases decomposition can be performed using methods for specific rule generation based on Boolean reasoning [12]. Moreover, we present the way the decomposition, stable with respect to information granule deviations, can be obtained.

First, let us start from some general remarks. Information granule decomposition methods are important components of methods for inducing AR schemes from data and background knowledge. Such methods are used to extract local decomposition schemes, called productions, from data [26]. The AR schemes are constructed by means of productions.

Decomposition methods are based on searching for the parts of information granules that can be used to construct relevant, higher level patterns that match, to a satisfactory degree, the target granule.

One can distinguish two kinds of parts (represented, e.g., by subformulas or subterms) of AR schemes. Parts of the first type are represented by expressions from a language, called the *domestic* language L_d, that has known semantics (consider, for example, semantics defined in a given information system [19]). Parts of the second type of AR scheme are from a language, called *foreign* language L_f (e.g., natural language), that has semantics definable only in an approximate way (e.g., by patterns extracted using rough, fuzzy, rough-fuzzy or other approaches). For example, the parts of the second kind of scheme can be interpreted as soft properties of sensor measurements [4].

For a given expression e, representing a given scheme that consists of subexpressions from L_f first it is necessary to search for relevant approximations in L_d of the foreign parts from L_f and next to derive global patterns from the whole expression after replacing the foreign parts by their approximations. This can be a multilevel process, i.e., we face problems of pattern propagation discovered through several domestic-foreign layers.

Productions from which AR schemes are built can be induced from data and background knowledge by pattern extraction strategies. Let us consider some such strategies. The first makes it possible to search for relevant approximations of parts by using the rough set approach. This means that each part from L_f can be replaced by its lower or upper approximation with respect to a set B of attributes. The approximation is constructed on the basis of a relevant data table [12,19]. With the second strategy, parts from L_f are partitioned into a number of subparts corresponding to cuts (or the set theoretical differences between cuts) of fuzzy sets representing vague concepts, and each subpart is approximated by rough set methods. The third strategy is based on searching for patterns sufficiently included in foreign parts. In all cases, the extracted approximations replace foreign parts in the scheme, and candidates for global patterns are derived from the scheme obtained after the replacement. Searching for relevant global patterns is a complex task because many parameters should be tuned, e.g., the set of relevant features used in approximation, relevant approximation operators, the number and distribution of objects from the universe of objects among different cuts, and so on. One can use evolutionary techniques [13] in searching for semioptimal patterns in the decomposition.

It has been shown that decomposition strategies can be based on rough set methods developed for decision rule generation in combination with the Boolean approach [16,23,24,35]. In particular, methods for decomposition based on background knowledge can be developed [31,34,36].

Now we can turn to some details of granule decomposition methods. We assume that a family of inclusion relations $\nu_p^i \subseteq G_i \times G_i$, $\nu_p^H \subseteq H \times H$ and a family of closeness relations $cl_p^1, \ldots, cl_p^k, cl_p^H$ for every $p \in [0,1]$ and $i = 1, \ldots, k$ are given [27]. Let us assume that two thresholds t, p are given. We define a relation

$$Q_{t,p}^{DT}(Pattern_1, \ldots, Pattern_k, \overline{v})$$

between granules called patterns $Pattern_1, \ldots, Pattern_k$ for arguments of f and the target pattern $\overline{v}$ representing the decision value vector in the following way:

$$Q_{t,p}^{DT}(Pattern_1, \ldots, Pattern_k, \overline{v}) \text{ if and only if} \tag{39}$$
$$\nu_p^H \left\{ f\left[Sem_{DT}(Pattern_1) \times \ldots \times Sem_{DT}(Pattern_k)\right], [\overline{v}]_{IND(d)} \right\} \text{ and}$$
$$card\left[Sem_{DT}(Pattern_1) \times \ldots \times Sem_{DT}(Pattern_k)\right] \geq t.$$

Let us now consider the following decomposition problem:

Granule decomposition problem
Input:

- Two thresholds t, p.
- A decision table $DT = (U, A, d)$ representing an operation $f : G_1 \times \ldots \times G_k \to H$ where $G_1, \ldots, G_k$ and H are given finite sets of information granules.

- A fixed decision value vector $\overline{v}$ represented by a value vector of decision attributes.

Output:

- A tuple $(Pattern_1, \ldots, Pattern_k)$ of patterns such that

$$Q^{DT}_{t,p}(Pattern_1, \ldots, Pattern_k, \overline{v}).$$

We consider a description given by decision rules extracted from the data table specifying the function f. Any left-hand side of a decision rule can be divided into parts corresponding to different arguments of the function f. The ith part, denoted by $Pattern_i$, specifies a condition that should be satisfied by the ith argument of f to obtain the function value specified by the decision attributes. For simplicity, we do not consider conditions specifying the relations between arguments. In this way, the left-hand sides of decision rules describe patterns, $Pattern_i$. The semantics of extracted patterns relevant for the target can be defined as the image with respect to f of the Cartesian product of sets $Sem_{DT}(Pattern_i)$, i.e., by $f\left[Sem_{DT}(Pattern_1) \times \ldots \times Sem_{DT}(Pattern_k)\right]$ (see Fig. 6). One can use one of the methods for decision rule generation, e.g., for generating of minimal rules or their approximations (e.g., in the form of association rules) [12] to obtain such decision rules.

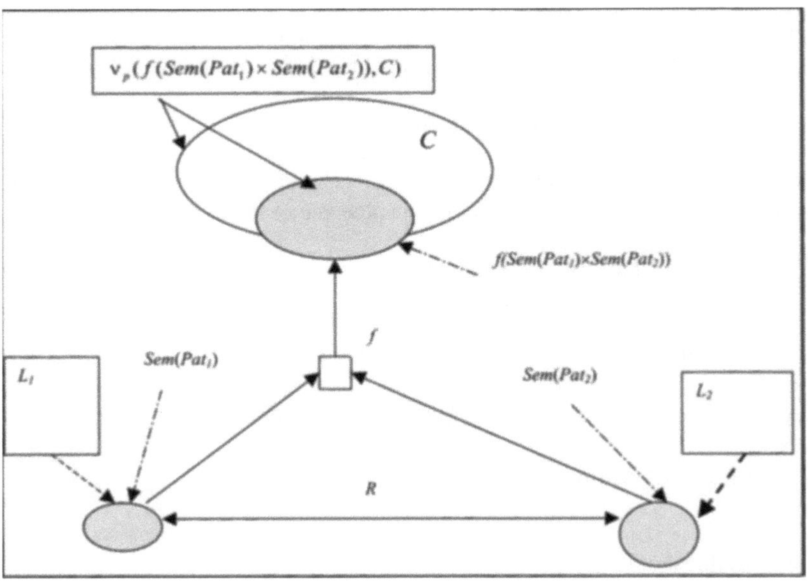

Fig. 6. Decomposition of information granule

In the former case, we receive the most general patterns for function arguments consistent with a given decision table, i.e., the information granules constructed by means of the function f from patterns extracted for arguments are included exactly in the information granule represented by a given decision value vector in the data table. In the latter case, we obtain more general patterns for function arguments having the following property: information granules constructed by means of f from such patterns will be included to a satisfactory degree in the information granule represented by a given decision value vector in the data table.

One of the very important properties of the operations on information granules discussed above is their robustness with respect to the deviations of arguments (see, e.g., [28]). This property can be formulated as follows: if an information granule constructed by means of f from the extracted patterns, $Pattern_1, \ldots, Pattern_k$, satisfies the target condition, then the information granule constructed from patterns, $Pattern'_1, \ldots, Pattern'_k$, sufficiently close to $Pattern_1, \ldots, Pattern_k$, respectively, satisfies the target condition, too. In this way, we obtain the following problem:

Robust decomposition problem (RD problem)
Input:

- Thresholds t, p.
- A decision table $DT = (U, A, d)$ representing an operation $f : G_1 \times \ldots \times G_k \to H$ where $G_1, \ldots, G_k$ and H are given finite sets of information granules.
- A fixed decision value vector $\bar{v}$ represented by a value vector of decision attributes.

Output:

- A tuple $(p_1, \ldots, p_k)$ of parameters.
- A tuple $(Pattern_1, \ldots, Pattern_k)$ of patterns such that

$$Q^{DT}_{t,p}(Pattern'_1, \ldots, Pattern'_k, \bar{v}),$$

if $cl^i_{p_i}(Sem_{DT}(Pattern_i), Sem_{DT}(Pattern'_i))$ for $i = 1, \ldots, k$.

It is possible to search for the solution of the RD problem by modifying the previous approach of decision rule generation. In the process of rule generation, one can impose a stronger discernibility condition by assuming that objects are discernible if their tolerance classes are disjoint. Certainly, one can tune parameters of tolerance relations to obtain rules of satisfactory quality. We would like to stress that efficient heuristics for solving these problems can be based on Boolean reasoning [12].

Searching for relevant patterns for information granule decomposition can be based on methods for tuning parameters of rough set approximations of fuzzy cuts or concepts defined by differences between cuts (see Sect. 8). In this case, pattern languages consist of parameterized expressions describing the rough set approximations of *parts* of fuzzy concepts as fuzzy cuts or differences between cuts. Hence, an interesting research direction related to the development of new hybrid rough-fuzzy methods arises aiming at developing algorithmic methods for rough set approximations of such parts of fuzzy sets relevant to information granule decomposition. An approach presented in this section can be extended to local granule decomposition based on background knowledge [36].

11 Basic Concepts Definable by Means of Rough Inclusions

In this section, we present several examples of basic notions for rough sets and granular computing definable by means of rough inclusion relation. The examples illustrate an important role of inclusion relations in rough set theory and granular computing.

11.1 Indiscernibility and Discernibility

Let us start from the fundamental notion of rough set theory, namely, indiscernibility of objects in a given information system. This notion, in granular computing, should be generalized to arbitrary information granules. Let us observe that indiscernibility is defined relative to a given information system that is, as shown, a special kind of information granule. We generalize the indiscernibility relation to arbitrary information granules and define it relative to given information granule and degree inclusion. Any information granules g_1, g_2 are indiscernible relative to a given information granule h and a degree p if for any exact part h' of h, granule g_1 is included to a degree at least p in h' if and only if g_2 is included to a degree at least p in h'.

More formally, granules g_1, g_2, h, p are given information granules. Granules g_1, g_2 are *h-indiscernible* (see Fig. 7) to a degree at least p, in symbols $g_1 IND_h^p g_2$, if and only if

$$\forall h' \left\{ \nu_1(h',h) \Rightarrow \left[\nu_p(g_1,h') \Leftrightarrow \nu_p(g_2,h')\right]\right\}. \tag{40}$$

Certainly, there are some other possibilities for introducing the indiscernibility of granules. For example, one could consider, instead of the rough inclusion relation, the closeness of granules to a degree (see Sect. 4.3) or only some parts of h, e.g., maximal proper parts of h.

When inclusion relations are set inclusions, $p = 1$, h is an information system, g_1, g_2 are objects in the information system and as exact parts of h are considered indiscernibility classes [19] of the information system, we obtain the definition of the indiscernibility considered in rough set theory.

Let us consider one more example of indiscernibility for decision rules. Intuitively, two such rules are indiscernible if they are matched by the same objects and they

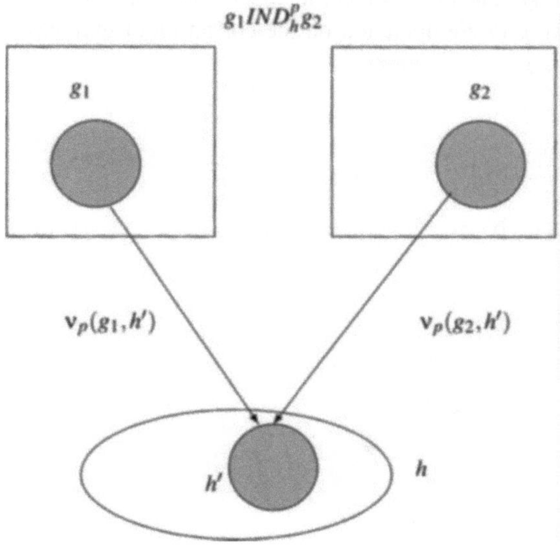

Fig. 7. Indiscernibility of information granules

predict the same decision. The indiscernibility of such information granules can be defined relative to a given information granule as a set of attribute value vectors corresponding to objects. Two decision rules r_1, r_2 are indiscernible relative to such a granule to a degree at least p if and only if any attribute value vector matches (i.e., is included to degree p) the left-hand side of rule r_1 if and only if it matches the left-hand side of rule r_2.

One direct method defining *discernibility* of information granules relative to a given information granule is to define such a relation as a complement of the indiscernibility relation, i.e.,

$$g_1 DIS_h^p g_2 \text{ if and only if } g_1 IND_h^p g_2 \text{ does not hold.} \tag{41}$$

Hence, two objects g_1, g_2 are discernible relative to h to a degree at least p if there exists a part h' of h discerning them, i.e., such that conditions $\nu_p(g_1, h')$ and $\nu_p(g_2, h')$ are not equivalent. Certainly, this is only the simplest case of discernibility. Discernibility can also be defined by not taking the complement of indiscernibility.

11.2 Semantics of Information Granules

Let us observe that *semantics of information granules* can be defined relative to preassumed inclusion relations. Let us consider an example illustrated in Fig. 8. For given information granules g,h, we define h-semantics of g, as a granule equal to the result of $Make\_granule$ operation on the collection of granules h' included to degree at least p in g.

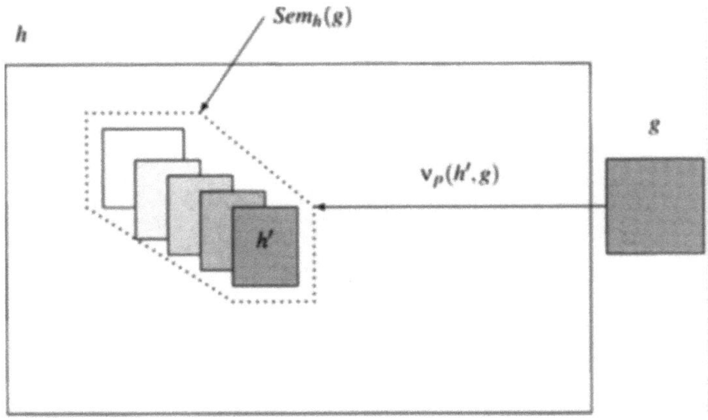

Fig. 8. Semantics of granule g defined relative to h

11.3 Approximation Spaces

Using rough inclusions, one can generalize the approximation operations for sets of objects, known in rough set theory, to arbitrary information granules. The idea is to consider a family $G = \{g_t\}_t$ of granules by means of which a given granule g should be approximated. We assume that for a given set $\{g_1,\ldots, g_k\}$ of information granules included to a degree at least p in g, there is a granule $Make\_granule(\{g_1,\ldots, g_k\})$ included to a degree at least $f(p)$ in g, representing in a sense a collection $\{g_1,\ldots, g_k\}$ where f is a function transforming inclusion degrees into inclusion degrees. A typical example of $Make\_granule$ is the set theoretical union used in rough set theory. Let us recall that inclusion degrees are partially ordered by a relation $\leq$.

Assume that p is an inclusion degree, $G = \{g_t\}_t$ is a given family of information granules, and g is a granule from a given information granule system S. The (G,p)-*approximation* of g, in symbols $APP_{G,p}(g)$, is an information granule defined by

$$Make\_granule(\{g_t : \nu_p(g_t,g)\}). \tag{42}$$

Now, assuming that, $p < q$, one can consider two approximations for a given information granule g by G. The (G,q)-*lower approximation* of g is defined by

$$LOW_{G,p,q}(g) = APP_{G,q}(g) \text{ (see Fig. 9).} \tag{43}$$

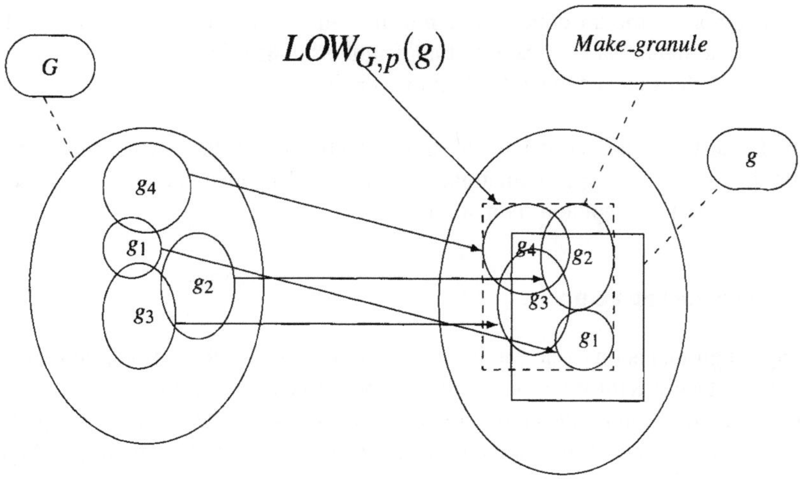

Fig. 9. (G,p)-lower approximation of granule g

The (G,q)-*upper approximation* of g is defined by

$$UPP_{G,p,q}(g) = Make\_granule(\{APP_{G,p'}(g) : p' > p\}). \tag{44}$$

Let us recall the general definition of an approximation space [33]. A *parameterized approximation space* is a system $AS_{\#,\$} = (U, I_{\#}, \nu_{\$})$, where

- U is a nonempty set of objects,
- $I_{\#} : U \rightarrow P(U)$, where $P(U)$ denotes the power set of U, is an uncertainty function, and
- $\nu_{\$} : P(U) \times P(U) \rightarrow [0,1]$ is a rough inclusion function.

If $p \in [0,1]$, then $\nu_p(X,Y)$ denotes that the condition, $\nu(X,Y) \geq p$, holds. The uncertainty function defines for every object x a set of similarly described objects, i.e., the neighborhood $I_{\#}(x)$ of x. A constructive definition of an uncertainty function can be based on the assumption that some metrics (distances) are given for attribute values.

A set $X \subseteq U$ is *definable in* $AS_{\#,\$}$, if it is a union of some values of the uncertainty function. The rough inclusion function defines the degree of inclusion between two subsets of U [33]. For example, if X is nonempty, then $\nu_p(X,Y)$ if and only if $p \leq \frac{card(X \cap Y)}{card(X)}$. If X is the empty set, we assume that $\nu_1(X,Y)$. For a parameterized approximation space $AS_{\#,\$} = (U, I_{\#}, \nu_{\$})$ and any subset $X \subseteq U$, the lower and the upper approximations are defined by

$$LOW\left(AS_{\#,\$}, X\right) = \{x \in U : \nu_{\$}\left(I_{\#}(x), X\right) = 1\} \text{ and} \tag{45}$$
$$UPP\left(AS_{\#,\$}, X\right) = \{x \in U : \nu_{\$}\left(I_{\#}(x), X\right) > 0\}, \text{ respectively.}$$

One can observe that the above definition of a parameterized approximation space is an example of the introduced notion of information granule approximation. It is enough to assume that G is the set of all neighborhoods $I_{\#}(x)$ for $x \in U$, $g \subseteq U$, and *Make_granule* is the set theoretical union, $p = 0$, $q = 1$.

It is useful to define parameterized approximations with parameters tuned in the searching process for approximations of concepts. This idea is crucial for methods of constructing concept approximations.

11.4 Granule Structures

Using rough inclusions, one can define the structures of information granules. The structure of a given information granule can be defined by means of a graph in which (directed) edges represent the relation as a part to a given degree between parts of an information granule. In searching for relevant patterns in data mining problems, one can look for clusters of information granule structures representing collections of granules with similar structures and having a required property, e.g., the majority of objects with such structure are in a given decision class. In the simplest case, two information structures are similar if they have the same graph structure and parts labeling nodes in the structure are sufficiently close. Such clusters of information granules are called *perception structures* (*terms*) (see Fig. 10).

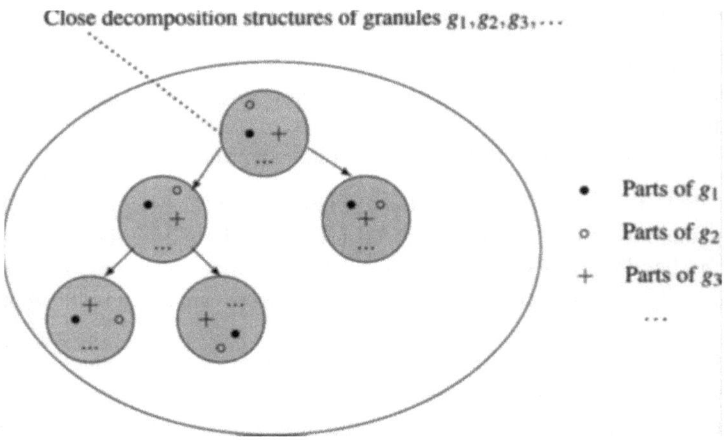

Fig. 10. Perception structure of information granules

12 Conclusions

We have outlined a methodology for approximate reasoning by means of RNC. Several research directions are related to AR schemes and rough-neural networks discussed. We enclose a list of such directions together with examples of problems.

1. *Developing foundations for information granule systems.* Certainly, still more work is needed to develop solid foundations for synthesizing and analyzing information granule systems. In particular, methods for constructing hierarchical information granule systems and methods for representing such systems should be developed.
2. *Algorithmic methods for inducing parameterized productions.* Some methods have already been reported, such as the discovery of rough mereological connectives from data [24] and methods based on decomposition [23,31,32,36]. However, these are only initial steps toward algorithmic methods for inducing parameterized productions from data. One interesting problem is to determine how such productions can be extracted from data and background knowledge. A method in this direction has been proposed in [4].
3. *Algorithmic methods for synthesizing of AR schemes.* It was observed [29,32] that problems of negotiation and conflict resolution are of great importance in synthesizing of AR schemes. The problem arises, e.g., when we are searching in a given set of agents for a granule sufficiently included or close to a given one. These agents, often working with different systems of information granules, can derive different granules, and their fusion will be necessary to obtain the relevant output granule. In the fusion process, negotiations and conflict resolutions are necessary. Much more work should be done in this direction by using the existing results on negotiations and conflict resolution. In particular, Boolean reasoning methods seem to be promising [32]. Another problem is related to the size of production sets. These sets can be large, and it is important to develop learning methods for extracting *small* candidate production sets in the process of extending of temporary derivations out of huge production sets. For solving this kind of problem, methods for clustering productions should be developed to reduce the size of production sets. Moreover, dialogue and cooperative strategies between agents can help to reduce the search space for necessary extension of the temporary derivations.
4. *Algorithmic methods for learning in rough-neural networks.* A basic problem in rough-neural networks is related to selecting relevant approximation spaces and to parameter tuning. One can also look up to what extent existing methods for classical neural methods can be used for learning in rough-neural networks. However, it seems that new approaches and methods for learning in rough-neural networks should be developed to deal with real-life applications. In particular, it is due to the fact that high-quality approximations of concepts can often be obtained only through dialogue and negotiation processes among agents in which the concept approximation is gradually constructed. Hence, for rough-neural networks, learning methods based on dialogue, negotiation and

conflict resolution should be developed. In some cases, one can use rough set and Boolean reasoning methods directly [35]. However, more advanced cases need new methods. In particular, hybrid methods based on rough and fuzzy approaches can bring new results [20].

5. *Fusion methods in rough-neural neurons.* A basic problem in rough neurons is fusion of the inputs (information) derived from information granules. This fusion makes it possible to contribute to the construction of new granules. When a granule constructed by a rough neuron consists of characteristic signal values made by relevant sensors, a step in the direction of solving the fusion problem can be found in [21].
6. *Adaptive methods.* Certainly, adaptive methods for discovering productions and for inducing AR schemes and rough-neural networks should be developed [13].
7. *Discovery of multiagent systems relevant to given problems.* Quite often, agents and communication methods among them are not given a priori with the problem specification, and the challenge is to develop methods for discovering multiagent system structures relevant to given problems, in particular, methods for discovering relevant communication protocols.
8. *Construction of multiagent systems for complex real-life problems.* Challenging problems are related to applying the methodology presented to real-life problems such as control of autonomous systems (see, e.g., www page of WITAS project [40]), web mining problems [10,31], sensor fusion [3,21,22], and spatial reasoning [6,8].
9. *Evolutionary methods.* For all of the above methods, it is necessary to develop evolutionary searching methods for semioptimal solutions [13].
10. *Parallel algorithms.* The problems discussed are of high computational complexity. Parallel algorithms searching for AR schemes and methods for their hardware implementation are one important research direction.

Acknowledgments

The research of Andrzej Skowron has been supported by the State Committee for Scientific Research of the Republic of Poland (KBN), research grant 8 T11C 025 19 and by the Wallenberg Foundation grant. The research of Jaroslaw Stepaniuk has been supported by the State Committee for Scientific Research of the Republic of Poland (KBN), research grant 8 T11C 025 19.

References

1. R. Agrawal, H. Mannila, R. Srikant, H. Toivonen, A. Verkano. Fast discovery of association rules. In U.M. Fayyad, G. Piatetsky-Shapiro, P. Smyth, R. Uthurusamy, editors. *Advances in Knowledge Discovery and Data Mining*, 307–328, AAAI/MIT Press, Cambridge, MA, 1996.
2. E. Bonabeau, M. Dorigo, G. Theraulaz, editors. *Swarm Intelligence: From Natural to Artificial Systems.* Oxford University Press, Oxford, 1999.

3. R.R. Brooks, S.S. Iyengar, editors. *Multi-Sensor Fusion*. Prentice-Hall, Upper Saddle River, NJ, 1998.
4. P. Doherty, W. Łukaszewicz, A. Skowron, A. Szałas. *Combining Rough and Crisp Knowledge in Deductive Databases* (this book).
5. G. Dong, J. Li. Interestingness of discovered association rules in terms of neighborhood-based unexpectedness. In X. Wu, K. Ramamohanarao, K.B. Korb, editors. *Proceedings of the 2nd Pacific-Asia Conference on Knowledge Discovery in Databases (PAKDD-98)*, LNAI 1384, 72–86, Springer, Berlin, 1998.
6. I. Düntsch, editor. Spatial reasoning (special issue). Vol. 45(1/2) of the *Fundamenta Informaticae*, 2001.
7. H-D. Ebbinghaus, H-D., J. Flum, W. Thomas. *Mathematical Logic*. Springer, Heidelberg, 1994.
8. M.T. Escrig, editor. *Qualitative Spatial Reasoning: Theory and Practice*. IOS Press, Amsterdam, 1998.
9. M.N. Huhns and M.P. Singh, editors. *Readings in Agents*. Morgan Kaufmann, San Mateo, CA, 1998.
10. H. Kargupta and P. Chan, editors. *Advances in Distributed and Parallel Knowledge Discovery*. MIT/AAAI Press, Cambridge, MA, 2001.
11. W. Kloesgen, J. Żytkow, editors. *Handbook of Knowledge Discovery and Data Mining*. Oxford University Press, Oxford, 2002.
12. J. Komorowski, Z. Pawlak, L. Polkowski, A. Skowron. Rough sets: A tutorial. In *[18]*, 3–98, 1999.
13. J.R. Koza. *Genetic Programming II: Automatic Discovery of Reusable Programs*. MIT Press, Cambridge, MA, 1994.
14. Y.Y. Lin, Y.Y. Yao, and L.A. Zadeh, editors. *Data Mining, Rough Sets and Granular Computing*. Physica, Heidelberg, 2002.
15. T.M. Mitchell. *Machine Learning*. McGraw-Hill, Portland, OR, 1997.
16. H.S. Nguyen, S.H. Nguyen, A. Skowron. Decomposition of task specification. LNCS 1609, 310–318, Springer, Berlin, 1999.
17. S.K. Pal, W. Pedrycz, A. Skowron, R. Swiniarski, editors. Rough-neuro computing (special issue). Vol. 36 of *Neurocomputing: An International Journal*, 2001.
18. S.K. Pal, A. Skowron, editors. *Rough-Fuzzy Hybridization: A New Trend in Decision Making*. Springer, Singapore, 1999.
19. Z. Pawlak. *Rough Sets: Theoretical Aspects of Reasoning about Data*. Kluwer, Dordrecht, 1991.
20. Z. Pawlak, A. Skowron. Rough set rudiments. *Bulletin of the International Rough Set Society*, 3/4: 181–185, 1999.
21. J. Peters, A. Skowron, S. Ramanna, M. Borkowski, Z. Suraj. Sensor, filter, and fusion models with rough Petri nets. *Fundamenta Informaticae*, 47: 307–323, 2001.
22. J.F. Peters, S. Ramanna, A. Skowron, J. Stepaniuk, Z. Suraj, M. Borkowski. Sensor fusion: A rough granular approach. In *Proceedings of the Joint 9th International Fuzzy Systems Association (IFSA) World Congress and 20th North American Fuzzy Information Processing Society (NAFIPS) International Conference*, 1367–1371, IEEE Press, Piscataway, NJ, 2001.
23. J.F. Peters, A. Skowron, J. Stepaniuk. Rough granules in spatial reasoning. In *Proceedings of the Joint 9th International Fuzzy Systems Association (IFSA) World Congress and 20th North American Fuzzy Information Processing Society (NAFIPS) International Conference, July 2001*, 1355–1361, IEEE Press, Piscataway, NJ, 2001.

24. L. Polkowski, A. Skowron. Rough mereological approach to knowledge-based distributed AI. In J.K. Lee, J. Liebowitz, and J.M. Chae, editors. *Critical Technology. Proceedings of the Third World Congress on Expert Systems*, 774–781, Cognizant Communications, New York, 1996.
25. L. Polkowski, A. Skowron. Rough mereology: A new paradigm for approximate reasoning. *International Journal of Approximate Reasoning*, 15(4): 333–365, 1996.
26. L. Polkowski, A. Skowron. Grammar systems for distributed synthesis of approximate solutions extracted from experience. In G. Paun, A. Salomaa, editors. *Grammar Models for Multiagent Systems*, 316–333, Gordon and Breach, Amsterdam, 1999.
27. L. Polkowski, A. Skowron. Towards adaptive calculus of granules. In *[44]*, 201–227, 1999.
28. L. Polkowski, A. Skowron. Rough mereological calculi of granules: A rough set approach to computation. *Computational Intelligence*, 17(3): 472–492, 2001.
29. L. Polkowski and A. Skowron. Rough-neuro computing. In W. Ziarko, Y.Y. Yao, editors. *Proceedings of the 2nd International Conference on Rough Sets and Current Trends in Computing (RSCTC 2000)*, LNAI 2005, 57–64, Springer, Berlin, 2001.
30. B.D. Ripley. *Pattern Recognition and Neural Networks*. Cambridge University Press, Cambridge, 1996.
31. A. Skowron. Toward intelligent systems: Calculi of information granules. *Bulletin International Rough Set Society*, 5(1/2): 9–30, 2001.
32. A. Skowron, L. Polkowski. Rough mereological foundations for design, analysis, synthesis, and control in distributive systems. *Information Sciences*, 104(1/2): 129–156, 1998.
33. A. Skowron, J. Stepaniuk. Tolerance approximation spaces. *Fundamenta Informaticae*, 27: 245–253, 1996.
34. A. Skowron, J. Stepaniuk. Information granule decomposition. *Fundamenta Informaticae*, 47(3/4): 337–350, 2001.
35. A. Skowron, J. Stepaniuk. Information granules: Towards foundations of granular computing. *International Journal of Intelligent Systems*, 16(1): 57–86, 2001.
36. A. Skowron, J. Stepaniuk, J.F. Peters. Extracting patterns using information granules. In *Proceedings of the JSAI International Workshop on Rough Set Theory and Granular Computing (RSTGC 2001)*, 135–142, Matsue, Japan, 2001.
37. P. Stone. *Layered Learning in Multi-Agent Systems: A Winning Approach to Robotic Soccer*. MIT Press, Cambridge, MA, 2000.
38. H. Toivonen, M. Klemettinen, P. Ronkainen, K. Hätönen, H. Mannila. Pruning and grouping discovered association rules. In *Proceedings of the MLnet Familiarization Workshop on Statistics, Machine Learning and Knowledge Discovery in Databases*, 47–52, Heraklion, Crete, Greece, 1995.
39. P.P Wang, editor. *Granular Computing*. Wiley, New York, 2001.
40. WITAS. available at http://www.ida.liu.se/ext/witas/eng.html. *Project web page*, 2001.
41. L.A. Zadeh. Fuzzy sets. *Information and Control*, 8: 333–353, 1965.
42. L.A. Zadeh. Fuzzy logic = computing with words. *IEEE Transactions on Fuzzy Systems*, 4: 103–111, 1996.
43. L.A. Zadeh. A new direction in AI: Toward a computational theory of perceptions. *AI Magazine*, 22(1): 73–84, 2001.
44. L.A. Zadeh, J. Kacprzyk, editors. *Computing with Words in Information/Intelligent Systems*, volumes 1, 2. Physica, Heidelberg, 1999.

Chapter 4
A Rough-Neural Computation Model Based on Rough Mereology

Lech Polkowski

Polish-Japanese Institute of Information Technology Research Center, Koszykowa 86, 02-008 Warsaw, Poland
and
Department of Mathematics and Computer Science, University of Warmia and Mazury, Żołnierska 14a, 10-561 Olsztyn, Poland
polkow@pjwstk.edu.pl

Summary. In this chapter, we propose a brief survey of rough mereology, i.e., approximate calculus of parts, an approach to reasoning under uncertainty based on the notion of an approximate part (part to a degree). Rough mereology is mentioned or applied in chapters by Gomolińska; Pal, Peters et al.; Skowron and Stepaniuk; Skowron and Swinarski; Peters, Ramanna et al.; and in some chapters in this book, ideas of approximations by parts are presented implicitly in the formal apparatus chosen there. Therefore, it seems desirable to give an account of this approach.

We exploit this approach as the basis for a model of rough-neural computations whose architecture is modeled on neural network architecture but whose computations are performed according to the ideas of rough set theory.

In the sections that follow, we review in a nutshell rough set theoretical notions (Sect. 1), ontology and mereology (Sect. 2), rough mereology (Sect. 3), and we present a rough-neural computation model based on rough mereological notions and inspired by a neural computing paradigm. As rough mereology may justly be regarded as a common generalization of rough and fuzzy set theories, this model of computations is a particular instance of a rough-fuzzy-neural computing paradigm (see chapter by Pal, Peters et al.).

1 Rough Sets: First Notions

Motivation for the approach called *rough mereology* comes from rough set theory as well as from fuzzy set theory. The two approaches to uncertainty and vagueness stem from a common inspiration: to describe inexact concepts by elaborating a language with which it would be possible to express properties of *boundary objects*, which every inexact concept must possess [14, 26].

Rough set theory approaches this problem with *descriptors* in an *attribute-value-based* language, viz., given a finite set U of objects and a finite set A of attributes, we call a *descriptor* [14] any pair of the form (a, v) where $a \in A$, $v \in V_a$, and V_a is the *value set* of the attribute $a : U \to V_a$ formally understood as a function.

Descriptors allow for a logical language: we call descriptors *elementary formulas* in *descriptor logic* (DL), and by formulas of (DL), we mean inscriptions formed from elementary formulas by means of propositional functors of disjunction $\vee$, conjunction $\wedge$, and negation $\neg$. Thus, any formula of logic DL may be written in the disjunctive normal form (DNF) as

$$\bigvee_i (a_{i_1}, v_{i_1}) \wedge \ldots \wedge (a_{i_k}, v_{i_k}), \tag{1}$$

where we observe that due to finiteness of U and A, we may dispense with negation. We may add a semantic ingredient to logic DL by interpreting the formulas in set U: for a formula α, the meaning $[\alpha]$ of α is defined by

$$[\alpha] = \{x \in U : x \models \alpha\}, \tag{2}$$

where the semantic satisfaction $\models$ is defined by recurrence as follows

$$\begin{gathered} x \models (a,v) \ iff \ a(x) = v, \\ x \models \gamma \vee \alpha \ iff \ x \models \gamma \ or \ x \models \alpha, \\ x \models \gamma \wedge \alpha \ iff \ x \models \gamma \ and \ x \models \alpha. \end{gathered} \tag{3}$$

A concept $X \subseteq U$ is *exact* when there exists a formula α of form (1) with the property that $X = [\alpha]$. Otherwise, the concept X is inexact; in this case, there exists a maximal formula $\underline{\alpha}_X$ with the properties,

$$\begin{gathered} (i) \ [\underline{\alpha}_X] \subseteq X, \\ (ii) \ [\beta] \subseteq X \Rightarrow [\beta] \subseteq [\underline{\alpha}_X], \end{gathered} \tag{4}$$

and a minimal formula $\overline{\alpha}_X$ with the properties,

$$\begin{gathered} (i) \ [\overline{\alpha}_X] \supseteq X, \\ (ii) \ [\beta] \supseteq X \Rightarrow [\beta] \supseteq [\overline{\alpha}_X], \end{gathered} \tag{5}$$

and

$$[\underline{\alpha}_X] \neq [\overline{\alpha}_X]. \tag{6}$$

The set,

$$[\neg \underline{\alpha}_X \wedge \overline{\alpha}_X] \neq \emptyset, \tag{7}$$

is the *boundary BdX of X*. By (4) and (5) along with (6), *BdX* consists of objects having logical counterparts in X as well as in $U \setminus X$, therefore, not uniquely admissible into X or $U \setminus X$.

The quality of approximation of X by its *lower approximation* $[\underline{\alpha}_X]$ and by its *upper approximation* $[\overline{\alpha}_X]$ can be measured [14] by the coefficients

$$\begin{gathered} \underline{\gamma}_X = \frac{|[\underline{\alpha}_X]|}{|X|}, \\ \overline{\gamma}_X = \frac{|X|}{|[\overline{\alpha}_X]|}, \end{gathered} \tag{8}$$

where $|A|$ is the cardinality (i.e., number of elements) of set A. These coefficients can be interpreted as follows in the language of *parts*:

1. $\underline{\gamma}_X$ is the degree to which $[\underline{\alpha}_X]$ is a part of X.
2. $\overline{\gamma}_X$ is the degree to which X is a part of $[\overline{\alpha}_X]$.

Thus, a proposition follows:

Proposition 1. *A set X is exact if and only if $\underline{\gamma}_X = \overline{\gamma}_X$.*

From now on, we use the following notation, where the formula α_X is written as $\bigvee_i (a_{i_1}, v_{i_1}) \wedge \ldots \wedge (a_{i_k}, v_{i_k})$ with all a_{i_j} exhausting the set $B \subseteq A$, i.e., it is written across the set B of attributes,

$$\begin{aligned} \underline{B}X &= [\underline{\alpha}_X], \\ \overline{B}X &= [\overline{\alpha}_X]. \end{aligned} \tag{9}$$

In particular, for $X = \{x\}$, we let $[x]_B = [\overline{\alpha}_{\{x\}}]$. The measure of X with respect to $\underline{B}X, \overline{B}X$ which takes into account local relations among those sets, is the *rough membership function* [15] defined as follows:

$$\mu_X^B(x) = \begin{cases} \frac{|[x]_B \cap X|}{|X|} & \text{if } X \neq \emptyset \\ 1 & \text{otherwise.} \end{cases} \tag{10}$$

This function measures the local degree in which $[x]_B$ is a part of X. Clearly,

Proposition 2.

1. $\mu_X^B(x) = 1$ if and only if $x \in \underline{B}X$.
2. $\mu_X^B(x) = 0$ if and only if $x \notin \overline{B}X$.
3. $\mu_X^B(x) \in (0,1)$ if and only if $x \in \overline{B}X \setminus \underline{B}X$.

We may observe that μ_X^B is a particular case of a *fuzzy membership function* induced from data.

Summing up, we have introduced in the universe of objects some notions of exactness as well as dual notions of roughness, and in the latter case, we have at our disposal some measures of partial containment.

A set description of the above notions is possible as well: letting

$$xIND_By \Leftrightarrow \forall a \in B.a(x) = a(y), \tag{11}$$

defines an equivalence relation of *B-indiscernibility*. Clearly, $[x]_B$ defined above is the equivalence class of this relation for x, and it may be used in a characterization of $\underline{B}X, \overline{B}X$, i.e.,

$$\begin{aligned} \underline{B}X &= \bigcup\{[x]_B : [x]_B \subseteq X\}, \\ \overline{B}X &= \bigcup\{[x]_B : [x]_B \cap X \neq \emptyset\}. \end{aligned} \tag{12}$$

2 The Leśniewski Ontology and Mereology

The reason for this pair of theories was the emergence of antinomies, i.e., paradoxes, notably the Russell paradox in naive set theory. This paradox was the result of an assumption made in the naive set theory (Frege) that there existed a one-to-one correspondence between sets and predicates, i.e., to each predicate $\Psi(x)$, there existed the set $s_\Psi = \{a : \Psi(a)\}$ and vice versa. Russell demonstrated the invalidity of this assumption by considering $\Psi(x) : x \notin x$ and observing that if $s = s_\Psi$ existed, we should have $s \in s \Leftrightarrow s \notin s$, a contradiction. The standard set theory has avoided this paradox by restricting the rules for set construction.

The approach taken by Leśniewski consisted of his entirely distinct method of analysis. In short, Leśniewski reexamined two basic notions of the naive set theory: the notion of an element [expressed by means of the primitive $\in$ (*ésti*)] and the notion of a set (class). He constructed *ontology*, i.e., a theory of the copula *is* : ε as a replacement for $\in$ as well as *mereology*, i.e., the theory of classes. In contrast to the set theoretical usage of a class in the *distributive* sense (i.e., as an aggregation of elements), he used the notion of a class in the *collective* sense, i.e., as an entity representing all objects in the class as a new object. A judicious selection of axioms for both theories allowed him to avoid the Russell antinomy as $s \varepsilon s$ holds for every individual s, and to build of a consistent theory.

We begin with an outline of ontology. This will be followed by an account of mereology. To differentiate these theories of sets from standard (naive) ones, we propose to use a distinct notation.

2.1 Ontology

Ontology was intended by Stanisław Leśniewski as a formulation of general principles of *being* [9]; cf. also [4–6, 21]. The only primitive notion of ontology is the copula *is* denoted by the symbol ε. Our exposition of ontology follows along the lines of [21], mostly with additions from papers cited above.

All well-formed expressions of ontology belong in classes called *semantic categories*. Categories are constructed starting with the two *basic semantic categories*: the *semantic category of names* and the *semantic category of propositions*. Either of those categories contains *constants* as well as *variables* of the given category. Higher order categories are constructed as functors under the agreement that each functor belongs either in the category of names (a *name-forming* functor) or in the category of propositions (a *proposition-forming* functor). However, a stratification of functors in the usual way is achieved by assigning functors to the same category if both are either name-forming or proposition-forming and if they have the same number of arguments falling into the same categories, respectively.

We will use standard symbols for propositional constants as well as quantifiers, parentheses, symbols, etc. Formulas of ontology will be constructed as in the standard predicate calculus. We begin with the axiom of ontology, i.e., a formula that introduces the copula ε. We denote names with capital letters $X, Y, Z, \ldots$.

Axiom of Ontology

$$\begin{gathered} X\varepsilon Y \Longleftrightarrow \exists Z.Z\varepsilon X \wedge \\ \forall U, W.(U\varepsilon X \wedge W\varepsilon X \Longrightarrow U\varepsilon W) \wedge \\ \forall Z.(Z\varepsilon X \Longrightarrow Z\varepsilon Y). \end{gathered} \tag{13}$$

In this axiom, the copula ε happens to occur in both sides of the equivalence, however, the definiendum $X\varepsilon Y$ belongs in the left side only, and we may understand the axiom as the definition of the meaning of $X\varepsilon Y$ via the meaning of terms of lower level $Z\varepsilon X$, $Z\varepsilon Y$, etc. According to this reading of the axiom, we gather that the proposition $X\varepsilon Y$ is true if and only if the conjunction of the following three propositions holds:

$$\begin{gathered} (i)\ \exists Z.Z\varepsilon X, \\ (ii)\ \forall U, W.(U\varepsilon X \wedge W\varepsilon X \Longrightarrow U\varepsilon W), \\ (iii)\ \forall Z.(Z\varepsilon X \Longrightarrow Z\varepsilon Y). \end{gathered} \tag{14}$$

Condition (i) asserts the existence of an object (name) Z which is X, and so X is not an empty name. Condition (ii) asserts that any two objects that are X are each other (a fortiori, they will be identified later on). This means that X is an individual name or that X is an individual entity, representable as a singleton. Condition (iii) asserts that every object that is X is Y as well or that X is contained in Y. The meaning of $X\varepsilon Y$ can be made clear now: X is an individual (name), and this individual is Y (i.e., belongs in Y, responds to the name of Y).

We apply ontology in what follows as a framework to discuss mereology and rough mereology: it will aid us in forming names of objects that are relations among simpler names or names of individuals that are collective classes of such names. The language of ontology seems to us a suitable vehicle to code approximate reasoning schemes as, e.g., we avoid deeper assumptions about the nature of sets, such as regularity axioms, which are particular to set theory.

Elementary Ontology Here, we present the basic ontological scheme for dealing with individuals and their distributive classes.

Rules for functor formation. Rules for new functor formation are as follows:

1. For proposition-forming functors: $f(X_1, X_2, \ldots, X_n) \Longleftrightarrow \alpha$ where α is a propositional expression of ontology; an example is the ontological axiom itself.
2. For name-forming functors: $X\varepsilon f(X_1, X_2, \ldots, X_n) \Longleftrightarrow X\varepsilon X \wedge \alpha(X_1, \ldots, X_n)$ where α is a propositional expression of ontology; examples of such functors are in this section (see Definition 6).

3. For nominal constants: $X\varepsilon C \Longleftrightarrow X\varepsilon X \wedge \alpha$
   where α is a propositional expression of ontology.

As basic examples, we introduce two nominal constants:

Definition 1.

1. $X\varepsilon\Lambda \Longleftrightarrow X\varepsilon X \wedge non(X\varepsilon X)$; constant Λ is an *empty name* [here $\alpha : non(X\varepsilon X)$].
2. $X\varepsilon V \Longleftrightarrow \exists Y.X\varepsilon Y$; constant V is a *universal name*. $X\varepsilon V$ reads as "X is an object" (as $\exists Y.X\varepsilon Y$ is equivalent to $X\varepsilon X$, we may take as α any tautology).

The basic consequences of the ontology axiom are

Proposition 3.

1. $X\varepsilon Y \Longrightarrow \exists Z.Z\varepsilon X$.
2. $X\varepsilon Y \Longrightarrow \forall U,W.(U\varepsilon X \wedge W\varepsilon X \Longrightarrow U\varepsilon W)$.
3. $X\varepsilon Y \Longrightarrow \forall Z.(Z\varepsilon X \Longrightarrow Z\varepsilon Y)$.

1,2,3 follow by the ontological axiom. We now introduce a name-forming functor $\subseteq$ via

Definition 2. $X\varepsilon \subseteq (Y) \Longleftrightarrow X\varepsilon X \wedge \forall Z.(Z\varepsilon X \Longrightarrow Z\varepsilon Y)$.

We will read $X\varepsilon \subseteq (Y)$ as "X is *contained* in Y." The basic properties of the functor $\subseteq$ are summarized below.

Proposition 4.

1. $X\varepsilon X \Longrightarrow X\varepsilon \subseteq (X)$.
2. $X\varepsilon \subseteq (Y) \wedge Y\varepsilon \subseteq (Z) \Longrightarrow X\varepsilon \subseteq (Z)$.
3. $X\varepsilon X \Longrightarrow X\varepsilon \subseteq (V)$.

Proof. 1 follows from the tautology $p \Longrightarrow p$ via instantiation $Z\varepsilon X \Longrightarrow Z\varepsilon X$ and universal quantification: $\forall Z.Z\varepsilon X \Longrightarrow Z\varepsilon X$. Similarly, 2 follows by omitting the general quantifier in the premises $X\varepsilon \subseteq (Y), Y\varepsilon \subseteq (Z)$, applying the inference rule

$$\frac{U\varepsilon X \Longrightarrow U\varepsilon Y, U\varepsilon Y \Longrightarrow U\varepsilon Z}{U\varepsilon X \Longrightarrow U\varepsilon Z},$$

and quantifying universally. For 3, we observe that $Z\varepsilon X \Longrightarrow \exists Y.Z\varepsilon Y \Longrightarrow Z\varepsilon V$. □

We introduce yet another name-forming functor via

Definition 3. $X\varepsilon = (Y) \Longleftrightarrow X\varepsilon X \wedge Y\varepsilon Y \wedge X\varepsilon Y \wedge Y\varepsilon X$.

The functor $=$ is the individual identity functor ("X is *identical* to Y"). With its help, we may write the individuality condition as

$$\forall U,W.(U\varepsilon X \wedge W\varepsilon X \Longrightarrow U = W).$$

This has to be discerned from the set identity functor defined for pairs of sets, not necessarily singletons. We state the basic properties of the existential statement $X\varepsilon Y$:

Proposition 5.

1. $X\varepsilon Y \wedge Z\varepsilon X \Longrightarrow Z\varepsilon Y$.
2. $X\varepsilon Y \Longrightarrow X\varepsilon X$.
3. $X\varepsilon Y \wedge Z\varepsilon X \Longrightarrow Z = X$.
4. $X\varepsilon X \Longleftrightarrow X\varepsilon V$.

Proof. Indeed, 1 follows from the ontological axiom; 2, 3 follow by virtue of definitions. Finally, $X\varepsilon X \Longrightarrow X\varepsilon V$ is obvious, and $X\varepsilon V \Longrightarrow X\varepsilon X$ patterns 2. □

On a higher level of generality, the counterpart of the identity functor $=$ is the scope equality functor $=_E$ defined as follows.

Definition 4. $X =_E Y \Longleftrightarrow \forall Z.(Z\varepsilon X \Longleftrightarrow Z\varepsilon Y)$.

The following are the properties of this notion:

Proposition 6.

1. $X =_E Y \Longleftrightarrow X\varepsilon \subseteq (Y) \wedge Y\varepsilon \subseteq (X)$.
2. $X = Y \Longleftrightarrow X\varepsilon Y \wedge X =_E Y$.
3. $X\varepsilon V \Longrightarrow (X = Y \Longleftrightarrow X =_E Y)$.

All of these properties follow immediately from definitions. Finally we give, following [21], a logical (extensional) content to individual identity. Let $\alpha(Z)$ be a propositional expression. We denote by $\alpha(Z/X)$ the expression formed from α by replacing Z with X. Then,

Proposition 7. $X = Y \Longrightarrow [\alpha(Z/X) \Longleftrightarrow \alpha(Z/Y)]$.

Proof. (Słupecki) Define a name-forming functor $f(Y,U,W,..)$ via

$$Z\varepsilon f(Y,U,W,..) \Longleftrightarrow Z\varepsilon Z \wedge \alpha(Z).$$

Assume that $X = Y$ and $\alpha(Z/X)$; hence, $Y\varepsilon X$ and $X\varepsilon\ f(Y,\ U,\ W\ ,\ldots)$, implying $Y\varepsilon f(Y,U,W,..)$ and $\alpha(Z/Y)$. The conclusion follows by symmetry. □

Equivalent Axiom Schemes One may choose axioms for ontology by means of other notions a fortiori by means of distinct schemes of axioms. The most important are axiomatic characterizations of ontology by means of ordering functors (relations) as they lead directly to well-known mathematical structures, viz., quasi-Boolean algebras; see also [4, 21]. We define a binary proposition-forming functor $ord(X,Y)$ as follows:

Definition 5. $ord(X,Y) \Longleftrightarrow \exists Z.Z\varepsilon X \wedge \forall T.(T\varepsilon X \Longrightarrow T\varepsilon Y)$.

Then,

Proposition 8.

$$X\varepsilon Y \iff ord(X,Y) \wedge \forall U,W.\left[ord(U,X) \wedge ord(W,X) \Longrightarrow ord(U,W)\right].$$

For better visualization of the formulas, we introduce a shortcut notation $X < Y$ for $ord(X,Y)$. This is, however, only a notational convenience, not a new type of functor.

Proof. Assume first that $X\varepsilon Y$; then clearly $X < Y$. Now assume that $U < X, W < X$. Observe that $U < X$ implies that U is an individual, i.e., $U\varepsilon U$, hence $U\varepsilon X$; similarly, $W\varepsilon X$ and as X is an individual, it follows that $U = W$. From $U\varepsilon V, W\varepsilon V$, it follows by Proposition 6 that $U =_E W$ and finally $U < W$.

Now assume that $ord(X,Y) \wedge \forall U,W.\left[ord(U,X) \wedge ord(W,X) \Longrightarrow ord(U,W)\right]$. It suffices to check that $U\varepsilon X \wedge W\varepsilon X \Longrightarrow U\varepsilon W$. Assume that $U\varepsilon X \wedge W\varepsilon X$; then $U < X$, $W < X$, and thus $U = W$, implying $X\varepsilon X$, hence $X\varepsilon Y$. □

Proposition 9. (Lejewski [6])

$$\exists Z.Z\varepsilon X \iff \exists Z.Z < X.$$

Proof. Clearly, $\exists Z.Z\varepsilon X$ implies $\exists Z.Z < X$. Conversely, $\exists Z.Z < X$ gives $A\varepsilon X$, for some A, hence, $\exists B.B\varepsilon A$ and $\forall T.(T\varepsilon A \Longrightarrow T\varepsilon X)$, so $B\varepsilon X$ and finally, $\exists Z.Z\varepsilon X$. □

The following proposition establishes a deeper parallelism between "is" and "entails."

Proposition 10. (Sobociński see [6])

$$\forall U,W.(U\varepsilon X \wedge W\varepsilon X \Longrightarrow U\varepsilon W) \iff \forall U,W.(U < X \wedge W < X \Longrightarrow U < W).$$

Proof. Assume $\forall U,W.(U\varepsilon X \wedge W\varepsilon X \Longrightarrow U\varepsilon W)$ and $U < X \wedge W < X$. For any P, it suffices to show that $P\varepsilon U \Longrightarrow P\varepsilon W$. Let $P\varepsilon U$; $Q\varepsilon W$; then $P\varepsilon X$, $Q\varepsilon X$, and by the premises, $P\varepsilon Q$, hence $P\varepsilon W$. Conversely, from $\forall U,W.(U < X \wedge W < X \Longrightarrow U < W)$ and $U\varepsilon X \wedge W\varepsilon X$, it follows that $U < X, W < X$, and so $U < W$; by Proposition 8, $U\varepsilon W$. □

We may now establish

Proposition 11. (Lejewski [6])

$$X < Y \iff \exists Z.Z < X \wedge \forall Z.\left[Z < X \Longrightarrow \exists W.(W\varepsilon Z \wedge W\varepsilon Y)\right].$$

Proof. First, let $X < Y$. Then, $\exists Z.Z < X$, hence $\forall T.(T\varepsilon Z \Longrightarrow T\varepsilon Y)$, and $\exists W.W\varepsilon Z$, which yields $A\varepsilon Z$; hence $A\varepsilon Y$ for some A, and finally $\exists W.W\varepsilon Z \wedge W\varepsilon Y$.
Now, let $\exists Z.Z < X \wedge \forall Z.\left[Z < X \Longrightarrow \exists W.(W\varepsilon Z \wedge W\varepsilon Y)\right]$; assume $Z\varepsilon X$, so $Z < X$. Then $W\varepsilon Z \wedge W\varepsilon Y$ for some W, hence $Z\varepsilon Y$. It follows that $\forall Z.(Z\varepsilon X \Longrightarrow Z\varepsilon Y)$, so $X < Y$. □

The following is an immediate consequence of the above propositions and Definition 5. We denote the formula below with the symbol (ATB) (atomic Boolean).

Proposition 12. (Lejewski [6])

$(ATB)\ X < Y \Longleftrightarrow \exists Z.Z < X \wedge \forall Z.[Z < X \Longrightarrow$
$\exists W.(W < Z \wedge W < Y) \wedge \forall P,Q.(P < W \wedge Q < W \Longrightarrow P < Q)]$.

Proposition 13. (Lejewski [6])

(ATB) is equivalent to the axiom of ontology.

Proof. First, we show that (ATB) implies the ontology axiom. It is sufficient to check that $(Z < X \Longrightarrow \exists W.W\varepsilon Z \wedge W\varepsilon Y) \Longleftrightarrow (Z\varepsilon X \Longrightarrow Z\varepsilon Y)$. Assume then that $(Z < X \Longrightarrow \exists W.W\varepsilon Z \wedge W\varepsilon Y)$ and $Z\varepsilon X$. Hence $Z < X$, and so $W\varepsilon Z, W\varepsilon Y$ for some W, and thus $Z\varepsilon Y$. Conversely, from $(Z\varepsilon X \Longrightarrow Z\varepsilon Y)$ and $Z < X$, it follows that $\exists W.W\varepsilon Z$ and $\forall W.(W\varepsilon Z \Longrightarrow W\varepsilon X)$; hence for some W, $W\varepsilon Z$; hence $W\varepsilon X$, and finally, $W\varepsilon Y$ proving $\exists W.W\varepsilon Z \wedge W\varepsilon Y$. That the ontology axiom implies (ATB) follows similarly. □

We now introduce a constant name AT.

Definition 6. $X\varepsilon AT \Longleftrightarrow X\varepsilon X \wedge \forall P,Q.(P < X \wedge Q < X \Longrightarrow P < Q)$.

The term $X\varepsilon AT$ reads "X is an atom." We may check that

Proposition 14. *(ATB) is an axiom for atomic Boolean algebra without a zero element.*

The proof consists in straightforward checking. Therefore,

Proposition 15. (cf. [4]) *Theorems of ontology are those that are true in every model for atomic Boolean algebra without a null element.*

This proposition reconciles ontology with naive set theory: ontological structures are equivalent to set-theoretical structures, viz., atomic Boolean algebras without a null element.

2.2 Mereology

Mereology was invented [11] as an alternative to naive set theory. Ontology defined and introduced in its basics in the previous section, we now may pass to mereology. Mereology, as already observed by us, is a theory of collective classes, i.e., individual entities representing general names as opposed to ontology that is a theory of distributive classes, i.e., general names. The distinction between a distributive class and its collective class counterpart is like the distinction between a family of sets and its union, i.e., a specific set. For instance, we may represent the United States through the list of its states (as a distributive class) or we may mean "United States" as the collective class of its states, e.g., their union (as the individual having spatial location in the Northern Hemisphere and containing as parts all of its states).

This explains the purpose of mereology, which is to express objects though their parts and to define some complex objects in terms of their parts. Mereology may be based on each of a few notions, such as those of a *part*, an *element (ingredient)*, a *class*, etc. Historically, it was conceived by Stanisław Leśniewski [8, 10, 11], cf. [3], [22–23], as a theory of the relation *part*. Here, we follow this line of development. In particular, we present the development of mereology within ontology: names of mereological constructs will be formed by means of ontological rules. We introduce the notion of the name-forming functor *pt* of *part*. Our presentation is based on [10] in the first place.

Mereology Axioms

(A1) $X \varepsilon pt(Y) \Longrightarrow X \varepsilon X \wedge Y \varepsilon Y$;
this means that the functor *pt* is defined only for *individual entities*.
(A2) $X \varepsilon pt(Y) \wedge Y \varepsilon pt(Z) \Longrightarrow X \varepsilon pt(Z)$;
this means that the functor *pt* is transitive, i.e., a part of a part is a part.
(A3) non $[X \varepsilon pt(X)]$;
this means that the functor *pt* is nonreflexive (or, equivalently, if $X \varepsilon pt(Y)$, then *non* $[Y \varepsilon pt(X)]$.

On the basis of the notion of a part, we define the notion of an *element* (possibly an improper part originally called an *ingredient*) as a name-forming functor *el*.

Definition 7. $X \varepsilon el(Y) \Longleftrightarrow X \varepsilon pt(Y) \vee X = Y$.

It is clearly possible to introduce mereology in terms of the functor *el* as a partial order functor (i.e., being consecutively *reflexive*: $X \varepsilon el(X)$, *transitive*: $X \varepsilon el(Y) \wedge Y \varepsilon el(Z) \Longrightarrow X \varepsilon el(Z)$, *weakly symmetrical*: $X \varepsilon el(Y) \wedge Y \varepsilon el(X) \Longrightarrow X = Y$) (see Proposition 35).

The remaining axioms of mereology are related to the class functor that converts distributive classes (general names) into individual entities: it may be used to represent the "United States" as an individual comprising all U.S. states. The class operator *Kl* is a principal tool in applications of rough mereology to problems of distributed systems, knowledge granulation, computing with words, where it does play the role of granulating (clustering) operator allowing for forming granules of knowledge and, subsequently, instrumental in calculi on them; see [16–20].

In this, *Kl* replaces the operators of a union of sets, acting on families of sets to produce a set (the union of the given family of sets). Here, we see a formal advantage of mereology: we have to deal only with objects, not with their families. We may now introduce the notion of a (collective) class via a name-forming functor *Kl*.

Definition 8.

$$
\begin{aligned}
X\varepsilon Kl(Y) \Longleftrightarrow \exists Z. Z\varepsilon Y \wedge \\
\forall Z. [Z\varepsilon Y \Longrightarrow Z\varepsilon el(X)] \wedge \\
\forall Z. \{Z\varepsilon el(X) \Longrightarrow \exists U, W. [U\varepsilon Y \wedge W\varepsilon el(U) \wedge W\varepsilon el(Z)]\}.
\end{aligned}
\tag{15}
$$

We will find the meaning of this definition. First, we may realize that the class operator Kl is intended as the operator converting names (general sets of entities) into individual entities, i.e., collective classes. Its role may be fully compared to the role of the union of sets operator in classical set theory. The analogy is not only functional but also formal. Let us look at the subsequent conjuncts in the defining formula above.

1. $\exists Z. Z\varepsilon Y$;
   this means that Y is a nonempty name (recall that the union of the empty family of sets is the empty set hence prohibited in ontology).
2. $\forall Z. [Z\varepsilon Y \Longrightarrow Z\varepsilon el(X)]$;
   meaning that any individual listed in Y is an element of $Kl(Y)$ (compare with: any element of the family of sets is a subset of the union of that family).
3. $\forall Z. \{Z\varepsilon el(X) \Longrightarrow \exists U, W. [U\varepsilon Y \wedge W\varepsilon el(U) \wedge W\varepsilon el(Z)]\}$
   this means that any element of $Kl(Y)$ has an element in common with an individual in Y (similarly, any element in the union of a family of sets is an element in at least one member of this family).

Thus, the class functor pastes together individuals in Y by means of their common elements The class functor is subject to the following postulates.

(A4) $X\varepsilon Kl(Y) \wedge Z\varepsilon Kl(Y) \Longrightarrow X\varepsilon Z$;
this means that $Kl(Y)$ is an individual name (entity) for any (nonempty) Y.
(A5) $\exists Z. Z\varepsilon Y \Longleftrightarrow \exists Z. Z\varepsilon Kl(Y)$;
meaning that $Kl(Y)$ exists (i.e., is a nonempty individual name) if and only if Y is a nonempty name.

First Consequences From the axioms above, we start a build-up of mereology. We begin with the simple consequences of axioms.

Proposition 16.

1. $X\varepsilon el(Y) \wedge Y\varepsilon el(Z) \Longrightarrow X\varepsilon el(Z)$.
2. $X\varepsilon el(Y) \wedge Y\varepsilon el(X) \Longrightarrow X = Y$.
3. $X\varepsilon el(X)$.

Proposition 17. $X\varepsilon X \Longrightarrow X\varepsilon Kl(elX)$ *where* $Z\varepsilon elX \Longleftrightarrow Z\varepsilon el(X)$.

Proposition 18. $X\varepsilon el(Y) \Longleftrightarrow \exists Z. X\varepsilon Z \wedge Y\varepsilon Kl(Z)$.

We define new name-forming functors:

Definition 9.

1. $X\varepsilon partY \Longleftrightarrow X\varepsilon pt(Y)$.
2. $X\varepsilon(\varepsilon Y) \Longleftrightarrow X\varepsilon Y$.

We have counterparts of Proposition 17.

Proposition 19. $X\varepsilon X \Longrightarrow X\varepsilon Kl(\varepsilon X)$.

Proof. Assume $X\varepsilon X$; then $\exists Z.Z\varepsilon X, \forall Z.[Z\varepsilon X \Longrightarrow Z\varepsilon el(X)]$, $\forall Z.\{Z\varepsilon el(X) \Longrightarrow \exists U,W.[U\varepsilon X \wedge W\varepsilon el(U) \wedge W\varepsilon el(Z)]\}$ are satisfied. □

Proposition 20. $\exists Z.Z\varepsilon pt(X) \Longrightarrow X\varepsilon Kl(partX)$.

Proposition 21. $X\varepsilon X \Longrightarrow X\varepsilon Kl(X)$.

We now define the notion of a set, weaker than that of a class. We may observe that a class is a set with the universality property, $\forall Z.\{Z\varepsilon Y \Longrightarrow Z\varepsilon el[Kl(Y)]\}$.

Definition 10. $X\varepsilon set(Y) \Longleftrightarrow \exists Z.Z\varepsilon Y \wedge \forall Z.\{Z\varepsilon el(X) \Longrightarrow \exists U,W.[U\varepsilon el(Z) \wedge U\varepsilon Y \wedge W\varepsilon el(U) \wedge W\varepsilon el(Z)]\}$.

We now recall, following [8, 10], some technical propositions leading to a thesis (Proposition 28) equivalent to (A4) and giving an inference rule about the functor *el*.

Proposition 22. $X\varepsilon set(Y) \wedge \forall Z.(Z\varepsilon Y \Longrightarrow Z\varepsilon W) \wedge T\varepsilon el(X) \Longrightarrow \exists P,R.[P\varepsilon el(T) \wedge P\varepsilon el(R) \wedge R\varepsilon W \wedge R\varepsilon el(X)]$.

Proposition 23. $X\varepsilon set(Y) \wedge \forall Z.(Z\varepsilon Y \Longrightarrow Z\varepsilon W) \Longrightarrow X\varepsilon set(W)$.

Proposition 24. $X\varepsilon Y \Longrightarrow X\varepsilon set(Y)$.

Proposition 25. $X\varepsilon Kl(Y) \Longrightarrow X\varepsilon set(Y)$.

The proofs are obvious.

Proposition 26. $X\varepsilon Kl[set(Y)] \wedge Z\varepsilon el(X) \Longrightarrow \exists U,W.[U\varepsilon el(Z) \wedge U\varepsilon el(W) \wedge W\varepsilon Y \wedge W\varepsilon el(X)]$.

Proof. Assume that $X\varepsilon Kl[set(Y)]$, $Z\varepsilon el(X)$. There exist U, W with the properties $U\varepsilon el(Z)$, $U\varepsilon el(W)$, $W\varepsilon set(Y)$, $W\varepsilon el(X)$; hence there exist P,Q with $P\varepsilon el(U)$, $P\varepsilon el(Q)$, $Q\varepsilon el(W)$, $Q\varepsilon Y$ implying $Q\varepsilon el(X)$. Thus P, Q satisfy the consequent. □

Corollary 1. $X\varepsilon Kl[set(Y)] \Longrightarrow X\varepsilon Kl(Y)$.

Proof. Assume that $X\varepsilon Kl[set(Y)]$. Then $Z\varepsilon Y$ implies $Z\varepsilon set(Y)$ by Proposition 2.22 and finally, by the assumption, $Z\varepsilon el(X)$. Now, for $Z\varepsilon el(X)$, there exist U,W with $U\varepsilon el(Z)$, $U\varepsilon el(W)$, $W\varepsilon set(Y)$, so by the definition of a set (Definition 2.10), we may assume that $\exists P,Q.P\varepsilon el(U), P\varepsilon el(Q), Q\varepsilon Y$. It follows that $X\varepsilon Kl(Y)$. □

Corollary 2. $X\varepsilon Kl[set(Y)] \Longleftrightarrow X\varepsilon Kl(Y)$.

Sets are elements of classes.

Proposition 27. $X\varepsilon set(Y) \Longrightarrow X\varepsilon el[Kl(Y)]$.

Proof. $X\varepsilon set(Y)$ implies $\exists U.U\varepsilon Y$; hence $\exists Z.Z\varepsilon Kl(Y)$ by (A5) and $Z\varepsilon Kl[set(Y)]$ by Corollary 2; so finally $X\varepsilon el(Z)$. □

Proposition 28. $X\varepsilon X \wedge \forall Z.[Z\varepsilon el(X) \Longrightarrow \exists T, T\varepsilon el(Z) \wedge T\varepsilon el(Y)] \Longrightarrow X\varepsilon el(Y)$.

Actually, one may prove that this proposition is equivalent to (A4). It may be regarded as an inference rule about the functor *el* and also as an alternative axiom.

Subset, Complement We define the notions of a subset and a complement. We define first the notion of a subset as a name-forming functor *sub* of an individual variable.

Definition 11. $X\varepsilon sub(Y) \Longleftrightarrow X\varepsilon X \wedge Y\varepsilon Y \wedge \forall Z[Z\varepsilon el(X) \Longrightarrow Z\varepsilon el(Y)]$.

Proposition 29. $X\varepsilon sub(Y) \Longrightarrow X\varepsilon el(Y)$.

Proof. As $X\varepsilon el(X)$ by Proposition 16, it follows that $X\varepsilon sub(Y)$ implies $X\varepsilon el(Y)$. □

Proposition 30. $X\varepsilon el(Y) \Longrightarrow X\varepsilon sub(Y)$.

Corollary 3. $X\varepsilon el(Y) \Longleftrightarrow X\varepsilon sub(Y)$.

We now define the notion of being external as a binary proposition-forming functor *ext* of individual variables.

Definition 12. $ext(X,Y) \Longleftrightarrow X\varepsilon X \wedge Y\varepsilon Y \wedge non[\exists Z.Z\varepsilon el(X) \wedge Z\varepsilon el(Y)]$.

Proposition 31.

1. $X\varepsilon X \Longrightarrow non[ext(X,X)]$.
2. $ext(X,Y) \Longleftrightarrow ext(Y,X)$.

The notion of a complement is rendered as a name-forming functor *comp* of two individual variables. We first define a new name Θ as follows: $U\varepsilon\Theta \Longleftrightarrow U\varepsilon el(Z) \wedge ext(U,Y)$.

Definition 13. $X\varepsilon comp(Y,Z) \Longleftrightarrow Y\varepsilon sub(Z) \wedge X\varepsilon Kl(\Theta)$.

Proposition 32. $X\varepsilon comp(Y,Z) \Longrightarrow ext(X,Y)$.

Proof. As $X\varepsilon comp(Y,Z)$, if $T\varepsilon el(X)$, then U,W with $U\varepsilon el(T)$, $U\varepsilon el(W)$, $W\varepsilon el(Z)$, $ext(W,Y)$ exist; hence $non[U\varepsilon el(Y)]$, and thus $non[U\varepsilon el(Y)]$. □

Corollary 4. $X\varepsilon X \Longrightarrow non[X\varepsilon comp(X,Z)]$.

Proposition 33. $X\varepsilon comp(Y,Z) \Longrightarrow X\varepsilon el(Z)$.

Proof. On lines of proof for Proposition 32. □

2.3 Other Axiomatics, Completeness

As with ontology, mereology may be axiomatized in terms of notions derived from the *part* functor. To acquaint the reader with this theme, we quote below some axioms, culminating in (CBA), below. We begin with the Sobociński axiom [22], formulated in terms of the functor *el*.

The Sobociński Axiom

(S) $X\varepsilon el(Y) \Longleftrightarrow Y\varepsilon Y \wedge \forall f,Z.\{\forall C.\{[C\varepsilon f(Z)] \Longleftrightarrow$
$\forall D.[D\varepsilon Z \Longrightarrow D\varepsilon el(C)] \wedge \forall D.[D\varepsilon el(C) \Longrightarrow \exists E,F.E\varepsilon Z \wedge F\varepsilon el(D) \wedge F\varepsilon el(E)]\}$
$\wedge Y\varepsilon el(Y) \wedge Y\varepsilon Z \Longrightarrow X\varepsilon el[f(Z)]\}$.

It is not difficult to see that (S) is a theorem of mereology.

Proposition 34. *(S) is a thesis of mereology.*

Proof. It is easily seen that $f(Z)$ is $Kl(Z)$, so (S) reads as the thesis

$$X\varepsilon el(Y) \Longleftrightarrow Y\varepsilon Y \wedge \forall Z.\{Y\varepsilon Z \Longrightarrow X\varepsilon el[Kl(Z)]\},$$

which is true in mereology. □

Proposition 35. *(S) implies axioms of mereology.*

Proof. Assume (S); then $\exists Z.Z\varepsilon el(Y)$ $(the\ left\ side)$ implies $Y\varepsilon el(Y)$. Similarly, taking into account that $f(Z)$ denotes $Kl(Z)$, we find that $X\varepsilon Kl(elX)$, and from this, we obtain that $X\varepsilon el(Y) \wedge Y\varepsilon el(Z) \Longrightarrow X\varepsilon el(Z)$. Letting $X\varepsilon pt(Y) \Longleftrightarrow X\varepsilon el(Y) \wedge non(X=Y)$, we arrive at (A1) and (A2). The uniqueness of $Kl(Z)$ follows from the subformula $C\varepsilon f(Z) \Longleftrightarrow$
$[\forall D.D\varepsilon Z \Longrightarrow D\varepsilon el(C)] \wedge [\forall D.D\varepsilon el(C) \Longrightarrow \exists E,F.E\varepsilon Z \wedge F\varepsilon el(D) \wedge F\varepsilon el(E)]$ of (S) and similarly, it does imply that $\exists E.E\varepsilon Z \Longleftrightarrow \exists C.C\varepsilon Kl(Z)$. □

A similar axiom has been proposed by Lejewski [7].

The Lejewski axiom.

(L) $X\varepsilon el(Y) \Longleftrightarrow Y\varepsilon Y \wedge \forall f,Z,C.\{\forall D.\{D\varepsilon f(Z) \Longleftrightarrow$
$\forall E.[\exists F.F\varepsilon el(D) \wedge F\varepsilon el(F)] \Longleftrightarrow [\exists G,H.G\varepsilon Z \wedge H\varepsilon el(E) \wedge H\varepsilon el(G))]\}$
$\wedge Y\varepsilon el(Y) \wedge Y\varepsilon el(C) \wedge Y\varepsilon Z \Longrightarrow X\varepsilon el[f(Z)]\}$.

One can show similarly that (L) and (S) are equivalent. A paraphrase of these axioms has been proposed by Clay [2].

The Clay axiom.

(Cl) $X\varepsilon el(Y) \Longleftrightarrow \{X\varepsilon X \wedge Y\varepsilon Y \wedge Y\varepsilon el(Y) \Longrightarrow \forall U,W.\{Y\varepsilon U \wedge$
$\forall C.\{C\varepsilon W \Longleftrightarrow \forall D.[D\varepsilon U \Longrightarrow D\varepsilon el(C)] \wedge \forall D.[D\varepsilon el(C) \Longrightarrow$
$\exists E,F.[E\varepsilon U \wedge F\varepsilon el(D) \wedge F\varepsilon el(E)]\} \Longrightarrow X\varepsilon el(W)\}\}$.

We may formally replace the functor *el* with a new name $\leq$, and we may read (and write) $X\varepsilon X$ as $X\varepsilon V$. Replacing U,W by small letters u,w symbolizing not necessarily individual names and denoting by f the field of $\leq$, we arrive at (Cl) in the form

(M) $X\varepsilon \leq (Y) \Longleftrightarrow \{X\varepsilon V \wedge Y\varepsilon V \wedge Y\varepsilon \leq (Y) \Longrightarrow \forall u,w.\{u \subset f \wedge w \subset f \wedge Y\varepsilon u \wedge$
$\forall C.\{C\varepsilon w \Longleftrightarrow \forall D.[D\varepsilon u \Longrightarrow D\varepsilon \leq\ (C)] \wedge \forall D.[D\varepsilon \leq (C) \Longrightarrow$
$\exists E,F.[E\varepsilon u \wedge F\varepsilon \leq (D) \wedge F\varepsilon \leq (E)]\} \Longrightarrow X\varepsilon \leq (w)\}\}$.

Substituting for V the universe U, neglecting the copula, and taking into account that w must be an individual name as pointed to by (M), one gets following [2]:

(CBA) $X \leq Y \Longleftrightarrow \{X \in U \wedge Y \in U \wedge Y \leq Y \Longrightarrow \forall u,w.\{u \subset U \wedge w \subset U \wedge Y \in u \wedge$
$(\forall C.\{C \in w \Longleftrightarrow \forall D.(D \in u \Longrightarrow D \leq C) \wedge \forall D.[D \leq C \Longrightarrow$
$\exists E,F.(E \in u \wedge F \leq D \wedge F \leq E)]\} \Longrightarrow \exists L.\ w = \{L\} \wedge X \leq L\}\}$.

It may be checked that (CBA) is the axiom for complete Boolean algebra without a null element. Therefore,

Proposition 36. (Tarski [25]). *Models of mereology are models of complete Boolean algebras without zero.*

Mereology is therefore complete with respect to algebraic structures, which are models for complete Boolean algebras without zero. In this way, mereology conceived as an alternative to naive set theory falls into the province of a most regular structure of set theory: complete Boolean algebras as an equivalent theory.

3 Rough Mereology: A Calculus of Approximate Parts

Rough mereology extends mereology by considering the functor μ_r of a *part to a degree r* for $r \in [0,1]$ [16, 17, 20]. The following is a list of basic postulates for rough mereology. We assume the notion of part defined so we discuss a mereological context. We introduce a family μ_r, where $r \in [0,1]$, called a *rough inclusion* that would satisfy

$$\begin{gathered}(RM1)\; x\mu_1 y \Leftrightarrow x\varepsilon el(y),\\ (RM2)\; x\mu_1 y \Rightarrow \forall z.(z\mu_r x \Rightarrow z\mu_r y),\\ (RM3)\; x = y \wedge x\mu_r z \Rightarrow y\mu_r z,\\ (RM4)\; x\mu_r y \wedge s \le r \Rightarrow x\mu_s y.\end{gathered} \tag{16}$$

The postulate (RM1) relates rough inclusion to mereology: $x\mu_1 y$ is equivalent to x as an element of y. In this way, an exact mereological structure is embedded into a rough mereological structure. It also follows that μ_r is defined on individual objects only. The postulate (RM2) does express the monotonicity of μ with respect to the relation *el*. By (RM3), μ_r is a congruence with respect to identity. (RM4) sets the meaning of μ_r: it means a degree at least r.

Example 1. A *generalized rough membership function*

$X\mu_r Y$ if and only if $\frac{|X\cap Y|}{|X|} \ge r$ in case X nonempty, 1 else;

see also [15] where X, Y are (either exact or rough) subsets (concepts) in the universe U of an information/decision system (U,A) is an example of a rough inclusion on concepts regarded as elements of the mereological universe.

It is evident that we cannot in general say more about the properties of μ_r: in particular, we lack the transitivity property. The class operator Kl may be recalled now. We use it in defining a sort of topology on the universe of individual objects. For given $r < 1$ and x, we let $g_r(x)$ denote the class $Kl(\Psi_r)$ where $\Psi_r(y) \Leftrightarrow y\mu_r x$. $g_r(x)$ collects in a single class-object all objects close to x in degree at least r.

From (RM1)–(RM4), the following properties may be deduced:

Proposition 37.

$$\begin{gathered}1.\; y\mu_r x \Rightarrow y\varepsilon el[g_r(x)].\\ 2.\; x\mu_r y \wedge y\varepsilon el(z) \Rightarrow x\varepsilon el[g_r(z)].\\ 3.\; \forall z.[z\varepsilon el(y) \Rightarrow \exists w,q.w\varepsilon el(z) \wedge w\varepsilon el(q) \wedge q\mu_r(x)] \Rightarrow y\varepsilon el[g_r(x)].\\ 4.\; y\varepsilon el[g_r(x)] \wedge z\varepsilon el(y) \Rightarrow z\varepsilon el[g_r(x)].\\ 5.\; s \le r \Rightarrow g_r(x)\varepsilon el[g_s(x)].\end{gathered} \tag{17}$$

Proof. 1 follows by definition of g_r, and 2 is implied by (RM2). For 3, consider a property Φ defined as $\Phi(y) \Leftrightarrow \forall z.[z\varepsilon el(y) \Rightarrow \exists w,q.w\varepsilon el(z) \wedge w\varepsilon el(q) \wedge q\mu_r(x)]$ and let $K\varepsilon = Kl(\Phi)$. Then, (i) $y\mu_r x \Rightarrow y\varepsilon el(K)$, (ii) $y\varepsilon el[g_r(x)] \Rightarrow y\varepsilon el(K)$, (iii) $y\varepsilon el(K) \Rightarrow y\varepsilon el[g_r(x)]$, which imply that $K\varepsilon =_E g_r(x)$ proving 3. 4 follows from 3 by the transitivity of el. 5 is a consequence of (RM4). □

The class $g_r(x)$ may be regarded as a neighborhood of x of radius r. Let us observe that $g_1(x) = x$ is the class of elements of x, hence x itself. We will single out a proposition for a rough inclusion in an information system $\mathcal{A} = (U,A)$. We define for $x,y \in U$, the set

$$DIS(x,y) = \{a \in A : a(x) \neq a(y)\}. \tag{18}$$

With help of $DIS(x,y)$, we define a *Gaussian rough inclusion* $\mu_r^{\mathcal{A}}$ by letting

$$x\mu_r^{\mathcal{A}}y \Leftrightarrow e^{-|\Sigma_{a\in DIS(x,y)} w_a|^2} \geq r, \tag{19}$$

where $w_a \in (0,\infty)$ is a *weight* associated with the attribute a for each $a \in A$. It remains to verify that (19) is a rough inclusion. We consider as individual objects the classes of indiscernibility relation IND_A, and we define the notion of an element as follows:

$$x\varepsilon el(y) \Leftrightarrow DIS(x,y) = \emptyset. \tag{20}$$

This notion of an element is then identical with = and corresponds to the empty part relation.

Proposition 38. *$\mu_r^{\mathcal{A}}$ satisfies (RM1)–(RM4) with the notion of an element as in (20).*

Proof. For (RM1), $x\mu_1^{\mathcal{A}}y$ if and only if $DIS(x,y) = \emptyset$ if and only if $x\varepsilon el(y)$. For (RM2), clearly, $DIS(x,y) = \emptyset$ implies $DIS(x,z) = DIS(y,z)$, and the same argument justifies (RM3). (RM4) follows by definition (19). □

Properties of Gaussian inclusions are collected in the next proposition. We denote with the symbol $g_r^{\mathcal{A}}$ the neighborhood induced by $\mu_r^{\mathcal{A}}$.

Proposition 39.

$$\begin{array}{c} 1.\ xIND_A y \Rightarrow x\mu_1^{\mathcal{A}}y. \\ 2.\exists\eta(r,s).x\mu_r^{\mathcal{A}}y \wedge y\mu_s^{\mathcal{A}}z \Rightarrow x\mu_{\eta(r,s)}^{\mathcal{A}}z. \end{array} \tag{21}$$

Proof. 1 follows directly from the proof of Proposition 38. To prove 2., we define

$$\eta(r,s) = r \cdot s \cdot e^{2\cdot(logr\cdot logs)^{\frac{1}{2}}}, \tag{22}$$

and we verify that $\eta(r,s)$ satisfies 2. We assume that $x\mu_r^{\mathcal{A}}y, y\mu_s^{\mathcal{A}}z$, so

$$\begin{array}{l} \Sigma_{a\in DIS(x,y)} w_a \leq (-logr)^{\frac{1}{2}}, \\ \Sigma_{a\in DIS(y,z)} w_a \leq (-logs)^{\frac{1}{2}}. \end{array} \tag{23}$$

As clearly, $DIS(x,z) \subseteq DIS(x,y) \cup DIS(y,z)$, assuming that t is the supremum of all u with $x\mu_u^{\mathfrak{A}} z$,

$$(-logt)^{\frac{1}{2}} \leq (-logr)^{\frac{1}{2}} + (-logs)^{\frac{1}{2}} \tag{24}$$

from which the formula

$$t \geq \eta(r,s) \tag{25}$$

follows. □

A look at (19) shows that $\mu_r^{\mathfrak{A}}$ is constant on indiscernibility classes of $IND_{\mathfrak{A}}$. In the sequel, we will use $\mu_r^{\mathfrak{A}}$ tacitly on objects as well as on indiscernibility classes. We apply Proposition 3.3 to show that $g_r(x)$'s induce a topology on classes of IND_A distinct from the standard clopen one see [13]:

Proposition 40.

$$\begin{gathered} x\varepsilon el[g_r(y)] \wedge x\varepsilon el[g_s(z)] \Rightarrow \\ g_t(x)\varepsilon el[g_r(y)] \wedge g_t(x)\varepsilon el[g_s(z)] \\ for\ t \geq max\{r^4, s^4\}. \end{gathered} \tag{26}$$

Proof. We need to solve the equation

$$w\mu_t^{\mathfrak{A}} x \Rightarrow w\mu_r^{\mathfrak{A}} y \tag{27}$$

with t. From (27), it follows that we need to satisfy the condition $\eta(r,t) \geq r$. Solving it, we get $t \geq r^4$, and similarly, $t \geq s^4$ implies $w\mu_t^{\mathfrak{A}} x \Rightarrow w\mu_s^{\mathfrak{A}} z$. □

Thus the Gaussian rough inclusion induces a regular structure into classes of the form $g_r(x)$. We have interpreted these classes in topological terms, but we may use them also as archetypical granules of knowledge.

Rough Mereological Component of Granulation The class $g_r(x)$ may be interpreted as a *granule of knowledge*, i.e., as a cluster of objects close to x in sufficient degree that the properties of these objects are satisfactorily close to properties of x, so there is no need to discern among x and them.

In the sequel, we discuss means for computing with granules of knowledge leading to networks of units organized similarly to neural networks but performing their computations in other ways. We consciously present this topic in a sterile form rid of application issues to present the reader with the basic ideas.

4 Rough-Neural Computation: A Case of Granular Computing

We define an intelligent unit *int-ag* modeled on a classical perceptron [1], and then we develop the notion of a network of intelligent units. The computation will consist of assigning a concept at the output unit to a collection of objects at input units, i.e., the result of a computation is classification of an input that assigns to the input a Gaussian granule of knowledge, i.e., an exact concept at the output unit. Thus the transition from the input to the output is a rough set theoretical operation, which is performed by a neural-like network operating on rough mereological principles and driven by gaussian rough inclusions at intelligent units in the network. Refer to [16–20] for a detailed analysis of those networks and to chapters in this volume by Pal, Peters et al., and Skowron and Stepaniuk for a presentation of this topic from the application point of view. Next, we begin our discussion with the simplest model of an intelligent unit.

4.1 Rough Mereological Perceptron

We exhibit the structure of a *rough mereological perceptron* (RMP). It consists of an intelligent unit *int-ag* denoted *ia* whose input is a finite tuple $\overline{x} = < x_1, x_2, \ldots, x_k >$ of objects. The symbol O_{ia} denotes the operation by which the input tuple $\overline{x}$ is converted into an object $x = O_{ia}(\overline{x}) \in U_{ia}$ where U_{ia} is the object universe of the information system $\mathcal{A}_{ia} = (U_{ia}, A_{ia})$ of *ia*. We endow *ia* with the Gaussian rough inclusion μ^{ia} at *ia*.

The unit *ia* is also equipped with a set of *target concepts* $T_{ia} \subset U_{ia}/IND_{A_{ia}}$. Each concept in T_{ia} is thus a class of indiscernibility $IND_{A_{ia}}$. The output of the RMP is the granule of knowledge $res_{ia}(\overline{x}) = g_{r_{res}}(x)$ with the property that

$$r_{res} = max\{r : \exists y \in T_{ia}.O_{ia}(\overline{x})\mu_r^{ia} y\}, \tag{28}$$

i.e., we essentially classify $\overline{x}$ as the exact concept being the class of indiscernibility classes as close to the indiscernibility class of x as the closest target class, i.e., in degree r_{res}. Let us observe that this classification depends on the weight system $\{w_a : a \in A_{ia}\}$ chosen. We may also observe that we could also present the result of computation $res_{ia}(\overline{x})$ as the tuple $< g_{r_i}(t_i) : t_i \in T_{ia} >$ where $r_i = sup\{r : O_{ia}(\overline{x})\mu_r^{ia} t_i\}$ for each $i \leq |T_{ia}|$.

4.2 Rough Mereological Network

Now we consider a network of intelligent agents (NIA) organized in the manner of a feedforward neural network, viz., we single out the following components in NIA:

1. The set *INPUT* of intelligent agents (*ias*) constituting the input layer of NIA. Each intelligent unit *ia* in *INPUT* acts as an RMP:

   - It receives a tuple $\overline{x}_{ia}$ of objects from the stock of primitive objects (signals, etc.).
   - It assembles $\overline{x}_{ia}$ into the object $O_{ia}(\overline{x}_{ia})$ in U_{ia}.
   - It classifies $O_{ia}(\overline{x}_{ia})$ with respect to its target concepts in T_{ia} and outputs the result $res_{ia}(\overline{x}_{ia})$.

2. Sets $LEV_1, \ldots, LEV_k$ constituting consecutive inner layers of NIA each $ia's$ in $\bigcup_i LEV_i$ acting as above.
3. The output RMP denoted ia_{out} acting as each ia except that its computation result is the collective computation result of the whole network.

Although, in general, the structure of neural networks is modeled on directed acyclic graphs, we assume for simplicity, that NIA is ordered by a relation ρ into a tree. Thus ia_{out} is the root of the tree, $INPUT$ is the leaf set of the tree, and each LEV_i constitutes the corresponding level of pairwise noncommunicating units in the tree.

Elementary Computations To analyze computation mechanisms in NIA, we begin with *elementary computations* performed by each subtree of the form $nia = \{ia, ia_0, \ldots, ia_m\}$ with $ia_j \rho ia$ for each $j \leq m$, i.e., ia is the root unit in nia, and $ia_0, \ldots, ia_m$ are its daughter units.

We make one essential assumption about NIA, viz., we presume that after preliminary training, the target sets T_{ia}, T_{ia_j} have been coordinated, i.e.,

1. For each $t \in T_{ia}$, there exist $t_1 \in T_{ia_1}, \ldots, t_m \in T_{ia_m}$ such that $t = O_{ia}(t_1, \cdots, t_m)$.
2. Each $t_j \in T_{ia_j}$ can be completed by $t_1 \in T_{ia_1}, \ldots, t_{j-1} \in T_{ia_{j-1}}, t_{j+1} \in T_{ia_{j+1}}, \ldots, t_m \in T_{ia_m}$ such that $t = O_{ia}(t_1, \ldots, t_m) \in T_{ia}$.

On the basis of assumptions 1, 2 above, we can describe elementary computations in terms of target sets. To this end, we introduce notions of *propagating functors* relative to sets of target concepts. The notion of a propagating functor was introduced in [16], and here we adopt it in essentially the same form.

We will say that a set $\sigma = \{t, t_1, \ldots, t_m\}$ of target concepts, where $t \in T_{ia}, t_1 \in T_{ia_1}, \ldots, t_m \in T_{ia_m}$, is *admissible* when $t = O_{ia}(t_1, \ldots, t_m)$. For an admissible set σ, the propagating functor $\Phi_{nia,\sigma}$ is defined as follows. For $\overline{x} =< x_1 \in U_{ia_1}, \ldots, x_m \in U_{ia_m} >$, we denote by $\overline{r}$ the tuple $< r_1, \ldots, r_m > \in \mathbf{R}^m$ defined by the condition

$$\forall i. r_i = sup\{r : x_i \mu_r^{ia_i} t_i\}, \tag{29}$$

and we let

$$\Phi_{nia,\sigma}(\overline{r}) = sup\{s : O_{ia}(x_1, \ldots, x_m) \mu_s^{ia} t\}. \tag{30}$$

The domain of $\Phi_{nia,\sigma}$ consists of a finite number of vectors in the cube $[0,1]^m$.
We order this set antilexicographically into a chain $\overline{r}_0 \prec \overline{r}_1 \prec \ldots \prec \overline{r}_n$. Thus, the ordering $\prec$ is the Pareto ordering, i.e.,

$$\overline{r}_i \prec \overline{r}_j \Rightarrow \forall u. r_{i_u} \leq r_{j_u}, \tag{31}$$

where $\overline{r} =< r_1, \ldots, r_m >$ denotes a vector with its coordinates.

Let us observe that $\overline{r}_n = \overline{1} =< 1,1,\ldots,1 >$. Given any vector $\overline{r} \in [0,1]^m$, we have the following cases:

$$(i)\ \overline{r} \prec \overline{r}_0; \quad (ii)\ \overline{r}_i \prec \overline{r} \prec \overline{r}_{i+1}. \tag{32}$$

In case (i),we set $\Phi_{nia,\sigma}(\overline{r}) = \overline{0} =< 0,0,\ldots,0 >$ and in case (ii), we set $\Phi_{nia,\sigma}(\overline{r}) = \Phi_{nia,\sigma}(\overline{r}_i)$. In this way, we extend $\Phi_{nia,\sigma}$ over the cube $[0,1]^m$. Let us also stress that $\Phi_{nia,\sigma}$ depends on weights w_a for $a \in \bigcup\{A_{ia} : ia \in \{ia, ia_1,\ldots, ia_m\}\}$ as on parameters; changing parameters w_a changes Φ. The direction of changes is indicated by the form of the function

$$f(x,y) = e^{-(\sum_{a\in DIS(x,y)} w_a)^2}, \tag{33}$$

which determines $\mu_r^{\mathfrak{A}}$ according to (19), i.e. via $x\mu_r^{\mathfrak{A}}y \Leftrightarrow f(x,y) \geq r$. Taking the gradient

$$\frac{\partial f}{\partial w} = f\cdot(-2\cdot\sum w_a), \tag{34}$$

we infer that the direction of maximizing f, hence $\mu_r^{\mathfrak{A}}$, is in minimizing $\sum w_a$.

We exploit the ordering introduced above to construct a piecewise linear hence almost everywhere differentiable approximation $\Psi_{nia,\sigma}$ of $\Phi_{nia,\sigma}$. To do this, we define the *indicator set* I_Φ as the maximal chain $\overline{r}_0 \prec \overline{r}_{i_1} \prec \ldots \prec \overline{r}_n$ with respect to the property that

$$\forall u. r_{i_u} < r_{i+1_u}. \tag{35}$$

Therefore, $\overline{r}_i, \overline{r}_{i+1} \in I_\Phi$ are antipodal vertices of the cube Q_i whose other vertices are vectors $\overline{r}$ with $\overline{r}_i \prec \overline{r} \prec \overline{r}_{i+1}$.

We define the mapping pr_{Δ_i} projecting vertices of Q_i onto the diagonal Δ_i of this cube. For each vertex $\overline{r} \in Q_i$, the coefficients α, β with $\alpha + \beta = 1$ such that $pr_{\Delta_i}(\overline{r}) = \alpha\cdot\overline{r}_i + \beta\cdot\overline{r}_{i+1}$ are defined, and we let

$$\Psi_{nia,\sigma}(\overline{r}) = \alpha\cdot\Phi_{nia,\sigma}(\overline{r}_i) + \beta\cdot\Phi_{nia,\sigma}(\overline{r}_{i+1}). \tag{36}$$

Clearly, $\Psi_{nia,\sigma}$ depends on weights w_a's as parameters. It is also obvious that

$$\Psi_{nia,\sigma}|I_\Phi = \Phi_{nia,\sigma}|I_\Phi.$$

The computation by *nia* may be described in terms of Ψ_{nia}. Given an input tuple $\overline{x}$ to leaf agents in *nia*, for each admissible σ, the vector $\overline{r}_{\overline{x}} \in Q_i$ for some i is defined, and $res'_{ia}(\overline{x}) = \Psi_{nia,\sigma}(\overline{r}_{\overline{x}})$ is calculated. The result of the computation is

$$res_{ia}(\overline{x}) = gr_{res'_{ia}(\overline{x})}(t). \tag{37}$$

This model of computing by *nia* may be called *linear Gaussian*.

Let us observe that computing acts by *nia* have a twofold aspect: on the *syntactic level*, the result is $O_{ia}(\overline{x})$, and on the *semantic level*, the result is $res_{ia}(\overline{x})$, which may be interpreted as the meaning of $O_{ia}(\overline{x})$.

Learning in nia We consider a learning problem.

Learning from positive examples.

Given:
a test sample $\mathbf{s} = (\overline{x}_1, \ldots, \overline{x}_k)$ of positive examples for
a target concept $g_r(t)$ for some $t \in T_{ia}$

Find: weights w_a such that

$$res_{ia}(\overline{x}_i)\varepsilon el[g_r(t)] \tag{38}$$

for each $i \leq k$.

To resolve the problem with respect to an admissible $\sigma = \{t_1, \cdots, t_m\}$, we proceed as in the back-propagation method [1] (see p.140 ff.), i.e., for each $\overline{x}_i$, we increment weights according to the rule,

$$w_a \rightarrow w_a + \theta \cdot \frac{\partial \psi}{\partial w_a}, \tag{39}$$

for each $a \in Dis[t, O_{ia}(\overline{x}_i)]$ with a parameter θ until $res_{ia}(\overline{x}_i)\varepsilon el[g_r(t)]$.

After successfully terminating the procedure, we have defined weights at *ia* at which all test tuples are classified into the target concept $g_r(t)$. From (37) it follows that with $\gamma = r - res'_{ia,\text{current}}(\overline{x})$ and a natural number k, θ_{current} should be of order

$$\frac{\gamma}{2kf^2 \cdot (\Sigma_a w_a)^2}$$

to make termination possible in k steps at each $i = 1, 2, \ldots, n$.

4.3 Computations in NIA

Any general NIA is a composition of local *nia*s according to the relation ρ. Each *ia* higher than leaf level receives objects from its daughters which it assembles into objects in its universe and classifies sending the assembled object up to its local head.

Accordingly, the global function Ψ is a superposition of local functions Ψ. Specifically, assuming that leaf units $ia_1, ia_2, \ldots, ia_q$ constituting $INPUT$ receive inputs $\overline{x}_1, \ldots, \overline{x}_q$, NIA produces from the joint vector $\overline{x}_{input} = <\overline{x}_1, \ldots, \overline{x}_q>$ the object $O_{NIA}(\overline{x}_{input}) \in U_{ia_{out}}$. For each $t \in T_{ia_{out}}$, we find, going top-down, target concepts t^*_{ia} for all *ia*s such that $O_{NIA}(t^*_{ia_1}, \ldots, t^*_{ia_q}) = t$ where $ia_1, ia_2, \ldots, ia_q$ constitute $INPUT$. We will say that t, t^*_{ia} form an admissible set of target concepts Σ. Given an admissible set Σ, we define the propagation functor $\Phi_{NIA,\Sigma}$ as in (29) and (30), and we apply the analogous procedure as with $\Psi_{nia,\sigma}$ to define $\Psi_{NIA,\Sigma}$ as a piecewise linear approximation of $\Phi_{NIA,\Sigma}$.

Now, in analogy to (37), we may define the result of a computation by NIA as follows:

$$res'_{NIA,\Sigma}(\overline{x}_{input}) = r_{res} = sup\{s : O_{NIA,\Sigma}(\overline{x}_{input})\mu_s^{ia_{out}} t\}, \tag{40}$$

and

$$res_{NIA,\Sigma}(\overline{x}_{input}) = g_{r_{res}}(t). \tag{41}$$

The learning problem may be stated for NIA in the same way it has been posed for *nia*. The learning algorithm is also similar with the difference that it goes top-down (or, back-propagating) from ia_{out} to leaf units and at each stage weights at the root of a current *nia* are adjusted incrementally.

Let us also observe that as rough mereology generalizes both rough and fuzzy set theories, the computation model presented above is a particular case of rough-fuzzy-neural computation, see the discussion in [12].

Acknowledgment This work was supported by the Grant No. 8T11C02417 from the State Committee for Scientific Research (KBN) of the Republic of Poland.

References

1. C. M. Bishop. *Neural Networks for Pattern Recognition.* Clarendon, Oxford, 1997.
2. R. Clay. Relation of Leśniewski's mereology to boolean algebra. *The Journal of Symbolic Logic*, 39: 638–648, 1974.
3. J. Srzednicki, S. J. Surma, D. Barnett, V. F. Rickey, editors. *Collected Works of Stanisław Leśniewski*, Kluwer, Dordrecht, 1992.
4. B. Iwanuś. On Leśniewski's elementary ontology. *Studia Logica*, 31: 73–119, 1973.
5. T. Kotarbiński. *Elements of the Theory of Knowledge, Formal Logic and Methodology of Science*. PWN, Warsaw, 1966.
6. Cz. Lejewski. On Leśniewski's ontology. *Ratio*, I(2): 150–176, 1958.
7. Cz. Lejewski. A contribution to Leśniewski's mereology. *Yearbook for 1954–55 of the Polish Society of Arts and Sciences Abroad V*, 43–50, London, 1954–55.
8. S. Leśniewski. Grundzüge eines neuen Systems der Grundlagen der Mathematik. *Fundamenta Mathematicae*, 14: 1–81, 1929.
9. S. Leśniewski. Über die Grundlegen der Ontologie. *Comptes Rendus des Séances de la Société des Sciences et des Lettres de Varsovie*, 3: 111–132, 1930.
10. S. Leśniewski. On the foundations of mathematics (in Polish). *Przegląd Filozoficzny*, 30: 164–206, 1927; 31: 261–291, 1928; 32: 60–101, 1929; 33: 77–105, 1930; 34: 142–170, 1931.
11. S. Leśniewski. On the foundations of mathematics. *Topoi*, 2: 7–52, 1982.
12. S. K. Pal, J. Peters, L. Polkowski, A. Skowron. Rough-neural computing: An introduction (this book).
13. Z. Pawlak. Rough sets, algebraic and topological approach. *International Journal of Computer and Information Sciences*, 11: 341–366, 1982.
14. Z. Pawlak. *Rough Sets: Theoretical Aspects of Reasoning about Data.* Kluwer, Dordrecht, 1992.

15. Z. Pawlak, A. Skowron. Rough membership functions. In R. R. Yager, M. Fedrizzi, J. Kacprzyk, editors, *Advances in the Dempster-Schafer Theory of Evidence*, 251–271, Wiley, New York, 1994.
16. L. Polkowski, A. Skowron. Rough mereology: a new paradigm for approximate reasoning. *International Journal for Approximate Reasoning*, 15(4): 333–365, 1997.
17. L. Polkowski, A. Skowron. Adaptive decision-making by systems of cooperative intelligent agents organized on rough mereological principles. *International Journal for Intelligent Automation and Soft Computing*, 2(2): 123–132, 1996.
18. L. Polkowski, A. Skowron. Grammar systems for distributed synthesis of approximate solutions extracted from experience. In G. Paun, A. Salomaa, editors, *Grammatical Models of Multi-Agent Systems*, 316–333, Gordon and Breach, Amsterdam, 1999.
19. L. Polkowski, A. Skowron. Towards an adaptive calculus of granules. In L. A. Zadeh, J. Kacprzyk, editors, *Computing with Words in Information/Intelligent Systems, Vol. 1*, 201–228, Physica, Heidelberg, 1999.
20. A. Skowron, L. Polkowski. Rough mereological foundations for design, analysis, synthesis and control in distributed systems. *Information Sciences. An Interntional Journal*, 104: 129–156, 1998.
21. J. Słupecki. S. Leśniewski's calculus of names. *Studia Logica*, 3: 7–72, 1955.
22. B. Sobociński. Studies in Leśniewski's mereology. In *Yearbook for 1954–55 of the Polish Society of Art and Sciences Abroad V*, 5: 34–43, London, 1954–55.
23. B. Sobociński. L'analyse de l'antinomie Russellienne par Leśniewski. *Methodos*, I: 94–107, 220–228, 308–316, 1949; II: 237–257, 1950.
24. A. Tarski. Appendix E. In J. H. Woodger, *The Axiomatic Method in Biology*, Cambridge University Press, Cambridge, 1937.
25. A. Tarski. Zur Grundlegung der Boolesche Algebra I. *Fundamenta Mathematicae*, 24: 177–198, 1935.
26. L. A. Zadeh. Fuzzy logic = computing with words. *IEEE Transactions on Fuzzy Systems*, 4: 103–111, 1996.

Chapter 5
Knowledge-Based Networking in Granular Worlds

Witold Pedrycz

Department of Electrical & Computer Engineering, University of Alberta, Edmonton, Alberta T6G 2M7, Canada
and
Systems Research Institute, Polish Academy of Sciences, Newelska 6, 01-447 Warsaw, Poland
pedrycz@ee.ualberta.ca

Summary. In this study, we develop models of collaborative clustering realized across a collection of databases, where each database can be treated as a granular world. The essence of a search for data structures carried out in this environment deals with a determination of crucial common relationships in databases. Depending upon the way in which databases are accessible and can collaborate, we distinguish between the vertical and horizontal collaboration. In the first case, the databases deal with objects defined in the same attribute (feature) space. Horizontal collaboration takes place when we deal with the same objects being defined in different attribute spaces and therefore giving rise to separate databases. We develop a new clustering architecture supporting the mechanisms of collaboration based on the standard fuzzy C-means (FCM) method. In the horizontal collaboration, the clustering algorithms interact by exchanging information about "local" partition matrices. In this sense, the required communication links are established at the level of information granules (more specifically, fuzzy sets or fuzzy relations forming the partition matrices) rather than patterns (data points) that are directly available in the databases. We discuss how this form of collaboration helps to meet requirements of data confidentiality. In vertical collaboration, the method operates at the level of the prototypes formed for each individual database or the induced partition matrices. Numerical examples are used to illustrate each of the methods.

1 Introductory Comments

Undoubtedly, the distributed nature of data is inherent in most information systems. Intelligent agents and their collaboration over the Internet are excellent testimony to such a claim [1, 9, 13, 14, 18]. In many areas of everyday activity, various databases are constructed, used, and maintained independently of each other. In each local environment, one tries to make sense of data by engaging in various activities of data mining and data analysis. The results obtained can be useful to a local community; yet they could be of significant interest to the others. This triggers interest in a collaborative effort where the data mining activities could exploit several databases and the ensuing results benefit a larger circle of users. Though it sounds appealing, one has to remember that sharing data, especially those of a more confidential nature, is a genuine obstacle. This matter has to be taken seriously in any collaborative pursuit in data analysis.

This collaboration-driven task of data mining calls for an orchestrated effort and implies the highly collaborative nature of search for dependencies in data so that such findings are common and relevant to all databases (as such, discoveries of global character are of genuine interest). To shed light on the spectrum of processing problems, we identify possible scenarios along with existing drawbacks, and we envision potential mechanisms of collaboration:

- **Search for a common structure in databases.** Within a given organizational structure (company, network of sales offices, etc.), there are several local databases of customers (e.g., each supermarket generates its own database, or a sales office maintains a local database of its customers). Generally, we can assume that all databases have the same attributes (features) whereas each database consists of different objects (patterns). To derive some global relationships that are common to all of these databases, we should allow the databases to collaborate at the level of patterns. Quite commonly, we may not be permitted to have access to all databases, but eventually we might be provided with some general aggregates (some synthetic indexes describing data or a mean value or median are good examples). Refer to Fig. 1, which illustrates the underlying concept. Bearing this in mind, we can talk about *vertical* (data-based) collaboration in the process of knowledge elicitation (that is, revealing a common structure in the data).

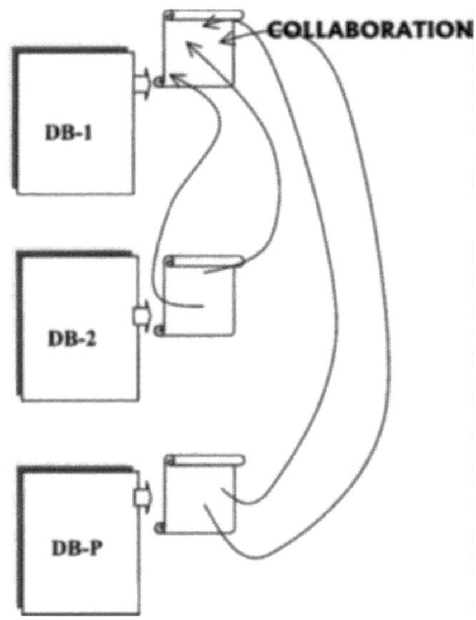

Fig. 1. Vertical collaboration between databases at a local level; in each database, objects are located in the same data space but deal with the different patterns

- **Security issues and discovery of data structures across different data sets.** Consider now that information about the same group of clients is collected in different databases where an individual company (bank, store, etc.) builds its own database. Because of confidentiality and security requirements, the companies cannot share information about clients directly. However, all of them are vitally interested in deriving some associations that help them learn about clients (namely, identifying their profiles and needs). As they are concerned with the same population of clients, we may anticipate that the basic structure of the population of such patterns, in spite of possible minor differences, should hold across all databases. The approach taken in this case would be to build clusters in each database and exchange information at the level of the clusters treated here as information granules. Subsequently, we allow all collaboration processes to be realized at this particular level. In this manner, the security issues are not compromised while a sound mechanism of collaboration/interaction between the databases becomes established. Graphically, we can envision the situation of such collaboration as that portrayed in Fig. 2. Evidently, in this case, we are concerned with *horizontal* (that is, feature-based) collaboration in the search for the data structure.

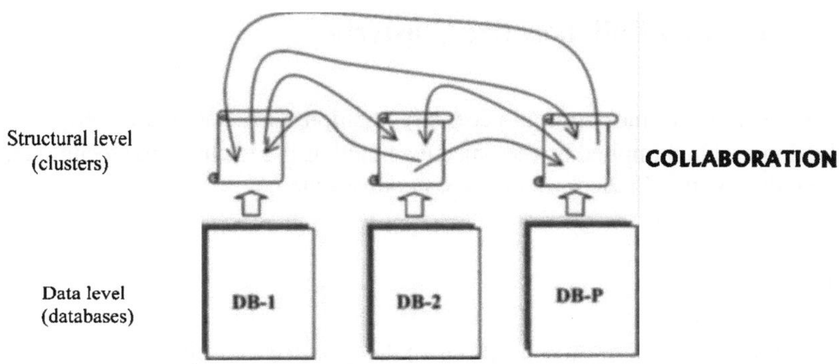

Fig. 2. Collaboration between databases at the level of local structures (clusters) discovered there. Note that no direct collaboration is allowed at the data level

As data structure elicitation is inherently user oriented and user friendly, we are interested in the collaborative clustering inasmuch as its results are information granules. In the sequel, this gives rise to a certain type of collaboration as indicated before, namely, vertical collaborative clustering that involves databases containing various objects and horizontal clustering where we are faced with the same objects being characterized by various attributes.

As far as the algorithmic issues are concerned, the underlying idea of collaboration dwells on well-known fuzzy C-means, see [3]. The reader may refer to pertinent details as to the generic method that is used as a canvass of the collaborative schemes developed in the study. In general, we can think of clustering [1, 4, 7, 8, 11, 12] as a vehicle for forming information granules. It is also worth stressing that fuzzy clustering arose as a fundamental and highly appealing technique in constructing fuzzy models; refer, e.g., to [6–7, 15–17]. Moreover, collaborative clustering can be cast in the realm of intelligent agents, see [14], whose activities may center around discovering and sharing knowledge.

The material is organized as follows. First, we proceed with horizontal collaborative clustering by introducing all necessary notations, formulating the problem itself, and discussing its algorithmic aspects. In the sequel, we use a number of numeric examples to illustrate the method. Second, we concentrate on vertical clustering following the same scheme of presentation as used in the first approach. Illustrative numeric examples are also covered. In Sect. 4, we contrast differences and similarities between vertical and horizontal clustering that are cast in terms of the space in which collaboration occurs. We also raise the underlying issues of data confidentiality and security, namely, that these terms are well-described in the language of fuzzy sets and become inherently non-Boolean (that is, non binary). Concluding remarks are covered in Sect. 5.

2 Horizontal Collaborative Clustering

In this section, we introduce all necessary notation, formulate the underlying optimization problem implied by the objective function-based clustering technique, and derive the solution in the form of some iterative scheme.

2.1 The Notation

In what follows, we consider p subsets of data located in different spaces (viz., the patterns there are described by different features), $ii = 1,2,\ldots,p$. As each subset has the same patterns (that is, each pattern results as a concatenation of corresponding subpatterns), the number of elements in each subset is the same and equal to N. We are interested in partitioning the data into c fuzzy clusters. The result of clustering completed for each subset of data comes in the form of a partition matrix and a collection of prototypes. We use bracket notation to identify the specific subset. Hence, we use the notation $U[ii]$ and $\mathbf{v}[ii]$ to denote the partition matrix and the ith prototype produced by the clustering realized for the iith set of data. Similarly, the dimensionality of the patterns (the number of their features) in each subset could be different. To underline this, we use a pertinent index, say $n[ii]$, $ii = 1,2,\ldots,p$. The distance function between the ith prototype and the kth pattern in the same set is denoted by $d_{ik}^2[ii]$, $i = 1,2,\ldots,c$, $k = 1,2,\ldots,N$. Again, the index used here (viz., ii) underlines the fact that we are dealing with a certain data space pertinent to the iith data set (database). Moreover, throughout the study, we confine ourselves to the weighted Euclidean distance of the form

$$d_{ik}^2[ii] = ||\mathbf{x}_k - \mathbf{v}_i[ii]||_{ii} = \sum_{j=1}^{n[ii]} \frac{(x_{kj} - v_{ij}[ii])^2}{\sigma_j^2[ii]}. \quad (1)$$

The objective function guiding the formation of the clusters that is completed for each subset assumes the well-known form encountered in the standard FCM algorithm

$$\sum_{k=1}^{N} \sum_{i=1}^{c} u_{ik}^2[ii] d_{ik}^2[ii], \quad (2)$$

$ii = 1, 2, \ldots, p$. Collaboration among subsets is established through a matrix of connections (interaction coefficients or interactions); see Fig. 3.

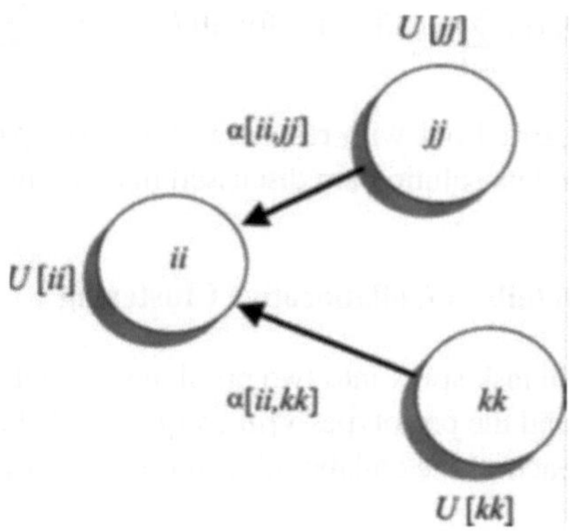

Fig. 3. Collaboration in the clustering scheme represented by the matrix of interactions among data sets

Each entry of the collaborative matrix describes an intensity of interaction. In general, $\alpha[ii, kk]$ assumes nonnegative values. The higher the value of the interaction coefficient, the stronger the collaboration among corresponding subsets. To accommodate the collaborative effect in the optimization process, the objective function is expanded into the form

$$Q[ii] = \sum_{k=1}^{N} \sum_{i=1}^{c} u_{ik}^2[ii] d_{ik}^2[ii] + \sum_{\substack{jj=1 \\ jj \neq ii}}^{P} \alpha[ii, jj] \sum_{k=1}^{N} \sum_{i=1}^{c} \{u_{ik}[ii] - u_{ik}[jj]\}^2 d_{ik}^2[ii], \quad (3)$$

$ii = 1, 2, \ldots, p$. The role of the second term in the above expression is to make the clustering based on the iith subset "aware" of the other partitions. It becomes obvious that if the structures in all data sets are similar, then the differences among the partition matrices tend to be lower. On the other hand, if we encounter higher differences, we anticipate that the collaboration will be able to address these needs.

The weight coefficient (α) helps control the effect of collaboration; the higher the value of α, the more impact comes from the collaborative side of the clustering.

As usual, we require that the partition matrix satisfies "standard" requirements of membership grades summing to 1 for each pattern and the membership grades contained in the unit interval. All in all, collaborative clustering converts into the following family of p optimization problems with membership constraints

$$Min\, Q[ii]$$

subject to

$$U[ii] \in \mathbb{U}[ii],$$

where $\mathbb{U}[ii]$ is a family of all fuzzy partition matrices, namely,

$$\mathbb{U}[ii] = \{u_{ik}[ii] \in [0,1] : \sum_{i=1}^{c} u_{ik}[ii] = 1 \quad \text{for all } k \text{ and } 0 < \sum_{k=1}^{N} u_i k[ii] < N \text{ for } i\}.$$

The minimization is carried out with respect to the fuzzy partition and the prototypes. This problem and its solution are discussed in detail in the ensuing section.

2.2 Optimization Details of Collaborative Clustering

The above optimization task splits into two problems, namely, determination of the partition matrix $U[ii]$ and the prototypes $\mathbf{v}_1[ii]$, $\mathbf{v}_2[ii]$, ..., $\mathbf{v}_c[ii]$. These problems are solved separately for each of the collaborating subsets of patterns.

To determine the partition matrix, we exploit the technique of Lagrange multipliers so that the constraint in the problem becomes integrated as a part of the objective function considered in the constraint-free optimization. The objective function $V[ii]$ that is considered for each k separately has the form

$$V[ii] = \sum_{i=1}^{c} u_{ik}^2[ii] d_{ik}^2[ii] + \sum_{\substack{jj=1 \\ jj \neq ii}}^{P} \alpha[ii,jj] \sum_{i=1}^{c} \{u_{ik}[ii] - u_{ik}[jj]\}^2 d_{ik}^2[ii] - \lambda\left(\sum_{i=1}^{c} u_{ik}[ii] - 1\right), \tag{4}$$

where λ denotes the Lagrange multiplier. The necessary conditions leading to the local minimum of $V[ii]$ read as follows:

$$\frac{\partial V[ii]}{\partial u_{st}[ii]} = 0, \qquad \frac{\partial V[ii]}{\partial \lambda} = 0, \tag{5}$$

$s = 1,2,\ldots,c$, $t = 1,2,\ldots,N$. Let us start with the explicit expression governing optimization of the partition matrix. Computing the derivative of V with respect to

u_{st} and zeroing it,

$$\frac{\partial V[ii]}{\partial u_{st}[ii]} = 2u_{st}[ii]d_{st}^2[ii] + 2\sum_{\substack{jj=1\\ jj\neq ii}}^{p} \alpha[ii,jj](u_{st}[ii] - u_{st}[jj])d_{st}^2[ii] - \lambda = 0. \quad (6)$$

To get $u_{st}[ii]$, we rewrite this expression as

$$u_{st}[ii] = \frac{\lambda + 2d_{st}^2[ii]\left(\sum\limits_{\substack{jj=1\\ jj\neq ii}}^{p} \alpha[ii,jj]u_{st}[jj]\right)}{2d_{st}^2[ii]\left(1 + \sum\limits_{\substack{jj=1\\ jj\neq ii}}^{p} \alpha[ii,jj]u_{st}[jj]\right)}. \quad (7)$$

To come up with a concise expression, we introduce some auxiliary notation,

$$\varphi_{st}[ii] = \sum_{\substack{jj=1\\ jj\neq ii}}^{P} \alpha[ii,jj]u_{st}[jj]$$

and

$$\psi[ii] = \sum_{\substack{jj=1\\ jj\neq ii}}^{P} \alpha[ii,jj].$$

By virtue of the normalization condition $\sum_{s=1}^{c} u_{st}[ii] = 1$, the Lagrange multiplier is

$$\lambda = \frac{1 - \sum\limits_{s=1}^{c} \frac{\varphi_{st}[ii]}{1+\psi[ii]}}{\sum\limits_{s=1}^{c} \frac{1}{2d_{st}^2[ii](1+\psi[ii])}},$$

which, taking into account (7), leads to the formula,

$$u_{st}[ii] = \frac{\varphi_{st}[ii]}{1+\psi[ii]} + \frac{1}{\sum\limits_{j=1}^{c} \frac{d_{st}^2}{d_{jt}^2}}\left[1 - \sum_{j=1}^{c} \frac{\varphi_{jt}[ii]}{1+\psi[ii]}\right]. \quad (8)$$

In calculations of the prototypes, we explicitly use the weighted Euclidean distance between the patterns and the prototypes. The necessary condition for the minimum of the objective function is of the form $\nabla_{\mathbf{v}[ii]}Q = 0$. The details are obvious, yet the calculations are somewhat tedious. Finally, the resulting prototypes are equal to

$$v_{st}[ii] = \frac{A_{st}[ii] + C_{st}[ii]}{B_s[ii] + D_s[ii]}, \quad (9)$$

$s = 1, 2, \ldots, c, t = 1, 2, \ldots, n[ii], ii = 1, 2, \ldots P$. The coefficients in the above expression are as follows:

$$A_{st}[ii] = \sum_{k=1}^{N} u_{sk}^2[ii] x_{kt}[ii], \tag{10}$$

$$B_s[ii] = \sum_{k=1}^{N} u_{sk}^2[ii], \tag{11}$$

$$C_{st}[ii] = \sum_{\substack{jj=1 \\ jj \neq ii}}^{P} \alpha[ii, jj] \sum_{k=1}^{N} (u_{sk}[ii] - u_{sk}[jj])^2 x_{kt}[ii], \tag{12}$$

$$D_s[ii] = \sum_{\substack{jj=1 \\ jj \neq ii}}^{P} \alpha[ii, jj] \sum_{k=1}^{N} (u_{sk}[ii] - u_{sk}[jj])^2 \tag{13}$$

(note that $\mathbf{x}_k[ii]$ denotes the kth pattern from the iith subset of patterns).

2.3 The Detailed Clustering Algorithm

The general clustering scheme consists of two phases:

- Generation of clusters without collaboration. This phase involves using the FCM algorithm applied individually to each subset of data. Obviously, the number of clusters needs to be the same for all of the data sets. During this phase, we independently seek a structure in each subset of data.
- Collaboration of the clusters. Here we start with the already computed partition matrices, set up the collaboration level (through the values of the interaction coefficients arranged in $\alpha[ii, jj]$), and proceed with simultaneous optimization of the partition matrices.

Moving on to the formal algorithm, the computational details are organized in the following way:

Given: subsets of patterns $X_1, X_2, \ldots, X_P$

Select: distance function, number of clusters (c), termination criterion, and collaboration matrix $\alpha[ii, jj]$.

Initiate randomly all partition matrices $U[1], U[2], \ldots, U[p]$.

Phase I

For each data set,

repeat

compute prototypes $\{\mathbf{v}_i[ii]\}, i = 1, 2, \ldots, c$ and partition matrices $U[ii]$ for all subsets of patterns

until a termination criterion has been satisfied.

Phase II

repeat

For the given matrix of collaborative links $\alpha[ii, jj]$, compute prototypes and partition matrices $U[ii]$ using (4) and (7)

until a termination criterion has been satisfied.

The termination criterion relies on the changes in the partition matrices obtained in successive iterations of the clustering method; for instance, a Tchebyschev distance could serve as a sound measure of changes in the partition matrices. Subsequently, when this distance is lower than an assumed threshold value ($\varepsilon > 0$), optimization is terminated.

2.4 Quantification of the Collaborative Phenomenon of Clustering

There are two levels of assessing a collaborative effect between clusters, namely, the level of data and the level of information granules (that is, fuzzy sets included in the partition matrix). In this latter quantification, we use the results of clustering without any collaboration as a point of reference.

The *level of data* involves a comparison carried out at the level of numeric representatives of clustering, that is, the prototypes (centroids). The impact of collaboration is then expressed in changes in the prototypes that result from the collaboration.

At the *level of information granules* (partitions and fuzzy sets), the effect of collaboration is expressed in two ways, as shown schematically in Fig. 4, where the collaboration involves two data sets (namely, $p = 2$) indicated by 1 and 2. Similarly, we denote by **1-ref** and **2-ref** the results (partition matrices) from clustering carried out without any collaboration. First, we express how close the two partition matrices are as a result of collaboration. The pertinent measure reads as an average distance between the partition matrices $U_1 = [u_{ik}[1]]$ and $U_2 = [u_{ik}[2]]$, that is,

$$\delta = \frac{1}{N * c} \sum_{k=1}^{N} \sum_{i=1}^{c} u_{ik}[1] - u_{ik}[2]|. \tag{14}$$

Evidently, the stronger the collaboration (higher values of the corresponding α), the lower the values of δ. In this sense, this index helps us translate the collaborative parameters (α) into the effective changes in membership grades (that are the apparent final result of such interaction). The plot of δ regarded as a function of α is useful in revealing how the collaborative phenomenon takes place. It tells how much the data subset is susceptible to the collaborative impact from the other subsets of patterns. For instance, no changes in the values of δ for increasing values of αs is an indicator of strong differences existing between the structures in the two data sets.
The second criterion takes into consideration the results of clustering obtained without any collaboration and treats this as a reference point. Using such partition matrices, we quantify how far collaboration affects the results of clustering. For instance,

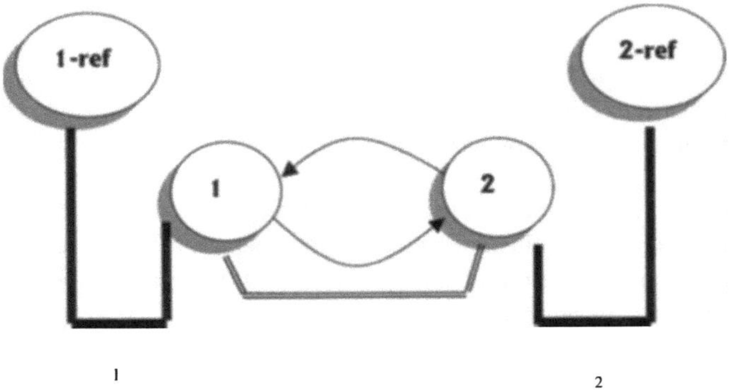

Fig. 4. Two ways of quantifying collaboration at the level of information granules (see the detailed description in the text)

for the first data set,

$$\Delta_1 = \frac{1}{N * c}\sum_{k=1}^{N}\sum_{i=1}^{c} u_{ik}[1] - u_{ik}[1 - ref]||. \tag{15}$$

For the second data subset, we obtain

$$\Delta_2 = \frac{1}{N * c}\sum_{k=1}^{N}\sum_{i=1}^{c} u_{ik}[2] - u_{ik}[2 - ref]||. \tag{16}$$

Although the above index exhibits a global character, one can investigate the changes at the level of the individual cluster and patterns. This local behavior of the collaboration is helpful in identifying elements whose membership grades are affected quite significantly as a result of collaboration and those whose structure is compatible across all data sets.

2.5 Experiments

In a series of numeric experiments, we used Boston housing data available on the Internet.[1] It consists of 506 patterns describing real estate in the Boston area. There are 14 features describing the patterns. These include crime rate, nitric acid concentration, median value of the house, just to name a few. We distinguish between two subsets of features where the first can be treated as descriptors of social aspects of the data:

A = {per capita crime rate by town, nitric oxide concentration (parts per 10 million), proportion of owner-occupied units built prior to 1940, weighted distances to five Boston employment centers, pupil–teacher ratio by town, % lower status of the population, median value of owner-occupied homes in \$1000s}
and

[1] see `ftp://ftp.ics.uci.edu/pub/machine-learning-databases/housing/`.

B = {proportion of residential land zoned for lots more than 25,000 sq.ft, proportion of nonretail business acres per town, Charles River dummy variable (equal to 1 if tract bounds river; 0, otherwise), average number of rooms per dwelling, index of accessibility to radial highways, full-value property-tax rate per \$10,000, 1000(Bk - 0.63)^2 where Bk is the proportion of blacks by town}

In the following experiments, we set up the number of the clusters to be equal to 5, $c = 5$. Several scenarios of collaboration are discussed; see Fig. 5 for the schematic notation. As only two subsets of data are involved, we drop indexes in the collaboration matrix; the meaning of collaboration becomes obvious from the context.

In all experiments, we start with clustering that takes place without any collaboration (it was found that 60 iterations was enough to assure no changes in the partition matrices, that is, the optimization process could be deemed complete). The collaboration takes place in the next phase.

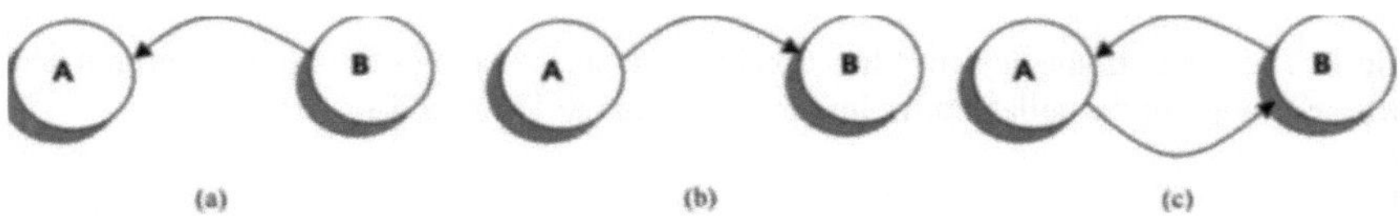

Fig. 5. Scenarios of collaborative clustering used in the experiments

a. There is a collaborative link originating from **B** and affecting **A**. The values of this link (α) are set successively to 0.05, 0.1, 0.5, and 1. The values of the objective function are shown in Fig. 6; as expected, the objective function assumes higher values for increasing levels of collaboration (this is not surprising when we note that the collaboration component contributes additively as part of this objective function). The drops in the values of the objective function at the beginning of the entire optimization are noticeable.

The resulting prototypes change once the collaboration assumes a different intensity, as shown below:

$$\alpha = 0.5$$

$\mathbf{v}_1$= [12.840698 0.675343 92.033089 1.984616 19.896008 21.301111 13.195952]
$\mathbf{v}_2$= [0.291484 0.437043 32.688419 6.474721 16.880703 6.680616 27.994970]
$\mathbf{v}_3$= [0.880803 0.528872 68.019592 3.802059 18.728121 12.382829 21.436878]
$\mathbf{v}_4$= [0.639216 0.500246 60.230255 4.270203 17.784111 8.671939 26.960485]
$\mathbf{v}_5$= [8.554115 0.661536 89.209915 2.277547 19.955612 17.302109 17.229208]

$$\alpha = 1.0$$

$\mathbf{v}_1$= [13.185606 0.673498 91.076195 2.016629 19.926291 20.879652 13.254309]

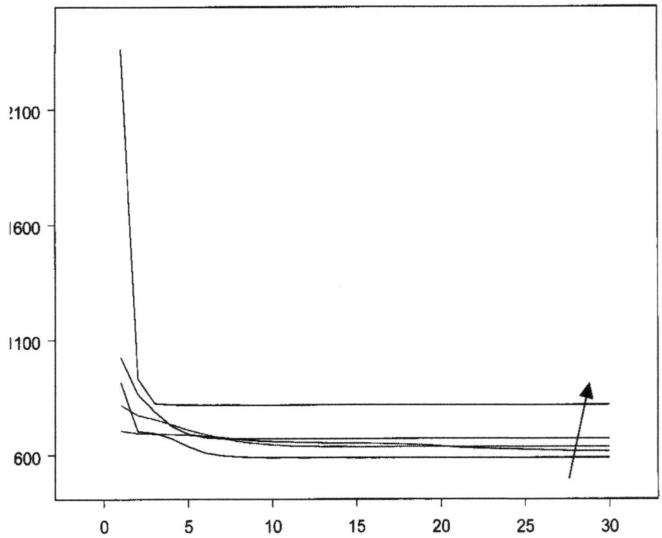

Fig. 6. The values of the objective function in successive iteration steps of the algorithm and selected values of the collaborative link (namely, 0.05, 0.1, 0.5, and 1)

$\mathbf{v}_2$= [0.264656 0.435453 33.055351 6.535963 16.788097 6.537128 28.447708]

$\mathbf{v}_3$= [0.759657 0.527652 67.316498 3.848079 18.747555 12.533010 21.190838]

$\mathbf{v}_4$= [0.561581 0.500323 59.888401 4.340404 17.798523 8.690695 26.753359]

$\mathbf{v}_5$= [9.424790 0.663229 88.969498 2.242356 19.995226 17.504406 17.110945]

For comparative reasons, the prototypes of the subset **A** without any collaboration are listed as follows:

$\mathbf{v}_1$= [11.491062 0.688633 94.221016 1.930663 19.940283 21.444845 13.103884]

$\mathbf{v}_2$= [0.394793 0.439771 31.897591 6.384313 17.012272 6.963907 27.159363]

$\mathbf{v}_3$= [0.860573 0.489002 52.550468 4.605259 18.520782 9.673503 24.041653]

$\mathbf{v}_4$= [1.307117 0.536930 75.181625 3.334807 17.237040 9.618564 27.298693]

$\mathbf{v}_5$= [3.288866 0.601333 86.465401 2.696251 19.858582 15.527621 18.926182]

One can note that higher values of α lead to more evident translations of the prototypes in comparison to their original location when no collaboration took place. The prototypes in each data set start resembling each other. The collaborative effect can be quantified in the language of membership functions (partition matrices). Following the notation introduced in Sect. 3, the values of the indexes δ and Δ_1 are illustrated in Fig. 7. As anticipated, the values of δ become lower as the collaboration level increases, whereas Δ_2 gets higher as we depart from the "local" partition matrix (namely, that computed without any

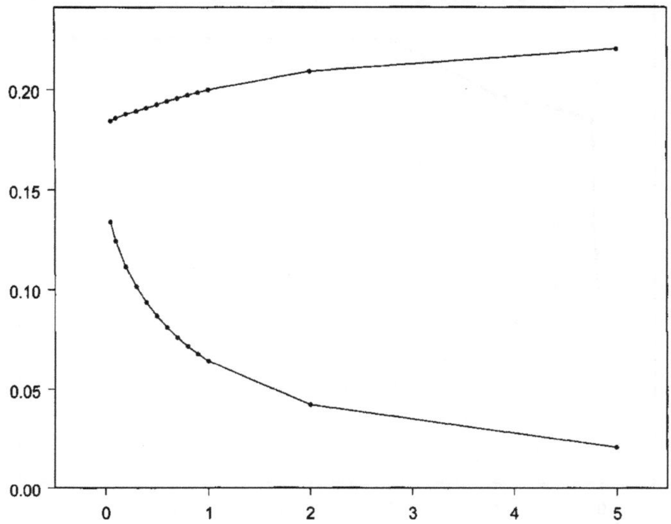

Fig. 7. Values of δ and Δ for selected values of α

collaboration) that is under collaborative pressure to accept some other sources of information about the overall data structure.

b. This experiment deals with the collaboration originating from the second group. The collaborative effect is quantified in Fig. 8. In comparison to the other collaborative scheme, there is quite a comparable level of changes in membership grades. The only significant jump is reported when the collaborative effect comes into play.

c. In this case, we allow reciprocal collaborative links, that is, **A** and **B** interact; see Fig. 9. The results are shown in the form of δ as well as Δ_1 and Δ_2. The values of δ decrease monotonically as values of α increase. An interesting effect occurs in terms of the collaboration: **A** tends to be more stiff in the collaborative interaction; the values of Δ_1, in spite of the increasing interaction (higher values of α), tend to remain constant. **B** is more flexible in that the collaboration more readily accepts collaborative signals that manifest in increasing values of Δ_2.

In the following experiments, we split the features into two groups: the first (**A**) includes all features but the price of real estate, which forms the second group (**B**). The collaborative link is activated by the first group (this group affects the clustering realized within **B**).

The prototypes in the median value of house change, depending on the values of collaborative feedback. Noticeably, with the increase in collaboration (denoted by α), the prototypes tend to occupy a narrower range in comparison to the situation

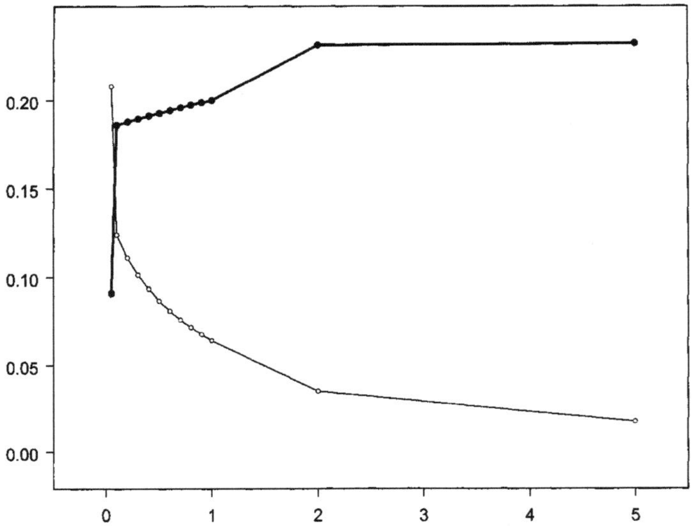

Fig. 8. Values of δ and Δ for selected values of α

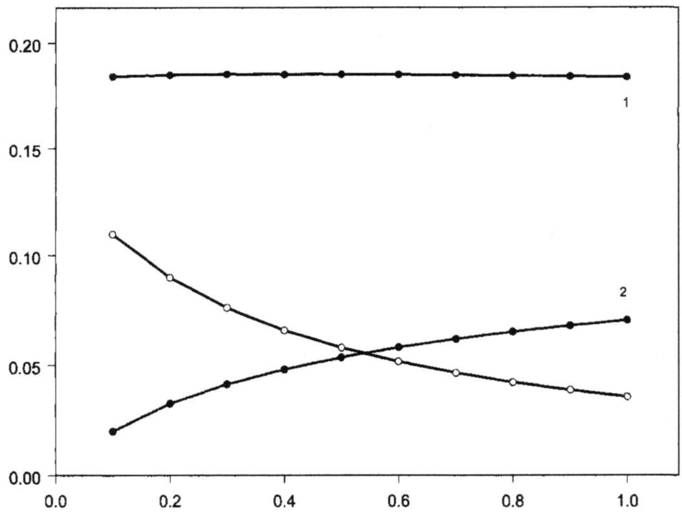

Fig. 9. Values of δ and Δ for selected values of α

where no interaction was present; see Fig. 10. Another way of investigating the way of visualizing the effects of collaboration is by looking at the changes in membership grades caused by collaboration. The changes in membership grades for the two selected levels of collaboration are shown in Fig. 11.

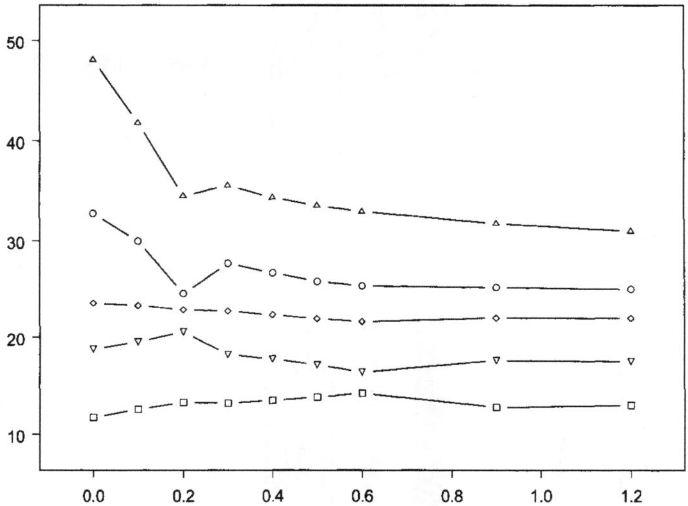

Fig. 10. Prototypes in the median value of real state as a function of collaborative linkage α

Now, we keep changing the number of clusters while retaining the same level of collaboration ($\alpha = 0.5$) to analyze how this affects the changes in δ and Δ. As expected, the values of δ decrease with an increasing number of clusters; see Fig. 12. The reason for this trend is obvious: we get more clusters, the individual membership grades decrease, and the differences become smaller. As to the second index, it is less monotonic, as with changes in the number of clusters, each data set has its own "plausible" number of clusters, and this could vary between them.

3 Vertical Collaborative Clustering

As already discussed, vertical collaborative clustering is concerned with a collection of databases involving different patterns defined in the same feature space, so that the patterns do not repeat across the databases. Inasmuch the feature space is common throughout the databases, we can use prototypes to facilitate collaboration between databases. The detailed algorithm discussed in the next section concentrates on this form of collaboration.

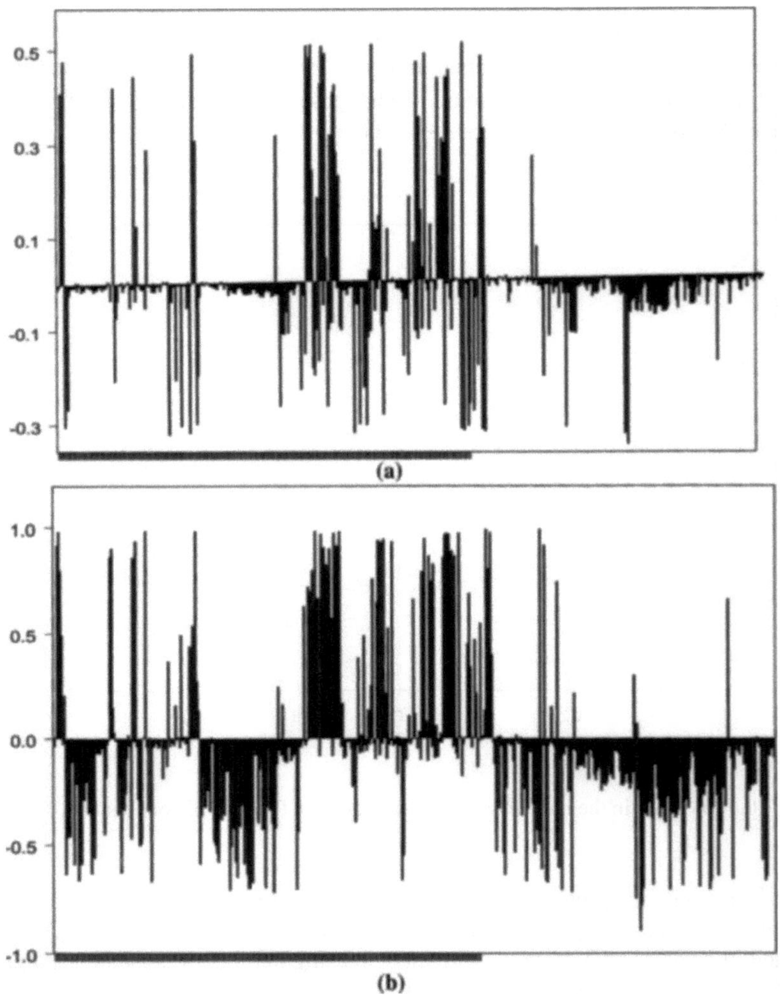

Fig. 11. Changes in membership grades for the first cluster for $\alpha = 0.2$ **(a)** and $\alpha = 0.5$ **(b)**

3.1 The Algorithm

We start by introducing the objective function that takes into account the vectors of prototypes specific for each database. With the same notation as before, the objective function is given as

$$Q[ii] = \sum_{i=1}^{c} \sum_{i=1}^{N[ii]} u_{ik}^2[ii] d_{ik}^2 + \sum_{\substack{jj=1 \\ jj \neq ii}}^{P} \beta[ii.jj] \sum_{k=1}^{N[ii]} \sum_{i=1}^{c} u_{ik}^2[ii] ||\mathbf{v}_i[ii] - \mathbf{v}_i[jj]||^2, \quad (17)$$

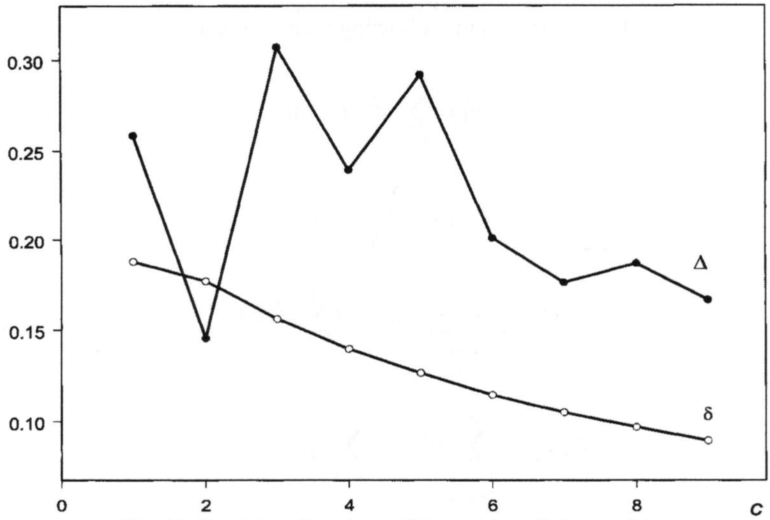

Fig. 12. δ and Δ as functions of the number of clusters (c)

where $\beta[ii,jj](>0)$ describes a level of collaboration between the data sets and $||\ \ ||$ denotes a distance function between the prototypes. The optimization of (15) is carried out for the partition matrix $U[ii]$ and the prototypes of the clusters $\mathbf{v}[ii]$. This implies two separate optimization problems; the first one involving the partition matrix is subject to constraints. Not including all computational details, the final expression governing computations of the partition matrix reads

$$u_{st} = \frac{1}{\sum_{j=1}^{c} \frac{D_{st}^2}{D_{jt}^2}} \tag{18}$$

$t = 1,2,\ldots,N[ii], s = 1,2,\ldots,c$, where D_{st} is computed as follows:

$$D_{st}^2 = d_{st}^2 + \sum_{\substack{jj=1\\ jj\neq ii}}^{P} \beta[ii,jj]||\mathbf{v}_s[ii] - \mathbf{v}_s[jj]||^2. \tag{19}$$

Proceeding with the optimization of the prototypes, we express a necessary condition for the minimum of Q to be in the form,

$$\frac{\partial Q}{\partial \mathbf{v}_s[kk]} = 0,\ kk = 1,2,\ldots,p. \tag{20}$$

This implies a system of linear equations with respect to v_{st}, that is,

$$v_{st}[ii] = \frac{F_{st}[ii] + A_{st}[ii]}{C_{st}[ii] + B_{st}[ii]} \tag{21}$$

$s = 1,2,\ldots,c$, $t = 1,2,\ldots,n$ with the following concise notation:

$$A_{st}[ii] = \sum_{k=1}^{N[ii]} u_{sk}^2[ii]x_{kt}[ii],$$

$$B_{st}[ii] = \sum_{k=1}^{N[ii]} u_{sk}^2[ii],$$

$$C_{st}[ii] = \sum_{\substack{jj=1 \\ jj \neq ii}}^{P} \beta[ii,jj] \sum_{k=1}^{N[ii]} u_{sk}^2[ii],$$

$$F_{st}[ii] = \sum_{\substack{jj=1 \\ jj \neq ii}}^{P} \beta[ii,jj] \sum_{k=1}^{N[ii]} u_{sk}^2[ii]v_{st}[jj].$$

The overall computing scheme can be presented in the following fashion:

Given: subsets of patterns $X_1, X_2, \ldots, X_P$ located in the same feature space
Select: distance function, number of clusters (c), termination criterion, and collaboration matrix $\beta[ii,jj]$
Initiate randomly all partition matrices $U[1], U[2], \ldots, U[p]$.
Phase I
For each data set,
repeat
compute prototypes $\{\mathbf{v}_i[ii]\}$, $i = 1,2,\ldots,c$ and partition matrices $U[ii]$ for all subsets of patterns
until a termination criterion has been satisfied.
Phase II
repeat
For the given matrix of collaborative links $\beta[ii,jj]$, compute prototypes and partition matrices $U[ii]$ using (19) and (16)
until a termination criterion has been satisfied.

Vertical collaboration can be realized not only by the prototypes, as discussed above, but also by establishing another vehicle of communication in the form of so-called induced partition matrices. The crux of this collaboration is as follows: Consider the prototypes of the clusters located in the data space of the jjth database, say $\mathbf{v}_s[jj]$, $s = 1,2,\ldots,c$. Now let us position these prototypes in the data space of the iith database. For any element $\mathbf{x}_t$ in this data space ($t = 1,2,\ldots,N[ii]$), we can compute *induced* membership grades (the grades induced by prototypes from a different space) in the form

$$u_{st}^{\sim}[ii][jj] = \frac{1}{\sum_{j=1}^{c} \frac{d_{st}^{\sim}[ii][jj]}{d_{jt}^{\sim}[ii][jj]}}, \tag{22}$$

where the distance $||.||_{ii}$ is computed in the iith space (and this is clearly identified by the corresponding subscript):

$$d_{st}^{\sim}[ii][jj] = ||\mathbf{x}_t - \mathbf{v}_s[jj]||_{ii}^2. \tag{23}$$

Now the objective function for the iith data set can be written in the form,

$$Q = \sum_{i=1}^{c}\sum_{k=1}^{N[ii]} u_{ik}^2[ii]d_{ik}^2[ii] + \sum_{\substack{jj=1\\ jj\neq ii}}^{p} \alpha[ii,jj]\sum_{i=1}^{c}\sum_{k=1}^{N[ii]} (u_{ik}[ii] - u_{ik}^{\sim}[ii][jj])^2 d_{ik}^2[ii], \tag{24}$$

with the notation already introduced above. The standard optimization requires two steps, that is, calculation of the partition matrix and the prototypes. Let us start with the partition matrix. Recalling that this implies constrained optimization, we use Lagrange multipliers that place the standard identity constraint as part of the objective function:

$$V = \sum_{i=1}^{c} u_{ik}^2[ii]d_{ik}^2[ii] + \sum_{\substack{jj=1\\ jj\neq ii}}^{p} \alpha[ii,jj]\sum_{i=1}^{c} (u_{ik}[ii] - u_{ik}^{\sim}[ii][jj])^2 d_{ik}^2[ii]$$

$$+\lambda(1 - \sum_{i=1}^{c} u_{ik}[ii]), \tag{25}$$

for all $k = 1, 2, \ldots, N[ii]$. The necessary condition for the minimum of (25) arises in the form

$$\frac{\partial V}{\partial u_{st}[ii]} = 2u_{st}[ii]d_{st}^2[ii] + 2\sum_{\substack{jj=1\\ jj\neq ii}}^{p} \alpha[ii,jj](u_{st}[ii] - u_{st}^{\sim}[ii][jj])d_{st}^2 - \lambda = 0.$$

Let us introduce the notation

$$D_{st}[ii] = 2d_{st}^2[ii](1 + \sum_{\substack{jj=1\\ jj\neq ii}}^{p} \alpha[ii,jj]),$$

$$F_{st}[ii] = 2d_{st}^2 \sum_{\substack{jj=1\\ jj\neq ii}}^{p} \alpha[ii,jj]u_{st}^{\sim}[ii][jj])d_{st}^2.$$

This yields the expressions,

$$u_{st}[ii]D_{st}[ii] - F_{st}[ii] - \lambda = 0$$

and

$$u_{it}[ii] = \frac{\lambda + F_{it}[ii]}{D_{it}[ii]},$$

that lead to the formula

$$\sum_{i=1}^{c} \frac{\lambda + F_{it}[ii]}{D_{it}[ii]} = 1.$$

Computing the Lagrange multiplier yields

$$\lambda = \frac{1 - \sum\limits_{i=1}^{c} \frac{F_{it}[ii]}{D_{it}[ii]}}{\sum\limits_{i=1}^{c} \frac{1}{D_{it}[ii]}}.$$

Finally, we obtain

$$u_{st}[ii] = \frac{1 - \sum\limits_{i=1}^{c} \frac{F_{it}[ii]}{D_{it}[ii]}}{\sum\limits_{i=1}^{c} \frac{D_{st}[ii]}{D_{it}[ii]}} + \frac{F_{st}[ii]}{D_{st}[ii]},$$

$s = 1, 2, \ldots, c, t = 1, 2, \ldots, N[ii]$. Proceeding with the computations of the prototypes, we determine them on the basis of the following conditions

$$\frac{\partial Q}{\partial v_{st}[ii]} = 0$$

that is,

$$\begin{aligned} \frac{\partial Q}{\partial v_{st}[ii]} &= \frac{\partial}{\partial v_{st}[ii]} \left\{ \sum_{i=1}^{c} \sum_{k=1}^{N[ii]} u_{ik}^2[ii] \sum_{j=1}^{n} \frac{(x_{kj}[ii] - v_{ij}[ii])^2}{\sigma_j^2[ii]} \right. \\ &\quad \left. + \sum_{\substack{jj=1 \\ jj \neq ii}}^{p} \alpha[ii, jj] \sum_{i=1}^{c} \sum_{k=1}^{N[ii]} (u_{ik}[ii] - u_{\widetilde{ik}}[ii][jj])^2 \sum_{j=1}^{n} \frac{(x_{kj}[ii] - v_{ij}[ii])^2}{\sigma_j^2[ii]} \right\} \\ &= 2 \sum_{k=1}^{N[ii]} u_{sk}^2[ii] \frac{x_{kt}[ii] - v_{st}[ii]}{\sigma_t^2[ii]} \\ &\quad + 2 \sum_{\substack{jj=1 \\ jj \neq ii}}^{p} \alpha[ii, jj] \sum_{k=1}^{N[ii]} (u_{sk}[ii] - u_{\widetilde{sk}}[ii][jj])^2 \frac{(x_{kt}[ii] - v_{st}[ii])}{\sigma_t^2[ii]}. \end{aligned}$$

To simplify the final formula, let us introduce the notation

$$A = \sum_{k=1}^{N[ii]} u_{sk}^2[ii] x_{kt}[ii],$$

$$B = \sum_{k=1}^{N[ii]} u_{sk}^2[ii],$$

$$C=\sum_{\substack{jj=1\\ jj\neq ii}}^{p}\alpha[ii,jj]\sum_{k=1}^{N[ii]}(u_{sk}[ii]-u_{sk}^{\sim}[ii][jj])^{2}x_{kt}[ii],$$

$$D=\sum_{\substack{jj=1\\ jj\neq ii}}^{p}\alpha[ii,jj]\sum_{k=1}^{N[ii]}(u_{sk}[ii]-u_{sk}^{\sim}[ii][jj])^{2},$$

that produces the final expression,

$$v_{st}[ii]=\frac{A+C}{B+D}.$$

Let us emphasize that this type of vertical collaboration occurs in the more abstract space of information granules (partition matrices) than the previous variant of collaboration.

3.2 Experiments

To illustrate how the method of this collaborative clustering works, we consider three collections of two-dimensional synthetic data collected in Table 1, where we identify the indexes of data points in the set. The elements that are substantially different from one data set to another are indicated in boldface. We partition the data into three clusters. The number of iterations that the clustering algorithm has been run is equal to 15 (practically, at this number, there are no further changes in the objective function).

Table 1. Three data sets used in the experiment of vertical clustering

No. of data	Data set #1	Data set #1	Data set #1
1	1.1 1.6	1.1 1.6	1.1 1.6
2	1.3 2.1	1.3 2.1	1.3 2.1
3	2.2 2.5	2.2 2.5	2.2 2.5
4	2.3 2.7	2.3 2.7	2.3 2.7
5	3.5 **6.7**	3.8 8.7	3.8 8.7
6	3.9 6.1	3.9 6.1	3.9 6.1
7	**3.3** 5.8	5.3 5.8	5.3 5.8
8	2.9 6.2	2.9 6.2	2.9 6.2
9	7.1 9.2	7.1 9.2	**5.1** 3.2
10	8.3 9.1	8.3 9.1	8.3 **3.1**
11	7.8 8.5	7.8 **5.5**	7.8 **3.5**
12	7.4 7.9	7.4 7.9	2.4 **3.9**

For comparative reasons, we start with a scenario in which there is no collaboration. The resulting partition matrices and prototypes are listed below:

Partition matrix; first data set

0.962885	0.029110	0.008005
0.987977	0.009657	0.002366
0.973869	0.021538	0.004593
0.948994	0.042514	0.008491
0.010942	0.977999	0.011059
0.013902	0.973516	0.012582
0.010637	0.983763	0.005600
0.015808	0.975499	0.008693
0.007394	0.024593	0.968013
0.006575	0.017805	0.975620
0.000675	0.002018	0.997307
0.009464	0.031015	0.959521

Partition matrix; second data set

0.025856	0.964681	0.009462
0.008925	0.988080	0.002995
0.014625	0.981004	0.004371
0.029609	0.961978	0.008414
0.708906	0.075178	0.215916
0.982872	0.008537	0.008591
0.780051	0.067610	0.152339
0.875552	0.078263	0.046185
0.054116	0.012098	0.933786
0.044340	0.012775	0.942884
0.284482	0.093742	0.621776
0.011821	0.002588	0.985591

Partition matrix; third data set

0.039863	0.032573	0.927565
0.017405	0.012870	0.969724
0.003631	0.002755	0.993615
0.009680	0.006951	0.983369
0.818173	0.091730	0.090097
0.975737	0.011179	0.013085
0.754797	0.161347	0.083856
0.913704	0.030640	0.055656
0.207793	0.492985	0.299222
0.013797	0.974664	0.011539
0.001866	0.996727	0.001407
0.183367	0.071699	0.744934

In all cases, we have several clearly visible clusters of data. The prototypes of the three data sets, tabulated in Table 2, are significantly different. In particular, the second and third prototype vary a lot across the data sets.

Table 2. Prototypes of data sets #1, #2, #3

Data set #1- prototypes	Data set #2 - prototypes	Data set #3- prototypes
[1.718665 2.222599]	[1.746825 2.253552]	[1.908616 2.492797]
[3.399974 6.196972]	[4.024643 6.495806]	[3.865419 6.565224]
[7.655201 8.676819]	[7.545924 8.293676]	[7.655816 3.344443]

Table 3. Prototypes of data sets #1, #2, #3 at level 1

Data set #1- prototypes	Data set #2 - prototypes	Data set #3- prototypes
[1.865954 2.338975]	[1.869684 2.344301]	[1.944257 2.400455]
[3.720608 6.258747]	[3.834615 6.340519]	[3.858967 6.242465]
[7.444662 7.334947]	[7.412808 7.155983]	[7.358653 6.138383]

Now, let us set up a collaborative level equal to 1; more specifically, $\beta[ii, jj] = 1.0$ for all $ii \neq jj$. The collaboration established in this way results in similar prototypes, as quantified in Table 3.

Noticeably, the prototypes start exhibiting a strong resemblance across the data, which is a visible indicator of the ongoing collaboration. The effect of collaboration driving the prototypes closer for each data set translates into changes in membership grades of the individual data points. Computationally, the change is taken as the sum of the absolute differences taken over all clusters:

$$\sum_{i=1}^{c} |u_{ik} - u_{ik}(no\_collaboration)|$$

with u_{ik}(no_collaboration) denoting the membership grade of the kth pattern in the ith cluster when collaboration is present. This effect of collaboration is shown in Fig. 13. Immediately, we recognize that some patterns are quite strongly affected by collaboration. Those are the patterns that are different between data sets. With the increasing values of β, the collaboration becomes more vigorous. Subsequently, the values of the changes in the membership grades are shown in Table 4. It can be seen that some of the patterns are heavily affected by the collaboration, meaning that at these points the structures are quite distinct, and any reconciliation between them requires a substantial level of effort. These particular patterns are indicated in boldface. These results correlate very evidently with the data; see Table 1. The method reveals that data points 9, 10, 11, and 1 are different — an observation included in Table 1. Interestingly, these patterns are the same as those that resulted in a substantial level of changes in membership during the process of collaboration. In the sequel, the total change in the membership (Δ) is determined as

$$\Delta = \sum_{k=1}^{N} \sum_{i=1}^{c} |u_{ik} - u_{ik}(no\_collaboration)|$$

and, now regarded as a function of β, is summarized in Fig. 13. Again, there is a strong monotonic relationship between the level of this collaboration and the mani-

Table 4. Changes in membership grades of the individual data points in three data sets for $\beta = 1.0$

Pattern no.	Change in membership (first data set)
1	0.032886
2	0.022677
3	0.027390
4	0.043216
5	0.011151
6	0.029918
7	0.064940
8	0.100724
9	**0.398732**
10	**0.323950**
11	**0.274974**
12	**0.144609**

Pattern no.	Change in membership (second data set)
1	0.035287
2	0.022392
3	0.014654
4	0.020496
5	0.024933
6	0.012801
7	**0.186115**
8	0.056257
9	**0.366174**
10	**0.287519**
11	**0.302670**
12	**0.186651**

Pattern no.	Change in membership (third data set)
1	0.032886
2	0.022677
3	0.027390
4	0.043216
5	0.011151
6	0.029918
7	0.064940
8	0.100724
9	**0.398732**
10	**0.323950**
11	**0.274974**
12	**0.144609**

festing changes in the partition matrix; the detailed relationships vary between data sets (groups of data). Next, we consider the same synthetic data, now using verti-

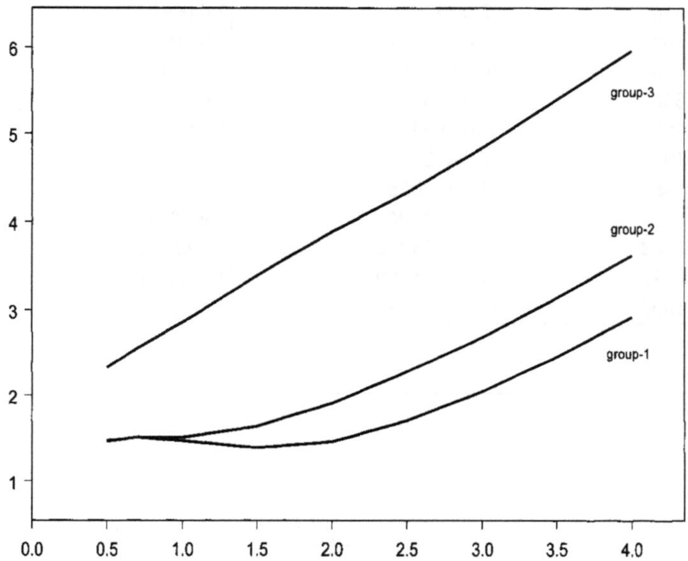

Fig. 13. Δ as a function of β for data sets used in the experiment

cal clustering where the methods collaborate at the level of membership functions (partition matrices). First the prototypes obtained at $\beta[ii][jj] = 1$ are summarized in Table 5. Subsequently, we show the changes in membership grades (Table 6). Finally, Fig. 14 depicts how the values of Δ are affected by the assumed level of collaboration.

Table 5. Prototypes of clusters for $\beta[ii][jj] = 1$

Data set 1	Data set 2	Data set 3
1.822371 2.348398	1.820462 2.341236	2.105824 2.529360
4.116829 6.610057	4.585076 6.841259	4.256230 6.166471
7.632895 8.640197	7.563589 8.204105	7.440859 3.623372

In general, the patterns that are identified as those requiring a high level of collaboration by the previous method are also highlighted as such by this approach. Comparing the plots of the changes in Δ (Figs. 13 and 14), we see that they are both monotonic, yet the type of monotonicity is not identical.

Table 6. Changes in membership grades for the three data sets. The most significant changes (values over 0.3) are indicated in boldface

Data set 1	Data set 2	Data set 3
1 0.041962	1 0.045628	1 0.030457
2 0.031656	2 0.032021	2 0.009927
3 0.037368	3 0.023017	3 0.002122
4 0.065780	4 0.039695	4 0.004779
5 0.105568	5 0.093528	5 0.233347
6 0.044639	6 0.063337	6 0.050150
7 0.296192	7 0.082744	7 0.148840
8 **0.340145**	8 0.142058	8 0.160036
9 **0.486945**	9 **0.417606**	9 **0.473936**
10 **0.424982**	10 **0.358432**	10 **0.871235**
11 **0.410413**	11 0.025409	11 **0.891050**
12 **0.371832**	12 **0.422575**	12 0.031298

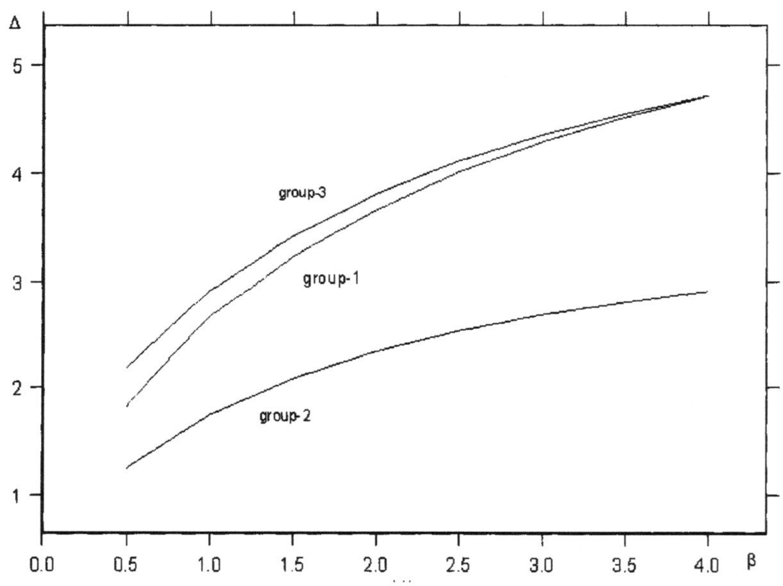

Fig. 14. Δ as a function of β

4 Vertical and Horizontal Clustering: Collaboration Space and Data Confidentiality and Security

The two types of collaboration between the clustering methods lead to an interesting taxonomy for the space of collaboration in horizontal collaboration: the methods exchange information in the space of information granules while not being involved in any communication occurring in the data space. We can say that the collaboration happens at a more abstract level, as the partition matrices provided to other collaborators are defined in the space of general data structures (partition matrices).

In vertical collaboration, we encounter two models of collaboration. The first is the same as outlined in horizontal collaboration, namely, the methods communicate at the level of their partition matrices. This collaboration at the level of information granules is more abstract than that in the data space itself. The difference lies in the fact that now such partition matrices used for communication are induced; we compute partition matrices on the basis of the prototypes available in the data space pertinent to the other data set. The second model of collaboration is realized in the data space and comes in the form of the data aggregates, such as prototypes.

Collaborative computing realized by clustering methods sheds light on the issue of data confidentiality and security. On one hand, collaboration calls for sharing data. On the other hand, there are some confidentiality requirements that prevent us from direct access to data in some other databases. The approach taken here is an interesting compromise between these two extremes. We do not share data (which is not feasible) but communicate at the level of information granules (and this collaboration does not violate the confidentiality requirement). Intuitively, we may note that the bigger the information granules, the less detailed information we share. This implies that the notion of data confidentiality (and alternatively, data security) is not a Boolean (two-valued) concept but one that can be represented as a fuzzy set. To maintain a certain level of data confidentiality, we may require that the information transactions (information sharing) occurs at a certain level of information granularity so that we do not get into too detailed information. In particular, we may require that the number of clusters does not exceed a certain maximal value.

5 Concluding Remarks

We have introduced the idea of collaborative processing in general, and collaborative clustering in particular. It has been shown that communication and collaboration between separate data sets can be effectively realized at the more abstract level of membership grades (partition matrices) and prototypes. Two types of collaboration (vertical and horizontal) were studied in detail. We provided a complete clustering algorithm by basing the method on the standard FCM method. Quantification of the collaborative effect can be realized either at the level of prototypes or partition matrices. An interesting expansion of the method discussed here involves partial (limited) collaboration where not all patterns are available to form collaborative links. This simply calls for an extra Boolean vector $\mathbf{b}= [b_1\, b_2 \ldots b_N]$ modifying the objective function in the form,

$$Q[ii] = \sum_{k=1}^{N} \sum_{i=1}^{c} u_{ik}^2[ii] d_{ik}^2[ii] + \sum_{\substack{jj=1 \\ jj \neq ii}}^{P} \alpha[ii, jj] \sum_{k=1}^{N} \sum_{i=1}^{c} \{u_{ik}[ii] - u_{ik}[jj]\}^2 b_k d_{ik}^2[ii],$$

where b_k assumes a value of 1 when the kth pattern is available for collaboration (otherwise, b_k is set to 0).

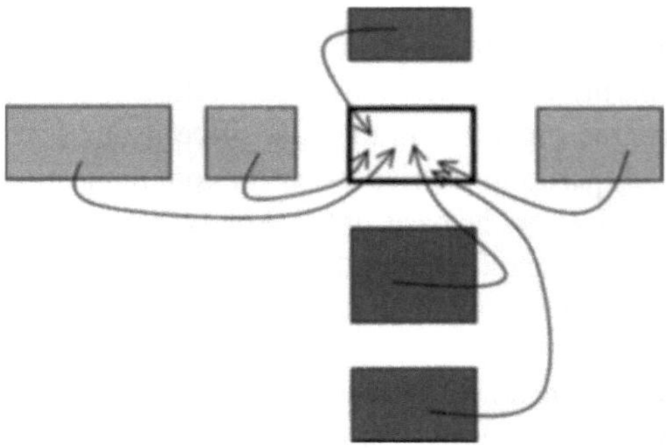

Fig. 15. Vertical and horizontal mode of collaboration between databases

In general, we can envision that the collaboration mechanism takes place both at the vertical (data) as well as the horizontal (feature) level; see Fig. 15. In terms of the objective function, this approach merges the two methods introduced before. As a matter of fact, we can write the following expression to emphasize that the collaboration mechanism is in place

$$U[ii] = F(U[jj], v[jj]),$$

where U and **v** are used to denote the information feedback of the other part of the system (both vertical and horizontal).

The approach presented here can easily be generalized to support more specific ideas such as rule-based systems. In this case, we are concerned with the reconciliation of rules in each subset of data. Obviously, the optimization details need to be refined, inasmuch as the specificity of the problem requires further in-depth investigation of a number of issues related to rules such as their specificity, consistency, and completeness.

References

1. A. Agah, K. Tanie. Fuzzy logic controller design utilizing multiple contending software agents. *Fuzzy Sets and Systems*, 106(2): 121–130, 1999.
2. M.R. Anderberg. *Cluster Analysis for Applications*. Academic Press, New York, 1973.
3. J.C. Bezdek. *Pattern Recognition with Fuzzy Objective Function Algorithms*. Plenum, New York, 1981.
4. R.N. Dave. Characterization and detection of noise in clustering. *Pattern Recognition Letters*, 12(11): 657–664, 1991.

5. M. Delgado, F. Gomez-Skarmeta, F. Martin. A fuzzy clustering-based prototyping for fuzzy rule-based modeling. *IEEE Transactions on Fuzzy Systems*, 5(2): 223–233, 1997.
6. M. Delgado, A.F. Gomez-Skarmeta, F. Martin. A methodology to model fuzzy systems using fuzzy clustering in a rapid-prototyping approach. *Fuzzy Sets and Systems*, 97(3): 287–302, 1998.
7. R.O. Duda, P.E. Hart, D.G. Stork. *Pattern Classification*, 2nd ed. Wiley, New York, 2001.
8. J.C. Dunn. A fuzzy relative of the ISODATA process and its use in detecting compact well-separated clusters. *Journal of Cybernetics*, 3(3): 32–57, 1973.
9. M.R. Genesereth, S.P. Ketchpel. Software agents. *Communications of the ACM*, 37(7): 48–53, 1994.
10. F. Hoppner et al. *Fuzzy Cluster Analysis*. Wiley, Chichester, 1999.
11. A. Kandel. *Fuzzy Mathematical Techniques with Applications*. Addison-Wesley, Reading, MA, 1986.
12. P.R. Kersten. Fuzzy order statistics and their application to fuzzy clustering. *IEEE Transactions on Fuzzy Systems*, 7(6): 708–712, 1999.
13. R. Kowalczyk, V. Bui. On constraint-based reasoning in e-negotiation agents. In F. Dignum, C. Sierra, editors, *Agent Mediated Electronic Commerce*, LNAI 1991, 31–46, Springer, Berlin, 2000.
14. V. Loia, S. Sessa. A soft computing framework for adaptive agents. In *Soft Computing Agents: New Trends for Designing Autonomous Systems*, Physica, Heidelberg (in press).
15. W. Pedrycz. *Fuzzy Sets Engineering*. CRC Press, Boca Raton, FL, 1995.
16. W. Pedrycz. Conditional fuzzy clustering in the design of radial basis function neural networks. *IEEE Transactions on Neural Networks*, 9(4): 601–612, 1999.
17. M. Setnes. Supervised fuzzy clustering for rule extraction. *IEEE Transactions on Fuzzy Systems*, 8(4): 416–424, 2000.
18. S. Zadrozny, J. Kacprzyk. Implementing fuzzy querying via the internet/www Java applets, activex controls and cookies. In *Proceedings of the 3rd International Conference on Flexible Query Answering Systems 1998*, LNCS 1495, 382–393, Springer, Berlin, 1999.

Chapter 6
Adaptive Aspects of Combining Approximation Spaces

Jakub Wróblewski

Polish-Japanese Institute of Information Technology, Koszykowa 86, 02-008 Warsaw
Poland
jakubw@mimuw.edu.pl, http://www.mimuw.edu.pl/~jakubw/

Summary. This chapter addresses issues concerning a problem of constructing an optimal classification algorithm. The notion of a parameterized approximation space is used to model the process of classifier construction. The process can be viewed as hierarchical searching for optimal information granulation to fit a concept described by empirical data. The problem of combining several parameterized information granules (given by classification algorithms) to obtain a global data description is described. Some solutions based on adaptive methods are presented.

1 Introduction

Many practical, complex problems cannot be solved efficiently (e.g., because of computational limitations) without decomposing them into easier subproblems. The hierarchical approach to problem solving is widely known and used, as in the case of a control problem (*layered learning* [32]) or decomposition of large databases in knowledge discovery in databases (KDD) [10]. Granular computing [12, 24, 36] (a new paradigm in computer science based on the notion of information granulation), when employed as machine learning, machine perception, and a KDD tool, also uses the advantages of a hierarchical structure.

This chapter addresses issues concerning the problem of constructing an optimal classification algorithm in KDD applications. Suppose that data is stored within *decision tables* [14], where each training case (elementary information granule) drops into one of predefined decision classes. By assumption, all available information about the universe of objects (cases) is collected in the decision table (or *information system*) $\mathbb{A} = (U, A, d)$, where each attribute $a \in A$ is identified with a function $a : U \longrightarrow V_a$ from the universe of objects U in the set V_a of all possible values of a and values $v_d \in V_d$ of $d \notin A$ (a distinguished decision attribute) correspond to mutually disjoint decision classes of objects. We will denote these classes by $D_1, \ldots, D_k$, where $D_i \subseteq U$.

The aim of data analysis is to construct an understandable description of data or a classifier (an algorithm that can classify previously unseen objects as members of appropriate decision classes). Methods of constructing of classifiers or descriptions can be regarded as tools for data generalization, i.e., tools that construct more and more general descriptions in terms of a hierarchy of information granules. Classifiers based on the rough set theory [14–17] are considered in this chapter.

The main notion of the rough set theory is the *indiscernibility relation*. Any two objects $u_1, u_2 \in U$ are indiscernible by a set of attributes $B \subseteq A$ [which is denoted by $(u_1, u_2) \in IND(B)$] iff there is no attribute $b \in B$ such that $b(u_1) \neq b(u_2)$. An indiscernibility class of object $u \in U$ is the set of objects (denoted as $[u]_B$) indiscernible with u:

$$[u]_B = \{u' \in U : \forall_{b \in B} b(u) = b(u')\}.$$

A *decision reduct* $B \subseteq A$ is the minimal (in terms of inclusion) set of attributes that is sufficient to discern any pair of objects from different decision classes, supposing that the whole set of attributes discerns the pair: $IND(B) \subseteq IND(\{d\}) \cup IND(A)$. Let us define the following rough set based notions:

Definition 1. Let indiscernibility relation $IND(B)$ be given. *The upper approximation* of a set X is defined as

$$\overline{X} = \{u \in U : X \cap [u]_B \neq \emptyset\}.$$

The lower approximation of a set X is defined by

$$\underline{X} = \{u \in U : [u]_B \subseteq X\}.$$

Definition 2. *The rough inclusion* of set Y in X is defined by

$$\mu(Y, X) = \begin{cases} \frac{|X \cap Y|}{|Y|} & \text{if } Y \neq \emptyset \\ 1 & \text{otherwise.} \end{cases}$$

The rough membership of object x in set X based on a set of attributes B is defined by

$$\mu_X^B(x) = \frac{|X \cap [x]_B|}{|[x]_B|}.$$

Indiscernibility classes are related to different levels of information granulation. Elementary granules correspond to $[u]_A$ classes (based on the whole set of attributes); every $B \subset A$ corresponds to a higher level granule, which may be used as a base for decision rule:

$$a_1(u) = v_1 \wedge \ldots \wedge a_j(u) = v_j \Longrightarrow d(u) = v_d, \tag{1}$$

for $B = \{a_1, \ldots, a_j\}$.

A notion of *approximation space*, a theoretical tool for data description with information granules is presented in the next sections of this chapter. A general composition scheme of data models (regarded as approximation spaces) into one classifier is presented as well.

The reader can find more details on the important role of approximation spaces in the process of information granule construction in Chap. 3.

2 Classification Algorithms

2.1 Approximation Spaces

The notion of an *approximation space* (see, e.g., [4, 15, 21–23, 25–27]) may be regarded as an extension of rough set theory. It is a tool for describing concepts not only in terms of their approximations but also in terms of the similarity of objects and concepts (see e.g., [15, 23, 25]). The notion of approximation space defined below is an extended form of definitions known from the literature (for more information see also Chap. 3 and [20]).

Definition 3. *An approximation space* is a tuple $AS = (U, I, \mathcal{R}, \nu)$, where
• U is a set of objects.
• $I : U \longrightarrow \mathcal{P}(U)$ is a function mapping every object from U into a subset (called a *neighborhood*), where $\forall_{u \in U}\ u \in I(u)$.
• $\mathcal{R} \subseteq \mathcal{P}(U)$ is a family of subsets of U (interpreted as a *set of templates*, or information granules, which are used to describe a concept).
•$\nu : \mathcal{P}(U) \times \mathcal{P}(U) \longrightarrow [0,1]$ is a function (interpreted as the *degree of inclusion* of subsets of U), where (see [23, 26])

1. $\forall_{A \subseteq U}\ \nu(A,A) = 1.$
2. $\forall_{A \subseteq U}\ \nu(\emptyset,A) = 1.$
3. $\forall_{A,B,C \subseteq U}\ \nu(A,B) = 1 \Rightarrow \nu(C,B) \geq \nu(C,A).$

An approximation space determines a language of describing concepts in U. It is useful especially in cases of vague, inaccurate, and incomplete descriptions of data. Function I expresses the idea of the indiscernibility of objects (a result of incompleteness of object descriptions), whereas family $\mathcal{R}$ determines a way of generalizing information about objects (which allows us to deal with inaccurate and vague data). $\mathcal{R}$ may be defined, e.g., by using language L of formulas based on descriptors $a_i(u) = v$ as atomic formulas (for $a \in A$, $v \in V_a$) and operation "$\wedge$". In this case [27],

$$\mathcal{R}_L = \{r_\alpha : \alpha \in L\}, \tag{2}$$

where $r_\alpha \subseteq U$ corresponds to the semantics of formula α in set U.

A goal of the KDD process in both a descriptive and predictive sense is to provide the best approximation of (one or more) concept $D \subset U$ based on known data by optimal information granulation. For a prediction task, the approximation takes the form of a *classification algorithm* — a function mapping vectors of values of conditional attributes into the set of decision classes $\{D_1, \ldots, D_k\}$. Selected decision class $D_i \subseteq U$ is described by AS as a rough set with upper and lower approximations given by

$$\overline{D_i} = \bigcup_{R \in \mathcal{R} :\, R \cap D_i \neq \emptyset} R; \qquad \underline{D_i} = \bigcup_{R \in \mathcal{R} :\, R \subseteq D_i} R.$$

Definition 4. Let $\mathbb{A}_1 = (U_1, A, d)$ be a decision table (training data set) and $AS = (U, I, \mathcal{R}, \nu)$ be an approximation space, where $U_1 \subseteq U$. Let $D \subseteq \mathcal{P}(U)$ be a partition of U onto disjoint decision classes $D = \{D_1, \ldots, D_k\}$, and let functions

$$\rho : \mathcal{R} \rightarrow \{\emptyset, 1, 2, \ldots, k\}$$

where $k = |D|$ and

$$\Phi : (\{\emptyset, 1, \ldots, k\} \times [0,1])^* \longrightarrow \{\emptyset, 1, \ldots, k\}$$

be given. *The classification algorithm* based on AS and ρ, Φ is a mapping

$$CA_{AS,D,\rho,\Phi} : U \longrightarrow \{\emptyset, D_1, D_2, \ldots, D_k\}$$

defined as

$$CA_{AS,D,\rho,\Phi}(u) = \Phi \left\{ \rho(R_1), \nu\left[I(u), R_1\right], \ldots, \rho(R_n), \nu\left[I(u), R_n\right] \right\}, \tag{3}$$

where $n = |\mathcal{R}|$. (We will omit subscripts AS, D, ρ, Φ for simplicity).

Typically, a given test object u is matched against templates from the family $\mathcal{R}$ (e.g., the left–hand sides of decision rules), and the best matching $R \in \mathcal{R}$ is selected. Then the most frequent decision class in R is taken as a result of the classification of u. In most cases, ρ is defined as

$$\rho(R) = \begin{cases} argmax_{i=1..k}[\nu(R, D_i)] & \text{for } \max_{i=1..k}[\nu(R, D_i)] > 0 \\ \emptyset & \text{otherwise.} \end{cases} \tag{4}$$

If an object can be matched to more than one template R, the final answer is selected by voting:

$$\Phi[(v_1, x_1), \ldots, (v_n, x_n)] = \begin{cases} argmax_{i=1..k}(\sum_{j \leq n: v_j = i} x_j) & \text{if } \exists_j x_j > 0 \\ \emptyset & \text{if } \forall_j x_j = 0, \end{cases} \tag{5}$$

for $n = |\mathcal{R}|$, i.e., given a set of partial answers v_i and corresponding coefficients x_i, one should select the most popular answer (in terms of the sum of x_i). The coefficients may be regarded as support of decision, credibility, or conviction factor, etc. For formula 3, it is the coefficient of relevancy of template R_i, i.e., the degree of inclusion of the test object in R_i.

Given template R may belong to the upper approximation of more than one decision class. The conflict is resolved by function ρ. Alternatively, the definition of the classification algorithm may be extended onto sets of decision classes or even onto probability distributions over them:

$$CA : U \longrightarrow \Delta^k,$$

where Δ^k denotes the k-dimensional simplex: $\Delta^k = \{x \in [0,1]^k : \sum_{i=1}^k x_i = 1\}$. In more general cases, the classification algorithm may take into account the degree of inclusion of an object u in the template R as well as the inclusion of R in decision classes.

2.2 Parameterized Approximation Spaces

The notion of a *parameterized approximation space* was introduced [18, 35] to provide more flexible, data-dependent description language of the set U. By AS_ξ, we will denote[1] an approximation space parameterized with a parameter vector $\xi \in \Xi$. The problem of optimal classifier construction is regarded as an optimization problem of finding optimal $\hat{\xi} \in \Xi$, i.e., of finding a vector of parameters such that $AS_{\hat{\xi}}$ generates an optimal (in the sense of, e.g., cross-validation results) classification algorithm. Parameter ξ is often used to maintain a balance between the generality of a model (classifier) and its accuracy.

Example 1 *An approximation space based on the set of attributes $B \subseteq A$ of information system $\mathbb{A} = (U, A, d)$ (see [26]). Let*

1. $I(u) = [u]_A$,
2. $\mathcal{R} = \{[u]_B : u \in U\}$,
3. $\nu(X_1, X_2) = \mu(X_1, X_2)$,

for $X_1, X_2 \subseteq U$, where μ is a rough inclusion function (Def. 2). Then $AS = (U, I, \mathcal{R}, \nu)$ is an approximation space related to a partition of the set U into indiscernibility classes of the relation $IND_{\mathbb{A}}(B)$. If we assume that B is a decision reduct of consistent data table $\mathbb{A}$, then family $\mathcal{R}$ corresponds to a set of consistent decision rules (i.e., for all $R \in \mathcal{R}$, there is a decision class D_i such that $R \subseteq D_i$). Every template $R \in \mathcal{R}$ corresponds to a decision rule r of the form of the conjunction of $a_i(u) = v_j$ descriptors, where $a_i \in B$, $v_j \in V_{a_i}$.

Now, let $AS_{B,\alpha}$, where $B \subseteq A$ and $\alpha \in [0,1]$, be a parameterized approximation space defined as follows (see [37, 39]):

1. $I(u) = [u]_A$,
2. $\mathcal{R} = \{[u]_B : u \in U\}$,
3. $\nu(X_1, X_2) = \begin{cases} \mu(X_1, X_2) & \text{if } \mu(X_1, X_2) \geq \alpha \\ 0 & \text{otherwise.} \end{cases}$

A classification algorithm based on $AS_{B,\alpha}$ works as follows: for any test object $u \in U$, find a template R matching it (i.e., a class of training objects identical to u with respect to attributes B), then check which is the most frequent decision class in a set R. If the most frequent decision class D_i covers at least α of R [i.e., $\mu(R, D_i) \geq \alpha$], object u is classified as a member of D_i [i.e., $\rho(R) = i$]. Otherwise, it is unclassified.

The goal of the above rough set based adaptive classification algorithm is to find such parameters (B, α) that the approximation space $AS_{B,\alpha}$ generates the best classifier. One can see that with parameter B, we adjust the generality of the model (the

[1] The notion of a parameterized approximation space is regarded in the literature as $AS_{\$,\sharp} = (U, I_\$, \nu_\sharp)$. The notation used in this chapter is an extension of the classical case.

smaller B is, the more general set of rules is generated, but also the less accurate rules we obtain). On the other hand, parameter α adjusts the degree of credibility of the model obtained: for $\alpha = 1$, there may be many unclassified objects, but only credible rules are taken into account; for small α, there may be no unclassified objects, but more objects are misclassified.

Example 2 *Let ρ be a metric on a set of objects U divided into disjoint decision classes $D = \{D_1, \ldots, D_m\}$. For each $u \in U$ and for test data set U_1, let $\sigma_{u,\rho}$ be a permutation of $\{1,..,|U_1|\}$, such that*

$$1 \leq i \leq j \leq |U_1| \Leftrightarrow \rho(u, u_{\sigma_{u,\rho}(i)}) \leq \rho(u, u_{\sigma_{u,\rho}(j)})$$

for $u_{\sigma_{u,\rho}(i)}, u_{\sigma_{u,\rho}(j)} \in U_1$.

Let $kNN_\rho : U \times \mathbb{N} \to 2^{U_1}$ be a function mapping each object u to a set of its k nearest neighbors according to metric ρ:

$$kNN_\rho(u,k) = \left\{u_{\sigma_u(1)}, \ldots, u_{\sigma_u(k)}\right\}.$$

Let $I_{k,\rho}(u) = kNN_\rho(u,k)$ for a given k; let $\mathcal{R} = \{R \subseteq U \ : \ |R| = k\}$ and $\nu(X_1,X_2) = \mu(X_1,X_2)$ (cf. Definition 2). Assume that ρ and Φ are defined by (4) and (5). Then $AS = (U, I_k, \mathcal{R}, \nu)$ is an approximation space, and $CA_{AS,D,\rho,\Phi}$ is a classification algorithm identical to the classical k-nearest neighbors algorithm. For each test object u, we check its distance (given by metric ρ) to all training objects from U_1. Then we find the k nearest neighbors [set $I_{k,\rho}(u)$] and define template $R = I_k(u)$. Object u is then classified into the most frequent decision class in R.

Let $n = |A|$ and $w \in \mathbb{R}^m$. Let ρ_w be the following metric:

$$\rho_w(u_1,u_2) = \sum_{i=1}^{n} w_i |a_i(u_1) - a_i(u_2)|.$$

The approximation space defined above may be regarded as the parameterized approximation space $AS_{k,w} = (U, I_{k,\rho_w}, \mathcal{R}, \nu)$, based on the k nearest neighbors and metric ρ_w. It is known that the proper selection of parameters (metric) is crucial for k-NN algorithm efficiency [2]).

3 Modeling Classifiers as Approximation Spaces

The efficiency of a classifier based on a given approximation space depends not only on domain-dependent information provided by values of attributes but also on its granularity, i.e., level of data generalization. Proper granularity of attribute values depends on the knowledge representation (data description language) and the generalization techniques used in the classification algorithm. In cases of data description by an approximation space $AS = (U, I, \mathcal{R}, \nu)$, the generalization is expressed by a family $\mathcal{R}$ of basic templates (granules) that form a final data model.

Some classification methods, especially these based on decision rules of the form (1), act better on discrete domains of attributes. Real-valued features are often transformed by discretization, hyperplanes, clustering, principal component analysis, etc. [6, 9, 11]. One can treat the analysis process on transformed data either as modeling of a new data table (extended by new attributes given as a function of original ones) or, equivalently, as an extension of model language. The latter means, e.g., change of metric definition in the k-NN algorithm (Example 2) or extension of descriptor language by interval descriptors "$a(u) \in [c_i, c_{i+1})$" in a rule based system.

An example of a new attribute construction method was presented by the author in [29]. A subset of attributes $B = b_1, \ldots, b_m \subseteq A$ is selected; then an optimal (in the sense of some quality measure) linear combination of them is constructed by an evolutionary strategy algorithm:

$$h(u) = \alpha_1 b_1(u) + \ldots + \alpha_m b_m(u),$$

where $\vec{\alpha} = (\alpha_1, \ldots, \alpha_m) \in \mathbb{R}^m$ is a vector of coefficients (assume $\|\vec{\alpha}\| = 1$). Note that every linear combination h corresponds to one vector of size $n = |A|$. An approximation space is based on a set of attributes containing a new one that is a discretization of h (see Fig. 1). If the process of constructing a classification system involves extension of $\mathbb{A}$ with k new attributes based on linear combinations, one may regard the process as optimization of an approximation space $AS_{\xi,\vec{\alpha_1},\ldots,\vec{\alpha_k}}$ parameterized by a set of parameters ξ (see Example 1) and a set of vectors $\vec{\alpha_1}, \ldots, \vec{\alpha_k}$ representing linear combinations of attributes.

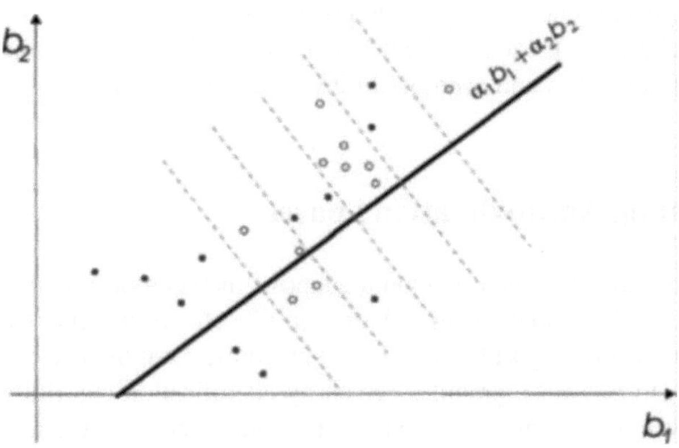

Fig. 1. A linear combination of two attributes and its discretization

The more general approach is presented in [35]. A model based on the notion of a relational information system [33], originally designed for relational database analy-

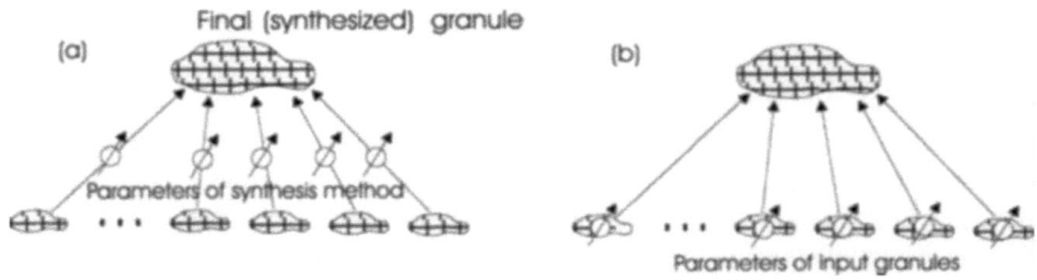

Fig. 2. Two general methods of adaptive combining granules: **(a)** by weights, **(b)** by adjusting model parameters on the lower level of a synthesis tree

sis, can be easily extended to cover virtually all possible transformations of existing data. The inductive closure $\mathbb{A}^*$ of an information system (or a relational information system) $\mathbb{A}$ is a decision table closed by an operation of adding (inequivalent) new attributes based on a given family of operations. Such a closure $\mathbb{A}^*$ is always finite since there is only a finite number of inequivalent attributes of any decision table $\mathbb{A}$. Hence, any classifying system based on transformed attributes may be modeled by a parameterized approximation space $AS_{\xi,B}$, where ξ is a set of parameters (influencing, e.g., a generalization level of rules) and $B \in A^*$ is a subset of attributes of inductive closure of $\mathbb{A}$.

When a final set of attributes (original, transformed, or created based on, e.g., relations and tables in relational database) is fixed, the next phase of classifier construction begins: data reduction and the model creation process. In rough set based data analysis, both steps are done by calculating reducts (exact or approximate) [28, 31, 35, 37] and a set of rules based on them. Unfortunately, a set of rules based on a reduct is not general enough to provide good classification results. A combination of rule sets (classifiers), each of them based on a different reduct, different transformations of attributes, and even on different subsets of training objects, must be performed.

4 Combining Approximation Spaces

One may distinguish between two main adaptive methods of granule combination (see Fig. 2). The first denoted (a) is based on a vector of weights (real numbers) used in a combination algorithm to adjust, somehow, the influence of a granule on a final model. In this case, granules (given by classification algorithms) are fixed, and the best vector of weights is used just to "mix" them (see the next section for more details). The second method denoted (b) consists of changing parameters of input granules, e.g., their generality, for a fixed combining method. In this section, we will consider one of the simplest adaptive combining methods: by zero–one weights, which is equivalent to choosing a subset of classifiers and combining them in a fixed way. We will refer to this subset as an ensemble of classifying agents (algorithms represented by an approximation space).

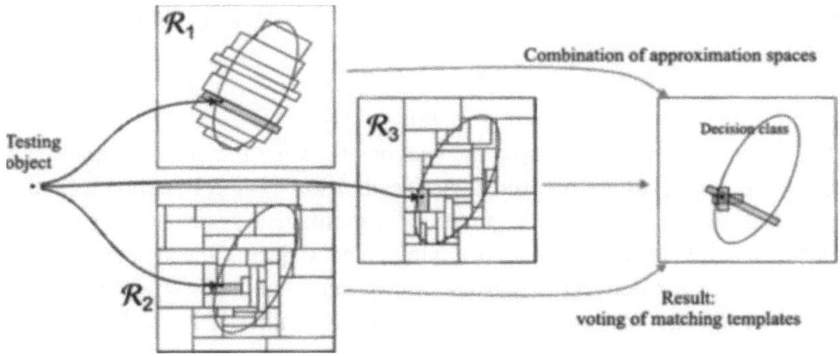

Fig. 3. A combination of approximation spaces (algorithms) and a new object classification

Assume that a classification system CA is composed of k classifying agents, each of them based on its own parameterized approximation space $AS_1,\ldots,AS_k$ and on its own subset of training examples $U_1,\ldots,U_k$ [using the same $I(u)$ and ν functions, limited to U_i]. Let us define an approximation space as a combination of $AS_1,\ldots,AS_k$.

Definition 5. *Operation of synthesis* of approximation spaces $AS_1, \ldots, AS_k$, where $AS_i = (U_i, I_i, \mathcal{R}_i, \nu_i)$ and $I_i = I|_{U_i}$, $\nu_i = \nu|_{U_i}$, is a mapping $S(AS_1,\ldots,AS_k) = AS'$, where $AS' = (U, I, \mathcal{R}, \nu)$ and

$$U = \bigcup_{i=1..k} U_i,$$
$$\mathcal{R} = \bigcup_{i=1..k} \mathcal{R}_i.$$

The classification of a new object u using AS' consists of finding all appropriate templates R [i.e., R such that $\nu[I(u), R]$ is large enough, see Definitions 3 and 4]. Then all values of $\rho(R)$ are collected, and the final answer is calculated by voting (function Φ).

Supposing that subsets U_i are significantly less than U, one can see that templates (in terms of subsets of objects matched) $R_{i,j} \in \mathcal{R}_i$ are relatively small as well. In practice, one should use a method of generalizing these templates onto the whole universe U.

If, for example, a family $\mathcal{R}_i$ is defined by a reduct $B \subseteq A$ (see Example 1)

$$\mathcal{R}_i = \{[u]_B \, : \, u \in U_i\},$$

then it will be generalized onto

$$\mathcal{R}_i' = \{[u]_B \, : \, u \in U\},$$

and a definition of synthesized $S(AS_1,\dots,AS_k) = AS'$ contains the following family $\mathcal{R}$:

$$\mathcal{R} = \bigcup_{i=1..k} \mathcal{R}_i'.$$

In [35], some remarks concerning connections between the above operations and rough mereology [16] are presented. A classification system based on a family of approximation spaces may be regarded as a multiagent system with one special agent for result synthesis. When classifying a new object u, the synthesizing agent sends a request for delivery of partial descriptions (templates R) of the object to subordinate agents. Then a complete description is synthesized based on Definition 5.

Note that a set of classifying agents may work on separate subsets $U_1,\dots,U_n$ of set U (e.g., in a distributed data mining system). Suppose that a set of approximation spaces $AS_1,\dots,AS_n$ was created based on reducts (see Example 1). Each AS_i is composed of a set of decision reducts, each of them related to one template $R \in \mathcal{R}_i$ (R is a set of objects matching the left–hand side of the rule) and a decision value $d = \rho(R)$. We tend to obtain the optimal synthesis of $AS_1,\dots,AS_n$, based on a measure Ψ of classification algorithm quality.

Let $S(AS_1,\dots,AS_n) = AS'$, where $AS' = (U, I, \mathcal{R}, \nu)$. Suppose that

$$U = \bigcup_{i=1..n} U_i,$$
$$\mathcal{R} = \bigcup_{i=1..n} \mathcal{R}_i,$$

for $AS_i = (U_i, I, \mathcal{R}_i, \nu)$. The space AS' is composed of all agents (approximation spaces) from the family $AS_1,\dots,AS_n$; our goal is to choose a subset $J = \{j_1,\dots,j_{|J|}\}$ that corresponds to the synthesized approximation space,

$$AS_J = S(AS_{j_1},\dots,AS_{j_{|J|}}), \tag{6}$$

providing optimal classification algorithm CA_{AS_J}. Let $Pos_{\mathbb{B}}(CA)$ and $Neg_{\mathbb{B}}(CA)$ denote a number of testing objects from table $\mathbb{B}$ properly and improperly (respectively) classified by CA. Let Ψ be a quality measure based on classification results on $\mathbb{B}$, satisfying the following conditions:

$$\begin{aligned} Pos_{\mathbb{B}}(CA_1) \subset Pos_{\mathbb{B}}(CA_2) \wedge Neg_{\mathbb{B}}(CA_1) = Neg_{\mathbb{B}}(CA_2) \Rightarrow \Psi(CA_1) < \Psi(CA_2), \\ Pos_{\mathbb{B}}(CA_1) = Pos_{\mathbb{B}}(CA_2) \wedge Neg_{\mathbb{B}}(CA_1) = Neg_{\mathbb{B}}(CA_2) \Rightarrow \\ \Rightarrow (\Psi(CA_1) < \Psi(CA_2) \Longleftrightarrow |J_1| > |J_2|), \end{aligned} \tag{7}$$

where $CA_1 = CA_{AS_{J_1}}$, $CA_2 = CA_{AS_{J_2}}$, and J_1, J_2 are subsets of agents. The above conditions mean that if two subsets of agents achieve the same results on a test table $\mathbb{B}$, we would prefer the smaller one.

Assume that CA_{AS_J} is based on a voting function Φ, such that

$$(\forall_i\, v_i = v \vee v_i = \emptyset) \wedge (\exists_i\, v_i = v) \Longrightarrow \Phi[(v_1,1),\dots,(v_k,1)] = v. \tag{8}$$

The following fact is true for families of classifying agents (see [35]):

Theorem 1. *Let a quality function Ψ (meeting conditions 7) be given. Suppose that $AS_1, \ldots, AS_n$ are approximation spaces (classifying agents) based on reducts. The problem of finding an optimal subset of agents (according to the function Ψ) is NP-hard.*

Proof. A similar result (for a problem formulated in a slightly different way) was presented in [34]. We will show that any minimal binary matrix column covering problem (known to be NP-hard) can be solved (in polynomial time) by selecting an optimal subset of agents for a certain data table and a set of classifying agents. Let $\mathbf{B} = \{b_{ij}\}$ be an $n \times m$ binary matrix to be covered by a minimal set of columns (suppose that there is at least one 1 in every row and column).

Let $\mathbb{A} = (U, A, d)$ be an information system, such that every row of matrix $\mathbf{B}$ corresponds to a pair of objects from U and every column of $\mathbf{B}$ corresponds to one attribute from A (hence $|A| = n$, $|U| = 2m$). Let attribute values be defined as follows:

$$a_i(u_{2j-1}) = 2 - b_{ij},$$
$$a_i(u_{2j}) = 2 - 2b_{ij},$$
$$d(u_j) = j \; mod \; 2,$$

where $j = 1..m$, $i = 1..n$. The set U of objects is partitioned into two decision classes D_0 and D_1.

Let us define a family of n approximation spaces based on subtables: $\mathbb{A}_i = (U_i, A, d)$, $i \in \{1,..,n\}$, where $U_i = \{u_{2j} \in U : b_{ij} = 1\} \cup \{u_{2j-1} \in U : b_{ij} = 1\}$. Let $AS_i = (U_i, I, \mathcal{R}_i, \nu)$ be an approximation space based on subtable $\mathbb{A}_i$ and the subset of attributes $B_i = \{a_i\}$ (which is a reduct of $\mathbb{A}_i$):

$$I(u) = [u]_A,$$
$$\mathcal{R}_i = \{[u]_{B_i} : u \in U_i\},$$
$$\nu(X_1, X_2) = \mu(X_1, X_2).$$

The set U_i contains pairs of objects u_{2j}, u_{2j-1} which correspond to rows $\mathbf{B}$ covered by column i. Let AS_J be an approximation space based on J (6). We will prove that classification algorithm CA_{AS_J} correctly classifies each object from U iff J corresponds to a column covering of $\mathbf{B}$. Let u_k be an object from U (suppose, without loss of generality, that k is even, $k = 2i$). Let $\mathcal{R}_J = \bigcup_{j \in J} \mathcal{R}_j$ be a family of templates of synthesized approximation space AS_J. Note that for any $R \in \mathcal{R}_j$,

$$u_{2i} \in R \in \mathcal{R}_j \iff b_{ij} = 1;$$

hence, as J corresponds to a covering of $\mathbf{B}$, there exists a template R that matches the object u_k. Note that for even numbers of objects,

$$[u_{2i}]_{B_j} = D_0,$$

where $u_{2i} \in U_j$. Hence,

$$u_{2i} \in R \in \mathcal{R}_j \implies \rho(R) = 0.$$

Every rule based on the template $R \in \mathcal{R}_J$ is deterministic. Therefore, for any voting function Φ (that meets condition 8), object u_k will be classified correctly. The same holds for odd k [in this case $\rho(R) = 1$].

Suppose that J corresponds to a set of columns which is not a covering of $\mathbf{B}$. In this case, there exists a row i not covered by any of the selected columns, and object u_{2i} is not contained by any U_j for $j \in J$. Object u_{2i} does not match any template from $\mathcal{R}_J$, so it will not be classified correctly.

It was proven that there exists a bijection between ensembles (subsets) of classifying agents (which classifies correctly all objects from $\mathbb{A}$) and coverings (subsets of columns) of matrix $\mathbf{B}$. Note that, by assumption (7), if there are many ensembles that classify every object in U, a function Ψ will prefer the smaller one. Hence, the optimal subset of agents corresponds to a minimal covering of $\mathbf{B}$. This completes the construction of the (polynomial) transformation of the matrix covering problem to the problem of selecting an optimal subset of agents, which proves the NP-hardness of the latter. □

5 Adaptive Strategies of Constructing Classifiers

The KDD process [5] consists of several stages; some of them may be performed automatically (some preprocessing steps, data reduction, method selection, data mining), whereas the others require expert knowledge (understanding the application domain, the goals of the analytic process, selecting an appropriate data set, interpreting and using results). One of the important fields in KDD research is seeking to develop methods of possibly automating many steps of the KDD process by using, e.g., automatic feature extraction, data reduction, or algorithm selection via parameterization. These methods are often based on an adaptation paradigm.

Let us consider an automatic classification system based on the KDD scheme. We will construct the classification algorithm step-by-step, by optimizing information granulation used at each level: feature extraction and preprocessing, data reduction and generalization, and synthesis of the final classifier (see Fig. 4). Some of these steps are known to be NP-hard, e.g., an optimal decomposition problem [11], optimal reduct finding (in the sense of its length or other measures, and also in cases of approximate or dynamic reducts [31, 35]), selection of optimal ensembles of agents (see above and [34, 35]). Approximate adaptive heuristics (e.g., based on evolutionary metaheuristics) should be used to optimize these steps.

A practical (partial) implementation of a classification system described in Fig. 4 was presented by the author in [35]. On the lower level, feature extraction evolu-

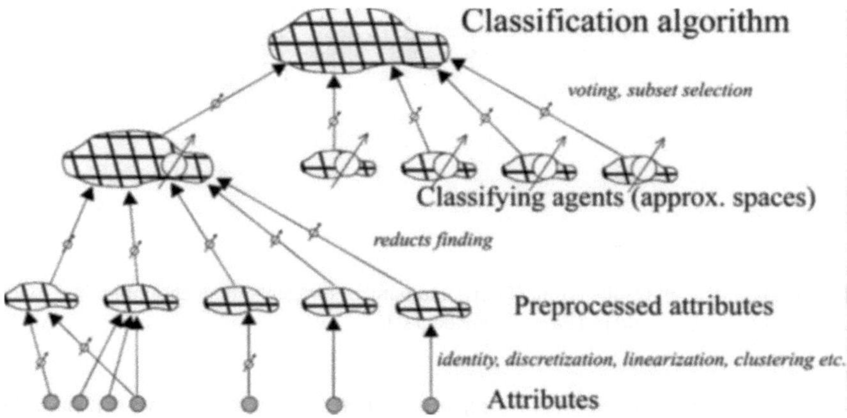

Fig. 4. Hierarchical construction of a classifying algorithm from granules (descriptors, approximation spaces); small circles with arrows denote adaptable parameters of information granules (or transforming/combining them)

tionary algorithms are used to create optimal linearization of attributes or new features based on a relational database (see Sect. 3). The process may be regarded as an optimization of weights in cases of linearization or as a selection (by 0–1 weights) of the best new attribute from the inductive closure of the database. There are other potential spaces of structures of new attributes, based on both a supervised and an unsupervised learning method. These spaces include clustering, PCA, discretization, and feature extraction methods used in cases of complex input objects (time series analyses, pattern recognition, etc.), which match the general scheme (Fig. 4).

The rough set based rule induction system is used at the generalization stage of an algorithm. A group of adaptation-based evolutionary (hybrid) algorithms for reduct finding creates a complete approximation space (by providing a set of rules as a source of templates forming the $\mathcal{R}$ family) parameterized by approximation coefficients in cases of approximate reducts [30]. The reduct finding process can be regarded as an optimization of 0–1 coefficients used in combining elementary granules (based on single attributes) into more complex ones (described by the approximation space).

The next step in the hierarchy depicted in Fig. 4 is concerned with creating optimal ensembles of classifying agents. The problem is NP-hard (see Theorem 1); the results of practical experiments confirm that increasing the number of agents in an ensemble does not necessarily lead to enhancing the classification results (see Fig. 5 and [34]). In [35], a genetic algorithm is used to find an optimal subset of agents. Chromosomes (binary coding) represent subsets of agents, and the fitness function is calculated based on classification results of an additional testing subtable.

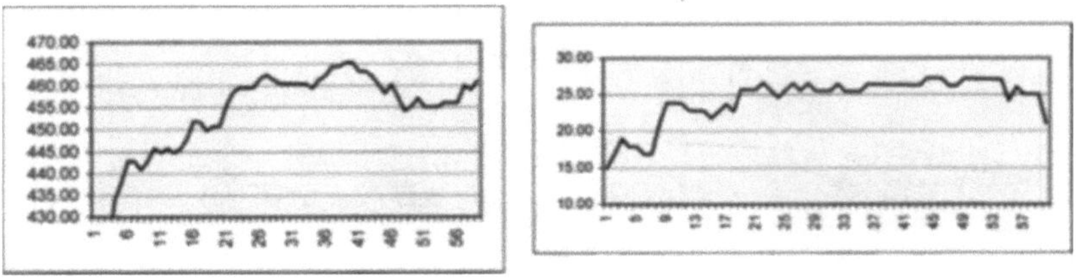

Fig. 5. Classification results (vertical axis) and number of agents in an ensemble (horizontal axis) – `DNA_splices` and `primary_tumor` data sets

There are two main conditions for regarding an algorithm as adaptive [1]: first, the algorithm should be parameterized (able to change itself); second, the criterion of parameter optimization should be based on the algorithm's efficiency. In the case of the adaptive scheme presented in Fig. 4, every level of the hierarchical granule combination process is parameterized either by weights (adjusting the method of combining granules) or by granule parameters. The optimization process for these parameters (e.g., the fitness function for genetic algorithms) at each level is based on an approximation (estimation) of the final classifier performance. In some cases, the estimate is based on results from an additional test sample (e.g., in optimization of an ensemble of agents [34]); at other levels, one should use more indirect approximation. In the adaptive system described in [35], both new features (e.g., given by linearization) and reducts are optimized by a probabilistic-based quality measure (a predictive measure [33]) estimating the final classifier quality indirectly. The popular criteria of the classifier optimization, based on the minimum description length principle [7], lead to an even more indirect approximation.

One may notice an interesting analogy between Fig. 4 and neural networks [13], [18]. In a multilayer feedforward artificial neural net, a model of input–output dependency is built as a combination of a number of linear (parameterized) and nonlinear functions. The adaptation process (implemented, e.g., as a back-propagation algorithm) is based on adjusting parameters (weights) based on the model prediction error, propagated down the net. There is no direct way to adapt this scheme to the general case of adaptive rule-based classifiers since there are no general methods of error propagation known in the discrete case (although some heuristics are used in this case). The most universal (but time-consuming) adaptation scheme is to collect new cases together with the correct answers and to rebuild the whole classification system or just a part of it (e.g., a new ensemble of agents) using the new data.

6 Results and Conclusions

This chapter describes a general scheme of modeling a process of classification system construction using the notion of an information granule. The process starts with a set of elementary information granules based on single attributes. The first level of the adaptive process of classifier construction is preprocessing of the initial attributes: discretization (generalizing several information granules into one), linearization (combining several attributes using an optimal, in some sense, linear combination of them where the final information granule is a combination of a set of granules based on a set of attributes), and other techniques. The next level of the hierarchical process is to combine information granules derived from the original attributes into approximation spaces (collections of information granules of a higher order). Rough set theory is a tool for generalizing descriptors (granules based on single attributes) onto the sets of rules.

The last level of the process described in the chapter is to combine a set of information granules (sets of rules, classifying agents) into one classification system and to resolve conflicts between them. The problem of selecting an optimal subset of agents is proven to be NP-hard, and a genetic algorithm is proposed to solve it approximately.

Since many of the problems concerning constructing and combining information granules are proven to be NP-hard, approximate heuristics should be used to obtain good results. The adaptive paradigm is the base of algorithms described in the chapter. All the steps of granule combination are parameterized, and some algorithms for parameter optimization are presented. Quality measures based on (estimated) efficiency of classifying new cases are proposed.

Table 1. Experimental results compared with two popular classifiers. The result column contains a number (percent) of properly classified test objects

Data	Size (training table)	k-NN	C4.5	Result
`Sat_image`	4435×37	90.6	85.0	91.05
`Letter`	15000×17	95.6	88.5	96.00
`Diabetes`	768×9	67.6	73.0	73.30
`Breast_cancer`	286×10	73.1	71.0	72.84
`Primary_tumor`	339×18	42.2	40.0	39.43
`Australian`	690×15	81.9	84.5	86.34
`Vehicle`	846×19	72.5	75.2	68.61
`DNA_splices`	2000×181	85.4	92.4	95.29
`Pendigits`	7494×16	97.8		98.28

The adaptive classification system described above was partially implemented by the author [29, 34, 35]. The results of experiments on some benchmark data tables are presented in Table 1.

Further research is needed in many detailed aspects of the process described. Regular examination of adaptive strategies of parameter optimization (especially when generalizing parameters, not only weights) should be performed. Although many parts of the process have been successfully implemented by the author, there are still no experimental results for the whole, fully adaptive algorithm. An integration of some methods described in the paper with RSES (rough set based data analysis system [19]) is to be done in the near future.

Acknowledgments

This work was supported by a grant of the Polish National Committee for Scientific Research (KBN), No. 8T11C02519.

References

1. Th. Bäck. An overview of parameter control methods by self-adaptation in evolutionary algorithms. *Fundamenta Informaticae*, 35(1): 51–66, 1998.
2. S.D. Bay. Combining nearest neighbor classifiers through multiple feature subsets. In *Proceedings of the 15th International Conference on Machine Learning (ICML'98)*, Morgan Kaufmann, San Mateo, CA, 1998.
3. J.G. Bazan, H.S. Nguyen, S.H. Nguyen, P. Synak, J. Wróblewski. Rough set algorithms in classification problem. In L. Polkowski, S. Tsumoto, and T.Y. Lin, editors, *Rough Set Methods and Applications: New Developments in Knowledge Discovery in Information Systems*, 49–88, Physica, Heidelberg, 2000.
4. I. Düntsch, G. Gediga. Uncertainty measures of rough set prediction. *Artificial Intelligence*, 106 : 77–107, 1998.
5. I. Düntsch, G. Gediga, H.S. Nguyen. Rough set data analysis in the KDD process. In *Proceedings of the 8th International Conference on Information Processing and Management under Uncertainty (IPMU 2000)*, 220–226, Madrid, Spain, 2000.
6. I.T. Jolliffe. *Principal Component Analysis*. Springer, Berlin, 1986.
7. M. Li, P. Vitanyi. *An Introduction to Kolmogorov Complexity and its Applications*. Springer, New York, 1993.
8. T.Y. Lin, A.M. Wildberger, editors. *Soft Computing: Rough Sets, Fuzzy Logic, Neural Networks, Uncertainty Management, Knowledge Discovery*. Simulation Councils, San Diego, CA, 1995.
9. H. Liu, H. Motoda, editors. *Feature Extraction, Construction and Selection: A Data Mining Perspective*. Kluwer, Dordrecht, 1998.
10. S.H. Nguyen, L. Polkowski, A. Skowron, P. Synak, J. Wrblewski. Searching for approximate description of decision classes. In *Proceedings of the 4th International Workshop on Rough Sets, Fuzzy Sets and Machine Discovery, (RSFD'96)*, 153–161, Tokyo, 1996.
11. H.S. Nguyen. *Discretization of Real Value Attributes: Boolean Reasoning Approach*. Ph.D. Dissertation, Faculty of Mathematics, Informatics and Mechanics, Warsaw University, 2002.
12. H.S. Nguyen, A.Skowron, J.Stepaniuk. Granular computing: A rough set approach. *Computational Intelligence*, 17(3): 514–544, 2001.
13. S.K. Pal, W. Pedrycz, A. Skowron, R. Swiniarski, editors. Rough-neuro computing (special issue). Vol. 36 of *Neurocomputing: An International Journal*, 2001.

14. Z. Pawlak. *Rough Sets: Theoretical Aspects of Reasoning about Data*. Kluwer, Dordrecht, 1991.
15. L. Polkowski, A. Skowron, J. Zytkow. Tolerance based rough sets. In *[18]*, 55–58, 1994.
16. L. Polkowski, A. Skowron. Rough mereological foundations for design, analysis, synthesis and control in distributed systems. *Information Sciences*, 104(1/2): 129–156, 1998.
17. L. Polkowski, A. Skowron, editors. *Rough Sets in Knowledge Discovery* Vols. 1, 2. Physica, Heidelberg, 1998.
18. L. Polkowski, A. Skowron, editors. Rough-Neuro Computing. In *Proceedings of the 2nd International Conference on Rough Sets and Current Trends in Computing (RSCTC 2000)*, LNAI 2005, 25–32, Springer, Heidelberg, 2001.
19. RSES homepage – rough set based data analysis system. Available at http://loic.mimuw.edu.pl/~rses/
20. A. Skowron. Approximation spaces in rough neurocomputing. In S. Hirano, M. Inuiguchi, S. Tsumoto, editors, *Rough Set Theory and Granular Computing*, Physica, Heidelberg, to appear.
21. A. Skowron, J. Stepaniuk. Approximation of relations. In *[38]*, 161–166, 1993.
22. A. Skowron, J. Stepaniuk. Generalized approximation spaces. In T.Y. Lin, A.M. Wildberger, editors, *Soft Computing: Rough Sets, Fuzzy Logic, Neural Networks, Uncertainty Management, Knowledge Discovery* 18–21, Simulation Councils, San Diego, CA, 1995.
23. A. Skowron, J. Stepaniuk. Tolerance approximation spaces. *Fundamenta Informaticae*, 27: 245–253, 1996.
24. A. Skowron, J. Stepaniuk. Information granule decomposition. *Fundamenta Informaticae*, 47: 337–350, 2001.
25. R. Słowiński, D. Vanderpooten. *Similarity Relation as a Basis for Rough Approximations*. Report number 53/95 of the Institute of Computer Science, Warsaw University of Technology, 1995; see also P.P. Wang, editor, *Advances in Machine Intelligence & Soft Computing*, 17–33, Bookwrights, Raleigh, NC, 1997.
26. J. Stepaniuk. Approximation spaces, reducts and representatives. In L. Polkowski, A. Skowron, editors, *Rough Sets in Knowledge Discovery, Vol. 2*, 109–126, Physica, Heidelberg, 1998.
27. J. Stepaniuk. Knowledge discovery by application of rough set methods. In L. Polkowski, T.Y. Lin, S. Tsumoto, editors, *Rough Sets: New Developments in Knowledge Discovery in Information Systems*, Physica, Heidelberg, 2000.
28. D. Ślęzak. Approximate reducts in decision tables. In *Proceedings of the 7th International Conference on Information Processing and Management under Uncertainty (IPMU'96)*, 1159–1164, Universidad da Granada, Granada, 1996.
29. D. Ślęzak, J.Wróblewski. Classification algorithms based on linear combinations of features. In *Proceedings of the 3rd European Conference on Principles and Practice of Knowledge Discovery in Databases (PKDD'99)*, LNAI 1704, 548–553, Springer, Berlin, 1999.
30. D. Ślęzak, J. Wróblewski. Application of normalized decision measures to the new case classification. In *Proceedings of the 2nd International Conference on Rough Sets and Current Trends in Computing (RSCTC 2000)*, LNAI 2005, 515–522, Springer, Berlin, 2000.
31. D. Ślęzak. *Approximate Decision Reducts*. Ph.D. Dissertation, Faculty of Mathematics, Informatics and Mechanics, Warsaw University, 2002 (in Polish).
32. P. Stone. *Layered Learning in Multi-Agent Systems: A Winning Approach to Robotic Soccer*. MIT Press, Cambridge, MA, 2000.

33. J. Wróblewski. Analyzing relational databases using rough set based methods. In *Proceedings of the 8th Conference on Information Processing and Management of Uncertainty in Knowledge-Based Systems (IPMU 2000)*, 256–262, Madrid, Spain, 2000.
34. J. Wróblewski. Ensembles of classifiers based on approximate reducts. In *Proceedings of the Workshop on Concurrency, Specification and Programming (CS&P 2000)*, volume 140(2) of *Informatik-Bericht*, 355–362, Humboldt-Universität, Berlin, 2000; also in *Fundamenta Informaticae*, 47(3/4): 351–360, 2001.
35. J. Wróblewski. *Adaptive Methods of Object Classification*. Ph.D. Dissertation, Faculty of Mathematics, Informatics and Mechanics, Warsaw University, 2002 (in Polish).
36. L.A. Zadeh, J.Kacprzyk, editors. *Computing with Words in Information/Intelligent Systems*, Vols. 1, 2. Physica, Heidelberg, 1999.
37. W. Ziarko. Variable precision rough set model. *Journal of Computer and System Sciences*, 46: 39–59, 1993.
38. W. Ziarko, editor. *Proceedings of the International Workshop on Rough Sets, Fuzzy Sets and Knowledge Discovery (RSKD'93)*, Workshops in Computing, Springer & British Computer Society, London, Berlin, 1994.
39. W. Ziarko. Approximation region-based decision tables. In *Proceedings of the 1st International Conference on Rough Sets and Current Trends in Computing (RSCTC'98)*, LNAI 1424, 178–185, Springer, Berlin, 1998.

Chapter 7
Algebras from Rough Sets

Mohua Banerjee,[1] Mihir K. Chakraborty[2]

[1] Department of Mathematics, Indian Institute of Technology, Kanpur 208 016, India
mohua@iitk.ac.in

[2] Department of Pure Mathematics, University of Kolkata, 35, Ballygunge Circular Road, Kolkata 700019, India
mihirc99@cal3.vsnl.net.in

Summary. Rough set theory has seen nearly two decades of research on both foundations and on diverse applications. A substantial part of the work done on the theory has been devoted to the study of its algebraic aspects. 'Rough algebras' now abound, and have been shown to be instances of various algebraic structures, both well-established and relatively new, e.g., quasi-Boolean, Stone, double Stone, Nelson, Łukasiewicz algebras, on the one hand, and topological quasi-Boolean, prerough and rough algebras, on the other. More interestingly and importantly, some of these latter algebras find a new dimension (interpretation) through representations as rough structures. An attempt is made here to present the various relationships and to discuss the representation results.

There have been other directions to the algebraic studies, too, especially in the development of relation and information algebras as well as rough substructures of algebraic structures, such as groups and semigroups, and structures on generalized approximation spaces and information systems. A brief overview of these approaches is also presented in this chapter.

1 Introduction

The theory of rough sets has gained steady ground since it was proposed in 1982 by Pawlak [28]. Quite a lot of research has been carried out on the theory itself, as well as on its various possible applications. Our focus in this chapter is on the algebraic aspects of the theory. From the beginning, a number of researchers have taken an interest in the algebraic forms of rough sets. They have brought into focus different features of these mathematical entities and have come up with various already known or new algebraic structures. All of these structures have rough sets as their instances (models). In addition to this, algebraic constructions have taken several other directions — all having roots in this theory. Sect. 9 of [19] is most probably the first attempt, though brief and incomplete, to point to some of these algebraic studies. The bibliography in [19] includes most of the work done in this area. Our present study may be considered a comprehensive extension of what was only initiated in [19].

A general feature in most of the algebraic studies that are under survey here has been to start with the formulation of a structure — a rough algebra that has rough sets as its elements, then investigate the properties of the structure, e.g., the kind of operations that might be defined in it, and then abstract these properties, thereby obtaining some new or known algebraic structure. If $\mathcal{A}$ denotes the class of structures thus evolved, a final step often taken is to establish a representation theorem for $\mathcal{A}$. By representation, we mean that there is a correspondence $c : \mathcal{A} \to \mathcal{R}$ (the class of rough algebras), such that any element $A \in \mathcal{A}$ is isomorphic to a subalgebra of $c(A)$. In some cases though, a representation of the rough structures is obtained in terms of other algebras. We shall see examples of both kinds of representation in the sequel.

A basic facet of these studies then, is the definition of a rough set that is chosen to be followed, and how the operations are defined in them (the rough sets). So far, we have found five definitions, three of which are equivalent to each other in a straightforward way. In fact, the equivalence could be extended to a general case but only under certain conditions. Let us start with these definitions.

As we know, rough set theory begins with the notion of an *approximation space*, which is a pair $< X,R >$; X is a nonempty set (the domain of discourse), and R an equivalence relation on it, representing indiscernibility at the object level. R is created by an *information system* $< X,AT,\{v_a : a \in AT\} >$, where AT is a set of attributes and for each $a \in AT$, v_a is a mapping from X to some value set. If $A \subseteq X$, the *lower approximation* $\underline{A}$ of A in the approximation space $< X,R >$ is the union of equivalence classes contained in A, and its *upper approximation* $\overline{A}$ in $< X,R >$ is the union of equivalence classes properly intersecting A. $\underline{A}$ ($\overline{A}$) is interpreted as the collection of those objects of the domain X that definitely (possibly) belong to A. The *boundary* BnA of A is given by $\overline{A} \setminus \underline{A}$, and so consists of elements possibly, but not definitely, in A.

The equivalence classes under R are called *elementary sets* of $< X,R >$. $A (\subseteq X)$ is said to be *definable* (or *exact*) in $< X,R >$ if and only if $\underline{A} = \overline{A}$, i.e., $BnA = \emptyset$. Equivalently, a definable set is a union of elementary sets. For any $A \subseteq X$, $\underline{A}$, $\overline{A}$, BnA are all definable sets, and it is important to note that an elementary set that is a singleton cannot be a constituent of (an elementary set in) the boundary of any set A.

Definition 1. [29] $A \subseteq X$ is a *rough set* in the approximation space $< X,R >$. (Actually A was termed rough in [29] provided $BnA \neq \emptyset$ — for generality, however, we could remove this restriction). Alternatively, to keep the context clear, we could call the triple $< X,R,A >$ a *rough set* [2].

Definition 2. (cf. [19]) The pair $< \underline{A},\overline{A} >$, for each $A \subseteq X$, is called a *rough set* in the approximation space $< X,R >$.

Definition 3. [27] The pair $< \underline{A},\overline{A}^c >$, for each $A \subseteq X$, is called a *rough set* in the approximation space $< X,R >$.

Definition 4. [25] Given an approximation space $<X,R>$, a *rough set* is an ordered quadruple $<X,R,L,B>$, where (i) L,B are disjoint subsets of X; (ii) both L and B are definable sets in $<X,R>$; and (iii) for each $x \in B$, there exists $y \in B$ such that $x \neq y$ and xRy (in other words, no equivalence class contained in B is a singleton).

A is said to be *roughly included* in B, $A,B \subseteq X$, if and only if $\underline{A} \subseteq \underline{B}$ and $\overline{A} \subseteq \overline{B}$. This brings us to the indiscernibility that rough set theory deals with at the concept level, viz., rough equality. Subsets A and B of X are *roughly equal sets* in $<X,R>$ (in notation $A \approx B$) if and only if each one is roughly included in the other, i.e., $\underline{A} = \underline{B}$ and $\overline{A} = \overline{B}$. This indicates that relative to the given partition of the domain, one cannot discern between the sets concerned. As before, this second indiscernibility relation also induces a partition, that of the power set $\mathcal{P}(X)$, so that we have the quotient set $\mathcal{P}(X)/\approx$.

Definition 5. [28] A *rough set* in the approximation space $<X,R>$ is an equivalence class of $\mathcal{P}(X)/\approx$.

It may be observed that Definitions 2, 3, and 5 are equivalent to each other for any given $<X,R>$ because, for any $A \subseteq X$, the entities $<\underline{A},\overline{A}>$, $<\underline{A},\overline{A}^c>$, and the equivalence class $[A]$ of A in $\mathcal{P}(X)/\approx$ are identifiable. Again, for a fixed $<X,R>$, a quadruple $<X,R,L,B>$ is essentially the pair $<L,B>$, and due to condition (iii) of Definition 4, one can always find a subset A of X, such that $\underline{A} = L$ and $BnA = B$. Hence, Definition 4 may be reformulated as follows : the pair $<\underline{A}, BnA>$ for each $A \subseteq X$ is a rough set so long as $<X,R>$ remains unchanged. So, via this interpretation, Definition 4 also becomes equivalent to 2, 3, and 5.

But interestingly, starting with these equivalent definitions, one arrives at different (though related) algebras by taking different definitions of union, intersection, complementation, and other algebraic operations. We shall encounter these in the sequel. We denote by $\mathcal{D}$ the collection of all definable sets of $<X,R>$. Also, let

1. $\mathcal{R} \equiv \{<D_1,D_2>,\ D_1 \subseteq D_2,\ D_1,D_2 \in \mathcal{D}\}$.
2. $\mathcal{RS} \equiv \{<\underline{A},\overline{A}>,\ A \subseteq X\}$.
3. $\mathcal{RS}' \equiv \{<\underline{A},\overline{A}^c>,\ A \subseteq X\}$.

Clearly, $\mathcal{RS} \subseteq \mathcal{R}$. Since $D_2 \setminus D_1$ may have singleton elementary classes among its constituents, $\mathcal{RS}$ may be a proper subset of $\mathcal{R}$. However, when there are no singleton elementary classes in $D_2 \setminus D_1$ for any $<D_1,D_2> \in \mathcal{R}$, the sets $\mathcal{R}$, $\mathcal{RS}$, $\mathcal{RS}'$ are in a one-to-one correspondence.

So, followers of Definitions 2, 3, and 5 have $\mathcal{RS}$, $\mathcal{RS}'$ and $\mathcal{P}(X)/\approx$ as domains (of rough sets). Some have taken $\mathcal{R}$ as the domain (in place of $\mathcal{RS}$). Once the domain is chosen, operations, such as join, meet, complementation, and others are defined suitably on it. These will be discussed while presenting the corresponding structures. However, in the case of $\mathcal{R}$, two natural operators of join ($\sqcup$) and meet ($\sqcap$) are defined as follows.

Definition 6. $< D_1, D_2 > \sqcup < D'_1, D'_2 > \equiv < D_1 \cup D'_1, D_2 \cup D'_2 >$,
$< D_1, D_2 > \sqcap < D'_1, D'_2 > \equiv < D_1 \cap D'_1, D_2 \cap D'_2 >$.

These definitions can be carried over to the subclass $\mathcal{RS}$ as follows:

Definition 7. $< \underline{A}, \overline{A} > \sqcup < \underline{B}, \overline{B} > \equiv < \underline{A} \cup \underline{B}, \overline{A} \cup \overline{B} >$,
$< \underline{A}, \overline{A} > \sqcap < \underline{B}, \overline{B} > \equiv < \underline{A} \cap \underline{B}, \overline{A} \cap \overline{B} >$.

That the right-hand entities in the above do belong to $\mathcal{RS}$, i.e. operations $\sqcup$ and $\sqcap$ are well-defined, can be shown, it is observed explicitly, e.g., in [4,17].

For $\mathcal{P}(X)/\approx$, one might be inclined to consider a join induced by the union operation in $\mathcal{P}(X)$, viz., $[A] \cup [B] \equiv [A \cup B]$, but this does not work. The inherent reason is that the lower approximation does not, in general, distribute over the union. A similar argument would apply for meet with respect to intersection. Thus we find two definitions for join and meet in $\mathcal{P}(X)/\approx$, in [8] and [4].

In the following section, we discuss in some detail the work of Iwiński [18], Obtułowicz [25], Pomykała [30,31], Comer [12,13], Bonikowski [8], Chuchro [10], Pagliani [27], Banerjee and Chakraborty [3,4], Wasilewska [37] and Iturrioz [17]. The results that are reported here have been chosen in compliance with the angle of presentation of this chapter, and constitute only a fragment of the work done by these authors.

Sect. 3 presents a comparative study of the above survey. In Sect. 4, we give a brief introduction to some other work, also done from the algebraic viewpoint, viz. on relation algebras, information algebras, rough substructures, and structures on generalized approximation spaces and information systems.

2 The Algebras and Their Representations

2.1 Iwiński

In [18], a rough algebra was presented for the first time. Definition 2 is followed, and the more general $\mathcal{R}$ (not $\mathcal{RS}$) is taken as the collection of rough sets. Operations of join ($\sqcup$) and meet ($\sqcap$) on $\mathcal{R}$ are given by Definition 6 of the previous section. It may be noted that any definable set A of $< X, R >$ is identifiable with the pair $< A, A >$ of $\mathcal{R}$. The following results are obtained:

Proposition 1. *$(\mathcal{R}, \sqcup, \sqcap, 0, 1)$ is a distributive complete lattice with 0 and 1, where $0 \equiv < \emptyset, \emptyset >$ and $1 \equiv < X, X >$. It is also atomic, with atoms of the form $< \emptyset, A >$, where A is an elementary set of $< X, R >$.*

Proposition 2. *The definable sets are the only Boolean elements of $(\mathcal{R}, \sqcup, \sqcap, 0, 1)$, and they form a maximal Boolean subalgebra of $(\mathcal{R}, \sqcup, \sqcap, 0, 1)$.*

It is important to observe that there is no Boolean complementation in $(\mathcal{R}, \sqcup, \sqcap, 0, 1)$. However, a *rough complement* may be associated with every element of $\mathcal{R}$:

Definition 8. $\neg < D_1, D_2 > \ \equiv \ < {D_2}^c, {D_1}^c >$.

It is easily seen that rough complementation satisfies the DeMorgan identities, and when restricted to definable sets, $\neg$ is the usual complement. In this context, we may recall the notion of a *quasi-Boolean algebra* (also known as a DeMorgan lattice) [32], it is a distributive lattice $(A, \leq, \vee, \wedge)$ with a unary operation $\neg$ that satisfies involution ($\neg\neg a = a$, for each $a \in A$) and makes the DeMorgan identities hold. So,

Proposition 3. $(\mathcal{R}, \sqcup, \sqcap, \neg, 0, 1)$ *is a complete atomic quasi-Boolean algebra.*

Now a basic finite quasi-Boolean algebra is $\mathcal{U}_0 \ \equiv \ (\{0, a, b, 1\}, \vee, \wedge, \neg, 1)$. It is a diamond as a lattice, viz.,

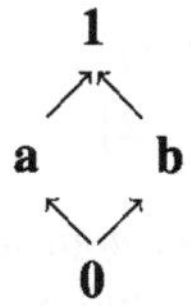

and $\neg$ is given by the equations

$$\neg 0 = 1,\ \neg 1 = 0,\ \neg a = a,\ \neg = b.$$

It is known [32] that any quasi-Boolean algebra is isomorphic to a subalgebra of the product $\prod_{i \in I} \mathcal{U}_i$, where I is a set of indexes, and $\mathcal{U}_i = \mathcal{U}_0$. Hence, to address the converse of Proposition 3, it seems natural to ask if $\mathcal{U}_0$ is isomorphic to some $\mathcal{R}$. The answer is negative, since for any member,

$$< D_1, D_2 > \text{ of } \mathcal{R} \text{ other than } < \emptyset, X >, \neg < D_1, D_2 > \neq < D_1, D_2 >,$$

whereas in $\mathcal{U}_0$, $\neg a = a$, $\neg b = b$ and $a \neq b$. So the class $\mathcal{R}$ is a proper subclass of the class of quasi-Boolean algebras. Iwiński calls the structure $(\mathcal{R}, \sqcup, \sqcap, \neg, 0, 1)$ a *rough algebra.*

2.2 Obtułowicz

[25] was published at the same time as [18] and was the first to hint at a three-valued nature of rough sets. The main result of interest to us here is that rough sets as given by Definition 4 can be essentially considered certain Heyting algebra valued sets [16].

Let $\mathcal{L}$-4 denote the set $\{0, 1, 2, 3\}$. This is a complete distributive lattice (equivalently a complete Heyting algebra), with meet $i \wedge j \ \equiv \ \min\{i, j\}$ and join $i \vee j \ \equiv \ \max\{i, j\}$.

An $\mathcal{L}$-4-*set* $\mathcal{L}$ is then a pair (X, δ), where X is a set, and $\delta : X \times X \to \mathcal{L}$-4 satisfies

$$\delta(x, y) = \delta(y, x), \ \text{ for all } x, y \in X, \ \text{ and}$$
$$\delta(x, y) \wedge \delta(y, z) \leq \delta(x, z), \ \text{ for all } x, y, z \in X.$$

A *rough* $\mathcal{L}$-4-*set* is an $\mathcal{L}$-4-set (X, δ) such that the following 'roughness' conditions hold :

1. $1 \leq \delta(x,x)$, for all $x \in X$.
2. if $2 \leq \delta(x,y)$, then $x = y$ for all $x,y \in X$.
3. if $\delta(x,y) = 1$, then $\delta(x,x) = \delta(y,y)$ for all $x,y \in X$.
4. for each $x \in X$ if $\delta(x,x) = 2$, then there exists $y \in X$ for which $\delta(x,y) = 1$.

We then have a characterization of rough $\mathcal{L}$-4-sets.

Theorem 1. *If* (X,δ) *is a rough* $\mathcal{L}$*-4-set, then* $< X,R,L,B >$ *is a rough set, where* R *is defined as*

$$xRy \text{ if and only if } \delta(x,y) \geq 1, L \equiv \{x \in X : \delta(x,x) = 3\}, \text{ and}$$
$$B \equiv \{x \in X : \delta(x,x) = 2\}.$$

Theorem 2. *If* $< X,R,L,B >$ *is a rough set, then the pair* (X,δ) *is a rough* $\mathcal{L}$*-4-set, where* $\delta : X \times X \to \mathcal{L}$*-4 is given as follows:*

$$\delta(x,y) = \begin{cases} 3 \text{ if } x = y \text{ and } x \in L \\ 2 \text{ if } x = y \text{ and } x \in B \\ 1 \text{ if } (xRy \text{ and } x \neq y) \text{ or } [x = y \text{ and } x \in X \setminus (L \cup B)] \\ 0 \text{ if it is not the case that } xRy. \end{cases}$$

The fact that, in a rough set $< X,R,L,B >$, B does not contain any singleton elementary set (see (iii), Definition 4), is used in the above characterization, viz., to verify condition 4 of the definition of a rough $\mathcal{L}$-4-set.

Thus, Theorem 1 gives a representation of any rough $\mathcal{L}$-4-set as a rough set. Further, Theorem 2 proposes a way of stratifying the R-relatedness of the elements in X relative to a pair $< L,B >$ and in turn assigns a degree of existence, viz., $\delta(x,x)$ of x in the rough set $< L,B >$. It is 3 (the highest) if $x \in L$, 2 if $x \in B$, and 1 if $x \in X \setminus (L \cup B)$. In essence, this points to a three-valued character in the ontology of rough sets.

An algebra of rough sets, however, is not defined in [25], but following the tradition of fuzzy set theory, one can define union, intersection, and complementation of $\mathcal{L}$-4-sets with the help of the operations in the Heyting algebra. This calculus is yet to be explored.

2.3 Pomykała

Definition 2 of rough sets is considered in [30], and the subset $\mathcal{RS}$ of $\mathcal{R}$ is taken. Pomykała came up with a number of algebraic structures that have $\mathcal{RS}$ as a domain and that differ from each other with respect to the complementation and implication operations chosen. We shall consider only one of the complementations (which is the same as the rough complement $\neg$ of Definition 8 in Sect. 2.1) and shall not be concerned with the implication operations. As would follow from Proposition 3 of Sect. 2.1 that $(\mathcal{RS}, \sqcup, \sqcap, \neg, 0, 1)$ is a quasi-Boolean algebra; the operations are restrictions of those in $\mathcal{R}$ (see, e.g., Definition 7 in Sect. 1). The author gives a representation of this algebra in terms of certain finite quasi-Boolean algebras. An important notion involved in this characterization is that of an individual atom — a singleton elementary class. Let us denote by S the collection of all individual atoms in the approximation space $< X, R >$.

Two subalgebras of the quasi-Boolean algebra $\mathcal{U}_0$ (see subsect. 2.1) are $\mathcal{B}_0$ and $\mathcal{C}_0$, where $\mathcal{B}_0 \equiv (\{0,1\}, \vee, \wedge, \sim, 1)$ and $\mathcal{C}_0 \equiv (\{0,a,1\}, \vee, \wedge, \sim, 1)$; the operations are restrictions of those in $\mathcal{U}_0$ to the sets $\{0,1\}$ and $\{0,a,1\}$, respectively. A representation theorem is obtained in terms of these two subalgebras.

Theorem 3. *Every rough algebra $(\mathcal{RS}, \sqcup, \sqcap, \neg, 0, 1)$ is isomorphic to a subalgebra of the product $\prod_{i \in I} \mathcal{U}_i$, where I is a set of indexes, and $\mathcal{U}_i = \mathcal{B}_0$ or $\mathcal{U}_i = \mathcal{C}_0$ for each $i \in I$.*

For the proof, I is considered to be the quotient set X/R, i.e. the family of all elementary sets in $< X, R >$. Further, if $i (\in I)$ is an individual atom, i.e., $i \in S$, then $\mathcal{U}_i = \mathcal{B}_0$, and $\mathcal{U}_i = \mathcal{C}_0$ otherwise. The isomorphism f between $\mathcal{RS}$ and $\prod_{i \in I} \mathcal{U}_i$ is defined as follows:

Let $< \underline{A}, \overline{A} > \in \mathcal{RS}$. Then,

$$f(< \underline{A}, \overline{A} >) \equiv (x_i)_{i \in I}, \ (x_i)_{i \in I} \in \prod_{i \in I} \mathcal{U}_i, \quad \text{if and only if}$$

1. $i \in S$ and $i \subseteq \underline{A}$ imply $x_i = 1$.
2. $i \in S$ and $i \not\subseteq \underline{A}$ imply $x_i = 0$.
3. $i \notin S$, $i \not\subseteq \underline{A}$ and $i \not\subseteq \overline{A}$ imply $x_i = 0$.
4. $i \notin S$, $i \not\subseteq \underline{A}$ and $i \subseteq \overline{A}$ imply $x_i = 1$.
5. $i \notin S$ and $i \subseteq \underline{A}$ imply $x_i = 1$.

In another paper [31], Pomykała defines a unary operation * on $\mathcal{RS}$ as $< \underline{A}, \overline{A} >^* \equiv < \overline{A}^c, \overline{A}^c >$, $< \underline{A}, \overline{A} > \in \mathcal{RS}$. Then,

Proposition 4. $(\mathcal{RS}, \sqcup, \sqcap, ^*, 0, 1)$ *is a Stone algebra, i.e., * is a pseudocomplementation on $\mathcal{RS}$ such that for all $< \underline{A}, \overline{A} > \in \mathcal{RS}$,*

$$< \underline{A}, \overline{A} >^* \sqcup < \underline{A}, \overline{A} >^{**} = 1 \ .$$

However, no representation is obtained.

2.4 Comer

In [13], it is seen that apart from the pseudocomplementation * defined above, a dual pseudocomplementation $^+$ may also be introduced in $\mathcal{RS}$. So the rough structure becomes a double Stone algebra (defined below), and a representation of certain double Stone algebras can also be obtained.

A *double Stone algebra* (DSA) is a structure $(L, \vee, \wedge, ^*, ^+, 0, 1)$ such that

1. $(L, \vee, \wedge, 0, 1)$ is a bounded distributive lattice.
2. $y \leq x^*$ if and only if $y \wedge x = 0$.
3. $y \geq x^+$ if and only if $y \vee x = 1$.
4. $x^* \vee x^{**} = 1$, $x^+ \vee x^{++} = 0$.

The DSA is *regular* if, in addition to the above, for all $x \in L$,

$$x \wedge x^+ \leq x \vee x^*$$

holds. This is equivalent to requiring that $x^* = y^*$, $x^+ = y^+$ imply $x = y$, for all $x, y \in L$.

Proposition 5. $(\mathcal{RS}, \sqcup, \sqcap, ^*, ^+, 0, 1)$, *for a given approximation space* $< X, R >$, *is a regular DSA, where the operations* $\sqcup, \sqcap$ *are as in Definition 7 (see Sect. 1),* $0, 1$ *as in Proposition 1 (see Sect. 2.1) and*

$$< \underline{A}, \overline{A} >^* \equiv < \overline{A}^c, \overline{A}^c >,$$
$$< \underline{A}, \overline{A} >^+ \equiv < \underline{A}^c, \underline{A}^c > .$$

In the converse direction, Comer first proves the following proposition.

Proposition 6. *Any complete, completely distributive, atomic regular DSA is isomorphic to* $\mathcal{RS}$ *for some approximation space* $< X, R >$.

He also establishes that any regular DSA is a subalgebra of a DSA of the type specified in Proposition 6 (with other properties as well), and hence obtains

Theorem 4. *Any regular DSA* $\mathcal{A}$ *is isomorphic to a subalgebra of* $\mathcal{RS}$ *for some approximation space* $< X, R >$.

In [12], however, the author considers a rough structure differently. The approximation algebras he comes up with, are representable by certain cylindrical algebras. He starts with an information system $S \equiv < X, AT, V, f >$, where X is a set, AT a finite set, V a function with domain AT, and $f : X \to \prod_{a \in AT} V_a$. One can easily notice that this is a slightly different, but equivalent formulation of an information system as presented in Sect. 1. Now each $P (\subseteq AT)$ induces an equivalence relation R_P on X, giving $< X, R_P >$, an *approximation space for knowledge P*. Each R_P in turn, induces an upper approximation operator $\overline{P} : \mathcal{P}(X) \to \mathcal{P}(X)$, i.e., $\overline{P}(A)$ is the union of equivalence classes under R_P of all elements of $A (\subseteq X)$.

Then, the structure $(\mathcal{P}(X), \cup, \cap, ^c, \overline{P}, \emptyset, X), P \subseteq AT$ is called a *knowledge approximation algebra of type AT derived from information system S*. For each $P \subseteq AT$, the structure $(\mathcal{P}(X), \cup, \cap, ^c, , \overline{P}, \emptyset, X)$ is called the (upper) *approximation closure algebra of P*. It may be noted that this is an instance of a monadic Boolean algebra [23] that was used in [17] later.

Now $(\mathcal{P}(X), \cup, \cap, ^c, \emptyset, X)$ is a complete atomic Boolean algebra. In fact, Comer observed that a knowledge approximation algebra of type AT is an instance of a general algebraic structure that consists of a complete atomic Boolean algebra $(B, \vee, \wedge, \sim, 0, 1)$ and a family of functions $K_P : B \to B$, $P \subseteq AT$, where AT is a finite set. Moreover, the functions satisfy the following, for $x, y \in B$ and $P, Q \subseteq AT$:

1. $K_P(0) = 0$.
2. $K_P(x) \geq x$.
3. $K_P[x \wedge K_P(y)] = K_P(x) \wedge K_P(y)$.
4. If $x \neq 0$, then $K_\emptyset(x) = 1$.
5. $K_{P \cup Q}(x) = K_P(x) \wedge K_Q(x)$, if x is an atom of B.

This leads to

Proposition 7. *Approximation closure algebras are complete atomic cylindrical algebras of dimension one.*

A representation theorem is subsequently obtained.

Theorem 5. *Every complete atomic cylindrical algebra of dimension one is isomorphic to an approximation closure algebra. In fact, every cylindrical algebra of dimension one can be embedded in an approximation closure algebra.*

However, one may say that the approximation closure algebras of Comer are not rough in our sense, as the elements of the domain considered are not rough sets. In fact, these have connections with information algebras [26] discussed in Sect. 4.

2.5 Bonikowski

This is an approach that adopts definition 5, i.e., a rough set in an approximation space $< X, R >$ is taken as an element of $\mathcal{P}(X)/\approx$.

Now, $(\mathcal{P}(X)/\approx, \leq)$ is clearly a partially ordered set, where $\leq$ is defined in terms of rough inclusion, i.e., $[A] \leq [B]$ if and only if A is roughly included in B, $[A], [B] \in \mathcal{P}(X)/\approx$. Bonikowski defines operations of join ($\cup_\approx$) and meet ($\cap_\approx$) (rough union and intersection) on $\mathcal{P}(X)/\approx$ to turn it into a lattice. As noted in the introduction, a join (meet) cannot be defined directly by the union (intersection) operations in $\mathcal{P}(X)$.

For a subset A of X, an *upper sample* P is such that $P \subseteq A$ and $\overline{P} = \overline{A}$. An upper sample P of A is *minimal*, if there is no upper sample Z of A with $Z \subsetneq P$. Then

$$[A] \cup_\approx [B] \equiv [\underline{A} \cup \underline{B} \cup P],$$

where P is a minimal upper sample of $\overline{A} \cup \overline{B}$.

Similarly,

$$[A] \cap_{\approx} [B] \equiv [(\underline{A} \cap \underline{B}) \cup P],$$

where P is a minimal upper sample of $\overline{A} \cap \overline{B}$. (Exterior) Complementation on $\mathcal{P}(X)/\approx$ is defined as

$$[A]^{\mathrm{ex}} \equiv [(\overline{A})^c].$$

The following propositions are then proved. The author includes $\emptyset$ among elementary sets. It may be noted that for Proposition 8, the domain of discourse X is assumed to be finite.

Proposition 8. $(\mathcal{P}(X)/\approx, \cup_{\approx}, \cap_{\approx}, ^{\mathrm{ex}}, [\emptyset], [X])$ *is a complete atomic Stone algebra, where the atoms are determined by proper subsets of the elementary sets or by singleton elementary sets in* $< X, R >$.

Proposition 9. $(\mathcal{D}/\approx, \cup_{\approx}, \cap_{\approx}, ^c, [\emptyset], [X])$ *is a complete atomic Boolean algebra, where* $\mathcal{D}$ *is the collection of all definable sets in the approximation space* $< X, R >$. *The atoms are determined by the elementary sets in* $< X, R >$.

Note that $\mathcal{D}/\approx$ is identifiable with $\mathcal{D}$, and the algebra $(\mathcal{D}/\approx, \cup_{\approx}, \cap_{\approx}, ^c, [\emptyset], [X])$ is just (up to isomorphism) the Boolean algebra formed by $\mathcal{D}$ with the usual set-theoretical operations.

2.6 Chuchro

This is another approach using Definition 5. However, the perspective is generalized in the following sense: Given an approximation space $< X, R >$, the lower approximation [treated as an operator on $\mathcal{P}(X)$] acts as an interior operation L (say) on $\mathcal{P}(X)$. The elementary sets, along with $\emptyset$ and X, form a basis of $\mathcal{P}(X)$ with respect to L. So, $(\mathcal{P}(X), \cup, \cap, ^c, L, \emptyset, X)$ is a topological Boolean algebra, given the special name, *topological field of sets* [32]. Note that a topological Boolean algebra is a Boolean algebra $(A, \vee, \wedge, \sim, 0, 1)$, equipped with a unary operation In that satisfies all the properties of an interior, viz., for each

$$a, b \in A, In(a) \vee a = a, In(a \wedge b) = In(a) \wedge In(b), In[In(a)] = In(a), \text{ and } In(1) = 1.$$

It is with this fact in the background that Chuchro considers an arbitrary topological Boolean algebra $(A, \vee, \wedge, \sim, In, 0, 1)$ and defines the relation of rough equality ($\approx$) on A:

$$x \approx y \text{ if and only if } In(x) = In(y) \text{ and } C(x) = C(y),$$

where $C(x) \equiv \sim In(\sim x), x, y \in A$ (i.e., C is the closure operation in A).

Now $\approx$ is an equivalence relation on A. The pair $< A, \approx >$ is called an *approximation algebra*, and rough sets are elements of $A/\approx$. In this work, the author characterizes certain rough sets in terms of the corresponding elements of A.

An element a of A is *regular open* in A, when $In[C(a)] = a$. For each regular open element a of A, a set $A(a)$ is defined as

$$A(a) \equiv \{x \in A : x = a \vee b, b \leq \sim a \wedge C(a), b \in A\}.$$

It is shown that

Proposition 10. $(A(a), \vee, \wedge, \neg, a, C(a))$ *is a Boolean algebra with a as zero,* $C(a)$ *as a unit, and complementation* $\neg$ *taken as*

$$\neg x \equiv a \vee \{\sim b \wedge [\sim a \wedge C(a)]\},$$

for $x \in A(a)$*, where* $\vee, \wedge$ *are, of course, restrictions of the corresponding operations in* A.

Then one obtains

Proposition 11. $A(a) = [a]$*, where* $[a] \in A/\approx$.

It is not difficult to observe that if the topological Boolean algebra $(A, \vee, \wedge, \sim, In, 0, 1)$ is zero-dimensional, i.e., $In[C(a)] = C(a)$ for all $a \in A$, the regular open elements of A would be precisely those that are clopen [$a = In(a) = C(a)$]. In that case, $A(a) = [a] = \{a\}$. In particular, as $(\mathcal{P}(X), \cup, \cap, ^c, L, \emptyset, X)$ is zero-dimensional, the characterization given by Proposition 11 works only for exact sets of $< X, R >$.

2.7 Pagliani

Pagliani has been exploring the algebraic and topological aspects of rough set theory intensively and extensively for quite some time now. However, a central structure that features in his studies is that of a Nelson algebra (also known as a quasi-pseudo-Boolean algebra)). He has been able to give a representation of certain Nelson algebras and three-valued Łukasiewicz algebras by rough set systems. In addition, rough set structures are shown to be instances of other algebras such as chain-based lattices and Post algebras. The main results depend on the assumption of a finite domain and/or involve a restriction about singleton elementary sets.

An algebra $(\mathcal{N}, \wedge, \vee, \neg, \sim, \rightarrow, 0, 1)$ is a Nelson algebra [32], provided

$$(\mathcal{N}, \wedge, \vee, \neg, 0, 1)$$

is a quasi-Boolean algebra, and for any $a, b, x \in \mathcal{N}$, it satisfies

1. $a \wedge \neg a \leq b \vee \neg b$.
2. $a \wedge x \leq \neg a \vee b$ if and only if $x \leq a \rightarrow b$.
3. $a \rightarrow (b \rightarrow c) = (a \wedge b) \rightarrow c$.
4. $\sim a = a \rightarrow \neg a = a \rightarrow 0$.

$\neg$ and $\sim$ are the strong and weak negations on $\mathcal{N}$, respectively. A Nelson algebra $\mathcal{N}$ is *semisimple*, if $a \vee \sim a = 1$ for all $a \in \mathcal{N}$.

Pagliani considers Definition 3 for rough sets in an approximation space $< X, R >$ with X finite and takes the set $\mathcal{RS}'$ of pairs $< \underline{A}, \overline{A}^c >, A \subseteq X$. The following is shown:

Proposition 12. $(\mathcal{RS}', \sqcap, \sqcup, \neg, \sim, \rightarrow, 0, 1)$ *is a semisimple Nelson algebra, whose operations are defined as*

$$
\begin{aligned}
< \underline{A_1}, \overline{A_1}^c > \sqcap < \underline{A_2}, \overline{A_2}^c > &\equiv < \underline{A_1} \cap \underline{A_2}, \overline{A_1}^c \cup \overline{A_2}^c >,\\
< \underline{A_1}, \overline{A_1}^c > \sqcup < \underline{A_2}, \overline{A_2}^c > &\equiv < \underline{A_1} \cup \underline{A_2}, \overline{A_1}^c \cap \overline{A_2}^c >,\\
< \underline{A_1}, \overline{A_1}^c > \rightarrow < \underline{A_2}, \overline{A_2}^c > &\equiv < \underline{A_1}^c \cup \underline{A_2}, \underline{A_1} \cap \overline{A_2}^c >,\\
\neg < \underline{A_1}, \overline{A_1}^c > &\equiv < \overline{A_1}^c, \underline{A_1} > \textit{ and}\\
\sim < \underline{A_1}, \overline{A_1}^c > &\equiv < \underline{A_1}^c, \underline{A_1} >.
\end{aligned}
$$

Further, a representation theorem is obtained.

Theorem 6. *Any finite semisimple Nelson algebra is isomorphic to* $\mathcal{RS}'$ *for some approximation space* $< X, R >$.

A three-valued Łukasiewicz (Moisil) algebra [22,23,7] $(A, \leq, \wedge, \vee, \neg, M, 0, 1)$ is such that $(A, \leq, \wedge, \vee, \neg, 0, 1)$ is a quasi-Boolean algebra and M is a unary operator on A satisfying, for all $a, b \in A$, the following.

M1. $M(a \wedge b) = Ma \wedge Mb$.

M2. $M(a \vee b) = Ma \vee Mb$.

M3. $Ma \wedge \neg Ma = 0$.

M4. $MMa = Ma$.

M5. $M\neg Ma = \neg Ma$.

M6. $\neg M \neg a \leq Ma$.

M7. $Ma = Mb, M\neg a = M\neg b$ imply $a = b$.

The author observes an equivalence between semisimple Nelson algebras and three-valued Łukasiewicz algebras via appropriate transformations, so that the preceding results can be carried over to the latter algebras. We note again that the results apply when the domain of the approximation space is finite.

Interestingly, in the special situation when the approximation space has no singleton elementary sets in it, the author shows that $\mathcal{RS}$ with its pairs in reverse order, viz., the collection of pairs $< \overline{A}, \underline{A} >, A \subseteq X$, turns out to be a Post algebra of order three [32]. In other words [11], it is a three-valued Łukasiewicz algebra with a center (i.e., an element c such that $\neg c = c$). In the general situation (with no restriction on the approximation space), the same structure can be made into an algebra that is a generalization of a Post algebra, viz., a certain chain-based lattice of order three [15].

$\mathcal{RS}'$ is also seen to be a Stone as well as a regular double Stone algebra with suitable operations — which is not surprising, considering Pomykała's and Comer's results about $\mathcal{RS}$ to the same effect (cf. Propositions 4 and 5). The operations in this case are derived from those that make $\mathcal{RS}'$ a Nelson algebra (cf. Proposition 12). $\sqcap, \sqcup$ remain the same, whereas the pseudocomplementation * is taken as $\neg \sim \neg$ and the dual pseudocomplementation $^+$ as $\sim$.

Pagliani relentlessly points at the local and global behavior of rough set systems caused by singleton and nonsingleton elementary sets, respectively. Let χ_A be the generalized characteristic function for the rough set A in $< X, R >$ defined by

$$\chi_A(x) = \begin{cases} 0 & \text{if } x \in \overline{A}^c \\ 1/2 & \text{if } x \in BnA \\ 1 & \text{if } x \in \underline{A}. \end{cases}$$

Then if $A \subseteq S \equiv$ the union of singleton elementary sets, χ_A takes only values 0 and 1, whereas if $A \subseteq P \equiv$ the union of nonsingleton elementary sets and A is not a definable set, χ_A is three-valued. Discussion of topological implications of this behavior is, however, outside the scope of this chapter.

2.8 Banerjee and Chakraborty

Initially, the endeavor in [3] was to look for a suitable logic, whose models would be rough sets, as in Definition 1. The modal system S_5 seemed to be an appropriate candidate, as a Kripke model, $< X, R >$ along with the interpretation of a well-formed formula α of S_5, $\pi(\alpha)$ is a rough set of Definition 1: here X is the domain, R the accessibility relation, an equivalence on X in case of S_5, and π the meaning function interpreting formulas as subsets of X. To address the notion of rough equality ($\approx$), a second implication $\Rightarrow$ (apart from that of S_5) was defined in the language of S_5 as $\alpha \Rightarrow \beta \equiv (L\alpha \rightarrow L\beta) \wedge (M\alpha \rightarrow M\beta)$. So, $\alpha \approx \beta \equiv (\alpha \Rightarrow \beta) \wedge (\beta \Rightarrow \alpha)$. Then, a Lindenbaum-like construction was carried out on the set of formulas using this $\Rightarrow$. Two formulas α, β were defined to be equivalent if and only if $\vdash_{S_5} \alpha \Rightarrow \beta$ and $\vdash_{S_5} \beta \Rightarrow \alpha$. Next, the logical operations were extended to the quotient space as

$$\begin{aligned}
&[\alpha] \sqcap [\beta] \equiv \{(\alpha \wedge \beta) \vee \alpha \wedge M\beta \wedge \neg M(\alpha \wedge \beta)\}, \\
&[\alpha] \sqcup [\beta] \equiv \{(\alpha \vee \beta) \wedge (\alpha \vee L\beta \vee \neg L(\alpha \vee \beta))\}, \\
&\neg[\alpha] \equiv [\neg\alpha], \\
&L[\alpha] \equiv [L\alpha]
\end{aligned}$$

for any $[\alpha], [\beta]$ in the quotient space.

The resulting structure turned out to be a quasi-Boolean algebra with an interior operation that satisfied the zero-dimensionality property. Thus, the notion of a *topological quasi-Boolean algebra* (tqBa) came up [3,36]. Formally, it is an algebra $(A, \leq, \sqcap, \sqcup, \neg, L, 0, 1)$ such that

A1. $(A, \leq, \sqcap, \sqcup)$ is a distributive lattice.
A2. $\neg\neg a = a$.
A3. $\neg(a \sqcup b) = \neg a \sqcap \neg b$.
A4. $La \leq a$.
A5. $L(a \sqcap b) = La \sqcap Lb$.
A6. $LLa = La$.
A7. $L1 = 1$.
A8. $MLa = La$,

where $Ma \equiv \neg L \neg a, a, b \in A$.

The semantic analogue of this exercise shows that $\{\mathcal{P}(X)/\approx, \sqcap, \sqcup, \neg, L, [\emptyset], [X]\}$ is a tqBa, such that

$$\begin{aligned} [S] \sqcap [T] &\equiv [(S \cap T) \cup (S \cap \overline{T} \cap \overline{S \cap T}^c)], \\ [S] \sqcup [T] &\equiv [(S \cup T) \cap (S \cup \underline{T} \cup \underline{S \cup T}^c)], \\ \neg[S] &\equiv [S^c], \\ L[S] &\equiv [\underline{S}], \end{aligned}$$

where $[S], [T] \in \mathcal{P}(X)/\approx$.

Wasilewska has generalized this construction in [37]. On the other hand, let us recall (cf. Sect. 1) that the set $\mathcal{R}$ of pairs of definable sets $< D_1, D_2 >$ in the approximation space $< X, R >$, $D_1 \subseteq D_2$. $(\mathcal{R}, \sqcup, \sqcap, \neg, L, 0, 1)$ is also seen to be a tqBa, with $\sqcup, \sqcap$ as in Definition 6 of Sect. 1, $\neg$ the rough complementation in Definition 8 of Sect. 2.1, and L defined as $L(< D_1, D_2 >) \equiv \; < D_1, D_1 >$. This algebra is called the *approximation space algebra* for $< X, R >$. It is no surprise that this algebra is isomorphic to $\{\mathcal{P}(X)/\approx, \sqcap, \sqcup, \neg, L, [\emptyset], [X]\}$ if and only if no definable set in $< X, R >$ is a singleton. Observing that $\mathcal{D}$, the collection of all definable sets in the approximation space $< X, R >$, forms a Boolean algebra, a general method of generating a tqBa is demonstrated in [4].

Given any Boolean algebra

$$\mathcal{B} \equiv (B, \leq, \wedge, \vee, \sim, 0, 1), \text{ the set } A \equiv \{< a, b > \in B \times B : a \leq b\}$$

is considered. Operations $\sqcap, \sqcup, \neg$ and L are defined on A just as in $\mathcal{R}$, and one finds that A, with these operations, is a tqBa. $< 0, 0 >$ and $< 1, 1 >$ are the zero and the unit, respectively. This special tqBa is called the *topological quasi-Boolean extension* of Boolean algebra $\mathcal{B}$, denoted as $TQ(\mathcal{B})$. Analogous to the observation of Iwiński (cf. Proposition 2), one finds that $\mathcal{B}$ is the maximal Boolean subalgebra of $TQ(\mathcal{B})$, up to isomorphism. It may be noticed that the approximation space algebra for $< X, R >$ is nothing but $TQ(\mathcal{D})$.

In fact, the tqBa of $\mathcal{P}(X)/\approx$ and $TQ(\mathcal{D})$ satisfy a few more properties. Abstracting these in two stages, one gets the notion of prerough and rough algebras. The following are added to the axioms of a tqBa, to get a *prerough algebra*.

A9. $\neg La \sqcup La = 1$.
A10. $L(a \sqcup b) = La \sqcup Lb$.
A11. $La \leq Lb, Ma \leq Mb$ imply $a \leq b$.
A12. $a \Rightarrow b = (\neg La \sqcup Lb) \sqcap (\neg Ma \sqcup Mb)$.

$TQ(\mathcal{B})$, for any Boolean algebra $\mathcal{B}$, is a prerough algebra.

A few intermediate algebras (tqBas, but not prerough algebras) are obtained when different proper subsets of the set of axioms A9–A12 are taken. These have been studied in [33].

A *rough algebra* $\mathcal{P} \equiv (A, \leq, \sqcap, \sqcup, \neg, L, \Rightarrow, 0, 1)$ is a prerough algebra such that the subalgebra $(L(A), \leq, \sqcap, \sqcup, \neg, 0, 1)$ of $\mathcal{P}$, where $L(A) \equiv \{La : a \in A\}$, is

A13. complete.
A14. completely distributive, i.e. $\sqcup_{i \in I} \sqcap_{j \in J} a_{i,j} = \sqcap_{f \in J^I} \sqcup_{i \in I} a_{i,f(i)}$

for any index sets I, J and elements $a_{i,j}, i \in I, j \in J$, of $L(A)$, where J^I is the set of maps of I into J.

Both the approximation space algebra $TQ(\mathcal{D})$ and $(\mathcal{P}(X)/\approx, \sqcap, \sqcup, \neg, L, [\emptyset], [X])$ are examples of rough algebras.

Prerough algebras have been investigated further [1], and one finds that

Proposition 13. *A prerough algebra is equivalent to a three-valued Lukasiewicz algebra, in the sense that the defining axioms of one are deducible from those of the other.*

Representation theorems for both prerough and rough algebras are obtained in [4]. The first allows us to view any prerough algebra as an algebra of pairs of Boolean elements — more specifically, as a subalgebra of $TQ(\mathcal{B})$ for some Boolean algebra $\mathcal{B}$. The second relates any rough algebra to a subalgebra of $TQ(\mathcal{D})$, and as a corollary, to $\mathcal{P}(X)/\approx$ for some approximation space $< X, R >$.

Theorem 7. *Any prerough algebra* $\mathcal{P} \equiv (A, \leq, \sqcap, \sqcup, \neg, L, \Rightarrow, 0, 1)$ *is isomorphic to the subalgebra of* $TQ[\mathcal{L}(A)]$ *formed by the set* $\{< La, Ma >: a \in A\}$.

Theorem 8. *Any rough algebra is isomorphic to a subalgebra of the approximation space algebra corresponding to some approximation space* $< X, R >$.

Proof. Let $\mathcal{P} \equiv (A, \leq, \sqcap, \sqcup, \neg, L, \Rightarrow, 0, 1)$ be a rough algebra.
Then, $\mathcal{L}(A) \equiv (L(A), \leq, \sqcap, \sqcup, \neg, 0, 1)$ is a complete and completely distributive Boolean subalgebra of $\mathcal{P}$ (by $A13$ and $A14$). Hence, $\mathcal{L}(A)$ is isomorphic to a complete

field of sets $C \equiv (C, \subseteq, \cap, \cup, ^c, \emptyset, 1)$ (say) [34].

C is atomic (and completely distributive). Let X denote the union of all of its atoms. The atoms induce a partition R (say) of X, such that any equivalence class due to R is an atom of C. Thus, we have an approximation space $< X, R >$.

As C is atomic, each of its elements is expressible as the union of some of its atoms. In particular, $C = \mathcal{D}$, the collection of all definable sets of $< X, R >$. So the isomorphism of $\mathcal{L}(A)$ and C implies the isomorphism of $TQ[\mathcal{L}(A)]$ and $TQ(\mathcal{D})$. By theorem 7, $\mathcal{P}$ has an isomorphic copy in $TQ[\mathcal{L}(A)]$, and therefore in $TQ(\mathcal{D})$.
This completes the proof. □

Corollary 1. *Any rough algebra is isomorphic to a subalgebra of $\mathcal{P}(X')/\approx$ for some approximation space $< X', R' >$.*

Proof. The copy of $\mathcal{P}$ in $TQ(\mathcal{D})$ may have an element $< D_1, D_2 >$, $D_1, D_2 \in \mathcal{D}$, for which there is no $S(\subseteq X)$ with $\underline{S} = D_1$, $\overline{S} = D_2$. As we have seen earlier, this problem occurs when there are singleton elementary classes in $< X, R >$. However, an approximation space $< X', R' >$ can always be obtained such that the following hold.

1. The Boolean algebras formed by the collections $\mathcal{D}$ and $\mathcal{D}'$ of definable sets in $< X, R >$ and $< X', R' >$, respectively, are isomorphic.
2. For any element $< D', D'' >$ of $TQ(\mathcal{D}')$, there is a rough set $< X', R', S >$ such that $\underline{S} = D'$ and $\overline{S} = D''$.
3. The approximation space algebras $TQ(\mathcal{D})$ and $TQ(\mathcal{D}')$ corresponding to $< X, R >$ and $< X', R' >$, respectively, are isomorphic.

Thus by the theorem, $\mathcal{P}$ has an isomorphic copy in $\mathcal{P}(X')/\approx$. □

2.9 Wasilewska

Wasilewska contributed to developing the formal definition of a topological quasi-Boolean algebra (tqBa) as given in [3]. She has worked with this notion further, and several properties of this structure were deduced by a theorem prover [36]. We shall, however, discuss another aspect of her work [37] in the given area.

Like [10] (cf. Sect. 2.6), in [37] too, an arbitrary topological Boolean algebra $(A, \vee, \wedge, \sim, In, 0, 1)$ is taken, and the same relation $\approx$ (called topological equality here) is defined on A. But the work takes a direction different from [10]. Operations of join ($\sqcup$) and meet ($\sqcap$) are defined on $A/\approx$ such that it becomes a lattice. Again, one might stress (as noted in the introduction) that the join operation $\vee$ in A does not induce a join in $A/\approx$, as In does not, in general, distribute across $\vee$. The construction that is adopted here is a direct generalization of that used to obtain $(\mathcal{P}(X)/\approx, \sqcap, \sqcup, \neg, L, [\emptyset], [X])$ from the topological Boolean algebra $(\mathcal{P}(X), \cup, \cap, ^c, L, \emptyset, X)$, cf. subsect. 2.8, viz.,

$$[a] \sqcup [b] \equiv [(a \vee b) \wedge (a \vee In(b) \vee \sim In(a \vee b))].$$

Similarly,

$$[a] \sqcap [b] \equiv \{(a \wedge b) \vee (a \wedge C(b) \wedge \sim C(a \wedge b))\},$$

where C is the closure operation on A. Complementation ($\neg$) and an interior (also denoted In) are induced on $A/\approx$:

$$\neg[a] \equiv [\sim a],\ In([a]) \equiv [In(a)],\ \text{for } a \in A.$$

The following proposition is then proved.

Proposition 14. $(A/\approx, \sqcup, \sqcap, \neg, [0], [1])$ *is a quasi-Boolean algebra with an interior operation In. If the topological Boolean algebra is zero-dimensional,* $A/\approx$ *is also zero-dimensional.*

Now, given any zero-dimensional topological space $< X, In >$ (In is the interior operation on X), there is an equivalence relation R on X such that the lower approximation operator L corresponding to R coincides with In [38]. This fact may justify calling a zero-dimensional topological Boolean algebra a generalized approximation space and elements of $A/\approx$ generalized rough sets. Wasilewska, however, is inclined to call any topological zero-dimensional space a generalized rough set.

2.10 Iturrioz

It is observed in [17] that, corresponding to an approximation space $< X, R >$, the structure $(\mathcal{P}(X), \cup, \cap, ^c, ^-, \emptyset, X)$ (the same considered by Comer, see Sect. 2.4) is a monadic Boolean algebra [$^-$ is the upper approximation operator on $\mathcal{P}(X)$]. It is a topological Boolean algebra with a closure operator.

Iturrioz takes Definition 2 for rough sets, and $\mathcal{RS} \equiv \{< \underline{A}, \overline{A} >, A \subseteq X\}$ as the domain. The universe X of discourse is assumed to be finite. (For further comments in this regard, see the following section).

With a unary operator M defined on $\mathcal{RS}$ as $M(< \underline{A}, \overline{A} >) \equiv < \overline{A}, \overline{A} >$, $A \subseteq X$, $\sqcup, \sqcap$ as in Definition 7 of Sect. 1, $\neg$ as in Definition 8 of Sect. 2.1,

$$0 \equiv < \emptyset, \emptyset > 1 \equiv < X, X >,$$

one finds that

Proposition 15. $(\mathcal{RS}, \sqcup, \sqcap, \neg, M, 0, 1)$ *is a three-valued Łukasiewicz algebra.*

A representation of three-valued Łukasiewicz algebras is then obtained.

Theorem 9. *Every three-valued Łukasiewicz algebra is isomorphic to a subalgebra of* $(\mathcal{RS}, \sqcup, \sqcap, \neg, M, 0, 1)$ *corresponding to some approximation space.*

3 A Comparative Study

Let us try to analyze the rough algebras and the representation results that were presented in the previous section. From the point of view of a mathematical description or representation of a structure, finiteness assumptions, if any, can play an important role. A representation of any finite three-valued Łukasiewicz algebra by a finite $\mathcal{RS}'$ is obtained in [27] (cf. Theorem 6). On the other hand, in [17], a representation of any (not necessarily finite) three-valued Łukasiewicz algebra by a subalgebra of $\mathcal{RS}$ for some approximation space $< X, R >$, is proved (cf. Theorem 9). However, the domain X considered in the proof is not necessarily finite, pointing to the irrelevance of the finiteness assumption in this context. ($\mathcal{RS}$, as noted earlier, is isomorphic to $\mathcal{RS}'$, so the above comparison is not unwarranted). Summing up the survey in the previous section, we find that the different algebras of rough sets are instances of

1. quasi-Boolean algebras [18,30].
2. regular double Stone algebras [13,27].
3. complete atomic Stone algebras [8].
4. semisimple Nelson algebras [27].
5. (a) topological quasi-Boolean algebras [3,36].
   (b) prerough algebras [4].
   (c) rough algebras [4].
6. three-valued Łukasiewicz algebras [27,17].

Now 1 is included in 5(a), which in turn is included in 5(b) and 5(c). Via suitable transformations 2, 4 and 6 are known to be equivalent to each other (see [27]). And if we restrict our discourse to the finitary properties of the collection of rough sets, its description as a prerough algebra is adequate. So we consider prerough algebras for comparison with others.

As observed in Proposition 13 of Sect. 2.8, prerough and three-valued Łukasiewicz algebras are equivalent. But the implication $\rightarrow$ in a three-valued Łukasiewicz algebra, defined as $a \rightarrow b \equiv (\neg La \vee b) \wedge (\neg a \vee Mb)$, and the implication $\Rightarrow$ in a prerough algebra differ. It is found that, in general, $a \Rightarrow b \leq a \rightarrow b$. In any case, in the light of Pagliani's result and this equivalence, it can be concluded that finite prerough algebras can be represented by the algebra of $\mathcal{RS}$. At the same time, through Iturrioz's result, one would obtain a representation of any prerough algebra by a subalgebra of $\mathcal{RS}$.

The equivalence of 2, 4, and 6 imply that prerough algebras are also equivalent to 2 and 4. Yet, it may be interesting to see the involved transformations explicitly. Given a semisimple Nelson algebra $\mathcal{N} \equiv (A, \wedge, \vee, \neg, \sim, \rightarrow, 0, 1)$, setting

$$La = \neg \sim a \text{ and } a \Rightarrow b = \neg \sim (a \hookrightarrow b),$$

where $a \hookrightarrow b \equiv (\sim a \wedge \sim \neg b) \vee (\neg \sim \neg a \vee b)$, a prerough algebra is obtained. Conversely, in a prerough algebra

$$(A, \wedge, \vee, \neg, L, \Rightarrow, 0, 1),$$

setting $\sim a = \neg La$ and $a \longrightarrow b = \neg La \vee b$, we obtain a semisimple Nelson algebra.

A prerough algebra $(A, \wedge, \vee, \neg, L, \Rightarrow, 0, 1)$ is also equivalent to a regular double Stone algebra $(L, \vee, \wedge, ^*, ^+, 0, 1)$, via the following transformations.

[DS-PR1.] $< a,b >^+ \equiv \neg L < a,b >$.
[DS-PR2.] $< a,b >^* \equiv L\neg < a,b >$.
[PR-DS.] $L(< a,b >) \equiv < \neg a,b >^+$.

Comer's representation result (Theorem 4) then vindicates our earlier observation that a prerough algebra is representable by a subalgebra of $\mathcal{RS}$.

There are some other algebraic structures to which prerough algebras become related, because of the preceding equivalences. For example, three-valued Łukasiewicz algebras are equivalent to Wajsberg algebras [7]. One finds that a Wajsberg algebra $(A, \longrightarrow, \neg, 1)$ becomes equivalent to a prerough algebra $(A, \wedge, \vee, \neg, L, \Rightarrow, 0, 1)$ via the transformations

[PR-W.] $a \longrightarrow b \equiv (M\neg a \vee b) \wedge (Mb \vee \neg a)$.
[W-PR1.] $a \vee b \equiv (a \longrightarrow b) \longrightarrow b$.
[W-PR2.] $a \wedge b \equiv \neg(\neg a \vee \neg b)$.
[W-PR3.] $Ma \equiv \neg a \longrightarrow a$.
[W-PR4.] $0 \equiv \neg 1$,

where $Ma \equiv \neg L \neg a$.

Further, a three-valued Łukasiewicz algebra is cryptoisomorphic to an MV_3-algebra [24] in the sense of Birkhoff [5]. Thus, there is a cryptoisomorphism between a prerough algebra and an MV_3-algebra as well.

Thus, we find that all of the different algebras just discussed and the rough algebras of Sect. 2.8 are representable by (subalgebras of) rough set algebras. The main theorems from which this conclusion is derived are Theorems 4, 6, 8, and 9. One may note that there is yet another representation — in the other direction — obtained by Pomykała (cf. Theorem 3).

Let us now turn to $\mathcal{P}(X)/\approx$, and the corresponding rough structures. As presented in the previous section, Bonikowski gives one of the versions of appropriate join ($\cup_\approx$) and meet ($\cap_\approx$) operations in $\mathcal{P}(X)/\approx$ to eventually form the complete atomic Stone algebra $(\mathcal{P}(X)/\approx, \cup_\approx, \cap_\approx, ^{ex}, [\emptyset], [X])$ (cf. Sect. 2.5). Another version of join ($\sqcup$) and meet ($\sqcap$) forms the tqBa $(\mathcal{P}(X)/\approx, \sqcap, \sqcup, \neg, L, [\emptyset], [X])$ (cf. Sect. 2.8). Let us denote the former algebra by $\mathcal{P}_b$, and the latter by $\mathcal{P}_{bc}$. It may be interesting to look at the relation between these two definitions.

First, it may be observed that a minimal upper sample P of a set $A(\subseteq X)$ in an approximation space $< X,R >$ can be formed by taking exactly one element from every equivalence class in $\overline{A}$. So, if an equivalence class $[x] \subseteq P$, then necessarily $[x] = \{x\}$, which means $x \in P$. This also implies that if P' is another minimal upper sample of A, $\underline{P} = \underline{P'}$. As $\overline{P} = \overline{P'}$, we conclude that any two minimal upper samples of A are roughly equal.

Now, it can easily be shown that $\overline{\underline{A} \cup \underline{B} \cup P} = \overline{A} \cup \overline{B}$, where P is a minimal upper sample of $\overline{A} \cup \overline{B}$. Further, $\underline{\underline{A} \cup \underline{B} \cup P} = \underline{A} \cup \underline{B}$:

$$[x] \subseteq \underline{\underline{A} \cup \underline{B} \cup P} \text{ implies } [x] \subseteq P,$$

so that, using the preceding observation,

$$\{x\} = [x] \subseteq P.$$

As $P \subseteq \overline{A} \cup \overline{B}$, $x \in \overline{A} \cup \overline{B}$, i.e.,

$$\{x\} = [x] \subseteq \underline{A} \cup \underline{B}.$$

The reverse direction is immediate. Referring to the definition of join $\sqcup$ in $\mathcal{P}_{bc}$, if

$$C \equiv A \sqcup B, \text{ i.e., } C \equiv (A \cup B) \cap (A \cup \underline{B} \cup \underline{A \cup B^c}),$$

we find that $\overline{C} = \overline{A} \cup \overline{B}$, $\underline{C} = \underline{A} \cup \underline{B}$.

(C is, clearly, an upper sample of $\overline{A} \cup \overline{B}$). Thus $C \approx \underline{A} \cup \underline{B} \cup P$, i.e.,

$$[A] \sqcup [B] \equiv [A \sqcup B] = [\underline{A} \cup \underline{B} \cup P] \equiv [A] \cup_{\approx} [B].$$

So the two definitions for join are the same. A similar case can be made for the definitions of meet. However, the complementations considered in the two cases vary, so that the final rough structures are different. For instance, it is seen that the rough structure $\mathcal{P}_{bc}$ is not necessarily a Stone algebra. In other words, a prerough algebra need not be a Stone algebra.

Of course, Bonikowski does not take into account any kind of interior operation, as done in $\mathcal{P}_{bc}$. What kind of structure one would obtain if the negation of $\mathcal{P}_b$ were incorporated in $\mathcal{P}_{bc}$, could be a separate matter of study in itself.

In [37], Wasilewska starts with a topological Boolean algebra $(A, \vee, \wedge, \sim, In, 0, 1)$, partitions A with the help of the topological equality $\approx$ on A, and obtains an algebra on $A/\approx$ (cf. Sect. 2.9). Further, the latter algebra is zero-dimensional, if the topological Boolean algebra on A is also such. As we saw, this is a direct generalization of the construction giving the tqBa $\mathcal{P}_{bc}$, starting from the zero-dimensional topological Boolean algebra $(\mathcal{P}(X), \cap, \cup, ^c, L, \emptyset, X)$. Now one may wonder if there is any relation between this tqBa and the tqBa $TQ(\mathcal{D})$ formed by the Boolean algebra $\mathcal{D}$ of definable subsets in $< X, R >$. (We may recall that $TQ(\mathcal{D})$ is the approximation space algebra $(\mathcal{R}, \sqcup, \sqcap, \neg, L, 0, 1)$, cf. Sect. 2.8). $\mathcal{D}$ is, of course, a subalgebra of the Boolean algebra $(\mathcal{P}(X),\cap,\cup,^c,L,\emptyset,X)$, and is $L[\mathcal{P}(X)]$. So $TQ(\mathcal{D})$ is just $TQ\{L[\mathcal{P}(X)]\}$. It turns out that there is a relation between $\mathcal{P}_{bc}$ and $TQ\{L[\mathcal{P}(X)]\}$ that can be generalized to the scenario in [37].

Proposition 16. *Let* $(A,\vee,\wedge,\sim,In,0,1)$ *be a zero-dimensional topological Boolean algebra. Then* $A/\approx$ *is isomorphic to a subalgebra of* $TQ[In(A)]$*, the topological quasi-Boolean extension of the Boolean algebra* $In(A)$.

Proof. $In(A)$ is a Boolean subalgebra of $(A,\vee,\wedge,\sim,In,0,1)$. Let us denote its topological quasi-Boolean extension $TQ(In(A))$ as A_0, i.e., A_0 consists of pairs

$$<In(a),\ In(b)>,\ In(a)\leq In(b).$$

Now we have a unary operator L on A_0, viz.,

$$L[<In(a),\ In(b)>]\ \equiv\ <In(a),In(a)>,\ a,b\in A.$$

So $In(A)$ clearly can be identified with $L(A_0)$. Let us take the subalgebra S of A_0 comprising pairs

$$<In(a),C(a)>=<In(a),In[C(a)]>,\ a\in A.\quad L(A_0)\subseteq S.$$

Then, it is easily seen that the algebra on $A/\approx$ is isomorphic to S by the correspondence assigning an equivalence class $[a]$ to the pair $<In(a),C(a)>, a\in A$. □

It is worth remarking upon the behavior of the operators In and L. Going back to the Boolean algebra on A, we can form its topological quasi-Boolean extension $TQ(A)$ and have a unary operator L on it as before, i.e., $L(<a,b>\equiv<a,a>,\ a\in A$. As evident, A is identifiable with $L[TQ(A)]$. $L(A_0)\subseteq L[TQ(A)]$. Then, on the one hand, L and In coincide across the isomorphic subsets $L(A_0)$ and $In(A)$ of $L[TQ(A)]$ and A respectively. On the other hand, L collapses into being the identity operator on $L[TQ(A)]$, while In is (generally) a nontrivial operator on A.

4 Some Other Directions

4.1 Relation Algebras

The set of all binary relations on a nonempty set X forms an algebra called the full algebra of relations. This is denoted by $Rel(X)$ and defined as

$$Rel(X)\ \equiv\ (\mathcal{P}(X\times X),\cap,\cup,\setminus,\emptyset,X\times X,\circ,^{-1},Id),$$

where $\cap,\cup,\setminus$ are set-theoretical operations, $\circ$ is the composition of relations, $^{-1}$ is the inverse operation, and Id the identity relation. Any subalgebra of $Rel(X)$ is called a binary relation algebra (BRA) on X.

Tarski generalized the idea to define a relation algebra. It is a structure

$$(A,+,\cdot,-,0,1,\circ,^{-1},in),$$

of type (2,2,1,0,0,2,1,0), satisfying the conditions

1. $(A,+,\cdot,-,0,1)$ is a Boolean algebra.
2. $(A,\circ,^{-1},in)$ is an involuted monoid.
3. $(a\circ b)\cdot c=0,\ (a^{-1}\circ c)\cdot b=0.\ (c\circ b^{-1})\cdot a=0$, for $a,b,c\in A$.

Now, instead of taking relations on X, one can take rough relations on an approximation space $<X,R>$. This is defined in the following way: $<X\times X,\approx>$ is an approximation space, where $(x,y)\approx(u,v)$ if and only if $xRu,\ yRv$. A rough relation on $<X,R>$ is a rough set in the approximation space $<X\times X,\approx>$, i.e., a pair $<\underline{R'},\overline{R'}>$, where $R'\subseteq X\times X$ and $\underline{R'}$, $\overline{R'}$ are the lower and upper approximations of R' in $<X\times X,\approx>$.

The set $Rough-Rel<X,R>$ of all rough relations can be given the structures of rough sets as discussed in Sect. 2. In particular, one can render the structure of a double Stone algebra (cf. Sect. 2.4). Some additional operations may now be defined by

$$\begin{aligned} <\underline{R}_1,\overline{R}_1>\circ<\underline{R}_2,\overline{R}_2> &\equiv <\underline{R}_1\circ\underline{R}_2,\overline{R}_1\circ\overline{R}_2>,\\ <\underline{R}_1,\overline{R}_1>^{-1} &\equiv <\underline{R}_1^{-1},\overline{R}_1^{-1}>,\\ in &\equiv <R,R>. \end{aligned}$$

The structure $(Rough-Rel<X,R>,\cap,\cup,\cdot,+,<\emptyset,\emptyset>,X\times X,\circ,^{-1},in)$ is called the full algebra of rough relations over $<X,R>$. Any subalgebra of this is called a rough relation algebra.

Following Tarski, a generalized notion of rough relation algebra may be defined. One such definition has been proposed by Comer and Düntsch (cf. [14]). Various properties of these algebras have been investigated, particularly those about their representability.

4.2 Information Algebras

The basic idea of information algebras is as follows: Let $<X,AT,\{v_a:a\in AT\}>$ be an information system. For any subset A of AT, the strong indiscernibility relation R_A defined by

$$xR_Ay \text{ if and only if } a(x)=a(y), \text{ for all } a\in A,$$

is an equivalence relation. Thus R_A gives rise to the approximation space $<X,R_A>$ that we have dealt with in the previous sections.

Now, a host of other binary relations may be defined on X in various ways, not all of which are equivalences. For example, a relation S_A called strong similarity may be defined thus:

$$xS_Ay \text{ if and only if } a(x)\cap a(y)\neq\emptyset, \text{ for all } a\in A.$$

This is not a transitive relation. All of these defined relations are categorized into two groups, those of indistinguishability and distinguishability relations.

For each such relation R_A, $A \subseteq AT$, four information operators may be defined on the power set $\mathcal{P}(X)$ of X:

1. $[R_A](S) \equiv \{x \in X : \text{for all y, (x,y)} \in \mathrm{R_A} \text{ implies y} \in \mathrm{S}\}$.
2. $< R_A > (S) \equiv \{x \in X : \text{there is y, such that (x,y)} \in \mathrm{R_A}\}$.
3. $[[R_A]](S) \equiv \{x \in X : \text{for all y,y} \in \mathrm{S} \text{ implies (x,y)} \in \mathrm{R_A}\}$.
4. $<< R_A >> (S) \equiv \{x \in X : \text{there is y} \notin \mathrm{S}, \text{ such that (x,y)} \notin \mathrm{R_A}\}$.

If R_A is the strong indiscernibility relation relative to the subset A of attributes, $[R_A](S) = \underline{S}$ and $< R_A > (S) = \overline{S}$, the lower and upper approximations of S, respectively.

Now, in the Boolean algebra of the subsets of X, some groups of information operators are added and what is thus obtained is an information algebra. For example, the following structure is an information algebra:

$$\{\mathcal{P}(X), \cap, \cup, ^c, \emptyset, X, \{< R_A >: A \subseteq AT\}\}.$$

Similarly, other such examples are found.

These typical information algebras are then generalized over any Boolean algebra. Let $(B, \wedge, \vee, \sim, 0, 1)$ be a complete atomic Boolean algebra, A a nonempty finite set, and f_P a unary operator on B for each $P \subseteq A$. Then the structure

$$(B, \cap, \cup, ^c, 0, 1, \{f_P : P \subseteq A\}),$$

is an information algebra. Studies on these algebras have opened up a new direction. For a brief introduction to this area, refer to [26].

4.3 Rough Substructures

In another approach, the so-called rough substructures of algebraic structures are being investigated. For instance, Biswas and Nanda [6] introduced the notion of rough subgroups of a group. Kuroki [20] has studied rough ideals in semigroups. The basic idea is to start with an algebraic structure having a congruence relation defined in it. The questions asked now are of the following type: which properties of a subset A of the algebra pass over to $\underline{A}$ and/or $\overline{A}$, and under which conditions? Some examples of results obtained are, if A is a subsemigroup of a semigroup, then so is $\overline{A}$, whereas $\underline{A}$ is also so, provided it is nonempty, and the congruence relation is complete. Thus, additional properties of the equivalence relation, e.g., completeness, find roles in this context. Multiple congruences are also considered.

4.4 Structures More Generalized

The approximation space as defined by Pawlak has been generalized by many and in different ways (cf. Sect. 8, [19]). Some have studied algebraic structures based on the resulting spaces; others have focused on applications of such spaces.

One idea is to replace the equivalence relation R of the approximation space $< X,R >$ by weaker relations. For instance, in [39], the author starts with a space $< X,R >$, where R does not have any conditions. In this case, the lower and upper approximations of a subset A of X are defined by

$$\begin{aligned} \underline{A} &\equiv \{x \in X : \text{ for all y, xRy implies y} \in \text{A}\}, \\ \overline{A} &\equiv \{x \in X : \text{ there exists y, such that xRy and y} \in \text{A}\}. \end{aligned}$$

With these definitions, by gradual addition of conditions on R, one arrives at various types of approximation spaces and corresponding algebras. This course of development has obvious connections with modal logics.

Considering tolerance (reflexive and symmetrical) relations in particular, leads to the notion of a generalized approximation space in [30]. Because tolerance classes form a cover of the domain X, an approximation space is considered here as any pair $< X,E >$, where E is a cover of X. Conjugate (dual) lower/upper approximation-like operations are then defined, and together with them, $\mathcal{P}(X)$ forms what is called an approximation algebra.

Axiomatizations of lower/upper approximation-like operators have also been attempted [40]. Two unary operators L and M on $\mathcal{P}(X)$ are taken such that

$$L(X) = X, M(\emptyset) = \emptyset$$

and L, M distribute over intersection, union, respectively. Then $(\mathcal{P}(X), {}^c, L, M, \cap, \cup)$ is called a rough set algebra.

In another direction, the domain X itself is given a structure [9], that of a partially ordered set (poset). An abstract approximation space comprises X and subposets of inner and outer definable elements of X satisfying certain properties. Lower and upper approximations are then specific maps between the domain and these subposets. Depending on how rich the poset is, one gets different approximation spaces. Carrying out this exercise on $\mathcal{P}(X)$, with the subposets as families of open and closed subsets, one gets a topological version of approximation spaces. The standard (Pawlak) approximation space is then the topological space of clopen sets.

In [21], applications of a definition of rough sets that is different from all those given so far, are explored. The notion of an information system (defined in Sect. 1) is modified to that of a decision system $< X, AT \cup \{d\} >$. Here, each member a of AT is regarded as a partial map from X to a value set V_a, and d, the decision attribute, is a partial map from X to $\{0,1\}$. It is possible that for some $x \in X$, all attribute values (including the value of d) are undefined. A rough set A is taken to be a pair $< A^+, A^- >$, where A^+ is the set of elements of X that may belong to A, whereas A^- contains those elements of X that may not belong to A. The Boolean attribute d is used to indicate the information about the membership of an object of X in A. Formally, let

$$AT \equiv \{a_1, \ldots, a_n\},\ AT(x) \equiv < a_1(x), \ldots, a_n(x) > \text{for each } x \in X,$$

and

$$AT^{-1}(t) \equiv \{x \in X : AT(x) = t\}, \text{for } t \in V_{a_1} \times \ldots \times V_{a_n}.$$

(Note that for some $x \in X$, $AT(x)$ could be undefined). Then,

$$A^+ \equiv \{x \in X : AT \text{ is defined for } x, \text{ and } d(x') = 1, \text{ for some } x' \in AT^{-1}[AT(x)]\},$$

and

$$A^- \equiv \{x \in X : AT \text{ is defined for } x, \text{ and } d(x') = 0, \text{ for some } x' \in AT^{-1}[AT(x)]\}.$$

This definition implies that A^+ and A^- may not be disjoint, allowing for the presence of conflicting (contradictory) decisions in the so-called decision table [29]. On the other hand, A^+ and A^- may not cover X either, allowing for the possibility that there is no available information about membership in A. Consideration of the latter case marks the significance of this particular approach to rough sets.

In this context, rough relations are considered. Standard relational database techniques, such as relational algebraic operations (e.g., union, complement, Cartesian product, projection) on crisp relations, are extended to rough relations. A declarative language for defining and querying these relations is introduced — pointing to a link of rough sets (as defined above) with logic programming.

The same scheme may be applied to the various definitions of rough sets (representing information granules) and the respective calculi presented earlier in this chapter. It may be interesting to investigate whether any differences regarding decisions emerge from this scheme due to the different ways of forming information granules.

5 Conclusions

A point that is clear and that has been made by most of the authors is that the collection $\mathcal{D}$ of all definable sets in an approximation space $< X, R >$, forms a complete atomic Boolean algebra. So we find that all rough algebras considered have a Boolean core constituted by $\mathcal{D}$ or an isomorphic copy of $\mathcal{D}$. This leads to an important observation about the algebraic structures discussed (e.g., quasi-Boolean, double Stone, three-valued Łukasiewicz or semisimple Nelson): these must have a Boolean core as well. The core is explicit in double Stone and quasi-Boolean algebras.

So, a general question may be raised whether these structures can be generated from this core in some manner — for instance, the way $TQ(\mathcal{B})$ is generated (cf. Sect. 2.8). (An example in this context would be the construction Vakarelov [35] works with to generate a semisimple Nelson algebra, though he starts with pseudo-Boolean algebra). Another question may be about the nature of this core, e.g., whether it would be atomic or complete. Further, one may inquire about the type of algebras constructed from the Boolean core when some of the properties (such as atomicity or completeness) are dropped.

It may be remarked that, though inadequate to capture the full import of rough set structures, topological quasi-Boolean algebra seems to be an interesting mathematical object. It has eluded description by a Hilbert-style formal system (logic), as it does not have an appropriate implication [4]. However, a Gentzen-style sequent calculus of which these algebras are models, has recently been proposed [33]. A proper search for a representation theorem for these algebras is also pending. At any rate, the world of algebras created out of Pawlak's rough sets has turned out to be extremely fascinating and is worth continued exploration.

Acknowledgments

The research of Mohua Banerjee has been supported by Project No. BS/YSP/29/2477 of the Indian National Science Academy.

References

1. M. Banerjee. Rough sets and three-valued Łukasiewicz logic. *Fundamenta Informaticae*, 32: 213–220, 1997.
2. M. Banerjee, M.K. Chakraborty. A category for rough sets. In R. Słowiński, J. Stefanowski, editors, *Proceedings of the 1st International Workshop on Rough Sets: State of the Art and Perspectives*, a volume of *Foundations of Computing and Decision Sciences* (special issue), 18(3/4): 167–180, 1993.
3. M. Banerjee, M.K. Chakraborty. Rough algebra. *Bulletin of the Polish Academy of Sciences. Mathematics*, 41(4): 293–297, 1993.
4. M. Banerjee, M.K. Chakraborty. Rough sets through algebraic logic. *Fundamenta Informaticae*, 28(3/4): 211–221, 1996.
5. G. Birkhoff. *Lattice Theory*. AMS Colloquium 25, Providence RI, 1967.
6. R. Biswas, S. Nanda. Rough groups and rough subgroups. *Bulletin of the Polish Academy of Sciences. Mathematics*, 42: 251–254, 1994.
7. V. Boicescu, A. Filipoiu, G. Georgescu, S. Rudeanu. *Łukasiewicz-Moisil Algebras*. North–Holland, Amsterdam, 1991.
8. Z. Bonikowski. A certain conception of the calculus of rough sets. *Notre Dame Journal of Formal Logic*, 33: 412–421, 1992.
9. G. Cattaneo. Abstract approximation spaces for rough theories. In L. Polkowski, A. Skowron, editors, *Rough Sets in Knowledge Discovery 1. Methodology and Applications*, 59–98, Physica, Heidelberg, 1998.
10. M. Chuchro. On rough sets in topological Boolean algebras. In W.P. Ziarko, editor, *Rough Sets, Fuzzy Sets and Knowledge Discovery. Proceedings of the International Workshop on Rough Sets and Knowledge Discovery (RSKD'94)*, 157–160, Springer, London, 1994.
11. R. Cignoli. Representation of Łukasiewicz and Post algebras by continuous functions. *Colloquium Mathematicum*, 24(2): 127–138, 1972.
12. S. Comer. An algebraic approach to the approximation of information. *Fundamenta Informaticae*, 14: 492–502, 1991.

13. S. Comer. Perfect extensions of regular double Stone algebras. *Algebra Universalis*, 34: 96–109, 1995.
14. I. Düntsch. Rough sets and algebras of relations. In E. Orłowska, editor, *Incomplete Information: Rough Set Analysis*, 95–108, Physica, Heidelberg, 1998.
15. G. Epstein, A. Horn. Chain based lattices. *Journal of Mathematics*, 55(1): 65–84, 1974. Reprinted in: D.C. Rine, editor, *Computer Science and Multiple-Valued Logic: Theory and Applications*, 58–76, North–Holland, Amsterdam, 1991.
16. D. Higgs. A category approach to Boolean-valued set theory. Preprint. University of Waterloo, 1973.
17. L. Iturrioz. Rough sets and three-valued structures. In E. Orłowska, editor, *Logic at Work: Essays Dedicated to the Memory of Helena Rasiowa*, 596–603, Physica, Heidelberg, 1999.
18. T.B. Iwiński. Algebraic approach to rough sets. *Bulletin of the Polish Academy of Sciences. Mathematics*, 35(9/10): 673–683, 1987.
19. J. Komorowski, Z. Pawlak, L. Polkowski, A. Skowron. Rough sets: A tutorial. In S.K. Pal, A. Skowron, editors, *Rough Fuzzy Hybridization: A New Trend in Decision-Making*, 3–98, Springer, Singapore, 1999.
20. N. Kuroki. Rough ideals in semigroups. *Information Sciences*, 100: 139–163, 1997.
21. J. Małuszyński, A. Vittoria. Towards rough datalog: Embedding rough sets in Prolog. (this book).
22. Gr. C. Moisil. Sur les idéaux des algébres łukasiewicziennes trivalentes. *Annales Universitatis C.I. Parhon, Acta Logica*, 3: 83–95, 1960.
23. A. Monteiro. Construction des algebres de Łukasiewicz trivalentes dans les algebres de Boole monadiques. *Mathematica Japonicae*, 12: 1–23, 1967.
24. D. Mundici. The C^*-algebras of three-valued logic. In R. Ferro, C. Bonotto, S. Valentini, A. Zanardo, editors, *Logic Colloquium'88*, 61–77, North–Holland, Amsterdam, 1989.
25. A. Obtułowicz. Rough sets and Heyting algebra valued sets. *Bulletin of the Polish Acaemy of Sciences. Mathematics*, 35(9/10): 667–671, 1987.
26. E. Orłowska. Introduction: what you always wanted to know about rough sets. In E. Orłowska, editor, *Incomplete Information: Rough Set Analysis*, 1–20, Physica, Heidelberg, 1998.
27. P. Pagliani. Rough set theory and logic-algebraic structures. In E. Orłowska, editor, *Incomplete Information: Rough Set Analysis*, 109–190, Physica, Heidelberg, 1998.
28. Z. Pawlak. Rough sets. *International Journal of Computer Information Sciences*, 11(5): 341–356, 1982.
29. Z. Pawlak. *Rough Sets: Theoretical Aspects of Reasoning about Data*. Kluwer, Dordrecht, 1991.
30. J.A. Pomykała. Approximation, similarity and rough construction. Report number CT-93-07 of *ILLC Prepublication Series*, University of Amsterdam, 1993.
31. J. Pomykała, J.A. Pomykała. The Stone algebra of rough sets. *Bulletin of the Polish Academy of Sciences. Mathematics*, 36: 495–508, 1988.
32. H. Rasiowa. *An Algebraic Approach to Non-Classical Logics*. North–Holland, Amsterdam, 1974.
33. J. Sen. *Some Embeddings in Linear Logic and Related Issues*. Ph.D. Dissertation, University of Calcutta, 2001.
34. R. Sikorski. *Boolean Algebras*. Springer, New York, 1969.
35. D. Vakarelov. Notes on $\mathcal{N}$-lattices and constructive logic with strong negation. *Studia Logica*, 36: 109–125, 1977.

36. A. Wasilewska, L. Vigneron. Rough equality algebras. In P.P. Wang, editor, *Proceedings of the International Workshop on Rough Sets and Soft Computing at 2nd Annual Joint Conference on Information Sciences (JCIS'95)*, 26–30, Raleigh, NC, 1995.
37. A. Wasilewska, L. Vigneron. On generalized rough sets. In *Proceedings of the 5th Workshop on Rough Sets and Soft Computing (RSSC'97) at the 3rd Joint Conference on Information Sciences (CJCIS'97)*, Durham, NC, 1997.
38. A. Wiweger. On topological rough sets. *Bulletin of the Polish Acaemy of Sciences. Mathematics*, 37: 51–62, 1988.
39. Y.Y. Yao. Constructive and algebraic methods of the theory of rough sets. *Information Sciences*, 109: 21–47, 1998.
40. Y.Y. Yao. Generalized rough set models. In L. Polkowski, A. Skowron, editors, *Rough Sets in Knowledge Discovery 1. Methodology and Applications*, 286–318, Physica, Heidelberg, 1998.

Part II

Hybrid Approaches

Introduction to Part II

Hybrid approaches to rough-neural networks (RNN) are presented in Part II. The chapters in Part II show how the rough set theoretical approach, in particular, the rough granulation paradigm, may be fused with other methodologies based on fuzzy sets, nonmonotonic reasoning, logic programming, database, neural networks, and self-adaptive maps.

The first chapter, *Approximation Transducers and Trees: A Technique for Combining Rough and Crisp Knowledge*, by Doherty, Łukaszewicz, Skowron, and Szałas introduces and studies approximation tranducers, devices that convert input approximate relations into output approximate ones by means of first-order theories. Different rough set techniques are applied here to produce approximations to relations. In defining inference mechanisms of approximate transducers, methods of relational databases are invoked and modified, which results in the notion of a rough relational database.

The motif of a rough relational database is pursued by Doherty, Kachniarz, and Szałas in their chapter, *Using Contextually Closed Queries for Local Closed-World Reasoning in Rough Knowledge Databases*. Here, the authors propose an architecture for an open-world querying system. To this end, the notion of a deductive database in a classical sense is recalled and generalized to that of a rough knowledge database in which relations are subject to rough set-like processing. Contextually closed queries are studied as an extension relative to a local closed-world (LCW) assumption.

Gediga and Düntsch discuss in *On Model Evaluation, Indexes of Importance, and Interaction Values in Rough Set Analysis* the problems of evaluating the quality of rough set data analysis. In terms of usefulness and significance of an approximation, they study some approximation quality indexes, based particularly on Choquet-type aggregations and the notion of entropy.

Rough sets are often combined with fuzzy sets in the literature. The chapter by Inuiguchi and Tanino, *New Fuzzy Rough Sets Based on Certainty Qualification*, offers a survey of existing rough-fuzzy approaches presenting a new one that is based on certainty qualifications. They study in detail the properties of this new construct showing that it offers a better approximation to fuzzy sets than the previous constructs.

Małuszyński and Vitória extend in *Toward Rough Datalog: Embedding Rough Sets in Prolog* the formalism of logic programming to cases where relations are known only roughly. They briefly recall the paradigm of logic programming and then embark on the construction of a rough datalog, a language for defining and querying rough relations. Syntax, semantics, and implementation of a rough datalog in Prolog are thoroughly examined.

The chapter *On Exploring of Soft Discretization of Continuous Attributes* by Nguyen addresses decision-tree construction problems in large data tables. He proposes the technique of soft discretization based on the notion of a soft cut, whose quality may be assessed by discernibility-based rough set methods. Discretization by soft cuts, searching for best cuts, and constructions of decision trees from data tables are examined in detail with references to practical problems.

The subject of rough-neural computing is the main concern of the last chapter by Pal, Dasgupta, and Mitra in Part II. In the chapter *Rough-SOM with Fuzzy Discretization*, self-organizing maps endowed with rough processing are studied. The traditional Kohonen network is augmented with fuzzy discretization of the input data set, making it into an information system, subject then to rough set reduct generation via a discernibility function. On the basis of implicants of this function, weights are assigned to a self-organizing map. A number of experiments bear out the superiority of this new approach.

Chapter 8
Approximation Transducers and Trees: A Technique for Combining Rough and Crisp Knowledge

Patrick Doherty,[1] Witold Łukaszewicz,[2] Andrzej Skowron,[3] Andrzej Szałas[2]

[1] Department of Computer and Information Science, Linköping University, 58183 Linköping, Sweden
`patdo@ida.liu.se`

[2] The College of Economics and Computer Science, Wyzwolenia 30, 10-106 Olsztyn, Poland
`witlu@ida.liu.se; andsz@ida.liu.se`

[3] Institute of Mathematics, Warsaw University, Banacha 2, 02-097 Warsaw, Poland
`skowron@mimuw.edu.pl`

Summary. This chapter proposes a framework for specifying, constructing, and managing a particular class of approximate knowledge structures for use with intelligent artifacts ranging from simpler devices such as personal digital assistants (PDAs) to more complex ones such as unmanned aerial vehicles (UAVs). This chapter introduces the notion of an *approximation transducer*, which takes approximate relations as input and generates a (possibly more abstract) approximate relation as output by combining the approximate input relations with a crisp local logical theory representing dependencies between input and output relations. Approximation transducers can be combined to produce *approximation trees*, which represent complex approximate knowledge structures characterized by the properties of elaboration tolerance, groundedness in the application domain, modularity, and context dependency. Approximation trees are grounded through the use of primitive concepts generated with supervised learning techniques. Changes in definitions of primitive concepts or in the local logical theories used by transducers result in changes in the knowledge stored in approximation trees by increasing or decreasing precision in the knowledge qualitatively. Intuitions and techniques from rough set theory are used to define approximate relations where each has an upper and a lower approximation. The constituent components in a rough set have correspondences in a logical language used to relate crisp and approximate knowledge. The inference mechanism associated with the use of approximation trees is based on a generalization of deductive databases that we call *rough relational databases*. Approximation trees and queries to them are characterized in terms of rough relational databases and queries to them. By placing certain syntactic restrictions on the local theories used in transducers, the computational processes used in the query/answering and generation mechanism for approximation trees remain in PTIME.

1 Introduction and Background

In this introductory section, we will set the context for the knowledge representation framework pursued in this chapter. We begin with a discussion of intelligent artifacts and a society of agent frameworks [7]. We proceed to a discussion of knowledge representation components for agents and consider the need for self-adapting knowledge representation structures and concept acquisition techniques. The core idea pursued in this chapter is to propose a framework for specifying, constructing, and managing of approximate knowledge structures for intelligent artifacts. The specific structures used are called approximation transducers and approximation trees. The specific implementation framework used is based on a generalization of deductive database technology. We describe the intuitions and basic ideas behind these concepts. We then conclude the introductory section with a brief description of the experimental platform from which these ideas arose and from which we plan to continue additional experimentation with the framework. The experimental platform is a deliberative/reactive software architecture for an unmanned aerial vehicle under development in the WITAS[1] Unmanned Aerial Vehicle Project at Linköping University, Sweden [1].

1.1 Intelligent Artifacts and Agents

The use of intelligent artifacts, both at the workplace and in the home, is becoming increasingly pervasive due to a number of factors, which include the accessibility of the Internet and the World Wide Web to the general public, the drop in price and increase in capacity of computer processors and memory, and the integration of computer technology with telecommunications. Intelligent artifacts are man-made physical systems containing computational equipment and software that provide them with capabilities for receiving and comprehending sensory data for reasoning and for performing rational action in their environments. The spectrum of capabilities and the sophistication of an artifact's ability to interface with its environment and to reason about it varies with the type of artifact, its intended tasks, the complexity of the environment in which it is embedded, and its ability to adapt its models of the environment at different levels of knowledge abstraction. Representative examples of intelligent artifacts ranging from less to more complex would be mobile telephones, mobile telephones with BLUETOOTH wireless technology,[2] personal digital assistants (PDAs), collections of distributed communicating artifacts that serve as components of smart homes, mobile robots, unmanned aerial vehicles, and many more.

One unifying conceptual framework for these increasingly complex integrated computer systems is to view them as societies of agents (virtually and/or physically embedded in their respective environments). These agents have the capacity to acquire information about their environments, structure the information and interpret it as knowledge, and use this knowledge rationally to enhance goal-directed behavior. Such behavior is used to achieve tasks and to function robustly in their dynamic and complex environments.

[1] WITAS (pronounced vee-tas) is an acronym for the Wallenberg Information Technology and Autonomous Systems Laboratory at Linköping University.

[2] BLUETOOTH is a trademark owned by Telefonaktiebolaget L M Ericsson, Sweden.

1.2 Knowledge Representation

An essential component in agent architectures is the agent's knowledge-representation component which includes a variety of knowledge and data repositories with associated inference mechanisms. The knowledge representation component is used by the agent to provide it with models of its embedding environment and of its own and other agent capabilities in addition to reasoning efficiently about them. It is becoming increasingly important to move away from the notion of a single knowledge representation mechanism with one knowledge source and inference method to multiple forms of knowledge representation with several inference methods. This viewpoint introduces an interesting set of complex research issues related to merging knowledge from disparate sources and using adjudication or conflict resolution policies to provide coherence of knowledge sources.

Due to the embedded nature of these agent societies in complex dynamic environments, it is also becoming increasingly important to take seriously the gap between access to low-level sensory data and its fusion and integration with more qualitative knowledge structures. These signal-to-symbol transformations should be viewed as an ongoing process with a great deal of feedback between the levels of processing. In addition, because the embedding environments are often as complex and dynamic as those faced by humans, the knowledge representations that are used as models of the environment must necessarily be partial, elaboration tolerant, and approximate.

Self-Adaptive Knowledge Structures A long-term goal of this research is developing a framework for specifying, implementing, and managing of self-adaptive knowledge structures containing both quantitative and qualitative components, where the knowledge structures are grounded in the embedding environments in which they are used. There has been very little work in traditional knowledge representation with the dynamics and management of knowledge structures. Some related work would include the development of belief revision and truth maintenance systems, in addition to the notion of *elaboration tolerant* knowledge representation and the use of *contexts* as first-class objects introduced by McCarthy [6]. In these cases, the view pertaining to properties and relations is still quite traditional with little emphasis on the approximate and contextual character of knowledge. The assumed granularity of the primitive components of these knowledge structures, in which these theories and techniques are grounded, is still that of classical properties and relations in a formal logical context. We will assume a finer granularity as a basis for concept acquisition, grounding, and knowledge structure design, which is the result of using intuitions from rough set theory.

Approximate Concept Acquisition and Management One important component related to the ontology used by agents is the acquisition-, integration-, and elaboration-tolerant update of concepts. Here, we will interpret the notion of *concept* in a broad sense. It will include both properties of the world and things in it and relations between them. The concepts and relations can be epistemic, and we will assume that the agent architectures contain syntactic components that correlate with these concepts and relations. The symbol-grounding problem, that of associating symbols with individuals, percepts, and concepts and managing these associations, will be discussed. The knowledge representation technique proposed will provide an interesting form of grounding, which we hope can contribute to an eventual solution of this important, complex, and yet unsolved problem. The symbol-grounding problem includes not only the association of symbols with concepts, but also ongoing concept acquisition and modification during the life of an agent. This aspect of the problem will involve the use of concept learning techniques and their direct integration in the knowledge structures generated.

To do this, we will assume that certain concepts, which we call *primitive concepts*, have been acquired through a learning process where learning samples are provided from sensor data and approximations of concepts are induced from the data. One particularly interesting approach to this is the use of rough set based supervised learning techniques. It is important to emphasize that the induced concepts are approximate and fluid in the sense that additional learning may modify them. In other words, concepts are inherently contextual and subject to elaboration and change in a number of ways. Primitive concepts may change as new sensor data are acquired and fused with existing data through diverse processes associated with particular sensory platforms. At some point, constraints associated with other more abstract concepts dependent on primitive concepts may influence the definition of the primitive concept.

As an example of these ideas, take a situation involving an unmanned aerial vehicle operating over a road and traffic environment. In this case, the meaning of concepts such as *fast* or *slow*, *small* or *large vehicle*, *near*, *far*, or *between*, will have meanings different from those in another application with other temporal and spatial constraints.

Assuming that these primitive concepts are given and that they are continually regrounded in changes in operational environment via additional learning or sensor fusion, we would then like to use these primitive concepts as the *ur*-elements in our knowledge representation structures. Since these *ur*-elements are inherently approximate, contextual, and elaboration tolerant, any knowledge structure containing these concepts should also inherit or be influenced by these characteristics. There are even more primitive *ur*-elements in the system we envision which can be used to define the primitive concepts themselves if a specific concept learning policy based on rough sets is used. These are the elementary sets used in rough set theory to define contextual approximations to sets.

1.3 Approximation Transducers and Trees

In the philosophical literature, W. V. O. Quine [11] has used the phrase *web of belief* to capture the intricate and complex dependencies and structures that make up human beliefs. In this chapter and in a companion chapter in this book [2], we lay the groundwork for what might properly be called *webs of approximate knowledge*. A better way to view this idea is starting with *webs of imprecise knowledge* and gradually incrementing these initial webs with additional approximate and sometimes crisp facts and knowledge. Through this process, a number of concepts, relations and dependencies among them become less imprecise and more approximate. There is a continual elastic process where precision in the meaning of concepts is continually modified in a change-tolerant manner. Approximate definitions of concepts will be the rule rather than the exception, even though crisp definitions of concepts are a special case included in the framework.

Specifically, webs of approximate knowledge will be recursively constructed from primitive concepts together with what we will call *approximation transducers*. An approximation transducer provides an approximate definition of one or more output concepts in terms of a set of input concepts and consists of three components:

1. an input consisting of one or more approximate concepts, some of which might be primitive;
2. an output consisting of one or more new and possibly more abstract concepts defined partly in terms of the input concepts;
3. a local logical theory specifying constraints or dependencies between the input concepts and the output concepts; the theory may also refer to other concepts not expressed in the input.

The local logical theory specifies dependencies or constraints that an expert in the application domain specify. Generally, the form of the constraints would be in terms of some necessary and some sufficient conditions for the output concept. The local theory is viewed as a set of *crisp* logical constraints specified in the language of first-order logic. The local theory serves as a logical template. During the generation of the approximate concept output by the transducer, the crisp relations mentioned in the local theory are substituted by the actual approximate definitions of the input. Either lower or upper approximations of the input concepts may be used in the substitution. The resulting output specifies the output concept in terms of newly generated lower and upper approximations. The resulting output relation may then be used as input to other transducers creating what we call *approximation trees*. The resulting tree represents a web of approximate knowledge capturing intricate and complex dependencies among an agent's conceptual vocabulary.

As an example of a transducer that might be used in the unmanned aerial vehicle (UAV) domain, we can imagine defining a transducer for the approximate concept of two vehicles *connected* in terms of *visible connection*, *small distance*, and *equal speed*. The latter three input concepts could be generated from supervised learning techniques where the data are acquired from a library of videos previously collected by the UAV on earlier traffic monitoring missions. As part of the local logical theory, an example of a constraint might state that "if two vehicles are *visibly connected*, are at a *small distance* from each other, and have *equal speeds*, then they are *connected*."

Observe that the resulting approximation trees are highly fluid, approximate, and elaboration tolerant. Changes in the definition of primitive concepts will trickle through the trees via the dependencies and connections, modifying some of the other concept definitions. Changes in local theories anywhere in the tree will modify those parts of the tree related to the respective output concepts for the local theories. This is a form of elaboration tolerance. These structures are approximate in three respects:

1. The primitive concepts themselves are approximate and generated through learning techniques. Rough learning techniques consist of upper and lower approximations induced from sample data.
2. The output concepts inherit or are influenced by the approximate aspects of the concepts input to their respective transducers.
3. The output concepts also inherit the incompletely specified sufficient and necessary conditions in the local logical theory specified in part by the input concepts.

Note that the transducers represent a technique for combining both approximate and crisp knowledge. The flow of knowledge through a transducer transforms the output concept from a less precise toward more approximate definition. The definition can continually be elaborated upon directly by modifying the local theory and indirectly via the modification of concept definitions on which it is recursively dependent or by retraining the primitive concepts through various learning techniques.

1.4 The WITAS UAV Experimental Platform

The ultimate goal of the research described in this chapter is using it as a basis for specifying, constructing, and managing a particular class of approximate knowledge structures in intelligent artifacts. In current research, the particular artifact we use as an experimental platform is an unmanned aerial vehicle flying over operational environments populated by traffic. In such a scenario, knowledge about both the environment below and the unmanned aerial vehicle agent's own epistemic state must be acquired in a timely manner for the knowledge to be of use to the agent while achieving its goals. Consequently, the result must provide for efficient implementation of both the knowledge structures themselves and the inference mechanisms used to query these structures for information.

The WITAS UAV Project is a long-term project whose goal is designing, specifying, and implementing the IT subsystem for an intelligent autonomous aircraft and embedding it in an actual platform [1]. We are using a Yamaha RMAX vertical take-off and landing system (VTOL) developed by the Yamaha Motor Company Ltd. An important part of the project is identifying core functionalities required for the successful development of such systems and doing basic research in the areas identified. The topic of this chapter is one such core functionality: approximate knowledge structures and their associated inference mechanisms.

The project encompasses the design of a command and control system for a UAV and integrating it in a suitable deliberative/reactive architecture; the design of high-level cognitive tasks, intermediate reactive behaviors, low-level control-based behaviors and integrating them with each other; integrating sensory capabilities with the command and control architecture, in particular, using an active vision system; the development of hybrid, mode-based low-level control systems to supervise and schedule control behaviors; signal-to-symbol conversions from sensory data to qualitative structures used in mediating a choice of actions and synthesizing plans to attain operational mission goals; and the development of the systems architecture for the physical UAV platform.

In addition, the project also encompasses the design and development of the necessary tools and research infrastructure required to achieve the goals of the project. This includes developing model-based distributed simulation tools and languages used in concurrent engineering to move incrementally from software emulation and simulation to the actual hardware components used in the final product.

The intended operational environment consists of widely varying geographical terrain with traffic networks and vehicle interaction of varying degrees of density. Possible applications are emergency service assistance, monitoring and surveillance, use of a UAV as a mobile sensory platform in an integrated real-time traffic control system, and photogrammetry application.

The UAV experimental platform offers an ideal environment for experimentation with the knowledge representation framework we propose because the system architecture is rich in different types of knowledge representation structures, the operational environment is quite complex and dynamic, and signal-to-symbol transformations of data are an integral part of the architecture. In addition, much of the knowledge acquired by the UAV will be necessarily approximate. In several of the sections in this chapter, we will use examples from this application domain to describe and motivate some of our techniques.

1.5 Generalized Deductive Databases

Due to the pragmatic constraints associated with deploying and executing systems such as the WITAS UAV platform, we will develop these ideas in the context of deductive database systems. Deductive database systems offer a reasonable compromise between expressiveness of the language used to model aspects of the embedding environment and the efficiency of the inference mechanisms required in the soft and sometimes hard real-time contexts in which the UAV operates. A deductive database concentric view of data and knowledge flow in a generic artificial intelligence application contains a number of components, as depicted in Fig. 1.

In the UAV architecture, there are a number of databases, or knowledge and data repositories. The dynamic object repository is a soft real-time database used to store preprocessed sensory data from the sensor platform and includes information about

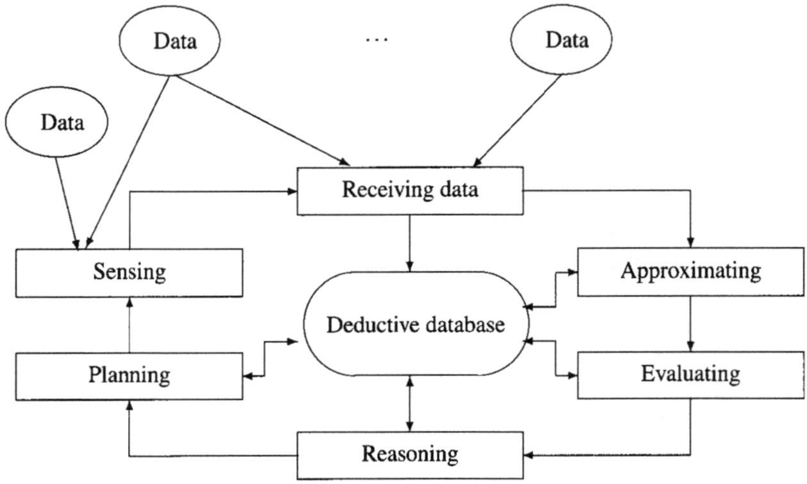

Fig. 1. Deductive database concentric view of data and knowledge flow

moving objects identified in the traffic scenarios observed by the UAV when flying over particular road systems. In addition there is an on-line geographic information repository containing data about the geographic area being flown over. This includes road data, elevation data, data about physical structures, etc.

The approximation trees described in the introduction are stored in part as relational database content and in part as additional specialized data structures. In this case though, the nature of the relational tables, queries to the database, and the associated inference mechanisms all have to be generalized. One reason for this is that all relations are assumed to be approximate, therefore, the upper and lower approximations have to be represented. This particular type of generalized deductive database, called a *rough relational database*. will be considered in Sect. 6. Rough relational databases, and the semantic and computational mechanisms associated with them provide us with an efficient means for implementing query/answering systems for approximation trees. How this is done, will be the main topic of the chapter.

2 Chapter Outline

In the remaining part of the chapter, we will provide details of the knowledge representation framework discussed in the introductory section. In Sect. 3, we begin with a brief introduction of some of the basics of rough set theory, primarily to keep the chapter self-contained for those not familiar with these concepts and techniques. In Sect. 4, we describe a logical language for referring to constituents of rough sets. The language is used as a bridge between more traditional rough set techniques and nomenclature and their use or reference in a logical language. We also define a subset of first-order logic that permits efficient computation of approximation transducers. In Sect. 5, we provide a more detailed description of approximation transducers and include an introductory example. In Sect. 6, we introduce a generalization of deductive databases that permits using and storing approximate or rough relations. Sect 7 contains the core formal results that provide semantics for approximation transducers and justifies the computational mechanisms used. The complexity of the approach is also considered. In Sect 8, we provide a more detailed example of the framework and techniques using traffic congestion as a concept to be modeled. In Section 9, we propose an interesting measure of the approximation quality of theories that can be used to compare approximate theories. In Sect. 10, we summarize the results of the chapter and provide a pointer to additional related work described in this volume.

3 Rough Set Theory

In the introductory section, we described a framework for self-adaptive and grounded knowledge structures in terms of approximation transducers and trees. One basic premise of the approach was the assumption that approximate primitive concepts could be generated by applying learning techniques. One particular approach to inducing approximations of concepts is by using rough set supervised learning techniques in which sample data is stored in tables and approximate concepts are learned. The result is a concept defined in terms of both a lower and an upper approximation. In this section, we will provide a short introduction to a small part of rough set theory and introduce terminology used in the remaining parts of the chapter. We will only briefly mention rough set learning techniques by describing *decision systems* that provide the basic structures for rough set learning techniques. Before providing formal definitions, we will first consider an intuitive example from a UAV traffic scenario application.

Example 1. Consider a UAV equipped with a sensor platform that includes a digital camera. Suppose that the UAV task is to recognize various situations on roads. It is assumed that the camera has a particular resolution. It follows that the precise shape of the road cannot be recognized if essential features of the road shape require a higher resolution than that provided by the camera. Figure 2 depicts a view from the UAV's camera, where a fragment of a road is shown together with three cars, c1, c2, and c3.

Observe that due to the camera resolution there are collections of points that should be interpreted as *indiscernible* from each other. The collections of indiscernible points are called *elementary sets*, using rough set terminology. In Fig. 2, elementary sets are illustrated by dashed squares and correspond to pixels. Any point in a pixel is not discernible from any other point in the pixel from the perspective of the UAV. Elementary sets are then used to approximate objects that cannot be precisely represented by (unions of) elementary sets. For instance, in Fig. 2, it can be observed that for some elementary sets, one part falls within and the other outside

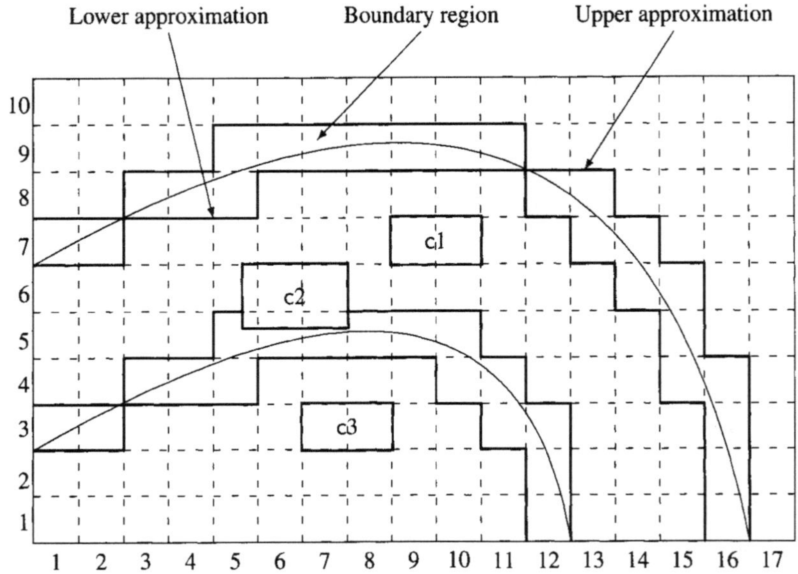

Fig. 2. Sensing the road considered in Example 1

the actual road boundaries (represented by curved lines) simultaneously.

Instead of precise characterization of the road and cars, using rough set techniques, one can obtain approximate characterizations, as depicted in Fig. 3. Observe that the road sequence is characterized only in terms of a lower and an upper approximation of the actual road. A boundary region, containing points that are not known to be inside or outside of the road's boundaries, is characterized by a collection of elementary sets marked with dots inside. Cars c1 and c3 are represented precisely, whereas car c2 is represented by its lower approximation (the thick box denoted by c2) and by its upper approximation (the lower approximation together with the region containing elementary sets marked by hollow dots inside). The region of elementary sets marked by hollow dots inside represents the boundary region of the car.

The lower approximation of a concept represents points that are known to be part of the concept, the boundary region represents points that might or might not be part of the concept, and the complement of the upper approximation represents points that are known not to be part of the concept. Consequently, car c1 is characterized as being completely on the road (inside the road's boundaries); it is unknown whether car c2 is completely on the road, and car c3 is known to be outside, or off the road.

As illustrated in Example 1, the rough set philosophy is founded on the assumption that we associate some information (data, knowledge) with every object of the

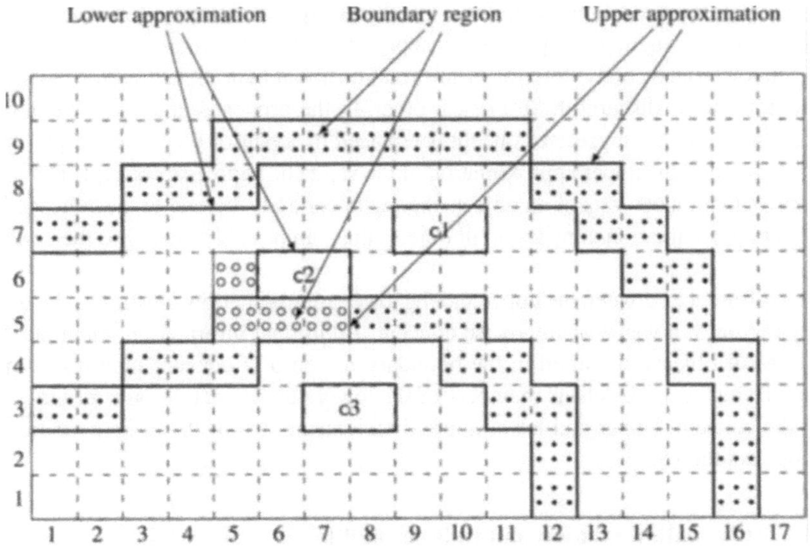

Fig. 3. An approximate view of the road considered in Example 1

universe of discourse. This information is often formulated in terms of attributes about objects. Objects characterized by the same information are interpreted as indiscernible (similar) in view of the available information about them. An indiscernibility relation, generated in this manner from the attribute/value pairs associated with objects, provides the mathematical basis of rough set theory.

Any set of all indiscernible (similar) objects is called an *elementary set* and forms a basic granule (atom) of knowledge about the universe. Any union of some elementary sets in a universe is referred to as a *crisp (precise) set*; otherwise, the set is referred to as a *rough (imprecise, vague) set*. In the latter case, two separate unions of elementary sets can be used to approximate the imprecise set, as we have seen in the example above.

Consequently, each rough set has what are called boundary-line cases, i.e., objects that cannot with certainty be classified either as members of the set or of its complement. Obviously, crisp sets have no boundary-line elements at all. This means that boundary-line cases cannot be properly classified by employing only the available information about objects.

The assumption that objects can be *observed* only through the information available about them leads to the view that knowledge about objects has a granular structure.

Due to this granularity, some objects of interest cannot always be discerned, given the information available; therefore, the objects appear the same (or similar). As a consequence, vague or imprecise concepts, in contrast to precise concepts, cannot be characterized solely in terms of information about their elements since elements are not always discernible from each other. In the proposed approach, we assume that any vague or imprecise concept is replaced by a pair of precise concepts called the lower and the upper approximation of the vague or imprecise concept. The lower approximation consists of all objects that, with certainty, belong to the concept. The upper approximation consists of all objects that can possibly belong to the concept.

The difference between the upper and the lower approximation constitutes the boundary region of a vague or imprecise concept. Additional information about attribute values of objects classified in the boundary region of a concept may result in reclassifying such objects as members of the lower approximation or as not included in the concept. Upper and lower approximations are two of the basic operations in rough set theory.

3.1 Information Systems and Indiscernibility

One of the basic concepts of rough set theory is the indiscernibility relation which is generated using information about particular objects of interest. Information about objects is represented in the form of a set of attributes and their associated values for each object. The indiscernibility relation is intended to express the fact that, due to lack of knowledge, we are unable to discern some objects from others simply by employing the available information about those objects. In general, this means that instead of dealing with each individual object, we often have to consider clusters of indiscernible objects as fundamental concepts of our theories.

Let us now present this intuitive picture about rough set theory more formally.

Definition 1. An *information system* is any pair $\mathcal{A} = \langle U, A \rangle$ where U is a nonempty finite set of *objects* called the *universe* and A is a nonempty finite set of *attributes* such that $a : U \rightarrow V_a$ for every $a \in A$. The set V_a is called the *value set* of a. By $Inf_B(x) = \{\langle a, a(x) \rangle : a \in B\}$, we denote the *information signature of x with respect to B*, where $B \subseteq A$ and $x \in U$.

Note that in this definition, attributes are treated as functions on objects, where $a(x)$ denotes the value the object x has for the attribute a.

Any subset B of A determines a binary relation $IND_{\mathcal{A}}(B) \subseteq U \times U$, called an indiscernibility relation, defined as follows:

Definition 2. Let $\mathcal{A} = \langle U, A \rangle$ be an information system, and let $B \subseteq A$. By the *indiscernibility relation determined by B*, denoted by $IND_{\mathcal{A}}(B)$, we understand the relation

$$IND_{\mathcal{A}}(B) = \{(x, x') \in U \times U : \forall a \in B.[a(x) = a(x')]\}.$$

If $(x,y) \in IND_{\mathcal{A}}(B)$, we say that x and y are *B-indiscernible*. Equivalence classes of the relation $IND_{\mathcal{A}}(B)$ (or blocks of the partition U/B) are referred to as *B-elementary sets*. The unions of B-elementary sets are called B-definable sets.

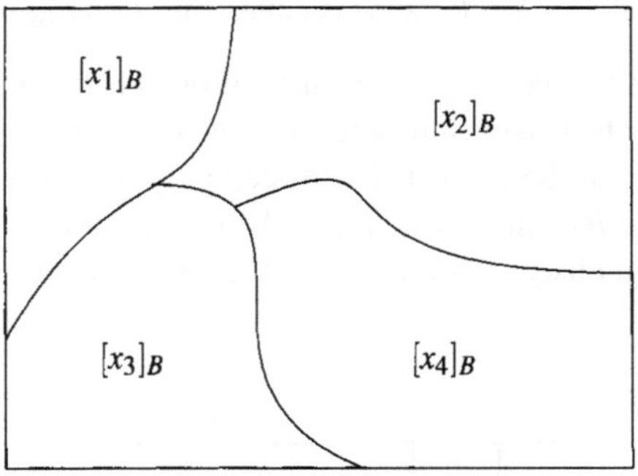

Fig. 4. A rough partition $IND_{\mathcal{A}}(B)$

Observe that $IND_{\mathcal{A}}(B)$ is an equivalence relation. Its classes are denoted by $[x]_B$. By U/B, we denote the partition of U defined by the indiscernibility relation $IND_{\mathcal{A}}(B)$. For example, in Fig. 4, the partition of U defined by an indiscernibility relation $IND_{\mathcal{A}}(B)$ contains four equivalence classes, $[x_1]_B, [x_2]_B, [x_3]_B$ and $[x_4]_B$. An example of a B-definable set is $[x_1]_B \cup [x_4]_B$, where $[x_1]_B$ and $[x_4]_B$ are B-elementary sets.

In Example 1, the indiscernibility relation is defined by a partition corresponding to pixels represented in Figs. 2 and 3 by squares with dashed borders. Each square represents an elementary set. In the rough set approach, the elementary sets are the basic building blocks (concepts) of our knowledge about reality.

The ability to discern between perceived objects is also important for constructing many entities like reducts, decision rules, or decision algorithms which are used in rough set based learning techniques. In the classical rough set approach, the *discernibility relation*, $DIS_{\mathcal{A}}(B)$, is defined as follows.

Definition 3. Let $\mathcal{A} = \langle U, A \rangle$ be an information system and $B \subseteq A$. The *discernibility relation* $DIS_{\mathcal{A}}(B) \subseteq U \times U$ is defined as $(x,y) \in DIS_{\mathcal{A}}(B)$ if and only if $(x,y) \notin IND_{\mathcal{A}}(B)$.

3.2 Approximations and Rough Sets

Let us now define approximations of sets in the context of information systems.

Definition 4. Let $\mathcal{A} = \langle U, A \rangle$ be an information system, $B \subseteq A$ and $X \subseteq U$. The *B-lower approximation* and *B-upper approximation* of X, denoted by $X_{+}B$ and $X^{+}B$ respectively, are defined by $X_{+}B = \{x : [x]_B \subseteq X\}$ and $X^{+}B = \{x : [x]_B \cap X \neq \emptyset\}$.

The B-lower approximation of X is the set of all objects that can be classified with certainty as belonging to X just using the attributes in B to discern distinctions.

Definition 5. The set consisting of objects in the *B-lower approximation* X_{B+} is also called the *B-positive region* of X. The set $X_{B^-} = U - X^+B$ is called the *B-negative region* of X. The set $X_{B\pm} = X^+B - X_+B$ is called the *B-boundary region* of X.

Observe that the positive region of X consists of objects that can be classified with certainty as belonging to X using attributes from B. The negative region of X consists of those objects that can be classified with certainty as not belonging to X using attributes from B. The B-boundary region of X consists of those objects that cannot be classified unambiguously as belonging to X using attributes from B.

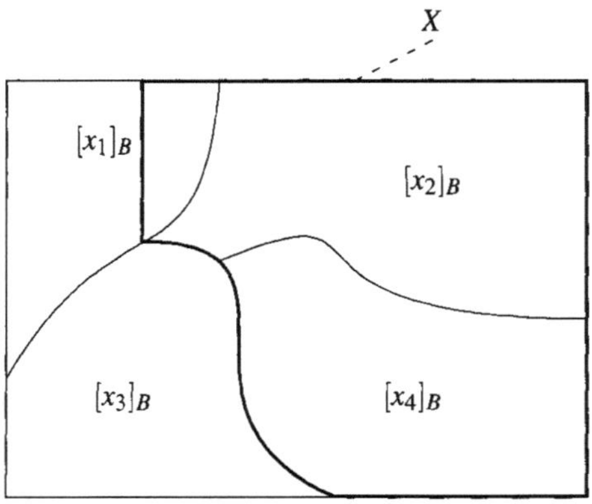

Fig. 5. A rough partition $IND_{\mathcal{A}}(B)$ and an imprecise set X

For example, in Fig. 5, the B-lower approximation of the set X, X_+B, is $[x_2]_B \cup [x_4]_B$. The B-upper approximation, X^+B, is $[x_1]_B \cup [x_2]_B \cup [x_4]_B \equiv [x_1]_B \cup X_+B$. The B-boundary region, $X_{B\pm}$, is $[x_1]_B$. The B-negative region of X, X_{B-}, is $[x_3]_B \equiv U - X^+B$.

3.3 Decision Systems and Supervised Learning

Rough set techniques are often used as a basis for supervised learning using tables of data. In many cases, the target of a classification task, that is, the family of concepts to be approximated, is represented by an additional attribute called a decision attribute. Information systems of this kind are called decision systems.

Definition 6. Let $\langle U, A \rangle$ be an information system. A *decision system* is any system of the form $\mathcal{A} = \langle U, A, d \rangle$, where $d \notin A$ is the *decision attribute* and A is a set of *conditional attributes*, or simply *conditions*.

Let $\mathcal{A} = \langle U, A, d \rangle$ be given, and let $V_d = \{v_1, \ldots, v_{r(d)}\}$. Decision d determines a partition $\{X_1, \ldots, X_{r(d)}\}$ of the universe U, where $X_k = \{x \in U \ : \ d(x) = v_k\}$ for

$1 \leq k \leq r(d)$. The set X_i is called the ith *decision class* of $\mathcal{A}$. By $X_{d(u)}$, we denote the decision class $\{x \in U : d(x) = d(u)\}$, for any $u \in U$.

One can generalize the above definition to decision systems of the form $\mathcal{A} = \langle U, A, D \rangle$ where the set $D = \{d_1, \ldots, d_k\}$ of decision attributes and A are assumed to be disjoint. Formally, this system can be treated as the decision system $\mathcal{A} = \langle U, C, d_D \rangle$ where $d_D(x) = (d_1(x), \ldots, d_k(x))$ for $x \in U$.

A decision table can be identified as a representation of raw data (or training samples in machine learning) that are used to induce concept approximations in a process known as supervised learning. Decision tables themselves are defined in terms of decision systems. Each row in a decision table represents one training sample. Each column in the table represents a particular attribute in A, with the exception of the first column which represents objects in U and selected columns representing the decision attribute(s).

A wide variety of techniques has been developed for inducing approximations of concepts relative to various subsets of attributes in decision systems. The methods are primarily based on viewing tables as a type of Boolean formula, generating reducts for these formulas, which are concise descriptions of tables with redundancies removed, and on generating decision rules from these formula descriptions. The decision rules can be used as classifiers or as representations of lower and upper approximations of the induced concepts. In this chapter, we will not pursue these techniques.

What is important for understanding our framework is that these techniques exist, they are competitive with other learning techniques, and often more efficient. Furthermore, given raw sample data, such as low-level feature data from an image processing system represented as tables, primitive concepts can be induced or learned. These concepts are characterized in terms of upper and lower approximations and represent grounded contextual approximations of concepts and relations from the application domain. This is all we need to assume to construct grounded approximation transducers and to recursively construct approximation trees.

4 A Logical Language for Rough Set Concepts

One final component that bridges the gap between more conventional rough set techniques and logical languages used to specify and compute with approximation transducers is a logical vocabulary for referring to constituent components of a rough set when viewed as a relation or property in a logical language. Note that this particular ontological policy provides the right syntactical characterization of the rough set concepts that we require for our framework. One could also envision a different ontological policy, with a higher level of granularity, for instance, that could be used for other purposes.

To construct a logical language for referring to constituent components of rough concepts, we introduce the following relation symbols for any rough relation R (see Fig. 6):

- R^+ represents the positive facts known about the relation. R^+ corresponds to the lower approximation of R. R^+ is called the *positive region (part)* of R.
- R^- represents the negative facts known about the relation. R^- corresponds to the complement of the upper approximation of R. R^- is called the *negative region (part)* of R.
- $R^\pm$ represents the unknown facts about the relation. $R^\pm$ corresponds to the set difference between the upper and lower approximations of R. $R^\pm$ is called the *boundary region (part)* of R.
- $R^\oplus$ represents the positive facts known about the relation together with the unknown facts. $R^\oplus$ corresponds to the upper approximation of R. $R^\oplus$ is called the *positive-boundary region (part)* of R.
- $R^\ominus$ represents the negative facts known about the relation together with the unknown facts. $R^\ominus$ corresponds to the upper approximation of the complement of R. $R^\ominus$ is called the *negative-boundary region (part)* of R.

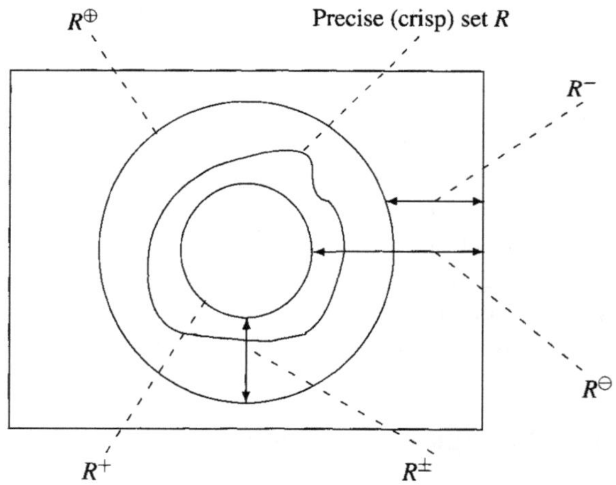

Fig. 6. Representation of a rough set in logic

For simplicity, in the rest of this chapter, we will assume that a theory defines only one intensional rough relation. We shall use the notation $Th(R;R_1,\dots,R_n)$ to indicate that R is approximated by Th, where the $R_1,\dots,R_n$ are the input concepts to a transducer. We also assume that negation occurs only directly before relation symbols.[3]

We write $Th^{+}(R;R_1,\ldots,R_n)$ (or Th^{+}, for short) to denote theory Th with all positive literals R_i substituted by R_i^{+} and all negative literals substituted by R_i^{-}. Similarly, we write $Th^{\oplus}(R;R_1,\ldots,R_n)$ (or $Th^{\oplus}$, in short) to denote theory Th with all positive literals R_i substituted by $R_i^{\oplus}$ and all negative literals substituted by $R_i^{\ominus}$. We often simplify the notation using the equivalences $\neg R^{-}(\bar{x}) \equiv R^{\oplus}(\bar{x})$ and $\neg R^{+}(\bar{x}) \equiv R^{\ominus}(\bar{x})$.

4.1 Additional Notation and Preliminaries

To guarantee that both the inference mechanism specified for querying approximation trees and the process used to compute approximate relations using approximation transducers are efficient, we will have to place a number of syntactic constraints on the local theories used in approximation transducers. The following definitions will be useful for that purpose.

Definition 7. A predicate variable R occurs *positively* (resp., *negatively*) in a formula Φ if the prenex and conjunctive normal form (i.e., the form with all quantifiers in the prefix of the formula and the quantifier-free part of the formula in the form of a conjunction of clauses) of Φ contains a literal of the form $R(\bar{t})$ (resp. $\neg R(\bar{t})$). A formula Φ is said to be *positive* (resp., *negative*) w.r.t. R iff all occurrences of R in Φ are positive (resp., negative).

Definition 8. A formula is called a *semi-Horn rule* (or *rule*, for short) w.r.t. relation symbol R provided that it is in one of the following forms:

$$\forall \bar{x}.[R(\bar{x}) \to \Psi(R,R_1,\ldots R_n)], \tag{1}$$

$$\forall \bar{x}.[\Psi(R,R_1,\ldots R_n) \to R(\bar{x})], \tag{2}$$

where Ψ is an arbitrary classical first-order formula positive w.r.t. R and $\bar{x}$ is an arbitrary vector of variable symbols. If formula Ψ of a rule does not contain R, the rule is called *nonrecursive w.r.t. R*.

Example 2. The first of the following formulas is a (recursive) semi-Horn rule w.r.t. R, whereas the second is a nonrecursive semi-Horn rule w.r.t. R:

$$\forall x,y.\,(\exists u.\,\{R(u,y) \vee \exists z.\,[S(z,x,z) \wedge \forall t.R(z,t)]\}) \to R(x,y),$$
$$\forall x,y.\,(\exists u.\,\{T(u,y) \vee \exists z.\,[S(z,x,z) \wedge \forall t.Q(z,t)]\}) \to R(x,y).$$

The following formula is not a semi-Horn rule, since R appears negatively in the left-hand side of the rule.

$$\forall x,y.\,(\exists u.\,\{\neg R(u,y) \vee \exists z.\,[S(z,x,z) \wedge \forall t.R(z,t)]\}) \to R(x,y).$$

[3] Any first- or second-order formula can be equivalently transformed into this form.

Observe that one could also deal with dual forms of the rules (1) and (2), obtained by replacing relation R by $\neg R$. It is sometimes more convenient to use rules of such a form. For instance, one often uses rules like "if an object on a highway is a car and is *not* abnormal, then it moves." Of course, the results we present can easily be adapted to such a situation.

We often write rules of the form (1) and (2) without initial universal quantifiers, understanding that the rules are always implicitly universally quantified.

5 Approximation Transducers

As stated in the introduction, an approximation transducer provides a means of generating or defining an approximate relation (the output) in terms of other approximate relations (the input) using various dependencies between the input and the output.[4] The set of dependencies is a logical theory where each dependency is represented as a logical formula in a first-order logical language. Syntactic restrictions can be placed on the logical theory to insure efficient generation of output.

Since we are dealing with approximate relations, both the inputs and output are defined in terms of upper and lower approximations. In Sect. 4, we introduced a logical language for referring to different constituents of a rough relation. It is not necessary to restrict the logical theory to just the relations specified in the input and output for a particular transducer. Other relations may be used since they are assumed to be defined or definitions can be generated simultaneously with the generation of the particular output in question. In other words, it is possible to define an approximation network rather than a tree, but for this presentation, we will stick to the tree-based approach. The network approach is particularly interesting because it allows limited forms of feedback across abstraction levels in the network.

The main idea is depicted in Fig. 7. Suppose, one would like to define an approximation of a relation R in terms of a number of other approximate relations $R_1, \ldots, R_k$. It is assumed that $R_1, \ldots, R_k$ consist of either primitive relations acquired via a learning phase or approximate relations that have been generated recursively via other transducers or combinations of transducers.

The local theory $Th(R; R_1, \ldots, R_k)$ is assumed to contain logical formulas relating the input to the output and can be acquired through a knowledge acquisition process with domain experts or even by using inductive logic programming techniques. Generally, the formulas in the logical theory are provided in the form of rules representing some sufficient and necessary conditions for the output relation in addition

[4] The technique also works for one or more approximate relations generated as output, but for clarity of presentation, we will describe the techniques using a single output relation.

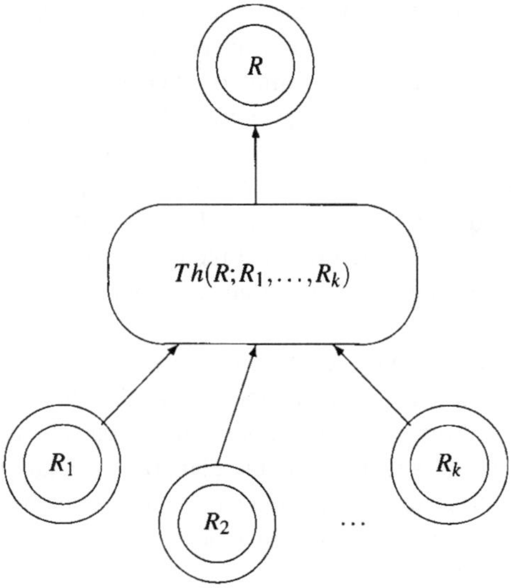

Fig. 7. Transformation of rough relations by first-order theories

possibly to other conditions. The local theory should be viewed as a logical template describing a dependency structure between relations.

The actual transduction process, which generates the approximate definition of relation R uses the logical template and contextualizes it with the actual contextual approximate relations provided as input. The result of the transduction process provides a definition of both the upper and lower approximation of R as follows:

- The lower approximation is defined as the least model for R w.r.t. the theory

$$Th^{+}(R;R_1,\ldots,R_k).$$

- The upper approximation is defined as the greatest model for R w.r.t. the theory

$$Th^{\oplus}(R;R_1,\ldots,R_k),$$

where Th^{+} and $Th^{\oplus}$ denote theories obtained from Th by replacing crisp relations with their corresponding approximations (see Sect. 4). As a result, one obtains an approximation of R defined as a rough relation. Note that appropriate syntactic restrictions are placed on the theory, so coherence conditions can be generated that guarantee the existence of the least and the greatest model of the theory and its consistency with the approximation tree in which its transducer is embedded. For details, see Sect. 7.

Implicit in the approach is the notion of abstraction hierarchies where one can recursively define more abstract approximate relations in terms of less abstract approximations by combining different transducers. The result is one or more approximation trees. This intuition has some similarity to the idea of layered learning (see, e.g., [12]). The technique also provides a great deal of locality and modularity in representation although it does not force this on the user, since networks violating locality can be constructed. If one starts to view an approximation transducer or subtree of approximation transducers as simple or complex agents responsible for the management of particular relations and their dependencies, this then has some correspondence with the methodology strongly advocated, e.g., in [7].

The ability to continually apply learning techniques to the primitive relations in the network and to continually modify the logical theories which are constituent parts of transducers provides a great deal of elaboration tolerance and elasticity in knowledge representation structures. If the elaboration is automated for an intelligent artifact using these structures, the claim can be made that these knowledge structures are self-adaptive, although a great deal more work would have to be done to realize this practically.

5.1 An Introductory Example

In this section, we provide an example of a single approximation transducer describing some simple relationships between objects on a road. Assume that we are provided with the following rough relations:

- $V(x,y)$ - there is a visible connection between objects x and y.
- $S(x,y)$ - the distance between objects x and y is small.
- $E(x,y)$ - objects x and y have equal speeds.

We can assume that these relations were acquired using a supervised learning technique where sample data were generated from video logs provided by the UAV when flying over a particular road system populated with traffic. Or we can assume that the relations were defined as part of an approximation tree using other approximation transducers.

Suppose that we would like to define a new relation C denoting that its arguments, two objects on the road, are connected. It is assumed that we, as knowledge engineers or domain experts, have some knowledge of this concept. Consider, for example, the following local theory $Th(C;V,S,E)$ approximating C:

$$\forall x,y.[V(x,y) \rightarrow C(x,y)], \tag{3}$$

$$\forall x,y.\{C(x,y) \rightarrow [S(x,y) \wedge E(x,y)]\}. \tag{4}$$

The former provides a sufficient condition for C, and the latter a necessary condition. Imprecision in the definition is caused by the following facts:

- The input relations V, S and E are imprecise (rough, noncrisp).
- The theory $Th(C;V,S,E)$ does not describe relation C precisely, as there are many possible models for C.

Then we accept the least model for C w.r.t. theory $Th(C;V^+,S^+,E^+)$ as the lower approximation of C and the upper approximation as the greatest model for C w.r.t. theory $Th(C;V^\oplus,S^\oplus,E^\oplus)$.

It can now easily be observed (and be computed efficiently) that one can obtain the following definitions of the lower and upper approximations of C:

$$\forall x,y.[C^+(x,y) \equiv V^+(x,y)], \tag{5}$$

$$\forall x,y.\left\{C^\oplus(x,y) \equiv \left[S^\oplus(x,y) \wedge E^\oplus(x,y)\right]\right\}. \tag{6}$$

Relation C can then be used, e.g., while querying the rough knowledge database containing this approximation tree or for defining new approximate concepts, provided that it is coherent with the database contents. In this case, the coherence conditions, which guarantee the consistency of the generated relation with the rest of the database (approximation tree), are expressed by the following formulas:

$$\forall x,y.\left\{V^+(x,y) \rightarrow \left[S^+(x,y) \wedge E^+(x,y)\right]\right\},$$

$$\forall x,y.\left\{V^\ominus(x,y) \rightarrow \left[S^\oplus(x,y) \wedge E^\oplus(x,y)\right]\right\}.$$

The coherence conditions can also be generated efficiently provided that certain syntactic constraints are applied to the local theories in an approximation transducer.

6 Rough Relational Databases

To compute the output of an approximation transducer, syntactic characterizations of both the upper and lower approximations of the output relation relative to the substituted local theories are generated. Depending on the expressiveness of the local theory used in a transducer, the results are either first-order formulas or fixed-point formulas. These formulas can then be used to query a rough relational database efficiently using a generalization of results from [4] and traditional relational database theory.

In this section, we define what rough relational databases are and consider their use in the context of approximation transducers and trees. In the following section, we provide the semantic and computational mechanisms used to generate output relations of approximation transducers, check the coherence of the output relations, and ask queries about any approximate relation in an approximation tree.

Definition 9. A *rough relational database B*, is a first-order structure $\langle U, r_1^{a_1}, \ldots, r_k^{a_k}, c_1, \ldots, c_l \rangle$, where

- U is a finite set.

- For $1 \leq i \leq k$, $r_i^{a_i}$ is an a_i-argument rough relation on U, i.e., $r_i^{a_i}$ is given by its lower approximation $r_i^{a_i+}$ and its upper approximation $r_i^{a_i\oplus}$.
- $c_1, \ldots, c_l \in U$ are constants.

By a *signature* of B, we mean a signature containing relation symbols $R_1^{a_1}, \ldots, R_k^{a_k}$ and constant symbols $c_1, \ldots, c_l$ together with equality symbol $=$.

According to the terminology accepted in the literature, a deductive database consists of two parts: an extensional and intensional database. The extensional database is usually equivalent to a traditional relational database, and the intensional database contains a set of definitions of relations in terms of rules that are not explicitly stored in the database. In what follows, we shall also use the terminology *extensional* (*intensional*) rough relation to indicate that the relation is stored or defined, respectively.

Note that intensional rough relations are defined in this framework by means of theories that do not directly provide us with explicit definitions of the relations. These are the local theories in approximation transducers. We apply the methodology developed in [4] which is based on the use of quantifier elimination applied to logical queries to conventional relational databases. The work in [4] is generalized here to rough relational databases.

According to [4], the computation process can be described in two stages. In the first stage, we provide a PTIME compilation process that computes explicit definitions of intensional rough relations. In our case, we would like to compute the explicit definitions of the upper and lower approximations of a relation output from an approximation transducer. In the second stage, we use the explicit definitions of the upper and lower approximations of the intensional relation generated in the first stage to compute suitable relations in the rough relational database that satisfy the local theories defining the relations.

We also have to check whether such relations exist relative to the rough relational database in question. This is done by checking so-called coherence conditions. It may be that the complex query to the rough relational database which includes the constraints in the local theory associated with a transducer is not consistent with relations already defined in the database itself. Assuming that the query is consistent, we know that the output relation for the approximation transducer in question exists and can now compute the answer.

Both checking that the theory is coherent and computing the output relation can be done efficiently because these tasks reduce to calculating fixed-point queries to relational databases over finite domains, a computation that is in PTIME (see, e.g., [5]). Observe that the notion of coherence conditions is adapted in this chapter to deal with rough relations rather than with precise relations, as done in [4].

7 Approximation Transducer Semantics and Computation Mechanisms

Our specific target is to define a new relation, say R, in terms of some additional relations $R_1,\ldots,R_n$ and a local logical theory $Th(R;R_1,\ldots,R_n)$ representing knowledge about R and its relation to $R_1,\ldots,R_n$. The output of the transduction process results in a definition of R^+, the lower approximation of R, as the least model of $Th^+(R;R_1,\ldots,R_n)$, and $R^\oplus$, the upper approximation of R, as the greatest model of $Th^\oplus(R;R_1,\ldots,R_n)$. The following problems must be addressed:

- Is $Th(R;R_1,\ldots,R_n)$ consistent with the database?
- Do a least and greatest relation R^+ and $R^\oplus$ exist that satisfy $Th^+(R;R_1,\ldots R_n)$ and $Th^\oplus(R;R_1,\ldots R_n)$, respectively?
- Is the complexity of the mechanisms used to answer the above questions and to calculate suitable approximations R^+ and $R^\oplus$ reasonable from a pragmatic perspective?

In general, consistency is not guaranteed. Moreover, the above problems are generally NPTIME-complete (across finite models). However, quite similar questions were addressed in [4], and a rich class of formulas has been isolated for which the consistency problem and the other problems can be resolved in PTIME. In what follows, we will use results from [4] and show that a subset of semi-Horn formulas from [4], which we call *semi-Horn rules* (or just rules, for short) and which are described in Sect. 4.1, guarantees the following:

- The coherence conditions for $Th(R;R_1,\ldots,R_n)$ can be computed and checked in polynomial time.
- The least and the greatest relations R^+ and $R^\oplus$, satisfying
$$Th^+(R;R_1,\ldots,R_n) \quad \text{and} \quad Th^\oplus(R;R_1,\ldots,R_n),$$
respectively, always exist provided that the coherence conditions are satisfied.
- The time and space complexity of calculating suitable approximations R^+ and $R^\oplus$ is polynomial w.r.t. the size of the database and that of calculating their symbolic definitions is polynomial in the size of $Th(R;R_1,\ldots,R_n)$.

In view of these positive results, we will restrict the set of formulas used in local theories in transducers to (finite) conjunctions of semi-Horn rules, as defined in Sect. 4.1. All theories considered in the rest of the chapter are assumed to be semi-Horn in the sense of Definition 8.

The following lemmas (Lemmas 1 and 2) provide us with a formal justification of Definition 10 which follows. Let us first deal with nonrecursive rules.[5] Lemma 1 is based on a lemma of Ackermann (see, e.g., [3]) and results of [4].

[5] Lemma 1 follows easily from Lemma 2 by observing that fixed-point formulas (14), (15), and (16) reduce in this case to first-order formulas (9), (10), and (11), respectively. However, reductions to classical first-order formulas are worth a separate treatment as these are less complex and easier to deal with.

Lemma 1. Assume that $Th(R;R_1,\ldots,R_n)$ consists of the following rules:

$$\forall\bar{x}.[R(\bar{x}) \rightarrow \Phi_i(R_1,\ldots,R_n)], \tag{7}$$

$$\forall\bar{x}.[\Psi_j(R_1,\ldots,R_n) \rightarrow R(\bar{x})], \tag{8}$$

for $i \in I$, $j \in J$, where I, J are finite, nonempty sets, and for all $i \in I$ and $j \in J$, formulas Φ_i and Ψ_j do not contain occurrences of R. Then, there exist the least and the greatest R satisfying (7) and (8). The least such R is defined by the formula,

$$R(\bar{x}) \equiv \bigvee_{j\in J} \Psi_j(R_1,\ldots,R_n), \tag{9}$$

and the greatest such R is defined by the formula,

$$R(\bar{x}) \equiv \bigwedge_{i\in I} \Phi_i(R_1,\ldots,R_n), \tag{10}$$

provided that the following coherence condition is satisfied in the database:

$$\forall\bar{x}.\left[\bigvee_{j\in J} \Psi_j(R_1,\ldots,R_n) \rightarrow \bigwedge_{i\in I} \Phi_i(R_1,\ldots,R_n)\right]. \tag{11}$$

Proof. Follows easily, e.g., from Theorem 5.3 of [4]. □

Denote by $\mu S.\alpha(S)$ the least and by $\nu S.\alpha(S)$ the greatest simultaneous fixed-point operator of $\alpha(S)$ (for the definition of fixed-points see, e.g., [5]). Then for recursive theories, we can prove the following lemma, based on the fixed-point theorem of [8] and the results of [4].

Lemma 2. Assume that $Th(R;R_1,\ldots,R_n)$ consists of the following rules:

$$\forall\bar{x}.[R(\bar{x}) \rightarrow \Phi_i(R,R_1,\ldots,R_n)], \tag{12}$$

$$\forall\bar{x}.[\Psi_j(R,R_1,\ldots,R_n) \rightarrow R(\bar{x})], \tag{13}$$

for $i \in I$, $j \in J$, where I, J are finite, nonempty sets. Then, there exist the least and the greatest R satisfying (12) and (13). The least such R is defined by the formula

$$R(\bar{x}) \equiv \mu R(\bar{x}).[\bigvee_{j\in J} \Psi_j(R,R_1,\ldots,R_n)], \tag{14}$$

and the greatest such R is defined by the formula

$$R(\bar{x}) \equiv \nu R(\bar{x}).[\bigwedge_{i\in I} \Phi_i(R,R_1,\ldots,R_n)], \tag{15}$$

provided that the following coherence condition holds:

$$\forall\bar{x}.\left\{\mu R(\bar{x}).[\bigvee_{j\in J} \Psi_j(R,R_1,\ldots,R_n)] \rightarrow \nu R(\bar{x}).[\bigwedge_{i\in I} \Phi_i(R,R_1,\ldots,R_n)]\right\}. \tag{16}$$

Proof. Follows easily, e.g., from Theorem 5.2 of [4]. □

The following definition provides us with a semantics of semi-Horn rules used as local theories in rough set transducers.

Definition 10. Let B be a rough relational database with extensional relation symbols $R_1, \ldots, R_n$, and let R be an intensional relation symbol.

By an *approximation transducer*, we intend the input to be $R_1, \ldots, R_n$, the output to be R, and the local transducer theory to be a first-order theory $Th(R; R_1, \ldots, R_n)$ expressed by rules of the form (12)/(13) or (7)/(8). Under these restrictions,

- The lower approximation of R is defined as the least relation R satisfying $Th(R; R_1, \ldots, R_n)$, i.e., the relation defined by $(9)^+$ or $(14)^+$, respectively, with $R_1, \ldots, R_n$ substituted as described in Sect. 4.
- The upper approximation of R is defined as the greatest relation R satisfying $Th(R; R_1, \ldots, R_n)$, i.e., the relation defined by $(10)^\oplus$ or $(15)^\oplus$, respectively, with $R_1, \ldots, R_n$ substituted as described in Sect. 4,

provided that the respective coherence conditions $(11)^+$ or $(16)^+$, for the lower approximation, and $(11)^\oplus$ or $(16)^\oplus$, for the upper approximation, are satisfied in database B.

Observe that we place a number of restrictions on this definition that can be relaxed, such as restricting the use of relation symbols in the local theory of the transducer to be crisp. This excludes the use of references to constituent components of other rough relations. In addition, since the output relation of a transducer can be represented explicitly in the rough relational database, approximation trees consisting of combinations of transducers are well defined.

7.1 The Complexity of the Approach

This framework is presented in the context of relational databases that have finite domains with some principled generalizations. In addition, both explicit definitions of approximations to relations and associated coherence conditions are expressed in terms of classical first-order or fixed-point formulas. Consequently, computing the approximations and checking coherence conditions can be done in time polynomial in the size of the database (see, e.g., [5]).

In addition, the size of explicit definitions of approximations and coherence conditions is linear in the size of the local theories defining the approximations. Consequently, the proposed framework is acceptable from the point of view of a formal complexity analysis. This serves as a useful starting point for efficient implementation of the techniques. It is clear though that, for very large databases of this type, additional optimization methods would be desirable.

8 A Congestion Example

In this section, we provide an example from the UAV-traffic domain that demonstrates one approach to the problem of defining the concept of traffic congestion using the proposed framework. We begin by assuming that the following relations and constants exist:

- l is a traffic lane on the road.
- $inFOA(l)$ denotes whether lane l is in the focus of the UAV camera.
- $inROI(x)$ denotes whether a vehicle x is in a region of interest.
- $Speed(x,z)$ denotes the approximate speed of x, where $z \in \{\mathsf{low}, \mathsf{medium}, \mathsf{high}, \mathsf{unknown}\}$.
- $Distance(x,y,z)$ denotes the approximate distance between vehicles x and y, where $z \in \{\mathsf{small}, \mathsf{medium}, \mathsf{large}, \mathsf{unknown}\}$.
- $Between(z,x,y)$ denotes whether vehicle z is between vehicles x and y.
- $Number(x,y,z)$ denotes the approximate number of vehicles between vehicles x and y in the region of interest, where $z \in \{\mathsf{small}, \mathsf{medium}, \mathsf{large}, \mathsf{unknown}\}$.
- $TrafficCong(l)$ denotes whether there is traffic congestion in lane l.

We define traffic congestion by the following formula:

$$\begin{aligned} TrafficCong(l) \equiv\ & inFOA(l) \wedge \\ & \exists x,y.\,(inROI(x) \wedge inROI(y) \wedge Number(x,y,\mathsf{large}) \wedge \\ & \forall z.\,[Between(z,x,y) \rightarrow Speed(z,\mathsf{low})] \wedge \\ & \forall z.\,\{Between(z,x,y) \rightarrow \exists t.\,[Distance(z,t,\mathsf{small})]\})\,. \end{aligned} \tag{17}$$

Observe that (17) contains concepts that are not defined precisely. However, for the example, we assume that the underlying database contains approximations of these concepts. We can then use the approximated concepts and replace (17) with the following two formulas representing the lower and upper approximation of the target concept:

$$\begin{aligned} & TrafficCong^{+}(l) \equiv \\ & \quad \exists x,y.\,\{inROI^{+}(x) \wedge inROI^{+}(y) \wedge Number^{+}(x,y,\mathsf{large}) \wedge \\ & \quad \forall z.\,[Between^{\oplus}(z,x,y) \rightarrow Speed^{+}(z,\mathsf{low})] \wedge \\ & \quad \forall z.\,[Between^{\oplus}(z,x,y) \rightarrow \exists t.Distance^{+}(z,t,\mathsf{small})]\} \end{aligned} \tag{18}$$

$$\begin{aligned} & TrafficCong^{\oplus}(L) \equiv \\ & \quad \exists x,y.\,\{inROI^{\oplus}(x) \wedge inROI^{\oplus}(y) \wedge Number^{\oplus}(x,y,\mathsf{large}) \wedge \\ & \quad \forall z.\,[Between^{+}(z,x,y) \rightarrow Speed^{\oplus}(z,\mathsf{low})] \wedge \\ & \quad \forall z.\,[Between^{+}(z,x,y) \rightarrow \exists t.Distance^{\oplus}(z,t,\mathsf{small})]\}\,. \end{aligned} \tag{19}$$

These formulas can be automatically generated using the techniques described previously.

It can now be observed that (17) defines a cluster of situations that can be considered traffic congestions. Small deviations of data do not have a substantial impact on the target concept. This is a consequence of the fact that in (17) we refer to values that are also approximated such as low, small, and large. Thus small deviations of vehicle speed or distance between vehicles usually do not change the qualitative classification of these notions.

Let us denote deviations of data by *dev* with suitable indexes. Now, assuming that the deviations satisfy the following properties:

$$\begin{aligned}
&x' \in dev_{inROI}(x) \equiv [inROI^{+}(x) \rightarrow inROI^{+}(x')], \\
&x' \in dev_{Speed}(x) \equiv [Speed^{+}(x, \mathsf{low}) \rightarrow Speed^{+}(x', \mathsf{low})], \\
&(x', y') \in dev_{Number}(x, y) \equiv \\
&\qquad [Number^{+}(x, y, \mathsf{large}) \rightarrow Number^{+}(x', y', \mathsf{large})], \\
&(x', y') \in dev_{Distance}(x, y) \equiv \\
&\qquad [Distance^{+}(x, y, \mathsf{small}) \rightarrow Distance^{+}(x', y', \mathsf{small})], \\
&(z', x', y') \in dev_{Between}(z, x, y) \equiv \\
&\qquad [Between^{+}(z, x, y) \rightarrow Between^{+}(z', x', y')],
\end{aligned} \tag{20}$$

one can conclude that

$$[TrafficCong^{+}(l) \wedge l' \in dev_{TrafficCong}(l)] \rightarrow TrafficCong^{+}(l'),$$

where $dev_{TrafficCong}(l)$ denotes the set of all situations obtained by deviations of l satisfying conditions expressed by (20).

The above reasoning schema is then robust w.r.t. small deviations of input concepts. Any approximation transducer, defined using purely logical means, enjoys this property since small deviations of data, by not changing basic properties, do not change the target concept.

A formal framework that includes the topics of robustness and stability of approximate reasoning schemas is presented, e.g., in [9, 10], where these notions have been considered in a rough mereological framework.

9 On the Approximation Quality of First-Order Theories

So far, we have focused on generating approximations to relations using local logical theories in approximation transducers and then building approximation trees from these basic building blocks. This immediately raises the interesting issue of viewing the approximate global theory itself as a conceptual unit. We can then ask what the approximation quality of a theory is and whether we can define qualitative or quantitative measures of the theory's approximation quality. If this is possible, then individual theories can be compared and assessed for their approximative value. One application of this measure would be to choose approximative theories for an application domain at the proper level of abstraction or detail, moving across the different levels of abstraction relative to the needs of the application. In this section, we provide a tentative proposal to compare the approximation quality of first-order theories.

9.1 Comparing Approximation Power of Semi-Horn Theories

Definition 11. We say that a *theory* $Th_2(R)$ *better approximates a theory* $Th_1(R)$ *relative to a database* B and denote this by $Th_1(R) \leq_B Th_2(R)$ provided that, in database B, we have $R_1^+ \subseteq R_2^+$ and $R_2^{\oplus} \subseteq R_1^{\oplus}$, where for $i = 1,2$, R_i^+ and $R_i^{\oplus}$ denote the lower and upper approximation of R defined by theory Th_i.

Observe that the notion of a better approximation has a correspondence to information orderings used in the model theory of a number of three-valued and partial logics.

Example 3. Let $CL(x,y)$ denote that objects x,y are close to each other, $SL(x,y)$ denote that x,y are on the same lane, $CH(x,y)$ denote that objects x,y can hit each other, and let $HR(x,y)$ denote that the relative speeds of x and y are high. We assume that lower and upper approximations of these relations can be extracted from data during learning acquisition or are already defined in a database, B. Consider the following two theories approximating the concept $D(x,y)$ that denotes a dangerous situation caused by objects x and y:

- $Th_1(D;CL,SL,CH)$ has two rules:

$$\begin{aligned} &\forall x,y.\{[CL(x,y) \wedge SL(x,y)] \rightarrow D(x,y)\}, \\ &\forall x,y.[D(x,y) \rightarrow CH(x,y)]. \end{aligned} \tag{21}$$

- $Th_2(D;CL,SL,HR)$ has two rules:

$$\begin{aligned} &\forall x,y.[CL(x,y) \rightarrow D(x,y)], \\ &\forall x,y.\{D(x,y) \rightarrow [HR(x,y) \wedge SL(x,y)]\}. \end{aligned} \tag{22}$$

Using Lemma 1, we can compute the following definitions of approximations of D:

- Relative to theory $Th_1(D;CL,SL,CH)$:

$$\begin{aligned} &\forall x,y.\left\{{D^{(1)}}^{+}(x,y) \equiv \left[CL^{+}(x,y) \wedge SL^{+}(x,y)\right]\right\}, \\ &\forall x,y.[{D^{(1)}}^{\oplus}(x,y) \equiv CH^{\oplus}(x,y)]. \end{aligned} \tag{23}$$

- Relative to theory $Th_2(D;CL,SL,HR)$:

$$\begin{aligned} &\forall x,y.[{D^{(2)}}^{+}(x,y) \equiv CL^{+}(x,y)], \\ &\forall x,y.\left\{{D^{(2)}}^{\oplus} \equiv \left[HR^{\oplus}(x,y) \wedge SL^{\oplus}(x,y)\right]\right\}. \end{aligned} \tag{24}$$

Obviously, $D^{(1)^+} \subseteq D^{(2)^+}$. If we also assume that, in our domain of discourse (and by implication in database B), $HR \cap SL \subseteq CH$ applies, we can also obtain the additional relation that $D^{(2)^\oplus} \subseteq D^{(1)^\oplus}$. Thus $Th_1 \leq_B Th_2$, which means that an agent possessing the knowledge implicit in Th_2 is better able to approximate concept D than an agent possessing knowledge implicit in Th_1.

These types of comparative relations between theories should prove to be very useful in cooperative agent architectures, but we leave this application for future work.

10 Conclusions and Related Work

In this chapter, we have presented a framework for generating, structuring, and reasoning about approximate relations having dependencies on each other. We began with a discussion of the subclass of approximate primitive concepts grounded in sensor or other data via the use of learning techniques. We then introduced the idea of an approximation transducer as a basic constituent in constructing more complex approximation trees consisting of combinations of a number of approximation transducers. An approximation transducer defines an approximate relation in terms of other approximate relations and a local transducer theory where dependencies among the relations are represented as logical formulas in a traditional manner. This combination of both approximate and crisp knowledge brings together techniques and concepts from two research disciplines. By providing syntactic characterizations of these ideas and techniques, we are able to propose a novel type of approximate knowledge structure which is elaboration tolerant, elastic, modular, and grounded in the particular contexts associated with various applications.

By restricting the syntax of local transducer theories, we can implement the approximation tree inference mechanism efficiently by using a slight generalization of deductive relational databases to include rough relations. Efficient reasoning mechanisms are important because experimentation is being done within the constraints of the WITAS UAV project where these techniques are intended to be used on-board the UAV as an integral part of its knowledge representation mechanisms. In the chapter we used a number of examples specific to the UAV domain to demonstrate the use and versatility of the techniques.

A richer and more complex type of generalized deductive database is proposed in a companion chapter in this volume [2]. In this case, an open-world assumption is assumed about relational information, rather than the standard closed-world assumption view. In addition, the query language used is logical and permits what we call *contextually closed queries*. These queries include the query itself, a local context in the form of integrity constraints and a minimization/maximization policy that permits locally closing parts of the database relative to a query. The latter technique replaces the global closed-world assumption and provides finer grained closure policies that are very useful for open-world planning applications. The use of approximate relations, contextually closed queries and a logical query language imply the use of a special inference mechanism. These generalized deductive databases, which we call *rough knowledge databases*, provide us with an even richer form of query/answering system that subsumes the idea of approximation trees and can also be used for many other applications involving queries on partial models.[6]

Acknowledgments

The research has been supported in part by the Wallenberg Foundation. Witold Łukaszewicz, Andrzej Skowron, and Andrzej Szałas have also been supported by the KBN grant 8 T11C 009 19 and the KBN grant 8T11C 025 19. We thank Jonas Kvarnström for thorough and valuable proofreading of this chapter.

References

1. P. Doherty, G. Granlund, K. Kuchcinski, K. Nordberg, E. Sandewall, E. Skarman, J. Wiklund. The WITAS unmanned aerial vehicle project. In *Proceedings of the 14th European Conference on Artificial Intelligence (ECAI 2000)*, 747–755, IOS, Amsterdam, 2000.
2. P. Doherty, J. Kachniarz, A. Szałas. Using contextually closed queries for local closed-world reasoning in rough knowledge databases. (this book).
3. P. Doherty, W. Łukaszewicz, A. Szałas. Computing circumscription revisited. *Journal of Automated Reasoning*, 18(3):297–336, 1997. See also: Report number LiTH-IDA-R-94-42 of Linköping University, 1994 and *Proceedings of the 14th International Joint Conference on AI (IJCAI'95)*, Morgan Kaufmann, San Francisco, 1995.
4. P. Doherty, W. Łukaszewicz, A. Szałas. Declarative PTIME queries for relational databases using quantifier elimination. *Journal of Logic and Computation*, 9(5):739–761, 1999. See also: Report number LiTH-IDA-R-96-34 of Linköping University, 1996.
5. H-D. Ebbinghaus, J. Flum. *Finite Model Theory*. Springer, Heidelberg, 1995.
6. J. McCarthy. Approximate objects and approximate theories. In A.G. Cohn, F. Giunchiglia, B. Selman, editors, *Proceedings of the 7th International Conference on Principles of Knowledge Representation and Reasoning, (KR 2000)*, 519–526, Morgan Kaufmann, San Francisco, 2000.
7. M. Minsky. *The Society of Mind*. Simon & Schuster, New York, 1986.
8. A. Nonnengart, A. Szałas. A fixpoint approach to second-order quantifier elimination with applications to correspondence theory. In E. Orłowska, editor, *Logic at Work: Essays Dedicated to the Memory of Helena Rasiowa*, 307–328, Physica, Heidelberg, 1998. See also Report number MPI-I-95-2-007 of Max-Planck-Institut fuer Informatik, Saarbruecken, 1995.
9. L. Polkowski, A. Skowron. Rough mereology: A new paradigm for approximate reasoning. *International Journal of Approximate Reasoning*, 15(4): 333–365, 1996.
10. L. Polkowski, A. Skowron. Towards adaptive calculus of granules. In L.A. Zadeh, J. Kacprzyk, editors, *Computing with Words in Information/Intelligent Systems 1*, 201–227, Physica, Heidelberg, 1999.
11. W.V.O. Quine. *The Web of Belief*, 2nd ed. McGraw-Hill, New York, 1978.
12. P. Stone. *Layered Learning in Multiagent Systems: A Winning Approach to Robotic Soccer*. MIT Press, Cambridge, MA, 2000.

[6] For a detailed presentation of rough knowledge databases and the use of contextually closed queries, we refer the reader to the companion chapter in this volume [2].

Chapter 9
Using Contextually Closed Queries for Local Closed-World Reasoning in Rough Knowledge Databases

Patrick Doherty,[1] Jarosław Kachniarz,[2] Andrzej Szałas[3]

[1] Department of Computer and Information Science, Linköping University, 58183 Linköping, Sweden
patdo@ida.liu.se

[2] Soft Computer Consultants, 34350 US19N, Palm Harbor, FL 34684, USA
jk@softcomputer.com

[3] The College of Economics and Computer Science, Wyzwolenia 30, 10-106 Olsztyn, Poland
andsz@ida.liu.se

Summary. Representing internal models of aspects of an autonomous agent's surrounding environment or of its own epistemic state and developing query mechanisms for these models based on efficient forms of inference are fundamental components in any deliberative/reactive system architecture used by an agent in achieving task goals. The problem is complicated by the fact that the models in question necessarily have to be incomplete due to the complexity of the environments in which such agents are intended to operate. Consequently, the querying mechanisms must be framed in the context of an *open-world assumption*. We propose an architecture for such a system that involves generalizing classical deductive databases to rough knowledge databases (RKDB), where relations in the database are defined as rough sets. We also propose the use of *contextually closed queries* (CCQs) where a context for a query and a local minimization policy are provided in terms of integrity constraints and techniques from circumscription. The concept of a contextually closed query is a generalization of querying in the context of a local closed-world assumption (LCW) previously proposed in the literature. CCQs have the effect of dynamically reducing the boundary regions of relations relative to a particular set of integrity constraints associated with the query before actually querying the RKDB. The general problem of querying the RKDB using CCQs is co-NPTIME complete, but we isolate a number of important practical cases where polynomial time and space complexity is achieved.

1 Introduction

Consider an autonomous system, such as a ground robot or an unmanned aerial vehicle (UAV), operating in a highly complex and dynamic environment. For systems of this sort to function intelligently and robustly, it is useful to have both deliberative and reactive capabilities. Such systems combine the use of reactive and deliberative capabilities in achieving task goals. Reactive capabilities are necessary so that the system can react to contingencies that arise unexpectedly and demand immediate response with little room for deliberation as to what the best response should be. Deliberative capabilities are useful in the sense that internal representations of aspects of the system's operational environment can be used to predict the course of events in the near or intermediate future. These predictions can then be used to determine more selective actions or better responses in the present, which will potentially save the system time, effort, and resources in the course of achieving task goals.

Due to the complexity of the operational environments in which such robotics systems generally operate and the inaccuracy of sensor data about the environment acquired through different combinations of sensors, these systems cannot be assumed to have complete information or models about their surrounding environment nor the effects of their actions on these environments. On the other hand, the deliberative component is dependent on synthesizing, managing, updating, and using of incomplete qualitative models of the operational environment represented internally in the system architecture. These internal models are used for reasoning about the system's environment and the effects of its actions on the environment while the system attempts to achieve task goals. In spite of the lack of complete information, such systems quite often have, or can acquire, additional information that can be used in certain contexts to assume additional knowledge about the incomplete parts of the specification. This information may be of a normative or default nature, may include rules of thumb particular to the operational domain in question, or may include knowledge implicit in the result of executing a sensing action.

One potentially useful approach that can be pursued in developing of on-line reasoning capabilities and representation of qualitative models of aspects of an autonomous system's operational environment is using traditional database technology combined with techniques originating from artificial intelligence research with knowledge-based systems. There are a number of different compositions of technologies that may be pursued, ranging from more homogeneous logic programming based deductive database systems to heterogeneous systems that combine the use of traditional relational database technology with specialized front-end reasoning engines.

The latter approach will be pursued in this chapter, but with a number of modifications of the standard deductive database framework. These modifications are made necessary by the requirement of representing and reasoning about incomplete qualitative models of the operational environments in which autonomous systems are embedded. A number of fundamental generalizations of standard semantic concepts used in the traditional deductive database approach will be made:

- The extensional database (EDB) which represents and stores base relations and properties about the external environment, or the system's internal environment, will be given formal semantics based on the use of rough sets [14, 19]. The extension of a database relation or property will contain explicit positive and negative information in addition to implicitly represented boundary information that is defined as the difference between upper and lower approximations of the individual relations and properties.

- The intensional database (IDB) will contain two rule sets generating implicit positive and negative information, respectively, via application of the rule sets in the context of the facts in the EDB. The *closed-world assumption* will *not* be applied to the resulting information generated from the EDB/IDB pair.
- An *open-world assumption* will be applied to the extensional and intensional database pair, which can be locally closed dynamically by using of *contextually closed queries*. A CCQ consists of the query itself, a context represented as a set of integrity constraints,[1] and a local closure policy specified in terms of the minimization/maximization of selected relations. The contextually closed query layer (CCQ layer) represents the closure mechanism and is used to answer individual CCQs.

In effect, the CCQ layer permits the representation of additional normative, default, or closure information associated with the operational environment at hand and the particular view of the environment currently used by the querying agent. Together with the rough set semantics for relations, a rough set knowledge base in this context represents an incompletely specified world model with dynamic policies that permit the local closure of parts of the world model when querying it for information.

The combination of the EDB, IDB, and CCQ layer will be called the rough knowledge database. The computational basis for the inference engine used to query the RKDB will be based on the use of circumscription, quantifier elimination, and the ability to automatically generate syntactic characterizations for the upper and lower bounds of rough relations in the RKDB.

1.1 Open-and Closed-World Reasoning

What is meant intuitively by open- and closed-world reasoning? In traditional databases, reasoning is often based on the assumption that information stored in a specific database contains a complete specification of the application environment at hand. If a tuple is not in a base relational table, it is assumed that it does not have that specific property. In deductive databases, if the tuple is not in a base relational table or any intensional relational tables generated implicitly by the application of intensional rules, it is again assumed that it does not have these properties. Under this assumption, an efficient means of representing negative information about the world depends on applying the *closed-world assumption* (CWA) [1, 20]. In this case, atomic information about the world, absent in a world model (represented as a database), is assumed to be false.

[1] We accept a paradigm, according to which integrity constraints are statements about database contents expressed as classical first-order formulas (see, e.g. [1]), that are to be satisfied by the instances of the database. However, since we also deal with incomplete information, we assume that the required satisfiability is restricted to tuples containing only complete information.

On the other hand, for many applications such as the autonomous systems applications already mentioned, the assumption of complete information is not feasible nor realistic, and the CWA cannot be used. In such cases, an *open-world assumption* (OWA), where information not known by an agent is assumed to be unknown, is often accepted, but this complicates both the representational and implementational aspects associated with inference mechanisms and the use of negative information. The CWA and OWA represent two ontological extremes. Quite often, a reasoning agent does have or acquires additional information that permits the application of the CWA *locally* in a particular context. In addition, if it does have knowledge of what it does not know, this information is valuable because it can be used in plan generation to acquire additional information by using of sensors.

In such a context, various forms of LCW assumptions have been defined (see, e.g., [8, 10]), and planning systems have been proposed (see, e.g., [11, 12]). The starting point for the approach proposed by in [8] several authors of this chapter is based on the approach to query answering using LCW assumptions described in [10], where the authors present a sound, but incomplete, tractable algorithm for LCW reasoning intended for use in the XII Planner [13]. The approach described in [10] was substantially strengthened in [8] by

- Providing formal semantics for the case where LCW assumptions and queries are expressed by arbitrary first-order formulas. The semantics is based on the use of formula circumscription and depends on minimizing formulas expressing LCW constraints.
- Isolating a more expressive language for LCW assumptions which subsumes that used in [10], permits limited use of negation and disjunction and still retains tractability.
- Providing a sound and complete, tractable deduction method for the more expressive language.

The semantics of LCW constraints, as defined in [8], depends on minimizing LCW constraints where it is specified that all relations in a constraint vary. The minimization process results in changing the varied database relations as a side effect of the process. Queries are then posed to the changed database. Initial practice in using the strengthened version of LCW assumptions showed that a finer granularity in the minimization policy for LCW assumptions was desirable as was a more intuitive methodology for expressing LCW policies to understand the results and provide intuitive semantics for the database changes. These desiderata have led to the proposal for the modifications and generalizations of deductive database technology described above.

1.2 A New Approach to Rough Set Based LCW Techniques

In the current chapter, we propose semantics and methodology for LCW reasoning that provides a more intuitive and general framework for integrating LCW reasoning in knowledge databases used by intelligent agents. The new approach differs from and subsumes that of [8] in the following manner:

- It generalizes to deductive databases, whereas the previous approach described in [8] is basically restricted to relational databases.
- Integrity constraints, absent in the previous approach, take on an important role in characterizing LCW assumptions in a principled manner. In most knowledge databases, the relationships between pieces of information are expressed by integrity constraints (e.g., defined by classical first-order formulas). When applying LCW policies locally to particular relations, one minimizes those relations. However, in such cases, the integrity constraints have to be preserved. This can result in implicit changes to some additional relations. However, the integrity constraints are still preserved, thus the knowledge structure represented continues to satisfy the desired properties. Such information was missing in the previous approach, thus it was much more difficult to understand the changes in the resulting database and to develop pragmatic implementation techniques for modifying and querying the knowledge database.
- Integrity constraints and local closure policies are decoupled from the knowledge database itself and associated dynamically with individual agent queries. The agents themselves possess local views and preferences about the world model that may or may not be shared by other agents or even the same agent using a different query.
- The formula-circumscription technique used in the previous approach is replaced by integrity constraints and standard circumscription. This modification permits selected fixing, varying, and minimizing of specific relations in integrity constraints, whereas the previous approach forced varying on all predicates in an LCW constraint. This provides the user with more flexibility in defining LCW constraints and brings the new approach closer to the methodology used in circumscription-based knowledge representation. It should be emphasized that the implementation is not always dependent on circumscription.
- At the semantic level we use rough sets to represent database information as a natural tool.[2] Rough sets contain information about tuples known to be in a relation (the lower approximation of the relation), tuples known not to be in the relation (the complement of the upper approximation of the relation) and tuples for which it is unknown whether they belong to the relation (the difference between the upper and lower approximation of the relation).

1.3 The Structure of This Chapter

In Sect. 2, we provide some notation and a number of definitions. In Sect. 3, we describe the basic architecture for rough knowledge databases consisting of the extensional, intensional, and contextual closure query layer. In Sect. 4, a detailed example from the domain of unmanned aerial vehicles is provided to demonstrate the need

[2] As discussed, e.g., in [14], rough relations appear in databases in many important contexts.

for the reasoning mechanisms we propose. In Sect. 5, we provide specifications of the languages used for the three rough knowledge database layers, and in Sect. 6, we provide the formal semantics for each of the three layers. In Sect. 7, we provide a high-level specification for an algorithm for computing queries to rough knowledge databases and consider complexity and expressiveness issues. In Sect. 8, we isolate a number of important special cases based on restrictions in using language at the three database layers which guarantee efficient mechanisms for computing queries for these cases. In Sect. 9, we conclude with a summary of results and some considerations on future work.

2 Preliminaries

We deal with the first-order language with equality, F_{I}, over a fixed vocabulary without function symbols, where Const is a finite set of *constant symbols*, V_{I} is a finite set of *first-order variables* and Rel is a finite set of *relation symbols*. Any rough relation $R()$ is defined by

- *The positive part of the relation*, containing positive information and denoted by $R^+()$ [it is simply the lower approximation of $R()$].
- *The negative part of the relation*, containing negative information and denoted by $R^-()$ [it is the complement of the upper approximation of $R()$].
- *The boundary region of the relation*, containing the unknown facts and denoted by $R^\pm()$ [it is the difference of the upper and the lower approximation of $R()$].

By F_{II} we denote the second-order language based on an alphabet whose symbols are those of Rel, together with a denumerable set V_{II} of n-ary predicate variables (for each $n \geq 0$). In the rest of the chapter, we shall use second-order circumscription. Our definition follows [15].

Definition 1. Let $\bar{P}$ be a tuple of distinct predicate constants, $\bar{S}$ be a tuple of distinct predicate constants disjoint with $\bar{P}$, and let $T(\bar{P},\bar{S})$ be a finite theory in the language F_{I}. The *second-order circumscription of $\bar{P}$ in $T(\bar{P},\bar{S})$ with variable $\bar{S}$*, written $CIRC(T(\bar{P},\bar{S});\bar{P};\bar{S})$, is the sentence (in the language F_{II})

$$T(\bar{P},\bar{S}) \wedge \forall\bar{\Phi}\forall\bar{\Psi}.\{[T(\bar{\Phi},\bar{\Psi}) \wedge \bar{\Phi} \leq \bar{P}] \rightarrow \bar{P} \leq \bar{\Phi}\}, \tag{1}$$

where $\bar{\Phi}$ and $\bar{\Psi}$ are tuples of predicate variables similar to $\bar{P}$ and $\bar{S}$, respectively,[3] $\bar{\Phi} \leq \bar{P}$ stands for

$$\bigwedge_{i=1}^{n}[\forall\bar{x}.\Phi_i(\bar{x}) \rightarrow P_i(\bar{x})] \text{ and } \bar{P} \leq \bar{\Phi} \text{ stands for } \bigwedge_{i=1}^{n}[\forall\bar{x}.P_i(\bar{x}) \rightarrow \Phi_i(\bar{x})].$$

[3] A tuple of predicate expressions $\bar{X}$ is said to be similar to a tuple of predicate constants $\bar{Y}$ iff $\bar{X} = (X_1,\ldots,X_n)$, $\bar{Y} = (Y_1,\ldots,Y_n)$ and, for all $1 \leq i \leq n$, X_i and Y_i are of the same arity.

In the following, we shall often write $CIRC(T;\bar{P};\bar{S})$ instead of $CIRC(T(\bar{P},\bar{S});\bar{P};\bar{S})$. We also allow minimization of negative literals. Definition 1 can easily be adjusted to this case, since minimizing a literal, say $\neg R$, means maximizing R. Thus it suffices to replace inequalities $\leq$ of (1) by inequalities $\geq$, i.e., to reverse the corresponding implication. We also require the following definition (see also [7,8]).

Definition 2. Let $\bar{x}$ be a tuple of first-order variables, $\bar{Q}$ be a tuple of relation symbols, $\Phi(\bar{Q})$ be a first-order formula positive w.r.t. all symbols in $\bar{Q}$, and $\Psi(\neg\bar{Q})$ be a first-order formula negative w.r.t. all symbols in $\bar{Q}$. Then

- By a *semi-Horn formula*, we shall understand any formula of the following form

$$[\forall\bar{x}.\Phi(\bar{Q}) \rightarrow \bar{Q}(\bar{x})] \wedge \Psi(\neg\bar{Q}).$$

- By a *weak semi-Horn formula*, we shall understand any formula of the following form:

$$[\Phi(\bar{Q}) \rightarrow Q_i(\bar{x})] \wedge \Psi(\neg\bar{Q}).$$

- By a *weak Ackermann formula* we shall understand a weak semi-Horn formula, in which Φ does not contain the relation Q_i.

Observe that in Definition 2, one can replace all occurrences of all relations of $\bar{Q}$ by their negations. This is useful for the application of dual forms of quantifier elimination techniques used in our algorithms.

3 The Architecture of Rough Knowledge Databases

Let us now discuss the architecture of rough knowledge databases as understood in this chapter. The kernel of the database is the so-called extensional database (see Fig. 1). We assume that the extensional database contains positive and negative facts. The facts that are not explicitly listed in the extensional database are assumed to be unknown in this layer of the database. Thus, in the extensional database layer we accept the open-world assumption. The intensional database layer provides rules that define some new relations, but also rules allowing one to extend the positive and negative parts of the extensional relations.[4] The outermost, most advanced layer, which we call the *contextual closure query layer* (CCQ layer), consists of the CCQ inference mechanism which represents the query/answer mechanism used by individual CCQs applied to the two lower layers of the RKDB.

The extensional database consists of rough relations. According to the methodology developed in [5], the rules of the intensional database function as rough set transducers (see also Sect. 6.3), transforming combinations of rough extensional relations

[4] We assume here that extensional and intensional databases are consistent. Let us note that the consistency condition, expressed in the language we consider, is tractable.

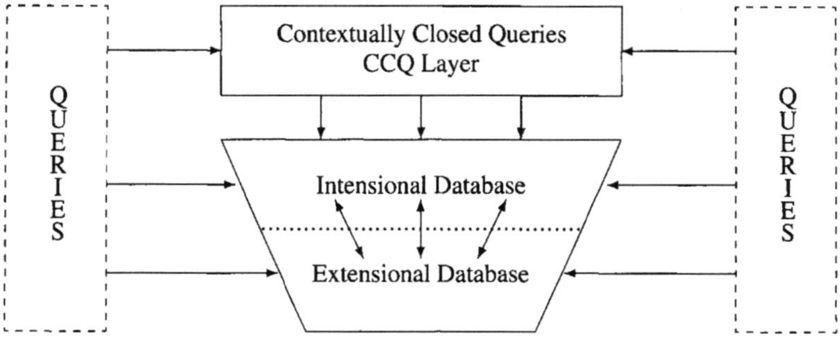

Fig. 1. The architecture of knowledge databases

into new relations that satisfy the constraints of the intensional rules. As in the extensional database, the open-world assumption is accepted in the intensional layer. Local closure context policies (LCC policies) allow us to minimize chosen relations (or their complements), while at the same time preserving the integrity constraints (ICs). Queries are posed via the outermost layer, but in some applications, it might be useful to submit queries to the intensional or even extensional layer. This feature can be provided in an obvious manner. We will focus on the CCQ layer.

4 A UAV Scenario Sensor Usage Example

The WITAS[5] Unmanned Aerial Vehicle Project [3] is a long-term basic research project located at Linköping University (LiU), Sweden, and the authors are participants in the project. The current work with rough knowledge databases and LCC reasoning is intended to be used in an on-line query-answering system which is part of the UAV's system architecture.

The long-term goal of the WITAS UAV Project is the development of the technologies and functionalities necessary for successfully deploying a fully autonomous UAV operating over road and traffic networks. While operating over such an environment, the UAV should be able to navigate autonomously at different altitudes (including autonomous takeoff and landing); plan for mission goals such as locating, identifying, tracking, and monitoring different vehicle types, and construct internal representations of its focus of attention for use in achieving its mission goals. Additionally, it should be able to identify complex patterns of behavior such as vehicle overtaking, traversing of intersections, parking lot activities, etc.

In the current project, we are using a Yamaha RMAX helicopter as the experimental physical platform on which to pursue our research. The helicopter is equipped with

[5] The Wallenberg Laboratory for Information Technology and Autonomous Systems (Pronounced *Vee-Tas*).

a sensor platform that includes a geographical positioning system (GPS), an inertial navigation system (INS), elevation sensors, and a magnetic compass, in addition to a video camera.

The system architecture for the UAV consists of both deliberative and reactive components, and the communication infrastructure for the components is based on the use of the standard object broker CORBA. There are a number of deliberative services such as task planners, trajectory planners, prediction mechanisms, and chronicle recognizers, that are dependent on internal qualitative representations of the environment over which the UAV operates. The knowledge representation components include a soft-real time database called the dynamic object repository, a standard relational database, a geographic information system containing road and geographic data, and a number of front-end query-answering systems that serve as inference engines and may be used by other components in the architecture. The research described in this chapter provides a basis for one of the inference engines. In addition to these components, there is an image processing module used for low and intermediate level vision tasks and a helicopter control module which is used to position the helicopter and camera dynamically and maintain positions during the execution of task goals which may include highly dynamic tasks such as tracking vehicles through a small village with building obstacles.
Let's examine a particular scenario from the UAV operational environment representative of the use of LCC reasoning in the UAV context.

Suppose the UAV receives the following mission goal from its ground control operator:

> Identify and track *all* moving vehicles in region X, and log the estimated velocities and positions of *all* small blue vehicles identified for the duration of their stay in region X, or until the UAV is low on fuel.

Achieving a mission goal such as this in a fully autonomous mode is extremely complex and would involve the concurrent use of many of the deliberative and reactive services in the architecture, in addition to a great deal of sophisticated reasoning about the operational environment. Both hard and soft real-time constraints must also be taken to consideration, particularly for query-answering during a plan execution phase. In this example, we will focus on a particular type of reasoning capability made possible by the combined use of LCC reasoning and rough knowledge databases.

The first step in achieving the mission goal would be to generate a task plan which would include the following steps:

1. Fly to a position that permits viewing region X and possibly an area surrounding the region.
2. Focus the camera on region X, and maintain position, focus, and coverage.

3. Initiate the proper image processing algorithms for identifying moving vehicles in region X.
4. Use the sensor data gathered in the previous step to produce knowledge as to what is seen or not seen by the UAV in region X.
5. Use the acquired knowledge to plan for the next series of actions which involve tracking, feature recognition, and logging.
6. Maintain execution of the necessary services and processes until the mission goal is completed.

We will concentrate on steps 3 and 4 whose successful completion is dependent on a combination of the open-world assumption, LCC reasoning, and rough knowledge database representation of relations and properties.

Observe that the mission goal above contains two universal statements, the first asks to "identify and track *all* moving vehicles in region X," and the second asks to "log the estimated velocities and positions of *all* small blue vehicles identified." The meaning of the second universal is naturally dependent on the meaning of the first universal. To achieve the mission goal, the inferencing mechanism used by the UAV during plan generation and plan execution must be able to circumscribe (in the intuitive sense) the meaning of "*all* moving vehicles in region X" and that of "*all* small blue vehicles identified."

What the UAV *can perceive as moving*, given the constraints under which it is operating, the character of the dynamics of its current operational environment, and the capabilities of its sensor and image processing functionalities in this context, is not necessarily the same thing as what *is actually moving* in region X. An additional problem, of course, is that the inferencing mechanism cannot appeal to the use of the closed-world assumption. If it could, it would register moving objects in region X and *assume* via application of the CWA that no other objects are moving. One cannot appeal to this mechanism because the open-world assumption is being used. Even if one could, this would be erroneous. Certainly, there may be vehicles in region X that are moving but can not be perceived due to limitations associated with the UAV's capabilities, and there may also be vehicles outside region X that are moving.

The key to solving this particular representational problem is to note that sensing actions, such as step 3 in the plan sketch above, implicitly generate local or contextual closure information (LCC policies) and that the UAV agent can query the rough knowledge database using the particular contextual closure that exists for the purpose at hand. For example, the sensing action in step 3 above not only generates information about specific moving individuals the UAV can perceive with its current sensor and image processing capabilities, but it also generates knowledge that this is all the UAV can see in the region of interest (ROI), region X. The nature of this information is that it is specific knowledge of what the UAV agent does not see rather than information derived via an assumption such as the CWA.
Of course, one has to (or more specifically, the UAV agent has to) supply the con-

textual closure information. This will be supplied in terms of one or more integrity constraints and an LCC policy consisting of particular LCC assumptions pertaining to the minimization, maximization, fixing, or varying of specific relations. The specific closure context for this situation could be paraphrased as follows:

> After sensing region X with a camera sensor, assume that all moving vehicles in the ROI (X) have been perceived except for those with a signature whose color feature is roadgray.

In the following example, we provide the particulars for representing the scenario above and reasoning about it using the proposed approach.

Example 1. [A UAV Scenario: Identify, Track, and Log]
Consider the situation where a UAV observes and classifies cars with different signatures based on color.[6] For the example, the domains considered consist of

- $Cars = \{\mathsf{c}_1, \mathsf{c}_2, \mathsf{c}_3, \mathsf{c}_4, \mathsf{c}_5, \mathsf{c}_6\}$.
- $Regions = \{\mathsf{r}_1, \mathsf{r}_2, \mathsf{r}_3, \mathsf{r}_4\}$.
- $Signatures = \{\mathsf{blue}, \mathsf{roadgray}, \mathsf{green}, \mathsf{yellow}\}$.

The following relations are also defined:[7]

- $Moving(\mathsf{c})$ the object c is moving.
- $InROI(\mathsf{r})$ the region r is in the region of interest.
- $See(\mathsf{c}, \mathsf{r})$ the object c is seen by the UAV in region r.
- $In(\mathsf{c}, \mathsf{r})$ the object c is in region r.
- $ContainedIn(\mathsf{r}, \mathsf{r}')$ region r is contained in region r′.
- $Sig(\mathsf{c}, \mathsf{s})$ the object c has signature s.

Suppose the actual situation in the operational environment over which the UAV is flying is as depicted in Fig. 2. For the mission goal, the UAV's initial region of interest (ROI) is region r_3.

At mission start, the following facts are in the UAV's on-line extensional database (EDB):

$$\{ContainedIn(\mathsf{r}_1, \mathsf{r}_2), ContainedIn(\mathsf{r}_2, \mathsf{r}_3)\}.$$

During mission preparation, the ground operator relays the following information to the UAV agent which is placed in the UAV's on-line EDB:

$$\{In(\mathsf{c}_2, \mathsf{r}_2), In(\mathsf{c}_3, \mathsf{r}_3), Moving(\mathsf{c}_2),$$
$$Sig(\mathsf{c}_2, \mathsf{roadgray}), Sig(\mathsf{c}_3, \mathsf{green}), Sig(\mathsf{c}_6, \mathsf{roadgray}), InROI(\mathsf{r}_3)\}.$$

[6] In an actual scenario, a vehicle signature would be more complex and contain features such as width, height, and length, or vehicle type.

[7] In addition, a number of type properties, such as *Car*(), *Region*(), etc. would also be defined.

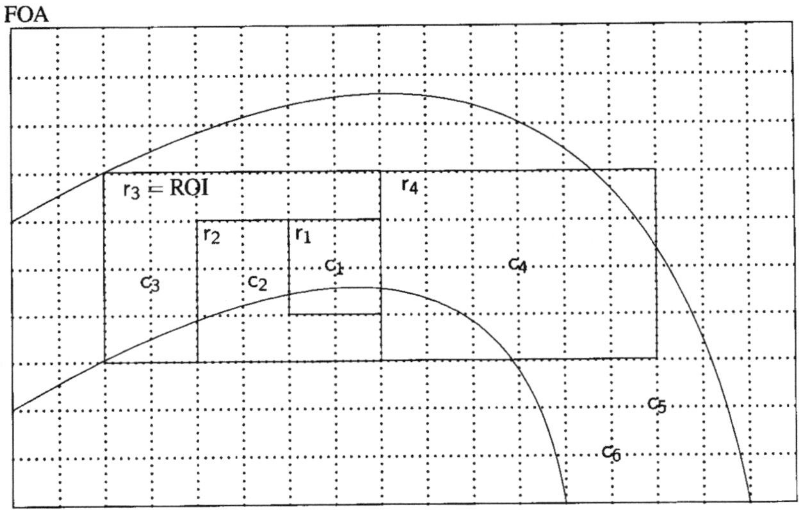

Fig. 2. The situation considered in Example 1.

The following rules are associated with the intensional database:

$$ContainedIn(r,s) \leftarrow \exists t.[ContainedIn(r,t) \wedge ContainedIn(t,s)], \quad (2)$$

$$InROI(s) \leftarrow \exists r.[ContainedIn(s,r) \wedge InROI(r)]. \quad (3)$$

The current EDB, together with the intensional database (IDB), would allow the UAV agent to infer the following additional facts:

$$\{ContainedIn(\mathsf{r}_1,\mathsf{r}_3), InROI(\mathsf{r}_2), InROI(\mathsf{r}_1)\}.$$

Observe that complete information about the *ContainedIn* and *InROI* relations is not yet assumed due to the application of an open-world assumption in the EDB and IDB.

Assume that the UAV generates a plan similar to that described at the beginning of this section and then executes steps 1–3 in this plan. Given its sensor capabilities under current weather conditions, suppose that the UAV agent can assert the following new set of facts in the EDB generated from its sensor actions and image processing facilities (step 3 in the plan):

$$\{In(\mathsf{c}_1,\mathsf{r}_1), In(\mathsf{c}_4,\mathsf{r}_4), Moving(\mathsf{c}_1), Moving(\mathsf{c}_4), Moving(\mathsf{c}_5)$$
$$Sig(\mathsf{c}_1,\mathsf{blue}), Sig(\mathsf{c}_4,\mathsf{yellow})\}.$$

After executing the sensor action in step 3, the UAV's on-line EDB contains the following facts:

$$\begin{aligned}&\{In(\mathsf{c}_1,\mathsf{r}_1), In(\mathsf{c}_2,\mathsf{r}_2), In(\mathsf{c}_3,\mathsf{r}_3), In(\mathsf{c}_4,\mathsf{r}_4),\\&Moving(\mathsf{c}_1), Moving(\mathsf{c}_2), Moving(\mathsf{c}_4), Moving(\mathsf{c}_5)\\&Sig(\mathsf{c}_1,\mathsf{blue}), Sig(\mathsf{c}_2,\mathsf{roadgray}), Sig(\mathsf{c}_3,\mathsf{green}), Sig(\mathsf{c}_4,\mathsf{yellow}),\\&Sig(\mathsf{c}_6,\mathsf{roadgray}), ContainedIn(\mathsf{r}_1,\mathsf{r}_2), ContainedIn(\mathsf{r}_2,\mathsf{r}_3),\\&InROI(\mathsf{r}_3)\}.\end{aligned} \tag{4}$$

At this point, observe that, due to the open-world assumption, it is unknown whether c_3 is moving and it is unknown what region c_5 is in or what color it is. Additionally, it is unknown what region c_6 is in or whether it is moving.

Before proceeding with the execution of the rest of the plan, the UAV must take stock of what it knows about the ROI, r_3. In other words, the UAV agent must query the RKDB with a particular policy of contextual closure to determine not only what it sees, but *all* that it sees under the current circumstances. The following closure context discussed above,

> after sensing region X with a camera sensor, assume that all moving vehicles in the ROI X have been perceived except for those with a signature whose color feature is roadgray,

can be represented as the following integrity constraint:

$$\begin{aligned}\forall x,r,z.[&Moving(x)\wedge In(x,r)\wedge InROI(r)\wedge Sig(x,z)\wedge\\&z\neq \mathsf{roadgray}]\rightarrow See(x,r),\end{aligned} \tag{5}$$

together with the following LCC policy:[8]

$$\mathsf{LCC}[See(x,r), ContainedIn(x,y); Moving()] : (5). \tag{6}$$

This combination states that relations $See(x,r)$ and $ContainedIn(x,y)$ are minimized, relation $Moving()$ is allowed to vary, and all other relations are fixed. The integrity constraint (5) is to be preserved. In essence, the UAV agent is assuming complete information locally about the $ContainedIn()$ and $See()$ relations by minimizing them. In addition, new information about moving may also be derived, but the only information about $Sig()$ and the other fixed relations that can be inferred is what is already in the EDB. That is the effect of *fixing* relations in this context.

Another way to view the integrity constraint (5) of the contextual closure is as the equivalent:

$$\begin{aligned}\forall x,r,z.[&In(x,r)\wedge InROI(r)\wedge Sig(x,z)\wedge z\neq \mathsf{roadgray}\wedge\\&\neg See(x,r)]\rightarrow \neg Moving(x),\end{aligned} \tag{7}$$

[8] The formal definition of an LCC policy is provided in Sect. 5.3, but we first treat it informally here.

which states that "if an object is in the ROI and it has a visible signature relative to the current capabilities of the UAV agent's sensors, if the UAV agent does not see it, then it is not moving." The integrity constraint is intended to represent strong coupling between moving and seeing due to the character of the sensor capabilities in this context.

After applying the LCC policy in the CCQ layer which includes integrity constraints, the resulting EDB/IDB combination would contain (explicitly and implicitly) the following facts:

$$\{In(\mathsf{c}_1,\mathsf{r}_1), In(\mathsf{c}_2,\mathsf{r}_2), In(\mathsf{c}_3,\mathsf{r}_3), In(\mathsf{c}_4,\mathsf{r}_4),$$
$$Moving(\mathsf{c}_1), Moving(\mathsf{c}_2), Moving(\mathsf{c}_4), Moving(\mathsf{c}_5), \neg Moving(\mathsf{c}_3),$$
$$Sig(\mathsf{c}_1,\mathsf{blue}), Sig(\mathsf{c}_2,\mathsf{roadgray}), Sig(\mathsf{c}_3,\mathsf{green}), Sig(\mathsf{c}_4,\mathsf{yellow}),$$
$$ContainedIn(\mathsf{r}_1,\mathsf{r}_2), ContainedIn(\mathsf{r}_2,\mathsf{r}_3), ContainedIn(\mathsf{r}_1,\mathsf{r}_3),$$
$$\neg ContainedIn(r,r') \text{ for all pairs } r,r' \text{ other than listed above},$$
$$InROI(\mathsf{r}_1), InROI(\mathsf{r}_2), InROI(\mathsf{r}_3)\}.$$

It is useful to note the following about the UAV agent's knowledge about the ROI, resulting from its sensing action in step 3 and subsequent reasoning about it. It still has incomplete information about the relations $In()$, $Sig()$, and $InROI()$. For example, it is unknown what signature object c_5 has or where it is. The UAV agent now knows that object c_3 is not moving and it does have complete information about the $ContainedIn()$ relation.

What about the relation $See()$, which has been minimized? One can now infer the following facts related to the relation $See()$ and the ROI, r_3:

- $See(\mathsf{c}_1,\mathsf{r}_3)$.
- $\neg See(\mathsf{c}_2,\mathsf{r}_3), \neg See(\mathsf{c}_3,\mathsf{r}_3), \neg See(\mathsf{c}_4,\mathsf{r}_3), \neg See(\mathsf{c}_6,\mathsf{r}_3)$.

c_3 is not seen because it is not moving. c_4 is not seen because it is not in the ROI, r_3. c_2 is not seen even though it is moving because of its signature. Most interestingly, it is unknown whether c_5 is seen because the UAV agent could not discern which region c_5 was in nor what its color signature was. In fact, since the UAV agent could identify c_5 as moving, the failure to discern a region for c_5 could be deduced as the reason for this, due to the tight coupling between moving and seeing. This could provide a reason for focusing on c_5 and trying to discern its region. c_6 is not seen because of its signature. What is interesting is that minimization of $See()$ does not change the status of c_6 w.r.t. $Moving()$, i.e., $Moving(\mathsf{c}_6)$ remains unknown.

The fact that $See(\mathsf{c}_5,\mathsf{r}_3)$ and $Moving(\mathsf{c}_6)$ remain unknown informs us of the subtlety of minimization in the context of rough sets. The minimization of a relation in the rough set context does not necessarily create a definition of the relation minimized. What it does do is move tuples in the boundaries of one or more relations into the positive or negative parts of the relation while meeting the conditions of the integrity constraints, whereas other tuples still remain in the boundaries. This is very important because it satisfies the ontological intuition associated with open-world reasoning.

It also worth emphasizing that if the integrity constraint (5) was defined as an intensional rule, like the following,

$$See(x) \leftarrow Moving(x) \wedge In(x,r) \wedge InROI(r) \wedge Sig(x,z) \wedge z \neq \mathsf{roadgray},$$

then the minimization of $See()$ would not result in changes in the relation $Moving()$ appearing in the body of the rule.

These subtle forms of inferencing are precisely what may be required for realistic inferencing mechanisms in fully autonomous systems, such as that described here, that may often be in situations where there is no communication for longer periods of time between the autonomous agent and ground operators. It remains to be seen whether such complex forms of inferencing can be implemented efficiently. This aspect will be considered in the remainder of the chapter.

5 The Languages of RKDBs

5.1 The Language of Extensional Databases

An extensional database consists of positive and negative facts. Thus, we assume that the language of the extensional database is a set of literals, i.e., formulas of the form $R(\bar{c})$ or $\neg R(\bar{c})$, where $R \in \mathsf{Rel}$ is a relation symbol and $\bar{c}$ is a tuple of constant symbols. It is assumed that the extensional database is consistent, i.e., it does not contain both $R(\bar{c})$ and $\neg R(\bar{c})$, for some relation $R()$ and tuple $\bar{c}$.

5.2 The Language of Intensional Databases

The intensional database is intended to infer new facts, both positive and negative, by applying intensional rules to the EDB. The process is similar to the approach using $\mathsf{Datalog}^{\neg\neg}$ (see, e.g., [1]). The rules have the form,

$$\pm P(\bar{x}) \leftarrow \pm P_1(\bar{x}_1), \ldots, \pm P_k(\bar{x}_k), \tag{8}$$

where $\pm$ is either the empty string or the negation symbol $\neg$ and any variable that appears in the head of a rule [i.e., any variable of $\bar{x}$ in a rule of the form (8)] appears also in the rule's body (i.e., among variables of $\bar{x}_1, \ldots, \bar{x}_k$ in the rule).
The rules can be divided into two layers, the first for inferring positive and the second for inferring negative facts. The first layer of rules (called the *positive* IDB *rule layer*), used for inferring positive facts, has the form,

$$P(\bar{x}) \leftarrow \pm P_1(\bar{x}_1), \ldots, \pm P_k(\bar{x}_k); \tag{9}$$

the second layer of rules (called the *negative* IDB *rule layer*), used for inferring negative facts, has the following form:

$$\neg P(\bar{x}) \leftarrow \pm P_1(\bar{x}_1), \ldots, \pm P_k(\bar{x}_k). \tag{10}$$

5.3 The Language of Integrity Constraints and LCC Policies

Integrity constraints are expressed as formulas of classical first-order logic. Intuitively, they can be considered implicit definitions of intensional relations that will be minimized or maximized by the LCC assumptions in a specific LCC policy. In the following sections, to obtain tractable instances of the general algorithm, we will impose some syntactic restrictions on the syntactic form of ICs together with the LCC assumptions in a specific LCC policy (see Sect. 8).

LCC *policies* are expressions of the form,

$$\mathsf{LCC}[L_1,\ldots,L_p;K_1,\ldots,K_r]\text{:IC}, \tag{11}$$

where $L_1,\ldots,L_p$ are (positive or negative) literals, $K_1,\ldots,K_r$ are relation symbols not appearing in L_i's, and IC is a set of integrity constraints. Literals $L_1,\ldots,L_p$ are minimized and relations $K_1,\ldots,K_r$ are allowed to vary. By an LCC assumption, we mean a minimization or maximization of a single literal from $L_1,\ldots,L_p$ in (11).
In the following sections, we often omit the part ":IC" of (11) if the corresponding integrity constraints are known from the context.

6 The Semantics of RKDBs

6.1 Notational Conventions

Let us denote the facts in the extensional database by EDB and the facts in the intensional database by IDB. Let $R_1,\ldots,R_n$ be all relations in the RKDB. For a specific relation R in the RKDB, we denote the positive atoms of R in the EDB by $EDB^+(R)$ and the negative atoms of R in the EDB by $EDB^-(R)$. Assume that

- By EDB^+, we denote the positive part of the EDB which is $\bigcup_{i=1}^{n} EDB^+(R_i)$.
- By EDB^-, we denote the negative part of the EDB which is $\bigcup_{i=1}^{n} EDB^-(R_i)$.

The EDB is then equivalent to $EDB^+ \cup EDB^-$.

For a specific relation R in the RKDB, we denote the positive atoms of R in the IDB generated by the positive intensional rules of form (9) by $IDB^+(R)$ (where it is assumed that $IDB^+(R) \equiv IDB^+(R) \setminus EDB^+(R)$) and the negative atoms of R in the IDB generated by the negative intensional rules of form (10) by $IDB^-(R)$ (where it is assumed that $IDB^-(R) \equiv IDB^-(R) \setminus EDB^-(R)$). Assume also that

- By IDB^+, we denote the positive part of the IDB which is $\bigcup_{i=1}^{n} IDB^+(R_i)$.
- By IDB^-, we denote the negative part of the IDB which is $\bigcup_{i=1}^{n} IDB^-(R_i)$.

The IDB is then equivalent to $IDB^+ \cup IDB^-$.

Let D be a finite set (a domain of the database). The semantics of constant symbols and variables is given by an assignment of domain values to constants and variables, called a *valuation*:

$$v : \mathsf{Const} \cup V_{\mathsf{I}} \longrightarrow D.$$

The valuation v is then extended to the vectors of constants and variables in the usual way. We also assume that the unique names assumption (UNA) holds, i.e., for all different constants c, c' in the RKDB, we assume that c and c' denote different objects. In other words, the formula $c_i \neq c_j$ is satisfied for each i, j, where $i \neq j$.

In the semantics defined in the following sections, all relations are interpreted as rough sets of tuples, where no form of domain closure is required. The symbol $\Vdash$ will denote the RKDB entailment relation and the symbol $\models$ will denote the classical two-valued entailment relation. By indexing relations with EDB, IDB, and LCC, we indicate that they are considered in the particular context as relations of the extensional, intensional, and CCQ layer of the RKDB, respectively.

6.2 The Semantics of Extensional Databases

The semantics of the extensional database is given by rough sets of tuples. Let $R()$ be a relational symbol appearing in the extensional database. Then $R()$ is interpreted as the rough set whose positive part contains all tuples $v(\bar{c})$ for which literal $R(\bar{c})$ is in the database, and the negative part contains all tuples $v(\bar{c})$ for which literal $\neg R(\bar{c})$ is in the database. All other tuples are in the boundary region of $R()$.

$$EDB \Vdash R(\bar{a}) \text{ iff } R(\bar{a}) \in EDB^{+}(R),$$
$$EDB \Vdash \neg R(\bar{a}) \text{ iff } \neg R(\bar{a}) \in EDB^{-}(R),$$

where $R()$ is a relation of the EDB and $\bar{a}$ is a tuple of constants.

Rough relations for the EDB are then defined as follows:

$$R^{+}_{EDB} = \{v(\bar{a}) : EDB \Vdash R(\bar{a})\},$$
$$R^{-}_{EDB} = \{v(\bar{a}) : EDB \Vdash \neg R(\bar{a})\},$$
$$R^{\pm}_{EDB} = \{v(\bar{a}) : EDB \nVdash R(\bar{a}) \text{ and } EDB \nVdash \neg R(\bar{a})\}.$$

6.3 The Semantics of Intensional Databases

The semantics of the intensional database is given by rough sets of tuples after application of the intensional rules to the extensional database. Intensional rules can be viewed as rough set transducers (see [5]). Basically, a rough set transducer takes rough sets as input and generates new or modified rough sets as output meeting the constraints of the transducer, a set of formulas.

To provide the semantics of the IDB, we will use the definition of the so-called Feferman–Gilmore translation (see, e.g., [2]) as a basis.

Definition 3. By a *Feferman–Gilmore translation of formula* α, denoted by $FG(\alpha)$, we mean the formula obtained from α by replacing all negative literals of the form $\neg R(\bar{y})$ by $R^-(\bar{y})$ and all positive literals of the form $R(\bar{y})$ by $R^+(\bar{y})$.

Let $\bar{S} = (S_1, \ldots, S_p)$ contain all relation symbols of the form R^+ and R^-, where R is a relation symbol in an IDB rule. For any relation S_i, all rules with S_i^+ (respectively, S_i^-) in their heads should be gathered into a single formula of the form,

$$\forall \bar{y}_i.[S_i^{\pm}(\bar{y}_i) \leftarrow \alpha_i(\bar{y}_i)],$$

where

$$\alpha_i(\bar{y}_i) \equiv \bigvee_j \exists \bar{z}_j.\beta_{ij}(\bar{z}_j)$$

where $\beta_{ij}(\bar{z}_j)$ denotes the bodies of the appropriate rules and $\pm$ stands for $+$ or $-$, respectively.

Denote by $\mu\bar{S}.[\alpha(\bar{S})]$ the least, and by $\nu\bar{S}.[\alpha(\bar{S})]$, the greatest simultaneous fixed-point operator of $\alpha(\bar{S})$ (for the definition of simultaneous fixed-points see, e.g. [9]). Define $\bar{S}_{IDB} \equiv \mu\bar{S}.[FG(\alpha_1), \ldots, FG(\alpha_p)]$. In some cases the IDB might appear inconsistent when there is a relation $R()$ such that $R^+ \cap R^- \neq \emptyset$. In what follows we require that the IDB is consistent, i.e., for all IDB relations $R()$, $R^+ \cap R^- = \emptyset$. This consistency criterion can be verified in time polynomial in the size of the database.

The semantics of IDB rules is then defined as follows:

$$IDB \Vdash\!\!\vdash R(\bar{a}) \text{ iff } \bar{a} \in EDB^+(R) \cup IDB^+(R),$$
$$IDB \Vdash\!\!\vdash \neg R(\bar{a}) \text{ iff } \bar{a} \in EDB^-(R) \cup IDB^-(R),$$

where $R()$ is a relation in the EDB or in the head of an intensional rule, $\bar{a}$ is a tuple of constants, and $IDB^+(R)$ and $IDB^-(R)$ are computed from the simultaneous fixed-point definition $\bar{S}_{IDB}$ defined above.

Rough relations for the IDB are then defined as follows:

$$R^+_{IDB} = \{v(\bar{a}) : IDB \Vdash\!\!\vdash R(\bar{a})\},$$
$$R^-_{IDB} = \{v(\bar{a}) : IDB \Vdash\!\!\vdash \neg R(\bar{a})\},$$
$$R^{\pm}_{IDB} = \{v(\bar{a}) : IDB \not\Vdash\!\!\vdash R(\bar{a}) \text{ and } IDB \not\Vdash\!\!\vdash \neg R(\bar{a})\}.$$

Observe that

$$EDB \Vdash\!\!\vdash R(\bar{a}) \text{ implies } IDB \Vdash\!\!\vdash R(\bar{a}),$$
$$EDB \Vdash\!\!\vdash \neg R(\bar{a}) \text{ implies } IDB \Vdash\!\!\vdash \neg R(\bar{a}).$$

Remark 1. If one wants to distinguish between facts entailed solely by application of intensional rules, this can be done in a straightforward manner, but as a rule, one is interested in querying both the EDB and IDB together, thus the choice of RKDB entailment from the IDB.

6.4 The Semantics of the CCQ Layer and LCC Policies

The inference mechanism associated with the CCQ layer is intended to provide a form of *contextual closure* relative to part of the EDB and IDB when querying the RKDB. A *contextually closed query* consists of

- The *query* itself, which can be any fixed-point or first-order query.
- The *context* represented as a set of one or more integrity constraints.
- A *local closure policy* representing the closure context and consisting of a minimization policy representing the local closure.

An LCC *policy* consists of a context and a local closure policy. LCC policies may also be viewed as rough set transducers with rough relations in the EDB and IDB as input, a transducer consisting of one or more integrity constraints and a minimization policy, and modified rough relations in the RKDB as output.

Let the EDB and IDB be defined as before, let IC denote a finite set of integrity constraints, and let RKDB:LCC$[\bar{L};\bar{K}]$:IC denote querying the three layers of the RKDB with a specific LCC policy LCC$[\bar{L};\bar{K}]$:IC. Then,[9]

$$\begin{aligned}
&\text{RKDB:LCC}[\bar{L};\bar{K}]\text{:IC} \Vdash R(\bar{a}) \text{ iff}\\
&\qquad\qquad CIRC(IC \cup IDB \cup EDB; \bar{L}; \bar{K}) \models R(\bar{a}),\\
&\text{RKDB:LCC}[\bar{L};\bar{K}]\text{:IC} \Vdash \neg R(\bar{a}) \text{ iff}\\
&\qquad\qquad CIRC(IC \cup IDB \cup EDB; \bar{L}; \bar{K}) \models \neg R(\bar{a}),
\end{aligned}$$

where the notation is, as in Sect. 6.1, under the assumption that the circumscriptive theory is consistent.

Thus, the CCQ layer has the purpose of dynamically redefining some relations to satisfy ICs in a particular query. A relation R which is minimized, maximized or allowed to vary is defined as the following rough relation:

$$\begin{aligned}
R^{+}_{\text{LCC}} &= \{v(\bar{a}) : \text{RKDB:LCC}[\bar{L};\bar{K}]\text{:IC} \Vdash R(\bar{a})\},\\
R^{-}_{\text{LCC}} &= \{v(\bar{a}) : \text{RKDB:LCC}[\bar{L};\bar{K}]\text{:IC} \Vdash \neg R(\bar{a})\},\\
R^{\pm}_{\text{LCC}} &= \{v(\bar{a}) : \text{RKDB:LCC}[\bar{L};\bar{K}]\text{:IC} \nVdash R(\bar{a}) \text{ and}\\
&\qquad \text{RKDB:LCC}[\bar{L};\bar{K}]\text{:IC} \nVdash \neg R(\bar{a})\}.
\end{aligned}$$

Intuitively, this means that the positive part of $R()$ contains tuples present in all extensions of $R()$ satisfying the ICs, the boundary part contains tuples present in some extensions of $R()$ satisfying the ICs, but not in all of them, and the negative part of $R()$ contains tuples not present in any extension of $R()$ satisfying the ICs.

The relations that are not minimized, maximized, or allowed to vary are not changed; thus their semantics is that given by the EDB and IDB layers of the RKDB.

[9] Observe that we abuse notation somewhat by using sets of literals, $\bar{L},\bar{K}$ for minimized and varied predicate constants in the circumscription formula $CIRC()$. Formally, we should use predicate constants contained in $\bar{L},\bar{K}$, respectively.

Remark 2. The inference mechanism associated with the CCQ layer is almost always used with both layers of the EDB and IDB. If one wants to apply LCC inference to just the EDB, this can also be done in a straightforward manner.

7 The Computation Method

7.1 The Pragmatics of Computing Contextual Queries

A contextual query in its simplest form involves the (implicit) generation of the extension of a relation R in the context of a set of integrity constraints and a minimization policy and asking whether one or more tuples is a member of that relation. Essentially, we are required to implicitly compute R^{+}_{LCC}, R^{-}_{LCC}, and $R^{\pm}_{\mathsf{LCC}}$ and determine whether the tuple or tuples are in any of the resulting rough set partitions of R. In Sect. 7.2, we will describe an algorithm to do this. Based on this specification, we will be able to show that in some cases, where the LCC policy associated with the query is restricted appropriately, querying the relation R can be done very efficiently. One of the more important results is that one can automatically generate syntactic characterizations of each of the partitions of a rough set relation without actually generating their explicit extensions. The syntactic characterizations can then be used to efficiently query the RKDB.

Since integrity constraints are not associated with the EDB/IDB pair, but with an agent posing a query, the integrity constraints associated with an agent are not necessarily satisfied together with the EDB/IDB. Checking satisfiability is tractable in this context, due to the first-order or fixed-point nature of the integrity constraints and the finiteness of the database. Under additional syntactic restrictions, the satisfiability of the circumscriptive theory can also be guaranteed. In the case of inconsistency, this would lead to the specification and computation of specific update policies which is a topic for future research.

7.2 The Algorithm

The algorithm presented below applies to the general case, i.e., to the problem which is co-NPTIME complete (see Sect. 7.4). However, in Sect. 8, we show specializations of the algorithm to some cases, where PTIME complexity is guaranteed. The inputs to the algorithm are

- An extensional database EDB
- An intensional database IDB
- A set of integrity constraints IC
- An LCC policy LCC$[\bar{L};\bar{K}]$:IC
- A relation symbol R.[10]

[10] The relation symbol R can be viewed as part of the query which consists of a number of relations that are required to compute the full query.

As output, the algorithm returns the definition of the relation $R()$ obtained by applying the LCC policy and preserving the ICs, according to the semantics defined in Sect. 6.

1. Construct $C \equiv CIRC(IC \cup IDB \cup EDB; \bar{L}; \bar{K})$ representing the given LCC policy applied to the IDB together with the EDB.
2. Eliminate second-order quantifiers from the formula obtained in step 1. In general, the elimination may fail, and the result is the initial second-order formula C, however, if certain restrictions concerning the form of IC are assumed, the elimination of second-order quantifiers is guaranteed (see Sect. 8).
3. Calculate the intersection of all extensions of R satisfying formula C. If there is not any relation $R()$ satisfying C, terminate and return the answer "unsatisfiable," meaning that either the EDB and IDB pair is inconsistent, or the ICs cannot be satisfied.
4. Calculate the union of all extensions of R satisfying formula C.
5. For any tuple $\bar{a}$:
   - if $v(\bar{a})$ is in the intersection calculated in step 3, add $v(\bar{a})$ to $R^+()$.
   - if $v(\bar{a})$ is not in the union calculated in step 4, add $v(\bar{a})$ to $R^-()$.
   - if none of the above two cases applies, then $v(\bar{a})$ is in $R^\pm()$.

In practice, one uses particular second-order quantifier elimination algorithms (see e.g., [6,16,17,21]) that may fail. Since second-order formulas are useless as results, it is reasonable to return the answer "unknown" when the elimination algorithm used in step 2 fails. This implies that the algorithm is only sound relative to the semantics provided in Sect. 6.4.

Observe also that, in practice, it is often better to calculate the definitions of new relations rather than calculating their extensions as in the above algorithm. To achieve this goal one can apply, e.g., techniques proposed in [4,7].

7.3 Expressiveness of the Approach

Noted that the approach we consider here subsumes that of [8], namely, consider a policy LCW[$\beta(\bar{R})$] of [8], meaning that formula $\beta(\bar{R})$ is to be minimized, whereas all relations in β, i.e. $\bar{R}$, are allowed to vary. This LCW policy is expressible by the following LCC policy:

$$\mathsf{LCC}[S;\bar{R}] : \{\forall \bar{x}.[\beta(\bar{R}) \rightarrow S(\bar{x})]\},$$

where $\bar{x}$ denotes all free variables of $\beta(\bar{R})$ and S is a fresh relation symbol, not appearing among symbols in $\bar{R}$.

An interesting question arises whether the current approach allows one to express all tractable LCC policies, where by a *tractable* LCC *policy*, we mean any LCC policy such that all minimized, maximized, and varied relations are PTIME-computable. The following characterization shows that the method presented is strong enough to express all tractable LCC policies. In other words, any tractable LCC policy can always be reformulated in the form used in Lemma 1 below. In Sect. 8, we provide additional syntactic characterizations of LCC policies that guarantee tractability.

Lemma 1. Given the LCC semantics for LCC policies provided in Sect. 6.4 and assuming that the database domain is ordered, all tractable LCC policies can be expressed as policies of the form,

$$\mathsf{LCC}[\bar{L};\bar{K}] : \{\beta_i(\bar{x}) \rightarrow L_i(\bar{x}) : L_i \in \bar{L}\},$$

where each $\beta_i(\bar{x})$ is a first-order formula positive w.r.t. L_i.

Proof. Any relation computable in PTIME can be expressed by means of the least fixed-point of a formula of the form,

$$\beta_i(\bar{x}) \rightarrow L_i(\bar{x}), \tag{12}$$

provided that the database domain is ordered (see, e.g., [9]). Since all minimized, maximized, and varied relations are assumed to be tractable, they can be expressed by the least fixed-points of formulas of the form (12), thus also by policy $\mathsf{LCC}[\bar{L};\bar{K}]$: $\{\beta_i(\bar{x}) \rightarrow L_i(\bar{x}) : L_i \in \bar{L}\}$. □

7.4 Complexity of the Approach

In general, the problem of querying the database in the presence of unrestricted ICs is co-NPTIME complete. On the other hand, some classes of LCC policies for which the computation mechanism is in PTIME can be isolated (see, e.g., [8] and also Sect. 8.3).

8 Important Particular Cases

In this section, we consider a number of restrictions on ICs that allow us to compute explicit definitions of the new relations as first-order and fixed-point formulas. In such cases, computing contextually closed queries is in PTIME.

Let $M(\bar{R})$ stand for $IC \cup EDB \cup IDB$, and assume that the ICs have the following form:

$$\forall \bar{x}.[\alpha(\bar{x}) \rightarrow \beta(\bar{x})], \tag{13}$$

where α and β are first-order formulas.

Definition 4. By a *marking* of relation symbols for the policy $\mathsf{LCC}[\bar{L};\bar{K}]$:IC, we understand a mapping assigning, to any relation symbol, both in the local closure policy $\mathsf{LCC}[\bar{L};\bar{K}]$ and IC in the LCC policy, the least subset of $\{\min, \max\}$ that is closed under the following rules:

1. For any relation symbol S appearing in $\bar{L}$ positively, min is in the set of marks of S.
2. For any relation symbol S appearing in $\bar{L}$ negatively, max is in the set of marks of S.
3. If $\alpha(R) \rightarrow \beta(S)$ is in IC, $R, S \in \bar{L} \cup \bar{K}$ and S occurs in β positively and is marked by min, or S occurs in β negatively and is marked by max; then,
   - If R occurs positively in α, min is in the set of marks of R.
   - If R occurs negatively in α, max is in the set of marks of R.
4. If $\alpha(R) \rightarrow \beta(S)$ is in IC, $R, S \in \bar{L} \cup \bar{K}$ and α contains a positive occurrence of R and R is marked by max, or α contains a negative occurrence of R and R is marked by min then:
   - if S occurs positively in β, then 'max' is in the set of marks of S.
   - if S occurs negatively in β, then 'min' is in the set of marks of S.

An $\mathsf{LCC}[\bar{L};\bar{K}]$:IC policy is called *uniform* if no relation symbol is marked by both max and min.

Example 2. Let us consider the following integrity constraint:

$$[Car(x) \wedge Red(x)] \rightarrow RedCar(x). \tag{14}$$

The marking for the policy,

$$\mathsf{LCC}[\{RedCar(x), Car(x)\}; \{Red(x)\}] : (14),$$

assigns the mark min to all the relation symbols. Thus the policy is uniform. On the other hand, the marking for the policy,

$$\mathsf{LCC}[\{RedCar(x), \neg Car(x)\}; \{Red(x)\}] : (14),$$

assigns the mark min to *Red* and the marks $\{\text{min}, \text{max}\}$ to *Car* and *RedCar*. Thus the latter policy is not uniform.

8.1 The Case of Universal LCC policies

Definition 5. By a *universal* LCC *policy*, we understand any uniform policy,

$$\mathsf{LCC}[\bar{L};\bar{K}]\text{: IC},$$

in which IC is a set of constraints of the following form,

$$\forall \bar{y}. \{[\pm P_1(\bar{x}_1) \wedge \ldots \wedge \pm P_k(\bar{x}_k)] \rightarrow \pm P(\bar{x})\}, \tag{15}$$

where $P_1, \ldots, P_k, P$ are relation symbols, $\bar{y}$ is the vector of all variables occurring in $\bar{x}_1, \ldots, \bar{x}_k, \bar{x}$, and $\bar{x} \subseteq \bar{x}_1 \cup \ldots \cup \bar{x}_k$.

For universal integrity constraints, we will have a computation method much more efficient than that described in Sect. 7.2. In the rest of this section, we will consider only universal LCC policies LCC$[\bar{L};\bar{K}]$:IC, for given sets of literals $\bar{L}$,$\bar{K}$ and a set IC of integrity constraints.

In the computation method for universal policies, we first construct minimal rough relations satisfying the EDB, IDB, and the integrity constraints, where minimality is defined w.r.t. the so-called *information ordering* considered by Fitting and van Benthem (see, e.g., [2]) in the context of three-valued logics. The definition of information ordering follows.

Definition 6. Let R and S be rough relations. We define *information ordering*, denoted by $R \sqsubseteq S$, as follows:

$$R \sqsubseteq S \stackrel{\text{def}}{\equiv} R^+ \subseteq S^+ \text{ and } R^- \subseteq S^- .$$

To find minimal w.r.t. $\sqsubseteq$ rough relations satisfying IC, EDB, and IDB, we will use the following tautologies of first-order logic:

$$\begin{array}{l} \forall \bar{x}. \{\alpha(\bar{R}) \rightarrow [\beta(\bar{R}) \vee M(\bar{y})]\} \equiv \forall \bar{x}. \{[\alpha(\bar{R}) \wedge \neg M(\bar{y})] \rightarrow \beta(\bar{R})\}, \\ \forall \bar{x}. \{[\alpha(\bar{R}) \wedge M(\bar{y})] \rightarrow \beta(\bar{R})\} \equiv \forall \bar{x}. \{\alpha(\bar{R}) \rightarrow [\beta(\bar{R}) \vee \neg M(\bar{y})]\}, \end{array} \tag{16}$$

where it is assumed that all double negations $\neg\neg$ are removed.

Definition 7. Let I be an integrity constraint of the form:

$$\forall \bar{x}. \{[\pm R_1(\bar{y}_1) \wedge \ldots \wedge \pm R_m(\bar{y}_m)] \rightarrow \pm S(\bar{z})\}. \tag{17}$$

Let $\mathcal{P} =$ LCC$[\bar{L};\bar{K}]$:I be an LCC policy. By the *expansion of I w.r.t. $\mathcal{P}$*, denoted by $Exp^{\mathcal{P}}(I)$, we understand the least set of constraints of the form,

$$\forall \bar{x}. \left\{ \left[\bigwedge_k L_k(\bar{x}_k) \right] \rightarrow \pm S(\bar{x}_S) \right\},$$

obtained from (17) by applying the tautologies (16), such that any (possibly negated) literal of (17) containing a relation symbol occurring in $\bar{L};\bar{K}$, is a consequent of exactly one constraint.

Example 3. Consider the integrity constraint

$$I \stackrel{\text{def}}{\equiv} \forall x,y. \{[\neg P(x) \wedge S(x,y)] \rightarrow P(y)\},$$

and the policy $\mathcal{P} =$ LCC$[P;S]$:I. The expansion of I w.r.t. $\mathcal{P}$ is defined as the following set of constraints:

$$\begin{aligned} Exp^{\mathcal{P}}(I) = \; & (\forall x,y. \{[\neg P(x) \wedge S(x,y)] \rightarrow P(y)\} \\ & \forall x,y. \{[\neg P(y) \wedge S(x,y)] \rightarrow P(x)\} \\ & \forall x,y. \{[\neg P(x) \wedge \neg P(y)] \rightarrow \neg S(x,y)\}). \end{aligned}$$

In the case of policy $\mathcal{P}'$ =LCC$[S;\emptyset]$:I, the expansion of I is defined as

$$Exp^{\mathcal{P}'}(I) = (\forall x,y.\{[\neg P(x) \wedge \neg P(y)] \rightarrow \neg S(x,y)\}).$$

Let us fix an LCC policy $\mathcal{P}$ =LCC$[\bar{L};\bar{K}]$:IC. To compute the definition of minimal w.r.t. $\sqsubseteq$ rough relations, satisfying the constraints IC, EDB, and IDB, we consider the following cases:

- If $S_i \notin \bar{L} \cup \bar{K}$, then the positive part of the resulting relation, $S_i^+()$, contains exactly the tuples present in $EDB^+(S_i) \cup IDB^+(S_i)$, and the negative part of the resulting relation, $S_i^-()$, contains exactly the tuples present in $EDB^-(S_i) \cup IDB^-(S_i)$.
- If $\bar{S} = \bar{L} \cup \bar{K}$, then we consider the set of integrity constraints:

$$\{FG(\alpha) : \alpha \in Exp^{\mathcal{P}}(I) \text{ and } I \in IC\},$$

where FG is the Feferman–Gilmore translation defined in Definition 3.

We assume that the following integrity constraints, reflecting the contents of EDB and IDB, are implicitly given:

$$\begin{aligned}
&\forall \bar{y}.\{EDB^+[S(\bar{y})] \rightarrow S^+(\bar{y})\},\\
&\forall \bar{y}.\{EDB^-[S(\bar{y})] \rightarrow S^-(\bar{y})\},\\
&\forall \bar{y}.\{IDB^+[S(\bar{y})] \rightarrow S^+(\bar{y})\},\\
&\forall \bar{y}.\{IDB^-[S(\bar{y})] \rightarrow S^-(\bar{y})\},
\end{aligned}$$

where the empty parts $EDB^+(S(\bar{y}))$, $EDB^-(S(\bar{y}))$, $IDB^+(S(\bar{y}))$, and $IDB^-(S(\bar{y}))$ are interpreted as false.

Now, for each $S_i \in \bar{S}$, gather all the ICs with S_i^+ as the consequent into the following single formula:

$$\forall \bar{y}.\left\{\left[\bigvee_{1\le k\le k_i} \exists \bar{z}_{ik}.\phi_{ik}(\bar{R}_k)\right] \rightarrow S_i^+(\bar{y})\right\}, \tag{18}$$

and all the ICs with S_i^- as the consequent into the following single formula:

$$\forall \bar{y}.\left\{\left[\bigvee_{1\le j\le j_i} \exists \bar{z}_{ij}.\psi_{ij}(\bar{R}_j)\right] \rightarrow S_i^-(\bar{y})\right\}. \tag{19}$$

The following definitions of the positive and the negative part of the required minimal rough relations wrt policy $\mathcal{P}$, indicated by the index $\mathcal{P}$, can now be derived:

$$\bar{S}^+_{\mathcal{P}}(\bar{y}) \equiv \mu\bar{S}(\bar{y}).\left[\bigvee_{1\le k\le k_1} \exists \bar{z}_{1k}.\phi_{1k}(\bar{R}_k), \ldots, \bigvee_{1\le k\le k_n} \exists \bar{z}_{nk}.\phi_{nk}(\bar{R}_k)\right] \tag{20}$$

$$\bar{S}^{-}_{\mathcal{P}}(\bar{y}) \equiv \mu\bar{S}(\bar{y}).\left[\bigvee_{1\le j\le j_1} \exists\bar{z}_{1j}.\psi_{1j}(\bar{R}_j),\ldots,\bigvee_{1\le j\le j_m} \exists\bar{z}_{mj}.\psi_{mj}(\bar{R}_j)\right]. \tag{21}$$

Observe that the syntactic restrictions placed on the ICs guarantee that the formulas under the fixed-point operators are positive, thus, the monotone w.r.t. S and consequently, the fixed-points exist. Observe also, that for nonrecursive universal LCC policies, the fixed-point operators can be removed, and the definitions obtained are classical first-order formulas.[11]

Having computed the suitable parts of the relations in all integrity constraints, one can easily perform a consistency check, indicating whether the ICs can be satisfied by the current contents of the $EDB \cup IDB$. For each relation $R()$, one needs to assure that $R^+ \cap R^- = \emptyset$.

Definition 8. Let $\mathcal{P}$ =LCC$[\bar{L};\bar{K}]$:IC be an LCC policy. The *rough negation for the policy* $\mathcal{P}$, denoted by $\sim_{\mathcal{P}}$, is defined as follows:

- $\sim_{\mathcal{P}}$ satisfies the usual DeMorgan laws for quantifiers, conjunction, and disjunction, and

$$\sim_{\mathcal{P}} \mu\bar{R}.\alpha(\bar{R}) \stackrel{\text{def}}{\equiv} \nu\bar{R}. \sim_{\mathcal{P}} \alpha(\bar{R}),$$
$$\sim_{\mathcal{P}} \nu\bar{R}.\alpha(\bar{R}) \stackrel{\text{def}}{\equiv} \mu\bar{R}. \sim_{\mathcal{P}} \alpha(\bar{R}).$$

- If $S \in \bar{L} \cup \bar{K}$, then,

$$\sim_{\mathcal{P}} S^+() \stackrel{\text{def}}{\equiv} \neg S^+(), \quad \sim_{\mathcal{P}} S^-() \stackrel{\text{def}}{\equiv} \neg S^-(),$$
$$\sim_{\mathcal{P}} \neg S^+() \stackrel{\text{def}}{\equiv} S^+(), \quad \sim_{\mathcal{P}} \neg S^-() \stackrel{\text{def}}{\equiv} S^-().$$

- If $S \notin \bar{L} \cup \bar{K}$, then,

$$\sim_{\mathcal{P}} S^+() \stackrel{\text{def}}{\equiv} S^-(), \qquad \sim_{\mathcal{P}} S^-() \stackrel{\text{def}}{\equiv} S^+(),$$
$$\sim_{\mathcal{P}} \neg S^+() \stackrel{\text{def}}{\equiv} \neg S^-(), \quad \sim_{\mathcal{P}} \neg S^-() \stackrel{\text{def}}{\equiv} \neg S^+().$$

If the ICs are consistent with $EDB \cup IDB$, then the definitions of minimal and maximal rough relations satisfying the ICs and reflecting the semantics introduced in Sect. 6.4 can be calculated as follows:[12]

$$S^+_{min}(\bar{y}) \equiv \bar{S}^+_{\mathcal{P}}, \tag{22}$$
$$S^-_{min}(\bar{y}) \equiv \sim_{\mathcal{P}} S^+_{min}(\bar{y}), \tag{23}$$
$$S^-_{max}(\bar{y}) \equiv \bar{S}^-_{\mathcal{P}}, \tag{24}$$
$$S^+_{max}(\bar{y}) \equiv \sim_{\mathcal{P}} S^-_{max}(\bar{y}). \tag{25}$$

[11] In both cases, however, computing the defined parts of relations can be done in time polynomial in the size of $EDB \cup IDB$.

[12] Observe that the LCC policies provide us with direct information about which relations are to be maximized, which are to be minimized, and which remain unchanged.

For nonrecursive universal policies, the fixed-point operators can be removed, as before.

Observe that definitions of varied predicates can now be computed by noticing that these are the minimal w.r.t. $\sqsubseteq$ rough relations satisfying the ICs in the new context of minimized and maximized relations. It then suffices to apply definitions (20) and (21) with minimized and maximized relations replaced by their definitions obtained as (22–25) , as appropriate.

Example 4. Consider the UAV sensing example introduced in Sect. 4. The definition of minimal *See*() is given by

$$\begin{aligned} See^+_{min}(x,r) \equiv\ & See^+(x,r) \vee \{\exists z.\ [Moving^+(x) \wedge In^+(x,r) \wedge \\ & InROI^+(r) \wedge Sig^+(x,z) \wedge z \neq \mathsf{roadgray}]\}, \\ See^-_{min}(x,r) \equiv\ & \neg See^+(x,r) \wedge \{\forall z.\ [\neg Moving^+(x) \vee In^-(x,r) \vee \\ & InROI^-(r) \vee Sig^-(x,z) \vee z = \mathsf{roadgray}]\}. \end{aligned}$$

The varied relation *Moving*() is defined by

$$\begin{aligned} Moving^+_{var}(x) \equiv\ & Moving^+(x), \\ Moving^-_{var}(x) \equiv\ & Moving^-(x) \vee \{\exists r \exists z.\ [In^+(x,r) \wedge InROI^+(r) \wedge \\ & Sig^+(x,z) \wedge z \neq \mathsf{roadgray} \wedge See^-_{min}(x,r)]\}. \end{aligned}$$

Example 5. Consider the problem of determining whether a given car on a road is seen. We assume that large cars are usually seen. Our database contains the following relations:

- *Car*() containing cars
- *Large*() containing large objects
- *See*() containing visible objects
- *Ab*() standing for abnormal objects, i.e., large but invisible objects.

Define the following integrity constraint IC:

$$\forall x.\ \{[Car(x) \wedge Large(x) \wedge \neg See(x)] \rightarrow Ab(x)\}.$$

We want to minimize abnormality, i.e., to minimize relation *Ab*, while keeping the relations *Car* and *Large* unchanged. The local closure policy is then

$$\mathsf{LCC}[\{Ab(x)\}; \{See(x)\}].$$

According to Lemma 2.4, we obtain the following characterizations of *Ab*() and *See*():

$$\begin{aligned} Ab^+_{min}(x) &\equiv Ab^+(x) \vee [Car^+(x) \wedge Large^+(x) \wedge See^-(x)], \\ Ab^-_{min}(x) &\equiv Ab^-(x) \vee [Car^-(x) \vee Large^-(x) \vee \neg See^-(x)], \\ See^+_{var}(x) &\equiv See^+(x) \vee [Car^+(x) \wedge Large^+(x) \wedge Ab^-_{min}(x)], \\ See^-_{var}(x) &\equiv See^-(x). \end{aligned}$$

8.2 The Case of Semi-Horn LCC Policies

Assume that

1. Any integrity constraint IC_j is expressed as a formula of the form
$$\forall \bar{x}.[\beta_j(\bar{x}) \rightarrow S_j(\bar{x})],$$
where for each $j = 1,\ldots,n$, S_j is a relation symbol and $\beta_j(\bar{x})$ is a first-order formula.
2. Any LCC assumption L_j, in the given LCC policy, has the form S_j or $\neg S_j$, where S_j is a relation symbol.

We now have the the following proposition.

Proposition 1. Under the above assumptions 1 and 2,

$$Circ(M(R);\bar{L};\emptyset) \equiv M(\bar{R}) \wedge \bigwedge_{j=1}^{n} \forall \bar{y}.\left[\neg L_j(\bar{y}) \vee \neg A_j(\bar{y})\right], \tag{26}$$

where $A_j(\bar{y})$ is the following second-order formula:

$$A_j(\bar{y}) \equiv \exists \bar{S}' \exists \bar{K}'. \left\{ M(\bar{R}') \wedge \neg L_j'(\bar{y}) \wedge \bigwedge_{1 \le i \le n, i \ne j} \forall \bar{x}. \left[\neg L_i'(\bar{x}) \vee L_i(\bar{x})\right] \right\}, \tag{27}$$

in which $\bar{R}' \equiv \bar{S}' \cup \bar{K}' \cup (\bar{R} - \bar{S} - \bar{K})$, L_j' stands for $L_j[\bar{S} \leftarrow \bar{S}', \bar{K} \leftarrow \bar{K}']$ and $M(\bar{R}')$ represents $IC' \cup IDB' \cup EDB'$.

The following lemma holds under assumptions 1 and 2.

Lemma 2.

1. If the LCC assumption L_j has the form $\neg S_j$ and β_j is a semi-Horn formula, then the elimination of second-order quantifiers is guaranteed in time polynomial in the size of formula β_j.
2. If β_j is expressed as a weak semi-Horn formula, then the elimination of second-order quantifiers is guaranteed in PTIME. The resulting formula $A_j(\bar{y})$ is a classical or fixed-point first-order formula. It is also guaranteed that elimination of second-order quantifiers from the formula $\exists \bar{K} Circ(\bar{K})$ succeeds in time polynomial in the size of formula β_j. However, necessary computations may require calculation of simultaneous fixed-points.
3. If β_j is expressed as a weak Ackermann formula, then the elimination of second-order quantifiers is guaranteed in PTIME, and $A_j(\bar{y})$ is a classical first-order formula. Elimination of second-order quantifiers from the formula $\exists \bar{K} Circ(\bar{K})$ is also guaranteed in time polynomial in the size of formula β_j.

4. For of a consistent EDB, a uniform LCC policy, and nonrecursive integrity constraints, $A_j(\bar{y})$ is expressed by the following formula:

$$A_j(\bar{y}) \equiv \neg L'_j(\bar{y})(\bar{K}' \leftarrow \neg\bar{K}^-, \neg\bar{K}' \leftarrow \neg\bar{K}^+),$$

in which all L'_j are replaced accordingly by $\beta_j(\bar{K}')$ or $\neg\beta_j(\bar{K}')$, $\neg K_i$ are replaced by $\neg K_i^+$, and K_i by $\neg K_i^-$.
5. If L_j is of the form S_j and $\neg\beta_j$ is a semi-Horn formula, then the elimination of second-order quantifiers is guaranteed. However, the resulting formula $A_j(\bar{y})$ may have an exponential size w.r.t. the size of β_j.

For semi-Horn formulas, the following lemma, simplifying the inference method, holds.

Lemma 3. If $\Psi(\bar{R})$ is a semi-Horn formula, then

$$\Psi(\bar{R}) \Vdash R(\bar{a}) \text{ iff } \bar{R}_{min} \models R(\bar{a}),$$
$$\Psi(\bar{R}) \Vdash \neg R(\bar{a}) \text{ iff } \bar{R}_{max} \models \neg R(\bar{a}),$$

where $\bar{R}_{min}$ (resp. $\bar{R}_{max}$) is a minimal (resp., maximal) relation satisfying $\Psi(\bar{R})$. In this case, both $\bar{R}_{min}$ and $\bar{R}_{max}$ can be computed in PTIME.

Thus, the general computation algorithm presented in Sect. 7.2 can be specialized in the following way: The inputs to the algorithm are the same as in Sect. 7.2, however, the LCC policy LCC$[\bar{L};\bar{K}]$:IC is assumed to satisfy syntactic restrictions as formulated in Lemma 2 by any of the points 1–4, accordingly. Note that assumptions 1 and 2 from the beginning of this section, as required by Lemma 2, should also be satisfied.

The specialized algorithm is formulated as follows:

1. Construct $C \equiv CIRC(IC \cup IDB \cup EDB; \bar{L}; \bar{K})$ representing the given LCC policy applied to the IDB together with the EDB.
2. Eliminate second-order quantifiers from the formula obtained in step 1. The elimination is guaranteed to succeed in PTIME. As a result, a first-order or fixed-point formula is obtained.
3. Calculate the minimal extension of R satisfying formula C. It can be done by computing the minimal R from the second-order formula, $\exists R.C(R)$. As a result, a definition of relation R, which is a definition of $R^+_{var}()$, and a coherence condition are obtained. If the coherence condition is not satisfiable, terminate and return the answer "unsatisfiable," meaning that either the EDB and IDB pair is inconsistent or the ICs cannot be satisfied. All of the above computations can be performed in PTIME.
4. Calculate the maximal extension of R satisfying formula C. It can be done by computing the maximal R from the second-order formula, $\exists R.C(R)$. As a result, a definition of the complement of R^-_{var} is obtained. All of the above computations can be performed in PTIME.

The algorithm presented above executes in PTIME and the necessary second-order quantifier elimination can be performed automatically (using the techniques described in [6,16,17,21]).

8.3 The Case of Nonuniform LCC policies

Observe that in the case of nonuniform LCC policies, we still might obtain tractable subcases of the general case of policies. One such large class is defined in [8]. The other classes are also discussed in Sect. 8.2, in particular, in Lemma 2.

One of the promising methods depends on first computing the corresponding circumscription (applying the Doherty, Łukaszewicz, Szałas (DLS) algorithm of [6]) and then on computing the definitions of the required minimal and maximal relations by using the methodology developed in [4] and [7].

9 Conclusions

We proposed the use of rough knowledge databases to represent incomplete models of aspects of an agent's operational environment or world model. Relations represented as tables were generalized to rough sets with partitions for positive, negative, and boundary information. Then, we introduced the idea of a contextually closed query consisting of a query, a context represented as a set of integrity constraints, and a local closure policy. The LCC policy, consisting of integrity constraints and local closure policy, was applied to the intensional and extensional database layers before actually querying the RKDB. The combination of a contextually closed query and a RKDB provided the basis for an inference mechanism that could be used under the open-world assumption.

The inference mechanism and modeling approach has many applications, particularly in the area of planning with an open-world assumption, where sensor actions and knowledge preconditions are essential components in a plan and an efficient query/answer system is used in both the plan generation and execution process. We demonstrated the idea with a scenario from an unmanned aerial vehicle project. In the general case, the problem of querying the RKDB using CCQs is co-NPTIME complete, but we could isolate a number of important practical cases where polynomial time and space complexity is achieved.

In the future, there are a number of interesting topics to pursue. The use of contextually closed queries in an open-world planner has already been mentioned. Another particularly interesting issue has to do with updating the RKDB. Since each querying agent "carries" its context with it, the issue of satisfiability of the integrity constraints relative to the EDB/IDB pair and satisfiability of the pursuant minimization policy are essential aspects of the approach. We have shown that satisfiability can be checked efficiently. Then, the research question is, "what should be done when a query is not satisfiable relative to the EDB/IDB pair?" This is a question posed and considered in the area of belief revision and update or in what is more traditionally called view update in the relational database area. One final pragmatic issue involves implementation of the techniques proposed in this chapter in an on-line query/answering system for the WITAS UAV project discussed in the chapter. Parts of a prototype system have already been implemented and empirical experiments in a real-time context are planned for the future.

Acknowledgments

This research has been supported in part by the Wallenberg Foundation, Sweden. Andrzej Szałas has additionally been supported by KBN grant 8T11C 025 19. We thank Jonas Kvarnström for thorough and valuable proofreading of this chapter.

References

1. S. Abiteboul, R. Hull, V. Vianu. *Foundations of Databases*. Addison-Wesley, Boston, 1995.
2. D. Busch. Sequent formalizations of three-valued logic. In P. Doherty, editor, *Partiality, Modality and Nonmonotonicity*, 45–75, CSLI Publications, Stanford, CA, 1996.
3. P. Doherty, G. Granlund, K. Kuchcinski, K. Nordberg, E. Sandewall, E. Skarman, J. Wiklund. The WITAS unmanned aerial vehicle project. In *Proceedings of the 14th European Conference on Artificial Intelligence (ECAI 2000)*, 747–755, IOS, Amsterdam, 2000.
4. P. Doherty, J. Kachniarz, A. Szałas. Meta-queries on deductive databases. *Fundamenta Informaticae*, 40(1): 17–30, 1999.
5. P. Doherty, W. Łukaszewicz, A. Skowron, A. Szałas. Approximation transducers and trees: A technique for combining rough and crisp knowledge (this book).
6. P. Doherty, W. Łukaszewicz, A. Szałas. Computing circumscription revisited. *Journal of Automted Reasoning*, 18(3): 297–336, 1997. See also Report number LiTH-IDA-R-94-42 of Linköping University, 1994 and *Proceedings of the 14th International Joint Conference on AI (IJCAI'95)*, Morgan Kaufmann, San Francisco, 1995.
7. P. Doherty, W. Łukaszewicz, A. Szałas. Declarative PTIME queries for relational databases using quantifier elimination. *Journal of Logic and Computation*, 9(5): 739–761, 1999. See also: Report number LiTH-IDA-R-96-34 of Linköping University, 1996.
8. P. Doherty, W. Łukaszewicz, A. Szałas. Efficient reasoning using the local closed-world assumption. In A. Cerri, D. Dochev, editors, *Proceedings of the 9th International Conference (AIMSA 2000)*, LNAI 1904, 49–58, Springer, Heidelberg, 2000.
9. H-D. Ebbinghaus, J. Flum. *Finite Model Theory*. Springer, Heidelberg, 1995.
10. O. Etzioni, K. Golden, D.S. Weld. Sound and efficient closed–world reasoning for planning. *Artificial Intelligence*, 89: 113–148, 1997.
11. O. Etzioni, S. Hanks, D.S. Weld, D. Draper, N. Lesh, M. Williamson. An approach to planning with incomplete information. In B. Nebel, C. Rich, W.R. Swartout, editors, *Proceedings of the 3rd International Conference on Principles of Knowledge Representation and Reasoning, (KR'92)*, 115–125, Morgan Kaufmann, San Francisco, 1992.
12. A. Finzi, F. Pirri, R. Reiter. Open world planning in the situation calculus. In *Proceedings of the 17th National Conference on Artificial Intelligence (NCAI 2000)*, MIT Press, Cambridge, MA, 2000.

13. K. Golden, O. Etzioni, D. Weld. XII: Planning for universal quantification and incomplete information. Technical Report of the University of Washington, Department of Computer Science and Engineering, Seattle, 1994.
14. J. Komorowski, Z. Pawlak, L. Polkowski, A. Skowron. Rough sets: A tutorial. In *[18]*, 3–98, 1999.
15. W. Lukaszewicz. *Non-Monotonic Reasoning - Formalization of Commonsense Reasoning*. Ellis Horwood, Chichester, 1990.
16. A. Nonnengart, H.J. Ohlbach, A. Szałas. Elimination of predicate quantifiers. In H.J. Ohlbach, U. Reyle, editors, *Logic, Language and Reasoning. Essays in Honor of Dov Gabbay, Part I*, 159–181, Kluwer, Dordrecht, 1999.
17. A. Nonnengart, A. Szałas. A fixpoint approach to second-order quantifier elimination with applications to correspondence theory. In E. Orłowska, editor, *Logic at Work: Essays Dedicated to the Memory of Helena Rasiowa*, 307–328, Physica, Heidelberg, 1998. See also Report number MPI-I-95-2-007 of Max-Planck-Institut fuer Informatik, Saarbruecken, 1995.
18. S.K. Pal, A. Skowron, editors. *Rough Fuzzy Hybridization: A New Trend in Decision-Making*. Springer, Singapore, 1999.
19. Z. Pawlak. *Rough Sets: Theoretical Aspects of Reasoning About Data*. Kluwer, Dordrecht, 1991.
20. R. Reiter. On closed world data bases. In H. Gallaire, J. Minker, editors, *Logic and Data Bases*, 55–76, Plenum, Dordrecht, 1978.
21. A. Szałas. On the correspondence between modal and classical logic: An automated approach. *Journal of Logic and Computation*, 3:605–620, 1993.

Chapter 10
On Model Evaluation, Indexes of Importance, and Interaction Values in Rough Set Analysis*

Günther Gediga,[1] Ivo Düntsch[2]

[1] Institut für Evaluation und Marktanalysen, Brinkstr. 19, 49143 Jeggen, Germany
gediga@eval-institut.de

[2] Department of Computer Science, Brock University, St. Catherines, Ontario L2S 3AI, Canada
duentsch@cosc.brocku.ca

Summary. As with most data models, "computing with words" uses a mix of methods to achieve its aims, including several measurement indexes. In this chapter, we discuss some proposals for such indexes in the context of rough set analysis, and we present some new ones.

In the first part, we investigate several classical approaches based on approximation quality and the drop of approximation quality when leaving out elements. We show that using the approximation quality index is sensible in terms of admissibility and present additional indexes for the usefulness and significance of an approximation. The analysis of a *drop* is reinterpreted in terms of a model comparison, and a general framework for all these concepts is presented.

In the second part of the chapter, we present an example showing how using similar nomenclature in the theory of Choquet-type aggregations of fuzzy measurements and rough set approximation quality, without regard to the fine structure of the underlying model assumptions, can suggest connections where there are none.

On a more positive note, we show that so-called qualitative power and interaction indexes, which are structurally similar to quantitative Choquet-type aggregations can be used in the context of rough set analysis. Furthermore, we propose an entropy-based measure which allows using qualitative power and interaction indexes as an approximation.

> "Mesmerized by a single-purpose, mechanised 'objective' ritual in which we convert numbers into other numbers and get a yes-no answer, we have come to neglect close scrutiny of where the numbers come from [3]."

---

* Equal authorship is implied

1 Introduction

Notwithstanding a widespread fascination with numbers, it is recognized that human behavior is often guided by imprecise concepts. In recognition of this fact, the direction "from computing with numbers to computing with words," put forward by [36], calls for a manipulation of perceptions instead of measurements:

> "The rationale for computing with words rests on two major imperatives: 1. computing with words is a necessity when the available information is too imprecise to justify the use of numbers and 2. when there is tolerance for imprecision which can be exploited to achieve tractability, robustness, low solution cost and better rapport with reality [36, p. 111]."

This is not to say that computing with words does not use numbers; for example, fuzzy events and constraint propagation assume a continuous scale of truth values for propositions. Leaving aside the question of where these initial parameters come from, one finds that subsequent processing may generate other numbers, such as averages or maxima, and the question arises in which sense are these measurements *meaningful*, and what, if anything, do these numbers actually measure? Wrong interpretations of numbers may lead to scaling artifacts — constructs that apply operations to a model which are not justified. A classical instance is a study by [20],[1] who, in an investigation of traffic deaths, assigned numbers to four groups of people (0 - white male, 1 - black male, 2 - white female, 3 - black female), and proceeded to take averages and variances. It turned out that across the population investigated, the average person to cause a fatal accident was a black male ($\overline{x} = 1$). Although this example is certainly extreme, the danger of falling into the trap so adequately described above by [3] is always present when relations among numbers are used for representing relations among objects.

The main topic of this chapter is the classical problem of building meaningful indexes from data or from other indexes. Both processes need a scaling theory, which enables the user to understand what the indexes mean in terms of the data and their context. It easy to declare a concept *useful* if it fits a few small examples, but it is usually quite hard to substantiate such a claim theoretically or experimentally. Even though some systems work "in spite of the erroneous assumptions that underlie them [1]," without such a theory, a sound basis for interpretation of numbers as measures is not given, and more often than not, such an interpretation will at some stage lead to wrong results. Data analysis methods that are not primarily quantitative (and consequently, do not have a built-in scaling theory) need to pay particular attention to this fact.

In this chapter, we will be concerned with a basic statistic of rough set data analysis, approximation quality, and its interpretation as an ordinal or interval scale. A main

[1] Reported by [34, p.64]

theme will be the distinction between *admissibility* and *usefulness* of a measure. In short, we call a measure *admissible* if it satisfies a given standard for describing a situation, and *useful*, if it can be used to give advice for decision making. We will see that these situations need different tools to cope with specific circumstances.

As an example, we will exhibit an instance where the application of one method (Choquet-like measures) to another (rough set approximation quality) leads to measurements without measure — numbers that cannot be meaningfully interpreted. We will focus on the influence of one or more features of a decision process as investigated in multicriteria decision making (MCDM), and the influence of one or more features of the approximation quality of rough set data analysis (RSDA) with respect to a decision attribute. We also suggest that capacities with a maximum operator instead of $\sum$, similar to the Sugeno integral, can be useful in the context of RSDA and that these measures can be applied to rough entropy in a meaningful way as well. It is not an aim of the chapter to criticize ad hoc methods for data analysis per se, which are certainly useful in specific cases. We also do not claim that the methods we suggest are universally better than others. The results of [35] show that no classification algorithm can always outperform another one. What we do want to show is that data modeling has to consider the scaling assumptions of the applied measures and indexes, and that things may be more complicated than they seem at first glance. It is necessary to guarantee the meaningfulness of measurements in the first place; the question of whether one is computationally better than another — or even more successful — can be posed after the first step. As expressed in the context of weighted voting in the seminal chapter by Banzhaf [1],

> "...its intent is only to explain the effects which necessarily follow once the mathematical model and the rules of its operation are established ..."

This chapter is organized as follows: We will first develop the necessary machinery from rough sets, followed by a discussion of several frequently used indexes and relations in RSDA, and we show how the model of a proportional error reduction helps to unify the different approaches. Afterward, we will recall various capacities and aggregation measures and show how they differ. In Sect. 4.3, we investigate the connection between the two contexts, based on a discussion of some examples. In Sect. 4.4, we present some thoughts on a reconciliation. Finally, we discuss some examples in which we show how the different techniques behave. We close the chapter with a future outlook.

2 Rough Set Theory and Approximation Quality

Rough set data analysis was introduced by Pawlak [22] and has since gained importance as an instrument for noninvasive data analysis. We invite the reader to consult [24] for a short introduction to traditional RSDA and [9] for a more detailed presentation.

Knowledge representation in rough set data analysis is done via *information systems*. These are structures of the form,

$$I = \langle U, \Omega, \{V_x : x \in \Omega\} \rangle, \tag{1}$$

where

- U is a finite set of objects.
- Ω is a finite set of mappings $x : U \to V_x$; each $x \in \Omega$ is called an *attribute*.
- V_x is the set of *attribute values* of attribute x.

Each set P of attributes defines an equivalence relation θ_P (and an associated partition) on U by

$$x\theta_P y \iff a(x) = a(y) \text{ for all } a \in P. \tag{2}$$

Note that $\theta_\emptyset = U \times U$.

Given an information system I, a basic construction of RSDA is the approximation of a given partition $\mathcal{R}$ of U by the partition generated by a set P of attributes. For example, one may want to reduce the number of attributes necessary to identify the class of an object $x \in U$ by its feature vector (i.e., looking for a key in a relational database) or generate rules from a decision system, which is an information system enhanced by a decision attribute.

By its very nature, RSDA operates on the lowest level of data modeling, on what is commonly called raw data, captured in an information table as defined above. Thus, in the language of measurement theory [32], RSDA resides on a *nominal scale* where meaningful operations are those that preserve equality. This parsimony[2] of model assumptions can be seen both as a strength and a weakness: Having only few model assumptions allows the researcher to investigate a greater variety of situations than, for example, data samples with a preassumed distribution such as normal or iid, which are required by many statistical methods. On the other hand, the results obtained by RSDA may be too weak to allow meaningful inference about the situation at hand. A case in point is the essentially logical (or algebraic) nature of RSDA: Its outcome is a set of if-then rules, which are logically true, and can be well used in a descriptive situation or for *deductive* knowledge discovery, i.e., analyzing the rules in a given data table. However, if one looks at *inductive* reasoning, such as prediction or classification of previously unseen objects, such rules are not necessarily useful. If, say, each rule is based on only one instance (i.e., each object determines a unique rule), then the rule set will not be helpful for classifying new

[2] It is sometimes claimed that RSDA has no prerequisites; a moment's reflection shows that this cannot be true: To be able to distinguish objects by their feature vector, one has to assume that the data contained in the vector are accurate.

elements. Therefore, in these situations, additional tools are required to supplement the results obtained by basic RSDA. As a first step, well within the RSDA approach, one uses information given by the data themselves, usually in the form of counting parameters. One of the first (and most frequent) to have been used is the *approximation quality* γ, which roughly speaking, measures the goodness of fit of expressing knowledge about the world, which is given by one set of features, by another set of features in the following way: Suppose that θ is an equivalence relation on a set U with the associated partition $\mathcal{P}$ and that $X \subseteq U$. Then, we first set

$$\pi_\theta(X) = \frac{|\bigcup\{Y \in \mathcal{P} : Y \subseteq X\}|}{|X|}.$$

This index measures the relative number of elements in X, which can be classified as certainly being in X, given the granularity provided by $\mathcal{P}$. If P is a set of attributes and $\mathcal{R}$ a fixed partition of U, then the *approximation quality of* P (with respect to $\mathcal{R}$) is defined as

$$\gamma_{\mathcal{R}}(P) = \sum_{Y \in \mathcal{R}} \frac{|Y|}{|U|} \cdot \pi_{\theta_P}(Y). \tag{3}$$

$\gamma_{\mathcal{R}}$ measures the relative cardinality of correctly classified elements of U as being in a class of $\mathcal{R}$ with respect to the indiscernability relation θ_P. In the situations that we are going to consider, the partition $\mathcal{R}$ arises from a decision attribute d; in the sequel, we assume that $\mathcal{R}$ is fixed, and we will just write γ instead of $\gamma_{\mathcal{R}}$. To avoid trivialities, we assume that $\mathcal{R}$ has at least two classes. By a *model*, we understand a set of attributes P along with the set of deterministic rules, which are generated by P with respect to the partition $\mathcal{R}$ generated by a decision attribute d. For reasons of brevity, we sometimes just call P a model.

3 Relations Based on Approximation Quality

Since $\gamma(P)$ is a real number between 0 and 1, one can, in principle, apply transformations and form relations with the γ values as real numbers. However, in order that such operations are meaningful and result in a valid interpretation, one has to have a theory, sometimes called a *scaling model*, which justifies using the transformation. In this section, we will discuss some "standard" approaches to handling γ; more complex transformations will be discussed later.

3.1 Comparing Approximation Qualities

One approach is the comparison of γ with a fixed number $0 \lneq c \leq 1$. Choosing the constant c is up to the user or investigator and is driven by practical necessities; having scanned the literature on applied RSDA, we have found that c is never chosen at less than 0.5 and is often close to 1.0. The usual interpretation says that any set of attributes P with $\gamma(P) \lneq c$ shows a "bad" approximation of $\mathcal{R}$, and any P with $\gamma(P) \geq c$ is "admissible" for the approximation of $\mathcal{R}$; this approach is used to define "isogamma" reducts and core [24, p. 51].

This is a straightforward and seemingly unequivocal interpretation of γ, and thus, this technique is frequently used. However, the question arises whether it is based on a meaningful interpretation of the approximation quality. The answer is, it depends on the context, and we have to distinguish between *admissibility* and *usefulness*: Suppose that $c \leq \gamma(P)$.

1. If P is claimed to be admissible for approximating $\mathcal{R}$, we do not run into problems, because the approximation quality induced by P is not less than the required relative number c of elements which are correctly classified with respect to $\mathcal{R}$; therefore, the approximation of $\mathcal{R}$ is admissible with respect to c. This interpretation is based purely on the algebraic structure of the equivalence relations and assumes that the data are correct *as given*. In other words, the approximation quality counts the relative number of elements that can be captured by the deterministic rules associated with P. In this sense, γ counts what is *logically true*.
2. If P is claimed to be useful (i.e., it can or should be used) for approximating $\mathcal{R}$, such as in decision support and medical diagnosis, the situation is more complicated. One has to take into account that attaining the standard c may have come about by random influences (noise, error) and that therefore, the application of P for approximating $\mathcal{R}$ is not necessarily useful [7]. A simple example of the divergence of admissibility and usefulness is an information system consisting of a running number and a decision attribute. Here, we have $\gamma = 1$, and the running number is helpful to identify any case without error. However, knowing the running number without knowing the value of the decision attribute does not help. For this purpose, $\gamma = 1$ is not useful.

A further example, shown in Table 1, will illustrate how random processes may influence the results for very low approximation qualities.

Table 1. A simple decision system [7]

U	p	q	d	U	p	q	d
1	0	0	0	5	1	0	1
2	0	2	0	6	1	2	1
3	0	2	0	7	1	2	1
4	1	1	0	8	0	1	1

If $\mathcal{R}$ is the partition associated with the decision attribute d, then

$$\gamma(p) = \gamma(q) = 0. \tag{4}$$

Whereas p is essential to predict d with only a class switch of 4 and 8 achieving perfect approximation quality, q is only required to *separate* 4 and 8 from their respective p-classes. In other words, a misclassification of 4 and 8 may well have taken place, owing to random influences in representing the data. The interpretation as usefulness needs an additional tool to cope with randomness. There are at least two different ways to evaluate the usefulness of P: cross-validation methods or statistical testing of *random admissibility*. We prefer the latter approach, because cross-validation methods require additional (non rough!) model assumptions for classifying of unseen cases in the learning sample, whereas testing usefulness, as random admissibility, does not.

The γ statistic is also used to define an ordinal relation among subsets of the attribute set by setting

$$P \preceq_\gamma Q \Longleftrightarrow \gamma(P) \leq \gamma(Q).$$

The interpretation "Q is at least as admissible as P" causes no difficulty because choosing a criterion c with $\gamma(P) \leq c \leq \gamma(Q)$ results in such an interpretation. On the other hand, interpreting $\preceq_\gamma$ by assigning the term *better than* (in the sense of more useful) to the defined relation is not meaningful. Since this interpretation makes sense only in a numerical system in which a $\leq$ relation is established,[3] the notation "better than" cannot be filled by an empirical interpretation.

3.2 Comparing Differences in Approximation Qualities

From its early days, identification of important features has been an integral part of RSDA:

> "The idea of attribute reduction can be generalised by an introduction of the concept of *significance*[4] *of attributes*, which enables an evaluation of attributes ... by associating with an attribute a real number from the $[0,1]$ closed interval; this number expresses the importance of the attribute in the information table [16]".

The first approach for analyzing the significance of an attribute with respect to a given attribute set was inspection of the *drop* in approximation quality when leaving

[3] Note that the numbers on the shirts of soccer players are $\leq$ related, but this does not induce a better than relation among soccer players.

[4] The terminology "significance" is somewhat unfortunate, because it has a fixed (and quite diverging) meaning in statistics. We think that the expression "importance of an attribute" would be a better choice.

out an attribute a from a set P defined by

$$d_1(P,a) = \gamma(P) - \gamma(P \setminus \{a\}) \tag{5}$$

(see, e.g., [23], p. 59). The analysis of the drop starts after P has been chosen as an admissible attribute set, for example, a reduct. Therefore, the difference cannot, in general, be interpreted in terms of admissibility because $\gamma(P \setminus \{a\})$ may have a smaller value than the chosen standard c, which sets the standard for admissibility. One can argue that d_1 is meaningful because γ forms an interval scale. In terms of admissibility, this is an additional condition which has as one consequence that a difference from $\gamma(P) = 0.1$ to $\gamma(P \setminus \{a\}) = 0.0$ is assumed to be identical (in terms of set approximation) to the difference among $\gamma(P) = 1.0$ and $\gamma(P \setminus \{a\}) = 0.9$. One has to be aware, however, that — contrary to the approximation quality comparison, which relies only on a monotone relationship among the numerical measures — the interpretation of differences needs much stronger assumptions than a comparison of numerical values. The problem that "differences ignore the base rate" occurs in other disciplines as well; in descriptive statistics, for instance, the odd ratio quotient of two probabilities is often used to describe the "difference" among probabilities in a meaningful way. Building the odd ratio of probabilities requires an extra scaling theory, which serves as a foundation for interpreting of the odd ratio. Obviously, something comparable would be needed in the context of RSDA as well. As far as we are aware, there has been no attempt to provide such a theory.

Simple differences should not be interpreted in terms of usefulness because the step from 0.0 to 0.1 can easily be explained by random processes, whereas it is much harder, given comparable distributions of attribute values, to result in a step from 0.9 to 1.0.

The problem of dealing with differences has been addressed by a newer approach for measuring the significance of an attribute [16,31]:

$$d_2(P,a) = \frac{\gamma(P) - \gamma(P \setminus \{a\})}{\gamma(P)} = 1 - \frac{\gamma(P \setminus \{a\})}{\gamma(P)}.$$

It is not at all clear how this function should be interpreted, and it has some peculiar properties: Given identical linear differences with different base rates, one can observe that the d_2 differences will be smaller, with an increasing base rate. For example,

$\gamma(P)$	$\gamma(P \setminus \{a\})$	$d_1(P,a)$	$d_2(P,a)$
1.0	0.9	0.1	0.1
0.6	0.5	0.1	0.167
0.2	0.1	0.1	0.5

As the preceding discussion shows, this property of $d_2(P,a)$ is not satisfactory because a constant gain should result in larger values if the base rate is at the upper end of the scale.

3.3 Averaging of Approximation Qualities

Another way of treating interval scale information is by forming averages. If this operation is applied to set approximation qualities, we again run into problems. Suppose that $\mathcal{A}$ is a collection of attribute sets. The average

$$E[\mathcal{A}] = \frac{1}{|\mathcal{A}|} \sum_{P \in \mathcal{A}} \gamma(P)$$

computes the expectation of the approximation quality when each of the sets $P \in \mathcal{A}$ has the same probability of being used for approximating a decision attribute. Unlike the case in which sampling properties can be described by the principle of indifference, forming attribute sets is not a random choice and is under the control of the researcher. Therefore, building an expectation value does not make much sense because the population for the sample cannot be properly defined. Furthermore, any researcher would agree that $\max\{\gamma(P) : P \in \mathcal{A}\}$ is a characteristic value for the set $\mathcal{A}$ and that the expectation may offer strange results, as Table 2 demonstrates.

Table 2. Strange results using expectations of approximation qualities

$\mathcal{A}$	γ	$\mathcal{B}$	γ
A_1	0.2	B_1	0.0
A_2	0.2	B_2	0.0
A_3	0.2	B_3	0.0
A_4	0.2	B_4	0.0
A_5	0.2	B_5	0.0
A_6	0.2	B_6	1.0
Maximum	0.2		1.0
Mean	0.2		0.167

There, the collection $\mathcal{A}$ consists of six non admissible attribute sets if $c \gtrsim 0.2$ for the approximation of $\mathcal{R}$, whereas one set in $\mathcal{B}$ is admissible for any c. The maximum of the approximation qualities will point to $\mathcal{B}$ as the *better set.* In contrast, using average values, one sees that the mean admissibility of $\mathcal{A}$ is higher than the mean admissibility of $\mathcal{B}$, although no element in $\mathcal{A}$ is admissible at all.

3.4 Indexes of Proportional Error Reduction as a General Concept

In the preceding section, both indexes d_1 and d_2 are based on transforming the γ index by assuming that the transformation somehow fits the semantics (the significance or importance). We have shown that d_1 and d_2 are not necessarily meaningful, either for admissibility or usefulness, due to the lack of a sound theory that guides the index building process.

In [10], we have introduced the proportional reduction of errors (PRE) approach of [15] into RSDA, which in the general case, describes the error reduction when a model is applied, based on the errors of a given benchmark model. In the context of RSDA, we say that an *error* is an object that cannot be explained with a deterministic rule. In line with this interpretation, the approximation quality becomes

$$\gamma(P) = 1 - \frac{1-\gamma(P)}{1-\gamma(\emptyset)},$$

which means that $\gamma(P)$ measures the proportional error reduction of a model using the attribute set P in comparison to the worst-case benchmark model in which every object is counted as an error.

Adapting the PRE approach to the importance problem, we find that

$$d_3(P,a) = \begin{cases} 0, & \text{if } \gamma(P \setminus \{a\}) = 1 \\ 1 - \frac{1-\gamma(P)}{1-\gamma(P\setminus\{a\})}, & \text{otherwise.} \end{cases} \tag{6}$$

is a suitable index for comparing a model using the attribute set P against a model using the attribute set $P \setminus \{a\}$. The index $d_3(P,a)$ measures the error reduction when using the set of attributes P compared with the benchmark model, which uses the set of attributes $P \setminus \{a\}$. This value can be compared to a threshold value c_g, and therefore, $d_3(P,a)$ can be interpreted as a measure of the admissibility of the gain.

Comparing the measures d_1, d_2, and d_3, we observe that the behavior of d_3 is as it should be:

$\gamma(P)$	$\gamma(P \setminus \{a\})$	$d_1(P,a)$	$d_2(P,a)$	$d_3(P,a)$
1.0	0.9	0.1	0.1	1.0
0.6	0.5	0.1	0.167	0.2
0.2	0.1	0.1	0.5	0.111

The evaluation of differences, given a small base rate, is lower than the same differences when given a high base rate.

Because admissibility gain does not take into account random influences, an index for a *usefulness gain* has to be defined as well. To this end, the results of [7] can be used to derive a descriptive measure for a usefulness gain.

Descriptive indexes, such as admissibility or usefulness, estimate the actual size of an effect, given a fixed set of model assumptions; therefore, such indexes are often called *effect size measures*. In statistical applications, one often considers one kind of effect size measure. However, this can be done only under rather restrictive assumptions. For instance, if it is assumed that two variables represent a bivariate normal distribution, the admissibility and usefulness of the correlation coefficient are identical. If this assumption is dropped, this identity does not hold in general.

Because effect sizes in terms of PRE measures are used in many contexts, there a rule of thumb exists for assessing effect sizes for expectation-based benchmark models [2]:

Effect size (ES)	Interpretation
$ES \lneq 0.1$	No effect
$0.10 \leq ES \lneq 0.3$	Small effect
$0.30 \leq ES \lneq 0.5$	Medium effect
$ES \geq 0.5$	Large effect

Effect size measures must rely on empirical data to estimate the range of effects in real-life data. This is just the way [2] arrived at the interpretation of effect sizes. But until a database exists for empirical studies that have been based on RSDA, the given rule of thumb can be used as a first approximation.

The PRE measures discussed in this section show peculiar behavior when $\gamma(P) = 1$. In this situation, any PRE measure will result only in 0 (if the benchmark model results in 1 as well) or 1. In terms of admissibility, this binary nature of the index cannot be resolved, but if a statistical benchmark model is used, it is easy to replace the descriptive PRE measure with a measure from inference statistics by computing the position of the observed error in the distribution of expected errors, given the benchmark model. This position is called statistical *significance*, and the value should be small (conventionally smaller than 5%) for a good model. Note that usability and significance are two different concepts that may dissociate, even though both use identical random processes. Whereas significance is changed when increasing the number of observations, usability remains unchanged.

Table 3 on the next page collects all approaches, PRE measures and significance, discussed so far, enhanced by *set gains*, which are a simple generalization of the preceding indexes by choosing $S = P$ or $S = \{a\}$, respectively.

4 Capacities, Power Indexes, and Values of Interaction

The influence and power of an attribute, as well as the interaction of several attributes, have been extensively studied in game theory and multicriteria decision analysis. For an overview of earlier work, we invite the reader to consult the collection of essays edited by Roth [26], and for more recent advances, the article by Grabisch [11]. Since the approximation quality has the same mathematical properties as a capacity or fuzzy measure (explained below), it has been claimed that

Table 3. PRE indexes as descriptive measures and significance values in RSDA

$\mathcal{P}^\sigma$:= Set of attribute sets that are constructed by random assignment of elements to the attributes P.
$\mathcal{S}^\sigma$:= Set of attribute sets that are constructed by random assignment of elements to the attributes S.
$\mathcal{A}^\sigma$:= Set of attribute sets that are constructed by random assignment of elements to the attribute a.

Error of the model	Error of the benchmark model	Interpretation	Source
$1-\gamma(P)$	$1-\gamma(\emptyset)=1$	Admissibility	[10]
$1-\gamma(P)$	$1-\mathcal{E}[\gamma(R)\|R \in \mathcal{P}^\sigma]$	Usefulness	[10]
$1-\gamma(P)$	$1-\gamma(P\setminus\{a\})=1-\gamma[(P\setminus\{a\})\cup\emptyset]$	Admissible gain	This text
$1-\gamma(P)$	$1-\mathcal{E}\{\gamma[(P\setminus\{a\})\cup R]\|R \in \mathcal{A}^\sigma\}$	Usable gain	[7]
$1-\gamma(P)$	$1-\gamma(P\setminus S)=1-\gamma[(P\setminus S)\cup\emptyset]$	Admissible set gain	This text
$1-\gamma(P)$	$1-\mathcal{E}\{\gamma[(P\setminus S)\cup R]\|R \in \mathcal{S}^\sigma\}$	Usable set gain	This text
Error of the model	Position of the error given the benchmark model	Interpretation	Source
$1-\gamma(P)$	$p[1-\gamma(R)\leq 1-\gamma(P)\|R \in \mathcal{P}^\sigma]$	Significance	[10]
$1-\gamma(P)$	$p\{1-\gamma[(P\setminus\{a\})\cup R]\leq 1-\gamma(P)\|R \in \mathcal{A}^\sigma\}$	Significant gain	[10]
$1-\gamma(P)$	$p\{1-\gamma[(P\setminus S)\cup R]\leq 1-\gamma(P)\|R \in \mathcal{S}^\sigma\}$	Significant set gain	This text

> "Due to this equivalence, it is possible to use different indices defined on fuzzy measures to assess the relative value of information supplied by each attribute and to analyze interaction between attributes [13]."

After introducing the necessary machinery, we shall show in this section that using γ as an interval scaled capacity leads to scaling artifacts which do not take into account the basic model assumptions of the indexes, and thus, they provide a *measurement without measure*. Alternatives that are more promising evaluation tools are presented as well.

4.1 Quantitative Indexes of Power and Interaction

In decision theory, the aggregation of criteria is usually done by weighted arithmetic means, or as they are sometimes called, discrete integrals. We will define below the most often used indexes. Throughout, we suppose that $U = \{1,...n\}$ is a finite set, which in the present context, can be interpreted as a set of criteria or a set of players.

A function $\mu : 2^U \to [0,1]$ is called a *capacity* or *fuzzy measure* if for all $X \subseteq U$,

$$\mu(\emptyset) = 0, \quad \mu(X) \leq 1, \tag{7}$$

$$A \subseteq B \subseteq X \text{ implies } \mu(A) \leq \mu(B). \tag{8}$$

We will usually identify singletons with the element they contain, e.g., we will write $\mu(p)$ instead of $\mu(\{p\})$.

The set function μ takes into account that the contribution of one criterion to a set S of criteria may vary, depending on the choice of S. In other words, μ is chosen so that it respects the interaction among criteria according to the belief of the investigator.

The common quantitative aggregation functions, resulting in power indexes, rely on a simple difference construction: If $K \subseteq U$ and $m \notin K$, we let

$$\Delta^{\mu}(K,m) = \mu(K \cup \{m\}) - \mu(K) \tag{9}$$

denote the (unweighted) marginal contribution of m to $\mu(K \cup \{m\})$. Two well-known power indexes are based on Δ^{μ}: The *Shapley value* [28] is defined by

$$\varphi_S^{\mu}(m) = \sum_{K \subseteq U \setminus \{m\}} \frac{(n-|K|-1)!|K|!}{n!} \Delta^{\mu}(K,m). \tag{10}$$

It is usually interpreted as a measure of the weighted marginal average contribution of m to sets of the form $\mu(K \cup \{m\})$ under the assumption that "all orders in which an individual enters any coalition are equiprobable [29]," and it is often called the *importance of m with respect to the weighting* μ.

Another value frequently considered is the *Banzhaf value* [1], given by

$$\varphi_B^{\mu}(m) = \frac{1}{2^{n-1}} \sum_{K \subseteq U \setminus \{m\}} \Delta^{\mu}(K,m). \tag{11}$$

Both indexes φ^{μ} make several model assumptions including the following:

A1. Taking averages of differences is meaningful.
A2. If μ is a capacity and σ a permutation of U, then, for all $m \in U$,

$$\varphi^{\mu}(m) = \varphi^{\sigma\mu}(\sigma(m)).$$

Here, $\sigma\mu(K) = \{\sigma(k) : k \in \mu(K)\}$.

We invite the reader to consult [4,5,17] for axiomatizations of the Shapley and Banzhaf values.

Apart from the *first-order* power indexes, more complicated indexes can be built as well. If $K \subseteq U$ and $i, j \notin K$, let

$$\Delta^{\mu}(K,\{i,j\}) = \mu(K \cup \{i,j\}) - \mu(K \cup \{i\}) - \mu(K \cup \{j\}) + \mu(K). \tag{12}$$

Weighted averages of these *second-order* differences, called *interaction values*, result in the Shapley interaction index,

$$\varphi_S^\mu(\{i,j\}) = \sum_{K \subseteq U \setminus \{i,j\}} \frac{(n-|K|-2)!|K|!}{(n-1)!} \Delta^\mu(K,\{i,j\}), \tag{13}$$

and the Banzhaf interaction index,

$$\varphi_B^\mu(\{i,j\}) = \frac{1}{2^{n-2}} \sum_{K \subseteq U \setminus \{i,j\}} \Delta^\mu(K,\{i,j\}), \tag{14}$$

respectively [21, 27]. Following [18], we say that $i,j \in U$ show a *negative interaction* if

$$\varphi^\mu(\{i,j\}) \lneqq 0, \tag{15}$$

they *do not interact* or are *uncorrelated*, if

$$\varphi^\mu(\{i,j\}) = 0, \tag{16}$$

and they *show a positive interaction*, if

$$\varphi^\mu(\{i,j\}) \gneqq 0. \tag{17}$$

4.2 Qualitative Indexes of Power and Interaction

As an alternative to the assumption of an interval scale for $\mu(K)$, Grabisch [12] and Dubois et al. [6] offer an index that assumes only ordinal scaling and uses neither differences nor the mean to calculate the influence of an element. Let $\bigoplus$ denote the maximum of an ordered set (if it exists), and set

$$a \ominus b = \begin{cases} a, & \text{if } a \gneqq b \\ 0, & \text{otherwise.} \end{cases}$$

The *qualitative power value of i* is defined by

$$\varphi_Q(i) = \bigoplus_{K \subseteq U \setminus \{i\}} \mu(K \cup \{i\}) \ominus \mu(K). \tag{18}$$

This expression can be handled formally as the indexes introduced in Sect. 4. The value $\varphi_Q(i)$ is the largest value of μ for a set containing i, which will drop, when i is left out. Based on investigations of [12], Dubois et al. [6] note that this value "seems to be the only reasonable definition for a qualitative Shapley value." It is based only on the ordinal scaling assumption because φ_Q is monotonically invariant with respect to μ, since for any monotonic mapping $T : \mathbb{R} \to \mathbb{R}$,

$$\varphi_Q^{T(\mu)}(m) = T\left[\varphi_Q^\mu(m)\right].$$

Therefore, φ_Q is an instance of the ordinal meaningful aggregation functions proposed by [19].

Although φ_Q shares some structural properties with φ_B and φ_S, the interpretation of φ_Q is quite different: Whereas φ_Q computes a maximum, the quantitative counterparts compute an average (based on different weights). Another difference is due to the monotonic invariance of φ_Q: Because any monotonic transformation of μ is admissible, there is no need to restrict μ to the interval $[0,1]$.

It is interesting to note that the application of φ_Q is not restricted to capacities but is meaningful with any set function: Even if we drop the assumption of monotonicity of μ with respect to $\subseteq$, the value of $\varphi_Q(i)$ is of interest because its value is the largest one for which i shows a nonredundant contribution to μ.

Let us note the following *relative commutativity* of operation $\ominus$:

Lemma 1. *$(a \ominus b) \ominus c = (a \ominus c) \ominus b$ for all a, b, c.*

Proof. The conclusion follows from the following table:

$a \ominus b$	$a \ominus c$	$(a \ominus b) \ominus c$	$(a \ominus c) \ominus b$
a	a	a	a
a	0	0	0
0	a	0	0
0	0	0	0

It is now straightforward to define a *qualitative value of interaction* as well, using the $\bigoplus$ and $\ominus$ operators: [5]

$$\varphi_Q(\{i,j\}) = \bigoplus_{K \subseteq U \setminus \{i,j\}} [\mu(K \cup \{i,j\}) \ominus \mu(K \cup \{i\})] \ominus \mu(K \cup \{j\}). \quad \square \tag{19}$$

By Lemma 1, $\varphi_Q(\{i,j\})$ is well defined, and it can be shown that this expression is formally the same as the quantitative interaction indexes. The index $\varphi_Q(\{i,j\})$ addresses the largest value of μ for which i, j truly interact in the sense that both elements contribute a part to μ that cannot be expressed by the other element. If $\varphi_Q(\{i,j\}) = 0$, no such interaction among i, j is observable. Because φ_Q uses monotonic invariant operators, it is monotonically invariant as well, and the name *qualitative value of interaction* is justified.

4.3 Applications of Capacities in Rough Set Data Analysis

Choquet-type aggregation measures have been considered in the context of RSDA:

[5] This construction may be known.

> "All these indices can be useful to study the informational dependence among the considered attributes and to choose the best reducts. ... the Shapley values ... can be interpreted as measures of importance of the corresponding attributes in the rough approximation [13, p. 102f]."

Below, we will investigate whether the aims addressed in the quotation can be fulfilled with the quantitative power indexes in question.

We start with a minor observation: If quantitative power indexes are used, one has to differentiate between the interpretations of the Shapley and the Banzhaf indexes since it is possible that $\varphi_S(p) \lneqq \varphi_S(q)$ and $\varphi_B(p) \gneqq \varphi_B(q)$ (for an example, see Table 4 in [33]). Therefore, it is not clear which of these indexes, if any, tells us something about the "informational dependence of the considered attributes [14]."

Table 4. A nonmonotonic relationship of Banzhaf and Shapley values

K	μ	$w_S(K)$	$\Delta(K,p)$	$\Delta(K,q)$	$w_S(K)\cdot\Delta(K,p)$	$w_S(K)\cdot\Delta(K,q)$
$\emptyset$	0	0.25	0.5	0	0.125	0
$\{r\}$	0	0.0833	0.5	0.8	0.0417	0.0667
$\{s\}$	0	0.0833	0.5	0.8	0.0417	0.0667
$\{p\}$	0.5	0.0833		0		0
$\{q\}$	0	0.0833	0.5		0.0417	
$\{r,s\}$	0	0.0833	1	1	0.0833	0.0833
$\{r,p\}$	0.5	0.0833		0.5		0.0417
$\{r,q\}$	0.8	0.0833	0.2		0.0167	
$\{s,p\}$	0.5	0.0833		0.5		0.0417
$\{s,q\}$	0.8	0.0833	0.2		0.0167	
$\{r,s,p\}$	1	0.0833		0		0
$\{r,s,q\}$	1	0.0833	0		0	
		Sum	3.4	3.6	0.3668	0.3

A more severe problem is the fact that comparing Banzhaf or Shapley values leads directly to a comparison of averages based on a collection of sets that may lead to "strange" results, as the example in Sect. 3.3 demonstrates. This can also be seen quite easily as follows: If we compare set functions of the type

$$\varphi^\mu(i) = \sum_{K\subseteq U\setminus\{i\}} w(|K|)\cdot[\mu(K\cup\{i\})-\mu(K)],$$

a straightforward calculation shows that

$$\varphi^\mu(i)-\varphi^\mu(j) = \sum_{K\subseteq U\setminus\{i,j\}} [w(|K|)+w(|K|+1)]\cdot[\mu(K\cup\{i\})-\mu(K\cup\{j\})].$$

Therefore, the difference in Banzhaf values can be rewritten as

$$\varphi_B^\mu(i) - \varphi_B^\mu(j) = \left[\frac{1}{2^{n-2}} \sum_{K \subseteq U \setminus \{i,j\}} \mu(K \cup \{i\})\right] - \left[\frac{1}{2^{n-2}} \sum_{K \subseteq U \setminus \{i,j\}} \mu(K \cup \{j\})\right].$$

This leads to a very simple interpretation of Banzhaf value differences: $i \prec_B j$ iff $\varphi_B^\mu(i) - \varphi_B^\mu(j) \lneq 0$, which means that the *average* of μ based on the sets $K \cup \{i\}$ is less than the *average* of μ based on the sets $K \cup \{j\}$.

For differences in Shapley values, we find

$$\varphi_S^\mu(i) - \varphi_S^\mu(j) = \sum_{K \subseteq U \setminus \{i,j\}} \frac{(n - |K| - 2)!|K|!}{(n-1)!} \left[\mu(K \cup \{i\}) - \mu(K \cup \{j\})\right],$$

which addresses the comparison of Shapley-weighted averages of μ values based on different sets.

Comparing Banzhaf or Shapley values leads to a comparison of mean values based on disjoint sets, which is exactly the situation we discussed in Sect. 3.3. One might argue that this cannot be observed in RSDA, but a more refined example can be constructed. Using the data in Table 5, we observe that $\gamma(\{p\}) = \gamma(\{a,p\}) = \gamma(\{b,p\}) = \gamma(\{c,p\}) = \gamma(\{a,b,p\}) = \gamma(\{a,c,p\}) = \gamma(\{b,c,p\}) = 0.125$, $\gamma(\{a,b,c,p\}) = 0.25$, $\gamma(\{q\}) = \gamma(\{a,q\}) = \gamma(\{b,q\}) = \gamma(\{c,q\}) = \gamma(\{a,b,q\}) = \gamma(\{a,c,q\}) = \gamma(\{b,c,q\}) = 0.0$, and $\gamma(\{a,b,c,q\}) = 1$. Now, it is easy to calculate that $0.140 = \varphi_B^\mu(p) \gneq \varphi_B^\mu(q) = 0.125$, and once again, a set of attribute sets with very low approximation qualities dominates another set with one perfect attribute combination.

Whichever power index will be used, it should be noted that this index cannot be interpreted in terms of usefulness. As an example, consider the case where $\Omega = \{p,q\}$, $\gamma(p) = c$ with $c \lneq 1$, and $\gamma(q) = 1$. The Banzhaf values and qualitative Shapley values are

$$\begin{aligned}
\varphi_B(p) &= \frac{1}{2}[(1-1)+(c-0)] = \frac{c}{2}, \\
\varphi_B(q) &= \frac{1}{2}[(1-c)+(1-0)] = 1 - \frac{c}{2}, \\
\varphi_Q(p) &= \max\{1 \ominus 1, c \ominus 0\} = c, \\
\varphi_Q(q) &= \max\{1 \ominus c, 1 \ominus 0\} = 1.
\end{aligned}$$

Now, $\varphi_x(p) \lneq \varphi_x(q)$ holds for every $x \in \{B,S,Q\}$ and any value of c strictly less than 1; regardless of the *quality* of the attribute q, it may be a running number or an attribute with low entropy. It is easy to construct situations in which attribute p is more useful than attribute q, but neither of the power indexes would detect this.

Table 5. Information system whose approximation qualities should not be averaged

Condition attributes					Decision attr.	Condition attributes					Decision attr.
a	b	c	p	q	d	a	b	c	p	q	d
1	1	0	0	1	0	1	1	1	1	1	4
2	1	0	0	2	0	2	1	1	1	2	4
1	1	0	1	2	1	1	1	1	1	2	5
2	2	0	1	1	1	2	2	1	1	1	5
1	2	0	1	1	2	1	2	1	1	1	6
2	2	0	1	2	2	2	2	1	1	2	6
1	2	0	1	2	3	1	2	1	1	2	7
2	1	0	1	1	3	2	1	1	1	1	7

Recall from (16) that in MCDM, two objects p,q are *uncorrelated* or *independent*, if $\mu(\{p,q\}) = \mu(p) + \mu(q)$. The application of the quantitative interaction values in RSDA are problematic as well, but the term "uncorrelated" should be used with caution: Assume an attribute q with $\gamma(q) = 0$, and use another attribute q', which generates the same partition on U. Then q and q' show no interaction with d and can be called uncorrelated, but their dependency is maximal.

4.4 A Qualitative Power Value Based on Rough Entropy

In [8], we noted that γ is a conditional measure, and therefore, comparisons of γ values are only valid in so-called *nested models*, which means that $\gamma(P)$ and $\gamma(Q)$ are comparable only in a meaningful manner if either $P \subseteq Q$ or $Q \subseteq P$ holds. To allow model selection from all possible attribute sets within RSDA, we have presented a measure called *entropy of deterministic rough approximation*, which is based on the maximum entropy principle as a worst case. Suppose that we have a fixed decision attribute d generating the equivalence relation θ_d on U. If Q is a set of attributes generating θ_Q, we define a new equivalence θ_Q^{det} by

$$x\theta_Q^{\text{det}} y \iff \begin{cases} x\theta_Q y, & \text{if } \theta_Q(x) = \theta_Q(y) \text{ and } \theta_Q(x) \subseteq \theta_d(z) \text{ for some } z \in U \\ x = y, & \text{otherwise.} \end{cases}$$

Its associated probability distribution, based on the principle of indifference, is given by $\{\hat{\psi}_K : K \in \mathcal{P}(\theta_Q^{\text{det}})\}$ with

$$\hat{\psi}(K) = \frac{|K|}{n}. \tag{20}$$

The *entropy of deterministic rough approximation* (with respect to Q and d) is now defined by

$$H^{\mathrm{det}}(Q) = \sum_K \hat{\psi}(K) \cdot \log_2 \frac{1}{\psi(K)}.$$

If

$$H(d) = \sum_{L \in \mathcal{P}_d} \frac{|L|}{n} \cdot \log_2 \frac{n}{|L|}$$

$H^{\mathrm{det}}(Q)$ can be standardized by

$$\mathrm{NRE}(Q) := 1 - \frac{H^{\mathrm{det}}(Q) - H(d)}{\log_2(n) - H(d)}, \tag{21}$$

assuming $H(d) \lesssim \log_2(n)$. We obtain a measure of approximation success within RSDA, which can be used to compare different models in terms of the combination of coding complexity and uncertainty outside the approximation in the sense that a perfect approximation results in $\mathrm{NRE}(Q) = 1$, and the worst case is at $\mathrm{NRE}(Q) = 0$. Unlike γ, NRE is an unconditional measure because both the complexity of the rules generated by the independent attributes and the uncertainty after approximation are merged into one measure.

We are now able to define the *qualitative power index of an attribute m using* NRE by

$$\varphi(m) = \max\{\mathrm{NRE}(K \cup \{m\}) \ominus \mathrm{NRE}(K) : K \subseteq U \setminus \{m\}\}. \tag{22}$$

This value is meaningful because it addresses the value of the maximum NRE (or minimum rough entropy) for which attribute m contributes a nonzero amount of additional information. It is easy to see that

$$\max_m \varphi(m) = \max_K \mathrm{NRE}(K),$$

so that the maximum of φ is also the highest NRE across all subsets of U.

5 An Example

One of the first published applications of RSDA was a study that describes patients after highly selective vagotomy for duodenal ulcer [25]. An enhanced data set of 122 patients was used in [30], and this data set will be used in the sequel.

The information system consisted of 11 condition attributes and a decision attribute, *Visick grading*. Comparing approximation qualities, it was decided that the attribute set P, consisting of

3: Duration of disease.
4: Complication.
5: Basic HCl concentration.
6: Basic volume of gastric juice.
9: Stimulated HCl concentration.
10: Stimulated volume of gastric juice.

is a good basis for approximating attribute *Visick grading* with an approximation quality $\gamma(P) = 0.795$. Inspecting the decline of the approximation quality (5) under the assumption of an admissibility threshold of $c = 0.55$, it was found that the attribute sets

$$A = \{4,5,6,9,10\},\ B = \{3,4,6,9,10\},\ C = \{3,4,5,6,9\}$$

are candidates for future research. These are presented in Table 6. It turns out that the attribute set $D^* = \{3,4,5,6,10\}$ with $\gamma = 0.631$ should have been included as well for further analysis.

Table 6. Analysis of the duodenal ulcer data, I

Attribute set	γ	Interpretation	d_3	Interpretation
3,4,5,6,9,10 (P)	0.795	Admissible		
·,4,5,6,9,10 (A)	0.590	Admissible	0.500	Nonadmissible gain
3,·,5,6,9,10	0.516	Not admissible	0.576	Admissible gain
3,4,·,6,9,10 (B)	0.680	Admissible	0.359	Nonadmissible gain
3,4,5,·,9,10	0.549	Not admissible	0.545	Admissible gain
3,4,5,6,·,10 (D*)	0.631	Admissible	0.444	Nonadmissible gain
3,4,5,6,9,· (C)	0.648	Admissible	0.418	Nonadmissible gain

The gain analysis with d_3 can be interpreted in the same way: Leaving out attributes 4 or 6 results in models that are not admissible, if we set $c_g = 0.5$. The analysis of gain complements the set admissibility: If a subset of P is labeled as admissible, the gain is labeled as nonadmissible and vice versa. This result is not a triviality, because the admissibility labels for sets and gains are driven by the different constants c and c_g.

As we have discussed above, the analysis of usefulness and significance requires a simulation frame: The results, based on 1000 simulated randomizations for each analysis, are gathered in Table 7. Column 1 shows the attributes under consideration, column 2 the observed approximation quality γ of this set, column 3 the expectation of γ given the benchmark model, column 4 the corresponding PREmeasure (usefulness), and column 5 the estimated position of γ in the distribution of the random matching assumption (significance).

Table 7. Analysis of the duodenal ulcer data, II (U: usefulness; S: significance)

	Analysis of the attribute set				
Attribute set	γ	$\mathcal{E}[\gamma]$	U	S	Interpretation
3,4,5,6,9,10	0.795	0.703	0.311	0.013	Significant, medium usefulness
·,4,5,6,9,10 (A)	0.590	0.554	0.081	0.153	Not significant , not useful
3,·,5,6,9,10	0.516	0.484	0.063	0.199	Not significant, not useful
3,4,·,6,9,10 (B)	0.680	0.579	0.241	0.018	Significant, small usefulness
3,4,5,·,9,10	0.549	0.487	0.121	0.084	Not significant, low usefulness
3,4,5,6,·,10 (D*)	0.631	0.515	0.240	0.008	Significant, small usefulness
3,4,5,6,9,· (C)	0.648	0.524	0.259	0.011	Significant, small usefulness

	Analysis of the gain within $\{3,4,5,6,9,10\}$				
Attribute	γ	$\mathcal{E}[\gamma]$	U	S	Interpretation
3	0.795	0.769	0.112	0.182	Not significant, low usefulness
4	0.795	0.751	0.178	0.099	Not significant, low usefulness
5	0.795	0.792	0.017	0.394	Not significant
6	0.795	0.760	0.145	0.107	Not significant, low usefulness
9	0.795	0.763	0.137	0.127	Not significant, low usefulness
10	0.795	0.786	0.044	0.310	Not significant

The first part of Table 7 presents the results of usefulness and significance for sets. The admissible sets (P, A, B, C, D^*) are significant as well. P offers a medium effect size, whereas the usefulness of the admissible subsets of P is smaller.

The analysis of gain within P (second part of Table 7) achieves an astonishing result: All attributes are conditional casual within P. This means that there are always only a few of the 122 observations that can be approximated additionally by introducing the attribute under study into the set. Thus, one can argue that the number of observations in the duodenal ulcer information system is too small, and the good results of P are pushed due to overfitting. Because rough entropy is helpful

Table 8. Analysis of the duodenal ulcer data, III

Attribute set	γ	*NRE*
3,4,5,6,9,10	0.795	0.063
·,4,5,6,9,10 (A)	0.590	0.046
3,·,5,6,9,10	0.516	0.070
3,4,·,6,9,10 (B)	0.680	0.079
3,4,5,·,9,10	0.549	0.064
3,4,5,6,·,10 (D*)	0.631	0.076
3,4,5,6,9,· (C)	0.648	0.092

in checking the complexity of the rule system based on the attributes and therefore, helpful in preventing overfitting, inspection of the normed rough entropy (NRE) values in Table 8 provides further insight: Among the given alternatives, set C has the highest NRE (or the lowest complexity), and from this point of view, can be regarded as a favorable model. Finally, we present the results of the qualitative power index analysis for this example in Table 9. In terms of NRE, the unique

Table 9. Analysis of the duodenal ulcer data, IV

Attribute	φ_Q	Due to attribute set
3	0.1006	3,4,6,10
4	0.1006	3,4,6,10
5	0.0982	3,4,5,9
6	0.1006	3,4,6,10
9	0.0947	3,4,9
10	0.1006	3,4,6,10

optimal set is $\{3,4,6,10\}$ and therefore, by construction of φ_Q in Sect. 4.2, we obtain $\varphi_Q(3) = \varphi_Q(4) = \varphi_Q(6) = \varphi_Q(10) = NRE(\{3,4,6,10\})$. For elements 5 and 9, there exist two further unique conditional optimal sets that are used to determine their qualitative power indexes.

To sum up, we conclude that set C seems to be the optimal choice: It shows an admissible approximation quality, its usefulness is near the optimal value, it is significant, and its complexity is close to the optimal value as well.

6 Conclusion

In this chapter, we have demonstrated that there is more than one way to evaluate a nonnumeric model, but we have also shown that an "anything goes" approach does not work: Forming simple differences or averages poses problems for interpretation, and the Choquet-type aggregation schemes will achieve strange results under certain circumstances. This does not mean that such approaches will not work most of the time, but there is no guarantee that they will. It is notable that [1] begins his analysis of voting schemes with exactly the same ideas. The replacement of quantitative indexes by their qualitative counterparts is a cure, but the results of these qualitative indexes are not overwhelming. They are simply pointers to maximal values of some basic evaluation function (γ, NRE). By reading out the results of the optimization of the basic evaluation, most of the results are achieved without the need for an extra theory.

Because there are several indexes, one has to find a guideline for when and how these indexes should be applied. The examples demonstrate that in RSDA a reasonable starting point for evaluation is inspection of the approximation quality because it is very easy to set a first restriction for a good model. A further restriction can be set by comparing the usefulness with the given standards of effect sizes. Models with very low effect sizes ($\lesssim 0.1$) have to be excluded, and the final model should be not too far away from the maximum effect size. The approximation quality of the final model must be significant, and its complexity (NRE) should be nearly optimal. The examples demonstrate as well that these four qualities need not be present in one model. Furthermore, there need not even be a successful combination at all. We have shown that the indexes may dissociate because they are looking only at partially overlapping features of a model. It may happen that approximation quality is high, but either usefulness is very low or significance is lacking. In such cases, the data do not vote for using RSDA, and this is a fair result as well.

Acknowledgment

Cooperation for this paper was supported by EU COST Action 274 "Theory and Applications of Relational Structures as Knowledge Instruments" (TARSKI).

References

1. J.F. Banzhaf. Weighted voting doesn't work: A mathematical analysis. *Rutgers Law Review*, 19: 317–343, 1965.
2. J. Cohen. *Statistical Power Analysis for the Behavioral Sciences*. Erlbaum, Hillsdale, NJ, 1988.
3. J. Cohen. Things I have learned (so far). *American Psychologist*, 45: 1304–1312, 1990.
4. P. Dubey. On the uniqueness of the Shapley value. *International Journal of Game Theory*, 4: 131–139, 1975.
5. P. Dubey, L. Shapley. Mathematical properties of the Banzhaf power index. *Mathematics of Operations Research*, 4(2): 99–131, 1979.
6. D. Dubois, M. Grabisch, F. Modave, H. Prade. Relating decision under uncertainty and multicriteria decision making models. Technical Report, IRIT-CNRS Université P. Sabatier, Toulouse, 1997.
7. I. Düntsch, G. Gediga. Statistical evaluation of rough set dependency analysis. *International Journal of Human–Computer Studies*, 46: 589–604, 1997.
8. I. Düntsch, G. Gediga. Uncertainty measures of rough set prediction. *Artificial Intelligence*, 106(1): 77–107, 1998.
9. I. Düntsch, G. Gediga. *Rough Set Data Analysis: A Road to Non-Invasive Knowledge Discovery*, Vol. 2 of *Methoδos Primers*. Methoδos Publishers (UK), Bangor, 2000.
10. G. Gediga, I. Düntsch. Rough approximation quality revisited. *Artificial Intelligence*, 132: 219–234, 2001.
11. M. Grabisch. The application of fuzzy integrals in multicriteria decision making. *European Journal of Operational Research*, 89: 445–456, 1996.

12. M. Grabisch. k-additive and k-decomposable measures. In *Proceedings of the Linz Seminar*, 1997.
13. S. Greco, B. Matarazzo, R. Słowinski. Fuzzy measure technique for rough set analysis. In H.-J. Zimmermann, editor, *Proceedings of the 6th European Congress on Intelligent Techniques and Soft Computing (EUFIT'98)*, 99–103, Verlag Mainz, Aachen, 1998.
14. S. Greco, B. Matarazzo, R. Słowinski. Rough sets theory for multicriteria decision analysis. *European Journal of Operational Research*, 129: 1–47, 2001.
15. D. Hildebrand, J. Laing, H. Rosenthal. Prediction logic and quasi-independence in empirical evaluation of formal theory. *Journal of Mathematical Sociology*, 3: 197–209, 1974.
16. J. Komorowski, Z. Pawlak, L. Polkowski, A. Skowron. Rough sets: A tutorial. In S.K. Pal, A. Skowron, editors, *Rough Fuzzy Hybridization: A New Trend in Decision–Making*, 3–98, Springer, Singapore, 1999.
17. A. Laruelle, F. Valenciano. Shapley–Shubik and Banzhaf indices revisited. *Mathematics of Operations Research* (in press).
18. J.-L. Marichal. An axiomatic approach of the discrete Choquet integral as a tool to aggregate interacting criteria. *IEEE Transactions on Fuzzy Systems* (submitted).
19. J.-L. Marichal, P. Mathonet. On comparison meaningfulness of aggregation function. *Journal of Mathematical Psychology*, 45: 213–223, 2000.
20. F. Miller. Computer study into the causes of 1965–1966 traffic deaths in Jacksonville, Florida (manuscript).
21. T. Murofushi, S. Soneda. Techniques for reading fuzzy measures iii: Interaction index. In *Proceedings of the 9th Fuzzy System Symposium*, 693–696, Sapporo, Japan, 1993 (in Japanese).
22. Z. Pawlak. Rough sets. *International Journal of Computer and Information Sciences*, 11: 341–356, 1982.
23. Z. Pawlak. *Rough Sets: Theoretical Aspects of Reasoning about Data*. Kluwer, Dordrecht, 1991.
24. Z. Pawlak. Rough set approach to knowledge-based decision support. *European Journal of Operational Research*, 99(1): 48–57, 1997.
25. Z. Pawlak, K. Słowiński, R. Słowiński. Rough classification of patients after highly selective vagotomy for duodenal ulcer. *International Journal of Man–Machine Studies*, 24: 413–433, 1986.
26. A.E. Roth, editor. *The Shapley Value — Essays in Honor of Lloyd S. Shapley*. Cambridge University Press, Cambridge, 1976.
27. M. Roubens. Interaction between criteria through the use of fuzzy measures. Technical Report number 96.007 of the Institute de Mathématique, Université de Liège, Liège, 1996.
28. L.S. Shapley. A value for *n*–person games. In H. W. Kuhn, A. W. Tucker, editors, *Contributions to the Theory of Games II*, 307–317, Princeton University Press, Princeton, NJ, 1953.
29. M. Shubik (1997). Game theory, complexity, and simplicity (manuscript). Available at citeseer.nj.nec.com/article/shubik97game.html.
30. K. Słowiński. Rough classification of HSV patients. In R. Słowiński, editor, *Intelligent Decision Support: Handbook of Applications and Advances of Rough Set Theory*, 77–94, Kluwer, Dordrecht, 1992.
31. J. Stepaniuk. Knowledge discovery by application of rough set models. In L. Polkowski, S. Tsumoto, S., T. Y. Lin, editors, *Rough Set Methods and Applications: New Developments in Knowledge Discovery in Information Systems*, 137–233, Physica, Heidelberg, 2000.

32. S.S. Stevens. Mathematics, measurement, and psychophysics. In S.S. Stevens, editor, *Handbook of Experimental Psychology*, Wiley, New York, 1951.
33. P.D. Straffin. The Shapley–Shubik and Banzhaf power indices as probabilities. In *[26]*, 71–81, 1976.
34. F. Vogel. *Probleme und Verfahren der numerischen Klassifikation*. Vandenhoeck & Ruprecht, Göttingen, 1975.
35. D.H. Wolpert, W.G. Macready. No free lunch theorems for search. Technical Report number SFI-TR-95-02-010 of the Santa Fe Institute, Santa Fe, NM, 1995.
36. L.A. Zadeh. From computing with numbers to computing with words: From manipulation of measurements to manipulation of perceptions. *IEEE Transactions in Circuits and Systems*, 45(1): 105–119, 1999.

Chapter 11
New Fuzzy Rough Sets Based on Certainty Qualification

Masahiro Inuiguchi, Tetsuzo Tanino

Department of Electronics and Information Systems, Graduate School of Engineering, Osaka University, 2-1 Yamadaoka, Suita, Osaka 565-0871, Japan
{inuiguti, tanino}@eie.eng.osaka-u.ac.jp

Summary. In this chapter, we propose a new definition of fuzzy rough sets based on certainty qualifications. First, rough sets and previously defined fuzzy rough sets are reviewed. Then, the certainty qualifications and the converse certainty qualifications are described. We show explicit representations of the smallest and the largest elements that satisfy given certainty and converse certainty qualifications, respectively. Based on the explicit representations of the smallest and largest elements, lower and upper approximations of a fuzzy subset are defined in two different ways. We show that those lower and upper approximations provide us with better approximations. We describe one of the lower (upper) approximations as requiring less information than the other. Using the lower and upper approximations, which require less information, we give a new definition of a fuzzy rough set. The fundamental properties of rough sets are examined in the case of the proposed fuzzy rough sets. It is shown that all but one of the properties are preserved in the proposed fuzzy rough sets.

1 Introduction

Rough sets were originally defined in the presence of an equivalence relation. This concept is applied to several cases where the equivalence relation, or equivalently, the partition, is replaced with several relations or with a covering. Słowinski and Vanderpooten [1] extended the rough set to a case in which an equivalence relation is replaced with a similarity relation (see also Greco et al. [2]). Yao and Lin [3] and Yao [4, 5] discussed rough sets under various kinds of extended equivalence relations. Bonikowski et al. [6] investigated rough sets under a covering which is an extension of a partition. Inuiguchi and Tanino [7] have also considered rough sets under a similarity relation and a covering. Greco et al. [2] defined rough sets under an ordering relation and showed the importance of their approach in multicriteria decision making problems.

Those studies show that different representations of the original rough sets produce distinct extensions of rough sets under a generalized setting. The selection from the distinct extensions under the same setting depends on the interpretation of rough sets, i.e., interpretations of lower and upper approximations; see [1, 7].

Departing from the crisp setting, we may discuss rough sets under a fuzzy setting. A fuzzy setting refers to a situation where a fuzzy relation instead of an equivalence relation, or a fuzzy partition instead of a partition, is given. The rough sets under fuzzy settings are called fuzzy rough sets. Fuzzy rough sets were originally proposed by Nakamura [8] and Dubois and Prade [9, 10].

Fuzzy rough sets may play a significant role in the treatment of continuous attribute values and vaguely categorized attributes in the reduction of information systems. In spite of their potential applicability, fuzzy rough sets have not yet been extensively investigated. Recently, Radzikowska and Kerre [11] have extended the definitions of fuzzy rough sets, introducing implication functions and t-norms. They have examined the preservation of basic properties of rough sets in the extended fuzzy rough set framework.

Fuzzy rough sets are defined by upper and lower approximations of fuzzy subsets under a fuzzy similarity relation. The upper and lower approximations are defined by using possibility and necessity measures, respectively. Moreover, in the presence of a fuzzy partition, a fuzzy rough set has also been defined [8, 9]. This fuzzy rough set is not a pair of fuzzy subsets but a pair of level-two fuzzy subsets [12], i.e., fuzzy subsets whose elements are fuzzy subsets.

As has already been demonstrated in many papers [1, 3–7] on generalizations of rough sets, there are several representations of original rough sets. Depending on the representation, we may produce a different extension. The fuzzy rough set described above is a fuzzy extension based on a representation of the original rough set. We may define new fuzzy rough sets as extensions of original rough sets using other representations.

In this chapter, after a brief review of original rough sets and previous fuzzy rough sets, we propose a new definition of a fuzzy rough set based on certainty qualifications. Before the discussion of the newly defined fuzzy rough sets, we investigate certainty qualifications. A certainty qualification depends on a necessity measure and a lot of necessity measures exist. In this chapter, we restrict ourselves to a discussion of a special but wide class of necessity measures that are composed of monotonic upper semicontinuous implication functions. It is shown that the smallest fuzzy subset satisfying given certainty qualification is represented explicitly. We discuss converse certainty qualifications and show that the largest fuzzy subset satisfying given converse certainty qualification is also represented explicitly.

Using explicit representations of the smallest and largest fuzzy subsets satisfying given certainty and converse certainty qualifications, the lower and upper approximations are defined in two different ways. We describe the relation of those definitions to representations of the original rough sets. Investigating the simple representation, one of the lower approximations and one of the upper approximations are selected for defining a fuzzy rough set. It is shown that the proposed fuzzy rough set provides better approximations than the previously proposed fuzzy rough set. Moreover, it is shown that a fuzzy rough set is defined as a pair of fuzzy subsets in the presence of a fuzzy covering, a more generalized setting. The basic properties of proposed fuzzy rough sets are investigated. The proposed fuzzy rough sets will be useful for extracting fuzzy rules for interpolated reasoning from information tables because they give better approximations than the previously used fuzzy rough sets.

2 Previous Fuzzy Rough Sets

2.1 The Original Rough Sets

Let Ω be a nonempty universal set, and let R be an equivalence relation on Ω. In rough set theory, a subset $A \subseteq \Omega$ is to be represented by elements of the quotient set Ω/R.

Given an equivalence relation R, lower and upper approximations of a subset $A \subseteq \Omega$ are, respectively, given as

$$R_*(A) = \{x \in \Omega \mid [x]_R \subseteq A\}, \tag{1}$$
$$R^*(A) = \{x \in \Omega \mid [x]_R \cap A \neq \emptyset\}, \tag{2}$$

where $[x]_R$ is the equivalence class of R with the representant x. A pair $[R_*(A), R^*(A)]$ is called a rough set of A with respect to R.

It is known that $R_*(A)$ and $R^*(A)$ can also be represented by

$$R_*(A) = \bigcup_{x\in\Omega:[x]_R\subseteq A} [x]_R, \tag{3}$$
$$R^*(A) = \bigcup_{x\in\Omega:[x]_R\cap A\neq\emptyset} [x]_R, \tag{4}$$
$$R_*(A) = \bigcap_{x\in\Omega:[x]_R\cap(\Omega-A)\neq\emptyset} (\Omega - [x]_R), \tag{5}$$
$$R^*(A) = \bigcap_{x\in\Omega:[x]_R\subseteq\Omega-A} (\Omega - [x]_R), \tag{6}$$
$$R_*(A) = \bigcup_{X\in\mathcal{D}(\Omega,R):X\subseteq A} X, \tag{7}$$
$$R^*(A) = \bigcap_{X\in\mathcal{D}(\Omega,R):X\supseteq A} X, \tag{8}$$

where $\mathcal{D}(\Omega, R)$ is a family of all definable sets and a definable set is a set X for which there exists a set $Y \subseteq \Omega$ such that $X = \bigcup_{x\in Y}[x]_R$ [1, 7].

The meaning of each representation is as follows. If we consider as $[x]_R$ the set of elements that we may misidentify with x, then (1) expresses a set of elements that we correctly judge as members of A, even if we misidentify them. Equation (2) expresses a set of elements that we possibly judge to be members of A considering the misidentification. On the other hand, (3) and (4) express inner and outer approximations of A by means of the union of elementary sets $[x]_R$, respectively. Representations (5) and (6) can be obtained from (3) and (4) via duality, $R_*(A) = \Omega - R^*(\Omega - A)$ and $R^*(A) = \Omega - R_*(\Omega - A)$, respectively. If we regard $\mathcal{D}(\Omega, R)$ as the set of all open sets, then (7) shows that $R_*(A)$ corresponds to the interior of A. If we regard $\mathcal{D}(\Omega, R)$ as the set of all closed sets, then (8) shows that $R^*(A)$ corresponds to the closure of A. From the topological viewpoint, definable sets can be regarded as *clopen* sets (closed and open sets).

Rough sets satisfy the fundamental properties listed in Table 1.

Table 1. Fundamental properties of rough sets

(i)	$R_*(A) \subseteq A \subseteq R^*(A)$
(ii)	$R_*(\emptyset) = R^*(\emptyset) = \emptyset$, $R_*(\Omega) = R^*(\Omega) = \Omega$
(iii)	$R_*(A \cap B) = R_*(A) \cap R_*(B)$, $R^*(A \cup B) = R^*(A) \cup R^*(B)$
(iv)	$A \subseteq B$ implies $R_*(A) \subseteq R_*(B)$, $A \subseteq B$ implies $R^*(A) \subseteq R^*(B)$
(v)	$R_*(A \cup B) \supseteq R_*(A) \cup R_*(B)$, $R^*(A \cap B) \subseteq R^*(A) \cap R^*(B)$
(vi)	$R_*(\Omega - A) = \Omega - R^*(A)$, $R^*(\Omega - A) = \Omega - R_*(A)$
(vii)	$R_*[R_*(A)] = R^*[R_*(A)] = R_*(A)$, $R^*[R^*(A)] = R_*[R^*(A)] = R^*(A)$

2.2 Previous Fuzzy Rough Sets

Several extensions of rough sets have been proposed. Dubois and Prade [9] discussed rough approximations of a fuzzy subset $A \subseteq \Omega$ which is called a *rough fuzzy set*. Yao and Lin [3] and Yao [4, 5] investigated rough sets under various relations. Słowinski and Vanderpooten [1] extended the rough set to a case in which an equivalence relation is replaced with a similarity relation. Bonikowski et al. [6] discussed rough sets when a covering is given. Nakamura [8] and Dubois and Prade [9, 10] proposed a fuzzy rough set that is a pair of lower and upper approximations of a fuzzy subset with respect to a similarity relation. Radzikowska and Kerre [11] extended the definition of the fuzzy rough set and investigated the fundamental properties.

Słowinski and Vanderpooten [1] treat a similarity relation R, which satisfies only reflexivity $\mu_R(x,x) = 1, \forall x \in \Omega$, where μ_R is a membership function of R. To avoid the complexity of explanation, we assume the symmetry, $\mu_R(x,y) = \mu_R(y,x), \forall x,y \in \Omega$. Thus an equivalence class $[x]_R$ can be defined by the following membership function:

$$\mu_{[x]_R}(y) = \mu_R(x,y) = \mu_R(y,x). \tag{9}$$

However, this assumption is not essential to defining fuzzy rough sets.

A fuzzy rough set [9–11] of a fuzzy subset $A \subseteq \Omega$ with respect to a similarity relation R is a pair $[R_*(A), R^*(A)]$ defined by the following membership functions:

$$\mu_{R_*(A)}(x) = \inf_y I[(\mu_{[x]_R}(y), \mu_A(y)], \tag{10}$$

$$\mu_{R^*(A)}(x) = \sup_y T[\mu_{[x]_R}(y), \mu_A(y)], \tag{11}$$

where μ_A is a membership function of A and $I : [0,1]^2 \to [0,1]$, $T : [0,1]^2 \to [0,1]$ are implication and conjunction functions such that

(I1) $I(1,0) = 0$, $I(0,0) = I(0,1) = I(1,1) = 1$.
(T1) $T(1,1) = 1$, $T(0,0) = T(0,1) = T(1,0) = 0$.

When T satisfies the following properties, T is called a t-norm:

(T1) $T(a,1) = T(1,a) = a$, $\forall a \in [0,1]$.
(T2) $T(a,b) \leq T(c,d)$, $0 \leq a \leq c \leq 1, 0 \leq b \leq d \leq 1$.
(T3) $T(a,b) = T(b,a)$, $\forall a,b \in [0,1]$.
(T4) $T[a, T(b,c)] = T[T(a,b), c]$, $\forall a,b,c \in [0,1]$.

Implication functions defined by $I(a,b) = n[T(a,n(b))]$, $I(a,b) = \sup_{0 \leq s \leq 1} \{s \mid T(a,s) \leq b\}$ and $I(a,b) = n\{T[a, n(T(a,b))]\}$ are called S-, R- and QL-implications, when T is a t-norm and $n : [0,1] \to [0,1]$ is a strong negation, i.e., a strictly decreasing function such that $n(0) = 1$ and $n[n(a)] = a$.

Fuzzy rough sets are defined by (10) and (11) with $I(a,b) = \max(1-a,b)$ and $T(a,b) = \min(a,b)$ in Dubois and Prade [9], with a t-norm T and S- or R- implication I in Dubois and Prade [10] and also in Radzikowska and Kerre [11] and with a t-norm T and QL-implication I in Radzikowska and Kerre [11]. Whereas Dubois and Prade [10] assumed the sup-T transitivity, Radzikowska and Kerre [11] assumed the sup-min transitivity. The fundamental properties are examined in those papers.

Let T be a continuous t-norm. Let R be a fuzzy relation satisfying reflexivity $\mu_R(x,x) = 1$, symmetry $\mu_R(x,y) = \mu_R(y,x)$, and t-transitivity $\mu_R(x,z) \geq t[\mu_R(x,y), \mu_R(y,z)]$. The results by Dubois and Prade [10] and Radzikowska and Kerre [11] are arranged in Table 2, where the union $A \cup B$, intersection $A \cap B$, complement $\Omega - A$ and inclusion relations $A \subseteq B$ are defined by

$$\mu_{A \cup B}(x) = \max[\mu_A(x), \mu_B(x)], \tag{12}$$

$$\mu_{A \cap B}(x) = \min[\mu_A(x), \mu_B(x)], \tag{13}$$

$$\mu_{\Omega - A}(x) = n[\mu_A(x)], \tag{14}$$

$$A \subseteq B \Leftrightarrow \mu_A(x) \leq \mu_B(x), \ \forall x \in \Omega. \tag{15}$$

Table 2. Fundamental properties of the previous fuzzy rough sets

I	S-implication	R-implication	QL-implication
(i)	Yes	Yes	Yes
(ii)	Yes	Yes	Part
(iii)	Yes	Yes	Yes
(iv)	Yes	Yes	Yes
(v)	Yes	Yes	Yes
(vi)	Yes	No	No
(vii-a)	Yes	Yes	No
(vii-b)	No	Yes	No

In Table 2, "Yes" means that the numbered property of Table 1 holds, and "No" means that the property is not always satisfied. "Part" in the (ii)-row and "QL-implication" column means a part of the property (ii), i.e., $R_*(\emptyset) = R^*(\emptyset) = \emptyset$ and $R^*(U) = U$, is satisfied. Rows (vii-a) and (vii-b) show the properties $R_*[R_*(X)] = R_*(X)$, $R^*[R^*(X)] = R^*(X)$ and $R^*[R_*(X)] = R_*(X)$, $R_*[R^*(X)] = R^*(X)$, respectively. A combination of (vii-a) and (vii-b) becomes (vii). From Table 2, we know that an arbitrary implication function does not always satisfy all properties, even when R satisfies reflexivity, symmetry, and t-transitivity conditions. However, an implication function that is an S-implication and at the same time an R-implication, such as the Łukasiewicz implication, satisfies all properties.

When a fuzzy subset A and a fuzzy relation R degenerate to a crisp subset A and a crisp relation R, (10) and (11) are reduced to (1) and (2), respectively. In other words, fuzzy rough sets defined by (10) and (11) are direct extensions of (1) and (2), respectively. Thus, $\mu_{R_*(A)}(x)$ shows the degree to which we judge with certainty x as a member of A under possible misidentification, whereas $\mu_{R^*(A)}(x)$ shows the degree to which we might judge x as a member of A under possible misidentification.

In the discussion above, we assume that a similarity relation is given. When a family of fuzzy subsets is given, a fuzzy rough set [9, 10] can also be defined. The family $\Phi = \{F_1, F_2, \ldots, F_n\}$ satisfies the following requirements:

(P1) $\inf_x \max_{i=1,2,\ldots,n} \mu_{F_i}(x) > 0$.

(P2) $\sup_x \min[\mu_{F_i}(x), \mu_{F_j}(x)] < 1$, for $i, j \in \{i = 1, 2, \ldots, n\}$, $i \neq j$.

Such a Φ can be regarded as a weak fuzzy partition so that F_i corresponds to $[x]_R$.

Given such a family Φ, lower and upper approximations of fuzzy subset A by Φ are defined as level-two fuzzy subsets [12], i.e., fuzzy subsets whose elements are fuzzy subsets by the following membership functions:

$$\mu_{\Phi_*(A)}(F_i) = \inf_x I[\mu_{F_i}(x), \mu_A(x)], \tag{16}$$

$$\mu_{\Phi^*(A)}(F_i) = \sup_x T[\mu_{F_i}(x), \mu_A(x)]. \tag{17}$$

$(\Phi_*(A), \Phi^*(A))$ is called a fuzzy rough set of A induced by a family Φ.

3 Certainty Qualifications

Possibility and necessity measures are significantly related to fuzzy rough sets. Possibility and necessity measures of a fuzzy subset B under a fuzzy subset A are defined by

$$\Pi_A(B) = \sup_{x \in \Omega} T[\mu_A(x), \mu_B(x)], \tag{18}$$

$$N_A(B) = \inf_{x \in \Omega} I[\mu_A(x), \mu_B(x)], \tag{19}$$

where μ_B is a membership function of B. $\Pi_A(B)$ evaluates the degree to which "an uncertain value u belongs to B" is possible when we know that u belongs to A, and $N_A(B)$ evaluates the degree to what extent "an uncertain value u belongs to B" is necessary (certain) when we know that u belongs to A.

As can be seen easily, $\mu_{R_*(A)}(x) = N_{[x]_R}(A)$, and $\mu_{R^*(A)}(x) = \Pi_{[x]_R}(A)$. To recover the original subset A by $R_*(A)$ and $R^*(A)$ when $[x]_R$ is a singleton $\{x\}$ for all $x \in \Omega$, we should assume some boundary conditions, i.e., $I(1,a) = a$, $I(0,a) = 1$, $T(1,a) = a$ and $T(0,a) = 0$. In this chapter, we define a new fuzzy rough set based on certainty qualifications. In new fuzzy rough set, we do not need to assume boundary conditions described above.

A certainty qualification [13] is a restriction of possible candidates for a fuzzy subset A by $N_A(B) \geq q$, where q and a fuzzy subset B is given. In many cases, the family of fuzzy subsets A which satisfy $N_A(B) \geq q$ is characterized by the greatest element in the sense of set-inclusion, i.e., $A_1 \subseteq A_2$ iff $\mu_{A_1}(x) \leq \mu_{A_2}(x)$, $\forall x \in \Omega$. To ensure this, we assume

(I2) $I(c,b) \leq I(a,d)$, $0 \leq a \leq c \leq 1$, $0 \leq b \leq d \leq 1$.
(I3) I is upper semicontinuous.

Note that S- and R-implications satisfy (I2), but QL-implication does not. Under assumptions (I2) and (I3), we obtain the greatest element $\hat{A}$ as

$$\mu_{\hat{A}}(x) = \sigma[I][q, \mu_B(x)], \tag{20}$$

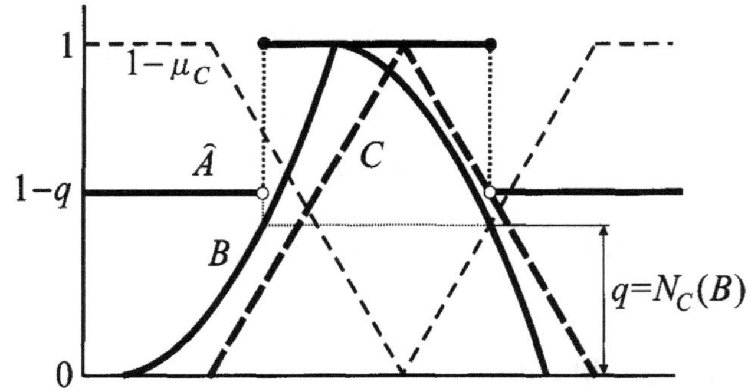

Fig. 1. $\hat{A}$ as an upper approximation of C

where the implication function I defines the necessity measure and a functional σ is defined by

$$\sigma[I](a,b) = \sup_{0 \le h \le 1} \{h \mid I(h,b) \ge a\}. \tag{21}$$

All fuzzy subsets A, such that $A \subseteq \hat{A}$, satisfy $N_A(B) \ge q$. Thus, we have the following note.

Note 1. Consider fuzzy subsets B and C. Let $q = N_C(B)$. Then, C satisfies $N_C(B) \ge q$. Therefore, $\hat{A}$ defined by (20) can be regarded as the upper approximation of C. For example, let us consider B and C in Fig. 1. Let N be a necessity measure defined by (19) with Dienes implication $I(a,b) = \max(1-a,b)$. The value $q = N_C(B)$ is depicted in Fig. 1. $\sigma[I]$ is obtained as

$$\sigma[I](a,b) = \begin{cases} 1, & \text{if } a \le b \\ 1-a, & \text{if } a > b. \end{cases}$$

Fuzzy subset $\hat{A}$ of (20) is obtained as shown in Fig. 1 and is an upper approximation of C.

Now let us consider a converse certainty qualification, i.e., a restriction of possible candidates for a fuzzy subset B by $N_A(B) \ge q$, where q and A are given. The family of fuzzy subsets Bs that satisfy $N_A(B) \ge q$ is often characterized by the smallest element in the sense of set-inclusion. Under assumptions (I2) and (I3), the smallest element $\check{B}$ is obtained as

$$\mu_{\check{B}}(x) = \xi[I][\mu_A(x), q], \tag{22}$$

where the implication function I defines the necessity measure and a functional ξ is defined by

$$\xi[I](a,b) = \inf_{0 \le h \le 1} \{h \mid I(a,h) \ge b\}. \tag{23}$$

All fuzzy subsets Bs such that $B \supseteq \check{B}$ satisfy $N_A(B) \ge q$.

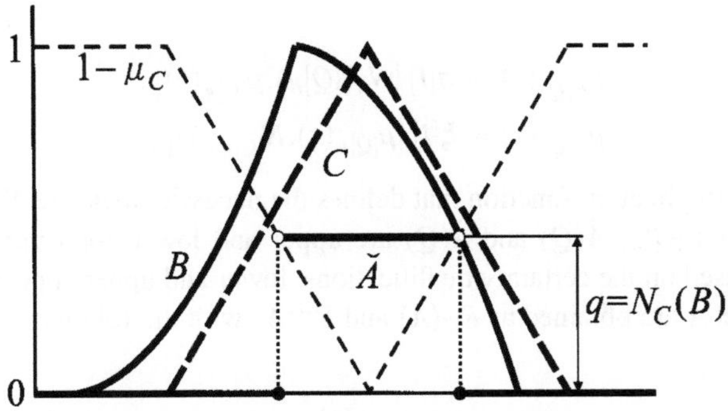

Fig. 2. $\check{A}$ as a lower approximation of B

Note 2. Again consider fuzzy subsets B and C. Let $q = N_C(B)$. Then, B satisfies $N_C(B) \geq q$. Therefore, $\check{A}$ defined by (22) can be regarded as the lower approximation of B. For example, let us consider B and C in Fig. 2. Let N be a necessity measure defined by (19) with Dienes implication $I(a,b) = \max(1-a,b)$. The value $q = N_C(B)$ is depicted in Fig. 2. $\xi[I]$ is obtained as

$$\xi[I](a,b) = \begin{cases} 0, & \text{if } 1-a \geq b \\ b, & \text{if } 1-a < b. \end{cases}$$

Fuzzy subset $\check{A}$ of (22) is obtained, as shown in Fig. 2, and is a lower approximation of B.

Before further discussion, we describe the properties of $\sigma[I]$ and $\xi[I]$. For $\sigma[I]$ and $\xi[I]$, from (I2) and (I3), the following equivalences hold, respectively:

$$\sigma[I](a,b) \geq c \Leftrightarrow I(c,b) \geq a, \tag{24}$$

$$\xi[I](a,b) \leq c \Leftrightarrow I(a,c) \geq b. \tag{25}$$

4 A New Definition of a Fuzzy Rough Set

Given a fuzzy similarity relation R, a fuzzy subset $[Q]_R$ composed of elements similar to elements of a crisp subset $Q \subset \Omega$ can be defined by the following membership function:

$$\mu_{[Q]_R}(x) = \begin{cases} \sup\limits_{y \in Q} \mu_{[y]_R}(x), & \text{if } Q \neq \emptyset \\ 0, & \text{if } Q = \emptyset. \end{cases} \tag{26}$$

$[Q]_R$ can be considered an extension of a definable set since it becomes a definable set when R is an equivalence relation.

Given a fuzzy subset A, we can obtain necessity degrees $N_{[Q]_R}(A)$ and $N_A([Q]_R)$ for each $[Q]_R$. We define fuzzy subsets $\hat{A}(Q)$ and $\check{A}(Q)$ by the following membership functions:

$$\mu_{\hat{A}(Q)}(x) = \sigma[I]\left[N_A([Q]_R), \mu_{[Q]_R}(x)\right], \tag{27}$$

$$\mu_{\check{A}(Q)}(x) = \xi[I]\left[\mu_{[Q]_R}(x), N_{[Q]_R}(A)\right], \tag{28}$$

where I is an implication function that defines the necessity measure N. By Notes 1 and 2, for each $[Q]_R$, $\hat{A}(Q)$ and $\check{A}(Q)$ are upper and lower approximations of A. Therefore, based on the certainty qualifications, lower and upper approximations of a fuzzy subset A are obtained as $R_{\Box}(A)$ and $R^{\Box}(A)$ with the following membership functions:

$$\begin{aligned}\mu_{R_{\Box}(A)}(x) &= \sup_{Q \subseteq \Omega} \mu_{\check{A}(Q)}(x) \\ &= \sup_{Q \subseteq \Omega} \xi[I]\left[\mu_{[Q]_R}(x), N_{[Q]_R}(A)\right], \end{aligned} \tag{29}$$

$$\begin{aligned}\mu_{R^{\Box}(A)}(x) &= \inf_{Q \subseteq \Omega} \mu_{\hat{A}(Q)}(x) \\ &= \inf_{Q \subseteq \Omega} \sigma[I]\left[N_A([Q]_R), \mu_{[Q]_R}(x)\right], \end{aligned} \tag{30}$$

where I is an implication function that defines the necessity measure N.

The complement of the lower (resp., upper) approximation of $\Omega - A$ is the upper (resp., lower) approximation of A. Let a strong negation n define the complement of a fuzzy subset. Then, other lower and upper approximations can be obtained as $R_{\diamond}(A)$ and $R^{\diamond}(A)$ with the following membership functions:

$$\mu_{R_{\diamond}(A)}(x) = \sup_{Q \subseteq \Omega} n\left\{\sigma[I]\left[N_{\Omega-A}([Q]_R), \mu_{[Q]_R}(x)\right]\right\}, \tag{31}$$

$$\mu_{R^{\diamond}(A)}(x) = \inf_{Q \subseteq \Omega} n\left\{\xi[I]\left[\mu_{[Q]_R}(x), N_{[Q]_R}(\Omega - A)\right]\right\}. \tag{32}$$

The fact that $R_{\diamond}(A)$ and $R^{\diamond}(A)$ are lower and upper approximations is confirmed in the following way: First

$$\begin{aligned}\mu_{R^{\Box}(\Omega-A)}(x) &= \inf_{Q \subseteq \Omega} \sigma[I]\left[N_{\Omega-A}([Q]_R), \mu_{[Q]_R}(x)\right] \\ &\geq \mu_{\Omega-A}(x) = n[\mu_A(x)], \\ \mu_{R_{\Box}(\Omega-A)}(x) &= \sup_{Q \subseteq \Omega} \xi[I]\left[\mu_{[Q]_R}(x), N_{[Q]_R}(\Omega - A)\right] \\ &\leq \mu_{\Omega-A}(x) = n[\mu_A(x)].\end{aligned}$$

Since n is strictly decreasing and $n[n(a)] = a$, we obtain

$$\begin{aligned}\mu_{R_{\diamond}(A)}(x) &= \sup_{Q \subseteq \Omega} n\left\{\sigma[I]\left[N_{\Omega-A}([Q]_R), \mu_{[Q]_R}(x)\right]\right\} \\ &\leq n\{n[\mu_A(x)]\} = \mu_A(x), \\ \mu_{R^{\diamond}(A)}(x) &= \inf_{Q \subseteq \Omega} n\left\{\xi[I]\left[\mu_{[Q]_R}(x), N_{[Q]_R}(\Omega - A)\right]\right\} \\ &\geq n\{n[\mu_A(x)]\} = \mu_A(x).\end{aligned}$$

From assumptions (I1)–(I3), we can prove the following theorem.

Theorem 1.

$$\mu_{R_\square(A)}(x) = \sup_{y\in\Omega}\xi[I]\left[\mu_{[y]_R}(x), N_{[y]_R}(A)\right], \tag{33}$$

$$\mu_{R^\diamond(A)}(x) = \inf_{y\in\Omega} n\left\{\xi[I]\left[\mu_{[y]_R}(x), N_{[y]_R}(\Omega - A)\right]\right\}. \tag{34}$$

Proof. When I satisfies (I1)–(I3), $\xi[I]$ is lower semicontinuous and nondecreasing with respect to both arguments. Thus $\xi[I](\cdot, b)$ is left continuous. Hence, we can prove (33) as

$$\begin{aligned}
\mu_{R_\square(A)}(x) &= \sup_{Q\subseteq\Omega}\xi[I]\left[\mu_{[Q]_R}(x), N_{[Q]_R}(A)\right] \\
&= \sup_{Q\subseteq\Omega}\xi[I]\left[\sup_{y\in Q}\mu_{[y]_R}(x), N_{[Q]_R}(A)\right] \\
&= \sup_{Q\subseteq\Omega}\sup_{y\in Q}\xi[I]\left[\mu_{[y]_R}(x), N_{[Q]_R}(A)\right] \\
&= \sup_{y\in\Omega}\xi[I]\left[\mu_{[y]_R}(x), N_{[y]_R}(A)\right].
\end{aligned}$$

The last equality comes from the following facts: (a) for any fuzzy subsets B and C such that $B \subseteq C$, $N_B(A) \geq N_C(A)$ holds because of (I2) and (b) for any $y \in Q$, $[y]_R \subseteq [Q]_R$ holds.

Equation (34) can be proved similarly. □

Though $R^\square(A)$ and $R_\diamond(A)$ require values of $N_A([Q]_R)$ and $N_{\Omega-A}([Q]_R)$ for every crisp subset $Q \subseteq \Omega$, from Theorem 1, $R_\square(A)$ and $R^\diamond(A)$ require values of $N_{[Q]_R}(A)$ and $N_{[Q]_R}(\Omega - A)$ only for every singleton $Q = \{x\}, x \in \Omega$. Thus, $R_\square(A)$ and $R^\diamond(A)$ are computed more easily. We define a fuzzy rough set of A as a pair $(R_\square(A), R^\diamond(A))$, where the implication functions of necessity measures that define $R_\square(A)$ and $R^\diamond(A)$ could be different. Let I_1 and I_2 be the implication functions that define $R_\square(A)$ and $R^\diamond(A)$, respectively. Thus we introduce the following definition.

Definition 1. Given a fuzzy subset $A \subseteq \Omega$, a fuzzy rough set of A based on certainty qualifications under a reflexive and symmetrical fuzzy relation A is defined by a pair of fuzzy subsets $[R_\square(A), R^\diamond(A)]$ with the following membership functions:

$$\mu_{R_\square(A)}(x) = \sup_{y\in\Omega}\xi[I_1]\left[\mu_{[y]_R}(x), N^1_{[y]_R}(A)\right], \tag{35}$$

$$\mu_{R^\diamond(A)}(x) = \inf_{y\in\Omega} n\left\{\xi[I_2]\left[\mu_{[y]_R}(x), N^2_{[y]_R}(\Omega - A)\right]\right\}, \tag{36}$$

where I_1 and I_2 are implication functions satisfying (I1)–(I3) and N^1 and N^2 are necessity measures defined by I_1 and I_2, respectively.

In Definition 1, the reflexivity and symmetry of R are not essential to the validity of

$$R_{\Box}(A) \subseteq A \subseteq R^{\Diamond}(A). \tag{37}$$

However, as we will see, the reflexivity is necessary for R to satisfy many fundamental properties.

Note 3. When R is not symmetrical, we can also define a fuzzy rough set of A by the following pairs of fuzzy subsets $(R_{\Box}^{-1}(A), R_{-1}^{\Diamond}(A))$ with the following membership functions:

$$\mu_{R_{\Box}^{-1}(A)}(x) = \sup_{y\in\Omega} \xi[I_1]\left[\mu_{[y]_{R^{-1}}}(x), N^1_{[y]_{R^{-1}}}(A)\right], \tag{38}$$

$$\mu_{R_{-1}^{\Diamond}(A)}(x) = \inf_{y\in\Omega} n\left\{\xi[I_2]\left[\mu_{[y]_{R^{-1}}}(x), N^2_{[y]_{R^{-1}}}(\Omega - A)\right]\right\}, \tag{39}$$

where $[y]_{R^{-1}}$ is defined by the membership function

$$\mu_{[y]_{R^{-1}}}(x) = \mu_R(x,y). \tag{40}$$

Consider the case when A and R are a crisp set and a relation. In this case, since $[Q]_R$ defined by (26) becomes a definable set, (29) and (30) degenerate to (7) and (8), respectively. On the other hand, (35) and (36) are reduced to (3) and (6), respectively. This fact shows that the new fuzzy rough set is defined as extensions of (3) and (6) whereas the previous fuzzy rough set is defined as those of (1) and (2).

We can prove that $R_{\Box}(A)$ and $R^{\Diamond}(A)$ are better approximations than $R_*(A)$ and $R^*(A)$, respectively.

Theorem 2. *Let R satisfy reflexivity, i.e., $\mu_R(x,x) = 1$ for all $x \in \Omega$. When I_j satisfies $I_j(1, b-\varepsilon) < b$, $\forall b \in [0,1]$, $\forall \varepsilon \in (0,1]$, $j = 1,2$, we obtain*

$$\begin{aligned} R_*(A) &\subseteq R_{\Box}(A), \\ R^{\Diamond}(A) &\subseteq R^*(A), \end{aligned} \tag{41}$$

where we assume that $R_*(A)$ is defined by using an implication function $I = I_1$ and $R^*(A)$ is defined by using a conjunction function $T(a,b) = n\{I_2[a, n(b)]\}$.

Proof. We prove $R^{\Diamond}(A) \subseteq R^*(A)$. $R_*(A) \subseteq R_{\Box}(A)$ can be proved in the same way.

Since I_2 satisfies $I_2(1, b-\varepsilon) < b$, $\forall b \in [0,1]$, $\forall \varepsilon \in (0,1]$, we obtain $\xi[I_2](1,b) \geq b$. Moreover, the reflexivity of R implies that $\mu_{[y]_R}(y) = 1$.

Hence,

$$\begin{aligned}
\mu_{R^{\diamond}(A)}(x) &= \inf_{y\in\Omega} n\left\{\xi[I_2]\left[\mu_{[y]_R}(x), N^2_{[y]_R}(\Omega - A)\right]\right\} \\
&\leq n\left\{\xi[I_2]\left[\mu_{[x]_R}(x), N^2_{[x]_R}(\Omega - A)\right]\right\} \\
&= n\left\{\xi[I_2]\left[1, N^2_{[x]_R}(\Omega - A)\right]\right\} \\
&\leq n\left(\inf_z I_2\left\{\mu_{[x]_R}(z), n[\mu_A(z)]\right\}\right) \\
&= \sup_z n\left(I_2\left\{\mu_{[x]_R}(z), n[\mu_A(z)]\right\}\right) \\
&= \sup_z T[\mu_{[x]_R}(z), \mu_A(z)] = \mu_{R^*(A)}(x). \quad \square
\end{aligned}$$

Observing that $I(1,b) = b$, $\forall b \in [0,1]$ holds for all S- and R-implications, we know that $I(1,b-\varepsilon) < b$, $\forall b \in [0,1]$, $\forall \varepsilon \in (0,1]$. Thus, (41) means that $R_{\square}(A)$ is a better lower approximation than $R_*(A)$ when we use S- and R-implications for I_1. Moreover, $R^{\diamond}(A)$ is a better upper approximation than $R^*(A)$ when we use an S-implication for I_2 since T defined by $T(a,b) = n\{I_2[a, n(b)]\}$ is a t-norm when I_2 is an S-implication [14].

Theorem 2 shows that new fuzzy rough sets are often composed of better lower and upper approximations. This is one of the main characteristics of the proposed fuzzy rough sets.

When a weak fuzzy partition Φ is given, the fuzzy rough set was a pair of level-two fuzzy subsets, fuzzy families of fuzzy subsets. In the proposed fuzzy rough set, the fuzzy rough set with respect to Φ can be defined as a pair of fuzzy subsets $\Phi_{\square}(A)$ and $\Phi^{\diamond}(A)$ with membership functions,

$$\mu_{\Phi_{\square}(A)}(x) = \max_{i=1,2,\ldots,n} \xi[I_1]\left[\mu_{F_i}(x), N^1_{F_i}(A)\right], \tag{42}$$

$$\mu_{\Phi^{\diamond}(A)}(x) = \min_{i=1,2,\ldots,n} n\left\{\xi[I_2]\left[\mu_{F_i}(x), N^2_{F_i}(\Omega - A)\right]\right\}. \tag{43}$$

Note that we can drop condition (P2) to define fuzzy subsets $\Phi_{\square}(A)$ and $\Phi^{\diamond}(A)$. In this case, the weak fuzzy partition is generalized to the weak fuzzy cover. In any case,

$$\Phi_{\square}(A) \subseteq A \subseteq \Phi^{\diamond}(A). \tag{44}$$

5 Fundamental Properties

In this section, let us examine the preservation of the fundamental properties of the rough sets listed in Table 1 in the newly defined fuzzy rough sets.

The first one in Table 1 has already been proven in the previous section. We discuss the preservation of (ii)–(vii).

Theorem 3. *We always have*

$$R_{\Box}(\emptyset) = \emptyset, \qquad R^{\Diamond}(\Omega) = \Omega. \tag{45}$$

When $I_j(1,b) < 1$, $\forall b \in [0,1)$, $j = 1,2$ is satisfied, under the reflexivity of R, we also have

$$R_{\Box}(\Omega) = \Omega, \qquad R^{\Diamond}(\emptyset) = \emptyset. \tag{46}$$

Proof. Equation (45) is straightforward from $\xi[I](a,0) = 0$, $\forall a \in [0,1]$ and for any implication function I. When $I(1,b) < 1$, $\forall b \in [0,1)$, $\xi[I](1,1) = 1$. From $\mu_{[x]_R}(x) = 1$ (reflexivity of R) and $N_{[x]_R}(\Omega) = 1$, (46) is obvious. □

Let I be an arbitrary implication function and N a necessity measure defined by I. For any fuzzy subsets A, B, and C such that $A \subseteq B$, $N_C(A) \leq N_C(B)$ holds, and for any $a,b,c \in [0,1]$ such that $a \leq b$, $\xi[I](c,a) \leq \xi[I](c,b)$. From those facts, we have the following theorem.

Theorem 4.

$$A \subseteq B \text{ implies } R_{\Box}(A) \subseteq R_{\Box}(B), \tag{47}$$

$$A \subseteq B \text{ implies } R^{\Diamond}(A) \subseteq R^{\Diamond}(B). \tag{48}$$

From Theorem 4, we have the following corollaries:

Corollary 1.

$$R_{\Box}(A \cap B) \subseteq R_{\Box}(A) \cap R_{\Box}(B), \tag{49}$$

$$R^{\Diamond}(A \cup B) \supseteq R^{\Diamond}(A) \cup R^{\Diamond}(B). \tag{50}$$

Corollary 2.

$$R_{\Box}(A \cup B) \supseteq R_{\Box}(A) \cup R_{\Box}(B), \tag{51}$$

$$R^{\Diamond}(A \cap B) \subseteq R^{\Diamond}(A) \cap R^{\Diamond}(B). \tag{52}$$

Theorem 4 and Corollary 2 correspond to properties (iv) and (v) of Table 1. Corollary 4 shows the half of property (iii) of Table 1. The other half is not always satisfied. A counterexample is given in the following example.

Example 1. Consider a case when $\Omega = \{1,2,3,4,5\}$, R defined by

$$\mu_R(x,y) = \begin{cases} 0, & \text{if } |x-y| > 2 \\ 0.4, & \text{if } 1 < |x-y| \leq 2 \\ 0.7, & \text{if } 0 < |x-y| \leq 1 \\ 1, & \text{if } |x-y| = 0, \end{cases}$$

and $I_1 = I_2 = I$ is the Gaines–Rescher implication, i.e., I is defined by

$$I(a,b) = \begin{cases} 1, \text{ if } a \leq b \\ 0, \text{ if } a > b. \end{cases}$$

Then,

$$\xi[I](a,b) = \begin{cases} a, \text{ if } b > 0 \\ 0, \text{ if } b = 0. \end{cases}$$

We select a strong negation $n(a) = 1 - a$ to define the complement of a fuzzy set.

Let A and B be fuzzy subsets defined by

$$\mu_A(1) = 1, \mu_A(2) = 0.8, \mu_A(3) = 0.5, \mu_A(4) = 0, \quad \mu_A(5) = 0,$$
$$\mu_B(1) = 0, \mu_B(2) = 0, \quad \mu_B(3) = 0.3, \mu_B(4) = 0.9, \mu_A(5) = 1.$$

Thus, $A \cap B$ and $A \cup B$ are obtained as

$$\mu_{A\cap B}(1) = 0, \mu_{A\cap B}(2) = 0, \quad \mu_{A\cap B}(3) = 0.3, \mu_{A\cap B}(4) = 0, \quad \mu_{A\cap B}(5) = 0,$$
$$\mu_{A\cup B}(1) = 1, \mu_{A\cup B}(2) = 0.8, \mu_{A\cup B}(3) = 0.5, \mu_{A\cup B}(4) = 0.9, \mu_{A\cup B}(5) = 1.$$

We have $N_{[1]_R}(A) = 1, N_{[x]_R}(A) = 0, x = 2,3,4,5, N_{[x]_R}(B) = 0, x = 1,2,3,4, N_{[5]_R}(B)$ $= 1$ and $N_{[x]_R}(A \cap B) = 0$, $x = 1,2,3,4,5$. Then, we obtain $R_\Box(A) = [1]_R$, $R_\Box(B) =$ $[5]_R$, and $R_\Box(A \cap B) = \emptyset$. $R_\Box(A) \cap R_\Box(B)$ is obtained as

$$\mu_{R_\Box(A)\cap R_\Box(B)}(3) = 0.4, \quad \mu_{R_\Box(A)\cap R_\Box(B)}(x) = 0, \ x = 1,2,4,5.$$

Hence, $R_\Box(A \cap B) \neq R_\Box(A) \cap R_\Box(B)$.

On the other hand, $N_{[x]_R}(\Omega - A) = 0, x = 1,2,3,4, N_{[5]_R}(\Omega - A) = 1, N_{[1]_R}(\Omega - B) =$ $1, N_{[x]_R}(\Omega - B) = 0, x = 2,3,4,5$ and $N_{[x]_R}[\Omega - (A \cup B)] = 0, x = 1,2,3,4,5$. Thus, we obtain $R^\Diamond(A) = \Omega - [5]_R$, $R^\Diamond(B) = \Omega - [1]_R$, and $R^\Diamond(A \cup B) = \Omega$. $R^\Diamond(A) \cup R^\Diamond(B)$ is obtained as

$$\mu_{R^\Diamond(A)\cup R^\Diamond(B)}(3) = 0.6, \quad \mu_{R^\Diamond(A)\cup R^\Diamond(B)}(x) = 1, \ x = 1,2,4,5.$$

Hence, $R^\Diamond(A \cup B) \neq R^\Diamond(A) \cup R^\Diamond(B)$.

The following theorem is straightforward from the definition.

Theorem 5. *When $I_1 = I_2$,*

$$R_\Box(\Omega - A) = \Omega - R^\Diamond(A), \quad R^\Diamond(\Omega - A) = \Omega - R_\Box(A). \tag{53}$$

Theorem 5 shows that we have (vi) of Table 1 when $I_1 = I_2$.

Because I satisfies (I2) and (I3), we have the following theorem related to (vii) of Table 1.

Theorem 6.

$$R_\Box[R_\Box(A)] = R_\Box(A), \quad R^\Diamond[R^\Diamond(A)] = R^\Diamond(A). \tag{54}$$

Proof. We prove only the first part, $R_\Box[R_\Box(A)] = R_\Box(A)$ since the second part can be proved similarly.

It suffices to prove that $N^1_{[y]_R}[R_\Box(A)] = N^1_{[y]_R}(A)$. From (I2),

$$\begin{aligned} N^1_{[y]_R}[R_\Box(A)] &= \inf_x I_1\left\{\mu_{[y]_R}(x), \sup_z \xi[I_1]\left[\mu_{[z]_R}(x), N^1_{[z]_R}(A)\right]\right\} \\ &\geq \inf_x I_1\left\{\mu_{[y]_R}(x), \xi[I_1]\left[\mu_{[y]_R}(x), N^1_{[y]_R}(A)\right]\right\}. \end{aligned}$$

Let $h(x) = \xi[I_1][\mu_{[y]_R}(x), N^1_{[y]_R}(A)]$. Then, from (25), $I_1[\mu_{[y]_R}(x), h(x)] \geq N^1_{[y]_R}(A)$, for all $x \in \Omega$. Hence,

$$N^1_{[y]_R}[R_\Box(A)] \geq \inf_x I_1\left[\mu_{[y]_R}(x), h(x)\right] \geq N^1_{[y]_R}(A).$$

On the other hand, from (I2), a necessity measure is nondecreasing with respect to set inclusion, i.e., $N^1_D(B) \leq N^1_D(C)$ if $B \subseteq C$. Because $R_\Box(A) \subseteq A$, we obtain $N^1_{[y]_R}[R_\Box(A)] \leq N^1_{[y]_R}(A)$. Hence, $N^1_{[y]_R}[R_\Box(A)] = N^1_{[y]_R}(A)$. □

When $I_1 = I_2$, we can prove the other part of (vii) of Table 1.

Theorem 7. *If $I_1 = I_2 = I$, then,*

$$R^\Diamond[R_\Box(A)] = R_\Box(A), \tag{55}$$

$$R_\Box[R^\Diamond(A)] = R^\Diamond(A). \tag{56}$$

Proof. From the assumption $I_1 = I_2 = I$, $N^1 = N^2 = N$. First, let us prove (55). It suffices to prove that $N_{[y]_R}[R_\Box(A)] = N_{[y]_R}(A)$. From (I2),

$$\begin{aligned} N_{[y]_R}[R_\Box(A)] &= \inf_x I\left\{\mu_{[y]_R}(x), \sup_z \xi[I]\left[\mu_{[z]_R}(x), N_{[z]_R}(A)\right]\right\} \\ &\geq \inf_x I\left\{\mu_{[y]_R}(x), \xi[I]\left[\mu_{[y]_R}(x), N_{[y]_R}(A)\right]\right\}. \end{aligned}$$

Let $h(x) = \xi[I][\mu_{[y]_R}(x), N_{[y]_R}(A)]$. Then, from (25),

$$I(\mu_{[y]_R}(x), h(x)) \geq N_{[y]_R}(A),$$

for all $x \in \Omega$. Therefore,

$$N_{[y]_R}[R_\Box(A)] \geq \inf_x I\left[\mu_{[y]_R}(x), h(x)\right] \geq N_{[y]_R}(A).$$

On the other hand, from $R_\Box(A) \subseteq A$ and the monotonicity of a necessity measure,

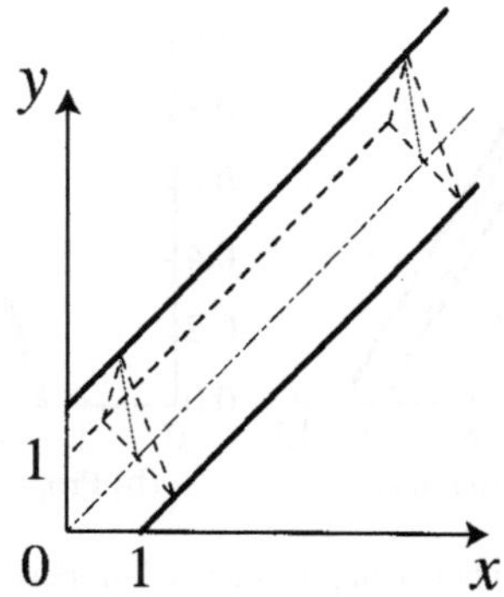

Fig. 3. The similarity relation in Example 2

we obtain $N_{[y]_R}[R_\Box(A)] \le N_{[y]_R}(A)$. Hence, $N_{[y]_R}[R_\Box(A)] = N_{[y]_R}(A)$.

To prove (56), it suffices to prove that $N_{[y]_R}[\Omega - R^\Diamond(A)] = N_{[y]_R}(\Omega - A)$. From Theorem 5, we have (53). Thus, we should prove that $N_{[y]_R}[R_\Box(\Omega - A)] = N_{[y]_R}(\Omega - A)$. Since A is arbitrary, this is equivalent to $N_{[y]_R}[R_\Box(A)] = N_{[y]_R}(A)$, which we have proven already. □

We have examined the fundamental properties of the proposed fuzzy rough sets. All but (iii) of them are preserved under weak requirements on implication functions I_1 and I_2. It is remarkable that all theorems in this chapter assume neither transitivity nor symmetry of R. Thus, all theorems are also valid for the fuzzy rough set $(\Phi_\Box(A), \Phi^\Diamond(A))$ defined by (42) and (43) in the presence of a fuzzy cover Φ as well as a fuzzy rough set defined by the nonsymmetrical fuzzy similarity relation R.

6 Example

To show the difference between previous and proposed fuzzy rough sets, let us consider the following example.

Example 2. Let $I_1(a,b) = I_2(a,b) = I(a,b) = \max(1-a,b)$ and $T(a,b) = \min(a,b)$. Let R be a similarity relation on **R** defined by

$$\mu_R(x,y) = \max(1 - |x-y|, 0). \tag{57}$$

R is depicted in Fig. 4. Obviously, R is reflexive and symmetrical but not T-transitive.

Let A be a fuzzy subset defined by the following membership function:

$$\mu_A(x) = \max\left[0, \min\left(1 - \frac{a^c - x}{\alpha}, 1 - \frac{x - a^c}{\beta}\right)\right]. \tag{58}$$

A is a so-called triangular fuzzy number.

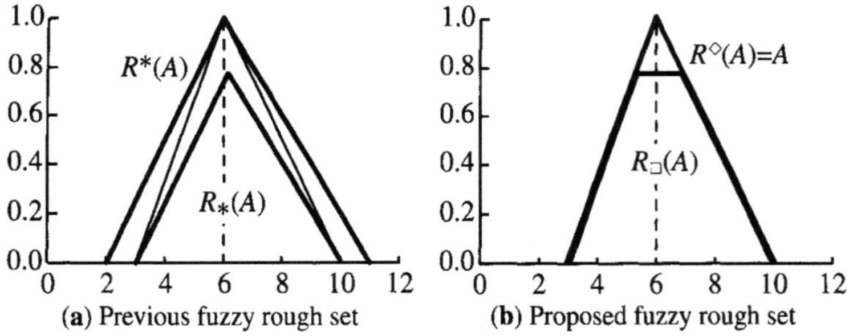

Fig. 4. Comparison between previous and proposed fuzzy rough sets

First, let us see the previous fuzzy rough set $(R_*(A), R^*(A))$. The membership functions of lower and upper approximations are obtained as

$$\mu_{R_*(A)}(x) = \max\left[0, 1 - \max\left(\frac{a^c - x + 1}{1+\alpha}, \frac{x - a^c + 1}{1+\beta}\right)\right], \tag{59}$$

$$\mu_{R^*(A)}(x) = \max\left[0, 1 - \max\left(\frac{a^c - x}{1+\alpha}, \frac{x - a^c}{1+\beta}\right)\right]. \tag{60}$$

When $a^c = 6$, $\alpha = 4$, and $\beta = 5$, $R_*(A)$ and $R^*(A)$ are depicted in Fig. 4(a).

On the other hand, the proposed fuzzy rough set $[R_\Box(A), R^\Diamond(A)]$ is obtained as

$$\mu_{R_\Box(A)}(x) = \min\left[(\mu_A(x), \frac{\alpha+\beta}{2+\alpha+\beta}\right], \tag{61}$$

$$\mu_{R^\Diamond(A)}(x) = \mu_A(x). \tag{62}$$

When $a^c = 6$, $\alpha = 4$, and $\beta = 5$, $R_\Box(A)$ and $R^\Diamond(A)$ are depicted in Fig. 4(b).

As depicted in Fig. 4, the proposed fuzzy rough set gives better approximations than the previous one.

7 Concluding Remarks

In this chapter, we have proposed a fuzzy rough set based on certainty qualifications. It is shown that the proposed fuzzy rough sets give better approximations of fuzzy subsets than the previously used fuzzy rough sets. The fundamental properties of rough sets are examined in the case of the proposed fuzzy rough sets. It is shown that all but one of the fundamental properties are preserved under a weak condition. The advantage of the proposed fuzzy rough sets over the previous ones lies not only in the fact that they provide better approximations but also in the fact that a fuzzy rough set can be defined as a pair of fuzzy subsets from an arbitrarily given fuzzy partition. The last fact will make an application of fuzzy rough sets easier since the fuzzy rough sets induced by a fuzzy partition are more tractable and require less information than fuzzy rough sets induced by a similarity relation.

Fuzzy rough sets are significantly related to necessity measures, i.e., the implication functions that define the necessity measure. A discussion about the way the approximation quality of fuzzy rough sets depends on the implication function that defines the necessity measure seems necessary. In this discussion, a level cut conditioning approach [15,16] to a necessity measure will be useful.

References

1. R. Słowinski, D. Vanderpooten. A generalized definition of rough approximations based on similarity. *IEEE Transactions on Data and Knowledge Engineering*, 12(2): 331–336, 2000.
2. S. Greco, B. Matarazzo, R. Słowinski. The use of rough sets and fuzzy sets in MCDM. In T. Gal, T. J. Stewart, T. Hanne, editors, *Multicriteria Decision Making: Advances in MCDM Models, Algorithms, Theory, and Applications*, 14.1–14.59, Kluwer, Boston, 1999.
3. Y.Y. Yao, T. Y. Lin. Generalization of rough sets using modal logics. *International Journal of Intelligent Automation and Soft Computing*, 2(2): 103–120, 1996.
4. Y.Y. Yao. Two views of the theory of rough sets in finite universes. *International Journal of Approximate Reasoning*, 15: 291–317, 1996.
5. Y.Y. Yao. Relational interpretations of neighborhood operators and rough set approximation operators. *Information Sciences*, 111: 239–259, 1998.
6. Z. Bonikowski, E. Bryniarski, and U. Wybraniec-Skardowska. Extensions and intensions in the rough set theory. *Information Sciences*, 107: 149–167, 1998.
7. M. Inuiguchi, T. Tanino. On rough sets under generalized equivalence relations. In *Proceedings of the International Workshop on Rough Sets Theory and Granular Computing (RSTGC 2001)*, vol. 5(1/2) of the *Bulletin of International Rough Set Society*, 167–171, 2001.
8. A. Nakamura. Fuzzy rough sets. *Notes on Multiple-Valued Logic in Japan*, 9(8): 1–8, 1998.
9. D. Dubois, H. Prade. Rough fuzzy sets and fuzzy rough sets. *International Journal of General Systems*, 17: 191–209, 1990.
10. D. Dubois, H. Prade. Putting rough sets and fuzzy sets together. In R. Słowinski, editor, *Intelligent Decision Support: Handbook of Applications and Advances of the Rough Sets Theory*, 203–232, Kluwer, Dordrecht, 1992.
11. A.M. Radzikowska, E.E. Kerre. Fuzzy rough sets revisited. In *Proceedings of the 5th European Conference on Fuzzy Information Technology (EUFIT'99)*, CD-ROM file cd 1-3-12679$_p$), Verlag Mainz, Aachen, 1999.
12. L.A. Zadeh. A fuzzy-set theoretic interpretation of linguistic hedges. *Journal of Cybernetics*, 2(2): 4–34, 1972.
13. D. Dubois, H.Prade. Fuzzy sets in approximate reasoning, part 1: Inference with possibility distributions. *Fuzzy Sets and Systems*, 40: 143–202, 1991.
14. M. Inuiguchi, M. Sakawa. On the closure of generation processes of implication functions from a conjunction function. In T. Yamakawa, G. Matsumoto, editors, *Methodologies for the Conception, Design, and Application of Intelligent Systems: Proceedings of*

the 4th International Conference on Soft Computing, Vol.1, 327–330, World Scientific, Singapore, 1996.

15. M. Inuiguchi, Y. Kume. Necessity measures defined by level set inclusions: Nine kinds of necessity measures and their properties. *International Journal of General Systems*, 22: 245–275, 1994.
16. M. Inuiguchi, T. Tanino. (2000) Necessity measures and parametric inclusion relations of fuzzy sets. *Fundamenta Informaticae*, 42(3/4): 279–302, 2000.

Chapter 12
Toward Rough Datalog: Embedding Rough Sets in Prolog

Jan Małuszyński,[1] Aida Vitória[2]

[1] Department of Computer and Information Science, Linköping University, 58183 Linköping, Sweden
janma@ida.liu.se
[2] Department of Science and Technology, Linköping University, 60174 Norrköping, Sweden
aidvi@itn.liu.se

Summary. This chapter extends the principles of logic programming for *rough relations*. The work is based on the formalism of *rough sets* [9, 16, 17], introduced for describing imprecise information. Its objective is to develop techniques that facilitate formal definition of imprecise concepts and reasoning about them.

The result is a logic programming language *Rough Datalog* that makes it possible to define systems of rough relations and to query them. We define the syntax and a fixed-point semantics of Rough Datalog and we show how to implement it in Prolog. Thus, we establish a link between rough set theory and logic programming that makes it possible to transfer expertise between these fields and to combine the techniques originating from both fields.

1 Introduction

This chapter presents a new language, called *Rough Datalog*, for defining and querying *rough relations*. The work is based on the formalism of *rough sets* [9, 16, 17] extended with explicit treatment of incomplete information. Rough relations can be seen as representations of imprecise concepts. An important aspect of Rough Datalog is that it makes it possible to combine such concepts into new ones and thus to develop systems of implicitly defined imprecise concepts. This is achieved by extending the principles and techniques of logic programming to rough relations.

The concept of rough set provides a formal basis for modeling imprecise data. In this formalism, a given data object may be classified as a sure or possible member of a considered set. This extends the usual concept of sets with sure membership, which in the rough set framework are called *crisp* sets. Recent publications show the usefulness of rough sets for representating and analyzing experimental results. In such applications, a collection of imprecise data is seen as a rough set. The research in this area has been mainly limited to rough set analysis and to algebraic aspects of rough sets (among others in [2,4,13,14]). In contrast to the previous research, our work aims at defining new rough sets from given ones and constructing a software system supporting reasoning with the rough sets defined. Moreover, the proposed formalism makes it possible to specify different abstract views on the underlying collections of imprecise data and consequently to reason about this data.

In our approach, a rough set S over an universe U is seen as a pair of not necessarily disjoint subsets (S^+, S^-) of U. Intuitively, S^+ consists of those elements of the universe that possibly belong to the rough set, and S^- of those elements that possibly do not belong to it. The remaining elements are those for which no membership information is known. We want to deal explicitly with the incomplete information, a common phenomenon in many applications. This is different from the usual approach to rough sets, where it is assumed that $U = S^+ \cup S^-$.

Syntactically, the programs of Rough Datalog resemble logic programs. However, predicates in Rough Datalog denote rough relations in contrast to the crisp relations denoted by the predicates of logic programs. Consequently, we have a three-valued semantics, which makes a clear distinction between positive, negative, and undefined information. The negative information is explicitly defined in the program. Therefore, in contrast to Prolog, negation as failure is not used to obtain negative information. Failure in Rough Datalog indicates the cases of undefined information.

We also define a language of queries, which syntactically resemble Prolog queries, but refer to rough relations. Ground queries make it possible to check whether given data may or may not belong to a rough set specified by a given program or whether the membership is undefined. Notice that the first two cases need not be exclusive. We can also use nonground queries to enumerate possible members of the relations specified in Rough Datalog.

A prototype implementation of Rough Datalog in Prolog was done by Govert Meuwis and is described in [11]. The implementation gives access to Prolog arithmetic and other built-ins, thus providing useful extensions to the basic idea.

The rest of the chapter is organized as follows. Sect. 2 surveys some basic concepts of rough sets and of logic programming. Sect. 3 defines the syntax, the fixed-point semantics of Rough Datalog, and introduces the notion of query. Sect. 4 discusses its implementation in Prolog and extensions with Prolog built-ins. Sect. 5 show how some operations on rough sets can be expressed in Rough Datalog. Finally, Sect. 6 includes discussion, conclusions, and future work.

2 Preliminaries

This section gives a brief and informal overview of rough sets and logic programming.

2.1 Rough Sets

We want to deal with the situation where there are conflicting judgments about classification of a given object. For example, two patients show identical results of clinical tests, but one of them has a certain disease, and the other does not have it, or the experts looking at the medical record of a patient may disagree on the diagnosis. The concept of a rough set makes it possible to describe such a situation. Intuitively, to define a rough set S, we will specify those elements of the universe that may belong to S and those that may not belong to S. Some elements may fall in both categories. They constitute the conflicting *boundary* cases. It is often assumed that the union of both categories covers the universe. It should be stressed that we do not make this assumption: for some elements of the universe the membership information concerning S may be missing. As this subset may be empty, our framework is more general.

We assume that each object of the universe is associated with a tuple of values. The positions of the tuples are called *attributes* and have predefined value domains. Therefore, information is usually represented as a table. Each row of the table corresponds to an object (e.g., a patient), and each column corresponds to an attribute (e.g., a symptom).

Formally, an attribute a can be seen as a partial function $a : U \rightarrow V_a$, where U is a universe of objects. The set V_a is called the *value domain* of a.,

We assume that tuples of values provide the only way of referring to objects. Two objects are *indiscernible* with respect to a selected set of attributes, if both have the same values for these attributes. As usual, with a set of attributes there is an associated equivalence relation, called an *indiscernibility relation.* Thus, in general, we are not able to refer to a unique object, and a tuple of attribute values is used to represent an equivalence class of indiscernible objects.

We specify a rough set S by saying which tuples represent objects that may belong to S (these form the set S^+) and that represent objects which may not belong to S (these form the set S^-). Therefore, a rough set S is represented by a pair of sets of tuples (S^+, S^-). For some tuples, both cases may hold and for some others no information may be provided. This is summarized by the following definition.

Definition 1. A *rough set* S is a pair (S^+, S^-) such that $S^+, S^- \subseteq V_{a_1} \times \ldots \times V_{a_n}$, for some nonempty set of attributes $\{a_1, \ldots, a_n\}$.

The components S^+ and S^- will be called the *positive region* (or the *positive information*) and the *negative region* (or the *negative information*) of S, respectively.

We will also use the following notions.

Definition 2. A *rough complement* of a rough set $S = (S^+, S^-)$ is the rough set (S^-, S^+). It will be denoted by $\neg S$.

Definition 3. Let $S = (S^+, S^-)$ be a rough set. Then, the *decomposition operations* $pos(S)$ and $neg(S)$ are defined as

$$pos(S) = (S^+, \emptyset),$$
$$neg(S) = (S^-, \emptyset).$$

The result of a decomposition operation on S is thus a rough set with an empty negative region. Its positive region is, depending on the operation, the positive or the negative region of S.

Given a rough set $S = (S^+, S^-)$, the sets S^+ and $(S^+ - S^-)$ are called the *upper approximation* and the *lower approximation* of S, respectively. Thus, the approximations of $\neg S$ are: S^- (the upper approximation) and $(S^- - S^+)$ (the lower approximation).

It is possible to represent a rough set S by a table. Technically, this can be done by adding a new Boolean attribute whose value indicates whether an object may belong to the set. This is summarized by the notions of decision system and rough set specified by a decision system. The next definition was adopted from [8].

Definition 4. A *(binary) decision system* is a pair $\mathcal{D} = (U, A \cup \{d\})$, where U is a universe of objects and $A \cup \{d\}$ is a nonempty finite set of *attributes*, such that $d : U \rightarrow \{true, false\}$. We allow that for some $u \in U$, all attribute values, including the value of d, are undefined.

Since decision systems are often represented in tabular form, they are also called decision tables.

In the sequel, we will assume that every attribute value is represented by a constant in some logical language. We also assume that there is some linear ordering on A such that all the attribute values of an object, if defined, can be represented by a tuple of constants. For a given $u \in U$, we denote by $A(u)$ the tuple $\langle a_1(u), \ldots, a_n(u) \rangle$. Recall that A may be undefined for some u; thus, A is a partial function on objects.

We can now use the decision attribute to define formally the notion of rough set specified by a decision system.

Definition 5. *A rough set D specified by a decision system* $\mathcal{D} = (U, A \cup \{d\})$ is a pair (D^+, D^-), where

$$D^+ = \{A(u) \mid u \in U \text{ and } d(u) = \text{true}\},$$
$$D^- = \{A(u) \mid u \in U \text{ and } d(u) = \text{false}\}.$$

Intuitively, the positive region of D represents all of the objects, which, according to the decision table, are known as possibly having the property d. Similarly, the negative region of D represents all of the objects that are known as possibly not having the property d.

The following example of a rough set specified by a decision table is a version of an example quoted in [16].

Example 1. The symptom attributes *fever, cough, headache, muscle pain* are used to decide whether a patient has flu. The decision system is represented by a table, which includes the symptom attributes and the Boolean decision attribute **flu**. The rows of the table represent the tuples specifying the rough set.

Temp	*Cough*	*Headache*	*Muscle pain*	**Flu**
Normal	No	No	No	No
Subfev	No	Yes	Yes	No
Subfev	No	Yes	Yes	Yes
Subfev	Yes	No	No	No
Subfev	Yes	No	No	Yes
High	No	No	No	No
High	Yes	No	No	No
High	Yes	No	No	Yes
High	Yes	Yes	Yes	Yes

The decision table above specifies the rough set $flu = (flu^+, flu^-)$,where [1]

$$\begin{aligned} flu^+ &= \{(\text{Subfev, No, Yes, Yes}), (\text{Subfev, Yes, No, No}), \\ &\quad (\text{High, Yes, No, No}), (\text{High, Yes, Yes, Yes})\} \\ flu^- &= \{(\text{Normal, No, No, No}), (\text{Subfev, No, Yes, Yes}), (\text{Subfev, Yes, No, No}), \\ &\quad (\text{High, No, No, No}), (\text{High, Yes, No, No})\}. \end{aligned}$$

The boundary of the rough set *flu* is defined by the patients who have subfever, or those who have high fever and cough, but do not have the remaining symptoms.

We conclude this example with the observation that the symbols flu^+ and flu^- represent sets of tuples, hence relations, and can thus be seen as predicates of a logical language. We will use this observation to relate rough sets and logic programming.

The example above illustrates the fact that the regions of the rough set defined by a decision system are relations. To emphasize this fact, the rough sets will often be called *rough relations*.

[1] We adopt the convention to name the rough set specified by a decision system by the name of the decision attribute, represented in print in italics.

2.2 Logic Programs

This section sets out an introduction to logic programming.

First, we recall briefly the syntax of logic programs. Second, we devote attention to the semantics of logic programs and show how fixed-point semantics makes it possible to view a logic program as the definition of a family of relations. Finally, the concept of query is introduced.

Notions are presented in a way facilitating extension of the formalism for defining families of rough relations. The discussion is restricted to *definite programs*, the basic class of logic programs. For a more general introduction to logic programming, the reader is referred to the literature, e.g, [10,12].

2.3 The Syntax

Definite programs are encoded in a formal language constructed over an alphabet, including *variables*, *function symbols*, and *predicate symbols*. Each function and predicate symbol has a specific *arity*. The zero-ary function symbols are called *constants*. We first define two kinds of expressions, terms and atomic formulas.

Definition 6. A *term* is a constant, a variable or an expression of the form

$$f(t_1,\ldots,t_n),$$

where f is an n-ary function symbol and $t_1,\ldots,t_n$ are terms. An *atomic formula* (or more simply, an *atom*) is an expression of the form $p(t_1,\ldots,t_m)$, where p is an m-ary predicate symbol and $t_1,\ldots,t_m$ are terms.

Next, we turn to the definitions of definite clause and definite program.

Definition 7. A *definite clause* is a formula of the form

$$h :- b_1,\ldots,b_n$$

where $n \geq 0$ and $h, b_1,\ldots,b_n$ are atomic formulas. If $n = 0$ the symbol $:-$ is omitted. Such clauses are sometimes called *facts* in contrast to the clauses with $n > 0$, which are called *rules*. A *definite program* $\mathcal{P}$ is a finite set of definite clauses.

Definite programs constitute a subclass of more general logic programs, which, however, are not used in this chapter. Therefore, in the sequel, when referring to a logic program, we have a definite logic program in mind.

A definite clause $h\ :-\ b_1,\ldots,b_n.$ can be seen as an alternative notation for the formula of first-order predicate logic,

$$\forall X(b_1 \wedge \ldots \wedge b_n \rightarrow h),$$

where $\forall$ is the universal quantifier over all variables X in the formula, and $\wedge, \rightarrow$ are the logical connectives conjunction and implication, respectively. A fact corresponds to the formula $\forall X(true \rightarrow h)$, where X represents all variables occurring in h.

Usually we assume that the alphabet of function symbols and predicates is specified implicitly, as consisting of those symbols that appear in a given program.

The set of all variable-free (or *ground*) terms that can be constructed using constant and function symbols in a logic program $\mathcal{P}$ is called the *Herbrand universe* for $\mathcal{P}$. The set of all ground atomic formulas whose predicate symbols appear in $\mathcal{P}$ and whose arguments belong to the Herbrand universe is called the *Herbrand base* for $\mathcal{P}$.

Let $\mathcal{P}$ be a definite logic program. By a *ground instance* of a clause $C \in \mathcal{P}$, we mean a clause C' obtained by replacing each occurrence of every variable v in C by a variable-free term t_v belonging to the Herbrand universe for $\mathcal{P}$. We denote by $g(\mathcal{P})$ the (possibly infinite) set of all ground instances of the clauses of $\mathcal{P}$.

In the examples, we follow Prolog notation, where variables are denoted by identifiers starting with a capital letter.

Example 2. In the following definite program $\mathcal{P}$, we use the unary predicates `list` and `elem`, the binary function symbol `c`, the constants `nil`, `0` and `1`, and the variables `E` and `L`.

```
elem(0).
elem(1).
list(nil).
list(c(E,L)) :- elem(E), list(L).
```

The program consists of three variable-free facts and one rule.

The Herbrand universe for this program is given by the infinite set

$$\{\texttt{0,1,nil,c(1,1),c(1,nil),c(nil,c(1,1))},\ldots\},$$

and the Herbrand base is

$$\{\texttt{elem(0),elem(1),elem(nil),elem(c(1,1)),list(c(1,nil)),}$$
$$\texttt{list(c(nil,c(1,1)))},\ldots\}.$$

The Herbrand universe is infinite due to the occurrence of the binary function symbol `c` in the program. Thus, both the Herbrand base and the set $g(P)$ are infinite, too.

The following are a few examples of ground instances of the rule.

```
list(c(0,nil)) :- elem(0), list(nil).
list(c(0,c(1,nil))) :- elem(0), list(c(1,nil)).
list(c(nil,c(1,1))) :- elem(0), list(c(1,1)).
```

2.4 The Semantics

Intuitively, the semantics assigns meanings to programs. This section outlines a commonly accepted logical semantics of definite programs.

We first recall some basic notions.

In logic, ground terms are used to represent values and ground atomic formulas are used to represent truth values of facts. To determine the represented values, one has to provide an *interpretation*.

Definition 8. Let $\mathcal{P}$ be a definite program. An *interpretation* I for $\mathcal{P}$, over a value domain D, is a mapping that to each n-ary function symbol f of $\mathcal{P}$ assigns an n-ary function $f^I : D^n \to D$, and to each m-ary predicate p of $\mathcal{P}$ assigns an m-ary relation p^I on D. Under the interpretation I, the term $f(t_1,\ldots,t_n)$ represents the value $[f(t_1,\ldots,t_n)]^I = f^I([t_1]^I,\ldots,[t_n]^I)$. Similarly, the atomic formula $p(t_1,\ldots,t_m)$ represents the truth value *true* if the tuple $([t_1]^I,\ldots,[t_n]^I)$ is in the relation p^I and the value *false*, otherwise.

A special kind of interpretation, where D is the Herbrand universe and each function symbol is interpreted as a term constructor, i.e., $[f(t_1,\ldots,t_n)]^I = f([t_1]^I,\ldots,[t_n]^I)$, is called a *Herbrand interpretation.* To define such an interpretation, it suffices to determine which ground atomic formulas take the truth value *true*. This justifies the following definition.

Definition 9. A *Herbrand interpretation* I for a definite program $\mathcal{P}$ is a subset of the Herbrand base for $\mathcal{P}$. A ground atomic formula a is true in I iff $a \in I$.

A Herbrand interpretation I assigns to each ground atomic formula one of the two truth values: *true* or *false*. A ground atom a is false if $a \notin I$. It is well known from logic programming that only Herbrand interpretations needed to be taken into account to define the semantics of logic programs.

We now give the fixed-point semantics of definite programs. With this semantics, we view each definite program as a kind of inductive definition of a unique Herbrand interpretation. Thus, each program specifies a finite family of relations over the Herbrand universe (i.e., each n-ary predicate denotes an n-ary relation over the Herbrand universe). But the Herbrand interpretation obtained by this construction is also a *model* of the program. This means that each clause of the program seen as a logic formula is true in this interpretation. In particular, it is the least Herbrand model with respect to set-inclusion ordering on Herbrand interpretations. Notice that a program may have several Herbrand models. For more details, see [10].

Before showing a formal definition of the semantics, we first discuss an example.

Example 3. Consider the program of Example 2. The intention is to associate with the unary predicate `list` a subset of the Herbrand universe, or intuitively, to say which terms are lists. The fact `list(nil)` is a base case. Intuitively, it says that the constant `nil` is used as the empty list. The inductive rule says that if E is an element and L is a list, then the term $c(E,L)$ is also a list. Thus, predicate `list` defines the least set *list* of terms including `nil` and closed under the construction of term $\texttt{c}(e,l)$, where e is `0` or `1` and $l \in list$.

In this way, we build a unique interpretation that assigns a relation over the Herbrand universe to each predicate. For `elem`, it is the relation {`elem(0), elem(1)`} and for `list`, it is the infinite set

{`list(nil),list(c(0,nil)),list(c(1,nil)),list(c(1,c(0,nil))),`...}.

The least Herbrand interpretation for the program is the union of these two sets.

We now describe how the technique illustrated in the example above applies to an arbitrary program $\mathcal{P}$. Following [1], we first associate with $\mathcal{P}$ a transformation T_P on Herbrand interpretations. For a given Herbrand interpretation I,

$$T_P(I) = \{h \mid h : -b_1, \ldots, b_n. \in g(\mathcal{P}),\ n \geq 0,\ b_1, \ldots, b_n \in I\}.$$

The set of all Herbrand interpretations is a complete lattice under set-inclusion. It is easy to check that T_P is monotonic. It follows by the Knaster–Tarski theorem [17] that T_P has a least fixed-point. This makes it possible to consider a given program to be a specification of this fixed-point.

Definition 10. The semantics M_P of a definite program $\mathcal{P}$ is the least Herbrand interpretation I of $\mathcal{P}$ such that

$$T_P(I) = I.$$

This interpretation is also the least Herbrand model of the program, considered as a set of logic formulas.

The model M_P of a definite program $\mathcal{P}$ is a two-valued model: the atoms not included in M_P are considered *false* in M_P. However, $\mathcal{P}$ does not define explicitly any negative information. Thus, we implicitly assume that $\mathcal{P}$ provides complete knowledge about its predicates. This assumption is known as the *closed-world assumption*. Instead, we could apply the *open-world assumption*, according to which the atoms not included in M_P are considered *undefined*.

The practical importance of the above definition comes from the fact that the relations defined in this way can be effectively queried in Prolog (for an explanation of the query-answering mechanism in Prolog, see [10]).

Queries are introduced in the next section.

2.5 Queries

A *query* Q is a sequence of atoms, possibly including variables.

Before we discuss the notion of an answer to a query, we need to introduce the definition of valuation.

Definition 11. A *valuation* ϑ is a finite set of the form $\{v_1/t_1,\ldots,v_n/t_n\}$, where each v_i is a variable, each t_i is a term distinct from v_i and the variables $v_1,\ldots,v_n$ are all distinct. Moreover, $\vartheta(Q)$ represents an *instance* of the query Q obtained from Q by simultaneously replacing each occurrence of the variable v_i in Q by the term t_i $(i = 1,\ldots,n)$.

Prolog answers a ground query to a program $\mathcal{P}$ by attempting to prove that each of its atoms belongs to the least Herbrand model of $\mathcal{P}$. If the attempt succeeds, then the answer `yes` is issued. But, if the attempt fails in a finite number of steps, then the answer is `no`. Notice that in some cases the attempt may not terminate. A query with variables is answered by showing valuations that instantiate the query to the atoms of the least Herbrand model. An answer `no` may also be obtained if such valuations do not exist.

Example 4. For the program of Example 2

- the query `? list(c(0,nil))` will result in the answer `yes`.
- `? list(0)` would be answered `no`.
- `? list(c(X,nil))` returns two answers: `X=0` and `X=1`.
- `? list(c(0,c(X, nil))),list(c(X,c(0,nil)))` will result in the answer `X=0`.

In practice, Prolog is also able to deduce negative information by using the closed-world assumption. For example, consider again Example 2, and suppose that we would like to check that `nil` *is not* an element. This can be formulated as the query `? \+elem(nil)` to the program (the symbol `\+` represents negation). According to the closed-world assumption, `0` and `1` are the only elements. Hence, we expect the answer `yes` to our query, and Prolog will indeed produce it. The answer to a negative query of the form `? \+p(a)` is constructed by Prolog by attempting to answer the query `? p(a)`. If this answer is `no` (`yes`), then, by the closed-world assumption, the answer to `? \+p(a)` is `yes` (`no`). This strategy for deducing negative information from a definite logic program is known as *negation as failure*.

Using negation as failure, it is possible to include negative atoms in the queries and also to extend definite clauses by allowing negative atoms in the bodies.

Example 5. For the program of Example 2, the query,

```
? list(c(0,c(X,nil))), \+list(c(0,c(0,X))),
```

returns the answers `X=0` and `X=1`.

Notice that negation as failure is less powerful than the closed-world assumption since the negation as failure requires execution to terminate. It may not be possible in a finite number of steps to determine that an atom does not belong to the semantics of the program. More details about this issue can be obtained from [10,12].

3 Kernel Rough Datalog

Rough relations are commonly specified in the form of decision systems. Our main objective is to provide a formalism for combining already defined rough relations into new ones and to support it by programming tools. To achieve this goal, we introduce a language for defining and querying rough relations, called *Rough Datalog*. As a matter of fact, the decision systems can also be encoded in Rough Datalog. Thus, we can define rough relations from scratch.

This section introduces the language. Syntactically, its programs resemble definite programs. However, we provide a different semantics, which associates a rough relation with each predicate of a Rough Datalog program.

We begin by presenting the syntax of the language. Then, we informally explain and illustrate by examples how rough relations can be defined in the language. This is followed by a formal definition of the semantics. Finally, we discuss how Rough Datalog programs can be queried.

For practical use, the syntax and the semantics presented in this section have to be extended with input-output and possibly with other features. These are not discussed in this section, which defines only the kernel of the language.

3.1 The Syntax

The basic lexical elements of Rough Datalog programs are *predicates, constants*, and *variables*. In concrete syntax, predicates are denoted by sequences of letters, digits, and signs $+$ or $-$, beginning with a lower case letter. Constants are denoted in the same way and also by standard Prolog [3] *numbers*. The variables are denoted by sequences of letters and digits beginning with an uppercase letter.

It is important to bear in mind that our alphabet does not include function symbols other than constants. Thus, the terms are restricted to variables and constants. Logic programs observing this restriction are called *datalog programs*. This explains the name of our language. The restriction conforms to the fact that attribute values in decision tables are represented by constants. Technically, it seems possible to extend the language for handling attributes ranging over tree domains. Whether such an extension is of practical interest is a matter of future studies.

Definition 12. A *literal* is an expression of the form a or $\neg a$, where a is an atomic formula (recall Definition 6). They are called *positive* and *negative* literals, respectively.

Definition 13. A *rough clause* is a formula of the form:

$$h : -B_1, \ldots, B_n,$$

where $n \geq 0$, h is an atom (called the *head*) and $B_1, \ldots, B_n$ are body elements. A *body element* B is defined by the following abstract syntax rule:

$$B \longrightarrow a \mid \mathtt{pos}(a) \mid \mathtt{neg}(a) \mid \neg\mathtt{pos}(a) \mid \neg\mathtt{neg}(a),$$

where a is an atom.

A *Rough Datalog program* $\mathcal{P}$ is a finite set of rough clauses.

Rough clauses with a nonempty body, i.e., at least one body element appears on the right of symbol $: -$, are called rules. Otherwise, they are called facts and are simply represented as $h.$. When it causes no confusion, we will often abbreviate the terms "rough clause" to "clause" and "Rough Datalog program" to "program."

The syntax presented is similar to that of definite programs. The only difference is that the body elements of definite clauses must be atomic formulas, whereas here they may include operators $\neg$, pos and neg. They refer to operations on rough sets discussed in Sect. 2.1.

The notion of *ground instance of a rough clause* is the same as that defined for definite programs in Sect. 2.3. The set of all ground instances of the clauses of a *Rough Datalog* program $\mathcal{P}$ is denoted by $g(\mathcal{P})$.

3.2 Defining Rough Relations in Rough Datalog: An Informal Introduction

Intuitively, each predicate p of a Rough Datalog program represents a rough relation P. A fact

$$\texttt{p}(t_1,\ldots,t_n)$$

states that the tuple $\langle t_1,\ldots,t_n\rangle$ possibly belongs to the rough relation denoted by predicate p (i.e., $\langle t_1,\ldots,t_n\rangle \in P^+$). Rules are used for defining new rough relations in terms of other rough relations. The predicate of the head of a rule denotes the new relation, which is defined in terms of the relations denoted by the predicates of the body. This will be formally described in the next section.

The operators pos and neg that can occur in the body of a rule are "decomposition" operators that extract the positive region and the negative region of a given rough relation, respectively. For example, consider the expression `pos[p(X,Y)]`. If `p(X,Y)` denotes the rough relation (P^+,P^-), then `pos[p(X,Y)]` denotes the rough set $(P^+,\emptyset)$. The operator $\neg$ simply exchanges the positive and negative regions of a rough relation denoted by a predicate (recall Definition 2).

Since a rough relation in our framework is defined by what might belong to the relation and by what might not belong to it, each rule has a dual role. It has to contribute to both the positive and the negative regions of the defined relation. Deciding how to define these contributions can be done in different ways, and it determines the semantics of the language. The rest of this section consists of two examples. The first of them motivates our decision in that matter. The other shows how rough relations specified by decision systems can be represented in Rough Datalog.

In the examples, we adopt the convention to denote the rough relation associated with the predicate of a program by the same identifier printed in italics.

Example 6. We consider the rough relation *flu* of Example 1 and a rough relation *patient* with the same attributes as *flu* extended with the new ones, *identification*, *age*, and *sex*. Intuitively, the universe of relation *patient* is a set of people who visited a doctor. Its decision attribute shows whether or not a person has to be treated for some disease and therefore, is considered a patient. The decision may be made independently by more than one expert. All decisions are recorded, which might make the relation rough. The example relation is defined by the following decision table.

Table 1. Example relation

Id	*Age*	*Sex*	*Temp*	*Cough*	*Headache*	*Muscle pain*	**Patient**
1	21	M	Normal	No	No	No	No
2	51	M	Subfev	No	Yes	Yes	Yes
3	18	F	Subfev	No	Yes	Yes	No
4	18	F	Subfev	No	Yes	Yes	Yes
5	18	M	High	Yes	Yes	Yes	Yes

To know who are the people to be treated for flu, we define a new rough relation *ft*. Intuitively, these are people possibly qualified as patients, who may have flu according to the decision table of Example 1. In Rough Datalog, we can express this intuition as the rule,

```
ft(Id,Age,Sex) :- patient(Id,Age,Sex,Fev,C,Ha,Mp),
                  flu(Fev,C,Ha,Mp).
```

Notice that, in contrast to logic programming, we deal here with rough relations (*flu* and *patient*), so that the new relation defined by the rule (*ft*) should also be rough. Thus, the rule has to express both the positive information, describing when a person may be treated for flu, and the negative, describing when a person may not be treated for flu.

Intuitively, concerning the positive information of *ft*, the rule above states that a person may be treated for flu if he/she has been qualified as a patient by at least one of the experts, and in addition, if he/she possibly has flu. We propose the following understanding for interpreting the negative information provided by the rule. A person may not be treated for flu in two cases:

- Either he/she was not qualified as a patient, and he/she has or has not flu-like symptoms (notice that definite information about flu symptoms is required).
- His/her symptoms may allow qualifying his/her disease as not flu, and he/she was or was not qualified as a patient (notice that a positive or negative qualification as a patient is required here).

It would be possible to define the meaning of the rule in other ways. For example, one could lift the requirement about completeness of flu symptoms and patient information. Alternatively, one could count contradictory expert opinions and use this information when defining regions of the new relation.

The above example illustrates our definition of the semantics stated formally in the next section. Roughly speaking, the positive regions of defined relations are constructed in a way similar to those of the least model of a definite program, using the positive regions of the relations referred to by the bodies. The construction of the negative regions is more complex and depends on both the negative and the positive regions of the relations occurring in the bodies. The motivation for this construction is that the positive and the negative regions of the new relation should be constructed from the defined information and must not refer to the missing information. The incompleteness of information dealt with is naturally captured by the three-valued semantics introduced formally in Sect. 3.3. This semantics is not based on the strong Kleene logic [7] since the conjunction of *undefined* truth value with *false* in our logic is still *undefined*. As explained in Sect. 4.1, the use of the strong Kleene logic would often lead to dramatic growth of the boundary regions of the defined relations.

We admit that different semantics of the rules may be more desirable in some applications. However, Rough Datalog is quite expressive, especially with the extensions discussed in Sect. 4. So, we hope that rather than changing the semantics of the rules for such applications, it will be possible to refine the rules, for example, by explicitly augmenting negative information, explicitly counting expert opinions, etc.

As the last example shows, our proposal allows for the possibility of defining a new rough relation in terms of other rough relations, some of which might be given by a finite decision table. We now show how a rough relation specified by a decision table $\mathcal{D}$ can be represented by rough clauses. Rules of the form,

$$p(X1,\ldots,Xn) :- \neg q(X1,\ldots,Xn),$$

play an important role in our representation.

First, we discuss informally the meaning of the rule above. Recall that if q denotes a rough relation $Q = (Q^+, Q^-)$, then $\neg q(X1,\ldots,Xn)$ denotes the rough complement of Q, i.e., $\neg Q = (Q^-, Q^+)$. Thus, the rule above states that p should represent the rough relation $\neg Q$. Alternatively, the rule can be understood as saying that *if* $q(c1,\ldots,cn)$ *might be true, then* $p(c1,\ldots,cn)$ *might be false*, for each ground instance $p(c1,\ldots,cn) :- \neg q(c1,\ldots,cn)$ of the clause.

Secondly, we introduce two auxiliary predicates $d+$ and $d-$ to represent, respectively, the regions D^+ and D^- of the rough relation D specified by the decision table (see Definition 5). Each row $\langle c1,\ldots,cn,d\rangle$ of the decision table, with $n+1$ attributes, is then represented as the fact `d+(c1, $\ldots$ ,cn)`, if the value of the decision attribute d is *true*, and as the fact `d-(c1,$\ldots$, cn)` if this value is *false*.

Finally, to represent the rough set D, we add the rules,

```
d(X1,$\ldots$,Xn) :- d+(X1,$\ldots$,Xn).
d(X1,$\ldots$,Xn) :- ¬d-(X1,$\ldots$,Xn).
```

This idea is illustrated by the following example.

Example 7. For the table of Example 1, we obtain the following program.

```
flu(X1,X2,X3,X4) :- flu+(X1,X2,X3,X4).
flu(X1,X2,X3,X4) :- ¬flu-(X1,X2,X3,X4).
flu-(Normal,No,No,No).
flu-(Subfev,No,Yes,Yes).
flu+(Subfev,No,Yes,Yes).
flu-(Subfev,Yes,No,No).
flu+(Subfev,Yes,No,No).
flu-(High,No,No,No).
flu-(High,Yes,No,No).
flu+(High,Yes,No,No).
flu+(High,Yes,Yes,Yes).
```

It is worth noting that the negative regions of the rough relations denoted by predicates `flu+` and `flu-` are empty because no negative information is provided for them (facts do not code any negative information about a relation). Therefore, the first rule does not introduce any negative information for the rough relation denoted by `flu`, and the second rule does not add any positive information.

3.3 The Semantics

Syntactically, Rough Datalog programs provide a simple extension of definite programs. The semantics of definite programs was discussed in Sect. 2.4. It was defined as the least fixed-point of the operator T_P on interpretations seen as sets of atomic formulas.

The concepts of the Herbrand universe and the Herbrand base are similar to those introduced for definite programs. Given a program $\mathcal{P}$, the *Herbrand universe* for $\mathcal{P}$ is the set of all constants occurring in $\mathcal{P}$. The *Herbrand base* for $\mathcal{P}$ is the set of all ground atoms $q(c_1,\ldots,c_n)$, such that q is a predicate symbol of $\mathcal{P}$ and $\{c_1,\ldots,c_n\}$ is a subset of the Herbrand universe.

Notice that our language supports some kind of negation represented in one of the forms $\neg a$, $\neg\mathsf{pos}(a)$ and $\neg\mathsf{neg}(a)$. Each of these body elements is associated with the rough complement of a rough set.

In the context of definite logic programs, an interpretation assigns to each ground atom either the truth value *true* or *false*. However, for logic programs with negation, there are good reasons to argue that an approach based on three-valued logic is more suitable (for an explanation, see [5]). Intuitively, the third truth value should be seen as *undefined* and is used to model incomplete information and non termination of programs. From this perspective, it seems natural to adopt three-valued logic for our framework, too.

In three-valued logic, the definition of the Herbrand interpretation has to be extended. For our purposes, we adopt the following definition extending to Rough Datalog programs the concept of the three-valued Herbrand interpretation used in logic programming.

Definition 14. Let $\mathcal{P}$ be a Rough Datalog program. A *rough Herbrand interpretation* I for $\mathcal{P}$ is a set of ground literals, such that for each ground literal a or $\neg a$ belonging to I, a is an element of the Herbrand base for $\mathcal{P}$. Given I, we distinguish two subsets I^+, I^- of the Herbrand base for $\mathcal{P}$:

$$I^+ = \{a \mid a \in I\},$$
$$I^- = \{a \mid \neg a \in I\}.$$

If $I^+ \cap I^- = \emptyset$, then I is a *three-valued Herbrand interpretation* of $\mathcal{P}$.

A rough Herbrand interpretation is a possibly infinite set of literals corresponding to a family of rough relations. Intuitively, the atoms in I^+ are (possibly) *true*, the atoms in I^- are (possibly) *false*, and those remaining are *undefined*. The definition of the three-valued Herbrand interpretation coincides with that used in logic programming. The notion of rough interpretation is more general since the same atom may be both in I^+ and in I^- and thus suits our purposes well. This corresponds to the situation where a tuple of values may be classified both as (possibly) belonging to a rough relation R and (possibly) not belonging to R (i.e., the tuple belongs to the boundary of the relation).

Alternatively, we could have given a four-valued interpretation, where each tuple falls in one of four disjoint categories: true, false, undefined, and contradictory.

It should be noticed that the set-theoretical operations of union and intersection make the partially ordered set of rough Herbrand interpretations of a program $\mathcal{P}$ into a complete lattice. The intuition of $I_1 \subseteq I_2$ in this ordering is that I_2 is more defined than I_1 in the sense that whenever an atom has a truth value *true* or *false* (or both) in I_1, it has also the same values in I_2, but not necessarily vice versa.

We extend the T_P operator of Sect. 2.4 to our programs, and we show that the extension is monotonic, thus guaranteeing the existence of the least fixed-point. Actually, we do not use the datalog restriction to obtain this result.

Before presenting the new definition of the T_P operator, we need to explain how a rough interpretation assigns truth values to ground body elements. Recall that a body element is either a literal or has the form $\mathsf{pos}(a)$, $\mathsf{neg}(a)$, $\neg\mathsf{pos}(a)$, or $\neg\mathsf{neg}(a)$, where a is an atom. The meaning of the expressions $\mathsf{pos}(a)$, $\mathsf{neg}(a)$ is associated with the decomposition operators already introduced.

Let b be a body element. We use the notation $I \nearrow b$ to denote that the truth value of b in I is *true* and $I \searrow b$ to denote that it is *false*. We allow I to assign both truth values to b, since we deal with rough relations.

Definition 15. Let b be a ground body element and I be a rough Herbrand interpretation.

- If b is an atom a, then
  - $I \nearrow b$ iff $a \in I$,
  - $I \searrow b$ iff $\neg a \in I$.
- If b is a negative literal $\neg a$, then
  - $I \nearrow b$ iff $\neg a \in I$,
  - $I \searrow b$ iff $a \in I$.
- If $b \equiv \mathsf{pos}(a)$, then
  - $I \nearrow b$ iff $a \in I$,
  - there is no atom a such that $I \searrow b$.
- If $b \equiv \neg\mathsf{pos}(a)$, then
  - $I \searrow b$ iff $a \in I$,
  - there is no atom a such that $I \nearrow b$.
- If $b \equiv \mathsf{neg}(a)$, then
  - $I \nearrow b$ iff $\neg a \in I$,
  - there is no atom a such that $I \searrow b$.
- If $b \equiv \neg\mathsf{neg}(a)$, then
  - $I \searrow b$ iff $\neg a \in I$,
  - there is no atom a such that $I \nearrow b$.

A body element b is said to be *defined* in I iff $I \nearrow b$ or $I \searrow b$. Otherwise, b is *undefined*.

From the definition above, we can easily prove the following:

Proposition 1. Let I_1 and I_2 be two rough Herbrand interpretations of a program $\mathcal{P}$ such that $I_1 \subseteq I_2$. Then,

- If $I_1 \nearrow b$, then $I_2 \nearrow b$.
- If $I_1 \searrow b$, then $I_2 \searrow b$.

We now extend the definition of the T_P operator to rough interpretations.

Definition 16. Let $\mathcal{P}$ be a Rough Datalog program, $\mathcal{R}_P$ be the set of all rough Herbrand interpretations of $\mathcal{P}$, and $I \in \mathcal{R}_P$. The mapping $T_P : \mathcal{R}_P \to \mathcal{R}_P$ is defined as follows:

$$\begin{aligned} T_P(I) = \{h \mid h \text{ :- } b_1,\ldots,b_n. \in g(\mathcal{P}),\ n \geq 0,\ \forall i(I \nearrow b_i\} \cup \\ \{\neg h \mid h \text{ :- } b_1,\ldots,b_n. \in g(\mathcal{P}),\ n > 0, \\ \exists k(I \searrow b_k) \text{ and } \forall i(I \nearrow b_i \text{ or } I \searrow b_i)\}. \end{aligned}$$

This definition consists of two parts. One of them defines all positive literals in the new interpretation; the other defines all negative literals. The intuition behind the first part is similar to that for definite programs. If I and the clauses of $\mathcal{P}$ include only positive literals, then the extended operator T_P works exactly as the original one (in Sect. 2.4), and the result contains only positive literals. However, the result is a three-valued interpretation, not two-valued. Thus, all literals not included in the interpretation have the truth value *undefined.*

The second part of the definition above describes our handling of negative information in the three-valued case. A negative literal is generated by a ground instance of a program clause, if at least one of its body elements is *false* in I and the others are defined (i.e., have truth values *true* or *false*). One point worth noting here is the requirement that all body elements have to be defined. The motivation for this will become clear in Sect. 4.1.

The next example shows the application of the T_P operator to a program.

Example 8. Consider the following program $\mathcal{P}$:

```
ft(Id,Age,Sex) :- patient(Id,Age,Sex,Fev,C,Ha,Mp),
                  flu(Fev,C,Ha,Mp).
flu(T,C,H,M) :- flu+(T,C,H,M).
flu(T,C,H,M) :- ¬flu-(T,C,H,M).
flu+(Subfev,No,Yes,Yes).
flu-(Subfev,No, Yes,Yes).
patient(2,51,M,Subfev,No,Yes,Yes).
```

Starting with $I_0 = \emptyset$, we get $I_1 = T_P(I_0)$:

$I_1 = \{$ `patient(2,51,M,Subfev,No,Yes,Yes),`
`flu+(Subfev,No,Yes,Yes),`
`flu-(Subfev,No,Yes,Yes)`$\}$.

Next, applying T_P to I_1, we obtain $I_2 = T_P(I_1)$:

$I_2 = I_1 \cup \{$`flu(Subfev,No,Yes,Yes)`, $\neg$`flu(Subfev,No,Yes,Yes)`$\}$.

Notice that the only negative literal of I_2 is obtained from the ground clause,

```
flu(Subfev,No,Yes,Yes) :- ¬flu-(Subfev,No,Yes,Yes).
```

Since $I_1 \nearrow$ `flu-(Subfev,No,Yes,Yes)`,
$I_1 \searrow$ `¬flu-(Subfev,No,Yes,Yes)`.
Hence, `¬flu(Subfev,No,Yes,Yes)` is included in I_2.

Finally, the next iteration gives $I_3 = T_P(I_2)$:

$I_3 = I_2 \cup \{$`ft(2,51,M)`, `¬ft(2,51,M)`$\}$.

Notice that I_3 is a fixed-point of T_P, i.e., $I_3 = T_P(I_3)$. Moreover, we can conclude that the tuple $\langle 2, 51, m \rangle$ belongs to the boundary region of the relation denoted by `ft`.

The example illustrates the fact that T_P is monotonic.

Proposition 2. Let I_1 and I_2 be two rough Herbrand interpretations of a program $\mathcal{P}$. If $I_1 \subseteq I_2$, then $T_P(I_1) \subseteq T_P(I_2)$.

This follows by Proposition 1. Each literal in $T_P(I_1)$ is produced by a ground clause C such that all elements in its body are defined in I_1. Since $I_1 \subseteq I_2$, the body literals of C are also defined in I_2 and, by Proposition 1, they admit at least the same truth values as in I_1. (They may admit more truth values inasmuch as the interpretations are rough and a body element may have both the value *true* and *false*.)

As the set of rough Herbrand interpretations form a complete lattice by the Knaster–Tarski theorem [17], there exists the least fixed-point of operator T_P.

Definition 17. The semantics M_P of a Rough Datalog program $\mathcal{P}$ is the least rough interpretation I such that

$$T_P(I) = I.$$

As Rough Datalog programs do not use function symbols other than constants, the use of recursion is rather limited. The least fixed-point of operator T_P (M_P) can then be obtained by a finite number of iterations, starting from the empty interpretation. Example 8 shows a finite sequence of iterations, the last of which gives the interpretation M_P of the example program.

3.4 Querying Rough Datalog Programs

The topic of this section is how to query Rough Datalog programs. We want to use similar queries as in Prolog.

We start by defining the syntax of queries.

Definition 18.

- A *query* $?Q$ is a nonempty sequence of query elements, separated by commas and ended with a point.
- A *query element* Q is defined by the following abstract syntax rule:

$$Q \longrightarrow a \mid \neg a \mid \texttt{pos}(a) \mid \texttt{neg}(a) \mid \texttt{lapos}(a) \mid \texttt{laneg}(a),$$

  where a is an atom.

Queries may include two new operators, `lapos` and `laneg`. The reasons that they are not allowed in the programs will be explained in the next section.

The predicates of the query refer to the rough relations defined by clauses of the program. The answers to an atomic query $?p(t_1,\ldots,t_n)$ concern the upper approximation of the rough relation P denoted by predicate p, whereas the query,

$$?\ \texttt{lapos}[p(t_1,\ldots,t_n)]$$

is about the lower approximation of P. The answers to a query $?\neg p(t_1,\ldots,t_n)$ concern the upper approximation of the rough relation $\neg P$ and the answers to the query $?\texttt{laneg}[p(t_1,\ldots,t_n)]$ are related to the lower approximation of $\neg P$.

Notice that both operators `lapos` and `laneg` cannot be otherwise expressed in Rough Datalog. However, they were easily implemented in our prototype by of Prolog negation as failure.

First, we give some examples of queries to a program and the corresponding answers.

Example 9. Consider the program obtained by coding in our language the decision tables of Examples 1 (see also Example 7) and 6, together with the clause defining the rough relation *ft*.

We illustrate the answers that should be obtained for some queries by referring to the rough relation *ft* of Example 6:

```
? ft(1,21,M).

no.
```

The tuple `(1,21,M)` does not belong to the upper approximation of the relation *ft*. Thus, person 1 may not be qualified for flu treatment.

```
? ¬ft(1,21,M).
```

```
yes.
```

The tuple (1,21,M) belongs to the upper approximation of the relation ¬*ft*. Thus, person 1 may not be qualified for flu treatment.

```
? ft(2,51,M).
```

```
yes.
```

The tuple (2,51,M) belongs to the upper approximation of the relation *ft*. Thus, person 2 may be qualified for flu treatment.

```
? ¬ft(2,51,M).
```

```
yes.
```

The tuple (2,51,M) belongs to the upper approximation of the relation ¬*ft*. Thus, person 2 may not be qualified for flu treatment.

```
? ¬ft(10,18,M).
```

```
undefined.
```

The tuple (10,18,M) belongs neither to the upper approximation of the relation *ft* nor to the upper approximation of the relation ¬*ft*.

```
? ft(I,A,S).

  I=2, A=51, S=M;
  I=4, A=18, S=F;
  I=5, A=18, S=M.
```

The answers lists all people in the database possibly qualified for flu treatment.

```
? ¬ft(I,A,S).

  I=1, A=21, S=M;
  I=2, A=51, S=M;
  I=3, A=18, S=F;
  I=4, A=18, S=F.
```

The answer lists all people in the database possibly not qualified for flu treatment.

```
? ft(I,A,S), ¬ft(I,A,S).

  I=2, A=51, S=M;
  I=4, A=18, S=F.
```

This is the query about the boundary region of the relation *ft*. The answer lists all people who may or may not be qualified for flu treatment.

```
? lapos(ft(I,A,S)).
```

```
I=5, A=18, S=M.
```

This is the query about the lower approximation of the relation *ft*. Thus, person 5 is, surely qualified for flu treatment.

```
? laneg(ft(I,A,S)).

I=1, A=21, S=M;
I=3, A=18, S=F.
```

This is the query about the lower approximation of the relation $\neg ft$. Thus, persons 1 and 3 certainly do not need a flu treatment.

The definition of answer uses the notion of valuation presented in Sect. 2.5 (see Definition 11).

For simplicity, we will consider one-element queries. As the example above shows, an answer to a query Q may be yes, no, undefined, or it may be a valuation of the variables of Q. The answer undefined means that there is no information.

For each form of a query element Q, we now summarize the answer that should be obtained.

- If Q is of the form $p(t_1,\ldots,t_n)$, then the answer should be
  - yes if $p(t_1,\ldots,t_n) \in M_P$,
  - no if $p(t_1,\ldots,t_n) \notin M_P$ and $\neg p(t_1,\ldots,t_n) \in M_P$,
  - a valuation ϑ if $\vartheta(Q) \in M_P$,
  - undefined in the other cases.
- If Q is of the form $\neg p(t_1,\ldots,t_n)$, then the answer should be
  - yes if $\neg p(t_1,\ldots,t_n) \in M_P$,
  - no if $\neg p(t_1,\ldots,t_n) \notin M_P$ and $p(t_1,\ldots,t_n) \in M_P$,
  - a valuation ϑ if $\vartheta(Q) \in M_P$,
  - undefined in the other cases.
- If Q is of the form pos$(p(t_1,\ldots,t_n))$, then the answer should be
  - yes if $p(t_1,\ldots,t_n) \in M_P$,
  - no is never obtained,
  - a valuation ϑ if $\vartheta(p(t_1,\ldots,t_n)) \in M_P$,
  - undefined in the other cases.
- If Q is of the form neg$(p(t_1,\ldots,t_n))$, then the answer should be
  - yes if $\neg p(t_1,\ldots,t_n) \in M_P$,
  - no is never obtained,
  - a valuation ϑ if $\vartheta(\neg p(t_1,\ldots,t_n)) \in M_P$,
  - undefined in the other cases.
- If Q is of the form lapos$(p(t_1,\ldots,t_n))$, then the answer should be
  - yes if $p(t_1,\ldots,t_n) \in M_P$ and $\neg p(t_1,\ldots,t_n) \notin M_P$,
  - a valuation ϑ if $\vartheta(p(t_1,\ldots,t_n)) \in M_P$ and $\vartheta(\neg p(t_1,\ldots,t_n)) \notin M_P$,

  - no if $\neg p(t_1,\ldots,t_n) \in M_P$,
  - `undefined` otherwise.
- If Q is of the form $\texttt{laneg}(p(t_1,\ldots,t_n))$, then the answer should be
  - yes if $\neg p(t_1,\ldots,t_n) \in M_P$, and $p(t_1,\ldots,t_n) \notin M_P$,
  - a valuation ϑ if $\vartheta(\neg p(t_1,\ldots,t_n)) \in M_P$ and $\vartheta(p(t_1,\ldots,t_n)) \notin M_P$,
  - no if $p(t_1,\ldots,t_n) \in M_P$,
  - `undefined` otherwise.

4 Implementing Rough Datalog in Prolog

In this chapter, we investigate how a Rough Datalog program $\mathcal{P}$ can be translated to a definite logic program whose semantics corresponds to the semantics of $\mathcal{P}$. This implies that programs in our language can be executed by a Prolog system.

First, we show how to compile each rule of a Rough Datalog program to definite clauses. Second, we concentrate on how to obtain answers for queries to our programs using a Prolog system. Third, we briefly describe a prototype. Finally, we examine how several Prolog built-ins enhance the expressive power of our language.

4.1 Compiling Rough Datalog to Definite Clauses

Let C be a clause of a Rough Datalog program. Remember that each rule encodes information about the positive and negative region of the rough relation denoted by the predicate defined by the rule. A fact specifies that a tuple belongs to the positive region of a rough relation and no information about the negative region can be extracted from it.

Each clause C is compiled in a nonempty set of definite clauses, denoted $\tau(C)$. Obviously, if C is a fact, then $\tau(C) = \{C\}$. Next, we explain how to obtain $\tau(C)$ when C is a rule.

In the compilation process, we extend the alphabet by adding for each predicate p a unique new predicate p'. This predicate p' accounts for the negative region of the rough relation denoted by p, and it will appear in some of the definite clauses generated by compilation.

The compilation of a rule C gives rise to two sets of definite clauses, $\tau^+(C)$ and $\tau^-(C)$, that represent the positive and negative information encoded in the rule, respectively. Then, $\tau(C) = \tau^+(C) \cup \tau^-(C)$. We start with $\tau^+(C)$, the easiest case.

Assume that C is the rule,

$$h :- b_1,\ldots,b_n.$$

If some b_i $(1 \leq i \leq n)$ is of the form $\neg\mathtt{pos}(a)$ or $\neg\mathtt{neg}(a)$ for an atom a, then no positive information is encoded by C and, consequently, $\tau^+(C) = \emptyset$. For example, recall that `pos(q(X1,...,Xm))` is interpreted as a decomposition operation on the rough relation $Q = (Q^+, Q^-)$ denoted by predicate q. Thus, `pos(q(X1,...,Xm))` denotes the rough relation $(Q^+, \emptyset)$ and $\neg$`pos(q(X1,...,Xm))` denotes the rough relation with an empty positive region $(\emptyset, Q^+)$. Similarly, $\neg$`neg(q(X1,...,Xm))` is also associated with a rough relation that has an empty positive region.

In the next definitions, we use the expression $E_1 \equiv E_2$ meaning that E_1 has the same syntactical form as E_2.

Definition 19. Let C be the rule,

$$h :- b_1, \ldots, b_n.$$

Then,

- If some b_i $(1 \leq i \leq n)$ is of the form $\neg\mathtt{pos}(a)$ or $\neg\mathtt{neg}(a)$ for an atom a, then $\tau^+(C) = \emptyset$.
- Otherwise,
$$\tau^+(C) = \{h :- \overline{b}_1, \ldots, \overline{b}_n. \mid \forall i \varphi^+(\overline{b}_i)\},$$
where the Boolean function φ^+ is defined as follows: $\varphi^+(\overline{b}) = true$ iff

- If b is a positive literal, then $\overline{b} = b$.
- If $b \equiv \neg p(t_1, \ldots, t_n)$, then $\overline{b} \equiv p'(t_1, \ldots, t_n)$.
- If $b \equiv pos\,[p(t_1, \ldots, t_n)]$, then $\overline{b} \equiv p(t_1, \ldots, t_n)$.
- If $b \equiv neg\,[p(t_1, \ldots, t_n)]$, then $\overline{b} \equiv p'(t_1, \ldots, t_n)$.

From the definition above, it is easy to see that $\tau^+(C)$ is either the empty set or a singleton.

Before defining set $\tau^-(C)$, we need to introduce some definitions. Consider once more the rule C, as above. Then,

- $\chi^C = \{b_1, \ldots, b_n\}$.
- $\chi_1^C = \{b \in \chi^C \mid b \equiv \neg\mathtt{pos}(a) \text{ or } b \equiv \neg\mathtt{neg}(a)\}$.
- $\chi_2^C = \{b \in \chi^C \mid b \equiv \mathtt{pos}(a) \text{ or } b \equiv \mathtt{neg}(a)\}$.
- $\chi_3^C = \chi^C - (\chi_1^C \cup \chi_2^C)$.

The clauses of $\tau^-(C)$ are constructed from C by transforming its head and some of its body elements, as described below. We first outline the main idea. We note that any body element $b \in \chi_1^C$ contributes only negative information and thus should be transformed in any clause of $\tau^-(C)$. On the other hand, any body element $b \in \chi_2^C$ contributes only positive information and thus should appear in unchanged form in any clause of $\tau^-(C)$. The remaining body elements contribute in general both kinds of information, and therefore will appear unchanged in some clauses of $\tau^-(C)$ and in transformed form in other ones. Thus, to construct a clause of $\tau^-(C)$, we choose a subset α of the remaining body elements ($\alpha \in \chi_3^C$), and we transform each $b \in \alpha$. This is repeated for all subsets α. However, if $\chi_1^C = \emptyset$, then the chosen set α cannot be the empty set. Otherwise, no negative information could be extracted from the rule. Notice that if $\chi^C = \chi_2^C$, then the rule does not give any negative information and, consequently, $\tau^-(C) = \emptyset$. Details are given in the next definition.

Definition 20. Let C be the rule,

$$q(t_1,\ldots,t_k) :- b_1,\ldots,b_n.$$

- If $\chi^C = \chi_2^C$, then $\tau^-(C) = \emptyset$.
- Otherwise,

$$\tau^-(C) = \{q'(t_1,\ldots,t_k) :- \overline{b}_1,\ldots,\overline{b}_m. \mid \exists\alpha \subseteq \chi_3^C[(\chi_1^C = \emptyset \rightarrow \alpha \neq \emptyset) \text{ and } \varphi^-(\{\overline{b}_1,\ldots,\overline{b}_m\},\alpha)]\},$$

where the Boolean function φ^- is defined as follows:
$\varphi^-(\{\overline{b}_1,\ldots,\overline{b}_m\},\alpha) = \mathit{true}$ iff all elements $\overline{b} \in \{\overline{b}_1,\ldots,\overline{b}_m\}$ satisfy the conditions:

- If $b \in \chi_1^C$ and $b \equiv \neg\mathsf{pos}\,[p(t_1,\ldots,t_m)]$, then $\overline{b} \equiv p(t_1,\ldots,t_m)$.
- If $b \in \chi_1^C$ and $b \equiv \neg\mathsf{neg}\,[p(t_1,\ldots,t_m)]$, then $\overline{b} \equiv p'(t_1,\ldots,t_m)$.
- If $b \in \chi_2^C$, then $\overline{b} = b$.
- If $b \in \chi_3^C \cap \alpha$ and $b \equiv p(t_1,\ldots,t_m)$, then $\overline{b} \equiv p'(t_1,\ldots,t_m)$.
- If $b \in \chi_3^C \cap \alpha$ and $b \equiv \neg p(t_1,\ldots,t_m)$, then $\overline{b} \equiv p(t_1,\ldots,t_m)$.
- If $b \in \chi_3^C - \alpha$, then $\overline{b} = b$.

Notice that the number of clauses in $\tau(C)$ may be exponential with respect to the number of body atoms of C. The reason for this is related to the number of subsets α that can be obtained from χ_3^C.

By $\tau(\mathcal{P})$, we mean the definite program obtained by the union of all sets $\tau(C)$ for $C \in \mathcal{P}$. Remember that the alphabet of $\tau(\mathcal{P})$ includes two predicates p and p' for each predicate p of $\mathcal{P}$.

Example 10. For the program $\mathcal{P}$ of Example 8, we obtain the following definite clauses:

```
ft(Id,Age,Sex)  :- patient(Id,Age,Sex,Fev,C,Ha,Mp),
                   flu(Fev,C,Ha,Mp).
ft'(Id,Age,Sex) :- patient(Id,Age,Sex,Fev,C,Ha,Mp),
```

```
                       flu'(Fev,C,Ha,Mp).
ft'(Id,Age,Sex)   :-  patient'(Id,Age,Sex,Fev,C,Ha,Mp),
                       flu(Fev,C,Ha,Mp).                     (*)
ft'(Id,Age,Sex)   :-  patient'(Id,Age,Sex,Fev,C,Ha,Mp),
                       flu'(Fev,C,Ha,Mp).                    (*)


flu(T,C,H,M)   :-  flu+(T,C,H,M).
flu'(T,C,H,M)  :-  flu+'(T,C,H,M).                           (*)


flu(T,C,H,M)   :-  flu-'(T,C,H,M).                           (*)
flu'(T,C,H,M)  :-  flu-(T,C,H,M).


flu+(Subfev,No,Yes,Yes).
flu-(Subfev,No, Yes,Yes).
patient(2,51,M,Subfev,No,Yes,Yes).
```

We can now use the T_P operator defined for definite logic programs (see Sect. 2.4) to compute the least fixed-point. We note that the clauses marked with `(*)` contribute no elements during this process. The interpretation obtained is

$M_{\tau(\mathcal{P})} = \{$`patient(2,51,M,Subfev,No,Yes,Yes),`
`flu+(Subfev,No,Yes,Yes),`
`flu-(Subfev,No,Yes,Yes), flu(Subfev,No,Yes,Yes),`
`flu'(Subfev,No,Yes,Yes),ft(2,51,M),ft'(2,51,M)`$\}$.

It can be seen as the encoding of the rough interpretation of Example 8, where all negative literals $\neg p(t_1,\ldots,t_n)$ in the former are represented by atoms with new predicates $p'(t_1,\ldots,t_n)$.

Let us look at our clauses as formulas of first-order predicate logic. A sequence of ground atomic formulas is then seen as a conjunction. Thus, b_1,b_2 is seen as $b_1 \wedge b_2$. A ground clause $h :- b_1,b_2$ defines the truth value of h as the conjunction of the truth values of b_1 and b_2. However, we deal here with three-valued interpretations. The question is then what is the result of the conjunction if one of the arguments is *undefined*. In our construction, the result of such a conjunction is *undefined*. As mentioned in Sect. 3.2, this is different from strong three-valued Kleene logic [7], often used in computer science, where the conjunction of *undefined* and *false* gives *false*. A version of our transformation function τ corresponding to this logic would define `ft'` by the definite clauses,

```
ft'(Id,Age,Sex)  :-  patient'(Id,Age,Sex,Fev,C,Ha,Mp),
ft'(Id,Age,Sex)  :-  flu'(Fev,C,Ha,Mp).
```

In this case the fixed-point semantics would define the relation ft′ as a full ternary relation over the Herbrand universe of $\tau(\mathcal{P})$, because the variables on the left-hand side in the second clause are disjoint from the variables on the right-hand side. To avoid this, our T_P operator for Rough Datalog produces a negative literal from a ground clause only if each body element of the clause has a truth value other than *undefined* (i.e., when all body elements are defined in a given interpretation).

The program transformation $\tau(\mathcal{P})$ reflects the definition of the T_P operator for a Rough Datalog program, as formally captured by the following proposition that follows directly from the definitions.

Proposition 3. Let $\mathcal{P}$ be a Rough Datalog program.

- A positive literal $l \in M_{\mathcal{P}}$ iff $l \in M_{\tau(\mathcal{P})}$.
- A negative literal $\neg p(t_1,\ldots,t_n) \in M_{\mathcal{P}}$ iff $p'(t_1,\ldots,t_n) \in M_{\tau(\mathcal{P})}$.

Thus, $\tau(\mathcal{P})$ is a definite program, whose semantics corresponds in a well-defined sense to the semantics of $\mathcal{P}$. The importance of this is that we can use Prolog implementations to run $\tau(\mathcal{P})$.

4.2 Answering Rough Datalog Queries in Prolog

Inasmuch $\tau(\mathcal{P})$ is a usual Prolog program, we may use it in a Prolog system to answer the Rough Datalog queries concerning $\mathcal{P}$. But then the query elements of a Rough Datalog query have to be compiled into Prolog queries, and Prolog answers are to be interpreted as Rough Datalog answers.

Sect. 3.4 introduces the concept of an answer to a query Q. We now show how these answers can be obtained by querying $\tau(\mathcal{P})$ in Prolog. But first, we need a couple of definitions.

Let Q be a literal. Then,

$$\overline{Q} = \begin{cases} p(t_1,\ldots,t_n) & \text{if } Q = p(t_1,\ldots,t_n) \\ p'(t_1,\ldots,t_n) & \text{if } Q = \neg p(t_1,\ldots,t_n), \end{cases}$$

and

$$\overline{\overline{Q}} = \begin{cases} p'(t_1,\ldots,t_n) & \text{if } Q = p(t_1,\ldots,t_n) \\ p(t_1,\ldots,t_n) & \text{if } Q = \neg p(t_1,\ldots,t_n). \end{cases}$$

For simplicity, we consider only one-element queries. Let Q be a query element and $\mathcal{P}$ be a program, both in Rough Datalog. It should be stressed that, in the algorithm below, the "otherwise" does not apply to nontermination because Prolog may not be able to find a correct answer to a query in a finite number of steps, even if one exists. We proceed by cases.

- Q is a literal. Then,
  - if the Prolog answer to $\tau(\mathcal{P})$ and $?\overline{Q}$ is yes or is a valuation ϑ, then it is the answer to $\mathcal{P}$ and $?Q$.
  - if the Prolog answer to $\tau(\mathcal{P})$ and $?\overline{Q}$ is no, and Q is nonground, then the answer to $\mathcal{P}$ and $?Q$ is undefined.
  - if the Prolog answer to $\tau(\mathcal{P})$ and $?\overline{Q}$ is no, and Q is ground, then two cases are possible.
    * If the Prolog answer to $\tau(\mathcal{P})$ and $?\overline{\overline{Q}}$ is yes, then the answer to $\mathcal{P}$ and $?Q$ is no.
    * Otherwise, the answer to $\mathcal{P}$ and $?Q$ is undefined.
- Q is of the form pos(a), where a is an atom. Then,
  - if the Prolog answer to $\tau(\mathcal{P})$ and $?a$ is yes or is a valuation ϑ, then it is the answer to $\mathcal{P}$ and $?Q$.
  - Otherwise, the answer to $\mathcal{P}$ and $?Q$ is undefined.
- Q is of the form neg(a), where a is an atom. Then,
  - if the Prolog answer to $\tau(\mathcal{P})$ and $?\overline{a}$ is yes or is a valuation ϑ, then it is the answer to $\mathcal{P}$ and $?Q$.
  - Otherwise, the answer to $\mathcal{P}$ and $?Q$ is undefined.
- Q is of the form lapos(a), where a is an atom. Then,
  - if the Prolog answer to $\tau(\mathcal{P})$ and $?a$, \+$\overline{\overline{a}}$ (where \+ is the Prolog negation as failure) is yes or is a valuation ϑ, then it is the answer to $\mathcal{P}$ and $?Q$.
  - If the Prolog answer to $\tau(\mathcal{P})$ and $?\overline{\overline{a}}$ is yes, then the answer to $\mathcal{P}$ and $?Q$ is no.
  - Otherwise, the answer to $\mathcal{P}$ and $?Q$ is undefined.
- Q is of the form laneg(a), where a is an atom. Then,
  - if the Prolog answer to $\tau(\mathcal{P})$ and $?\overline{\overline{a}}$, \+a is yes or is a valuation ϑ, then it is the answer to $\mathcal{P}$ and $?Q$.
  - if the Prolog answer to $\tau(\mathcal{P})$ and $?a$ is yes, then the answer to $\mathcal{P}$ and $?Q$ is no.
  - Otherwise, the answer to $\mathcal{P}$ and $?Q$ is undefined.

4.3 The Prototype Implementation

A prototype implementation of Rough Datalog in Prolog is described in [11]. Recall that the exponential size of $\tau(\mathcal{P})$, with respect to the size of $\mathcal{P}$, is caused by the number of definite clauses, obtained by compilation, defining the negative region of the rough relations. These clauses are necessary for answering negative queries and lower bound queries. It seems that positive queries concerning upper approximations may be more common than negative queries. In that case, if negation is not extensively used in program clauses of $\mathcal{P}$, the majority of the clauses of $\tau(\mathcal{P})$ will be unused during query answering. Therefore, in this implementation, $\mathcal{P}$ is not compiled into $\tau(\mathcal{P})$. Instead, the system is an interpreter of Rough Datalog written in Prolog. If a negative literal is queried, then the interpreter simulates execution of the relevant clauses of $\tau(\mathcal{P})$. For a more detailed description, refer to [11]. This first prototype is a pilot project. For combining rough relations described by large decision tables common in practical applications, a different implementation technique will probably be needed similar to those used in deductive databases.

4.4 Extending Rough Datalog with Prolog Built-Ins

Prolog built-ins introduce several extensions to pure logic programming, such as cut, negation as failure, arithmetic, and many others. We have shown how Rough Datalog programs can be implemented in Prolog. This makes it possible to extend Rough Datalog with Prolog built-ins. This section shows how introducing arithmetic and negation as failure enhances the expressive power of Rough Datalog.

Our prototype allows using of arithmetic constraints, which are handled by Prolog arithmetic, so that a ground arithmetic expression can be evaluated. An arithmetic constraint has the form,

$$\langle arithmetic\ term\rangle \otimes \langle arithmetic\ term\rangle\,,$$

where $\otimes$ is a predefined arithmetic predicate. There is a finite number of predefined arithmetic predicates, such as equality, inequality, disequality, etc. with a predefined interpretation on a domain of numbers (floating-point numbers or integers).

Arithmetic terms are built from usual arithmetic operators and constants. The arithmetic constraints can be placed in clause bodies. Let C be a Rough Datalog clause with an arithmetic constraint added in the body and C' be the same clause but without the arithmetic constraint. Then, C has the same effect as the set $\tau(C')$ of Prolog clauses, with the constraint added in each of them. This makes it possible to define rough relations with an infinite number of elements.

Example 11. The following program defines an infinite binary rough relation.

```
r(X,Y):- r+(X,Y).
r(X,Y):- ¬r-(X,Y).
r+(X,Y):- X>Y, X+2<Y, X>0.
r-(X,Y):- X^2 + Y^2 < 4.
```

Its positive region includes pairs of positive numbers, which can be geometrically represented as points between two parallel lines. Its negative region can be depicted as a circle of radius 2 centered at $(0,0)$.

Definitions of infinite rough relations may be used for stating hypotheses that are expected to generalize the results of experiments. Such hypotheses may be stated by experts or generated automatically in some data mining frameworks, such as inductive logic programming, (see, e.g., a recent paper [9] on rough set based ILP). Given a (finite) n-ary rough relation e based on a experimental results and a possibly infinite relation h describing the hypothesis, both defined in Rough Datalog, we may use the definitions to check how well the hypothesis conforms to the experimental data. For example, the valuations obtained by the query,

```
? lapos[e(X1,...,Xn)], laneg[h(X1,...,Xn)],
```

indicate the (certainly) positive outcomes of the experiment that are (certainly) classified as negative by the hypothesis. Similarly, the query,

```
? laneg[e(X1,...,Xn)], lapos[h(X1,...,Xn)],
```

will identify negative outcomes of the experiments classified as positive by the hypothesis. A similar check can be done for the upper approximations.

One can consider using negation as failure explicitly in the body of program clauses. The potential usefulness of such an extension is illustrated by the following example.

Example 12. Assume that a number of n-ary rough relations $r_1, \ldots, r_m$ is defined in Rough Datalog. We want to define a new n-ary relation r such that $c = \langle c_1, \ldots, c_n \rangle$ is in the positive region of r iff c is in the positive region of the majority of the relations r_i. A similar condition defines the negative region of r. We propose the following solution, based on negation as failure and Prolog arithmetic. First, we define a new $n+1$-ary predicate rp_i, for each $i = 1, \ldots, m$. The additional argument of i is a 0 or 1 attribute which indicates whether or not a given tuple is in the positive region of r_i. Using negation as failure, we define `rpi` as follows:

```
rpi(X1,...,Xn,1) :- pos[ri(X1,...,Xn)]
rpi(X1,...,Xn,0) :- \+ ri(X1,...,Xn).
```

Now the positive region `rpos` of the relation r can be defined.

```
rpos(X1,...,Xn) :- rp1(X1,...,Xn,C1),...,rpm(X1,...,Xn,Cm),
                   C is C1+...+Cm, C>m/2 ,
```

where m is the number of "voting" predicates. We use here upper approximations of the predicates r_i. If so desired, we could instead refer to lower approximations since the extended language allows Prolog's negation as failure.

The negative region `rneg` of r can be defined in a similar way.

```
1. rni(X1,...,Xn,1) :- neg[ri(X1,...,Xn)].
2. rni(X1,...,Xn,0) :- \+ neg(ri(X1,...,Xn)).

3. rneg(X1,...,Xn) :- rn1(X1,...,Xn,C1),...,rnm(X1,...,Xn,Cm),
                      C is C1+...+Cm, C>m/2.
```

The final definition of `r` combines both regions.

```
r(X1,...,Xn) :- rpos(X1,...,Xn).
r(X1,...,Xn) :- ¬rneg(X1,...,Xn).
```

5 Operations on Rough Relations

Rough Datalog makes it possible to define and to query families of rough relations. This resembles techniques used in (deductive) relational databases (see [18]). The latter are based on relational algebras, thus on some specific operations on (crisp) relations, such as union, complement, Cartesian product, projection, etc. To extend relational database techniques to rough relations, one has to extend the relational operations. The question is then how to express the suitable operations in Rough Datalog.

Extensions of set operations to rough sets have been studied by many authors (among others [2,4,6,13,14]). However, they do not capture the phenomenon of incomplete information. This section presents a preliminary proposal for a few operations on rough relations with incomplete information and shows how they can be expressed in Rough Datalog.

5.1 Expressing Relational Operations in Prolog

We first recall how crisp relational operations can be expressed in definite clauses. The argument relations are assumed to be represented by predicates. We represent each operation by a new predicate, for which we provide definite clauses. We restrict discussion to the union, intersection, Cartesian product, and projection.

- *Union* $\cup$
  The union u of two n-ary relations p and r is the set of all tuples that belong to one or both relations. It can be defined by the clauses,
  ```
  u(X1,...,Xn) :- p(X1,...,Xn),
  u(X1,...,Xn) :- r(X1,...,Xn).
  ```
- *Intersection* $\cap$
  The intersection s of two n-ary relations p and r is the set of all tuples that belong to both relations. It can be defined by the clause,
  ```
  s(X1,...,Xn) :- p(X1,...,Xn), r(X1,...,Xm).
  ```
- *Cartesian product* $\times$
  The Cartesian product cp of an n-ary relation p and an m-ary relation r is a $(n+m)$-ary relation consisting of all $(n+m)$-tuples, such that the first n components form a tuple in p and the remaining m components form a tuple in r. It can be defined by the clause,
  ```
  cp(X1,...,Xn,Y1,..., Ym) :- p(X1,...,Xn), r(Y1,...,Ym).
  ```
- *Projection* $\pi_i(r)$, with $1 \leq i \leq n$
  The projection p_i selects the ith component from each n-tuple in r and makes them into a set (i.e., into a unary relation). It can be defined by the clause,
  ```
  pi(Xi) :- r(X1,...,Xi,...,Xn).
  ```

The complement of an n-ary relation p is a relation *comp* including all n-tuples of the universe that are not in p. This is not expressible in definite clauses, but the following clause using Prolog negation as failure may be used to check whether a given tuple in the universe belongs to the complement of p:

```
comp(X1,...,Xn) :- \+p(X1,...,Xn).
```

5.2 Defining Operations on Rough Relations

As discussed in Sect. 2.1, a rough relation R is represented by a pair of crisp relations (R^+, R^-). The question is then how to define the extensions of the above mentioned operations such that they can be represented in our framework. The extensions have to satisfy the condition that if the arguments are crisp and the information is complete (i.e., $R^+ \cap R^- = \emptyset$ and $R^+ \cup R^-$ is the universe), the result is as defined by the original operations.

There are several proposals for algebraic operations on rough sets (among others, in [4,14]). Based on those ideas, we present the following extensions:

- *Union* $\sqcup$
  $R_1 \sqcup R_2 = (R_1^+ \cup R_2^+, R_1^- \cap R_2^-)$.
- *Intersection* $\sqcap$
  $R_1 \sqcap R_2 = (R_1^+ \cap R_2^+, R_1^- \cup R_2^-)$.
- *Cartesian product* $\times$
  $R_1 \times R_2 = (R_1^+ \times R_2^+, (R_1^- \times R_2^-) \cup (R_1^+ \times R_2^-) \cup (R_1^- \times R_2^+))$.
- *Projection* Π_i
  $\Pi_i(r) = (\pi_i(R^+), \pi_i(R^-))$.

It can be checked that they satisfy the requirement.

The operations above rely on the usual set operations on the (positive and negative) regions of the arguments. We can thus express them directly in Rough Datalog by using the primitives pos and neg. For example, given the binary rough relations P and Q, the relation $S = P \sqcap Q$ can be described as follows:

```
sp(X1,X2)  :- pos[p(X1,X2)], pos[q(X1,X2)].
sn(X1,X2)  :- neg[p(X1,X2)].
sn(X1,X2)  :- neg[p(X1,X2)].
s(X1,X2)   :- sp(X1,X2).
s(X1,X2)   :- ¬sn(X1,X2).
```

It should be stressed that operations on rough relations corresponding to intersection and union are in general not defined in a unique way. In different applications, the user may prefer to define some application-specific versions more suitable for the problem at hand. The above definitions should rather be considered as examples. The examples of Sect. 4.4 demonstrate clearly that possible user preferences, such as voting techniques, can be easily encoded in Rough Datalog and used for defining application-specific versions of operations on rough relations. A possible topic of future work could be the development of a library of Rough Datalog definitions of commonly used operations on rough relations.

6 Conclusions

The notion of a rough relation emerges naturally as an abstraction of incomplete and contradictory decision tables. We introduced a new declarative language for defining and querying rough relations, inspired by the concepts of logic programming. We presented a fixed-point semantics of the language, and we outlined an approach to its implementation in Prolog. Several examples were presented to illustrate its potential usefulness. We are not aware of any similar work done previously.

One of the most important contributions of this study is to establish a formal link between the theory of rough sets and logic programming, thus opening the way for transferring techniques between these fields and for generalizing of both formalisms. For example, this chapter shows how the rough relations defined by decision systems can be combined into new ones using techniques originating from logic programming. The rough relations defined in that way can be effectively queried by using query answering techniques similar to those used in Prolog or from deductive databases. In contrast to the difficulties in defining semantics of logic programs with negation, the fixed-point semantics of Rough Datalog is very simple, although negative information may be explicitly represented in our programs. This is due to the fact that the rough interpretations allow inconsistencies forbidden in logic. A natural extension of Rough Datalog would allow general terms in clauses. This would open the way for more advanced recursive definitions. The present restriction comes from the usual form of the decision tables. Fixed-point semantics extends smoothly for general terms.

Possible topics of future work include

- Development of a Rough Datalog library of commonly used operations on rough relations. This would provide a practical assessment of the language and may suggest further extensions. Moreover, it may also contribute to the methodology of rough set applications.
- Development of a more realistic implementation of Rough Datalog in Prolog, or in other languages, and investigation of other implementation techniques like

those used in the deductive databases. The implementation should make possible clean integration of Rough Datalog programs with software written in other languages. In particular, the existing rough sets software, such as tools computing reducts, finding discretization, etc. should be made accessible in Rough Datalog.

- Practical case studies and development of applications.
- Investigation of further extensions to Rough Datalog, such as the use of general terms, and different kinds of constraints (e.g., finite-domain constraints).

Rough sets are sometimes viewed as information granules [8]. Using extensions of Sect. 4.4, one can define advanced operations on rough sets, thus implementing calculi of information granules. In particular, it seems possible to implement *rough inclusion* [8] tests using Prolog built-ins such as `findall`. Investigation of the usefulness of Rough Datalog in rough mereology is yet another topic of future research.

Acknowledgments

The authors gratefully acknowledge stimulating discussions with Jan Komorowski and Andrzej Skowron. A prototype implementation of Rough Datalog was developed by Govert Meuwis and is described in his Master's Thesis [11]. Discussions with Govert Meuvis contributed substantially to the development of this chapter, which extends the preliminary ideas described in [11] by a formal definition of the fixed-point semantics of Rough Datalog.

References

1. K.R. Apt , M. van Emden. Contributions to the theory of logic programming. *Journal of the ACM*, 29: 841–862, 1982.
2. M. Banerjee , M.K. Chakraborty. Algebras from rough sets (this book).
3. P. Deransart, A. Ed-Bali, L. Cervoni. *Prolog: The Standard, Reference Manual*. Springer, Berlin, 1996.
4. I. Duentsch I. Rough sets algebras of relations. In E. Orłowska, editor, *Incomplete Information: Rough Set Analysis*, 95–108, Physica, Heidelberg, 1998.
5. M.C. Fitting. A Kripke/Kleene semantics for logic programs. *Jornal of Logic Programming*, 2: 295–312, 1985.
6. T. Iwiński. Algebraic approach to rough sets. *Bulletin of the Polish Academy of Sciences. Mathematics*, 35: 673–683, 1987.
7. S.C. Kleene. *Introduction to Metamathematics*. Van Nostrand, Princeton, NJ, 1950.
8. J. Komorowski, Z. Pawlak, L.Polkowski, A. Skowron. Rough sets: A tutorial. In S. K. Pal, A. Skowron, editors, *Rough Fuzzy Hybridization: A New Trend in Decision Making*, Springer, Singapore, 1999.
9. C. Liu, N. Zhong. Dealing with imperfect data by RS-ILP. In S. Hirano, M. Inuiguchi, S.Tsumoto, editors, *Proceedings of the International Workshop on Rough Set Theory and Granular Computing (RSTGC 2001)*, Vol. 5(1/2) of the *Bulletin of International Rough Set Society*, 2001.

10. J.W. Lloyd. *Foundations of Logic Programming*, (2nd ed.) Springer, Berlin, 1987.
11. G. Meuvis. *Developing a Prolog System for Rough Set Reasoning*, Master's Thesis, Linköping University, 2001.
12. U. Nilsson, J. Maluszynski. *Logic, Programming and Prolog*, (2nd ed.) Wiley, New York, 1995. Available at http://www.ida.liu.se/ ulfni/lpp/copyright.html.
13. A. Obtułowicz. Rough sets and Heyting algebra valued sets. *Bulletin of the Polish Academy of Sciences. Mathematics*, 35(9/10): 667–671, 1987.
14. P. Pagliani. Rough set theory and logic-algebraic structures. In E. Orłowska, editor, *Incomplete Information: Rough Set Analysis*, 109–190, Physica, Heidelberg, 1998.
15. Z. Pawlak. Rough sets. *International Journal of Computer and Information Sciences*, 11(5): 341–356, 1982.
16. Z. Pawlak. *Rough Sets: Theoretical Aspects of Reasoning about Data*, Kluwer, Dordrecht, 1991.
17. A. Tarski. A lattice-theoretical theorem and its applications. *Pacific Journal of Mathematics*, 5: 285–309, 1955.
18. J.D. Ullman. *Principles of Database and Knowledge-Base Systems*. Computer Science Press, Rockville, MD, 1988.

Chapter 13
On Exploring Soft Discretization of Continuous Attributes

Hung Son Nguyen

Institute of Mathematics, Warsaw University, Banacha 2, 02-097 Warsaw , Poland
son@mimuw.edu.pl

Summary. Searching for a binary partition of attribute domains is an important task in data mining. It is present in both decision tree construction and discretization. The most important advantages of decision tree methods are compactness and clearness of knowledge representation as well as high accuracy of classification. Decision tree algorithms also have some drawbacks. In cases of large data tables, existing decision tree induction methods are often inefficient in both computation and description aspects. Another disadvantage of standard decision tree methods is their instability, i.e., small data deviations may require a significant reconstruction of the decision tree. We present novel *soft discretization* methods using *soft cuts* instead of traditional *crisp* (or sharp) cuts. This new concept makes it possible to generate more compact and stable decision trees with high accuracy of classification. We also present an efficient method for soft cut generation from large databases.

1 Introduction

Classification and description of target concepts are among the most important tasks in knowledge discovery in database (KDD) processes. There are many efficient classification techniques, but only approaches based on a decision rule set and a decision tree can be applied to perform both classification and description tasks. The other approaches such as case base reasoning (e.g., the nearest neighbor method) or artificial neural networks are not suitable for description tasks. In practice, one can notice the unpleasant fact that although rule based methods and decision tree methods are more complex than, e.g., the nearest neighbor method, their classification accuracy is not better for some data sets. Let us explain the reason for this phenomenon.

Usually, the existing (traditional) classification methods based on rule set generation require a data preprocessing step called discretization of continuous attributes, which divides the attribute domain into intervals. Decision tree methods can be used for the best partition of attribute domain extraction. The problem of searching for optimal partitions of real value attributes, defined by so-called cuts, has been studied by many authors ([1–3, 5, 12, 22]), where optimization criteria are defined by, e.g., the height of the decision tree obtained, the number of cuts, or the classification accuracy of the decision tree on new unseen objects. In general, all of those problems are hard from the computational point of view. Hence, numerous heuristics have been investigated to develop approximate solutions of these problems. One of the major tasks of these heuristics is to define some approximate measures estimating the quality of extracted cuts. Hence, in both discretization and decision tree construction methods, it is necessary to use crisp conditions expressed by cuts for object discerning. In our opinion, this approach can lead to misclassification of new objects which are, e.g., close to the separating boundary between decision classes and provide low quality of new object classification. Furthermore, in some data the values of attributes are measured by sensors, and we know that these values are not perfect with regard to device errors. In such cases, low classification accuracy can be caused by the impossibility of analyzing noisy data by using crisp cuts.

We propose a novel approach based on *soft cuts*, which makes it possible to overcome this difficulty. Our methods are based on the main approach of *rough set data analysis* methods, i.e., based on *handling the discernibility between objects* [20, 24]. Using rough set and Boolean reasoning-based methods, one can define the quality of cuts by the number of pairs of objects discerned by the partition (called *the discernibility measure*).

In this chapter, we consider a discretization problem defined by *soft cuts* and the rough-fuzzy reasoning scheme. We also propose some modifications of existing rule induction methods by soft cuts.

We also present efficient strategies of searching for the best cuts (both soft and crisp cuts). In this chapter, we discuss two strategies called *local* and *global searching strategies*. These strategies allow us to implement the proposed method for large data tables stored in databases.

There are two strategies for solving the problem of searching for an optimal decision tree or discretization for real value data, assuming that a large data table is represented in the relational database. The most popular strategy is based on the sampling technique, i.e., on building a decision tree for a small, randomly chosen subset of data and then evaluating the quality of the decision tree for all of the data. If the quality of a generated decision tree is not sufficient, we have to repeat this step for a new sample. In this chapter, we propose new methods using all of the data. Using a straightforward approach to optimal partition selection (with respect to a given measure), the number of necessary queries is of order $O(N)$, where N is the number of preassumed partitions of the searching space. For large databases, even linear complexity is not acceptable because of the time necessary for one step. The critical factor for time complexity of algorithms solving the problem discussed is the number of simple structured query language (SQL) queries such as, *select count from ... where attribute between ...*, (related to some interval of attribute values) necessary to construct such partitions. We assume that the answering time for such queries does not depend on the interval length. We show some properties of considered optimization measures allowing us to reduce the searching space size. Moreover, we prove that using only $O(\log N)$ simple queries, one can construct a partition very close to optimal. We have shown that the main part of the formula estimating the quality of the best cut for independent variables from [18] is the same for fully dependent variables.

This chapter is organized as follows: in Sect. 2 we present the main notations related to rough set theory, discretization, and the decision tree problem. The definition and application of soft cuts in classification problems are presented in Sect. 3. In Sect. 4 we present efficient searching methods for the best soft and crisp cuts. Conclusions and remarks are presented in Sect. 5.

2 Basic Notions

An *information system* [20] is a pair $\mathbb{A} = (U, A)$, where U is a nonempty, finite set called the *universe* and A is a nonempty finite set of *attributes* (or *features*), i.e., $a : U \to V_a$ for $a \in A$, where V_a is called *the value set of* a. Elements of U are called *objects* or *records*. Two objects $x, y \in U$ are said to be *discernible* by attributes from A if there exists an attribute $a \in A$ such that $a(x) \neq a(y)$.

Any information system of the form $\mathbb{A} = (U, A \cup \{dec\})$ is called a *decision table* where $dec \notin A$ is called a *decision attribute*. Without loss of generality, we assume that $V_{\text{dec}} = \{1, \ldots, d\}$. Then the set $DEC_k = \{x \in U : dec(x) = k\}$ will be called the kth *decision class of* $\mathbb{A}$ for $1 \leq k \leq d$. Any pair (a, c), where a is an attribute and c is a real value, is called *a cut*. We say that *cut* (a, c) *discerns a pair of objects* x, y if either $a(x) < c \leq a(y)$ or $a(y) < c \leq a(x)$.

2.1 Rough Set Based Discretization Method

Discretization is a process of determining the partition of attribute domains into intervals. Such a partition can be defined uniquely by some set of cuts.

Definition 1. For a given decision table $\mathbb{A} = (U, A \cup \{dec\})$, the set of cuts $\mathbf{C}$ is called $\mathbb{A}$-*consistent*, if for any pair of objects (x, y) such that $dec(x) \neq dec(y)$ and x, y are *discernible* by A, there exists a cut $(a, c) \in \mathbf{C}$ discerning x from y. The set of cuts $\mathbf{P}^{irr}$ is $\mathbb{A}$-*irreducible* if $\mathbf{P}$ is not $\mathbb{A}$-consistent for any $\mathbf{P} \subset \mathbf{P}^{irr}$. The set of cuts $\mathbf{P}^{opt}$ is $\mathbb{A}$-*optimal* if $card\,(\mathbf{P}^{opt}) \leq card\,(\mathbf{P})$ for any $\mathbb{A}$-consistent set of cuts $\mathbf{P}$.

Rough set based discretization methods are oriented to searching for an optimal set of cuts. In previous papers, we have shown the following theorems:

Theorem 1. *[12] For a given decision table $\mathbb{A}$ and an integer k: The decision problem of checking if there exists an $\mathbb{A}$-irreducible set of cuts $\mathbf{P}$ such that card$(\mathbf{P}) < k$ is NP-complete. The problem of searching for an $\mathbb{A}$-optimal set of cuts is NP-hard.*

To prove this theorem, one can construct efficient heuristics using a so-called Boolean reasoning approach.

For a given decision table $\mathbb{A} = (U, A \cup \{d\})$, a new decision table,

$$\mathbb{A}^* = (U^*, A^* \cup \{d^*\}),$$

is constructed as follows:

- $U^* = \{(u,v) \in U^2 : d(u) \neq d(v)\} \cup \{\perp\}$.
- $A^* = \{c : c \text{ is a cut on } \mathbb{A}\};\ c(\perp) = 0$.
  $$c[(u_i,u_j)] = \begin{cases} 1 & \text{if cut } c \text{ discerns } u_i, u_j \\ 0 & \text{otherwise.} \end{cases}$$
- $d(\perp) = 0; d^*(u_i,u_j) = 1$.

$\mathbb{A}^*$	c_1	c_2	...	c	...	d^*
(u_1,u_2)	1	0	...	...	...	1
⋮						
(u_i,u_j)	0			1		1
⋮			...			
$\perp$	0	0	...	0	...	0

It has been shown that any set of cuts is $\mathbb{A}$-irreducible if and only if it is a reduct of A^*.

Our maximal discernibility heuristic (MD) based on searching for cuts with a maximal number of object pairs discerned by this cut [12] is described as follows:

> *From the set of all possible cuts A^*, the cut discerning the maximal number of pairs of objects from different decision classes is chosen, and this step is repeated until no two objects from different decision classes discerned by some cut can be found.*

This heuristic can be realized quite efficiently. In [15], we have shown that the total time of discretization is $O(nk \cdot |\mathbf{C}|)$, where $\mathbf{C}$ is a final set of cuts for discretization and n and k are numbers of objects and attributes in the decision table.

2.2 Decision Tree Construction from Decision Tables

The decision tree for a given decision table is (in the simplest case) a binary directed tree with *test functions* (i.e., Boolean functions defined on the information vectors of objects) labeling internal nodes and decision values labeling leaves. In this chapter, we consider decision trees using cuts to represent test functions. Any cut (a,c) is associated with a test function $f_{(a,c)}$ such that for any object $u \in U$, the value of $f_{(a,c)}(u)$ is equal to 1 (true) if and only if $a(u) > c$. The typical algorithm for decision tree induction can be described as follows:

1. For a given set of objects U, select a cut (a, c_{Best}) of high quality among all possible cuts and all attributes.
2. Induce a partition U_1, U_2 of U by (a, c_{Best}).
3. Recursively apply Step 1 to both sets U_1, U_2 of objects until some stopping condition is satisfied.

In developing some decision tree induction methods [5,22] and some supervised discretization methods [1,3,12,15], it is often necessary to solve the following problem:

> *For a given real value attribute a and set of candidate cuts $\{c_1, \ldots, c_N\}$, find a cut (a, c_i) belonging to the set of optimal cuts with the highest probability.*

Usually, we use some *measure* (or *quality function*) $F : \{c_1, \ldots, c_N\} \to \mathbb{R}$ to estimate the quality of cuts. For a given measure F, any *straightforward algorithm* should compute the values of F for all cuts: $F(c_1), \ldots, F(c_N)$. The cut c_{Best} that maximizes or minimizes the value of function F is selected as the result of the searching process. Let us consider the attribute a and the set of all relevant cuts $\mathbf{C}_a = \{c_1, \ldots, c_N\}$ on a.

Definition 2. The d-tuple of integers $\langle x_1, \ldots, x_d \rangle$ is called the class distribution of the set of objects $X \subset U$ iff $x_k = card(X \cap DEC_k)$ for $k \in \{1, \ldots, d\}$. If the set of objects X is defined by $X = \{u \in U : p \leq a(u) < q\}$ for some $p, q \in \mathbb{R}$, then the class distribution of X can be called the class distribution in $[p; q)$.

Any cut $c \in \mathbf{C}_a$ splits the domain $V_a = (l_a, r_a)$ of the attribute a into two intervals: $I_L = (l_a, c); I_R = (c, r_a)$. We will use the following notation:

- U_{L_j}, U_{R_j} — the sets of objects from the jth class in I_L and I_R where $j \in \{1, \ldots, d\}$.
- $U_L = \bigcup_j U_{L_j}$, and $U_R = \bigcup_j U_{R_j}$.
- $\langle L_1, \ldots, L_d \rangle$ and $\langle R_1, \ldots, R_d \rangle$ — class distributions in U_L and U_R.
- $L = \sum_{j=1}^{d} L_j$, and $R = \sum_{j=1}^{d} R_j$.
- $C_j = L_j + R_j$ — number of objects in the jth class.
- $n = \sum_{i=1}^{d} C_j = L + R$ — the total number of objects.

In the following sections, we recall the most frequently used measures for decision tree induction such as *"entropy function" and "discernibility measure."*

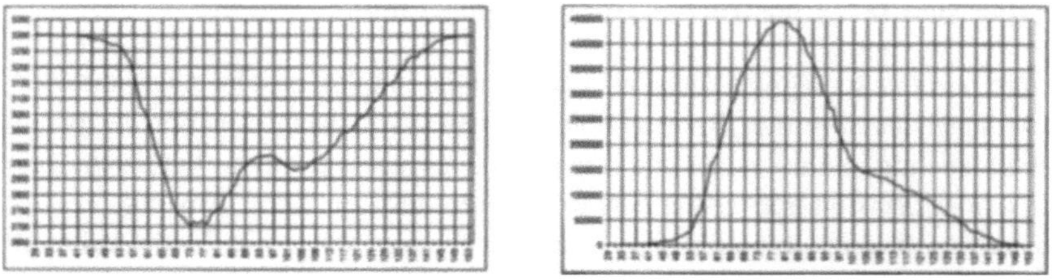

Fig. 1. Illustration of entropy measure (left) and discernibility measure (right)

Entropy methods A number of methods based on entropy measure have been developed in the domain of decision tree induction and discretization. These methods use class entropy as a criterion to evaluate the list of best cuts, which together with the attribute domain induce the relevant intervals. The class information entropy of the set of N objects X with class distribution $\langle N_1, \ldots, N_d \rangle$, where $N_1 + \ldots + N_d = N$, is defined by $Ent(X) = -\sum_{j=1}^{d} \frac{N_j}{N} \log \frac{N_j}{N}$. Hence, the entropy of the partition induced by a cut point c on attribute a is defined by

$$E(a,c;U) = \frac{|U_L|}{n} Ent(U_L) + \frac{|U_R|}{n} Ent(U_R),$$

where $\{U_L, U_R\}$ is a partition of U defined by c. For a given feature a, the cut $c_{\min}$, which minimizes the entropy function over all possible cuts is selected, see Fig. 1. The methods based on information entropy are reported in [1, 2, 6, 22].

Maximal discernibility principle Cuts in Boolean reasoning methods are treated as Boolean variables, and the searching problem for an optimal set of cuts can be characterized by a Boolean function $f_{\mathbb{A}}$ (where $\mathbb{A}$ is a given decision table). Any set of cuts is $\mathbb{A}$-consistent if and only if the corresponding evaluation of variables in $f_{\mathbb{A}}$ returns the value *True* [12]. We have shown that the quality of cuts can be measured by their *discernibility properties*. Intuitively, the energy of the set of objects $X \subset U$ can be defined by the number, called $conflict(X)$, of pairs of objects from X to be discerned. Let $\langle N_1, \ldots, N_d \rangle$ be a class distribution of X, then $conflict(X)$ can be computed by

$$conflict(X) = \sum_{i<j} N_i N_j.$$

The cut c, which divides the set of objects U into U_1 and U_2 is evaluated by

$$W(c) = conflict(U) - conflict(U_1) - conflict(U_2),$$

i.e., the higher the number of pairs of objects discerned by the cut (a,c), the larger is the chance that c can be added to the optimal set of cuts. Hence, the decision tree induction algorithms based on the rough set and the Boolean reasoning approach use the quality of a given cut c defined by

$$W(c) = \sum_{i \neq j}^{d} L_i R_j = \sum_{i=1}^{d} L_i \sum_{i=1}^{d} R_i - \sum_{i=1}^{d} L_i R_i. \tag{1}$$

The algorithm based on such a measure is called maximal-discernibility heuristics (MD-heuristics) for decision tree construction. Figure 1 illustrates changes of values of entropy and discernibility functions over the set of possible cuts on one of the attributes of SatImage data. One can see that the cuts preferred by both measures are quite similar. The high accuracy of decision trees constructed by using discernibility measures and their comparison with entropy-based decision methods are reported in [16, 17].

3 Discretization by Soft Cuts

So far, we have presented discretization methods working with sharp partitions defined by cuts, i.e., domains of real values are partitioned by them into disjoint intervals. One can observe that in some situations, similar (class) objects can be treated by cuts as very different. In this section, we introduce *soft cuts* discerning two given values if those values are far enough from the cut. The formal definition of soft cuts follows:

> *A soft cut is any triple $p = \langle a, l, r \rangle$, where $a \in A$ is an attribute, $l, r \in \mathbb{R}$ are called the left and right bounds of p ($l \leq r$); and the value $\varepsilon = \frac{r-l}{2}$ is called the uncertainty radius of p. We say that a soft cut p discerns the pair of objects x_1, x_2 if $a(x_1) < l$ and $a(x_2) > r$.*

The intuitive meaning of $p = \langle a, l, r \rangle$ is such that there is a real cut somewhere between l and r. So we are not sure where one can place the real cut in the interval $[l, r]$. Hence, for any value $v \in [l, r]$, we are not able to check whether v is on the left side or on the right side of the real cut. Then we say that the interval $[l, r]$ is an uncertain interval of the soft cut p. One can see that any normal cut is a soft cut of radius equal to zero.

Any set of soft cuts splits the real axis into intervals of two categories: the intervals corresponding to new nominal values and the intervals of uncertain values called boundary regions.

The problem of searching for a minimal set of soft cuts with a given uncertainty radius can be solved in a way similar to the case of sharp cuts. We propose some heuristic for this problem in the last section of the chapter. The problem becomes more complicated if we want to obtain the smallest set of soft cuts with a radius as large as possible. We will discuss this problem later. Now, we recall some existing rule induction methods for real value attribute data and their modifications using soft cuts.

Instead of sharp cuts (see previous sections), soft cuts determine additionally some uncertainty regions. Assume that $\mathbf{P} = \{p_1, p_2, \ldots, p_k\}$ is a set of soft cuts on attribute $a \in A$, where $p_i = (a, l_i, r_i); l_i \leq r_i$ and $r_i < l_{i+1}$for $i = 1, \ldots, k-1$. The set of soft cuts $\mathbf{P}$ defines on $\mathbb{R}$ a partition,

$$\mathbb{R} = (-\infty, l_1) \cup [l_1, r_1] \cup (r_1, l_2) \cup \ldots \cup [l_k, r_k] \cup (r_k, +\infty),$$

and at the same time defines a new nominal attribute $a^{\mathbf{P}} : U \to \{0, 1, \ldots, k\}$, such that $a^{\mathbf{P}}(x) = i$ if and only if $a(x) \in (r_i, l_{i+1})$; $i = 1, \ldots, k$. We are proposing some possible classification methods using soft discretization. These methods are based on the fuzzy set approach, rough set approach, clustering approach, and decision tree approach.

3.1 Fuzzy Set Approach

In the fuzzy set approach, one can treat the interval $[l_i, r_i]$ for any $i \in \{1, \ldots, k\}$ as a kernel of some fuzzy set Δ_i. The membership function $f_{\Delta_i} : \mathbb{R} \to [0, 1]$ is defined as follows:

1. $f_{\Delta_i}(x) = 0$ for $x < l_i$ or $x > r_{i+1}$.
2. $f_{\Delta_i}(x)$ increases from 0 to 1 for $x \in [l_i, r_i]$.
3. $f_{\Delta_i}(x)$ decreases from 1 to 0 for $x \in [l_{i+1}, r_{i+1}]$.
4. $f_{\Delta_i}(x) = 1$ for $x \in (r_i, l_{i+1})$.

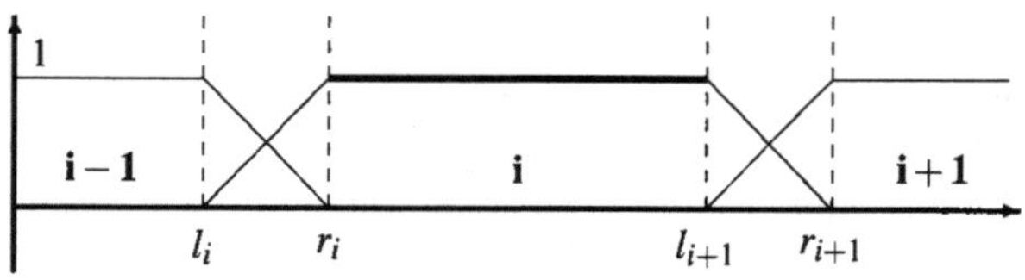

Fig. 2. Membership functions of intervals

Having defined membership function, one can use the idea of a *fuzzy graph* [4] to represent the knowledge discovered.

3.2 Rough Set Approach

The boundary interval $[l_i, r_i]$ can be treated as an uncertainty region for a real sharp cut. Hence, using the rough set approach, the intervals (r_i, l_{i+1}) and $[l_i, r_{i+1}]$ are treated as the lower and the upper approximations of any set X. Hence, we use the following notation: $\mathbf{L}_a(X_i) = (r_i, l_{i+1})$ and $\mathbf{U}_a(X_i) = [l_i, r_{i+1}]$, such that $(r_i, l_{i+1}) \subseteq X \subseteq [l_i, r_{i+1}]$.

Having approximations of nominal values of all attributes, we can generate an upper and a lower approximation of decision classes by taking the Cartesian product of rough sets. For instance, let the set X be given by its rough representation $[\mathbf{L}_B(X), \mathbf{U}_B(X)]$ and the set Y by $[\mathbf{L}_C(Y), \mathbf{U}_C(Y)]$, and let $B \cap C = \emptyset$. One can define a rough representation of $X \times Y$ by $[\mathbf{L}_{B \cup C}(X \times Y), \mathbf{U}_{B \cup C}(X \times Y)]$, where

$$\mathbf{L}_{B \cup C}(X \times Y) = \mathbf{L}_B(X) \times \mathbf{L}_C(Y)$$

and

$$\mathbf{U}_{B \cup C}(X \times Y) = \mathbf{U}_B(X) \times \mathbf{U}_C(Y).$$

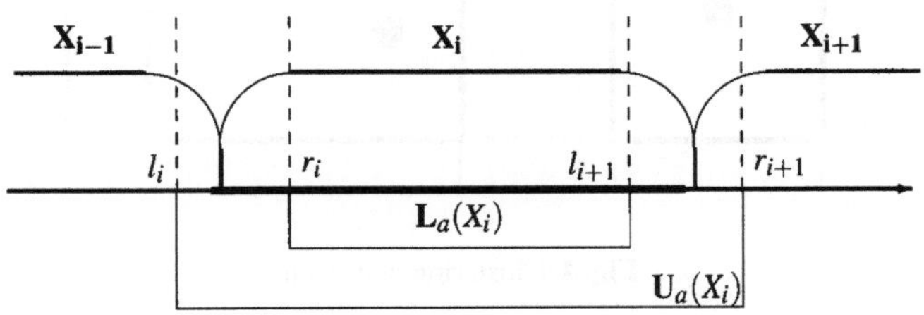

Fig. 3. Illustration of soft cuts

3.3 Clustering Approach

Any set of soft cuts $\mathbf{P}$ defines a partition of real values of attributes into disjoint intervals, which determine a natural equivalence relation $IND(\mathbf{P})$ over the set of objects. New objects belonging to the boundary regions can be classified by applying the rough set membership function to test the hypothesis that the new object belongs to a certain decision class.

One can also apply the idea of clustering. Any set of soft cuts defines a partition of $\mathbb{R}^k$ into k-dimensional cubes. Using the rough set approach one can classify some of those cubes in the lower approximation of a certain set, and they can be treated as clusters. To classify a new object belonging to any boundary cube one can compare distances from this object to the centers of adjacent clusters (see Fig. 4).

3.4 Decision Tree with Soft Cuts

In [13], we have presented some methods for decision tree construction from cuts (or oblique hyperplanes). Here, we propose two strategies that are modifications of that method using soft cuts (fuzzy separated cuts) described above. They are called *fuzzy decision tree* and *rough decision tree*.

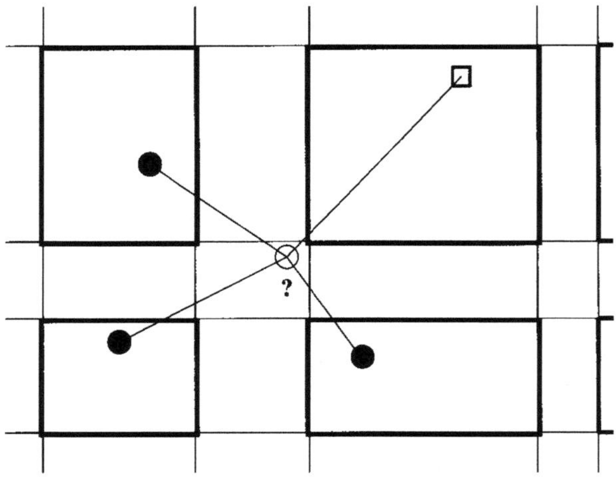

Fig. 4. Clustering approach

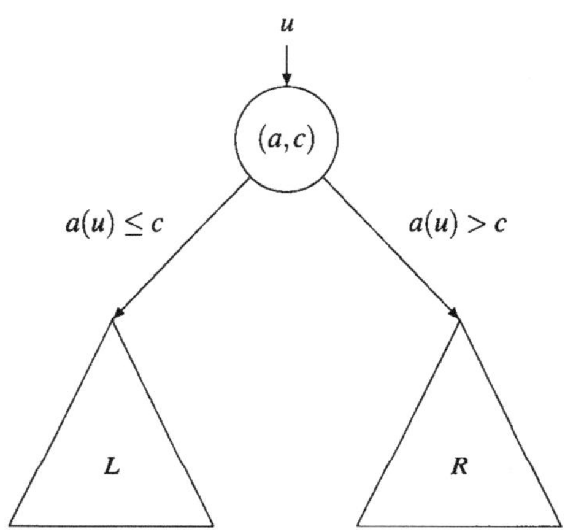

Fig. 5. Standard decision tree approach

The new object $u \in U$ can be classified by a given (traditional) decision tree as follows:

> *We start from the root of the decision tree. Let (a,c) be a cut labeling the root. If $a(u) > c$, we go to the right subtree, and if $a(u) \leq c$, we go to the left subtree of the decision tree. The process is continued for any node until we reach any external node.*

If the fuzzy decision tree method is used, then instead of checking the condition $a(u) > 0$, we have to check the strength of the hypothesis that u is on the left or right side of the cut (a,c). This condition can be expressed by $\mu_L(u)$ and $\mu_R(u)$, where μ_L and μ_R are membership functions of left and right intervals, respectively. The values of those membership functions can be treated as a probability distribution of u in the node labeled by the soft cut $(a, c-\varepsilon, c+\varepsilon)$. Then, one can compute the probability of the event that object u reaches a leaf. The decision for u is equal to decision labeling the leaf with the largest probability.

If the rough decision tree is used and we are not able to decide whether to turn left or right, we do not distribute the probabilities to children of the node considered. We have to compare the children's answers, taking into account the numbers of objects supported by them. The answer with the largest number of supported objects determines the decision for a given object.

4 Searching for Best Cuts in Large Data Tables

In this section, we present an efficient approach to searching for optimal cuts in large data tables. We consider some modifications of our MD-heuristic described in previous sections. The next sections describe some techniques that have been presented in previous papers (see [18]).

4.1 Tail Cuts Can Be Eliminated

First, let us consider two cuts $c_L < c_R$. Let $\langle L_1, \ldots, L_d\rangle$ be the class distribution in $(-\infty; c_L)$, $\langle M_1, \ldots, M_d\rangle$ the class distribution in $[c_L; c_R)$, and $\langle R_1, \ldots, R_d\rangle$ the class distribution in $[c_R; \infty)$.

Lemma 1. *The following equation holds:*

$$W(c_R) - W(c_L) = \sum_{i=1}^{d} \left[(R_i - L_i) \sum_{j \neq i} M_j \right]. \tag{2}$$

This lemma implies the following techniques for eliminating irrelevant cuts:

For a given set of cuts $\mathbf{C}_a = \{c_1, \ldots, c_N\}$ on a, by the median of the kth decision class we mean the cut $c \in \mathbf{C}_a$, which minimizes the value $|L_k - R_k|$. The median of the kth decision class will be denoted by $Median(k)$. Let $c_1 < c_2 \ldots < c_N$ be the set of candidate cuts, and let

$$c_{\min} = \min_i \{Median(i)\} \text{ and } c_{\max} = \max_i \{Median(i)\}.$$

This property is formulated in the following theorem:

Theorem 2. *The quality function $W : \{c_1, \ldots, c_N\} \to \mathbb{N}$ defined over the set of cuts increases in $\{c_1, \ldots, c_{\min}\}$ and decreases in $\{c_{\max}, \ldots, c_N\}$. Hence,*

$$c_{\text{Best}} \in \{c_{\min}, \ldots, c_{\max}\}.$$

This property is interesting because one can use only $O(d \log N)$ queries to determine the medians of decision classes by using the binary search algorithm. Hence, the tail cuts can be eliminated by using using $O(d \log N)$ SQL queries. Let us also observe that if all decision classes have similar medians, then almost all cuts can be eliminated.

Example. We consider a data table consisting of 12,000 records. Objects are classified into three decision classes with the distribution $\langle 5000, 5600, 1400 \rangle$, respectively. One real value attribute has been selected, and $N = 500$ cuts on its domain have generated class distributions as shown in Fig. 6. The medians of the three decision classes are c_{166}, c_{414} and c_{189}, respectively. The median of every decision class has been determined by a *binary search algorithm* using $\log N = 9$ simple queries. Applying Theorem 2, we conclude that it is enough to consider only cuts from $\{c_{166}, \ldots, c_{414}\}$. In this way, 251 cuts have been eliminated by using only 27 simple queries.

4.2 Divide and Conquer Strategy

The main idea is to apply the *divide and conquer* strategy to determine the best cut $c_{\text{Best}} \in \{c_1, \ldots, c_n\}$ with respect to a given quality function.

First, we divide the set of possible cuts into k intervals, where k is a predefined parameter ($k \geq 2$). Then, we choose the interval to which the best cut may belong with the highest probability. We will use some approximating measures to predict the interval that probably contains the best cut with respect to the discernibility measure. This process is repeated until the interval considered consists of one cut. Then, the best cut can be chosen among all visited cuts.

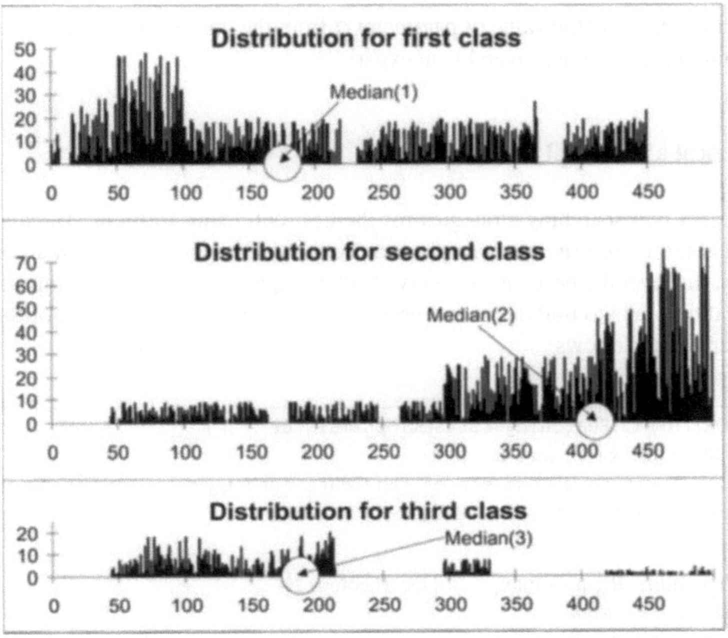

Fig. 6. Distributions for decision classes 1, 2, and 3

The problem arises how to define the measure evaluating the quality of the interval $[c_L;c_R]$ having class distributions: $\langle L_1,\ldots,L_d\rangle$ in $(-\infty;c_L)$, $\langle M_1,\ldots,M_d\rangle$ in $[c_L;c_R)$, and $\langle R_1,\ldots,R_d\rangle$ in $[c_R;\infty)$. This measure should estimate the quality of the best cut among those belonging to the interval $[c_L;c_R]$. In the next section, we present some theoretical considerations about the quality of the best cut in $[c_L;c_R]$. These results will be used to construct the relevant measure to estimate the quality of the whole interval.

Evaluation Measures In our previous papers (see [18]), we have proposed the following measures to estimate the quality of the best cut in $[c_L;c_R]$:

$$Eval\left([c_L;c_R]\right) = \frac{W(c_L)+W(c_R)+conflict([c_L;c_R])}{2} + \Delta, \tag{3}$$

where the value of Δ is defined by

$$\Delta = \frac{[W(c_R)-W(c_L)]^2}{8\cdot conflict([c_L;c_R])} \quad \text{(in the dependent model),}$$

$$\Delta = \alpha\cdot\sqrt{D^2(W(c))} \qquad \text{for some } \alpha\in[0;1] \quad \text{(in the independent model).}$$

The choice of Δ and the value of parameter α from the interval $[0;1]$ can be tuned in a learning process or are given by an expert.

4.3 Local and Global Search

We present two searching strategies for the best cut, using formula (3), called the *local* and the *global search* strategies, respectively. Using a local search algorithm, first, we discover the best cuts on every attribute separately. Next, we compare all of the local best cuts to find the global best one. The details of the local algorithm can be described as follows:

ALGORITHM 1: Searching for semioptimal cut
PARAMETERS: $k \in \mathbb{N}$ and $\alpha \in [0;1]$.
INPUT: attribute a; the set of candidate cuts $\mathbf{C}_a = \{c_1, \ldots, c_N\}$ on a;
OUTPUT: The optimal cut $c \in \mathbf{C}_a$

begin
  $Left \leftarrow \min$; $Right \leftarrow \max$; {see Theorem 2}
  while $(Left < Right)$
    1.Divide $[Left; Right]$ into k intervals of equal length defined by $(k+1)$ boundary points

$$p_i = Left + i * \frac{Right - Left}{k};$$

    for $i = 0, \ldots, k$.
    2.For $i = 1, \ldots, k$ compute $Eval([c_{p_{i-1}}; c_{p_i}], \alpha)$ using Formula (3). Let $[p_{j-1}; p_j]$ be the interval with the maximal value of $Eval(.)$;
    3.$Left \leftarrow p_{j-1}$; $Right \leftarrow p_j$;
  endwhile;
  return the cut c_{Left};
end

One can see that to determine the value $Eval([c_L; c_R])$, we need only $O(d)$ simple SQL queries of the form:

```
SELECT COUNT
FROM data_table
WHERE attribute BETWEEN c_L AND c_R.
```

Hence the number of SQL queries necessary for running our algorithm is of the order $O(dk \log_k N)$. We can set $k = 3$ to minimize the number of queries, because the function $f(k) = dk \log_k N$ takes the minimum over positive integers for $k = 3$. For $k > 2$, instead of choosing the best interval $[p_{i-1}; p_i]$, one can select the best union $[p_{i-m}; p_i]$ of m consecutive intervals in every step for the predefined parameter $m < k$. The modified algorithm needs more — but still of order $O(\log N)$ —

simple questions only.

The global strategy is searching for the best cut over all attributes. At the beginning, the best cut can belong to every attribute; hence, for each attribute, we keep the interval in which the best cut can be found (see Theorem 2), i.e., we have a collection of all potential intervals:

$$\textbf{Interval\_List} = \{(a_1, l_1, r_1), (a_2, l_2, r_2), \ldots, (a_k, l_k, r_k)\}.$$

Next, we iteratively run the following procedure:

- Remove the interval $I = (a, c_L, c_R)$ having the highest probability of containing the best cut [using formula (3)].
- Divide interval I into smaller ones $I = I_1 \cup I_2 \ldots \cup I_k$.
- Insert $I_1, I_2, \ldots, I_k$ into **Interval_List.**

This iterative step can be continued until we have a one-element interval or the time limit of the searching algorithm is exhausted. This strategy can be simply implemented by using a priority queue to store the set of all intervals, where the priority of intervals is defined by formula (3).

4.4 Example

In Fig. 7, we show the graph of $W(c_i)$ for $i \in \{166, \ldots, 414\}$, and we illustrate the outcome of the application of our algorithm to the reduced set of cuts for $k = 2$ and $\Delta = 0$.

First, the cut c_{290} is chosen, and it is necessary to determine to which of the intervals $[c_{166}, c_{290}]$ and $[c_{290}, c_{414}]$ the best cut belongs. The values of function $Eval$ on these intervals is computed:

$$Eval([c_{166}, c_{290}]) = 23927102, \qquad Eval([c_{290}, c_{414}]) = 24374685.$$

Hence, the best cut, it is predicted, belongs to $[c_{290}, c_{414}]$, and the search process is reduced to the interval $[c_{290}, c_{414}]$. The above procedure is repeated recursively until the selected interval consists only of a single cut. For our example, the best cut c_{296} has been successfully selected by our algorithm. In general, the cut selected by the algorithm is not necessarily the best. However, numerous experiments on different large data sets have shown that the cut c^* returned by the algorithm is close to the best cut c_{Best} (i.e., $\frac{W(c^*)}{W(c_{\text{Best}})} \cdot 100\%$ is about 99.9%).

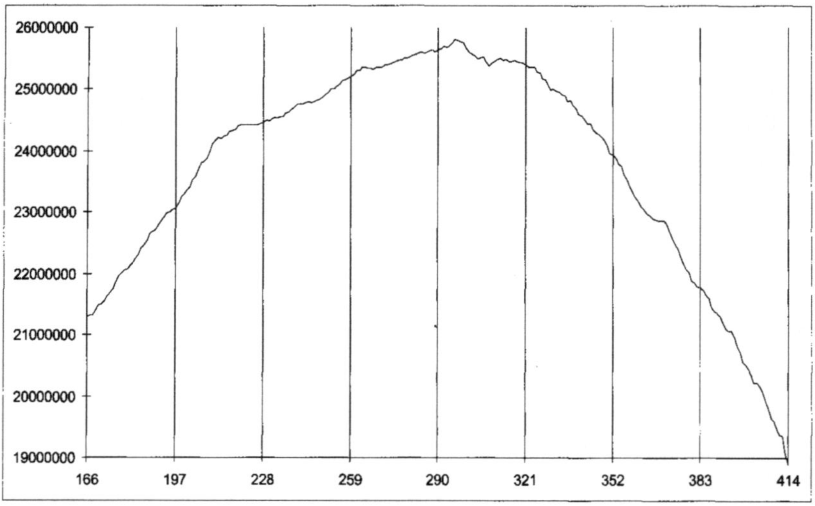

Fig. 7. Graph of $W(c_i)$ for $i \in \{166,\ldots,414\}$

4.5 Searching for Soft Cuts

One can modify Algorithm 1 presented in the previous section to determine soft cuts in large databases. The modification is based on changing the stop condition. In every iteration of Algorithm 1, the current interval $[Left; Right]$ is divided equally into k smaller intervals, and the best smaller interval is chosen as the current interval. In the modified algorithm, one can either select one of smaller intervals as the current interval or stop the algorithm and return the current interval as a result.

Intuitively, the divide and conquer algorithm is stopped and returns the interval $[c_L; c_R]$ if the following conditions hold:

- The class distribution in $[c_L; c_R]$ is stable, i.e., there is no subinterval of $[c_L; c_R]$ that is considerably better than $[c_L; c_R]$ itself.
- The interval $[c_L; c_R]$ is sufficiently small, i.e., it contains a small number of cuts.
- The interval $[c_L; c_R]$ does not contain too many objects (because the large number of uncertain objects can result in a larger decision tree that prolongs the time of decision tree construction).

These conditions can be controlled by some parameters. Now we are describing the stability measure for intervals.

5 Conclusions

The problem of optimal binary partition of a continuous attribute domain for large data sets stored in *relational databases* has been investigated. We show that one can reduce the number of simple queries from $O(N)$ to $O(\log N)$ to construct a partition very close to the optimal one. We plan to extend these results to other measures.

Acknowledgments

This paper has been partially supported by Polish State Committee of Research (KBN) grant No 8T11C02519, a grant of the Wallenberg Foundation, and a British Council/KBN grant "Uncertainty Management in Information Systems: Foundations and Applications of Non-Invasive Methods (1999–2001)."

References

1. J. Catlett. On changing continuous attributes into ordered discrete attributes. In Y. Kodratoff, editor, *Proceedings of the European Working Session on Learning (EWSL'91)*, LNAI 482, 164–178, Springer, Berlin, 1991.
2. M.R. Chmielewski, J.W. Grzymala–Busse. Global discretization of attributes as preprocessing for machine learning. In T. Y. Lin, A. M. Wildberger, editors, *Soft Computing: Rough Sets, Fuzzy Logic Neural Networks, Uncertainty Management, Knowledge Discovery*, 294–297, Simulation Councils, San Diego, CA, 1995.
3. J. Dougherty, R. Kohavi, M. Sahami. Supervised and unsupervised discretization of continuous features. In *Proceedings of the 12th International Conference on Machine Learning (ICML'95)*, 194–202, Morgan Kaufmann, San Francisco, 1995.
4. D. Dubois, H. Prade, R.R. Yager, editors. *Readings in Fuzzy Sets for Intelligent Systems*. Morgan Kaufmann, San Francisco, 1993.
5. U.M. Fayyad, K.B. Irani. On the handling of continuous-valued attributes in decision tree generation. *Machine Learning*, 8: 87–102, 1992.
6. U.M. Fayyad, K.B. Irani. The attribute selection problem in decision tree generation. In *Proceedings of the 10th National Conference on Artificial Intelligence (AAAI 1992)*, 104–110, MIT Press, Cambridge, MA, 1992.
7. S.J. Hong. Use of contextual information for feature ranking and discretization. *IEEE Transactions on Knowledge and Data Engineering*, 9(5): 718–730, 1997.
8. G.H. John, P. Langley. Static vs. dynamic sampling for data mining. In *Proceedings of the 2nd International Conference on Knowledge Discovery and Data Mining (KDDM'96)*, 367–370, AAAI Press, Menlo Park, CA, 1996.
9. R. Kerber. Chimerge. Discretization of numeric attributes. In *Proceedings of the 10th National Conference on Artificial Intelligence (AAAI'92)*, 123–128, MIT Press, Cambridge, MA, 1992.
10. H. Liu, R. Setiono. Chi2: Feature selection and discretization of numeric attributes. In *Proceedings of the 7th IEEE International Conference on Tools with Artificial Intelligence (TAI'95)*, 388–391, IEEE Press, Washington, DC, 1995.
11. S. Murthy, S. Kasif, S. Saltzberg, R. Beigel. OC1: Randomized induction of oblique decision trees. In *Proceedings of the 11th National Conference on Artificial Intelligence (AAAI'93)*, 322–327, MIT Press, Cambridge, MA, 1993.
12. H.S. Nguyen, A. Skowron. Quantization of real value attributes. In P. P. Wang, editor, *Proceedings of the 2nd Annual Joint Conference on Information Sciences (JCIS'95)*, 34–37, Wrightsville Beach, NC, 1995.

13. S.H. Nguyen, H.S. Nguyen. From optimal hyperplanes to optimal decision tree. In *Proceedings of the 4th International Workshop on Rough Sets, Fuzzy Sets and Machine Discovery (RSFD'96)*, 82–88, Tokyo, 1998.
14. H.S. Nguyen, S.H. Nguyen, A. Skowron. Searching for features defined by hyperplanes. In Z. W. Raś, M. Michalewicz, editors, *Proceedings of the 9th International Symposium on Methodologies for Information Systems (ISMIS'96)*, LNAI 1079, 366–375, Springer, Berlin, 1996.
15. H.S. Nguyen. Discretization methods in data mining. In L. Polkowski, A. Skowron, editors, *Rough Sets in Knowledge Discovery 1*, 451–482, Physica, Heidelberg, 1998.
16. H.S. Nguyen, A. Skowron. Boolean reasoning for feature extraction problems. In Z. W. Raś, A. Skowron, editors, *Proceedings of 10th International Symposium on the Foundations of Intelligent Systems (ISMIS'97)*, LNAI 1325, 117–126, Springer, Heidelberg, 1997.
17. H.S. Nguyen, S.H. Nguyen. From optimal hyperplanes to optimal decision trees. *Fundamenta Informaticae*, 34(1/2): 145–174, 1998.
18. H.S. Nguyen. Efficient SQL-querying method for data mining in large databases. In *Proceedings of the 16th International Joint Conference on Artificial Intelligence (IJCAI'99)*, 806–811, Morgan Kaufmann, San Francisco, 1999.
19. H.S. Nguyen. On efficient construction of decision tree from large databases. In *Proceedings of the 2nd International Conference on Rough Sets and Current Trends in Computing (RSCTC 2000)*, LNAI 2005, 316–323, Springer, Berlin, 2000.
20. Z. Pawlak. *Rough Sets: Theoretical Aspects of Reasoning about Data.* Kluwer, Dordrecht, 1991.
21. L. Polkowski, A. Skowron, editors. *Rough Sets in Knowledge Discovery*, Vol. 1,2. Physica, Heidelberg, 1998.
22. J.R. Quinlan. C4.5: Programs for Machine Learning. Morgan Kaufmann, San Francisco, 1993.
23. A. Skowron, C. Rauszer. The discernibility matrices and functions in information systems. In R. Słowiński, editor, *Intelligent Decision Support: Handbook of Applications and Advances of the Rough Sets Theory*, 311–362, Kluwer, Dordrecht, 1992.
24. J. Komorowski, Z. Pawlak, L. Polkowski, A. Skowron. Rough sets: A tutorial. In S. K. Pal and A. Skowron, editors, *Rough–Fuzzy Hybridization: A New Trend in Decision Making*, 3–98, Springer, Singapore, 1998.
25. W. Ziarko. Rough sets as a methodology in data mining. In L. Polkowski, A.Skowron, editors, *Rough Sets in Knowledge Discovery 1*, 554–576, Physica, Heidelberg, 1998.

Chapter 14
Rough-SOM with Fuzzy Discretization

Sankar K. Pal, Biswarup Dasgupta, and Pabitra Mitra

Machine Intelligence Unit, Indian Statistical Institute, 203 B.T. Road, Kolkata 700108, India
sankar@isical.ac.in, biswarupdg@yahoo.com, pabitra_r@isical.ac.in

Summary. A rough self-organizing map (RSOM) with fuzzy discretization of feature space is described here. Discernibility reducts obtained using rough set theory are used to extract domain knowledge in an unsupervised framework. Reducts are then used to determine the initial weights of the network, which are further refined using competitive learning. The superiority of this network in terms of the quality of clusters, learning time, and representation of data is demonstrated quantitatively through experiments across the conventional SOM.

1 Introduction

Rough set theory [1] provides an effective means for classificatory analysis of data tables. The main goal of rough set theoretical analysis is to synthesize or construct approximations (upper and lower) of concepts from the acquired data. The key concepts here are those of "information granule" and "reducts." The information granule formalizes the concept of finite precision representation of objects in real life situations, and the reducts represent the *core* of an information system (both in terms of objects and features) in a granular universe. An important use of rough set theory has been in generating logical rules for classification and association [2]. These logical rules correspond to different important granulated regions of the feature space, which represent data clusters.

Recently, rough sets have been integrated with neural networks [3]. In the framework of rough-neural integration, research has mainly been done in using of rough sets for encoding the weights of knowledge-based networks. However, mainly layered networks in a supervised learning framework have been considered so far [4]. This chapter is an attempt to incorporate rough set methodology in the framework of unsupervised networks.

A self-organizing map (SOM) (see [5]) is an unsupervised network, which has been recently popular for unsupervised mining of large data sets. However, SOM suffers from the problem of local minima and slow convergence. Here, rough set theory offers a fast and robust solution to the initialization and local minima problem. In this chapter, we integrate rough set theory with SOM by designing what is called a rough-SOM (RSOM). Here rough set theoretical knowledge is used to encode the weights as well as to determine the network size. Fuzzy set theory is used for discretization of the feature space. The performance of networks is measured by using quantization error, entropy, and a fuzzy feature evaluation index. The compactness of the network and the convergence time are also computed. All of these characteristics have been demonstrated with two artificially generated linearly and nonlinearly separable data sets and compared with that of the conventional SOM.

2 Rough Sets

2.1 Definitions

Here, we present some preliminaries of rough set theory that are relevant to this chapter.

An information system is a pair $S = <U, A>$, where U is a nonempty finite set called the universe and A is a nonempty finite set of attributes. An attribute a can be regarded as a function from the domain U to some value set V_a.

An information system may be represented as an attribute-value table, in which rows are labeled by objects of the universe and columns by attributes.

With every subset of attributes $B \subseteq A$, one can easily associate an equivalence relation I_B on U:

$$I_B = \{(x,y) \in U : \text{forverya} \in \text{B}, \text{a(x)} = \text{a(y)}\}.$$

Then, $I_B = \cap_{a \in B} I_a$.

If $X \subseteq U$, the sets $\{ x \in U : [x]_B \subseteq X \}$ and $\{x \in U : [x]_B \cap X \neq \emptyset\}$, where $[x]_B$ denotes the equivalence class of the object $x \in U$ relative to I_B, are called the B-lower and the B-upper approximation of X in S and denoted by $\underline{B}X$ and $\overline{B}X$, respectively.

$X (\subseteq U)$ is B-exact or B-definable in S if $\underline{B}X = \overline{B}X$. It may be observed that $\underline{B}X$ is the greatest B-definable set contained in X and $\overline{B}X$ is the smallest B-definable set containing X.

We now define the notions relevant to knowledge reduction. The aim is to obtain irreducible but essential parts of the knowledge encoded by the given information system; these would constitute reducts of the system. So one is, in effect, looking for the maximal sets of attributes taken from the initial set (A, say) that induce the same partition on the domain as A. In other words, the essence of the information remains intact, and superfluous attributes are removed. Reducts have been nicely characterized in [2] by discernibility matrices and discernibility functions. Consider $U = \{x_1, \ldots, x_n\}$ and $A = \{a_1, \ldots, a_m\}$ in the information system $S = <U, A>$. By the discernibility matrix $M(S)$ of S is meant an $n \times n$ matrix such that

$$c_{ij} = \{a \in A : a(x_i) \neq a(x_j)\}. \tag{1}$$

A discernibility function f_S is a function of m Boolean variables $a_1, \ldots, a_m$ corresponding to attributes $a_1, \ldots, a_m$, respectively, and defined as follows:

$$f_S(a_1, \ldots, a_m) = \wedge\{\vee(c_{ij}) : 1 \leq i, j \leq n, j < i, c_{ij} \neq \emptyset\}, \tag{2}$$

where $\vee(c_{ij})$ is the disjunction of all variables a with $a \in c_{ij}$. It is seen in [2] that $\{a_{i_1}, \ldots, a_{i_p}\}$ is a reduct of S if and only if $a_{i_1} \wedge \ldots \wedge a_{ip}$ is a prime implicant (constituent of the disjunctive normal form) of f_S.

2.2 Indiscernibility of Patterns and Fuzzy Discretization of the Feature Space

A primary notion of rough set theory is that of an indiscernibility relation. For continuous valued attributes, the feature space needs to be discretized for defining indiscernibility relations and equivalence classes. Discretization is a widely studied problem in rough set theory, and in this chapter we use fuzzy set theory for effective discretization. Use of fuzzy sets has several advantages over hard discretization, such as modeling of overlapped clusters and linguistic representation of data. We discretize each feature into three levels: low, medium, and high; finer discretizations may lead to better accuracy at the cost of a higher computational load.

Each feature of a pattern is described in terms of its fuzzy membership values in the linguistic property sets *low* (L), *medium* (M), and *high* (H). Let these be represented by L_j, M_j, and H_j, respectively. The features of the ith pattern $\mathbf{F}_i$ are mapped to the corresponding three-dimensional feature space of $\mu_{low(F_{i_j})}(\mathbf{F}_i)$, $\mu_{medium(F_{i_j})}(\mathbf{F}_i)$ and $\mu_{high(F_{i_j})}(\mathbf{F}_i)$ by (3). An n-dimensional pattern $\mathbf{F}_i = (F_{i_1}, F_{i_2}, \ldots, F_{i_n})$ is represented as an $3n$-dimensional vector [6,7]

$$\mathbf{F}_i = (\mu_{low(F_{i_1})}(\mathbf{F}_i), \ldots, \mu_{high(F_{i_n})}(\mathbf{F}_i)), \tag{3}$$

where the μ values indicate the membership functions of the corresponding linguistic Π-sets, low, medium, and high along each feature axis. This effectively discretizes each feature into three levels.

Then consider only those attributes that have a numerical value greater than some threshold TH (= 0.5, say). This implies clamping only those features demonstrating high membership values with unity, whereas the others are fixed at zero. An attribute-value table is constructed comprising the above binary valued $3n$-dimensional feature vectors.

We use the Π-fuzzy sets (in one-dimensional form) with a range [0,1], represented as

$$\Pi(F_j;c,\lambda) = \begin{cases} 2(1-||F_j-c||/\lambda)^2 & \text{for} \lambda/2 \leq \quad ||\mathrm{F_j}-\mathrm{c}|| \leq \lambda \\ 1-2(||F_j-c||/\lambda)^2 & \text{for} 0 \leq ||\mathrm{F_j}-\mathrm{c}|| \leq \lambda/2 \\ 0 & \text{otherwise,} \end{cases} \tag{4}$$

where $\lambda(>0)$ is the radius of the Π function with c as the central point. The details of the above method may be found in [6].

Let us now explain the procedure for selecting centers (c) and radii (λ) of the overlapping Π sets. Let m_j be the mean of the pattern points along the jth axis. Then m_{j_l} and m_{j_h} are defined as the mean (along the jth axis) of the pattern points having coordinate values in the range $[F_{j_{\min}}, m_j)$ and $(m_j, F_{j_{\max}}]$ respectively, where $F_{j_{\max}}$ and $F_{j_{\min}}$ denote the upper and lower bounds of the dynamic range of feature F_j (for the training set) considering only numerical values. For the three linguistic property sets along the jth axis, the centers and the corresponding radii of the corresponding Π functions are defined as

$$\begin{aligned} c_{low(F_j)} &= m_{j_l}, \\ c_{medium(F_j)} &= m_j, \\ c_{high(F_j)} &= m_{j_h}, \\ \lambda_{low(F_j)} &= c_{medium(F_j)} - c_{low(F_j)}, \\ \lambda_{high(F_j)} &= c_{high(F_j)} - c_{medium(F_j)}, \\ \lambda_{medium(F_j)} &= c_{high(F_j)} - c_{low(Fj)}, \end{aligned} \tag{5}$$

respectively. Here, we take into account the distribution of pattern points along each feature axis while choosing the corresponding centers and radii of the linguistic properties. The nature of membership functions is illustrated in Fig. 1.

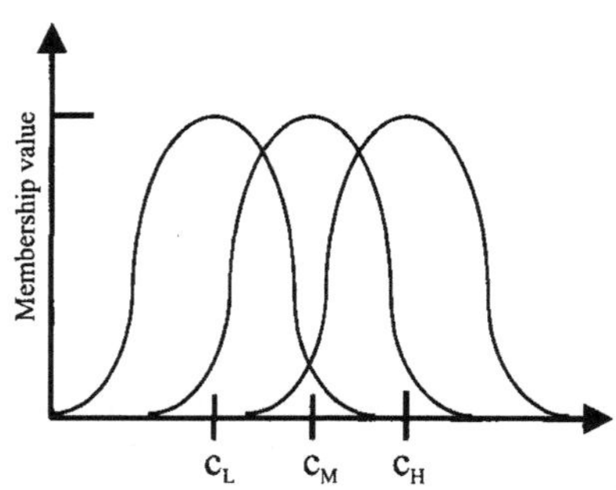

Fig. 1. Π membership functions for linguistic property sets, *low* (L), *medium* (M), and *high* (H) for each feature axis

2.3 Methodology of Generating Reducts

Let there be m sets $O_1,\ldots,O_m$ of objects in the attribute-value table (obtained by the procedure described in the previous section) having identical attribute values, and $card(O_i) = n_{ki}$, $i = 1,\ldots,m$ such that $n_{k_1} > n_{k_2} > \ldots > n_{k_m}$ and

$$\sum_{i=1}^{m} n_{k_i} = n_k.$$

The attribute-value table can now be represented as an $m \times 3n$ array. Let $n'_{k_1}, n'_{k_2},\ldots,$ n'_{k_m} denote the distinct elements among n_{k_1}, $n_{k_2},\ldots,$ n_{k_m} such that $n'_{k_1} > n'_{k_2} > \ldots > n'_{k_m}$.

Let a heuristic threshold be defined as [4]

$$Tr = \left\lceil \frac{\sum_{i=1}^{m} \frac{1}{n'_{k_i} - n'_{k_{i+1}}}}{\mathrm{TH}} \right\rceil, \tag{6}$$

so that all entries having frequency less than Tr are eliminated from the table, resulting in the reduced attribute-value table S. Note that the main motive of introducing this threshold function lies in reducing the size of the model. One attempts to eliminate noisy pattern representatives (having lower values of n_{k_i}) from the reduced attribute-value table. From the reduced attribute-value table obtained, reducts are determined using the methodology described below.

Let $\{x_{i_1},\ldots,x_{i_p}\}$ be the set of those objects of U that occur in S. Now a discernibility matrix [denoted $M(B)$] is defined as follows:

$$c_{ij} = \{a \in B : a(x_i) \neq a(x_j)\} \text{ for } i,j = 1,\ldots,n. \tag{7}$$

For each object $x_j \in \{x_{i_1},\ldots,x_{i_p}\}$, the discernibility function f_{x_j} is defined as

$$f_{x_j} = \wedge\{\vee(c_{ij}) : 1 \leq i,j \leq n, j < i, c_{ij} \neq \emptyset\}, \tag{8}$$

where $\vee(c_{ij})$ is the disjunction of all members of c_{ij}. One thus obtains a rule r_i, viz., $P_i \rightarrow cluster_i$, where P_i is the disjunctive normal form (dnf) of f_{x_j}, $j \in \{i_1,\ldots i_p\}$.

3 Rough-SOM

3.1 Self-Organizing Maps

The Kohonen feature map is a two-layered network. The first layer of the network is the input layer, and the second layer, called the competitive layer, is usually organized as a two-dimensional grid. All interconnections go from the first layer to the

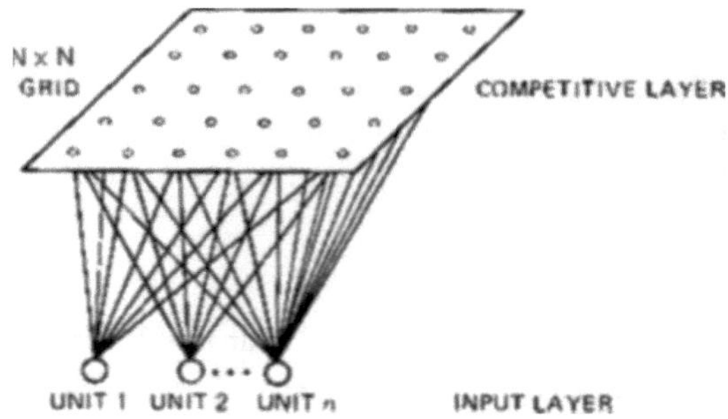

Fig. 2. The basic network structure for the Kohonen feature map

second; the two layers are fully connected to all of the units in the competitive layer (Fig. 2).

All nodes in the competitive layer compare the inputs with their weights and compete with each other to become the winning unit having the lowest difference. The basic idea underlying what is called competitive learning is roughly as follows: Assume a sequence of input vectors [$x = x(t) \in R^n$, where t is the time coordinate] and a set of variable reference vectors [$m_i(t) : m_i \in R^n$, $i = 1, 2, \ldots, k$, where k is the number of units in the competitive layer]. Initially, the values of the reference vectors (also called weight vectors) are set randomly. At each successive instant of time t, an input pattern $x(t)$ is presented to the network. The input pattern $x(t)$ is then compared with each $m_i(t)$, and the best matching $m_i(t)$ is updated to match even more closely the current $x(t)$.

If the comparison is based on some distance measure $d(x, m_i)$, altering m_i must be such that, if $i = c$, the index of the best-matching reference vector, then $d(x, m_c)$ is reduced, and all the other reference vectors m_i, with $i \neq c$, are left intact. In this way, the different reference vectors tend to become specifically "tuned" to different domains of the input variable x.

Learning The first step in the operation of a Kohonen network is to compute a matching value for each unit in the competitive layer. This value measures the extent to which the weights or reference vectors of each unit match the corresponding values of the input pattern. The matching value for each unit i is $|| \mathrm{x} - m_i ||$, which is the distance between vectors x and m_i and is computed by

$$\sqrt{\sum_j (x_j - m_{ij})^2} \quad \mathrm{for j} = 1, 2, \ldots, \mathrm{n}. \tag{9}$$

The unit with the lowest matching value (the best match) wins the competition. In other words, the unit c is said to be the best matched unit if

$$||x - m_c|| = \min_i \{||x - m_i||\}, \tag{10}$$

where the minimum is taken over all units i in the competitive layer. If two units have the same matching value, then by convention, the unit with the lower index value i is chosen.

The next step is to self-organize a two-dimensional map that reflects the distribution of input patterns. In biophysically inspired neural network models, correlated learning by spatially neighboring cells can be implemented by using various kinds of lateral feedback connections and other lateral interactions. Here, the lateral interaction is enforced directly in a general form for arbitrary underlying network structures by defining a neighborhood set N_c around the winning cell. At each learning step, all cells within N_c are updated, whereas cells outside N_c are left intact. The update equation is

$$\Delta m_{ij} = \begin{cases} \alpha(x_j - m_{ij}) & \text{if unit i is in the neighborhood } N_c \\ 0 & \text{otherwise} \end{cases} \tag{11}$$

and

$$m_{ij}^{new} = m_{ij}^{old} + \Delta m_{ij}. \tag{12}$$

Here, α is the learning parameter. This adjustment modifies weights of the winning unit and its neighbors, so they become more like the input pattern. The winner then becomes more likely to win the competition should the same or a similar input pattern be presented subsequently.

Effect of neighborhood The width or radius of N_c can be time-variable; for good global ordering, it has experimentally turned out to be advantageous to let N_c be very wide at the beginning and shrink monotonically with time (Fig. 3) because a wide initial N_c, corresponding to a coarse spatial resolution in the learning process, first induces a rough global order in the m_i values, after which narrowing of N_c improves the spatial resolution of the map. The acquired global order, however, is not destroyed later on. This allows forming the topological order of the map.

3.2 Incorporation of Rough Sets in SOM

As described in Sect. 2.1, the dependency rules generated by using rough set theory from an information system are used to discern objects with respect to their attributes. However, the dependency rules generated by a rough set are coarse and therefore need to be fine-tuned. Here, we have used the dependency rules to get

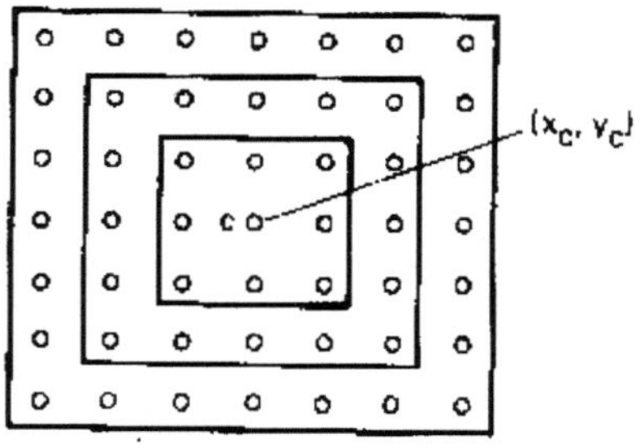

Fig. 3. Neighborhood N_c, centered on unit c (x_c, y_c). Three different neighborhoods are shown at distances $d = 1, 2$, and 3

a crude knowledge of the cluster boundaries of the input patterns to be fed to a self-organizing map. This crude knowledge is used to encode the initial weights of the nodes of the map, which is then trained using the usual learning process (Sect. 3.1.1). Since initial knowledge about the cluster boundaries is encoded into the network, the learning time is reduced greatly with improved performance.

The steps involved in the process are summarized here:

1. From the initial data set, use the fuzzy discretization process to create the information system.
2. For each object in the information table, generate the discernibility function,
$$\mathcal{F}_A(\overline{a_1}, \overline{a_2}, \ldots, \overline{a_{3n}}) = \bigwedge\{\bigvee c_{ij} \quad |1 \leq j \leq i \leq n, \quad c_{ij} \neq \emptyset\}, \tag{13}$$
where $\overline{a_1}$, $\overline{a_2}$, ..., $\overline{a_{3n}}$ are the $3n$ Boolean variables corresponding to the attributes $a_1, a_2, \ldots, a_{3n}$ of each object in the information system. The expression $\mathcal{F}_A$ is reduced to the set of all prime implicants of $\mathcal{F}_A$ that determines the set of all reducts of A.
3. The self-organizing map is created with $3n$ inputs (Sect. 2.2), which correspond to the attributes of the information table, and a competitive layer of $N \times N$ grid of units where N is the total number of implicants present in the discernibility functions of all objects of the information table.
4. Each implicant of the function $\mathcal{F}_A$ is mapped to a unit in the competitive layer of the network, and high weights are given to those links that come from the attributes, which occur in the implicant expression. The idea behind this is that when an input pattern belonging to an object, say O_i, is applied to the inputs of the network, one of the implicants of the discernibility function of O_i will be satisfied, and the corresponding unit in the competitive layer will fire and emerge as the winning unit. All of the implicants of an object O_i are placed in the same layer, whereas the implicants of different objects are placed in different layers separated by the maximum neighborhood distance. In this way, the initial knowledge obtained with rough set methodology is used to train the SOM. This is explained with the following example.

Let the reduct of an object O_i be

$$O_i : (F_{1\text{low}} \wedge F_{2\text{medium}}) \vee (F_{1\text{high}} \wedge F_{2\text{high}})$$

where $F_{(.)\text{low}}$, $F_{(.)\text{medium}}$ and $F_{(.)\text{high}}$ represent the low, medium, and high values of the corresponding features.

Then the implicants are mapped to the nodes of the layer in the following manner: Here high weights (H) are given only to those links that come from the features present in the implicant expression. Other links are given low weights.

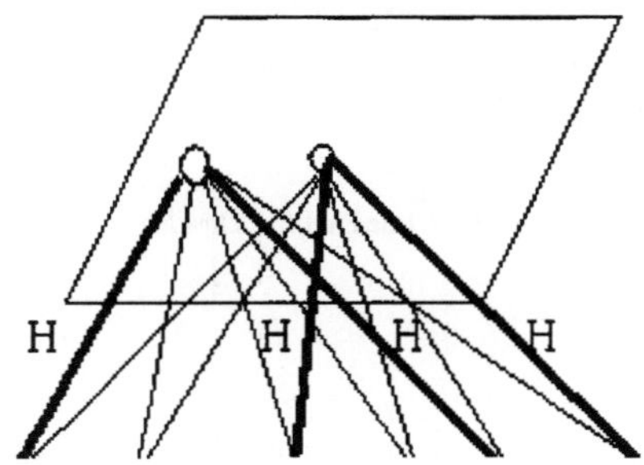

Fig. 4. Mapping of reducts in the competitive layer of an RSOM

4 Experimental Results

4.1 Data Sets Used

We have considered two artificially generated data sets for our experiment. The first data set (Fig. 5) consists of two features containing 1000 data points from two well-separated circles. The second data set (Fig. 6) consists of two features containing 417 points from two horseshoe-shaped clusters. The parameters of the fuzzy membership functions and the rough set reducts obtained from the two data sets are as follows:

Circle data:

$$\begin{aligned} c_{\text{low}(F1)} &= 4.948250 & c_{\text{low}(F2)} &= 4.954297 \\ c_{\text{medium}(F1)} &= 8.550404 & c_{\text{medium}(F2)} &= 10.093961 \\ c_{\text{high}(F1)} &= 11.956331 & c_{\text{high}(F2)} &= 14.953643 \\ \lambda_{\text{low}(F1)} &= 3.602154 & \lambda_{\text{low}(F2)} &= 5.139664 \\ \lambda_{\text{high}(F1)} &= 3.504041 & \lambda_{\text{high}(F2)} &= 4.999673 \\ \lambda_{\text{medium}(F1)} &= 3.405928 & \lambda_{\text{medium}(F2)} &= 4.859682 \end{aligned}$$

$$O_1 : (F_{1\text{low}} \wedge F_{2\text{low}}) \vee (F_{1\text{high}} \wedge F_{2\text{high}}) \vee (F_{1\text{medium}} \wedge F_{2\text{high}})$$
$$O_2 : (F_{1\text{medium}} \wedge F_{2\text{low}}) \vee (F_{1\text{high}} \wedge F_{2\text{medium}}) \vee (F_{1\text{low}} \wedge F_{2\text{medium}}).$$

Shoe data:

$$\begin{aligned} c_{\text{low}(F1)} &= 0.223095 & c_{\text{low}(F2)} &= 0.263265 \\ c_{\text{medium}(F1)} &= 0.499258 & c_{\text{medium}(F2)} &= 0.511283 \\ c_{\text{high}(F1)} &= 0.753786 & c_{\text{high}(F2)} &= 0.744306 \\ \lambda_{\text{low}(F1)} &= 0.276163 & \lambda_{\text{low}(F2)} &= 0.248019 \\ \lambda_{\text{high}(F1)} &= 0.265345 & \lambda_{\text{high}(F2)} &= 0.240521 \\ \lambda_{\text{medium}(F1)} &= 0.254528 & \lambda_{\text{medium}(F2)} &= 0.233022 \end{aligned}$$

$$O_1 : (F_{1\text{low}} \wedge F_{2\text{medium}}) \vee (F_{1\text{high}} \wedge F_{2\text{medium}})$$
$$O_2 : (F_{1\text{low}} \wedge F_{2\text{high}})$$
$$O_3 : (F_{1\text{high}} \wedge F_{2\text{low}})$$
$$O_4 : (F_{1\text{medium}} \wedge F_{2\text{high}}) \vee (F_{1\text{medium}} \wedge F_{2\text{low}}).$$

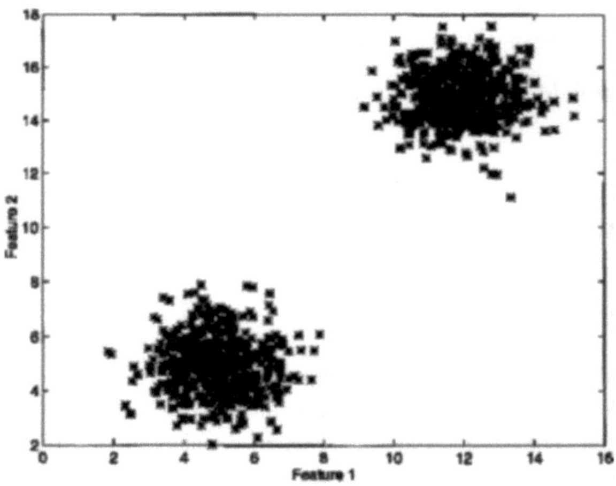

Fig. 5. Nature of the first data set (circle)

4.2 Results

To demonstrate the effectiveness of the proposed knowledge-encoding scheme (RSOM), its comparison with a randomly initialized self-organized map is presented. The following quantities are considered for comparison.

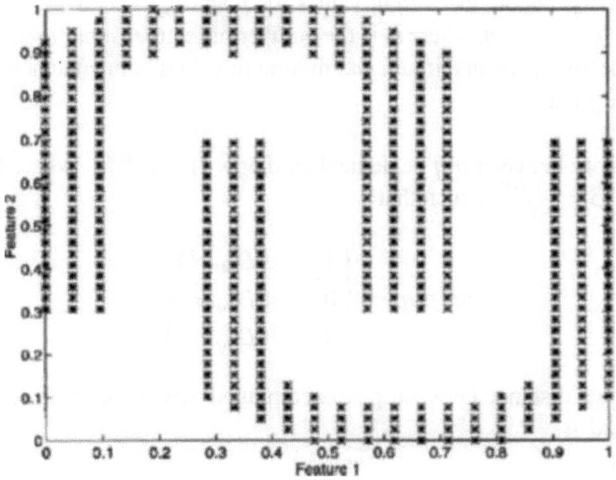

Fig. 6. Nature of the second data set (horseshoe)

A. Quantization Error

The quantization error (q_E) measures how fast the weight vectors of the winning node in the competitive layer align themselves with the input vectors presented during training. The quantization error is calculated by the following equation:

$$q_E = \frac{\sum\limits_{\text{all input patterns}} \left[\sum\limits_{\text{all wining nodes}} \sqrt{\sum\limits_j (x_j - m_j)^2} \right]}{\text{number of patterns}} \tag{14}$$

where $j = 1, \ldots, m$, where m is the number of input features to the net. For the circle data $m = n$, and for the shoe data, $m = 3n$. Hence, the higher the quantization error (q_E), the more the difference between the reference vectors and the input vectors of the nodes in the competitive layer.

B. Entropy and Fuzzy Feature Evaluation Index

To measure the quality of cluster structure, we have used an entropy measure and fuzzy feature evaluation index (FFEI) [7]. These are defined below.

Entropy: Let the distance between two weight vectors p, q be

$$D_{pq} = \sqrt{\sum_j \left(\frac{x_{p_j} - x_{q_j}}{\max_j - \min_j} \right)^2}, \tag{15}$$

where x_{p_j} and x_{q_j} denote the weight values for p and q, respectively along the jth direction and $j = 1,\ldots,m$, where m is the number of features input to the net. $\max_j$, $\min_j$ are respectively the maximum and minimum values computed across all samples along the jth axis.

Let the similarity between p,q be defined as $sim(p,q) = e^{-\beta D_{pq}}$, where β is a positive constant ($\beta = \frac{-\ln 0.5}{\overline{D}}$), such that

$$sim(p,q) = \begin{cases} 1 & \text{if } D_{pq} = 0 \\ 0 & \text{if } D_{pq} = \infty \\ 0.5 & \text{if } D_{pq} = \overline{D}. \end{cases} \tag{16}$$

$\overline{D}$ is the average distance between points computed across the entire data set. Entropy is defined as

$$E = -\sum_{p=1}^{l}\sum_{q=1}^{l}\{sim(p,q) \times \log sim(p,q) + [1 - sim(p,q]) \times \log[1 - sim(p,q)]\}. \tag{17}$$

If the data are uniformly distributed in the feature space, entropy is maximum. When the data have well-formed clusters, uncertainty is low and so is entropy.

Fuzzy Feature Evaluation Index: The fuzzy feature evaluation index is defined as [7]

$$FFEI = \frac{2}{lL}\sum_{p}\sum_{q\neq p}\frac{1}{2}\left[\mu_{pq}^{R}(1-\mu_{pq}^{O}) + \mu_{pq}^{O}(1-\mu_{pq}^{R})\right], \tag{18}$$

where l is the number of nodes, L is the number of input data points, and μ_{pq}^{O} and μ_{pq}^{R} are the degrees that both patterns p and q belong to the same cluster in the feature spaces Ω_O and Ω_R, respectively. Ω_O is the space of input patterns, and Ω_R is the weight space. Membership function μ_{pq} may be defined as

$$\mu_{pq} = \begin{cases} 1 - \frac{D_{pq}}{D_{\max}} & \text{if } D_{pq} \leq D_{\max} \\ 0 & \text{otherwise.} \end{cases} \tag{19}$$

D_{pq} is the distance between patterns p and q, and $D_{\max}$ is the maximum separation between patterns in the respective feature spaces.

The value of FFEI decreases as the intercluster/intracluster distances increase or decrease. Hence, the lower the value of FFEI, the more crisp the cluster structure.

C. Frequency of Winning Nodes (f_k)

Here, we have used the number of winning top k nodes (f_k) in the competitive layer, where k is the number of rules (characterizing the clusters) obtained using rough sets. $k = 2$ for circle data and $k = 4$ for shoe data. f_k reflects the error if all but k nodes would have been pruned. In other words, it measures the number of sample points correctly represented by these nodes.

D. Number of Iterations

We compute the number of iterations at which the error does not change much.

The comparative results are presented in Table 1.

Table 1. Comparison of RSOM with SOM

Data	Initialization	Quantization error	Iteration at which error converged	Entropy	FFEI	f_k
Circle	Random	0.9	5000	0.5701	0.2841	157
	Rough	0.1	10	0.1672	0.0354	248
Shoe	Random	0.038	5000	0.7557	0.5206	83
	Rough	0.022	50	0.6255	0.3912	112

The following conclusions can be derived from the results obtained:

1. Better cluster quality: As seen from Table 1, the RSOM has lower values of both entropy and FFEI thus implying lower intracluster distance and higher intercluster distance in the clustered space compared to the conventional SOM. The quantization error of RSOM is also far less than that of SOM.
2. Less learning time: The number of iterations required to achieve the error level is far less in RSOM compared to SOM. The convergence curves of the quantization errors are presented in Figs. 7 and 8 for circle and shoe data, respectively. It is seen that RSOM starts from a very low value of quantization error compared to SOM.
3. Compact representation of data: It is seen that in the RSOM, fewer nodes in the competitive layer dominate, i.e., they win for most of the samples in the training set. On the other hand, in conventional SOM, this number is higher. This is quantified by the frequency of winning for the top k nodes. It is observed that this value is much higher for the RSOM, thus signifying less error if all but k nodes would have been pruned. In other words, the RSOM achieves a more compact representation of the data. The distribution of the frequency of winning nodes is shown in Figs. 9–12.

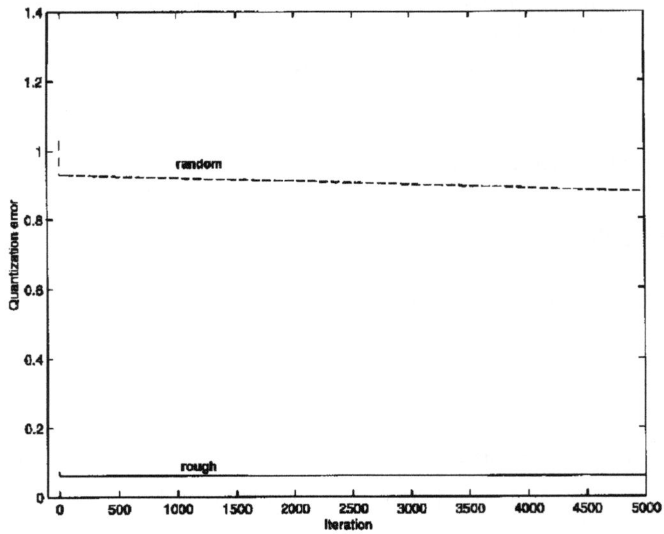

Fig. 7. Variation of quantization error with iteration for the circle data

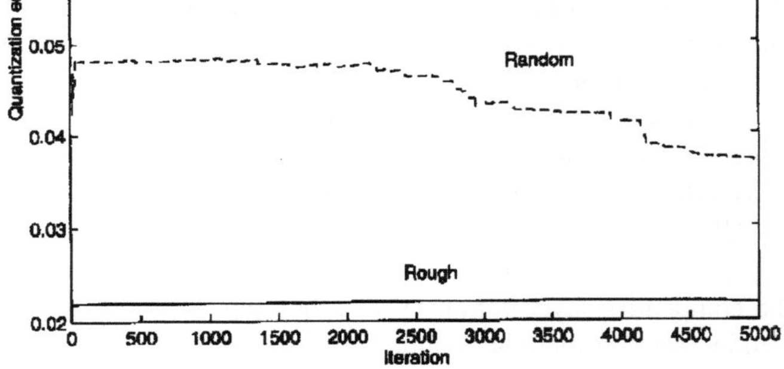

Fig. 8. Variation of quantization error with iteration for the shoe data

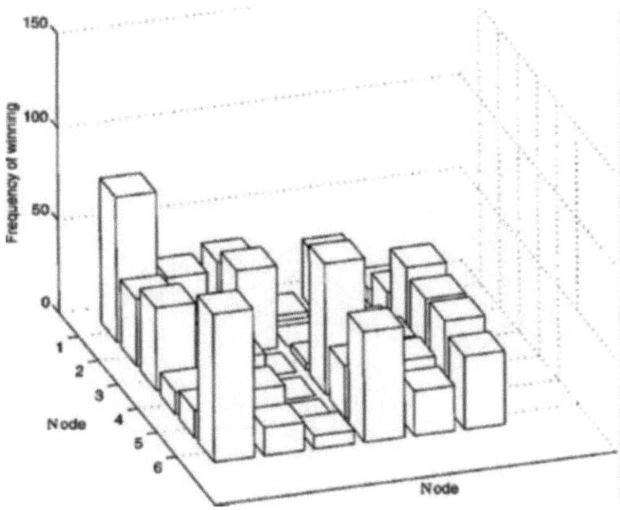

Fig. 9. Plot showing the frequency of winning nodes using random weights for the circle data

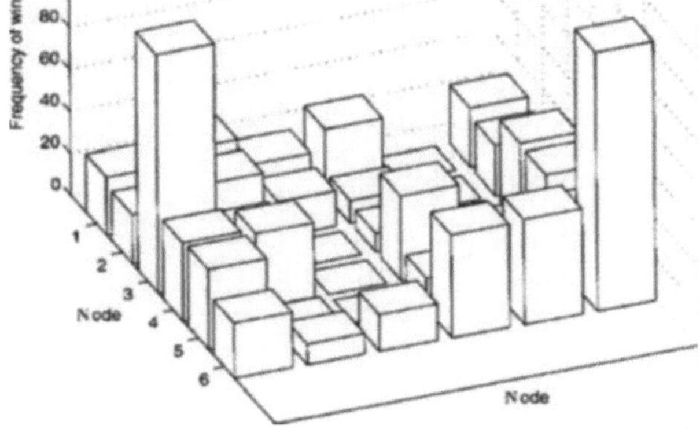

Fig. 10. Plot showing the frequency of winning nodes using rough set knowledge for the circle data

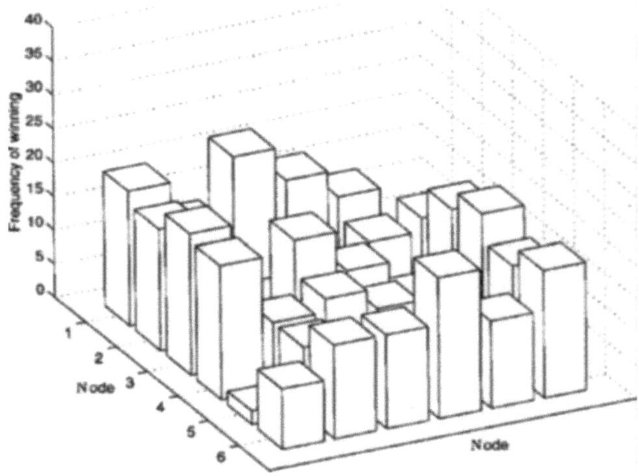

Fig. 11. Plot showing the frequency of winning nodes using random weights for the shoe data

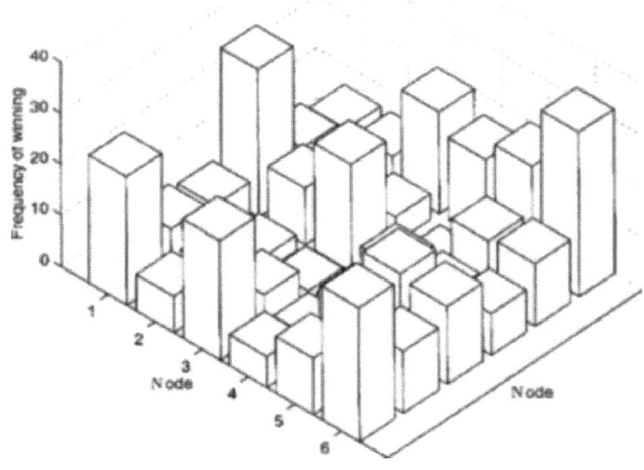

Fig. 12. Plot showing the frequency of winning nodes using rough set knowledge for the shoe data

5 Conclusions

A self-organizing map incorporating the theory of rough sets with fuzzy discretization is designed. Rough set theory is used to encode domain knowledge in the form of crude rules, which are mapped for initialization of the weights of the SOM. The superiority of the model, compared with random initialization of the weights of the SOM, is demonstrated both for linearly and nonlinearly separable data sets in terms of learning time, quality of clusters, and representation of data.

Since the RSOM achieves compact clusters, this will enable one to extract nonambiguous rules. Its significance in mining large data sets is evident.

References

1. Z. Pawlak. *Rough Sets: Theoretical Aspects of Reasoning About Data.* Kluwer, Dordrecht, 1991.
2. A. Skowron, C. Rauszer. The discernibility matrices and functions in information systems. In R. Słowiński, editor, *Intelligent Decision Support: Handbook of Applications and Advances of the Rough Sets Theory*, 331–363, Kluwer, Dordrecht, 1992.
3. S.K. Pal, W. Pedrycz, A. Skowron, R. Swiniarski, editors. Rough-neurocomputing (special issue). Vol. 36 of *Neurocomputing: An International Journal*, 2001.
4. M. Banerjee, S. Mitra, S.K. Pal. Rough fuzzy MLP: Knowledge encoding and classification. *IEEE Transactions on Neural Networks*, 9: 1203–1216, 1998.
5. T. Kohonen. *Self-Organizing Maps.* Springer, Heidelberg, 2001.
6. S.K. Pal, S. Mitra. Multi-layer perceptron, fuzzy sets and classification. *IEEE Transactions on Neural Networks*, 3: 683–697, 1992.
7. S.K. Pal, S. Mitra. *Neuro-Fuzzy Pattern Recognition: Methods in Soft Computing.* Wiley, New York, 1999.

Part III

Exemplary Application Areas

Introduction to Part III

Part III, *Exemplary Application Areas*, consists of chapters that deal with granulation of knowledge (based on rough sets) in biomedical reasoning, socially embedded games, risk theory, and decision theory.

The chapter, *A Semantic Model of Biomedical Cognition*, by Doroszewski examines the structure of the intricate process of solving biomedical problems from the cognitive point of view. A structural and semantic analysis of statements leading to extraction of their components such as causal structure, causal paraphrase, and propositional paraphrase is presented along with an analysis of temporal and uncertainty factors. As a result of in-depth analysis, a hypothesis is suggested that an important role in biomedical reasoning is played by the logical type of probability. An analysis of a textbook case as well as a clinical case is made as exemplary cases of biomedical reasoning.

In *Fundamental Mathematical Notions of the Theory of Socially Embedded Games: A Granular Computing Perspective*, Gomolińska studies the problems of social games in the light of granular computing. She uses mereological notions (see Part I chapter on a rough mereological basis for RNN by Polkowski) to define partial containment of complexes of points. Then, she discusses rule complexes as granules of information and eventually models social actors by means of rule complexes.

The topic of social games is pursued by Burns and Roszkowska in their article, *Fuzzy Game Theory: The Perspective of the General Theory of Games on Nash and Normative Equilibria*. This chapter presents a survey of fuzzy game theory, including a discussion of Nash equilibria. It provides a valuable reference to the subject.

In *Rough Neurons: Petri Net Models and Applications*, Peters, Ramanna, Suraj, and Borkowski discuss Petri net models for rough-neural computing in the forms of training set production and approximate reasoning schemes, considered in the framework of parameterized approximation spaces. A Petri net model of a neural network classifier with optimal feature selection is presented, and Petri net models of elementary approximation neurons are discussed in detail. In addition, the design of three sample elementary approximation neurons is given. Training of elementary approximation neurons is also discussed.

The subject of information granulation and approximation is discussed by Yao in *Information Granulation and Approximation in a Decision-Theoretical Model of Rough Sets*. The author studies the rough set mechanism of granulation and approximation, keeping the decision theoretical issues in mind. After a careful survey of various granulation and approximation methodologies, the author models the Bayesian decision theory with rough set classifiers.

Chapter 15
Biomedical Inference: A Semantic Model

Jan Doroszewski

Department of Biophysics and Biomathematics, Medical Center of Postgraduate Education, Marymoncka 99, 01-813 Warszawa, Poland
jandoro@cmkp.edu.pl

Summary. This chapter presents a model of the structure of a typical fragment of pathophysiological and diagnostic knowledge and of simple steps for biomedical problem solving. The main method used in this study consists of a structural and semantic analysis of the biomedical statements that appear in original clinical textbooks and documents. Various transformations of these statements reveal their different aspects as causal structure and meaning (causal paraphrase) and a simplified logical skeleton (propositional paraphrase), as well as temporal and uncertainty factors. The possibilities of performing some inference-making procedures based on the results of this analysis are shown. Biomedical types of understanding and use of the notions of causal relationship and probability are discussed from the point of view of their roles in biology and medicine. It is suggested that in a biomedical context, an especially important role is played by the logical type of probability, i.e., estimation of the degree of certainty of statements based on their contextual justification. The process of biomedical hypotheses verification is based, to a considerable degree, on the estimation of changes in their probabilities (confirmation) in the light of increasing evidence. This approach, called a semantic model of biomedical cognition, is exemplified by the analysis of numerous original statements and a study of a handbook text fragment, as well as by a discussion of an authentic clinical case.

1 Introduction

This chapter describes the results of a model approach to the study of some aspects of biomedical knowledge, namely, those that are especially important from the point of view of solving diagnostic problems. It is a continuation of previous research in which biomedical knowledge was studied in the systems' and semantic framework [15–17]. The approach on which this study is based consists mainly in analyzing natural-language medical texts as reflecting the fragments of natural medical knowledge. It is supposed that the results of such a study may contribute to better understanding of biomedical reasoning, perhaps also to the development of a more formal representation of biomedical knowledge in the framework of artificial intelligence.

Although the literature on various aspects of medical knowledge and problem solving, theoretical as well as practical, methodological, psychological, and others, is vast and manifold [3, 10, 12, 20, 22, 31, 43, 44], the works directly related to semantic biomedical problems are not so numerous [2, 9, 14, 25, 37, 40, 46]. This study was partly inspired by research on qualitative reasoning [26, 27] and some aspects of AI knowledge representation [36, 45]. The analytical model character of this approach should be firmly stressed since neither the methods applied nor the results obtained are directly connected with research on the psychology of thinking, in which considerable progress constantly occurs (e.g., [21, 23, 28, 39, 41]).

Biomedical knowledge is considered here as a system of statements describing features of the human organism and its parts. Due to the extreme complexity, as well as the dynamic and variable character of biological and pathological objects, the statements describing them are, on the one hand, interconnected in a multiple way and, on the other hand, denote, in an overt or a hidden way, the temporal aspects of phenomena and the degree of certainty of judgments.

One of the most specific features of practical biomedical knowledge and reasoning is its close connection with natural language and relatively limited use of formal methods of inference making in everyday problem solving, despite the rapid development of applications of mathematical methods. However, to better understand the meanings of biomedical statements, they should be extricated, as it were, from their natural linguistic surroundings.

In the search for the basic elements of biomedical statements, the method of simplifying but sense-preserving graphs and paraphrases representing linguistic expressions is useful. It resembles the logical approach to the analysis of natural language (e.g., [30, 32, 34]), but its purpose is limited here only to medical language. Theoretical structures revealed in this way may be used as a base for schemes of reasoning and applied to simplified inference making.

Scientific domains, such as medicine as a whole, specialties, for example, cardiology, problems or fragments of knowledge (e.g., description of a disease), and the like are systems of laws and other general statements that describe objects as characterized by their features. Such features are first-order statements [17], e.g., anatomical and relations connecting the features of objects and their values, which are second-order statements [17], e.g., physiological. In this work, the latter category of scientific statements is studied, especially those that describe relations linking particular values of features.

1.1 General Description of the Disease

This study is centered on the medical cognition of the patient, based on a general model description of the disease, which is composed of statements representing pathological and other phenomena and their interrelations. The reasoning of the doctor aims at making a mental reconstruction of unobserved facts (signs) on the grounds of those that have been observed. A typical set of pathological phenomena (composed of a model of a disease, or nosological unit) that are the correlates of the biomedical statements may be represented as in Fig. 1.

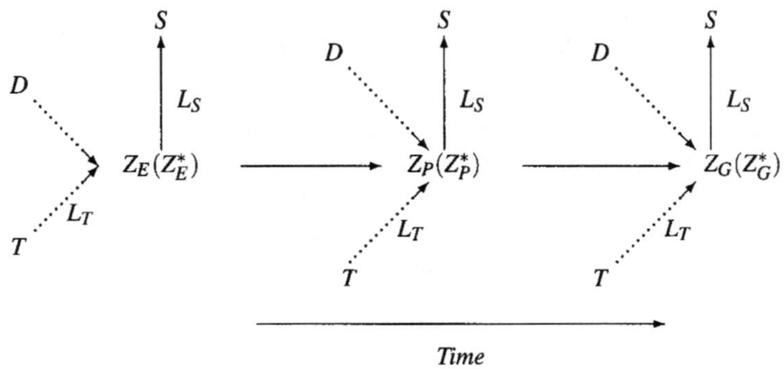

Fig. 1. A scheme of disease in general terms (a fragment of biomedical knowledge). The symbols may be interpreted as denoting the phenomena or corresponding names. Z_E: etiologic factors; Z_P: main (specific) pathological phenomena (direct and indirect effects of etiologic factors); Z_G: effects of specific pathological phenomena (Z_P) concerning the organism as a whole; S: observable phenomena (signs). Arrows: relations linking the phenomena (asymmetrical, mostly causal). D: diagnostic actions causing the appearance of signs; T: therapeutic actions potentially transforming undesired pathological phenomena into desired ones (Z_E^*, Z_P^*, Z_G^*).

1.2 A Biomedical Statement

A specific feature of pathological (clinical and others) statements is that they describe the relations between pathological values of the human organism and its particular parts. It should be remembered, however, that their explicit meaning is only a small part of overall, mostly normal physiological and anatomical knowledge.

The basic type of statements that are analyzed in this work may be regarded either as sentences composed of the grammatical object and a complex predicate or as statements containing two or more clauses linked by a relation; one of the clauses denotes the main object, and the other(s) describe(s) secondary ones; still other(s) represent the relation(s). The statements of this kind describe the relations between two or more values of the features exhibited by the objects; their extremely simplified general structure (neglecting the modal auxiliaries) may be summarized as follows: (1) *Functional form*: "for every object x (some object x, $f\%$ objects x) in conditions W, the values of feature E_1 of the object E are linked with the values of the feature E_2 of this object by the relation R." (2) *Particular-value form*: "for every object x (some object x, $f\%$ objects x) in conditions W, the feature E^1 of the object x is related to the feature E^2 of this object by the relation R_{12}." As examples, the statements analyzed below may be cited: (3) "*the oxygen supply* (feature E_1) *in the tissue (object* x_1*) depends (functional relation R) on the blood flow* (feature E_2) *in the artery* (x_2); " (4) "*the decrease in the blood flow* (E_2^1) *in the artery* (x_2) *is the cause* (R_{12}) *of the decrease in the oxygen supply* (E_1^1) *in the tissue* (x_1)." The objects x_1 and x_2 are linked by a certain spatial relation.

The professional-language statements, according to widely accepted usage, contain the phrases signifying the time and uncertainty factors in special situations rather than as a general rule. The reason probably is that these notions are so deeply embedded in medical science and the way of reasoning that they frequently seem self-evident and are automatically supplemented, as it were, by the participants in the communication process. To consider these factors in an explicit way, the above statement describing the relation between particular values (form 2), should be reformulated as follows: "for every object x (some object x, $f\%$ objects x) in conditions W, the value E_i^1 of the feature E^1 of the object E occurring in time t is connected by the relation R_{12} with the value E_j^2 of the feature E^2 of this object occurring in time t' with probability π." Besides the artificiality and complexity of such a statement, its precision and clarity are rather doubtful. Perhaps the difficulty of formulating too complex a linguistic expression is the reason that various levels of their sense are separated in practice: the basic, "core" meaning, on the one hand, and modifying factors, on the other hand. Both of them are, to a certain extent, though in a different manner, contained in the significance of a special kind of relation that may link the phenomena as cause and effect. The time sequence is one of the constitutive conditions of the biomedical causal relations (see Sect. 2) and therefore, may not be treated separately. On the other hand, the probability of the occurrence of the phenomena, even if connected with the notion of causality, is a factor that should be analyzed apart (see Sect. 2).

In comparison with many natural professional-language expressions, the above schemes of a sentence are, of course, an obvious simplification, possibly even an oversimplification. Take, e.g., an authentic sentence: "*Experimental studies suggest that the endothelial barrier, broken by mechanical or chemical injury, is associated with a tissue response that includes local platelet adhesion and aggregation* [5]." It would be impossible to render the sense of this or a similar sentence in the above scheme. A sequence, however, of several sentences (paraphrases) could convey its meaning in a relatively exact way. The above approach seems to be especially suitable for analyzing the description of pathological phenomena contained, e.g., in clinical textbooks, which present fragments of biomedical knowledge in a manner, to some extent, prepared for practical use (see Sect. 4). The remarks suggest that to make a biomedical statement clear and unambiguous (formulated, as normally, in natural linguistic shape), i.e., to express it in an overt form, it is necessary to enrich it with certain elements or to modify its wording, or both. It would be, however, an impossible task to attempt to express the whole meaning of a statement, since it would be equivalent to setting forth a fragment of knowledge of considerable size. As concerns the specific purposes of this analysis, it is sufficient, however, to show more clearly, i.e., to extract from a deeper context, only certain aspects of the sense of the statement, e.g., the time factor involved, the degree of certainty of the judgment, etc. or to modify its structure so that it could be more suitable for use in a certain type of inference. This purpose may be achieved by making various paraphrases of the analyzed statement the main types of which either express the kind of relation or simplify it. Inasmuch as one of the especially common and important types of relations described by biomedical statements is the cause-effect connection, the former transformation may be called a *causal paraphrase*; on the other hand, to show a certain type of model inference, a change based on the propositional (sentential) scheme, *propositional paraphrase*, may be useful. Besides the linguistic transformation, the use of a graphic representation is useful for semantic analysis.

1.3 A Fragment of Biomedical Knowledge

Below we present an analysis of quasi-natural statements in the frame of semantic model (a study of an authentic text is presented in Sect. 4). To be closer to real biomedical knowledge, the statements comprise more than two arguments.

Let us consider the following statements (uppercase and lowercase letters are assigned to the components of the sentences for further use).

1. *Quasi-natural statements in functional form* (a description of the relation linking the features): Oxygen content in a tissue (Q_1) depends on the blood flow in the supplying artery (P_1), arterial blood oxygen saturation (P_2), oxygen consumption in the tissue (P_3), and other factors. The supply of glucose (Q_2), the action of hormones on this tissue (Q_3), etc. depend also on arterial blood flow (P_1).

2. *Quasi-natural statements in a causal form* (a description of the relation linking particular values of features: A decrease in the oxygen supply in a tissue (q_1) is caused by a decrease in the blood flow in the supplying artery (p_1) and/or a decrease in the arterial oxygen content (p_2) and/or a decrease in oxygen consumption in this tissue (p_3) and/or by other factors. The decrease in blood flow in an artery (p_1) causes a decrease in the oxygen supply in a tissue supplying this tissue (q_1) and/or a decrease in the glucose supply (q_2) and/or a decrease in the action of hormones on this tissue (q_3), etc. 3. *Propositional form* (a description of the relation linking particular values of features by conditional sentences): If the blood flow in an artery (p_1) and/or arterial oxygen content (p_2) and/or oxygen consumption in this tissue

(p_3), etc. are decreased, then the oxygen supply in the tissue supplied by this artery (q_1) is decreased. If the blood flow in an artery (p_1) is decreased, then in the tissue supplied by this artery, the oxygen supply (q_1) and/or glucose supply (q_2) and the hormone action (q_3), etc. are decreased. In an abbreviated form:

1. If p_1 and/or p_2 and/or p_3 and/or ..., then q_1.
2. If p_1 then q_1 and/or q_2 and /or q_3 and/or ...,

where the functor and/or may be interpreted as alternative, disjunction, or conjunction.

The above sets of statements (a knowledge fragment in various linguistic shapes) may be represented as directed graphs (see Fig. 2; see also Figs. 1 and 3).

These formulations do not contain expressions indicating the time in which the phenomena occur and their degree of certainty. The time factor is implicitly assumed in the causal relation (see Sect. 2); in some cases, however, time requires an apparent expression. The same may regard the probability of the statement (see Sect. 3); more frequently, however, this factor is represented by a modal part of the sentence, e.g., "the phenomenon p_1 is a *probable cause* of the phenomenon q_1" and "the phenomenon q_1 is a *probable effect* of the phenomenon p_1." This problem will be discussed later.

The propositional representation of the statements may be useful for revealing the basic scheme (a skeleton, as it were) of various inferences; it may also show more clearly the temporal and probabilistic aspects of sentences. The statements presented as propositional paraphrases may constitute grounds for the following inference: (1) p_1 occurs; therefore q_1 and/or q_2 and/or q_3 and/or ... occur, (2) p_1 and/or p_2 and/or p_3 and/or ... occur; therefore q_1 occurs. In practice, the usefulness and effectiveness, even feasibility, of such complex reasoning is limited, and it is seldom performed (in any case, in an overt form). To arrive at a valuable conclusion, various assumptions should be adopted not only to introduce a simplification but also to make clearer some aspects of the statements, especially the meaning of the functor and/or in a given context.

An extreme simplification would consist in assuming that the influence of most of elements may be disregarded because either their insignificance is known or because they are unknown, etc. This assumption reduces the whole fragment of knowledge to the sentence: "p_i is the only cause of q_j" or "q_j is the only effect of p_i." In a propositional paraphrase, it would run: "p_i if and only if q_j" that would make possible a two-sided inference, namely, "p_i is accepted; therefore q_j," and "q_j is accepted; therefore p_i." In most situations, however, such a strong assumption would be as unjustified as taking into account all possibly relevant factors is unrealistic.

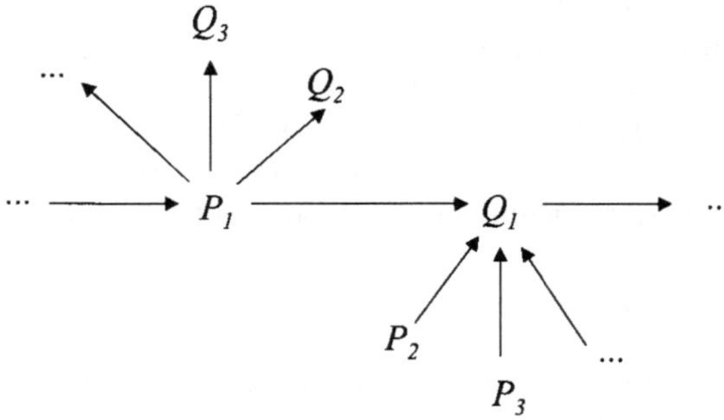

(a)

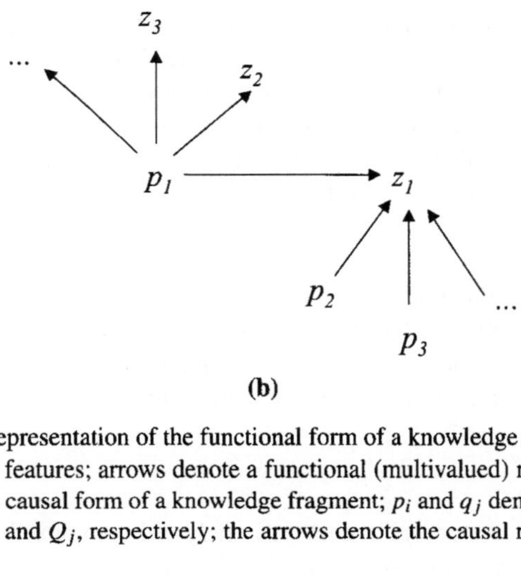

(b)

Fig. 2. (a) Graphic representation of the functional form of a knowledge fragment. P_i and Q_j denote the names of features; arrows denote a functional (multivalued) relation. **(b)** Graphic representation of the causal form of a knowledge fragment; p_i and q_j denote the names of the values of features P_i and Q_j, respectively; the arrows denote the causal relationship between these values.

The solution seems to be, in practice, to perform the reasoning without deciding in an explicit manner which elements are considered and which are disregarded. It is a kind of counterfactual thinking, of pondering "as if," or of drawing a conclusion with a certain doubt, being ready to modify it if new evidence appears. This hypothesis is corroborated by the common use of statements that are not clear in their semantic range. For example, the sentence "the phenomenon p is the cause of the phenomenon q" ("a decrease in the oxygen supply in the tissue is caused by a decrease in the blood flow in the artery") or "the phenomenon q is the effect of the phenomenon p" ("a decrease in the blood flow in the artery causes a decrease in the oxygen supply in the tissue") may be understood as signifying that p is the unique cause (effect) of q or that p is one of the causes (effects) of q. Similarly, in natural language, the statement equivalent to the conditional sentence "if p, then q" may mean "only if p, then q", "if p, then only q," etc. A conclusion based on such statements, "accepted p, therefore q" or otherwise, may be considered correct only with the above-mentioned reservation, i.e., on condition of its provisional character. The possibility of modifying the result of reasoning is assured by the fact that each statement constitutes a part of a broader fragment of knowledge that may be a source of new pieces of evidence. It seems that reasoning surrounded by a "penumbra of doubt," understood as probability or otherwise, is a normal procedure in medical problem solving.

Among the many assumptions that may constitute more or less firm grounds for accepting a conclusion in complex reasoning, one is especially important and frequently used in practice. It consists in accepting tacitly a supposition concerning the values of other features, which are components of a set of premises appearing in a given inference. For example, the conclusion that the oxygen supply in a tissue is decreased when the arterial blood flow is decreased is correct, only if it is assumed that either the causes are independent or that other features that (jointly or alternatively) are the causes of this condition display certain values (are normal, are not increased). Such an assumption would constitute the condition of the validity of the reverse inference, i.e., that the arterial blood flow is decreased based on the fact that the oxygen supply in this tissue is decreased.

In natural discourse, the modal auxiliaries are expressed in an almost infinite manner (see Sect. 3). In many cases, they are taken as self-evident and are not specified at all. On the other hand, the notions of temporal sequence and uncertainty or probability are, to a certain extent, implied in the meaning of the concept of causal relationship (see Sect. 2).

The full meaning of a causal statement is rich and complex. For the time being, it may be assumed that the statement "the phenomenon p is the cause of the phenomenon q" means that p and q occur in certain times t and t' and that the occurrence of q as the effect of p is not necessary. The paraphrase may be formulated as follows: *if in time t the phenomenon p occurs, then in time t′ the phenomenon q occurs with probability* π (probably occurs), where $t = t'$ or $t < t'$ or $t > t'$; π is a qualitative estimate or quantitative measure of probability. Another formulation expresses the notion of the probability of statements rather than that of events: *if in time t, the phenomenon p occurs, then in time t′, the phenomenon q occurs is probable (justified) in grade* π (see Sect. 3). According to the context, the expressions "*probable*" or "*in grade* π" may have a quantitative or a qualitative interpretation, i.e., may be equivalent to a certain number, to an adjective in a basic or comparative form, or may indicate the revision of an estimate in the light of new evidence. The inference based on the sentence *if* $p(t)$, *then* $q(t')$ *with probability* π, $t < t'$ runs: $p(t)$ is accepted, therefore q occurs (simultaneously as or later, or earlier than q) with probability π, where the phrase "probability π" may be understood in each of the above manners.

Propositional paraphrases, whose meanings are enriched by the time and probability aspects, i.e., in fully developed form, are the following:

1. If $p_1(t_1)$, and/or $p_2(t_2)$, and/or $p_3(t_3)$, and/or ..., then $q_1(t')$ with π, where the relations of times depend on the interpretation of the functor and/or.
2. If $p_1(t)$, then $q_1(t')$ with π_1 and/or $q_2(t_2)$ with π_2 and/or $q_3(t_3)$ with π_3 ... where the relations of times are as before.

A fully developed paraphrase, as well as the paraphrases that don't contain modal auxiliaries (see above) are sometimes reduced to quasi-deterministic conditional sentences: "if $p_i(t)$, then $q_j(t')$ with π_1" and "if $q_j(t')$, then $p_i(t)$ with π_2"; as before, a "provisional" conclusion may be drawn, namely, $q_j(t')$ with π_1, and $q_i(t')$ with π_2.

1.4 Inference Model

The above presented analysis may be summarized as follows:

An example of a rather typical, though not very precise, formulation of a biomedical (pathophysiological) statement may be put in words in the following manner: *a decrease in the blood flow in the artery supplying a tissue* (p_1) *is accompanied by a decrease in the oxygen supply in this tissue* (q_1). It could be used as the general premise in two kinds of inferences: (a) the blood flow in the artery is decreased (p_1), therefore probably the oxygen content in the tissue is decreased (q_1); (b) the oxygen content in the tissue is decreased (q_1), therefore probably the blood flow in the artery is decreased (p_1). In natural reasoning, both inferences would probably be accepted. The correctness of the above conclusions depends, however, on other connected statements and may be verified in various ways. Let us transform the initial statement into causal and propositional forms (paraphrases).

Effect-cause paraphrase
The decrease in the oxygen supply in a tissue (q_1) *is caused by a decrease in the blood flow in the supplying artery* (p_1).

Paraphrase of the above form:
If p_1 *occurs, then* q_1 *occurs with probability* π.

Inference:
p_1 occurs; therefore q_1 occurs with probability π; in words, *the blood flow in the*

artery is decreased (p_1)*; therefore probably (with probability* π*) the oxygen content in the tissue is decreased* (q_1).

This conclusion is arrived at without taking into consideration other than p_1 possible causes (see Fig. 2). Whether it is true or false depends on other phenomena, namely, the decrease in the oxygen saturation of the blood (p_2) and the increased oxygen consumption in the tissue (p_3). If the relation linking all these phenomena is interpreted as a weak alternative, i.e., if p_1 or p_2 or p_3 or ..., then q_1; probability π of q_1 increases when p_1 and p_2, when p_1 and p_2 and p_3 ... then the conclusion (a) is correct; taking into consideration the possible or actual appearance of p_2 and/or p_3 would only modify (enhance) the probability of the conclusion (see later).

Cause-effect paraphrase:
The decrease in the blood flow in the supplying artery (p_1) *is the effect of the decrease in the oxygen supply in a tissue* (q_1).

Paraphrase of the above form:
If q_1 *occurs, then* p_1 *occurs with probability* π'.

Inference:
q_1 occurs; therefore p_1 occurs with probability π'; in words, *the oxygen content in the tissue is decreased* (q_1)*; therefore, probably (with probability* π'*) the blood flow in the artery is decreased* (p_1).

Whether this conclusion is true or false depends on the interpretation of the functor and/or. For the general assumption, as before, if p_1 and $not(p_2)$ and $not(p_3)$, therefore p_1; the probability of p_1 is modified (decreases) when not only p_1 appears but also p_2, and p_1, together with p_2 and p_3.

2 Biomedical Aspects of the Notion of Causality

Causality constitutes one of the most important philosophical and logical problems of the general methodology of science (see, e.g., [5,6,29]), including the theory of computer science and artificial intelligence [33]. The analysis presented in this work concerns only limited and rather special aspects of the problem of a causal relation, namely, the manner in which the statements formulated in a causal convention are used and understood in the medical context, i.e., in connection with natural, oral, and written communication between physicians, biomedical scientists, etc. Although such an analysis cannot reveal the philosophical essence of causality, it may contribute to a better understanding of the meaning of statements falling under the scheme *phenomenon p is the cause of phenomenon q* and *phenomenon p is the effect of phenomenon q* and the role they play in biomedical discourse.

2.1 Meaning of Biomedical Causal Statements

The majority of biomedical causal statements exhibit the following features (a similar list of various aspects of the causal relation from a general viewpoint is presented by Bunge [6]):

1. Causal statements describe relations between particular values of features of objects (in contradistinction to functional laws denoting the relation linking the features, i.e., the sets of values).
2. A relation described by a causal statement is asymmetrical, i.e., if p, then q and not otherwise, where p is the cause and q is the effect.
3. Grammatical objects (object clauses) of causal statements denote dynamic phenomena, i.e., changes of values or states remaining after the occurrence of the change; in other words, in principle, the cause-effect relation links changes of the state of affairs and not the states themselves; when a statement describes the causal relation between two states, it is tacitly understood that a given change did occur earlier than the change constituting its effect.
4. The event that is the cause occurs earlier than the event that is the effect and, inversely, the effect (resulting change) occurs later than the cause (initial change); the difference in time between the occurrence of the initial change and the resulting change may be insignificant but is always present. This time sequence is related to (or is the essence of) the asymmetry of the relation.
5. The relation described by the causal statement may be either deterministic or uncertain, i.e., probabilistic or possibilistic (see Sect. 3). To underline the probabilistic character, the causal statements are formulated, e.g., *p is a probable (possible) cause of q* or similar.
6. The connection between phenomena described by causal statements is not a single relation but embraces numerous other links connecting many phenomena. In other words, every statement considered causal may be proved by and/or derived from other accepted general statements that is a condition sine qua non of belonging to this category.
7. From the above conditions follows the most important aspect of the meaning of the causal relation in a biomedical context, namely, the belief that some changes artificially provoked in a certain phenomenon (the cause) may provoke the desired changes in another phenomenon. This conviction constitutes the foundation of the majority of actions whose goal is to produce a purposeful modification of a certain state of affairs. The famous saying *felix qui potuit rerum cognoscere causas* not only reflects the intellectual satisfaction that gives the cognition of the causes of things, i.e., of the relations linking the phenomena, but also is connected with the practical value of this type of knowledge. In medical practice, acquaintance with the cause-effect relations is the basis of "causal therapy" whose essence consists in provoking the appearance of phenomena that bring about the desired effects.

As an example, let us take the statement, *impaired myocardial perfusion is caused by a decrease in coronary blood flow* [11].

1. The phrases *impaired myocardial perfusion* and *decrease in coronary blood flow* signify the values of features *myocardial perfusion* and *coronary blood flow* displayed by the object *myocardium (heart muscle)* and the complex object *blood flowing in the coronary artery.*
2. It is assumed that if coronary blood flow is decreased, then myocardial perfusion is impaired and not otherwise around.
3. The phrase *decrease in coronary blood flow* should be understood as indicating that blood flow has decreased at a certain time, whereas the phrase *impairment of myocardial perfusion* means that this state has once begun and afterward remains.
4. The beginning of the decrease in coronary blood flow preceded the impairment of myocardial perfusion.
5. The supposition that impaired myocardial perfusion is connected with the decrease in coronary blood flow is very probable, although not certain.
6. The statement under consideration may be explained on the grounds of anatomical conditions (connection of the coronary artery with the myocardial capillaries), continuity of the flow of blood in the artery and capillaries and the physical relations from which follow the direction of change of coronary blood flow and myocardial perfusion.
7. It is known that when coronary blood flow increases, myocardial perfusion will augment; therefore, if the decrease in blood flow in the coronary artery would be artificially changed to a normal value, myocardial perfusion would probably become normal; this expectation is the basis of the therapeutic management of ischemic heart disease.

To sum up, the superficial as well as the deeper meaning of the sentence formulated according to the scheme, *p is the cause of q*, may be presented as follows:

1. p and q are values of features P and Q of an object (see below).
2. The relation between p and q is asymmetrical.
3. p and q are dynamic phenomena.
4. p occurs or begins earlier than q.
5. The occurrence of q on the condition that p occurs (did occur) is usually not certain.
6. p and q are connected by other (mostly numerous) phenomena and relations.
7. An influence exerted on p may be followed by a modification of q.

Mutatis mutandis, the meaning of the statement "q is the effect of p" may be explicated in a similar way.

2.2 Causal Inference

The major premise of a causal inference is either the statement, "p is the cause of q" (identification of the cause), or the statement, "q is the effect of p" (identification of the effect), the minor premises "q occurs" or "p occurs," and the conclusions, neglecting the time factor, "p occurs" or "q occurs," respectively. The formal bases of these inferences are the following schemes:

1. If p, then q (propositional paraphrase of the cause-indicating statement); p occurs, therefore, q occurs.
2. If q, then p (paraphrase of the effect-indicating statement); q occurs, therefore, p occurs.

In comparison with the whole, very rich meaning of causal statements (see above), the propositional paraphrases, as well as the conclusions they may generate, are extremely poor. To represent a "semantic minimum," they should be enriched at least by introducing an indication of the time sequence:

1. If $p(t)$, then $q(t')$, $t' < t$ (if p, then earlier q).
2. If $q(t)$, then $p(t')$, $t < t'$ (if q, then later p).

It should be stressed that the inferences based on cause-indicating and effect-indicating statements are not interchangeable (or are such only on certain assumptions), since q may have more causes than only p, and p may have more effects then only q (see below).

Another step enriching a propositional paraphrase is to include not only time, but also an uncertainty factor, as expressed, e.g., by an estimate of the probability:

1. If $p(t)$, then $q(t')$ with probability π, $t' < t$ (if p, then probably earlier q).
2. If $q(t)$, then $p(t')$ with probability π', $t < t'$ (if q, then probably later p).

The probabilities of the causes and effects may be different due to the same reasons mentioned before.

The above aspects of the meaning of causal statements, namely, time and probability, constitute only a small part of their whole sense, namely, only conditions 4 and 5 from the above list. A statement that could explicitly take into account the whole meaning of the sentence, "p is the cause of q", and of the sentence, "q occurs as the effect of p," besides formulating the above conditions, should embrace many other factors.

2.3 Uncertainty of Causal Statements

In most cases, inferences based on causal statements are subjectively self-evident. Usually, however, the formulation of a detailed proof of the correctness of conclusions based on premises expressed as causal propositions could be a difficult task since it would involve a more or less rich and complex fragment of biomedical knowledge. For example, a conclusion based on the general statement *impaired myocardial perfusion is caused by a decrease in coronary blood flow* (see above) contains a tacit assumption that the impairment of myocardial perfusion occurs as the cause of decrease in coronary blood flow, and such a statement is an abbreviation for a considerable number of interrelated statements. In an "unfolded" form, such a knowledge fragment would describe various possible causes, a temporal relation, and an estimate of the degree of certainty, etc. Thus, the full meaning of the causal inference is composed of explicit and implicit parts:

1. *p is the cause of q, p occurs, therefore, q occurs (will occur) as the effect of p* (where q is connected with p by the relation of being the effect).
2. *q is the effect of p, q occurs, therefore p occurs (has occurred) as the cause of q* (where p is connected with q by the relation of being the cause).

Causal statements are usually formulated either in a pseudodeterministic wording or in a manner that reveals their temporal and/or certainty factors. A typical pseudodeterministic wording of a causal statement is exemplified by the above cited statement: *impaired myocardial perfusion is caused by a decrease in coronary blood flow*. The precise characteristic of the time factor depends on numerous conditions and may be expressed accordingly in various ways. In principle, however, it is contained (as an enthymematic component) in the causal relation itself (see above). A few remarks presented below concern the degree of certainty of causal statements; a more detailed analysis of the problems connected with the notion of probability in a biomedical context is presented in Sect. 3.

The necessary condition of the validity of a pseudodeterministic form of a causal statement is the tacit assumption of isolation of the phenomena described by it. When the variable influences of other phenomena are to be taken into consideration in an overt form, the statement should be supplemented with a modal qualifier, indicating that the effect or the cause of an event probably occurs (is probable) or may occur (is possible).

The degree of certainty of causal propositions is expressed in genuine texts by a lot of phrases. For example, various causes of ischemic heart disease are described in [11] as follows (with minor modifications, partly enumerated in a box): *Although atherosclerotic obstructive coronary artery disease is the most common cause of chronic myocardial ischemia, it can result also from coronary artery spasm, congenital coronary artery anomalies, coronary artery embolism. ...Systemic collagen vascular disease, extrinsic compression of the coronary arteries by tumors can infrequently be a cause of myocardial ischemia.* The same author describes possible effects in the form of manifestations and complications in the following manner: *Acute myocardial infarction may be the first evidence of ischemic heart disease.*

Infrequently, arrhythmias and congestive heart failure can be the predominant consequence of ischemic heart disease. However, angina pectoris is by far the most common clinical manifestation. Some authors mention a quantitative estimation, e.g., *atherosclerosis is the most frequent cause of coronary insufficiency; it appears in 90% of patients, in the remaining 10% it may be the consequence of abnormal blood flow in coronary arteries, coronary embolism or other non-atherosclerotic changes in coronary arteries* [24].

Various modal expressions appear in the above phrases. They denote the probability of the cause or effect (in a qualitative or an approximative quantitative manner) or possibility, e.g., p is the most common (frequent) cause of q, p is the cause of q in 90% of cases, p can infrequently be a cause of q, q can infrequently be the consequence of q, q is by far the most common manifestation (effect) of p, q can result from p.

The set of above statements signifying causal relations partly enriched by estimation of probability (by means of linguistic or numerical variables) may be considered a fragment of knowledge describing the connections of ischemic heart disease with other pathological phenomena. Its graphic representation is given in Fig. 3.

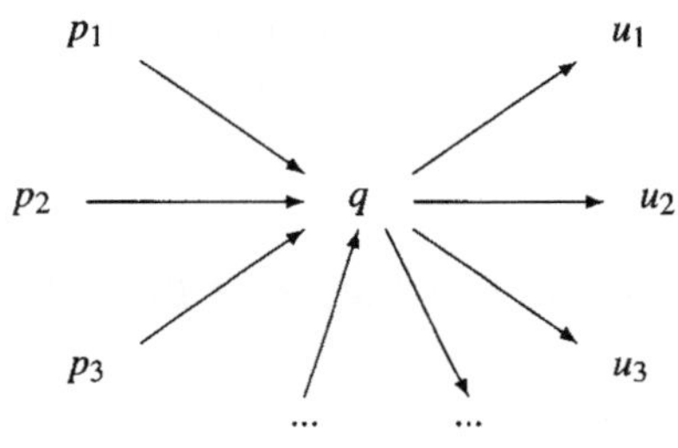

Fig. 3. Plurality of causes and effects (see text). q: ischemic heart disease; p_1: atherosclerotic obstructive coronary artery disease; p_2: coronary spasm; p_3: congenital coronary artery anomalies; u_1: angina pectoris (retrosternal pain); u_2: arrhythmias; u_3: myocardial infarction

The relations represented in Fig. 3 may be described as follows:

Causal form:

q is caused by p_1 and/or $p_2 \ldots$ and/or p_m.
s_1 and/or $s_2 \ldots$ and/or s_n are the effects of q.

Causal form enriched by probability estimation:

q is caused by p with probability π_i.
u_j is the effect of q with probability π_j.

Propositional form:

If p_1 and/or p_2 ... and/or p_m, then q.
If q, then earlier p_1 and/or p_2 ... and/or p_m.
If q, then u_1 and/or u_2 ... and/or u_n.
If u_1 and/or u_2 ... and/or u_n, then earlier q.

Propositional form enriched by probability estimation:

If p, then q with probability π_i.
If q, then u_j with probability π_j.

The degree of certainty (π) is estimated by means of an adjective (e.g., high, low), in a comparative manner (e.g., most probable), by a numerical value (e.g., 90%), etc. or by mentioning it as a possibility (without probabilistic qualification). The time factor is represented above by the adverb "earlier," and it may also be expressed more precisely.

The range of possible elementary problem solving processes based on the above (or similar) set of statements is great; the kind of particular inferences, however, depends in the first place on the type of links symbolized by the "and/or" functor that may be equivalent to a weak or strong alternative or conjunction.

In the above example, most of the elements of the sets of causes and effects are linked by weak alternatives which, however, for practical purposes are sometimes interpreted disjunctively.

3 Biomedical Aspects of the Notion of Probability

From a general point of view, the notion of probability is an even more complex and multifaceted problem in comparison with that of causality and since antiquity has been treated by philosophers, mathematicians, logicians, and thinkers from other fields. The problem of probability compared with that of causality, has greater direct impact on the development of the majority of sciences, from physics to sociology, and is more important from the viewpoint of practical applications. In biomedical sciences, various branches of mathematics related to probability are the cornerstones of research and clinical activity, and their role is rapidly increasing with the progress of applications of computers in medicine and the development of artificial intelligence (for a review, see [4]).

3.1 Logical Aspects of Probability in a Biomedical Context

A considerable part of biomedical knowledge is based on the results of statistical research and quantitative measurements. They constitute the very foundation of modern biological and medical sciences. For practical medical problem solving, however, probability based on estimation of relative frequencies, even if available, is not expressed mostly in numbers, but by means of linguistic variables expressed as modal qualifiers. Such is the usual manner in which the numerical estimates are introduced into the processes of natural thinking. It is in this way that the statements such as *if coronary arterial resistance is increased, then in 90% of patients myocardial perfusion is impaired,* are transformed into their qualitative counterparts, *if coronary arterial resistance is increased, then very probably myocardial perfusion is impaired.*

The remarks preceding the discussion of the notion of causality apply also to this analysis of the concept of probability in biomedical sciences. The following analysis is centered on those aspects of probability that are typical for biological and medical natural knowledge and reasoning (perhaps, however, some of these views might have also a more general bearing). In other words, this study concerns the problems of uncertainty as expressed in the professional medical language of everyday communication. It is based mainly on the analysis of biomedical texts that are easily accessible and are a reliable source of information, enriched by the results of behavioral observation. It is evident that this study, whose aim and methods are restricted by such research material, could contribute to a psychological or formal theory of medical problem solving only in an indirect way.

The majority of biological and biomedical laws describe events that are time-related and whose appearance is not certain or not quite determined, although in many scientific statements, time and uncertainty factors are not indicated in an explicit way. Another characteristic of typical descriptions of biological phenomena and biomedical problems is the fact that they are composed, to a considerable extent, of statements that link, in a general way, concrete values of features instead of describing the relations between sets of values, as, e.g., in physical sciences. Among various reasons for such an approach is the fact that in biomedical sciences an important role is played by qualitative rather than numerical scales. Hence, concrete statements (concerning various objects) formulated as conditional sentences, such as *if the value p of the feature P of the object O occurs, then the value q of the feature Q occurs* appear more frequently than the general laws stating that *the values of the feature Q of the object O depend in such and such a manner on the value of the feature P of this object.* The concrete character of biomedical knowledge is one of the principal causes for the necessity of using the concept of probability (frequently linked with the temporal aspects of phenomena) as a basic component of biomedical laws as well as the statements describing particular facts.

Similarly, as in other sciences and in everyday life, in medicine the notion of probability denotes, apart from a strictly formal way of understanding, the degree of anticipation of the appearance (presence) of an event (phenomenon) founded on past experience (frequency) and/or rational justification. The sentence, *if p occurs, then q occurs (or will probably occur),* means (in an extreme simplification) that on the grounds of the occurrence of p, we expect the occurrence of q rather than $not(q)$. In the context of this work, the expressions denoted by p and q mean either purely qualitative states (nominal values) or intervals of magnitudes distinguished on a numerical scale, as expressed in the form of linguistic variables. They represent either a qualitative description of basically quantitative values or an attempt to make basically qualitative ones more exact. In both cases, in every probabilistic statement (in the above meaning,) appears an exact "core" surrounded by a "shadow" of uncertainty (see below).

Sometimes, we have at our disposal relevant statistical evidence on whose grounds it is possible to assess probability quantitatively. For, however, a biomedical context, such a possibility is far from common not only due to the lack of appropriate data, but also because the relevant conditions of applicability frequently are not fulfilled. Also, in most cases, the quantitative probability does not represent the essence of biomedical problems and only partly meets the needs of reasoning processes.

The second qualitative aspect of the notion of probability is related to the logical grounds on which it is estimated, i.e., on the evidence furnished by a certain set of general, already accepted statements or a fragment of general biomedical knowledge as a whole. In medicine, as well as in biology and perhaps in other sciences, such as psychology and sociology, the notion of the probability of a statement (hypothesis) is related to the totality of evidence that speaks in favor or against it. Thus, the notion of probability connected with biomedical sciences seems to be closer to the logical than to purely numerical interpretation. According to this approach, *probability measures the degree of confidence that would be rationally justified by the available evidence* (Salmon [38]). The definition given by Ajdukiewicz [1] runs, *the logical probability of a statement A relative to a statement B is the highest degree of certainty of acceptance of the statement A to which we are entitled by a fully certain and valid acceptance of the statement B* and stresses that *the condition of conclusiveness of an inference is relative to the premises and the body of knowledge*. It should be underlined that in this interpretation, probability is an attribute of the statements describing the events and not of the event described by these statements. Although Carnap distinguished various kinds of probability [7], according to Ajdukiewicz [1] and Reichenbach [35], it is possible that logical and mathematical probability have a common basis; this problem, however, exceeds by far the question of probability as used in biomedicine. The psychological aspects of probability are outside the scope of this chapter; in this context, however, note that the qualitative assessment of probability was studied in connection with probabilistic networks [42].

3.2 Justification of Uncertain Biomedical Statements

The probability of a statement describing biological or pathological phenomena, understood as the degree of its general justification, depends on various factors. The most important are (1) the quantitative characteristic of the description of the phenomena, (2) the semantic aspects of statements, (3) the interconnections of various phenomena with other ones, and (4) the place of a given statement in a certain fragment of the system of science.

1. The quantitative characteristics of phenomena are related to the distributions of continuous or discrete values that represent the direct or indirect results of measurements and observations performed in the frame of various experiments. Due to the variability inseparably connected with biological and pathological phenomena, the classes of values and objects are unsharp and overlap more or less considerably. Therefore, it is commonly impossible to ascribe a particular case (value or object) to a given class with absolute certainty, but only with a certain probability, which may be expressed as a numerical magnitude or as a linguistic variable. The latter approach is useful because natural reasoning (in which it is normally involved) is not performed on numbers. Moreover, the statistical data describe external data rather than the "inside" of phenomena and represent only one of many facets of the reality, frequently not the most significant. The numerical probability, when available and reliable, is an important component of the estimation of the degree of certainty of judgments only when it is integrated into the general bulk of knowledge, i.e., regarded as one of the pieces of evidence.
2. Typical general statements that are considered in this work are conditional sentences composed of objects (in a grammatical sense) and predicates containing relational and adjectival parts. The range of denotation of the majority of these expressions is frequently broad, i.e., they describe classes composed of more or less numerous kinds of objects, relations, and features. Moreover, their meaning is usually ambiguous and unsharp, and the sentences are mostly formulated so that an important part of their meaning is not expressed in an explicit way. It is evident that the statements exhibiting such a semantic characteristic cannot contain precise expressions denoting the uncertainty factor.
3. Biomedical phenomena appear as elements of other phenomena and, accordingly, the statements that describe them constitute the components of complex systems of laws. Their probability, therefore, depends to a considerable extent on their connections with various parts of a greater whole. Although a "core probability" of a statement, i.e., its probability considered in abstraction from the broader environment, is relatively stable, it is surrounded by a shadow of uncertainty due to the fact that in different situations, the phenomena are subject to various influences. Obviously, the variability of phenomena, e.g., of the values of features of the organism or of its different parts, is more conspicuous in the state of disease than in normal conditions. Therefore, reasoning, whose aim is estimation of the probability of pathological phenomena, i.e., their identification, including the degree of certainty, is especially typical for the process of diagnosis.
4. From the statistical point of view, the justification of a statement consists in performing mathematical operations on numerical magnitudes. On the other hand, in a biomedical context, the majority of premises are expressed not by numerical values but by linguistic variables, and the way of arriving at the conclusions

is not a calculus but operations on the judgments of another kind. On the other hand, a strictly formal approach to qualitative (linguistic) expressions, which are the basic units of biomedical knowledge and reasoning, may be used only with considerable limitations. The logical structure of biomedical statements, general as well as particular, constitutes a "deterministic skeleton" of their meaning that is surrounded by an informal "flesh." The answer to the question whether, or with what certainty, a given statement should be accepted, may be arrived at by proving or justifying it by using various methods of inference making, i.e., making a logical analysis on various levels of knowledge. By "logical approach," I mean here an analysis as detailed, ordered, and precise as possible and useful, rather than the application of a formal system.

3.3 Typical Expressions Denoting Probability and Their Interpretation

In natural, professional medical language, various levels of probability are designated by a great variety of expressions, e.g., "in most (some) cases," "sometimes," "frequently (seldom, exceptionally)," "probably," "very likely," "something may (can) occur," "something is a probable cause (effect)," "something is supposed to be," etc. The meaning of these and similar expressions is ambiguous and unsharp to the same extent as those that are used in everyday language. However, experience with the verbal usage of doctors and the study of professional texts and other sources suggest that the following main types of medical subjective probability may be distinguished: formal certainty, quasi-certainty, high probability, intermediate probability, low probability, and possibility.

The above types may be illustrated by the following sentences; it should be remembered, however, that their sense overlaps to a considerable extent.

Formal certainty: blood flow is either normal or decreased (still, these ranges are unsharp).

Quasi-certainty: if the artery is constricted (and blood pressure is not increased and so on), then almost certainly blood flow in this artery is decreased.

High probability: if the patient complains of retrosternal pain, then very probably this patient suffers from ischemic heart disease.

Intermediate probability: if the patient suffers from atherosclerosis, this patient probably displays an elevated level of a certain fraction of cholesterol.

Low probability: the probability that the patient suffering from mild myocardial ischemia will develop severe heart failure is small.

Possibility: the patient suffering from myocardial ischemia may display a normal electrocardiographic recording.

Although the majority of doctors will probably agree with the above examples, nobody would be inclined to consider them exact. Sometimes, however, similar statements, or rather their meanings, appear as important elements in the decision making process.

The above classification embraces qualitative degrees of probability, i.e., is related to a discrete scale. Although it may be considered a "strong" type of probability estimation (probabilistic verification), not less important is a "weaker" type of the assessment of probability, namely, (1) comparison of the probability of various statements and (2) comparison of the probability of the same statement in the light of different pieces of evidence, i.e., its confirmation.

1. The former type, i.e., *verification of hypotheses*, consists of indicating which of the statements describing the two values of the same feature is more probable. For example, the judgment stating that the patient suffering from atherosclerosis probably displays an elevated blood level of a certain fraction of cholesterol is more probable than the statement suggesting that this patient will display a normal or diminished level of this compound.
2. The notion of *confirmation* of statements is based on the assumption that if new (supplementary), previously unknown or neglected, evidence is taken into account, then the probability of a given statement increases or decreases in comparison with that estimated before this information has been obtained or considered. For example, the probability of the statement, *if the patient suffers from atherosclerosis and retrosternal pain, then this patient suffers from myocardial ischemia,* is higher in comparison with the statement, *if the patient suffers from atherosclerosis, then this patient suffers from myocardial ischemia.* The problems related to the confirmation of hypotheses play especially important roles in the approach presented in this work and are analyzed in more detail in the next paragraph.

The above considerations may be summarized as follows:

The notion of probability has several meanings that depend on the context in which it is used by the physicians and/or appears in biomedical texts. On the grounds of general usage, several qualitative degrees of uncertainty (probability) may be distinguished. In practice, they are well understood and discriminated by the doctors such a categorization, however, may serve mainly to introduce a certain order (of a rather soft kind) rather than using them as a basis for a precise scheme of inference. The interpretation of the way of understanding the notion of the probability of a statement (of a phenomenon) may be either strong or weak. The strong meaning corresponds to the qualitative interval scale or a nominal scale and is expressed by adjectival expressions such as high, medium, low, almost equal to certainty (positive or negative), etc. The weak interpretation is based on a comparative scale; this category embraces the confirmation of hypotheses.

The probability of a statement depends on its justification on the grounds of certain knowledge (general and particular), i.e., on the kind, strength, and amount of evidence. In some cases the evidence exists that may constitute sufficient basis for a strong kind of probability assessment (including practical certainty), in other situations, only a weak assessment is justified on the grounds of existing evidence. In the latter case, a certain number of "weak" pieces of evidence may be sufficient for a strong assessment. This is important point since probability is one of the factors on which the process of decision making is based, and for a decision concerning an action, frequently, a strong type of its assessment, instead of a weak one, is necessary. Commonly, in the absence of a single strong piece of evidence, the decision making subject is compelled to use a set of weak pieces of evidence to arrive at a sufficient estimate of probability.

The appearance of a phenomenon or the acceptance of a statement may have a different influence on the probability of a given phenomenon (statement); in other words, the revision of probability may be greater or smaller. In clinical practice we say that the symptoms and signs are deciding, pathognomic, strong, weak etc. and a similar "strength" of every evidence may be distinguished. Thus, when a series of observations is performed, the probability of a given phenomenon (statement) undergoes a sequence of changes depending on the number of pieces of evidence as well as on their confirmatory strength.

3.4 Confirmation of Hypotheses: A Theoretical Model

The assignment of a certain value of probability to a hypothesis may be called *probabilistic verification*, whereas the estimation of the direction and approximative magnitude of the change (revision) of probability of a hypothesis is called confirmation [8]. As the latter type of probability evaluation seems to be especially important for a biomedical problem solving theory, especially in a model approach, it deserves a more detailed presentation.

In research on the fundamental problems of the methodology of sciences, the authors [1,8,38] frequently use a notation that shows the connection of the probability of statements with general knowledge. It is assumed that an analogy exists between the estimate of numerical and logical probability that justifies the similarity of these approaches. According to this usage, the probability of the hypothesis h in the light of general knowledge k may be designated by symbol $p(h/k)$, and its probability in the light of general knowledge k and evidence e by the symbol $p(h/k \text{ and } e)$. Usually, the symbol p may be interpreted as denoting either quantitative or qualitative probability. In the present analysis, I use the symbol π for probability based on various kinds of justification (logical as well as quantitative), and a piece of evidence here is designated by the letter s (from sign), h and k denoting the hypothesis and general knowledge, respectively.

According to my previous work [13] and in a way similar to the usage of this term by above mentioned authors, the *confirmation* or *degree of confirmation* $c(h,k,s)$ of the hypothesis h is equal to its probability, as estimated in the light of general knowledge k and observation (acceptance) of novel evidence (e.g., sign s) divided by the probability of this hypothesis h estimated without considering this evidence. Hence,

$$c(h,k,s) = \frac{\pi(h/k \text{ and } s)}{\pi(h/k)}.$$

For two alternative and mutually exclusive hypotheses h_1 and h_2 [i.e., h_2 is $not(h_1)$], it may be shown (on the grounds of the Bayes formula) that if we accept the evidence s, whose probability is higher on the grounds of hypothesis h_1 than on the grounds of hypothesis h_2 (the evidence s speaks in favor of hypothesis h_1 and against hypothesis h_2 or, using statistical terminology, the likelihood of h_1 in the light of s is greater than the likelihood of h_2 in the light of s, or), then the probability of hypothesis h_1 increases, and that of the hypothesis h_2 decreases.

The following conditions are equivalent:

1. $\pi(s/k \text{ and } h_1) > \pi(s/k \text{ and } h_2)$.
2. $c(h_1,k,s) > c(h_2,k,s)$.
3. $\pi(h_1/k \text{ and } s) > \pi(h_1/k)$.
4. $c(h_1,k,s) > 1$.
5. $\pi(h_2/k \text{ and } s) < \pi(h_2/k)$.
6. $c(h_2,k,s) < 1$.

Hence, in particular, the following implications hold:

- If $\pi(s/k \text{ and } h_1) > \pi(s/k \text{ and } h_2)$ then $c(h_1,k,s) > c(h_2,k,s)$.

- If $\pi(s/k \text{ and } h_1) > \pi(s/k \text{ and } h_2)$ then $\pi(h_1/k \text{ and } s) > \pi(h_1/k)$.

- If $\pi(s/k \text{ and } h_1) > \pi(s/k \text{ and } h_2)$ then $c(h_1,k,s) > 1$. In other words, in the situation described by any of conditions (1)–(6), confirmation of hypothesis h_1 is greater than one, and the confirmation of the hypothesis h_2 is less than one (evidence s strengthens hypothesis h_1 and weakens hypothesis h_2). The above conclusion is valid irrespective of the initial probability of hypotheses h_1 and h_2. In other words, to estimate the direction of the change (revision) in the probability of a hypothesis, we do not need to know its initial probability. The magnitude, however, of the change in probabilities in the light of evidence s depends on the initial probabilities of the hypotheses (for hypothesis h_i it is the greater, the smaller the initial probability of this hypothesis), as well as on the magnitude of likelihood values.

Analogously, one can obtain:

$$\text{if } \pi(s/k \text{ and } h_1) < \pi(s/k \text{ and } h_2), \text{ then } c(h_2,k,s) < 1.$$

Every empirical examination, scientific as well as practical, consists of gathering successively the pieces of evidence (e.g., signs s_1, s_2, and so on) that are relevant, novel, and independent (or at least are assumed to be independent). The greater the number of observed pieces of evidence (signs) that speak in favor of (suggest) a given hypothesis, the greater the degree of its confirmation. Theoretical aspects of this problem are discussed by Ajdukiewicz [1] and Caws [8].

Assuming, s_1, s_2 are conditionally independent (with respect to conditions described by k and h) and $\pi(h/k \text{ and } s_1) > \pi(h/k)$ as well as $\pi(h/k \text{ and } s_2) > \pi(h/k)$, one can show that the probability of hypothesis h in the light of general knowledge k and the pieces of evidence s_1 and s_2 satisfies the following inequality:

$$\pi[h/(k \text{ and } s_1 \text{ and } s_2)] > max[\pi(h/k \text{ and } s_1), \pi(h/k \text{ and } s_2)].$$

The above concept of probability revision in the light of growing evidence constitutes the basis of the semantic model of hypothesis verification presented in this work.

3.5 An Example of the Semantic Model (a General Presentation)

The subject (e.g., a doctor) verifies (in a broad sense) the hypothesis h_1 versus an alternative hypothesis h_2 [or *not*(h)] in the light of a set of accessible pieces of evidence or signs s_1, s_2, and so on. On the grounds of relevant general knowledge, the hypotheses are a priori neither excluded nor certain. Some of the signs confirm hypothesis h_1 (and disconfirm h_2), others confirm hypothesis h_2 (and disconfirm h_1). The signs have various confirmational strength, i.e., the degree of confirmation of each hypothesis in the light of a given sign may be weak or strong. In other words, every sign confirms hypothesis h_1 or h_2 weakly or strongly. It is assumed that the weak and strong confirmation mean that the change in probability is equal to one or two confirmational steps, respectively. The subject performs the appropriate examination, i.e., successively observes (or accepts on theoretical grounds) the occurrence of signs and intends to assess the direction of revision of the probability of hypotheses to decide which hypothesis is more probable than its counterpart (to make a weak verification), or to assign to the hypotheses certain values of probability (to verify them in a strong sense). If all observed signs or their majority confirm the same hypothesis and the number of signs as well as their strength are considerable, then the probability of this hypothesis unequivocally increases, and one of them may be verified or even considered practically certain in the light of a given set of signs. If, however, the observed signs don't confirm (weakly or strongly) the same hypothesis, the examination permits only (in the best case) choosing the hypothesis whose probability is higher in the light of the set of signs (weak verification).

Let us consider two possible situations: at first, two signs s_1 and s_2 having equal strength, secondly, signs of various strengths.

1. *Signs of equal strength:*
   (a) Both signs confirm hypothesis h_1 and disconfirm h_2, i.e.,
   $\pi(s_1/k \text{ and } h_1) > \pi(s_1/k \text{ and } h_2)$,
   $\pi(s_2/k \text{ and } h_1) > \pi(s_2/k \text{ and } h_2)$,
   where k is the relevant general knowledge.
   Therefore,
   $\pi(h_1/k \text{ and } s_1) > \pi(h_1/k)$,
   $\pi(h_1/k \text{ and } s_2) > \pi(h_1/k)$,
   $\pi(h_1/k \text{ and } s_1 \text{ and } s_2) > max\,[\pi(h_1/k \text{ and } s1), \pi(h_1/k \text{ and } s1)]$.
   (b) s_1 confirms h_1, s_2 confirms h_2, i.e.,
   $\pi(s_1/k \text{ and } h_1) > \pi(s_1/k \text{ and } h_2), \pi(s_2/k \text{ and } h_1) < \pi(s_2/k \text{ and } h_2)$.
   Therefore,
   $\pi(h_1/k \text{ and } s_1) > \pi(h_1/k)$,
   $\pi(h_1/k \text{ and } s_2) < \pi(h_1/k)$,
   in view of various directions of confirmation of hypotheses in the light of $s_1 ands_2$, the overall confirmation cannot be determined.
2. *Signs of various strengths:*
   We assume that the confirmation of h_1 in the light of s_1 is weak and the confirmation of h_1 in the light of s_2 is strong:
   $\pi(s_1/k \text{ and } h_1) > \pi(s_1/k \text{ and } h_2), \pi(s_2/k \text{ and } h_1) >> \pi(s_2/k \text{ and } h_2)$.
   Therefore,
   $\pi(h_1/k \text{ and } s_1) > \pi(h_1/k)$,
   $\pi(h_1/k \text{ and } s_2) >> \pi(h_2/k)$.
   $\pi(h_1/k \text{ and } s_1 \text{ and } s_2) >> \pi(h1/k \text{ and } s_1 \text{ or } s_2)$.

Similarly, the observation of a greater number of signs possibly exhibiting various degrees of strength may be interpreted from the point of view of the confirmation and verification of hypotheses. For some purposes (see Sect. 5) an alternative notation of the direction and strength of confirmation may be useful, namely, $\uparrow$ or $\uparrow\uparrow$, etc., for a weak and strong positive confirmation and $\downarrow$ or $\downarrow\downarrow$, etc. for a weak and strong negative confirmation, respectively.

The problem of the qualitative influence of signs in the frame of probabilistic networks was discussed by Wellman [42] and Druzdzel [18, 19].

The application of the above described model to the analysis of an authentic clinical case is presented in Sect. 5.

4 Analysis of an Original Text (Ischemic Heart Disease) Using the Semantic Model

The application of the above described approach to the analysis of a clinical text is presented here using as an example a fragment of the chapter entitled "Ischemic heart disease" (K. Chatterjee) from the textbook edited by J.H. Stein [11]. The statements form partly a continuous text, are partly cited from different places, and are only slightly abbreviated. Here and there the wording is insignificantly simplified (numbering is inserted for the use in the discussion).

1. *Obstructive coronary artery disease caused by atherosclerosis is the most common cause of chronic ischemic heart disease.*
2. *Ischemic heart disease may be totally silent or may manifest in angina, arrhythmias, and heart failure.*
3. *Myocardial ischemia stems from the imbalance between myocardial oxygen requirements and oxygen supply which can occur from a primary decrease in coronary blood flow or from an increase in myocardial oxygen requirements or their combination.*
4. *Impaired myocardial perfusion, caused by a decrease in coronary blood flow resulting from an increase in the coronary arterial resistance and/or abnormalities of the coronary vascular autoregulatory mechanisms, appears to be the principal cause for myocardial ischemia in other clinical syndromes.*
5. *Angina pectoris in by far the most common clinical manifestation of ischemic heart disease.*
6. *Downsloping electrocardiographic S-T segments are highly specific for coronary artery disease.*
7. *The definitive diagnosis of coronary artery disease can be made only by coronary angiography.*

4.1 Causal Paraphrases of the Original Statements

In the sentences below, the names of phenomena (complexes object, feature, value) are distinguished by italics and marked by uppercase letters for use in Fig. 4a.

1. The most probable cause of *chronic ischemic heart disease* (E) is *atherosclerotic obstructive coronary artery disease* (L).
2. *Angina* (H), *arrhythmias* (J), and *heart failure* (K) may be the effects of *ischemic heart disease* (E).
3. *Myocardial ischemia* (E) may be caused by a *decrease in coronary blood flow* (B) and/or from an *increase in myocardial oxygen requirements* (F).
4. *Myocardial ischemia* (E) is caused by *impaired myocardial perfusion* (A).
5. *Impaired myocardial perfusion* (A) is caused by a *decrease in coronary blood flow* (B).

6. *Decrease in coronary blood flow* (B) is caused by an *increase in coronary arterial resistance* (C) and/or *abnormalities of coronary vascular autoregulatory mechanisms* (D).
7. *Angina pectoris* (M) is the most probable effect of *ischemic heart disease* (E).
8. Appearance of *downsloping electrocardiographic S–T segments* (N) is a very probable effect of *coronary artery disease* (E).
9. *Coronary artery disease* (E) causes the *appearance of a specific image in coronary angiography* (T).

The fragment of knowledge composed of the above causal statements may be represented as a directed graph, as in Fig. 4a.

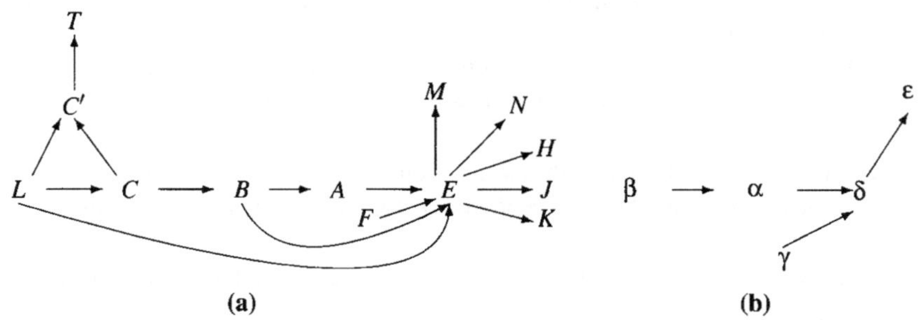

Fig. 4. (a) Graphic representation of a causally paraphrased fragment of knowledge (see text). Letters correspond to phrases designating the value of a feature of an enthymematic object; e.g., E : "myocardial ischemia," i.e., the phrase meaning a decreased value of the blood supply (feature) to the myocardium (heart muscle, object). Arrows denote causal relations described by the phrase "is the cause (effect) of" in which time and probability factors are enthymematically comprised. Long arrows represent shortcut relations mentioned in the original text. The element C' means "atheromatic constriction of the coronary arteries" absent from the original text and introduced here to ensure appropriate links among other elements.
(b) Arrows denote functional relations (phrase "depends on", see text).

4.2 Paraphrases of Functional Laws

According to the natural understanding of the sentences contained in the original text, they constitute the concretization of more comprehensive (general, functional) laws describing the relation linking whole features instead of their particular values. The following statements represent the general laws that embraced some of the concrete statements analyzed above (Greek letters are used in Fig. 4b).

1. Myocardial perfusion (α) depends on coronary blood flow (β).

2. Myocardial oxygen supply (δ) depends on myocardial perfusion (α) and/or myocardial oxygen requirements (γ).
3. Retrosternal pain sensations (ε) depend on myocardial oxygen supply (ε).

4.3 Propositional Paraphrases (Conditional Sentences)

To present the propositional form, three statements are represented below in the form of conditional sentences; in this case, the time and probability factors are added or preserved (in the original texts, they are commonly omitted).

1. If in time t coronary blood flow is decreased (B), then in time $t' \geqslant t$, myocardial perfusion is probably impaired (A).
2. If in time t, myocardial perfusion is impaired (A) and/or in time t, myocardial oxygen requirements are increased (F), then in time $t' \geqslant t$, myocardial ischemia (E) probably occurs.
3. If in time t, myocardial blood supply is decreased (i.e., myocardial ischemia occurs, E), then in time $t = t'$, retrosternal pain occurs (H).

The regressive form based on the assumption of isolation of the above statements is the following (the time factor is rendered by typical linguistic expressions):

1. If myocardial perfusion is impaired (A), coronary blood flow has probably been decreased (B).
2. If myocardial ischemia (E) occurs, then myocardial perfusion has probably been impaired (A), and/or myocardial oxygen requirements have probably been increased (F).
3. If retrosternal pain occurs (H), then myocardial blood supplies probably are decreased (i.e., myocardial ischemia (E) probably occurs).

The following sentences are complementary for the above statements (1)–(3):

1. If coronary blood flow is normal (B′), then myocardial perfusion is probably normal (A′).
2. If myocardial perfusion is normal (A′) and/or myocardial oxygen requirements are increased (F′), then myocardial blood supply is normal (i.e., myocardial ischemia does not occur, E′).
3. If myocardial blood supply is normal (i.e., myocardial ischemia does not occur, E′), retrosternal pain does not occur (H′).

Propositional paraphrases may be used as a basis for a pathophysiological inference.

Progressive inference:

1. In time t, coronary blood flow is decreased (B); therefore, in time $t' \geqslant t$, myocardial perfusion is probably impaired (A).

2. In time t, myocardial perfusion is impaired (A); therefore, in time $t' \geqslant t$, myocardial ischemia (E) probably occurs.
3. In time t, myocardial oxygen requirements are increased (F); therefore, in time $t' \geqslant t$, myocardial ischemia (E) probably occurs.

Regressive inference (with the assumption of isolation):

1. Myocardial perfusion is impaired (A); therefore, coronary blood flow has probably decreased (B).
2. Myocardial ischemia (E) occurs; therefore, myocardial perfusion has probably been impaired (A) and/or myocardial oxygen requirements have probably increased (F).

The diagnostic inferences based on propositional paraphrases may be exemplified as follows.

1. *Progressive form:*
   In time t, myocardial blood supply is decreased (i.e., myocardial ischemia occurs, E); then in time $t = t'$, retrosternal pain (will) occurs (H).
2. *Regressive:*
   Retrosternal pain occurs (H); therefore, myocardial blood supply probably is decreased (i.e., myocardial ischemia probably occurs, E).

5 Analysis of an Authentic Clinical Case Using the Semantic Model

The present clinical example is based on the following text extracted from a document concerning the treatment of a patient during his stay in the clinical department (epicrisis).

"66-year-old patient was admitted to the Coronary Care Unit with a severe retrosternal pain of three hour duration. Acute ischemic heart disease with a possibility of myocardial infarction was diagnosed. The chest pain resolved after morphine and the treatment with subcutaneous low molecular heparine, aspirine and intravenous nitrates was administrated. The diagnosis of myocardial infarction was ruled out based on a creatine kinase level estimation and subsequent electrocardiographic recordings. Urgent coronary angiogram was performed which showed total occlusion of the left anterior descending artery and two lesions of the proximal left circumflex artery (stenosis of 60 and 90%). Percutaneous transluminar coronary angioplasty was performed."

The above document contains the description of signs exhibited by the patient, diagnostic hypotheses (statements), and therapeutic procedures. The last component is presented only to underline the inseparable connection between medical cognition and decision making. The signs (partly completed on the grounds of the context) are the following: retrosternal pain, normal blood level of creatine kinase, ECG recordings not characteristic for myocardial infarction, and a coronarographic image characteristic of coronary stenosis. The diagnostic hypotheses embraced acute ischemic heart disease and myocardial infarct. In other words, the epicrisis contains (apart from therapeutic information) the premises and conclusions of the doctor's hypothetical reasoning.

According to the model presented in this work, in the process of solving a biomedical problem, e.g., a diagnostic one, the final steps are based on relatively simple propositions (working statements) derived from statements belonging to general knowledge that have various forms and exhibit different levels of generality and degree of precision. In a model approach, the connection of the general (theoretical) statements with the working propositions may be represented as a sequence of paraphrases of original, e.g., textbook, sentences in which they are transformed into typical causal, propositional, and working statements (see Sect. 1). For example, an original textbook formulation: "angina pectoris (retrosternal pain) is by far the most common manifestation of ischemic heart disease" [11], may be expressed as a causal proposition: "the most probable cause of retrosternal pain is ischemic heart disease," which in its turn is approximately equivalent to the propositional paraphrase "if myocardial ischemia occurs, then probably retrosternal pain occurs." Finally as a working proposition, "if retrosternal pain occurs, then probably myocardial ischemia occurs."

A general working proposition in conjunction with the particular statement "retrosternal pain occurs," is the basis for the conclusion "therefore, probably myocardial ischemia occurs."

Similarly other working statements may be derived and applied in the inference-making process.

If myocardial ischemia occurs, then, myocardial infarction is possible (the same as a propositional paraphrase).

If blood CK is normal, then, very probably myocardial infarction does not occur.

If the ECG curve does not exhibit typical features, then, myocardial infarction very probable does not appear.

If myocardial ischemia occurs, then, very probably the coronary artery is constricted.

If a coronary angiogram exhibit typical features, then, the coronary artery is almost certainly constricted.

On the grounds of the above working statements and the statements describing the signs presented by the patient, the following conclusions are drawn:

1. Severe retrosternal pain occurs; therefore, very probably the patient suffers from ischemic heart disease, possibly with myocardial infarction.
2. Assuming that ischemic heart disease occurs, myocardial infarction is possible.
3. Creatine kinase (CK) level is normal; therefore, probably myocardial infarction does not occur.
4. Electrocardiographic (ECG) recordings are normal; therefore, probably myocardial infarction does not occur.
5. Assuming that ischemic heart disease is present, obstruction of the coronary artery is very probable.
6. Occlusion and stenosis of coronary arteries appear in the angiogram; therefore, obstruction of the coronary artery is practically certain.
7. Considering that the obstruction of the coronary artery is almost certain, therefore, ischemic heart disease is very probable.

The final solution of the problem may be formulated as follows: the hypothesis of ischemic heart disease is accepted (ischemic heart disease is certain), and the hypothesis of myocardial infarction is rejected (myocardial infarction is improbable).

It should be stressed that each of both diagnostic hypotheses, i.e., that of ischemic heart disease, as well as that of the absence of myocardial infarction, is supported by two signs: "retrosternal pain" and "occlusion of coronary arteries in the angiogram" as concerns the former, and "normal CK level" and "normal ECG recordings," as concerns the latter. Therefore, it seems justified to estimate the resultant probability of these statements as "almost certain," despite the fact that in the light of the signs taken separately, the probability of these hypotheses would rather be "very high."

The clinical document, the extract of which is being analyzed here, presents the results of the patient's examination without mentioning the obvious fact that the appropriate tests were previously designed. It seems worthwhile, however, to complete the present study by indicating the logical grounds for the choice of tests.

The reasoning connected with the planning of the diagnostic examination is approximately the following. For example, the doctor suspects myocardial infarction and wishes to verify this hypothesis. He or she knows that if the patient suffers from this condition, the CK blood level would very probably be elevated. If myocardial infarction does not appear, the CK blood level would very probably be normal. In the former case, he or she would accept (with appropriate probability) the hypothesis of myocardial infarction, in the latter case he or she would reject it. Thus, the doctor takes into consideration two statements: "if myocardial infarction occurs, then the blood level of CK is very probably elevated" and "if myocardial does not occur, then probably the blood CK level is normal," which play the role of working hypotheses. The conclusion is based on hypothetical reasoning: assuming that myocardial infarction is present, it should be supposed that the blood CK level is elevated, on the contrary, assuming that myocardial infarction is absent, it should be supposed that the blood CK level is normal. It is easy to see that in the process of planning examinations, the working statements are equivalent to the propositional paraphrase; in other words, as concerns the planning reasoning, the statements' transformation may be simpler in comparison with the interpretation of already observed signs. On the other hand, to decide which test is necessary for proving a given hypothesis, at least two statements, instead of only one, should be considered.

Below is a list of general statements (and their interrelations) on which the process of solving the problem under consideration could be based, together with some examples of inferences. General statements, as well as particular premises and conclusions, are artificially reconstructed on the grounds of the semantic model described in this work. The list embraces functional laws and statements formulated in causal and propositional form; the model inference is presented according to the confirmational approach (see Sects. 3.4 and 3.5).

Functional Laws (see Fig. 5)

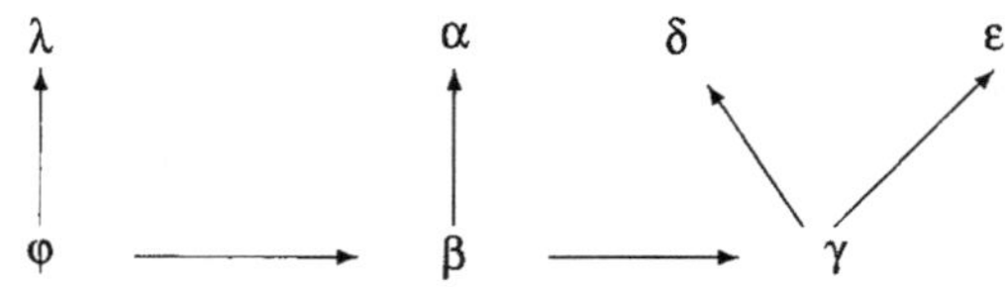

Fig. 5. Scheme of functional laws (see text)

1. Retrosternal pain sensations (α) depend on myocardial oxygen supply (β).
2. The morphological state of the heart muscle cells (γ) depends on the myocardial oxygen supply (β).
3. The level of blood CK (δ) depends on the morphological state of the heart muscle cells (γ).
4. Some features of the ECG curve (ε) depend on the state of the heart muscle cells (γ).
5. The myocardial oxygen supply (β) depends on the morphological features of the coronary arteries (φ).
6. The features of the coronary angiogram (λ) depend on the morphological features of the coronary arteries (φ).

Causal Statements

1. Myocardial ischemia (E) is the (a probable) cause of retrosternal pain (M).
2. Myocardial infarction (P) is the effect (a probable, possible effect) of myocardial ischemia (E).
3. Myocardial infarction (P) is the cause of the elevated blood CK level (U).
4. Myocardial infarction (P) is the cause of the appearance on the ECG curve of features typical of this condition (V).
5. Constriction of coronary arteries (W) is the (a probable) cause of myocardial ischemia (E).
6. Constriction of coronary arteries (W) is the cause of characteristic features in the coronary angiogram (Z).

The fragment of knowledge (general functional laws) on which the above problem and its solution are based may be represented in graphic form (Fig. 5).

A list of propositional probabilistic paraphrases follows:

1. If myocardial ischemia occurs (E), then, probably retrosternal pain occurs (M).
2. If retrosternal pain (M) occurs, then, probably myocardial ischemia occurs (E).
3. If myocardial ischemia occurs (E), then, myocardial infarction is possible (P).
4. If myocardial infarction occurs (P), then, myocardial ischemia certainly occurs (E).
5. If myocardial infarction occurs (P), then, probably the blood CK level is elevated (U).
6. If the blood CK level is elevated (U), then, probably myocardial infarction occurs (P).
7. If myocardial infarction does not occur (P′), then, probably the blood CK level is normal (U).
8. If blood CK level is normal (U′), then, probably myocardial infarction does not occur (P′)
9. If myocardial infarction occurs (P), then, almost certainly the ECG curve exhibits features (V) typical of myocardial infarction.
10. If the ECG curve exhibits typical features (V), then, myocardial infarction very probably occurs (P′).
11. If myocardial infarction does not occur (P′), then almost certainly the ECG curve typical (of myocardial infarction) does not appear (V′).
12. If the ECG curve does not exhibit typical features (V′), then, myocardial infarction very probably does not occur (P′).
13. If constriction of coronary arteries occurs (W), then, probably myocardial ischemia occurs (E).
14. If myocardial ischemia occurs (E), then, constriction of coronary arteries probably occurs (W).
15. If constriction of coronary arteries occurs (W), then, almost certainly characteristic features appear in the coronary angiogram (Z).

16. If characteristic features appear in the coronary angiogram (Z), then, almost certainly constriction of coronary arteries occurs (W).

Diagnostic conclusions (interpretation of signs) based on the above list of propositional probabilistic paraphrases:

17. Retrosternal pain (M) occurs; therefore, probably myocardial ischemia occurs (E).
18. Myocardial ischemia occurs (E); therefore, myocardial infarction is possible (P).
19. The blood CK level is normal (U′); therefore, probably myocardial infarction does not occur (P′).
20. The ECG curve does not exhibit typical features (V′); therefore, myocardial infarction very probably does not occur (P′).
21. Characteristic features appear in the coronary angiogram (Z); therefore, almost certainly constriction of coronary arteries occurs (W).

Pathophysiological inference based on the above list of propositional probabilistic paraphrases 13, 14:

22. Regressive conclusion: assuming that myocardial ischemia occurs (E); therefore, probably constriction of coronary arteries occurs (W).
23. Progressive conclusion: assuming that constriction of coronary arteries occurs (W); therefore, myocardial ischemia occurs (E).

Confirmational model based on the above list of propositional probabilistic paraphrases.

24. Retrosternal pain (M) strongly suggests myocardial ischemia (E)
    If M, then, $\uparrow\uparrow E$ or $\downarrow\downarrow$E′.
25. Myocardial ischemia (E) suggests the possibility of myocardial infarction (P).
    If E, then, $\uparrow$P or $\downarrow$P′.
26. The normal blood level of CK (U′) strongly suggests the absence of myocardial infarction (P′).
    If U′, then $\uparrow\uparrow$P′ or $\downarrow\downarrow$P.
27. The absence of features typical of the myocardial infarction in the ECG curve (V′) strongly suggests absence of myocardial infarction (P′).
    If V′, then, $\uparrow$P′ or $\downarrow$P.
28. Constriction of coronary arteries (W) strongly suggests myocardial ischemia (E).
    If W, then, $\uparrow\uparrow$E or $\downarrow\downarrow$E′.
29. Myocardial ischemia (E) strongly suggests constriction of coronary arteries (W).
    If E, then, $\uparrow\uparrow$W or $\downarrow\downarrow$W′.

30. The appearance of characteristic features in the coronary angiogram (Z) proves the constriction of coronary arteries (W).
    If Z, then W and W′.

Conclusions

31. (E and U′ and V′); therefore, P.
32. If Z, then, W; if W, then ↑↑E or ↓↓E′; Z; therefore, ↓↓↓E or ↓↓↓E′.

Acknowledgment

I deeply appreciate the kind help and criticism of Professor Piotr Kułakowski of the Department of Cardiology of the Medical Center of Postgraduate Education in Warsaw. Needless to say, deficiencies in this work are my sole responsibility. The research has been partially supported by the State Committee for Scientific Research of the Republic of Poland (KBN) research grant 8 T11C 025 19.

References

1. K. Ajdukiewicz. *Pragmatic Logic*. PWN, Warsaw, 1974.
2. R. Baud. Present and future trends with nlp (natural language processing). *International Journal of Medical Informatics*, 52: 133–139, 1998.
3. M.S. Blois. *Information and Medicine. The Nature of Medical Descriptions*. University of California Press, Berkeley, CA, 1984.
4. G.W. Bradley. *Disease, Diagnosis and Decisions*. Wiley, Chichester, 1993.
5. M. Bunge. *Causality – the Place of the Causal Principle in Modern Science*. Harward University Press, Cambridge, MA, 1959.
6. M. Bunge. Conjonction, succession, determination, causalité. In *Les Theories de Causalité*, 112–132, University of Paris Press, Paris, 1971.
7. R. Carnap. *Logical Foundations of Probability*, (3rd ed.). University of Chicago Press, Chicago, 1950.
8. P. Caws. *The Philosophy of Science. A Systematic Account*. Van Nostrand, Princeton, NJ, 1965.
9. W. Ceuster, P. Spyns, G. de Moor. From syntactic–semantic tagging to knowledge discovery in medical texts. *International Journal of. Medical Informatics*, 52: 149–157, 1998.
10. T. Chałubiński. *The Method of Arriving at Medical Indications*. Gebethner and Wolff, Warsaw, 1874 (in Polish).
11. K. Chatterjee. Ischemic heart disease. In J.H. Stein, editor, *Internal Medicine*, (4th ed.), 149–169, Mosby, St. Louis, 1994.
12. P. Cutler. *Problem Solving in Clinical Medicine*. Williams and Wilkins, Baltimore, MD, 1979.
13. J. Doroszewski. Evaluation of probability in the process of verification of diagnostic hypotheses. *Studia Filozoficzne*, 10: 121–142, 1972 (in Polish).

14. J. Doroszewski. Polish medical language. In W. Pisarek, editor, *Polszczyzna 2000*, Jagiellonian University Press, Kraków, 1999.
15. J. Doroszewski. An analysis of medical knowledge and reasoning. In A. Tymieniecka, Z. Zalewski, editors, *Analecta Husserliana*, 57–66, Kluwer, Dordrecht, 2000.
16. J. Doroszewski. Solving pathophysiological problems. *Med. Sci. Mon.*, 6: 1–7, 2000.
17. J. Doroszewski. A model of medical knowledge based on systems' and semiotic approach. *Studia Semiotyczne*, 23, 2001 (in Polish).
18. M.J. Druzdzel, M. Henrion. Intercausal reasoning with uninstantiated ancestor nodes. In *Proceedings of the 9th Annual Conference on Uncertainty in AI*, 317–325, Washington, DC, 1993.
19. M.J. Druzdzel. Qualitative verbal explanations in Bayesian belief networks. *AISB Q.*, 94: 43–54, 1996.
20. B. Eiseman, R. Wotkyns. *Surgical Decision Making*. Saunders, Philadelphia, 1978.
21. H.T. Engelhardt, S.F. Spicker, B. Towers. *Clinical Judgment: A Critical Appraisal*. Riedel, Dordrecht, 1977.
22. A.R. Feinstein. *Clinical Judgment*. Williams and Wilkins, Baltimore, MD, 1957.
23. M. Henle. On the relation between logic and thinking. *Psych. Rev.*, 69: 366–578, 1962.
24. M. Hoffman. Diseases of heart and blood vessels. In R. Brzozowski, editor, *Vademecum of Diagnosis and Theraphy*, 97–184, Wyd. Lek. PZWL, Warszawa, 1993 (in Polish).
25. G. Hripcsak. Writing arden syntax medical logic modules. *Comp. Biol. Med.*, 24: 331–363, 1994.
26. K.D. Forbus Qualitative process theory. In D.G. Bobrow, editor, *Qualitative Reasoning about Physical Systems*, MIT Press, Cambridge, MA, 1995.
27. B. Kuipers. Qualitative simulation. *Artificial Intelligence*, 29: 289–338, 1986.
28. H. Llewelyn, A. Hopkins, editors. *Analysing How We Reach Clinical Decisions*. Royal College of Physicians, London, 1993.
29. J.L. Mackie. *The Cement of the Universe. A Study of Causation*. Oxford University Press, Oxford, 1974.
30. W. Marciszewski. *Methods of Analysis of a Scientific Text*. PWN, Warsaw, 1977 (in Polish).
31. E.A. Murphy. *The Logic of Medicine*, (2nd ed.) Johns Hopkins University Press, Baltimore, MD, 1978.
32. J. Packard, J.E. Faulconer. *Introduction to Logic*. Van Nostrand, New York, 1980.
33. J. Pearl. *Causality — Models, Reasoning, and Inference*. Cambridge University Press, Cambridge, 2000.
34. H. Reichenbach. *Elements of Symbolic Logic*. Macmillan, New York, 1948.
35. H. Reichenbach. *The Theory of Probability*. University of California Press, Berkeley, CA, 1949.
36. C. Rieger. An organization of knowledge representation. *Artificial Intelligence*, 7: 89–127, 1976.
37. N. Sager, C. Friedman, M.S. Lyman. *Medical Language Processing. Computer Management of Narrative Data*. Addison-Wesley, Reading, MA, 1987.
38. W.C. Salmon. *The Foundations of Scientific Inference*. University of Pittsburgh Press, Pittsburgh, PA, 1967.
39. S. Schwartz, T.M. Griffin. *Medical Thinking. The Psychology of Medical Judgment and Decision Making*. Springer, New York, 1986.
40. J. Stausberg, M. Person. A process model of diagnostic reasoning in medicine. *Journal of Medical Informatics*, 54: 9–23, 1999.
41. P. Szolowitz, S.G. Pauker. An organization of knowledge representation. *Artificial Intelligence*, 11: 115–144, 1978.

42. M.P.Wellman. Fundamental concepts of qualitative probabilistic networks. *Artificial Intelligence*, 44(3): 257–303, 1990.
43. H.J. Wright, D.B. Macadam. *Clinical Thinking and Practice*. Churchill Livingstone, Edinburgh, 1979.
44. H.R. Wulff, S.A. Pedersen, R. Rosenberg. *Philosophy of Medicine. An Introduction*. Blackwell, Oxford, 1990.
45. L.A. Zadeh. A new direction in AI: Toward a computational theory of perceptions. *AI Magazine*, 22(1): 73–84, 2001.
46. D. Zigmond, L.A. Lenert. Monitoring free-text data using medical language processing. *Computers and Biomedical Research*, 26: 467–481, 1993.

Chapter 16
Fundamental Mathematical Notions of the Theory of Socially Embedded Games: A Granular Computing Perspective

Anna Gomolińska

University of Białystok, Department of Mathematics, Akademicka 2, 15-267 Białystok, Poland
anna.gom@math.uwb.edu.pl

Summary. The aim of this article is to present the key mathematical notions of the theory of socially embedded (or generalized) games (GGT) such as rule, rule application, and rule complex from the standpoint of granularity of information. In GGT, social actors as well as social games and, more generally, interactions, are represented by granules of information called rule complexes. Rule complexes are special cases of complexes of points of some space where the points are rules over a considered language. In many cases, rule complexes may be viewed as granules of information just because the rules constituting the rule complexes are drawn together on the base of similarity and/or functionality. On the other hand, rules represent granules of information. Therefore, rule complexes are multilevel structures representing granules of information as well. In this chapter, we also discuss mereological questions, and, in particular, we define three kinds of "crisp" ingredient-whole relationships with respect to complexes of points (viz., being of a g-element, being of a g-subset, and being of a subcomplex of a complex of points) and present some ideas on how to define the notion of a part (ingredient) to a degree in cases of complexes of points.

To my Parents Emilia and Kazimierz

1 Introduction

The theory of socially embedded (or generalized) games (GGT) [1–5] is a sociological, yet mathematically grounded framework for analyzing various forms of social games and, more generally, interactions[1] that actually may not fulfill the assumptions of classical game theory, i.e., the game theory created by von Neumann and Morgenstern [28]. In their approach, intended to formalize the economic behavior of fully rational agents, a game is specified by a collection of predetermined rules, a set of players, and possible moves and strategies for each and every player. Players

[1] Roughly speaking, a game is an interaction where the participating actors are conscious of being involved in the interaction.

have to follow the rules of the game and are not allowed to change them, unless stated otherwise. Nevertheless, they may choose a strategy and decide which of the possible moves to take.

The theory of socially embedded games extends classical game theory in several aspects: (1) Rules of the game may be underspecified, imprecise, and tolerate exceptions. (2) Participating actors (agents, players) may be neither strict rule followers nor pure rationalists maximizing a value. In fact, several pure action modalities (and a number of mixed cases) may be distinguished: instrumental rationality, normatively oriented action, procedural modality, ritual and communication, play [2]. Real actors actually may not know strategies for themselves and/or for other participants of the interaction. Nevertheless, they can modify, change, or even refuse to follow rules of the game, construct or plan actions, fabricate rules. (3) Actors' social roles and, in particular, values and norms are important factors having an impact on the behavior of the actors. (4) Information may be not only incomplete but also vague, and this imprecision is an additional source of uncertainty.

The key mathematical notions of GGT are *rule* and *complex of rules* or, in other words, *rule complex*.[2] Actors are social beings trying to realize their relationships and cultural forms. With every actor, we associate values, norms, actions and action modalities, judgment rules and algorithms, beliefs, knowledge, and roles specific to the actor. All the constituents that may be expressed in a language of representation are formalized as rules or rule complexes. In this way, we can speak of (1) value complexes consisting of evaluative rules and, in particular, norms; (2) action complexes collecting various action rules and procedures; (3) modality complexes, including rules and procedures describing possible modes of acting; (4) models formed of beliefs and knowledge; and (5) role complexes collecting all rules and complexes of rules relevant for actors' social roles in interactions. Value, action, and modality complexes as well as models are parts (more precisely, subcomplexes) of appropriate role complexes. On the other hand, a social interaction, and a game in particular, may be given the form of a rule complex as well. Such a rule complex specifies more or less precisely who the actors are; what their roles, rights, and obligations are; what the interaction is; what the action opportunities, resources, goals, procedures, and payoffs of the game are; etc. Thus, social actors, their systems as well as social interactions considered in GGT are uniformly represented by rule complexes. Rule complexes seem to be a flexible and powerful tool not only for representing actors and their interactions but also for analyzing various living problems in the area of social interactions.

Informally speaking, a rule complex is a set built of rules of a given language and/or an empty set, according to some formation rules. More precisely, it is a multilevel structure consisting of rules and other rule complexes, where a single rule or rule

---

[2] Primarily, rule complexes were coined by the present author as finitary multilevel structures built of rules [1]. The definition proposed in this chapter is new.

complex may occur many times at various levels. A prototype of a rule complex is an algorithm with embedded procedures. Rules occurring in the rule complex correspond to single instructions, and rule complexes occurring in it correspond to procedures of the algorithm.

The notion of granulation in the context of information, language, and reasoning was introduced by Zadeh, the founder of fuzzy sets and fuzzy logic [29], in his paper [30] from 1973. According to Zadeh's definition of granule, a *granule* is a clump of objects (or points) of some class, drawn together by indistinguishability, similarity, or functionality. Lin [9] proposes to replace the phrase "drawn together," as suggesting symmetry among objects, by the expression "drawn towards an object." It seems to be reasonable when single points may be distinguished as centers of accumulation in granules and/or where symmetry does not need to take place. We feel free to use both expressions, depending on the situation. Thus, we can say that *granulation of information* is a collection of granules (e.g., fuzzy granules, rough granules) where a granule is a clump of objects (points) drawn together and/or toward an object. Formation of granules from objects of a space (universe) is an important stage in granular computing (GC) [8–10, 17, 19, 24–26, 31–34]. As briefly characterized by Lin [9],

> "The primary goal of granular computing is to elevate the lower level data processing to a high level knowledge processing. Such an elevation is achieved by granulating the data space into a concept space. Each granule represents a certain primitive concept, and the granulation as a whole represents knowledge."

Apart from the granulation step, GC includes a representation step, consisting in naming granules and hence representing knowledge by words, and also the word-computing step.

The idea of computing with words or word computing (CW) was proposed by Zadeh in his seminal paper [32] from 1996, but the very foundations were laid many years earlier, starting with Zadeh's paper [30] from 1973. The reader interested in more recent developments is referred to [33, 34]. CW is a methodology where words are used instead of numbers for computing and reasoning. In Zadeh's original formulation, CW involves a fusion of natural languages and computation with fuzzy variables. A word is viewed as a label of a fuzzy granule. On the other hand, the granule is a denotation of the word and works as a fuzzy constraint on a variable. Under the assumption that information may be conveyed by constraining values of variables, the main role in CW is played by fuzzy constraint propagation from premises to conclusions. As pointed out by Zadeh [32],

> "There are two major imperatives for computing with words. First, computing with words is a necessity when the available information is too imprecise to justify the use of numbers. And second, when there is a tolerance for

> imprecision which can be exploited to achieve tractability, robustness, low solution costs and better rapport with reality. Exploitation of the tolerance for imprecision is an issue of central importance in CW."

Rough sets, invented by Pawlak (a good reference in English is [12]) and later developed and extended by a number of scientists (see, e.g., [6, 11, 16]), is a methodology for handling uncertainty arising from granularity of information in the domain of discourse. The form of granulation of information primarily considered in rough sets was based on indiscernibility among objects in a set [17]. The approach was later extended to capture similarity-based granulation [10, 24–26]. The very idea of rough sets is deep yet simple. Any concept that may be represented in the form of a set x of objects of a space is approximated by a pair of exact sets of objects, called the lower and upper rough approximations of x. More precisely, knowledge about a domain is represented in the form of an *information system*,[3] formalized as a pair $\mathcal{A} = (U,A)$ where U is a space of objects, A is a set of attributes, each attribute $a \in A$ is a mapping $a : U \mapsto V_a$, and V_a is the set of values of a. If an attribute $d : U \mapsto V_d$, the *decision attribute*, is added to A, $\mathcal{A}_d = (U, A \cup \{d\})$ is called a *decision system*. It is assumed that objects having identical descriptions in an information system are indiscernible to the users of the system. With each $B \subseteq A$, we can associate an equivalence relation $\mathrm{Ind}_B \subseteq U^2$, the *indiscernibility relation* on U relative to B. Intuitively, elements of the quotient structure U/Ind_B are appropriate candidates for elementary granules of information given $\mathcal{A} = (U,A)$ and $B \subseteq A$. In Polkowski and Skowron's paper [17], they are called *elementary B-pregranules* and are used to generate a Boolean algebra of B-pregranules by means of the set-theoretical operations of union, intersection, and complement. Given nonempty subsets $B_i \subseteq A$ $(i = 1,2)$, by a (B_1,B_2)-*granule of information* relative to $\mathcal{A}$ and B_1, B_2, as above, we mean any pair (x_1,x_2) where x_i is a B_i-pregranule $(i = 1,2)$. This construction may easily be adapted to the decision system case. The notion of granule just described was later refined and generalized [10, 24–26]. It is worth mentioning that decision rules obtained from decision systems represent granules of information.

Basic relations between granules of information are closeness (i.e., how close or similar two granules are) and inclusion (i.e., what it means that a granule x is a part or, more generally, an ingredient of a granule y). The part–whole relation is one of the main concepts of Leśniewski's mereology [7, 27]. Polkowski and Skowron extended the original system of Leśniewski to the system of rough mereology [13, 14, 18, 20] where an intuitive notion of a *part to a degree* was formalized and studied.

As mentioned above, rules in information systems represent granules of information. This observation may be generalized to arbitrary rules, as well as those considered in GGT. As built of rules, rule complexes represent compound granules of information. On the other hand, rule complexes modeling social actors and their interactions may be viewed themselves as granules of information. To see this, let us

[3] A historical name given by Pawlak, nowadays used also in other contexts.

recall that rules in such rule complexes are drawn together on the base of similarity and/or functionality. As already said, evaluative rules are collected to form value complexes; action rules and procedures are drawn together to build action complexes; beliefs and knowledge form models, etc. Role complexes, built of various rules describing actors' social roles in interactions and including the just mentioned complexes as parts, are next drawn together with some other complexes of rules (e.g., complexes of control and inference rules) to form rule complexes representing actors and their interactions.

The main theoretical issues addressed in GGT are application and transformation of rule complexes. Actors apply their rule complexes in situations of action or interaction to achieve private or group objectives, to plan and implement necessary activities, and to solve problems. Rules and rule complexes are subject to modifications or more serious transformations whenever needed and possible. For instance, application of a rule complex in a given situation s possibly changes the situation to a new situation s^*. If an actor is conscious of the transformation of s into s^*, then the actor will typically try to update his/her model of s to obtain a model of s^*. Here, the term "situation s" is a label for an imprecise concept, approximated by the actor by means of a rule complex, consisting of the actor's beliefs and knowledge about the situation s considered and called "the model of s." If we add that rules in GGT may be formulated in natural or seminatural language(s), then it becomes clear that GGT may substantially benefit by employing fuzzy-rough GC and related methods. On the other hand, GGT offers a vast area of application, hardly exploited by knowledge discovery methodologies.

In Sect. 2 of this chapter, we present a general notion of a complex of points. In Sect. 3, the notions of a g-(general)element, a g-(general) subset, and a subcomplex of a complex are defined and studied. From the mereological perspective, g-elements are proper parts, whereas g-subsets and subcomplexes are ingredients of complexes. Section 4 is devoted to the notions of a rule and a rule complex viewed from the standpoint of granularity of information. In the next section, we briefly describe the GGT model of a social actor, based on the notion of a rule complex. Section 6 contains a brief summary. Let us note that the present notions of a rule and a rule complex differ from those used in [1–5].

For any set x, we denote the cardinality of x by $\#x$, the power set by $\wp(x)$, and the complement by $-x$. The set of natural numbers (with zero) will be denoted by $\mathbf{N}$. Given sets x, x_0, y, y_0 such that $x_0 \subseteq x$ and $y_0 \subseteq y$, and a mapping $f : x \mapsto y$, by $f^{\rightarrow}(x_0)$ and $f^{\leftarrow}(y_0)$, we denote the image of x_0 and the inverse image of y_0 given by f, respectively. We abbreviate $z_0 \in z_1 \wedge z_1 \in z_2 \wedge \ldots \wedge z_n \in z_{n+1}$, where $n \in \mathbf{N}$, by $z_0 \in z_1 \in \ldots \in z_n \in z_{n+1}$.

2 The Notion of a Complex of Points

From the set-theoretical point of view, complexes of points of a space are sets, built of points of the space and/or the empty set according to some formation rules. In general, complexes of points are multilevel structures resembling algorithms with embedded procedures which, by the way, were prototypes of complexes of rules. Where points represent granules of information as in the case of rules, complexes of such points also represent some (compound) granules of information. Complexes of points can be granules of information themselves, e.g., role complexes, value complexes, action complexes. Components of these complexes are drawn together on the base of similarity and/or functionality. Thus, constituents of role complexes are linked together as they are relevant to the topics "social roles," components of value complexes are relevant to the topics "values and norms," and ingredients of action complexes are relevant to the topics "action and interaction." The notion of a complex emerged during the author's work on formalization of GGT. Primarily only complexes of rules, i.e., rule complexes were considered [1–5]. As far as application in GGT is concerned, rule complexes seem to be a convenient and sufficient tool for representing both social actors as well as their interactions in a uniform way and for analyzing various aspects of social interactions. In our opinion, the general notion of a complex of points is not only interesting in itself, but it can be useful as well.

Given a set of points U, called the space, from elements of U and the empty set, we build particular sets called complexes of points of U.

Definition 1. The class of *complexes of points of U* (or simply *complexes* if U is known from the context), written $C(U)$, is the least class C of sets that is closed under the formation rules (cpl1)–(cpl4):

(cpl1) Every subset of U belongs to C.
(cpl2) If x is a set of members of C, then $\bigcup x \in C$.
(cpl3) If $x \subseteq y$ and $y \in C$, then $x \in C$ as well.
(cpl4) If $x \in C$, then the power set of x, $\wp(x)$, is a member of C.

We say that a complex x is finite if $\#x < \aleph_0$, the cardinality of $\mathbf{N}$.

Proposition 1. *Let C be a class of sets satisfying (cpl3). Then, (a) for every nonempty set x of members of C,* $\bigcap x \in C$ *and (b) for each* $x \in C$ *and a set y,* $x - y \in C$.

Hence the set-theoretical intersection of a nonempty set of complexes and the set-theoretical difference of a complex and a set are complexes.

Example 1. Let $x_i \in U$ $(i = 0, \ldots, 3)$ be different points. Sets $x_4 = \{x_0, x_2\}$, $x_5 = \{x_1, x_6\}$, $x_6 = \{x_0, x_2, x_3\}$, $x_7 = \{x_0, x_1, x_4, x_5\}$, and

$$x = \{\underbrace{\{\ldots\{}_{n} x_0 \underbrace{\}\ldots\}}_{n} \mid n \in \mathbf{N}\}$$

are complexes. Unlike x, the remaining complexes are finite. x_0 occurs in x_7 three times: as an element of x_4, x_6, x_7 where x_6 is an element of x_5. On the other hand, x_0 occurs infinitely many times in x.

Given a class of sets C and a set x, let us define

$$\varphi(U,C,x) \text{ iff } \forall y \in x.(y \in U \vee y \in C) \tag{1}$$

and consider the following conditions:

(cpl5) Arbitrary sets of elements of C belong to C.
(cpl6) For every set x, $x \in C$ iff $\varphi(U,C,x)$.

Theorem 1.

(a) If C satisfies (cpl3), (cpl5), *then* (cpl4) *holds as well.*

(b) Conditions (cpl2)–(cpl4) *imply* (cpl5).

(c) Condition (cpl6) *implies* (cpli) *for $i = 1, \ldots 5$.*

(d) The class $C(U)$ satisfies (cpl6).

Proof. We prove only (d) and leave the rest as an exercise. Let x be any set. In fact, the right-to-left part of (cpl6) holds for every class of sets C satisfying (cpl1)–(cpl4). To this end, assume that $\varphi(U,C,x)$ holds. Then $x = (x \cap U) \cup \{y \in x \mid y \in C\}$. Clearly, $x \cap U \in C$ by (cpl1). On the other hand, $\{y \in x \mid y \in C\} \in C$ by (cpl5) and (b). Hence by (cpl2), $x \in C$. Finally, notice that $C(U)$ satisfies (cpl1)–(cpl4) by definition. For the left-to-right part assume that $x \in C(U)$. Let us observe that the rules (cpl1)–(cpl4) are the only formation rules for building complexes of points of U. (i) If $x \subseteq U$, then for each $y \in x$, $y \in U$ as well, that is, $\varphi(U, C(U), x)$ holds. (ii) Assume that (A) $x = \bigcup y$ where (B) y is a set of complexes z such that $\varphi(U, C(U), z)$. Consider any $u \in x$. By (A), there is $z \in y$ such that $u \in z$. By (B), $u \in U$ or $u \in C(U)$. Hence, $\varphi(U, C(U), x)$. (iii) Now assume $x \subseteq y$ where $\varphi(U, C(U), y)$ holds. Suppose $z \in x$. By the assumption, $z \in y$, and subsequently $z \in U$ or $z \in C(U)$. Hence, $\varphi(U, C(U), x)$. Finally, consider the case that (iv) $x = \wp(y)$ where $\varphi(U, C(U), y)$. Let $z \in x$. By the assumption, $z \subseteq y$. Again by the assumption and (iii), $\varphi(U, C(U), z)$. By virtue of the right-to-left part of (cpl6), $z \in C(U)$. Hence, $\varphi(U, C(U), x)$, as required. □

As a consequence, complexes may be described as follows:

Corollary 1. *A set x is a complex of points of U iff for each $y \in x$, it holds that $y \in U$ or y is a complex of points of U.*

One can easily see that $\wp(U)$ is the least class of sets satisfying (cpl1)–(cpl3). However, $\wp(U)$ is not closed under (cpl4) since $U \in \wp(U)$, but $\wp(U) \notin \wp(U)$. Thus, the conditions (cpl1)–(cpl3) alone are not sufficient to imply (cpl4). Next, one can wonder whether $C(U)$ is the set of all subsets of

$$U \cup \wp(U) \cup \wp[U \cup \wp(U)] \cup \wp\{U \cup \wp(U) \cup \wp[U \cup \wp(U)]\} \ldots \tag{2}$$

It is not the case, as shown below.

Example 2. Consider $U = \{x\}$ and a complex of points of U

$$y = \{\underbrace{\{\ldots\{}_{n} x \underbrace{\}\ldots\}}_{n} \mid n \in \mathbf{N}\}.$$

The set $z = \{y\}$ is a complex of points of U as well. However, z is not a subset of the set given by (2) since $y \notin U$, $y \notin \wp(U)$, $y \notin \wp[U \cup \wp(U)]$, etc.

The class of complexes of points of U is proper, i.e., complexes of points of U do not form a set. Suppose, to the contrary, that $C(U)$ is a set. By Theorem 1, $C(U)$ must be a complex consisting of all complexes. Since each $x \subseteq C(U)$ is a complex by (cpl3), $x \in C(U)$. Hence $\wp[C(U)] \subseteq C(U)$. [Notice also that $\wp[C(U)] \in C(U)$.] Then $\#\wp[C(U)] \leq \#C(U)$ contrary to Cantor's theorem.

If points of U are granules of information, the corresponding complexes may be viewed as compound granules of information having a multilevel structure. The set-theoretical union and intersection of granules as well as the complement of a granule are granules of information themselves, according to [10]. We may also view a set of granules, the power set of a set of granules, pairs and, more generally, tuples of granules, and Cartesian products of sets of granules as granules of information. Finally, complexes of granules may be treated as granules of information as well.

3 Mereological Functors Associated with Complexes of Points

In this section, we discuss mereological questions in our complex-based framework. Concisely speaking, mereology is a theory of part–whole relations. This theory, invented by Leśniewski [7] in 1916, may be described after Sobociński [27] as

> "a deductive theory which inquires into the most general relations that may hold among objects [...]. Mereology can be regarded as the theory of collective classes in contradistinction to Ontology which is the theory of distributive classes."

In Leśniewski's mereology, a "part" means a proper part, i.e., it never holds that x is a part of itself. If x is a part of y or x, y are identical, then we say that x is an *ingredient of* y. Apart from elements and subsets, we distinguish three other kinds of "crisp" ingredients associated with the notion of a complex: g-elements, g-subsets, and subcomplexes of a complex. At the end of this section, we present some preliminary ideas for defining the notion of a part (or ingredient) of a complex of points to a degree.

Informally speaking, a point or complex of points x of a space U is a g-element of a complex y of points of U if x occurs in y at some level. A complex x is a g-subset of a complex y if all g-elements of x are g-elements of y as well. Thus g-subsets of y are arbitrary complexes built of g-elements of y. Subcomplexes of a complex y are also formed of g-elements of y, but the idea is different. A complex $x \neq y$ is a subcomplex of y if x can be obtained from y by deleting some occurrences of g-elements of y and possibly some parentheses. Subcomplexes of y are not g-subsets of y in general. The functor of being of a g-element of a complex, $\in_g$, is a part–whole relationship, generalizing the set-theoretical functor of being of an element of a set. The functors of being of a g-subset and being of a subcomplex of a complex, $\subseteq_g$ and $\sqsubseteq$, respectively, are ingredient–whole relationships, generalizing the set-theoretical functor of being of a subset of a set. Our aim here is not to give a formal theory in the spirit of Leśniewski's mereology but rather to investigate properties of the functors $\in_g$, $\subseteq_g$, and $\sqsubseteq$. Let us emphasize that the notion of a subcomplex is of particular interest and importance from the perspective of GGT.

Definition 2. Given a complex y of points of U, we say that x is a *g-element* of y, written $x \in_g y$, iff

$$x \in y \vee \exists n \in \mathbf{N}. \exists z_0, \ldots, z_n. x \in z_0 \in \ldots \in z_n \in y. \tag{3}$$

Observe that x is a point of U or a complex of points of U, whereas $z_0, \ldots, z_n$ must be complexes of points of U.

Let us note a few basic properties.

Proposition 2. *For every x and any complexes y, z:*

1. *If $x \in y$, then, $x \in_g y$.*
2. *If $x \in_g y$ and $y \in_g z$, then, $x \in_g z$.*
3. *If $x \in_g y$ and $y \subseteq z$, then, $x \in_g z$.*

Proof. We prove only 2. The remaining cases are left as exercises. Thus, assume that $x \in_g y$ and $y \in_g z$. The following cases hold by the definition: (i) $x \in y \in z$ or (ii) $x \in y$, and there are $n \in \mathbf{N}$ and complexes $y_0, \ldots, y_n$ such that $y \in y_0 \in \ldots \in y_n \in z$, or (iii) there are $m \in \mathbf{N}$ and complexes $x_0, \ldots, x_m$ such that $x \in x_0 \in \ldots \in x_m \in y \in z$, or (iv) there are $m, n \in \mathbf{N}$ and complexes $x_0, \ldots, x_m, y_0, \ldots, y_n$ such that $x \in x_0 \in \ldots \in x_m \in y$ and $y \in y_0 \in \ldots \in y_n \in z$. In each case, $x \in_g z$ follows by the definition. □

Two different complexes may have the same g-elements, i.e., g-elements determine complexes only in part.

Example 3. Consider $x = \{z_0, z_1\}$ and $y = \{z_1, z_2\}$ where $z_0 = \{z_2, z_3\}$, $z_1 = \{z_0, z_4\}$, and z_2, z_3, z_4 are different elements of U. Complexes x and y are different but have the same g-elements.

All g-elements of x that are points of U form the *point base* of x, $\mathrm{pb}(x)$. Similarly, all g-elements of x being complexes form the *complex base* of x, $\mathrm{cb}(x)$:

$$\begin{aligned} \mathrm{pb}(x) &\stackrel{\mathrm{def}}{=} \{y \in_g x \mid y \in U\}, \\ \mathrm{cb}(x) &\stackrel{\mathrm{def}}{=} \{y \in_g x \mid y \in C(U)\}. \end{aligned} \tag{4}$$

For convenience, let us denote the set of all elements of x that are complexes by $\mathrm{cp}(x)$.

Example 4. Consider the complexes x, y from Example 3. Obviously, their point bases are equal, and the same holds for their complex bases. Thus, $\mathrm{pb}(x) = \mathrm{pb}(y) = \{z_2, z_3, z_4\}$ and $\mathrm{cb}(x) = \mathrm{cb}(y) = x$.

Example 5. Infinite complexes may have finite point bases, whereas complexes with infinite point bases may be finite. Let U be an infinite set, $x \in U$, and y be an infinite subset of U. Define $z_1 = \{\underbrace{\{\ldots\{}_{n} x \underbrace{\}\ldots\}}_{n} \mid n \in \mathbf{N}\}$ and $z_2 = \{y\}$. Then, $\mathrm{pb}(z_1) = \{x\}$ and $\mathrm{pb}(z_2) = y$.

Theorem 2. *For any complexes x, y and $\tau \in \{\mathrm{pb}, \mathrm{cb}\}$:*

1. $\mathrm{pb}(x) \cap \mathrm{cb}(x) = \emptyset$.
2. *If* $x \in \mathrm{cb}(y)$ *or* $x \subseteq y$, *then,* $\tau(x) \subseteq \tau(y)$.
3. $\mathrm{cb}(x) = \mathrm{cp}(x) \cup \bigcup\{\mathrm{cb}(y) \mid y \in \mathrm{cp}(x)\}$.
4. $\mathrm{pb}(x) = [x \cup \bigcup \mathrm{cb}(x)] \cap U$.
5. $\mathrm{cb}(x) = \emptyset$ *iff* $\mathrm{cp}(x) = \emptyset$.
6. $\mathrm{pb}(x) = \emptyset$ *iff* $x \cap U = \emptyset$ *and* $\forall y \in \mathrm{cb}(x) y \cap U = \emptyset$.
7. *If* $x \cap U = \emptyset$, *then,* $\mathrm{cb}(\bigcup x) = \bigcup\{\mathrm{cb}(y) \mid y \in \mathrm{cp}(x)\}$.
8. *If* $x \cap U = \emptyset$, *then,* $\mathrm{pb}(\bigcup x) = \mathrm{pb}(x)$.
9. $\mathrm{cb}[\wp(x)] = \wp(x) \cup \mathrm{cb}(x)$.
10. $\mathrm{pb}[\wp(x)] = \mathrm{pb}(x)$.

Proof. We prove only 3, leaving the remaining cases as exercises. To this end, assume that $z \in \mathrm{cb}(x)$ first. By definition, z is a complex such that $z \in_g x$. Hence, $z \in x$, or there are $n \in \mathbf{N}$ and complexes $z_0, \ldots, z_n$ such that $z \in z_0 \in \ldots \in z_n \in x$. In the former case, $z \in \mathrm{cp}(x)$, and we are done. In the latter, $z \in \mathrm{cb}(z_n)$ since $z \in_g z_n$ by definition. Moreover, $z_n \in \mathrm{cp}(x)$. As a consequence, $z \in \bigcup\{\mathrm{cb}(y) \mid y \in \mathrm{cp}(x)\}$. To prove the remaining part, consider a complex z such that $(*)$ $z \in x$ or $(**)$ there is a complex $y \in x$ such that $z \in \mathrm{cb}(y)$. From $(*)$, it directly follows that $z \in \mathrm{cb}(x)$. From $(**)$ and the definition, $y \in x$ and $z \in_g y$. Hence by the properties of $\in_g$, $z \in_g x$. Finally, $z \in \mathrm{cb}(x)$. □

Let us observe that complexes are well-founded sets. There is no infinite sequence of complexes $x_0, x_1, x_2, \ldots$ such that $(*) \ldots \in x_2 \in x_1 \in x_0$. By Theorem 2, the only operation responsible for an increase in the "complexity" of a complex is the power set operation. According to our definition, complexes are formed by application of (cpl1)–(cpl4) a finite, even if very large number of times. Therefore, every sequence of the form $(*)$ must be finite. There is no complex x that $x \in_g x$, either.

Recall that for any complexes (and sets in general) x and y, $x \subseteq y$ iff $\forall z.(z \in x \rightarrow z \in y)$. There arises a question what ingredient–whole relation can be obtained if we replace $\in$ by its generalization $\in_g$. We say that x is a *g-subset* of y, written $x \subseteq_g y$, iff $\forall z.(z \in_g x \rightarrow z \in_g y)$. By the definition of $\in_g$, it is easy to see that

$$x \subseteq_g y \text{ iff } \mathrm{pb}(x) \subseteq \mathrm{pb}(y) \text{ and } \mathrm{cb}(x) \subseteq \mathrm{cb}(y). \tag{5}$$

Thus, x is a g-subset of y if all of the points and complexes that form x are also points and complexes constituting y, respectively.

Example 6. Consider complexes

$$x = \{\{x_1, x_2\}, \{x_3, \{x_1, x_2\}\}\} \text{ and } y = \{x_0, \{x_3, \{x_1, x_2\}\}\}$$

where $x_i \in U$ $(i = 0, \ldots, 3)$ are different. In this case $x \subseteq_g y$ and neither $x \subseteq y$ nor $y \subseteq x$.

The fundamental properties of the notion of a g-subset are stated below.

Theorem 3. *For any complexes x, y, z:*

1. $x \subseteq_g x$.
2. *If $x \subseteq y$, then, $x \subseteq_g y$.*
3. *If $x \in_g y$, then, $x \subseteq_g y$.*
4. *If $x \subseteq_g y$ and $y \subseteq_g z$, then, $x \subseteq_g z$.*
5. *If $x \subseteq_g y$ and $y \in_g z$, then, $x \subseteq_g z$.*

Clearly, a g-subset of complex y may or may not be a g-element of y. Notice that $x \subseteq_g z$ in 5 cannot be replaced by $x \in_g z$.

Example 7. Let $x_i \in U$ $(i = 0,1,2)$ be different points, $x_3 = \{x_0\}$, $x_4 = \{x_1, x_2, x_3\}$, $x_5 = \{x_1, x_2\}$, and $x = \{x_0, x_4\}$. Observe that $x_5 \subseteq x_4$ (and hence $x_5 \subseteq_g x_4$) and $x_4 \in_g x$. However, $x_5 \notin_g x$ since $x_5 \notin \mathrm{cb}(x) = \{x_3, x_4\}$.

When complexes may be represented by collections of their g-elements, $\subseteq_g$ plays the role of $\subseteq$. Then, we would view complexes having the same g-elements as equivalent. Formally, let us define $x \equiv_g y$, where x, y are complexes, as follows:

$$x \equiv_g y \text{ iff } x \subseteq_g y \text{ and } y \subseteq_g x. \tag{6}$$

Example 8. Let $x_i \in U$ $(i = 0,\ldots,3)$ be different points. Consider two complexes $x = \{x_0, \{x_1, x_2\}, \{x_3, \{x_1, x_2\}\}\}$ and $y = \{x_0, \{x_3, \{x_1, x_2\}\}\}$. Clearly, $x \equiv_g y$.

The following properties of $\equiv_g$ easily follow from Theorem 3.

Proposition 3. *For any complexes x, y, z, (a) $x \equiv_g x$; (b) $x \equiv_g y$ implies $y \equiv_g x$; and (c) if $x \equiv_g y$ and $y \equiv_g z$, then $x \equiv_g z$.*

Thus, $C(U)$ is divided by $\equiv_g$ into classes of complexes with the same g-elements.

Recall briefly that a complex is a g-subset of a complex y if it is built of some (or all) g-elements of y. Another idea underlies the notion of a subcomplex. Assume that x is a complex, and consider a complex $y = \underbrace{\{\ldots\{}_{n} x \underbrace{\}\ldots\}}_{n}$ where $n \in \mathbf{N} - \{0\}$. From the GGT perspective, these parentheses are redundant[4] since application of y will resolve itself into application of x. Let us define an operation of removing parentheses, $\odot$, as follows:

$$\odot(x) \stackrel{\text{def}}{=} \begin{cases} y & \text{if } x = \{y\} \text{ and } y \in C(U) \\ x & \text{otherwise.} \end{cases} \tag{7}$$

Multiple parentheses may be removed by iteration of $\odot$.

Definition 3. A complex x is a *subcomplex* of a complex y, $x \sqsubseteq y$, if $x = y$, or x may be obtained from y by deleting some occurrences of g-elements of y and/or redundant parentheses.

A few properties of subcomplexes are given below.

Theorem 4. *For any complexes x, y, z:*

1. $x \sqsubseteq x$.
2. *If $x \subseteq y$, then, $x \sqsubseteq y$.*
3. *If $x \in_g y$, then, $x \sqsubseteq y$.*
4. *If $x \sqsubseteq y$ and $y \sqsubseteq z$, then, $x \sqsubseteq z$.*
5. *If $x \sqsubseteq y$ and $y \in_g z$, then, $x \sqsubseteq z$.*

[4] This is a relatively simple redundancy. In the future, more complicated forms of redundancy should be taken into account as well.

Proof. We prove only 3 and leave the rest as an exercise. Assume that $x \in_g y$. By definition, $x \in y$, or there are $n \in \mathbf{N}$ and complexes $x_0, \ldots, x_n$ such that $x \in x_0 \in \ldots \in x_n \in y$. In the former case, $x = \odot[y - (y - \{x\})]$. Hence $x \sqsubseteq y$ by the definition. In the latter case, let $x_{-1} = x$ and $x_{n+1} = y$. Then for $i = -1, 0, \ldots, n$, $x_i = \odot[x_{i+1} - (x_{i+1} - \{x_i\})]$. Thus, x may be obtained from y according to the definition of $\sqsubseteq$, i.e., $x \sqsubseteq y$. □

As stated in 2, all subsets of a complex y are subcomplexes of y. The converse does not hold in general. Obviously a subcomplex of y may or may not be a g-element of y. The notions of g-subset and subcomplex are different as well.

Example 9. Let $x_i \in U$ $(i = 0, \ldots, 3)$ be different points, $x_4 = \{x_1\}$, $x_5 = \{x_2, x_3, x_4\}$, $x_6 = \{x_0, x_5\}$, and $y = \{x_0, x_1, x_5, x_6\}$. Define $x_7 = \{x_2, x_4\}$, $x_8 = \{x_2, x_3\}$, $x_9 = \{x_0, x_8\}$, and $x = \{x_1, x_7, x_9\}$, that is, $x_7 = x_5 - \{x_3\}$, $x_8 = x_5 - \{x_4\}$, $x_9 = (x_6 - \{x_5\}) \cup \{x_8\}$, and $x = (y - \{x_0, x_5, x_6\}) \cup \{x_7, x_9\}$. Thus, $x \sqsubseteq y$. Notice that $\mathrm{pb}(x) = \mathrm{pb}(y) = \{x_0, \ldots, x_3\}$, $\mathrm{cb}(x) = \{x_4, x_7, x_8, x_9\}$, and $\mathrm{cb}(y) = \{x_4, x_5, x_6\}$. Hence, x is neither a subset nor a g-subset of y.

Example 10. Let $x_i \in U$ $(i = 0, 1, 2)$ be different points, $x_3 = \{x_1\}$, $x_4 = \{x_2, x_3\}$, $x = \{x_3, x_4\}$, and $y = \{x_0, x_4\}$. In this case, $\mathrm{pb}(x) = \{x_1, x_2\}$, $\mathrm{pb}(y) = \{x_0, x_1, x_2\}$, and $\mathrm{cb}(x) = \mathrm{cb}(y) = \{x_3, x_4\}$. Thus, x is a g-subset of y but not a subcomplex of y.

The following observations hold by Theorem 4.

Proposition 4. *For any complexes x, z and a set y, (a) $x \cap y \sqsubseteq x$; (b) $x - y \sqsubseteq x$; (c) $x \sqsubseteq x \cup z$; (d) if $x \cap U = \emptyset$, then, $\forall z \in x. \bigcap x \sqsubseteq z$; and (e) $x \sqsubseteq \wp(x)$.*

It can be that $x \sqsubseteq y$ and $y \sqsubseteq x$, while nevertheless $x \neq y$.

Example 11. Let $z \in U$, and consider two different complexes:
$x = \{\underbrace{\{\ldots\{}_{n} z \underbrace{\}\ldots\}}_{n} \mid n \in \mathbf{N} - \{0,1\}\}$ and $y = \{\underbrace{\{\ldots\{}_{n} z \underbrace{\}\ldots\}}_{n} \mid n \in \mathbf{N} - \{0\}\}$. Clearly $x \sqsubseteq y$ since $x \subseteq y$. On the other hand, we can obtain y from x by applying $\odot$ to each and every element of x, that is, $y \sqsubseteq x$.

As for g-subsets, we may draw together complexes that are subcomplexes of one another. For this purpose, let us define $x \cong y$, where x, y are complexes, as follows:

$$x \cong y \text{ iff } x \sqsubseteq y \text{ and } y \sqsubseteq x. \tag{8}$$

The basic properties of $\cong$ follow from Theorem 4.

Proposition 5. *For any complexes x, y, z, (a) $x \cong x$; (b) if $x \cong y$, then, $y \cong x$ as well; (c) if $x \cong y$ and $y \cong z$, then, $x \cong z$.*

Thus, by means of $\cong$, the class of complexes $C(U)$ is divided into classes of complexes that are subcomplexes of one another.

Polkowski and Skowron extended Leśniewski's mereology to the system of rough mereology [13,14,18,20]. The key concept of their framework is the notion of a *part to a degree*. We briefly present it here in keeping with the terminology of approximation spaces [10,23]. An *approximation space* $\mathcal{U}$ is a triple $\mathcal{U} = (U, I, \kappa)$ where U is the universe, $I : U \mapsto \wp(U)$ is an uncertainty mapping, and $\kappa : [\wp(U)]^2 \mapsto [0,1]$ is a rough inclusion function.[5] Let us assume that $u \in I(u)$ for each object $u \in U$. Hence $I^{\rightarrow}(U)$ is a covering of U. The mapping I works as a granulation function which associates with every $u \in U$ a clump of objects of U that are similar to u in some respect. Sets $I(u)$ may be called *basic granules* of information in $\mathcal{U}$. For any pair of sets of objects (x,y), the rough inclusion function κ determines the degree of inclusion of x in y. Along standard lines, we may assume that

$$\begin{aligned} &\kappa(x,y) = 1 \text{ iff } x \subseteq y, \\ &\kappa(x,y) > 0 \text{ iff } x = \emptyset \text{ or } x \cap y \neq \emptyset. \end{aligned} \tag{9}$$

Thus, x is included in y to the (highest) degree 1 iff x is a subset of y, and x is included in y to the (lowest) degree 0 iff x and y are disjoint. When x is finite, κ may be defined as follows [6]:

$$\kappa(x,y) \stackrel{\text{def}}{=} \begin{cases} \frac{\#(x \cap y)}{\#x} & \text{if } x \neq \emptyset \\ 1 & \text{otherwise.} \end{cases} \tag{10}$$

If $\kappa(x,y) = k$, then, we say that x is a *part* (or more precisely, an *ingredient*) of y *to the degree k*. Now, consider an object $u \in U$ such that $I(u)$ is finite and a set of objects $x \subseteq U$. In the crisp approach, u is an element of x or not. According to the rough approach, u may be a member of x to a degree, being a real number of the unit interval $[0,1]$. Thus, u is said to be a *part* of x *to a degree k* when $\kappa[I(u), x] = k$, i.e., if

$$\frac{\#[I(u) \cap x]}{\#I(u)} = k. \tag{11}$$

There arises a question, *how is the notion of a part to a degree defined for complexes of points?* For the time being, we are able to answer this question only partially. First, consider a complex of objects x and an object $u \in U$ such that $\#I(u) < \aleph_0$. As earlier, $\mathrm{pb}(x)$ and $\mathrm{cb}(x)$ denote the point base and the complex base of x, respectively. We say that u is a *part* of x *to a degree* $k \in [0,1]$ just in case u is a part of $\mathrm{pb}(x)$ to the degree k, with the latter notion defined as above. The definition (10) of a rough inclusion function may be extended to complexes of points. In what follows, we generalize the crisp notion of a g-subset to the rough case. Consider complexes x, y of points of U such that the point and complex bases of x are finite. We define a mapping κ_g on such pairs of complexes into $[0,1] \times [0,1]$ as follows:

$$\kappa_g(x,y) = (\kappa[\mathrm{pb}(x), \mathrm{pb}(y)], \kappa[\mathrm{cb}(x), \mathrm{cb}(y)]). \tag{12}$$

[5] For simplicity, we omit the lists of parameters occurring in the original definition.

Then, we say that x is an *ingredient* of y *to the degrees* (k_1,k_2) $(k_1,k_2 \in [0,1])$ iff $\kappa_g(x,y) = (k_1,k_2)$. According to this definition, complexes having the same point bases and complex bases are equivalent. In other words, complexes x,y such that $x \equiv_g y$ are treated as the same. In our opinion, the above definition is unfortunately not adequate for the rough version of a subcomplex.

4 Rules, Rule Complexes, and Granules of Information

A rule is a fundamental concept of GGT. Rules are major components of games and interactions. Instructions of algorithms may be seen as rules. Values and norms as well as beliefs and knowledge of social actors may also be represented in the form of rules. Let us mention action rules specifying pre- and postconditions of various actions and interactions, rules of logical inference, generative rules and specific situational rules, control rules and in particular judgment rules, strict rules and rules with exceptions, precise and vague rules, and last but not least, metarules of various kinds.

Below, we introduce a formal notion of a rule[6] over a given language $\mathcal{L}$. At the present stage, we do not specify $\mathcal{L}$ totally, assuming that, e.g., it is rich enough to formally express rules of various games considered in GGT or to mention and use the names of rules and rule complexes within $\mathcal{L}$. To denote formulas of $\mathcal{L}$, we use lowercase Greek letters with subscripts whenever needed.

By a *rule* r over $\mathcal{L}$, we mean a triple $r = (x,y,\alpha)$ where x, y are finite sets of formulas of $\mathcal{L}$ called the sets of *premises* and *justifications* of r, respectively, and α is a formula of $\mathcal{L}$ called the *conclusion* of r. Premises and justifications have to be declarative statements, whereas conclusions may be declarative or imperative but not interrogative statements. Rules without premises and justifications are called *axiomatic*. There is a one-to-one correspondence between the set of all formulas of $\mathcal{L}$ and the set of all axiomatic rules over $\mathcal{L}$. Rules without justifications are the usual if-then rules. Since a pair (a,b) is defined as the set $\{\{a\},\{a,b\}\}$ and a triple (a,b,c) is simply $((a,b),c)$, we can easily conclude that a rule r, as above, is a complex of formulas. Thus, rule complexes over $\mathcal{L}$ are complexes of formulas of $\mathcal{L}$. The converse does not hold in general. Nevertheless, every complex of formulas may be transformed into a rule complex.

The informal meaning of $r = (x,y,\alpha)$ is that if all formulas of x hold and all formulas of y possibly hold, then one may conclude α. Thus, premises are stronger preconditions than justifications. A justification $\beta \in y$ may actually not hold, but it suffices for the sake of application of r that we do not know surely that it does not hold. The name "justification" is adopted from the formalism introduced by Reiter and widely known as *default logic* [22]. Our rules make it possible to reason by

[6] The present notion of a rule differs from that used in [1,2,5].

default and to deal with exceptions that can be particularly useful when formalizing commonsense reasoning. Needless to say, such a form of reasoning is common in social life and hence, in social actions and interactions. Of course, not all rules tolerate exceptions, i.e., in many cases, the set of justifications will be empty.

In rough set methodology, decision rules obtained from decision systems represent (or more precisely, define) corresponding rough granules of information [10, 17]. Rules considered in our framework are built of formulas of $\mathcal{L}$, formulas are formed of words of $\mathcal{L}$, whereas words are labels for fuzzy granules of information (see [32]). We argue that although they are more complicated than decision rules, rules in GGT represent granules of information in the spirit of rough set methodology as well.

Let S denote a nonempty set of all situations considered. Situations that are similar and, even more, indiscernible[7] to an actor or a collective of actors form basic granules of information. More formally, let us consider an approximation space [10, 23] $\mathcal{S} = (S, I, \kappa)$ where S is the universe, $I : S \mapsto \wp(S)$ is an uncertainty mapping, and $\kappa : [\wp(S)]^2 \mapsto [0,1]$ is a rough inclusion function.[8] We assume, as earlier, that for each $s \in S$, $s \in I(s)$. Thus, $I^{\rightarrow}(S)$ is a covering of S. Basic granules of information in $\mathcal{S}$ are of the form $I(s)$ where $s \in S$. As earlier, the rough inclusion function κ satisfies (9). For finite sets of situations, κ may be defined, e.g., as in (10).

Starting with basic granules of information in $\mathcal{S}$, we can construct more compound, yet still quite simple granules of information by means of the set-theoretical operations of union, intersection, and complement. Such granules will be referred to as *simple*. We may also view sets of granules of information, the power sets of sets of granules, pairs and, more generally, tuples of granules, and Cartesian products of sets of granules as compound granules of information. Finally, complexes of granules are granules of information as well.

According to [10], a set of situations $x \subseteq S$ is *definable* in $\mathcal{S}$ if it is a union of some values of I, i.e., if there is a set $y \subseteq S$ such that $x = \bigcup I^{\rightarrow}(y)$. Consider the classical case where $I^{\rightarrow}(S)$ is a partition of S. Then an arbitrary set of situations x may be approximated by a pair of definable sets of situations

$$(\mathrm{LOW}_{\mathcal{S}}(x), \mathrm{UPP}_{\mathcal{S}}(x)), \tag{13}$$

where $\mathrm{LOW}_{\mathcal{S}}(x)$ and $\mathrm{UPP}_{\mathcal{S}}(x)$ are the *lower* and *upper rough approximations* of x in $\mathcal{S}$, respectively, defined as follows [6,12]:

$$\begin{aligned} \mathrm{LOW}_{\mathcal{S}}(x) &\stackrel{\mathrm{def}}{=} \{s \in S \mid \kappa[I(s), x] = 1\}, \\ \mathrm{UPP}_{\mathcal{S}}(x) &\stackrel{\mathrm{def}}{=} \{s \in S \mid \kappa[I(s), x] > 0\}. \end{aligned} \tag{14}$$

[7] Two situations may be treated as indiscernible not only because they cannot be distinguished. Another motivation is that the observable differences are negligible.

[8] As in the preceding section, we omit parameters for simplicity.

Thus,

$$\begin{aligned} \mathrm{LOW}_{\mathcal{S}}(x) &= \{s \in S \mid I(s) \subseteq x\}, \\ \mathrm{UPP}_{\mathcal{S}}(x) &= \{s \in S \mid I(s) \cap x \neq \emptyset\}. \end{aligned} \tag{15}$$

It is easy to prove that the lower and upper rough approximations of x may be characterized by the following equations as well:

$$\begin{aligned} \mathrm{LOW}_{\mathcal{S}}(x) &= \bigcup\{I(s) \mid s \in S \wedge I(s) \subseteq x\} \\ \mathrm{UPP}_{\mathcal{S}}(x) &= \bigcup\{I(s) \mid s \in S \wedge I(s) \cap x \neq \emptyset\}. \end{aligned} \tag{16}$$

Elements of $\mathrm{LOW}_{\mathcal{S}}(x)$ surely belong to x, whereas elements of $\mathrm{UPP}_{\mathcal{S}}(x)$ possibly belong to x, relative to $\mathcal{S}$. One can see that x is definable in $\mathcal{S}$ iff

$$\mathrm{LOW}_{\mathcal{S}}(x) = \mathrm{UPP}_{\mathcal{S}}(x). \tag{17}$$

In a more general case where $I^{\rightarrow}(S)$ is a covering but not a partition of S, the above observation is not valid. Equations (16) do not hold in general, either. The problem of finding of the best candidates for generalized lower and upper rough approximations is discussed in detail in a separate paper. Let us note only that in case I satisfies the condition

$$\forall s, s^* \in S.[s \in I(s^*) \rightarrow s^* \in I(s)], \tag{18}$$

$\mathrm{UPP}_{\mathcal{S}}(x)$ given by (14) is still a reasonable candidate for a generalized upper rough approximation of x. On the other hand,

$$\mathrm{LOW}^*{}_{\mathcal{S}}(x) = \bigcup\{I(s) \mid s \in S \wedge \forall t \in I(s).\kappa[I(t), x] = 1\} \tag{19}$$

seems to be a better candidate than $\mathrm{LOW}_{\mathcal{S}}(x)$ for a generalized lower rough approximation of x. $\mathrm{LOW}^*{}_{\mathcal{S}}(x)$ is definable in $\mathcal{S}$, whereas $\mathrm{LOW}_{\mathcal{S}}(x)$ may actually be undefinable in our sense.

With every formula α of $\mathcal{L}$, we associate the set of situations of S, where α holds. Postponing to another occasion the discussion on how to understand that a formula holds in a situation, let us define the *meaning* of α, $||\alpha||$, as the set of situations of S, where α holds. Formulas having the same meaning are semantically equivalent. If the meaning of α is definable in $\mathcal{S}$, then $||\alpha||$ is a simple granule of information in $\mathcal{S}$ represented by α. If $||\alpha||$ is not definable in $\mathcal{S}$, we may approximate it by a pair of definable sets of situations. Then α may be viewed as a symbolic representation of a pair of granules of information[9] in $\mathcal{S}$. If $I^{\rightarrow}(S)$ is a partition of S, α may be viewed as a representation of the lower and upper rough approximations of $||\alpha||$, i.e.,

$$(\mathrm{LOW}_{\mathcal{S}}(||\alpha||), \mathrm{UPP}_{\mathcal{S}}(||\alpha||)). \tag{20}$$

[9] According to the previous remarks, a pair of granules of information is a compound granule of information.

If $I^{\rightarrow}(S)$ is not a partition of S, then $\mathrm{LOW}_{\mathcal{S}}$ and $\mathrm{UPP}_{\mathcal{S}}$ should be replaced by appropriate generalizations. The meaning of a set of formulas x, $||x||$, is defined as

$$||x|| = \{||\alpha|| \mid \alpha \in x\}. \tag{21}$$

Clearly, $||x||$ is a set of granules of information in $\mathcal{S}$ and hence, may be seen as a granule of information in $\mathcal{S}$. Thus, x represents a granule of information. The meaning of a rule $r = (x, y, \alpha)$, $||r||$, may be defined as the triple $||r|| = (||x||, ||y||, ||\alpha||)$. Thus, $||r||$ is a granule of information in $\mathcal{S}$ as a triple of granules of information in $\mathcal{S}$, and hence, r represents a granule of information. As a consequence, transformations of rules (e.g., composition and decomposition) may be viewed as granular computing problems. Along the same lines, the meaning of a set of rules x, $||x||$, may be defined as

$$||x|| = \{||r|| \mid r \in x\}. \tag{22}$$

Arguing as before, we can draw a conclusion that x represents a granule of information. Finally, the meaning of a rule complex x, $||x||$, may be defined as

$$||x|| = \{||y|| \mid y \in x\}. \tag{23}$$

Thus, rule complexes represent compound granules of information.

Another key issue is application of rules. A necessary but usually insufficient condition for applying a rule $r = (x, y, \alpha)$ in a situation s is that r is activated in s. Given an approximation space $\mathcal{S}$, as above, we say that r is activated in a situation s if each premise of r certainly holds in s and each justification of r possibly holds in s. If $I^{\rightarrow}(S)$ is a partition of S, then r is said to be *activated* in s iff

$$\forall \beta \in x. s \in \mathrm{LOW}_{\mathcal{S}}(||\beta||) \text{ and } \forall \beta \in y. s \in \mathrm{UPP}_{\mathcal{S}}(||\beta||). \tag{24}$$

If $I^{\rightarrow}(S)$ is merely a covering of S, then $\mathrm{LOW}_{\mathcal{S}}$ and $\mathrm{UPP}_{\mathcal{S}}$ in the formula above should be replaced by appropriate generalized rough approximation mappings.

Consider situations $s \in S$ such that $I(s)$ is finite. Then we define that a formula α *holds* in a situation s *to a degree* $k \in [0, 1]$ when

$$\frac{\#[I(s) \cap ||\alpha||]}{\#I(s)} = k \tag{25}$$

by a straightforward adaptation of (11), that is, α holds in a situation s to the degree k if s belongs, to the degree k, to the set of situations where α holds.

The granular computing approach sheds a new light on the problems of what we mean by "a formula holds in a situation," application of rules and rule complexes, and many others.

5 GGT Model of Social Actors in Terms of Rule Complexes

In GGT, social actors as well as the games played and, more generally, social interactions are modeled by rule complexes. These complexes represent compound granules of information, as argued in the preceding section. On the other hand, rules constituting actors' rule complexes are drawn together on the base of similarity and/or functionality, that is, such rule complexes are themselves granules of information.

In this section, we consider the actor case, but social interactions may also be represented by appropriate rule complexes. The GGT model of a social actor is presented here in a very general way. We aim more at showing the structure of the model than describing a particular actor participating in a social interaction. A more detailed presentation of GGT models of social actors and interactions will be postponed to another occasion. We do not discuss such questions as the acquisition of information/knowledge, the formation of rules and rule complexes, or communication of actors, either. With the help of the figures below, we illustrate some ideas of our modeling of social actors. The broad boundaries of the ellipses suggest the vague nature of the concepts presented, whereas the arrows point at the existence of relationships among complexes.

The totality of rules, associated with an actor i in a situation s, is organized into a rule complex called i's *actor complex* in s, written $\mathrm{ACTOR}(i,s)$. This rule complex is usually a huge granule of information, consisting of various rules and rule complexes that themselves are granules of information as well. Two main subcomplexes of $\mathrm{ACTOR}(i,s)$ are distinguished (Fig. 1): i's *role complex* in s, $\mathrm{ROLE}(i,s)$ and i's *control complex* in s, $\mathrm{CTRL}(i,s)$. The first rule complex describes the actor i's social roles in the situation s in terms of rules and rule complexes. The notion of a social role is one of the key notions of sociology. An actor may have many different and often mutually incompatible social roles in a given interaction, e.g., family roles, roles played in the workplace, the role of customer, the role of church member, etc. $\mathrm{ROLE}(i,s)$ is obtained from $\mathrm{ACTOR}(i,s)$ by neglecting all the rules of $\mathrm{ACTOR}(i,s)$ that are irrelevant to the topics "social roles." On the other hand, $\mathrm{CTRL}(i,s)$ consists of management rules and procedures that control functioning of the whole complex $\mathrm{ACTOR}(i,s)$ and its parts, describe how to derive new rules from primary ones, to draw conclusions, to make judgments, and to transform rules and rule complexes.

To play their social roles, actors are equipped with systems of norms and values, telling them what is good, bad, worth striving for, what ought to be done, and what is forbidden. In GGT, norms and values are modeled by rules and rule complexes, and hence they represent some granules of information, as argued in the preceding section. Systems of norms and values of the actor i in s are represented by a rule complex $\mathrm{VALUE}(i,s)$, referred to as i's *value complex* in s (Fig. 2). $\mathrm{VALUE}(i,s)$ is

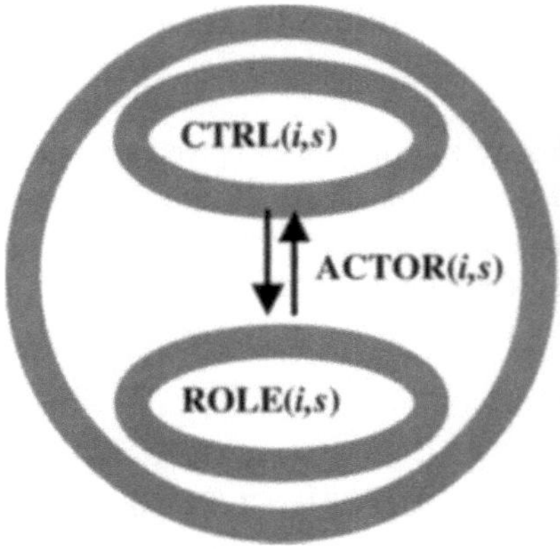

Fig. 1. An actor i's rule complex and its main subcomplexes in a situation s

a subcomplex of ROLE(i,s). Actually, it is a granule of information obtained from the role complex by removing all rules and rule complexes irrelevant to the topics "values and norms."

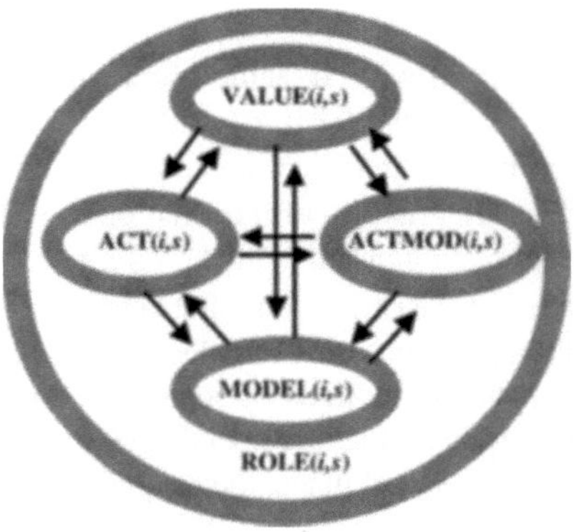

Fig. 2. Main subcomplexes of i's role complex in s

Actors have beliefs and knowledge about themselves, about other actors involved in the interaction, and about the situation. The actor i's beliefs and knowledge in s are represented by rules and rule complexes that form a rule complex MODEL(i,s), called i's *model* in s. MODEL(i,s) is a granule of information, a subcomplex of ROLE(i,s) (see Fig. 2). Being to some extent provided with possible action repertoires, actors can also plan and construct appropriate actions if needed and possible. In GGT, actions are modeled by rules and rule complexes. Modes of acting are described by procedures called *action modalities*. Several pure action modalities are distinguished, viz., instrumental rationality (goal- or payoff-oriented action), normatively oriented action, procedural modality (e.g., medical procedures, bureaucratic orders), ritual and communication, and play [2]. Rules and rule complexes, representing possible actions associated with i in s, are composed into a rule complex $\mathrm{ACT}(i,s)$ called i's *action complex* in s. Action modalities of i in s are represented in the form of rule complexes that are next composed into a rule complex $\mathrm{ACTMOD}(i,s)$ referred to as i's *modality complex* in s. Both $\mathrm{ACT}(i,s)$ and $\mathrm{ACTMOD}(i,s)$ are subcomplexes of $\mathrm{ROLE}(i,s)$. They are information granules formed from $\mathrm{ROLE}(i,s)$ by neglecting all of its parts that are irrelevant to the topics "action and interaction" (Fig. 2).

Given a situation s, suppose that the actor i participates in a game or interaction G. All rules specifying and describing i's role as a player in the game G in the situation s form a subcomplex $\mathrm{ROLE}(i,s,G)$ of i's role complex $\mathrm{ROLE}(i,s)$ (see Fig. 3). Similarly, evaluative rules (i.e., rules representing values and norms) associated with G constitute a subcomplex $\mathrm{VALUE}(i,s,G)$ of i's value complex $\mathrm{VALUE}(i,s)$; action rules (i.e., rules describing actions) relevant to G form a subcomplex $\mathrm{ACT}(i,s,G)$ of $\mathrm{ACT}(i,s)$; rules and rule complexes representing action modalities appropriate for G form a subcomplex $\mathrm{ACTMOD}(i,s,G)$ of $\mathrm{ACT}(i,s)$; and finally, rules and rule complexes representing beliefs and knowledge of i in the game G in s constitute a subcomplex $\mathrm{MODEL}(i,s,G)$ of i's model $\mathrm{MODEL}(i,s)$. On the other hand, the rule complexes $\mathrm{VALUE}(i,s,G)$, $\mathrm{ACT}(i,s,G)$, $\mathrm{ACTMOD}(i,s,G)$, and $\mathrm{MODEL}(i,s,G)$ just mentioned are subcomplexes of i's role complex in the game G and the situation s, $\mathrm{ROLE}(i,s,G)$. These relationships may be depicted, as in Fig. 2, by adding a new parameter G.

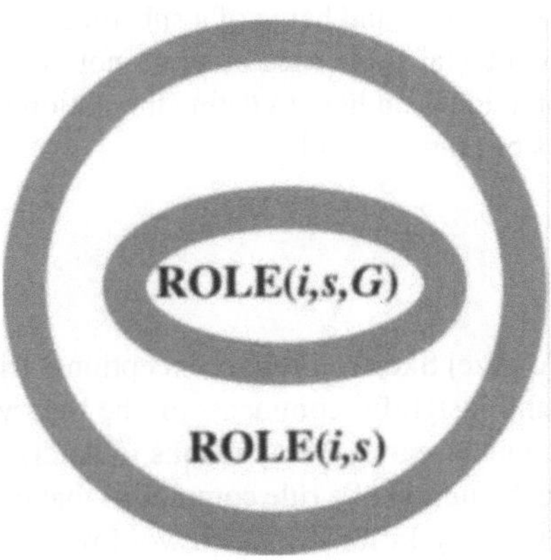

Fig. 3. An actor i's role complex relevant to a game G as a subcomplex of i's (general) role complex in s

It is hardly an exaggeration to say that, as for single rules, the main questions concerning rule complexes are application and transformation. The application of a rule complex resolves itself into the application of some rules. The transformation of a complex of points (in particular, rules) includes the transformation of points of the complex as a special case. The main, yet not necessarily elementary, kinds of transformations are composition and decomposition. Considering our previous remarks on rule complexes versus the granules of information,[10] it becomes clear that the application and transformation of rule complexes may be treated as GC problems.

6 Summary

In this chapter, we presented the mathematical foundations of the theory of socially embedded games from the perspective of granular computing. Rules in GGT represent granules of information; so do rule complexes as multilevel structures built of rules. Rule complexes are special cases of complexes of points of a space where the points are just rules over a language considered. Some rule complexes may be viewed as granules of information themselves since their constituents are drawn together on the base of similarity and/or functionality. As a consequence, the application and transformation of rules and rule complexes may be treated as GC problems. This creates new prospects for further developments of both GGT and GC. GC methods may help to solve some key problems addressed in GGT. On the other hand, those very problems may stimulate innovations and improvements of GC tools.

In this chapter, we also discussed mereological questions with respect to complexes of points. Three kinds of crisp ingredient–whole relationships were defined: being of a g-element, being of a g-subset, and being of a subcomplex of a complex of points. We presented the fundamental properties of these notions and several illustrative examples. We gave some ideas on how to define the notion of a part (ingredient) to a degree for complexes of points as well.

Acknowledgments

I am very grateful to Andrzej Skowron for his exceptional friendliness and support, valuable discussions, and insightful comments on the theory of complexes (still in progress) and the previous research. Many thanks to Andrzej Wieczorek who suggested that I generalize the notion of a rule complex to that of a complex of arbitrary objects, and to Adam Grabowski for help with word processing. Special thanks to Wojciech Gomoliński for invaluable help with graphics. Last, but not least, I thank the anonymous referees for the constructive criticism and useful remarks that helped to improve the final version of this chapter. The research was supported in part by the Polish National Research Committee (KBN), grant # 8T11C 02519.

[10] Rule complexes may be viewed as structures representing granules of information, and, moreover, some rule complexes are granules of information themselves.

References

1. T.R. Burns, A. Gomolińska. Modelling social game systems by rule complexes. In *[15]*, 581–584, 1998.
2. T.R. Burns, A. Gomolińska. The theory of socially embedded games: The mathematics of social relationships, rule complexes, and action modalities. *Quality and Quantity: International Journal of Methodology*, 34(4): 379–406, 2000.
3. T.R. Burns, A. Gomolińska. Socio-cognitive mechanisms of belief change: Applications of generalized game theory to belief revision, social fabrication, and self-fulfilling prophesy. *Cognitive Systems Research*, 2(1): 39–54, 2001. Available at http://www.elsevier.com/locate/cogsys.
4. T.R. Burns, A. Gomolińska, L.D. Meeker. The theory of socially embedded games: Applications and extensions to open and closed games. *Quality and Quantity: International Journal of Methodology*, 35(1): 1–32, 2001.
5. A. Gomolińska. Rule complexes for representing social actors and interactions. *Studies in Logic, Grammar and Rhetoric*, 3(16): 95–108, 1999.
6. J. Komorowski, Z. Pawlak, L. Polkowski, A. Skowron. Rough sets: A tutorial. In *[11]*, 3–98, 1999.
7. S. Leśniewski. *Foundations of the General Set Theory I (in Polish)*. Works of the Polish Scientific Circle, Vol. 2, Moscow, 1916. See also S.J. Surma et al., editors, *Stanisław Leśniewski. Collected Works*, 128–173, Kluwer, Dordrecht, 1992.
8. T.Y. Lin. Granular computing on binary relations 1. Data mining and neighborhood systems. In *[16]*, Vol. 1, 107–121, 1998.
9. T.Y. Lin. Granular computing: Fuzzy logic and rough sets. In *[34]*, Vol. 1, 183–200, 1999.
10. S.H. Nguyen, A. Skowron, J. Stepaniuk. Granular computing. A rough set approach. *Computational Intelligence*, 17(3): 514–544, 2001.
11. S.K. Pal, A. Skowron, editors. *Rough-Fuzzy Hybridization: A New Trend in Decision Making*. Springer, Singapore, 1999.
12. Z. Pawlak. *Rough Sets: Theoretical Aspects of Reasoning about Data*. Kluwer, Dordrecht, 1991.
13. L. Polkowski, A. Skowron. Adaptive decision-making by systems of cooperative intelligent agents organized on rough mereological principles. *International Journal of Intelligent Automation and Soft Computing*, 2(2): 121–132, 1996.
14. L. Polkowski and A. Skowron. Rough mereology: A new paradigm for approximate reasoning. *International Journal of Approximated Reasoning*, 15(4): 333–365, 1996.
15. L. Polkowski, A. Skowron, editors. *Proceedings of the 1st International Conference on Rough Sets and Current Trends in Computing (RSCTC'1998)*, LNAI 1424, Springer, Berlin, 1998.
16. L. Polkowski, A. Skowron, editors. *Rough Sets in Knowledge Discovery*, Vols. 1, 2. Physica, Heidelberg, 1998.
17. L. Polkowski, A. Skowron. Towards adaptive calculus of granules. In *[34]*, Vol. 1, 201–228, 1999.
18. L. Polkowski, A. Skowron. Rough mereology in information systems. A case study: Qualitative spatial reasoning. In *[21]*, 89–135, 2000.
19. L. Polkowski, A. Skowron. Rough-neuro computing. In W. Ziarko, Y. Y. Yao, editors, *Proceedings of the 2nd International Conference on Rough Sets and Current Trends in Computing (RSCTC 2000)*, LNAI 2005, 25–32, Springer, Berlin, 2001.

20. L. Polkowski, A. Skowron, J. Komorowski. Towards a rough mereology-based logic for approximate solution synthesis 1. *Studia Logica*, 58(1): 143–184, 1997.
21. L. Polkowski, S. Tsumoto, T. Y. Lin, editors. *Rough Set Methods and Applications: New Developments in Knowledge Discovery in Information Systems*. Physica, Heidelberg, 2000.
22. R. Reiter. A logic for default reasoning. *Artificial Intelligence*, 13: 81–132, 1980.
23. A. Skowron, J. Stepaniuk. Tolerance approximation spaces. *Fundamenta Informaticae*, 27: 245–253, 1996.
24. A. Skowron, J. Stepaniuk. Information granules in distributed environment. In *Proceedings of the Conference on New Directions in Rough Sets, Data Mining, and Granular Soft Computing (RSDGrC'99)*, LNAI 1711, 357–365, Springer, Berlin, 1999.
25. A. Skowron, J. Stepaniuk. Towards discovery of information granules. In *Proceedings of the 3rd European Conference on Principles and Practice of Knowledge Discovery in Databases (PKDD'99)*, LNAI 1704, 542–547, Springer, Berlin, 1999.
26. A. Skowron, J. Stepaniuk, S. Tsumoto. Information granules for spatial reasoning. *Bulletin of the International Rough Set Society*, 3(4): 147–154, 1999.
27. B. Sobociński. Studies in Leśniewski's mereology. In T. J. Srzednicki, V. F. Rickey, J. Czelakowski editors, *Leśniewski's Systems. Ontology and Mereology*, 217–227, Nijhoff/Ossolineum, The Hague, 1984. See also *Yearbook for 1954–55 of the Polish Society of Arts and Sciences Abroad V*, 34–48, London, 1954–55.
28. J. von Neumann, O. Morgenstern. *Theory of Games and Economic Behaviour*. Princeton University Press, Princeton, NJ, 1944.
29. L.A. Zadeh. Fuzzy sets. *Information and Control*, 8: 338–353, 1965.
30. L.A. Zadeh. Outline of a new approach to the analysis of complex system and decision processes. *IEEE Transactions on Systems, Man, and Cybernetics*, 3: 28–44, 1973.
31. L.A. Zadeh. Fuzzy sets and information granularity. In M. Gupta, R. Ragade, R. Yager, editors, *Advances in Fuzzy Set Theory and Applications*, 3–18, North-Holland, Amsterdam, 1979.
32. L.A. Zadeh. Fuzzy logic = computing with words. *IEEE Transactions on Fuzzy Systems*, 4(2): 103–111, 1996.
33. L.A. Zadeh. Toward a theory of fuzzy information granulation and its certainty in human reasoning and fuzzy logic. *Fuzzy Sets and Systems*, 90: 111–127, 1997.
34. L.A. Zadeh and J. Kacprzyk, editors. *Computing with Words in Information/Intelligent Systems*, Vols. 1, 2. Physica, Heidelberg, 1999.

Chapter 17
Fuzzy Games and Equilibria: The Perspective of the General Theory of Games on Nash and Normative Equilibria

Tom R. Burns,[1] Ewa Roszkowska[2]

[1] Uppsala Theory Circle, Department of Sociology, University of Uppsala, Box 821, 75108 Uppsala, Sweden
tom.burns@soc.uu.se

[2] University of Białystok, Faculty of Economics, Warszawska 63, 15-062 Białystok, Poland and
Białystok School of Economics, Choroszczańska 31, 15-732 Białystok, Poland
erosz@w3cache.uwb.edu.pl

Summary. Extending the general theory of games (GGT), this paper fuzzifies judgment, decision making, and game equilibria. It models the ways in which players use approximate reasoning and deal with imprecise information in making decisions, interacting, and generating equilibria. The conceptualization of fuzzy judgment entails a two process model: (1) the judgment of similarity and dissimilarity (where threshold functions) are provided; (2) the judgment of fit or degree of membership formulated as a fuzzy set M taking on values between $[0,1]$. The paper goes on to provide a theory of equilibria. In particular, a generalized Nash equilibrium is formulated in fuzzy set terminology. Moreover, normative and nonnormative Nash equilibria are distinguished. Normative equilibria satisfy or realize an appropriate norm or value r in the game (not all games have normative equilibria; not all normative equilibria are Nash equilibria, and vice versa). This fuzzified GGT can be applied to classical closed games as well as to open games. The prisoners' dilemma game is considered for the closed game analysis, and it is shown how social relationships between players ("the social embeddedness" of games) impact players' judgments, internation patterns, and equilibria. Application to open games is illustrated by considering bilateral bargaining games in market exchange.

1 Introduction to the General Theory of Games

In their classic work [28], Von Neumann and Morgenstern defined a game as simply the totality of the rules that describe it. They did not, however, elaborate a theory of rules. Such considerations lead to conceptualizing rules and rule configurations as mathematical objects, specifying the principles for combining rules, developing the theory of revising, replacing, and, in general transforming rules and rule complexes. In the general theory of games (GGT) ([4–8,12]), games are conceptualized in a uniform and general way as rule complexes.[1]

A well-specified game at time t is a particular interaction where the participating actors (players) have defined roles and role relationships [although not all games are necessarily well-defined with, for instance, clearly specified and consistent roles and role relationship(s)]. A social role is a particular rule complex, operating as the basis of an incumbent's judgments and actions in relation to other actors (players) in their roles in the defined game. The notion of a *situation* is a primitive. Situations are denoted by S with subscripts, if needed. We use the lowercase t, possibly with subscripts, to denote points of time (or reference). Thus, S_t denotes a situation at time t. Given a concrete situation S at time t, a *general game structure* is represented as a particular rule complex $G(t)$. This complex includes roles as rule subcomplexes along with norms and other rules. Suppose that a group or collective $I = \{1, \ldots, m\}$ of players is involved in a game $G(t)$. ROLE(i,t,G) denotes actor i's *role complex* in the game $G(t)$, and ROLE(I,t,G) denotes the *role configuration* of all players in I engaged in $G(t)$.[2]

Role relationships provide complexes of appropriate rules, including values and norms, particular ways to classify and judge action, and "internal" interpretations and meanings [3]. "Noncooperation" in, for instance, a prisoners' dilemma (PD) situation will not be merely "defection" if the players are friends or relatives in a solidary relationship, but a form of "disloyalty" or "betrayal" and subject to harsh social judgment and sanction. In the case of enemies, "defection" in the PD game would be fully expected and considered "natural" — neither shameful nor contemptible, but right and proper rejection of or damage to the other, and, hence, not a matter of "defection" at all. Such a sociological perspective on games enables us to systematically identify and analyze their symbolic and moral aspects associated with established social relationships.

An actor's (player's) role is specified in GGT in terms of a few basic cognitive and normative components (formalized as mathematical objects in [4–6,8,12]). A rule complex VALUE(i,t,G) represents actor (player) i's value complex and is the

[1] A major concept in GGT is the rule complex. The motivation behind the development of this concept has been to consider repertoires of rules in all their complexity with complex interdependencies among the rules and, hence, not to consider them merely as sets of rules. The organization of rules in rule complexes provides us with a powerful tool for investigating and describing various sorts of rules with respect to their functions, such as values, norms, judgment rules, prescriptive rules, and metarules as well as more complex objects consisting of rules such as roles, routines, algorithms, action modalities, models of reality as well as social relationships and games. Informally speaking, a rule complex is a set consisting of rules and/or other rule complexes (see Gomolińska, this volume).

[2] In a game $G(t)$, the individual role complex, ROLE(i,t,G), for every actor (player) $i \in I = \{1, \ldots, m\}$ is a subcomplex of the role configuration ROLE(I,t,G), and the latter complex is a subcomplex of $G(t)$ in the formalism of rule complex theory. These relationships are represented as follows ([5,8]): ROLE$(i,t,G) \subseteq_g$ ROLE$(I,t,G) \subseteq_g G(t)$.

basis for generating evaluations and preferences through judgment processes; it contains norms and values relating to what to do and what not to do and what is good or bad. MODEL(i,t,G) represents the actor's (player's) belief structure about herself and her environment, relevant conditions, and constraints in the situation S_t. This is a complex of beliefs or rules that frame and define the reality of relevant interaction situations. ACT(i,t,G) represents the repertoire of acts, strategies, routines, programs, and action available to actor (player) i in her particular role in the game situation S_t and MODALITY(i,t,G) is her mode or modes of generating or determining action in the situation S_t; they are the particular ways to organize the determination of actions in a given role in relation to particular other actors (players).[3]

The point of departure of GGT has been to explore fruitful ways to extend and develop classical game theory (GT). In GGT there are several parallels as well as extensions of concepts of rational choice and GT. The concept of value complex, VALUE(i,t,G), has parallels to, but also encompasses and extends, the concept of a utility function. Utility is a unidimensional value satisfying artificial rules of rationality such as consistency and complete ordering of preferences. The value complex VALUE(i,t,G) concept makes no such claim of universal consistency and well-orderedness in human judgments. A complex, however, is the basis on which players make evaluative judgments and generate preferences, although it may be characterized by inconsistencies, gaps, and fuzziness. Value complexes are closely associated with particular roles, social relationships, and institutional arrangements. The model complex MODEL(i,t,G) describes belief structures and cognitive frames and provides a particular perspective on, and a basis for understanding, the perceived reality of an interaction in the context of defined social relationships. In classical GT, all game players are assumed to have the same representation of the situation, and, moreover, this representation is assumed to be identical to that of the game theorist. The GGT complex MODEL(i,t,G), corresponding to the rational choice notion of "information" (perfect or less than perfect), conceptualizes a player's bounded, possibly distorted, or even false beliefs and cognitive frames (as in the case of collective ignorance or in negotiation processes where participants employ deception [5]). The next important component of $G(t)$ is a MODALITY(i,t,G) complex that describes the kind of modalities involved in action in a game. Classical GT assumes that players operate with a common universal modality, namely, that of instrumental rationality (for instance, maximizing expected gain or utility). In GGT, a variety of socially embedded judgment rules and algorithms replaces a given fixed algorithm, such as maximization of utility. In related articles [5,6], GGT formulates the concept of multimodal social action, which generalizes a number of other approaches. For instance, the modality of instrumental rationality, the routine or habitual modality,

---

[3] For each actor (player) i, a role is specified as follows:
MODEL(i,t,G), VALUE(i,t,G), ACT(i,t,G), MODALITY$(i,t,G) \subseteq_g$ ROLE$(i,t,G) \subseteq_g G(t)$, that is, the role components are subcomplexes of the role, and the roles (along with other roles) are subcomplexes of the game.

the normative modality, "communicative/dramaturgy," "aesthetic," and "play." Role conceptions of action may combine modalities such as instrumental and normative. Modalities for determining action encompass processes of "choice" or "decision" but also account for determination processes such as habitual action devoid of deliberation and choice. Finally, GGT conceptions of action alternative and strategy differ from those of game theory. The complex $\mathrm{ACT}(i,t,G)$ in a role or action complex includes "action alternatives" in a decision or game situation. But it also includes routines, programs, and habits with which players may respond automatically (i.e., without deliberation or choice) to particular problems or conditions in that situation. It includes as well operations that relate to stopping (or pausing) in a program or routine as well as acts making up a reflexive mode with respect to rule complexes. Of particular importance is that in the GGT conceptualization, $\mathrm{ACT}(i,t,G)$ may be subject to attempts by the players themselves to manipulate and control. In other words, $\mathrm{ACT}(i,t,G)$ may become an object players try to constrain or transform, that is a policy variable. Obviously, particular social, institutional, and other factors constrain the complex of options, limiting the extent it can be manipulated or transformed by the participants. The useful distinction between *open and closed games* (see later in this chapter) derives from this conceptualization of action repertoires and the capacity of agents to construct and transform them.

Our generalization of game theory (GT) implies that there are many game theories or models reflecting or referring to diverse social relationships and corresponding rationalities or action logics. GT is, therefore, a general model limited in its scope and applicable to a particular type of social relationship, namely, that between unrelated or anomic agents acting and interacting in accordance with rationality rules and modalities. The players lack sentiments, either for or against one another. They are purely neutral and egoistic in their relationship. Moreover, their games are closed. The players may not change the rules such as the number and qualities of participants, the specific action alternatives and outcomes, the modality of action, or the particular social relationships obtaining between them. GGT reconceptualizes a major result of classical GT, namely, the Nash equilibrium. It also introduces the concept of normative equilibrium which may or may not be a Nash equilibrium. In GGT, a normative equilibrium is an interaction pattern, set of consequences, or payoff that participants in a game judge to realize or satisfy a norm or value appropriate for the situation. Whether a game has or does not have such an equilibrium depends on the type of relationship among participating players, the relevant norms or values, and the situational conditions obtaining. Table 1 presents distinctions between classical GT, on the one hand, and GGT, on the other.

This chapter, after briefly introducing the GGT concepts of judgment, action determination, and Nash as well as normative equilibria, reformulates these and other concepts in fuzzy set terms. For instance, fuzzy sets methods prove useful for describing Nash as well as normative equilibria and also imply particular factors generating and explaining patterns of game equilibria [1,2,10,16,24].

Table 1. Comparison of the general theory of games and classical game theory

The General Theory of Games	Classical Game Theory
Game rule complex, $G(t)$. Together with physical and ecological constraints, it structures and regulates action and interaction.	**Game constraints** ("rules").
VALUE(i,t,G) **complex.** A player's values and evaluative structure derives from the social context of the game (institutional arrangements, social relationships, and roles).	**Utility function** or preference ordering given or exogenously determined.
MODEL(i,t,G) **complex.** A player's model of the game situation is to a greater or lesser extent, incomplete, fuzzy, or even false. Reasoning processes may or not follow standard logic.	**Perfect or imperfect information** about the game players, their options, payoffs, and preference structures.
Imprecise (or fuzzy/rough) entities. Data and rules, strategies, and judgments.	**Crisp** information, strategies, and decisions.
ACT(i,t,G) **complex.** It represents the repertoire of acts, strategies, routines, programs, and actions available to player i in her particular role and role relationships in the game situation.	**Set of strategies.**
MODALITY(i,t,G) **complex.** There are multiple modalities of action determination including instrumental, normative, and habitual modes of action determination, among others.	**Singular modality.** Rational choice or instrumental rationality.
Value or norm realization. The universal motivational factor is the drive to realize or achieve a value or norm.	**Maximization of expected utility** as universal principle.
Normative and nonnormative equilibria.	**Nash equilibrium** (which conflates different types of socially meaningful equilibria).
Open and closed games.	**Closed games.**

2 Fuzzification of Decision Making, Games, and Equilibria

2.1 Judgment of Similarity and Dissimilarity

Judgment is a central concept in GGT. The major basis of judgment is matching, that is, a process of comparing and determining similarity or dissimilarity ([5,6,27]). Judgment like many other notions in GGT is contextualized to *particular actor(s)* [*player(s)*], *situation, time,* etc.

Let $J(i,t)$ be a rule complex, possibly in the form of an algorithm, with which a player i makes a comparison and judges the degree of similarity of goodness of fit of two objects (x,y) under consideration in situation S at time t. The result of application of $J(i,t)$ to a pair objects $(x,y), J(i,t)(x,y)$, is an expression describing in qualitative or quantitative terms whether or not x,y are similar and to what extent. It can take the form: "similar," "dissimilar," "not decided," "almost similar," "sufficiently similar," "highly similar," or "dissimilar to the degree d," etc. Associated with the judgment complex J, there are typically thresholds for dissimilarity and similarity, as discussed below. (Henceforth, we drop the context-specific defining indexes i and t unless necessary to distinguish players and contexts, or we refer to a list of parameters p possibly containing a player's name, time, situation, "topics," etc.).

The capacity of players to judge similarity or likeness (that is, up to some threshold specified by a metarule or norm of stringency) is the foundation for rule following or rule application and plays a major part in belief change and action processes generally [5,12]. It is also the point of departure for conceptualizing fuzzy judgment.

2.2 Fuzzy Judgment

GGT models the judgment process, and in particular, the judgment of similarity or dissimilarity and the use of approximate reasoning and imprecise information in making decisions, interacting, and stabilizing situations. The conceptualization of fuzzy judgment entails a two process model:

1. The judgment of similarity and dissimilarity (where thresholds threshold functions are provided).
2. The judgment of fit or degree of membership formulated as a fuzzy set M taking on values between $[0,1]$.

In the next section, the judgment expression $J(i,t)$ and its threshold functions are transformed into a fuzzy function M. M does two things. First, it normalizes judgment with a minimum (0) for sufficiently dissimilar and a maximum (1) for sufficiently similar. Second, when one has less than perfect similarity (or dissimilarity), it distinguishes degree of fit or membership, which may be based on fuzzy verbal distinctions such as "moderately fitting," "borderline," and "not fitting very well," that is, it represents fuzzy judgments between the maximum and minimum with breakpoints or thresholds based on distinctions, such as *moderately fitting* and *borderline,* or other semantic distinctions. Rather than judgment as a matter of yes or no, it may express a degree of, for instance, preference, consensus, compliance, rule matching, or normative equilibrium. In general, GGT is able to use key social concepts which are imprecise and ambiguous: definition of the situation, the game or type of game, role, norm, value, particular types of action such as "cooperation" and "noncooperation" or "compliance" and "noncompliance."

Let U be a universe of all objects to be compared with one another. Given a list of parameters p (the list may contain a player's name, time, situation, "topics," etc.), suppose there is a function $M_p : U \times U \rightarrow [0,1]$, measuring in real numbers a player's judgment of *the degree of membership of pairs* (x,y) *of objects of* U to the set of all pairs of objects similar to each other relative to p. In other words, M_p measures the degree of similarity of objects. Now, suppose the player has threshold functions $f_{p,1}$ and $f_{p,2}$ for dissimilarity and similarity, respectively. Assume that for all $x,y \in U$, $f_{p,1}(x,y) < f_{p,2}(x,y)$. Then for any pair (x,y) of objects, we define that x and y are dissimilar relative to p, written $dissim_p(x,y)$, iff $M_p(x,y) \leq f_{p,1}(x,y)$, and x and y are similar relative to p, written $sim_p(x,y)$, iff $f_{p,2}(x,y) \leq M_p(x,y)$. We have also $M_p(x,x) = 1$.

Definition 1. [11,14] A fuzzy set A in a universe of discourse $U = \{u_1, u_2, \ldots, u_i, \ldots, u_n\}$ will be represented by a set of pairs $(M_A(u), u)\ \forall u \in U$, where $M_A : U \rightarrow [0,1]$ is a player's *judgment of the fit or degree of membership degree of u in A*:[4] from full membership (= 1) to full nonmembership (= 0) through all intermediate values.

Consider that one of the objects in the universe U is a rule or standard r. We are interested in judgments about the degree of fit with, or membership degree of, some condition or action x with respect to a norm or value r that is, $J(i,t)(x,r)$. If that norms and values as well as actions and outcomes are completely crisp, we have the classical case. The major principles of GGT are discussed in the following section in fuzzy set terminology, which presents several GGT results, including classical forms as special cases.

[4] In a finite universe of discourse $U = \{u_1, u_2, \ldots, u_i, \ldots, u_n\}$ assumed here, the fuzzy set A represented by the set of pairs $(M_A(u), u)\ \forall u \in U = \{u_1, u_2, \ldots, u_i, \ldots, u_n\}$ will be denoted as $A = M_A(u_1)/u_1 + \cdots + M_A(u_n)/u_n$, where $M_A(u_1)/u_1$ is the pair "degree of membership/element" and '+' is in the set-theoretical sense. The basic operations on fuzzy sets are defined in a standard way, that is
complementation: $M_{\neg A}(u) = 1 - M_A(u),\ \forall u \in U$,
union: $M_{A+B}(u) = M_A(u) \cup M_B(u) = max[M_A(u), M_B(u)],\ \forall u \in U$,
intersection: $M_{A-B}(u) = M_A(u) \cap M_B(u) = min[M_A(u), M_B(u)],\ \forall u \in U$.

2.3 Action Determination in Fuzzy Terms

In determining or deciding action, a player compares and judges the similarity between an action or action alternatives and the salient or primary value or goal that the player is oriented to realize or achieve in the situation. More precisely, the player tries to determine if a finite set of expected or predicted consequences of an action a, $Con(a,t)$, in situation S at time t is *sufficiently similar* as measured by $M_{J(i,t)}(a,r)$ to the set of those consequences $Con(r)$ prescribed by the norm or value r. Assume that a player i is oriented to a value r, activated at t, which is contained in VALUE(i,t,G) and which specifies consequences that are to be realized or achieved through action in the situation S_t. Suppose that the player i considers prescribed or expected consequences of r, which are represented by formulas and form a finite set $Con(r,i,t)$. The player tries to identify or construct an action or a sequence of actions a which has anticipated consequences $Con(a,i,t)$ like those specified or directed in $Con(r,i,t)$. More precisely, the consequences of the constructed action(s) a are believed or expected to match or correspond to those specified or prescribed by r. Suppose that $\gamma_1,\gamma_2, \ldots, \gamma_k \in Con(r,i,t)$ are the consequences prescribed or specified by r and $\gamma_1^*,\gamma_2^*,\ldots, \gamma_k^* \in Con_{exp}(a,i,t)$ are the actual expected consequences of i's action or activity. If for each consequence $\gamma_j \in Con(r,i,t)$, there is an actual or anticipated consequence $\gamma_j^* \in Con_{exp}(a,i,t)$ judged by player i at t as sufficiently similar to γ_j, then, the sets $Con_{exp}(a,i,t)$ and $Con(r,i,t)$ are seen as sufficiently similar as well:

$$J_{sim}(i,t)(Con_{exp}(a,i,t),Con(r,i,t)) = \text{sufficiently similar}. \tag{1}$$

This similarity or correspondence is the aim of constructing an action or a sequence of actions a according to or realizing a given value r. This is following or applying a rule in one of its senses (see footnote 8).

This basis for choosing or determining action is formulated in terms of the following principle.

The Principle of Action Determination. Construct, or find and select, an action a in situation S at time t which satisfies rule r; we say then that $sim_{J(i,t)}(a,r)$ and write

$$J(i,t)[Con(a,t),Con(r)] = sim_{J(i,t)}(a,r). \tag{2}$$

Recall that using fuzzy set terminology, we have the following:

$$sim_{J(i,t)}(a,r) \text{ if and only if } f_{J(i,t),2}(a,r) \le M_{J(i,t)}(a,r), \tag{3}$$

where $f_{J(i,t),2}$ is a threshold function, as introduced earlier.

The player may fail to find or to construct an ideal action, for instance, in a closed game or under time or other constraints. A metanorm may allow the actor to relax or compromise with the principle: he may choose that action, whose consequences, although not fully satisfying r, are judged closest possible to those indicated or specified by the norm or value r (see [5] on maximizing the degree of similarity). [5]

For player i, the fuzzy judgment of degree of fit or membership of action a with respect to the norm or value r is given by the function $M_{J(i,t)}(a,r)$, which may be represented, for instance, as follows:

$$M_{J(i,t)}(a,r) = \begin{cases} 1 & \text{for actions that are judged totally consistent with the norm or value } r \\ c & \text{for actions that moderately satisfy } r \text{ but are not outstanding, that is, they are judged relatively close to } r \\ 0.5 & \text{for actions that just minimally fit or satisfy } r \\ 0 & \text{for actions or expressions that fail to satisfy minimum standards or criteria of the norm or value } r, \end{cases}$$

where $c \in (0.5, 1)$.

A player is motivated or driven to maximize the fit or degree of membership in her value complex or goal set — maximize $M_{J(i,t)}(a,r)$ — through her construction and/or choice of action a. This assumes that she knows or hypothesizes the connection between a and its consequences, $Con(a)$. The GGT principle of action determination corresponds to the principle of maximizing utility in rational choice and classical game theories. Its conception of comparison and judgment of similarity and dissimilarity is a major distinguishing feature.

2.4 Generalized Nash Equilibria in Fuzzy Terms from the GGT Perspective

The notion of an *equilibrium point* was the basic component in Nash's classic paper [18]. The equilibrium point is also known as a *noncooperative equilibrium, the Nash noncooperative equilibrium,* and *the Nash–Cournot equilibrium.* It is also the major equilibrium concept in noncooperative game theory. Since agreements in noncooperative games are not enforceable, "rational players" are expected to choose a strategy combination that is self-stabilizing in the sense that the players will have some incentive to abide by a strategy combination if they expect all other players to abide by it. That means that they always choose a strategy combination with the property that every player's strategy is a best countering strategy to all other players' strategies. A strategy combination with this property is called an *equilibrium*

[5] This entails following or applying a rule in one of its senses (see footnote 11). The basic process is one of comparison-test and is universal in constructing, selecting, and assessing action.

(point). Nash ([18], Theorem 1) has also shown that every finite game with complete information has at least one equilibrium point (in pure strategies or sometimes only in mixed strategies).

The Nash result requires that each and every player can compare and rank order their options (or has a utility function over her options). Formulated in fuzzy terms, the judgments with respect to options would take the following form [11,14]:

For player i, given a set of alternatives $\mathrm{ACT}(i,t) = \{a_1^i, a_2^i, \ldots, a_{n_i}^i\}, i \in I$, actor i's fuzzy judgment function or fuzzy preference relation $J(i,t)(a_j^i, a_k^i)$ at time or context t is given by its membership function $M_{J(i,t)} : \mathrm{ACT}(i,t) \times \mathrm{ACT}(i,t) \to [0,1]$, that is, a matrix $J(i,t)(a_j^i, a_k^i) = M_{J(i,t)}(a_j^i, a_k^i)$, or simply $J(i,t)_{jk} = M_{J(i,t)}(a_j^i, a_k^i)$, where

$$M_{J(i,t)}(a_j^i, a_k^i) = \begin{cases} 1 & \text{if } a_j^i = a_k^i \text{ or } a_j^i \text{ is judged as definitely} \\ & \text{better or preferred to } a_k^i \\ c & \text{if } a_j^i \text{ is judged somewhat better} \\ & \text{or preferred to } a_k^i \\ 0.5 & \text{if there is no preference (i.e., indifference)} \\ & \text{between } a_j^i \text{ and } a_k^i \text{ where } a_j^i \neq a_k^i \\ d & \text{if } a_k^i \text{ is judged somewhat better} \\ & \text{or preferred to } a_j^i \\ 0 & \text{if } a_k^i \text{ is judged definitely better} \\ & \text{or preferred to } a_j^i, \end{cases} \tag{4}$$

where $a_j^i, a_k^i \in \mathrm{ACT}(i,t), c \in (0.5, 1), d \in (0, 0.5)$, $J(i,t)(a_j^i, a_k^i) + J(i,t)(a_k^i, a_j^i) = 1$ for $j \neq k$.

For a given set of m players, $I = \{1, \ldots, m\}$, and a set of options (strategies)

$$\mathrm{ACT}\ (i,t) = \{a_1^i, a_2^i, \ldots, a_{n_i}^i\},$$

for each player $i \in I$, where $n_i \in N$, consider that each player i at time or context t has a fuzzy evaluative judgment function or preference relation as expressed in (4).

A generalized Nash equilibrium can now be reformulated in terms of fuzzy judgment functions. Let x denote i-player option (strategy) $a_{j_i}^i$, that is, $x = (a_{j_i}^i)$ and y options (strategies) of other players,

$$y = (a_{j_1}^1, a_{j_2}^2, \ldots a_{j_{i-1}}^{i-1}, a_{j_{i+1}}^{i+1}, \ldots, a_{j_m}^m);$$

then, $J(i,t)(x,y)$ represents player i's evaluative judgments of, or preferences with respect to, the outcomes of her option $a_{j_i}^i$ in situation S_t, given the choices,

$$(a_{j_1}^1, a_{j_2}^2, \ldots, a_{j_{i-1}}^{i-1}, a_{j_{i+1}}^{i+1}, \ldots, a_{j_m}^m),$$

of all other participants. Instead of $J(i,t)(x,y)$, we write $J(i,t)[(a^1_{j_1},a^2_{j_2},\dots,a^i_{j_i},\dots,a^m_{j_m})]$ or $J(i,t)(a)$, where $a=(a^1_{j_1},a^2_{j_2},\dots,a^i_{j_i},\dots,a^m_{j_m})$. Let us denote by

$$\mathrm{ACT}(I,t)=\{a : a=(a^1_{j_1},a^2_{j_2},\dots,a^i_{j_i},\dots,a^m_{j_m}),\ \text{where } a^i_{j_i}\in ACT(i,t),\ \text{and } i\in I\}.$$

In the analysis of behavior, it is often necessary to look at one player's choices while holding the choices of all others fixed. To facilitate this with a convenient notation, $a\backslash a^i_k \equiv (a^1_{j_1},a^2_{j_2},\dots,a^i_k,\dots,a^m_{j_m})$ will be used. Thus, $a\backslash a^i_k$ is

$$a=(a^1_{j_1},a^2_{j_2},\dots,a^i_{j_i},\dots,a^m_{j_m})$$

with its ith coordinate replaced by a^i_k.

Definition 2. An m-tuple of strategies $a_I=(a^1_{j_1},a^2_{j_2},\dots,a^i_{j_i},\dots,a^m_{j_m})$ is a *generalized Nash equilibrium* in pure strategies if the following evaluative judgment or preference below holds:

$$J(i,t)(a_I)\geq J(i,t)(a_I\backslash a^i_s), \tag{5}$$

for all $a^i_s\in\mathrm{ACT}(i,t)$, and all $i\in I$.

A generalized Nash equilibrium obtains when each player in a game is predisposed to choose a strategy in a way that maximizes her expected payoff, *given* that all of the other players act in accordance with the equilibrium. All players are assumed to make their choices independently and without social coordination among them. This is the usual condition for a "noncooperative game" (i.e., they are not communicating or coordinating with one another in an organized manner).

Remark 1. Observe that if $J(i,t)(a_I)$ denotes i-player's payoff function in crisp terms, then we obtain the classical definition of the Nash equilibrium in pure strategies [15].

Definition 3. A strategy a^i_k from $\mathrm{ACT}(i,t)$ is *the best countering strategy or response for player i in pure strategies against the strategy* $a\in\mathrm{ACT}(I,t)$ if the following evaluative judgment or preference relation holds:

$$J(i,t)(a\backslash a^i_k)\geq J(i,t)(a\backslash a^i_s), \tag{6}$$

for all strategies $a^i_s\in\mathrm{ACT}(i,t)$.

Let us denote the set of all best countering strategies for player i against the strategy $a\in\mathrm{ACT}(I,t)$ by $\mathrm{BCS}(i,t)(a)$, that is,

$$\mathrm{BCS}(i,t)(a)=\{a^i_k\in\mathrm{ACT}(i,t) : J(i,t)(a\backslash a^i_k)\geq J(i,t)(a\backslash a^i_s)\ \text{for all } a^i_s\in\mathrm{ACT}(i,t)\}.$$

The strategy a^i_k is the best countering strategy for player i against the strategy $a\in\mathrm{ACT}(I,t)$ if it maximizes i's evaluative judgement, given the strategy choices of the other.

Definition 4. An m-tuple of strategies $a_I = (a^1_{j_1}, \ldots, a^i_{j_i}, \ldots, a^m_{j_m})$ is *the best countering strategy or response in pure strategies against the strategy* $s \in \mathrm{ACT}(I,t)$ if and only if $a^i_{j_i} \in \mathrm{BCS}(i,t)(s)$, for all $i \in I$.

Let us denote the set of the best countering strategies in pure strategies against the strategy $s \in \mathrm{ACT}(I,t)$ by $\mathrm{BCA}(I,t)(s)$. Thus,

$$a_I = (a^1_{j_1}, a^2_{j_2}, \ldots, a^i_{j_i}, \ldots, a^m_{j_m}) \in \mathrm{BCA(I,t)}(s) \text{ iff } a^i_{j_i} \in \mathrm{BCS}(i,t)(s), \text{ for all } i \in I.$$

The best countering strategy provides a natural way to think about generalized Nash equilibrium states because all such states satisfy the condition that

$$a_I = (a^1_{j_1}, a^2_{j_2}, \ldots, a^i_{j_i}, \ldots, a^m_{j_m})$$

is a generalized Nash equilibrium if and only if

$$a_I = (a^1_{j_1}, a^2_{j_2}, \ldots, a^i_{j_i}, \ldots, a^m_{j_m}) \in \mathrm{BCA(I,t)}(a_I).$$

In other words, the Nash equilibrium is a best countering strategy to itself, and conversely any m-tuple of strategies which is a best countering strategy to itself is a generalized Nash equilibrium.

Lemma 1. An m-tuple of strategies $a_I = (a^1_{j_1}, a^2_{j_2}, \ldots, a^i_{j_i}, \ldots, a^m_{j_m})$ is a generalized Nash equilibrium in pure strategies if and only if

$$a_I = (a^1_{j_1}, a^2_{j_2}, \ldots, a^i_{j_i}, \ldots, a^m_{j_m}) \in \mathrm{BCA}(I,t)(a_I).$$

Proof. From the definition of the generalized Nash equilibrium state, if a_I is an equilibrium state, then $a_I = (a^1_{j_1}, a^2_{j_2}, \ldots, a^i_{j_i}, \ldots, a^m_{j_m}) \in \mathrm{BCA}(I,t)(a_I)$.

Now suppose that $a_I = (a^1_{j_1}, a^2_{j_2}, \ldots, a^i_{j_i}, \ldots, a^m_{j_m}) \in \mathrm{BCA}(I,t)(a_I)$. To see that a_I must be a generalized Nash equilibrium state, consider the definition of the best countering strategy: $a^i_{j_i} \in \mathrm{BCS}(i,t)(a_I)$ for $i \in I$. But the latter condition means that no player could achieve a greater payoff by using a different strategy, given the strategies of the other players. Thus, a_I is a generalized Nash equilibrium state. □

Remark 2. From Lemma 1, it follows that *the best countering strategy for player i may also be referred to as the generalized Nash best countering strategy for player i.*

Lemma 2. A strategy $a^i_{j_i}$ from $\mathrm{ACT}(i,t)$ is *the best countering strategy for player i in pure strategies against the strategy* $a \in \mathrm{ACT}(I,t)$ if and only if

$$M_{J(i,t)}(a^i_{j_i}, a^i_s) \geq 0.5 \tag{7}$$

for every strategy $a^i_s \in \mathrm{ACT}(i,t)$.

Proof. First we show that if a strategy $a^i_{j_i}$ from $\mathrm{ACT}(i,t)$ is *the best countering strategy for player i in pure strategies against the strategy* $a \in \mathrm{ACT}(I,t)$, then,

$$M_{J(i,t)}(a^i_{j_i}, a^i_s) \geq 0.5$$

for every strategy $a^i_s \in \mathrm{ACT}(i,t)$.

Recall the definition of the best countering strategy for player i against the strategy $a \in \mathrm{ACT}(I,t)$. A strategy $a^i_{j_i} \in \mathrm{ACT}(i,t)$ is the best countering strategy for player i against the strategy $a \in \mathrm{ACT}(I,t)$ if the following holds:

$$J(i,t)(a \backslash a^i_{j_i}) \geq J(i,t)(a \backslash a^i_s)$$

for all $a^i_s \in \mathrm{ACT}(i,t)$.

This inequality means that player i always chooses a strategy with the property that his/her strategy is the best countering strategy (given players' conditions, rules, norms, etc.) to the strategy a. Thus, it follows that player i judges that his/her strategy $a^i_{j_i}$ is better than (preferred to) a^i_s or there is no preference (i.e., indifference) between $a^i_{j_i}$ and a^i_s. Hence,

$$M_{J(i,t)}(a^i_{j_i}, a^i_s) \geq 0.5$$

holds for every strategy $a^i_s \in \mathrm{ACT}(i,t)$, such that $a^i_{j_i} \neq a^i_s$.

Moreover, for $a^i_{j_i} = a^i_s$, $M_{J(i,t)}(a^i_{j_i}, a^i_s) = 1$. Thus

$$M_{J(i,t)}(a^i_{j_i}, a^i_s) \geq 0.5$$

holds for every strategy $a^i_s \in \mathrm{ACT}(i,t)$.

Let $M_{J(i,t)}(a^i_{j_i}, a^i_s) \geq 0.5$ for every strategy $a^i_s \in \mathrm{ACT}(i,t)$.

If $a^i_{j_i} = a^i_s$,

$$J(i,t)(a \backslash a^i_{j_i}) = J(i,t)(a \backslash a^i_s) \text{ for all } a \in \mathrm{ACT}(I,t).$$

If $a^i_{j_i} \neq a^i_s$, then player i judges that strategy $a^i_{j_i}$ is better than (preferred to) a^i_s or there is no preference (i.e., indifference) between $a^i_{j_i}$ and a^i_s. This means that strategy $a^i_{j_i}$ is the best countering strategy (given players' conditions, rules, norms, etc.) to all other players' strategies. So, the following holds:

$$J(i,t)(a \backslash a^i_{j_i}) \geq J(i,t)(a \backslash a^i_s) \text{ for all } a^i_s \in \mathrm{ACT}(i,t), \text{ and all } a \in \mathrm{ACT}(I,t). \quad \square$$

From Lemma 1 and 2, we have the following:

Corollary 1. An m-tuple of strategies $a_I = (a^1_{j_1}, a^2_{j_2}, \ldots, a^i_{j_i}, \ldots, a^m_{j_m})$ is a Nash equilibrium in pure strategies if and only if

$$M_{J(i,t)}(a^i_{j_i}, a^i_s) \geq 0.5 \tag{8}$$

for all $a^i_s \in \mathrm{ACT}(i,t)$, and all $i \in I$.

Observe that Corollary 1 describes the generalized Nash equilibrium state in fuzzy terms (from a different theoretical perspective, [24] also arrives at Nash fuzzy equilibria).

Theorem 1. Let $\mathrm{ACT}(i,t) = \{a^i_1, a^i_2, \ldots, a^i_{n_i}\}$ be a finite set of pure strategies of player i and let R_i be a relation defined in the following way: for strategies $a^i_k, a^i_s \in \mathrm{ACT}(i,t)$ $a^i_k R_i a^i_s$ if and only if strategy a^i_k is judged better (or preferred to a^i_s), or $a^i_k = a^i_s$, or there is no difference (indifference) between a^i_k and a^i_s.

i) If relation R_i is transitive, then there exists a strategy from $\mathrm{ACT}(i,t)$ which is *the best countering strategy for player i* in pure strategies against every strategy $a \in \mathrm{ACT}(I,t)$.
ii) If relation R_i is transitive for every $i \in I$, then there exists an m-tuple of strategies $a_I = (a^1_{j_1}, a^2_{j_2}, \ldots, a^i_{j_i}, \ldots, a^m_{j_m})$ which is a generalized Nash equilibrium in pure strategies.

Proof.

i) Observe first that if M is a function defined as (3), then $a^i_k R a^i_s$ if and only if $M(a^i_k, a^i_s) \geq 0.5$. Moreover, for every $i \in I$, relation R_i has the following properties: aRa (reflexive); aRb or bRa (coherent). Hence, because set $\mathrm{ACT}(i,t)$ is finite, we can order strategies in the following way: $a^i_{j_1} R_i a^i_{j2}, \ldots, a^i_{jn_i-1} R_i a^i_{jn_i}$. Thus, if relation R_i is transitive, for strategy $a^i_{j_1}$, we have $a^i_{j_1} R a^i_s$ for every strategy $a^i_s \in \mathrm{ACT}(i,t)$. Hence, $M(a^i_{j_1}, a^i_s) \geq 0.5$ for every strategy $a^i_s \in \mathrm{ACT}(i,t)$. Thus from Lemma 2, it follows that strategy $a^i_{j_1}$ is the best countering strategy for player i in pure strategies against every strategy $a \in \mathrm{ACT}(I,t)$.
ii) Follows immediately from i) and from Corollary 1. □

Remark 3. Theorem 1 is equivalent to the original Nash theorem for a finite set of pure strategies.[6]

Remark 4. The Nash equilibrium would not hold for player i under some conditions, such as fuzzy intransitivity (discussed below). Consider the following voting situation. The player has three candidates of choice: A, B, and C, and she has three

[6] Observe that this more general formulation reduces to the original case under crisp conditions.

strategies:
a_1: voting for candidate A,
a_2: voting for candidate B,
a_3: voting for candidate C.

But the following intransitivity may obtain:

She definitively prefers candidate A over B, so

$$M_{J(i,t)}(a_1,a_2) = 1, M_{J(i,t)}(a_2,a_1) = 0.$$

To a considerable degree, she prefers candidate B over C, so,

$$M_{J(i,t)}(a_2,a_3) = 0.7, M_{J(i,t)}(a_3,a_2) = 0.3.$$

But she also slightly prefers candidate C to A, so,

$$M_{J(i,t)}(a_3,a_1) = 0.6, M_{J(i,t)}(a_1,a_3) = 0.4.$$

The relation preference is not transitive in this case, and we obtain the following fuzzy matrix:

$$[J(i,t)_{jk}]_{3\times 3} = \begin{bmatrix} 1 & 1 & 0.4 \\ 0 & 1 & 0.7 \\ 0.6 & 0.3 & 1 \end{bmatrix}.$$

Moreover, there is no strategy a_i such that $M_{J(i,t)}(a_i,a_j) \geq 0.5$ for every a_j. Therefore, the Nash equilibrium would not hold for this player, and, of course, this is because in the relation $J(i,t)(a^i_j,a^i_k) = M_{J(i,t)}(a^i_j,a^i_k)$, strategy a is "best" or "equivalent" is not transitive. Such a situation would not arise under classical assumptions of crisp, transitive preference relations.

2.5 Normative Equilibria in Fuzzy Terms

In GGT, an activity, program, outcome, condition, or state of the world is in a normative equilibrium if it is judged to realize or satisfy an appropriate norm or value r in the situation S. This may be based on *ex ante* judgments [that is, as expressed in (2)], *and/or ex post (that is, experienced)* judgments.[7] Note: In a Nash equilibrium

[7] What has been formulated here entails following or applying a rule in a certain sense [5] (see footnote 11). This may not be a trivial matter, as Wittgenstein [30] and Winch [29] have pointed out. We limit ourselves to the following observations. Some of the actors in I may allege a violation of the norm. This may not be a dispute over the norm itself, but over its application, an issue of fact. Related problems may arise: some of the actors have conflicting interpretations of the meanings of the norm or of its application in the situation S. Or the participants, although adhering to the common norm, introduce different (and possibly incompatible) rules of other sorts, potentially affecting the scope of the norm and the equilibrium in the situation.

([6,15]), the player's strategies are in *an expected or anticipated equilibrium.*

If the player judges that expected or actual consequences, $Con(a,t+1)$, of a planned or implemented action a in S at time $t+1$, are sufficiently similar to the set of consequences prescribed by the salient or primary norm or value r, $Con(r)$, then we denote this by $sim_{J(i,t)}(a,r)$ and write

$$J(i,t)[Con(a,t+1),Con(r)] = sim_{J(i,t)}(a,r). \tag{9}$$

Definition 5. If the judgment specified in (9) holds, $Con(a,t+1)$ is a *normative i-player equilibrium* with respect to norm r in situation S_t.[8]

Using a fuzzy set conceptualization, we have the following:

Corollary 2. $Con(a,t+1)$ is a *normative i-player equilibrium* with respect to norm r in situation S_t if the following holds:

$$f_{J(i,t),2}(a,r) \leq M_{J(i,t)}(a,r). \tag{10}$$

When an action determination results in a normative equilibrium, the determination is stabilized. Normative equilibria are a function of (1) the value or norm r appropriate or activated in the situation S at a given time t, (2) the action possibilities found or constructed in the situation and the consequences attributed or associated with the action(s), and (3) the player's judgment processes in the situation.[9]

[8] Normative equilibria refer to consequences associated with an action or program a, where for our purposes here, time $t+1$ refers either (1) to the set of consequences expected at a later time $t+1$ that is envisioned or imagined prior to implementing the action or (2) to an actual or experienced set of consequences at $t+1$ after the execution of the action. Thus, in the case of action determination (*ex ante*), a normative equilibrium with respect to the norm or value r obtains if the actor judges that she has found or constructed an action — and can execute it — with consequences Con(a,t) that are *expected or planned* at time $t+1$ to sufficiently realize or satisfy r in situation S. In the *ex post* situation, after action a has been performed, the *actual or experienced consequences*, $Con(a,t+1)$, in S at time $t+1$ are judged "sufficiently similar" to the set of consequences prescribed by the norm or value r. In both cases, the player would not feel motivated or disposed to change the consequences, $Con(a,t+1)$, other things being equal. Rather, she is confident and feels that she is doing (or has done) the right and proper thing: for instance, successfully performing an appropriate role pattern or norm, maximizing utility, or experiencing a suitable emotional "high" in connection with activity a. Other types of equilibria, based on opportunism or coercion, may take place but are potentially unstable [5].

[9] Obviously, the concrete realizations or rule-governed activities – practices – are not identical to the rules, nor to the intended or planned action a. Though normative equilibria are generated or supported by rules, norms, and institutional arrangements, there is no one-to-

The normative equilibria associated with performances of roles, norms, and institutional arrangements make for *social facts* and *"focal points"* [6,26] to which participants orient — even in cases of deviance.[10] In any institutionalized setting or domain, there are multiple normative equilibria in each of the interaction settings (some of which may also be contradictory). Also, there are typically such equilibria on different levels. In general, one finds an *ecology or infrastructure of normative equilibria* [6].

Although the concept of normative equilibria may be applied to role performances and to individuals implementing norms, we are particularly interested here in *global normative equilibria*, for instance, normative equilibria for a particular game or institutional setting. This means that r_I specifies a *normative procedure* or *institu-*

---

one relationship. A situation consists of concrete conditions, constraints, other agents, and rules that may make the realization of a norm or value problematic, in particular:

1. Player i cannot find or construct a suitable "realizer" a in S, that is, an option that is judged to satisfy or realize r. Or, there appear to be several, but she cannot determine which to choose.
2. The player lacks the resources or capability to carry out a.
3. Other relevant, core norms or values contradict r. There is no action a that can be constructed or found that can satisfy both of these norms in S in accordance with (9). For instance, there is an active core norm $r' = (X', Y', O\beta')$ such that β and β' cannot hold simultaneously in any situation, nor to put it in other words, relative to any and every list of parameters. Here, β' can be, for example, the complement β^c of β. If r is realized, then r' is not satisfied, and vice versa. Thus, the realization of r in S is constrained by another norm or value r'; r' implies some degree of proscription of r (or its potential realizer a).
4. Conditions (including players who apply selective sanctions) produce (or construct) a subset of consequences of a, such that at least one of these consequences exceeds some prescribed maximum or limit (or minimum), for instance, one or more dimensions specified by r', therefore, making a unacceptable. This maximum (or minimum) represents a limit of commitment or level of sacrifice for r in the context of the relevance and applicability of r'. This constrains, of course, the scope of r. Consider, for instance, an interaction process that generates unintended consequences, which some participants experience as unfair results that exceed a maximum threshold of tolerance. This defines "enough is enough." Similarly, a development may lead to consequences that are above a "minimum level of security" or of "efficiency." These cases entail the activation and application of values that normally are in the background, taken for granted. Risky or actual negative developments bring the background values into play in the context given by parameters p, including S, and evoke a transformation of the "operative value complex" in the situation (see [4] and [7]).

[10] Schelling (see [26] pp. 57–58) refers also to "clues," "coordinators" that have "some kind of prominence or conspicuousness." From a conceptual point of view, his characterization is vague. For instance, "But it is a prominence that depends on time and place and who people are. Ordinary folk lost on a plane circular area may naturally go to the center to meet each other.... But in the final analysis we are dealing with imagination as much as with logic...Poets may do better than logicians at this game."

tional arrangement, and the participants (or their leaders) judge an m-tuple $a_I = (a_1, a_2, \ldots, a_i, \ldots, a_m)$ in terms of whether it realizes or satisfies r_I. Examples of such procedures are the democratic process, adjudication, and bilateral negotiation [6,9]. These are ways to produce normative equilibria under conditions of conflict and contentiousness.

For such a rule or rule complex, (9) becomes a case of collective judgment about whether the collective action is sufficiently similar to the norm or institutionalized pattern, $sim_p(a_I, r_I)$:

Definition 6. $Con(a_I, t+1)$ is a *collective normative equilibrium* for a set of players I with respect to norm r_I in situation S_t if the following holds:

$$J(I,t)[Con(a_I,t+1), Con(r_I)] = sim_{J(I,t)}(a_I, r_I). \tag{11}$$

Using fuzzy set terminology, we have the following.

Corollary 3. $Con(a_I, t+1)$ is a *collective normative equilibrium* for a set of players I with respect to norm r in situation S_t if the following holds:

$$f_{J(I,t),2}(a_I, r_I) \leq M_{J(I,t)}(a_I, r_I). \tag{12}$$

Consideration of diverse normative equilibria enables us to compare and analyze similarities and differences between GGT and Nash approaches to game equilibria [9].

2.6 Normative and Nonnormative Nash Equilibria

It is apparent in social life that many interactions have no equilibrium, and many social equilibria in reality are not Nash equilibria [9]; so, any serious consideration of the Nash equilibrium theorem as an empirical proposition is highly dubious. GGT is a scientific theory as opposed to a mathematical ideal. It may be used to specify the properties of normative equilibria, their scope, and the conditions under which they do or do not occur. Also, normative equilibria may be distinguished from nonnormative equilibria, that is, equilibria that fail to satisfy the relevant norm or value applying in the particular game situation.

Game equilibria may or may not be normatively grounded. The Nash theory does not distinguish between normative and nonnormative equilibria — and, of course, was never designed to make such a distinction. GGT can be used to systematically distinguish between, for instance, normative and nonnormative Nash equilibria. Briefly,

Definition 7. Strategy $a^i_{j_i} \in \mathrm{ACT}(i,t)$ is a *normative equilibrium with respect to rule r for player i* if the following holds:

$$J(i,t)[Con(a^i_{j_i},t+1),Con(r)] = sim_{J(i,t)}(a^i_{j_i},r). \tag{13}$$

Definition 8. The m-tuple of strategies $a_I = (a^1_{j_1}, a^2_{j_2}, \ldots, a^i_{j_i}, \ldots, a^m_{j_m})$ is a *normative equilibrium with respect to rule r* if the following holds:

$$J(i,t)[Con(a^i_{j_i},t+1),Con(r)] = sim_{J(i,t)}(a^i_{j_i},r) \tag{14}$$

for all $a^i_{j_i} \in \mathrm{ACT}(i,t)$, and all $i \in I$.

This means that a_I is a *normative equilibrium with respect to rule r* for each and every player $i \in I$.

Definition 9. An m-tuple of strategies $a_I = (a^1_{j_1}, a^2_{j_2}, \ldots, a^i_{j_i}, \ldots, a^m_{j_m})$ is *a normative Nash equilibrium with respect to rule r* if

i) $J(i,t)(a_I) \geq J(i,t)(a_I \backslash a^i_s)$
for all $a^i_s \in \mathrm{ACT}(i,t)$, and all $i \in I$.
ii) $J(i,t)[Con(a^i_{j_i},t+1),Con(r)] = sim_{J(i,t)}(a^i_{j_i},r)$,
for all $a^i_{j_i} \in \mathrm{ACT}(i,t)$, and all $i \in I$.

Similarly we could define *a normative best countering strategy against the strategy a with respect to rule r for player i.*

Definition 10. A strategy $a^i_{j_i}$ is *a normative best countering strategy against strategy a with respect to rule r for player i* if

i) $J(i,t)(a \backslash a^i_{j_i}) \geq J(i,t)(a \backslash a^i_s)$ for all strategies $a^i_s \in \mathrm{ACT}(i,t)$.
ii) $J(i,t)[Con(a^i_{j_i},t+1),Con(r)] = sim_{J(i,t)}(a^i_{j_i},r)$.

Then, an m-tuple of strategies $a_I = (a^1_{j_1}, a^2_{j_2}, \ldots, a^i_{j_i}, \ldots, a^m_{j_m})$ is a normative Nash equilibrium with respect to rule r, if $a^i_{j_i} \in \mathrm{ACT}(i,t)$ is the best countering strategy against the strategy a_I as well as a normative equilibrium with respect to rule r for each and every player i.

Using fuzzy set conceptions, we obtain the following.

Corollary 4. An m-tuple of strategies $a_I = (a^1_{j_1}, a^2_{j_2}, \ldots, a^i_{j_i}, \ldots, a^m_{j_m})$ is a normative Nash equilibrium with respect to rule r if

i) $M_{J(i,t)}(a^i_{j_i}, a^i_s) \geq 0.5$ for all $a^i_s \in \mathrm{ACT}(i,t)$, and all $i \in I$.

ii) $f_{J(i,t),2}(a^i_{j_i}, r) \geq M_{J(i,t)}(a^i_{j_i}, r)$ for all $a^i_{j_i} \in \mathrm{ACT}(i,t)$ and for all $i \in I$.

Corollary 5. A strategy $a^i_{j_i} \in \mathrm{ACT}(i,t)$ is a normative best countering strategy against the strategy a with respect to rule r for player i if

i) $M_{J(i,t)}(a^i_{j_i}, a^i_s) \geq 0.5$ for all $a^i_s \in \mathrm{ACT}(i,t)$.
ii) $f_{J(i,t),2}(a^i_{j_i}, r) \leq M_{J(i,t)}(a^i_{j_i}, r)$.

Definition 11. An m-tuple of strategies,

$$a_I = (a^1_{j_1}, a^2_{j_2}, \ldots, a^i_{j_i}, \ldots, a^m_{j_m}),$$

is *a nonnormative Nash equilibrium with respect to rule r* if

i) $J(i,t)(a_I) \geq J(i,t)(a_I \backslash a^i_s)$ for all $a^i_s \in \mathrm{ACT}(i,t)$ and $i \in I$.
ii) $J(i,t)[Con(a^i_{j_i}, t+1), Con(r)] = dissim_p(a^i_{j_i}, r)$ for some $a^i_{j_i} \in \mathrm{ACT}(i,t)$, where $i \in I$.

Similarly, we can define *a nonnormative best countering strategy against strategy a with respect to rule r for player i.*

Definition 12. A strategy $a^i_{j_i} \in \mathrm{ACT}(i,t)$, is *a nonnormative best countering strategy against strategy a with respect to rule r for player i* if

i) $J(i,t)(a \backslash a^i_{j_i}) \geq J(i,t)(a \backslash a^i_s)$ for all strategies $a^i_s \in \mathrm{ACT}(i,t)$.
ii) $J(i,t)[Con(a^i_{j_i}, t+1), Con(r)] = dissim_{J(i,t)}(a^i_{j_i}, r)$.

Then, an m-tuple of strategies $a_I = (a^1_{j_1}, a^2_{j_2}, \ldots, a^i_{j_i}, \ldots, a^m_{j_m})$ is a nonnormative Nash equilibrium with respect to rule r, if $a^i_{j_i} \in \mathrm{ACT}(i,t)$ and is the best countering strategy against strategy a_I for each and every player i; yet, $a^i_{j_i}$ is not a normative equilibrium for player $i \in I$ with respect to rule (norm, value) r (i.e., it is unsatisfactory with respect to the norm or value r).

Using a fuzzy set conceptualization, we obtain the following.

Corollary 6. An m-tuple of strategies,

$$a_I = (a^1_{j_1}, a^2_{j_2}, \ldots, a^i_{j_i}, \ldots, a^m_{j_m}),$$

is a nonnormative Nash equilibrium with respect to rule r if

i) $M_{J(i,t)}(a^i_{j_i}, a^i_s) \geq 0.5$ for all $a^i_s \in \mathrm{ACT}(i,t)$, and all $i \in I$.
ii) $M_{J(i,t)}(a^i_{j_i}, r) \leq f_{J(i,t),1}(a^i_{j_i}, r)$ for some $a^i_{j_i} \in \mathrm{ACT}(i,t)$, where $i \in I$.

In sum, a *normative Nash equilibrium* is one that satisfies the Nash formula (5) and the equation for normative equilibrium (10). It means that the Nash equilibrium or preferred strategy a_i (within the constraints of the game) fully realizes from i's perspective the value or norm r relevant and applicable in the game situation. A *nonnormative Nash equilibrium* would satisfy the Nash formula (5) but *not* the equations specifying normative equilibria [for either role (5) or collective norms (11)]. Thus, although $a_{I(\text{Nash})}$ consisting of the m-tuple of strategies is the option closest to or most similar to the goal or value r — and, therefore, at least a Nash equilibrium — it would not be a normative equilibrium since it would fail to fully realize or satisfy the value or norm r_i to a sufficient degree. On the other hand, if each and every player i in I believes that she must play the game and has no other viable options, she would accept the nonnormative equilibrium and be ill-disposed to shift her position, although it does not satisfy her value or goal in the situation. A normative non-Nash equilibrium is one that satisfies the equation for normative equilibrium (10) and does not satisfy the Nash formula (5).

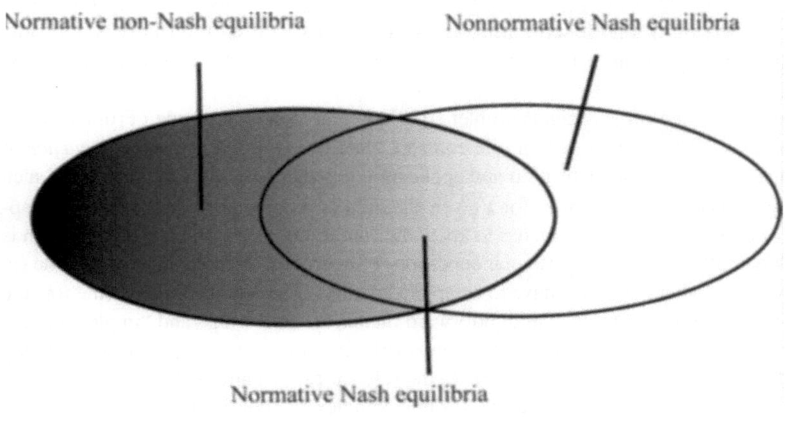

Fig. 1. Relationship between normative and Nash equilibria

The failure to achieve normative equilibria may have multiple causes, among others, the general conditions of constrained or even coerced equilibria: *for instance, people's strong commitment to a primary value or norm in the situation — which they refuse to opportunistically adjust to the circumstances — is confronted with concrete game conditions that constrain or frustrate the realization or achievement of the value or norm.* They may be forced to accept a state of affairs, for instance, strategy choices, interaction patterns, and outcomes, that fail to realize or satisfy their primary norm or value in the situation.

3 Applications of Fuzzy Game Theory

3.1 Introduction

Interactions or games taking place under well-defined conditions entail the application of or implementation of relevant rule complexes.[11] This is usually *not* a mechanical process, nor is it crisp. Players conduct situational analyses; the rules have to be interpreted, filled in, and adapted to the specific circumstances. Some interaction processes may be interrupted or blocked because of application problems such as contradictions among rules, situational constraints or barriers, etc. (see later). Above all, the players make fuzzy judgments about such conditions as the definition of the situation, the type of game, their roles (for a particularly innovative application of fuzzy logic to role, see [17]), particular values, beliefs, and types of action such as "cooperation" and "noncooperation."

Game situations may be distinguished in terms of whether or not the actors are able to construct or change their options. Classical games are *closed games* with the options given. Such closed game situations with specified alternatives and outcomes are distinguishable from *open game situations*.[12] In the following sections,

[11] This relates to the Wittgensteinian problem ([29],[30]) of "following a rule." From the GGT perspective, "following a rule" (or rule complex) entails *several phases and a sequence* of judgments: in particular, activation and application together with relevant judgments such as judgments of appropriateness for a given situation or judgments of applicability. To apply a rule (or rule complex), one has to know the conditions under which the application is possible or allowed and the particular conditions of execution or application of the rule (in part, whether other rules may have to be applied earlier). The application of a rule (or rule complex) is not then simply a straightforward matter of "following" and "implementing" it: the conditions of execution may be problematic; the situation (or situational data) may not fit or be fully coherent with respect to the rule (or rule complexes); actors may reject or refuse to seriously implement a rule (or rule complex); a rule (or rule complex) may be incompatible or inconsistent with another rule that is to be applied in the same situation. In general, actors may experience ambiguity, contradiction, dilemmas, and predicaments in connection with "following a rule" making for a problematic situation and possibly the unwillingness or inability to "follow a particular rule."

[12] Open and closed games can be distinguished more precisely in terms of the properties of the action complex, $\mathrm{ACT}(I,t,G)$ for the group of players I at time t in the game $G(t)$ (see [5,6]). In closed game conditions, $\mathrm{ACT}(i,t,G)$ is specified and invariant for each player i in I, situation S_t, and game $G(t)$. Such closure is characteristic of classical games (as well as parlor games), whereas most real human games are open. In open games, the players participating in $G(t)$ construct or "fill in" $\mathrm{ACT}(I,t,G)$, as, e.g., in a bargaining process where they alter their strategies or introduce new strategies during the course of their negotiations. In such bargaining processes, social relationships among the players involved — whether relations of anomie, rivalry, or solidarity — guide the construction of options and the patterning of interaction and outcomes. In general, for each player i in I, her repertoire of actions, $\mathrm{ACT}(i,t,G)$, is constructed and reconstructed by her (possibly with the

we consider these types of games and the implications of fuzzy judgment processes for interaction patterns and equilibria.

3.2 Closed Games: The Socially Embedded Prisoners' Dilemma Game

Classical games are *closed games* with given, fixed players, fixed action alternatives and outcomes, and particular anomic type relationships among players [8]. According to GGT, the players' defined social relationships shape and influence their value judgments of actions and outcomes in the interaction situation, lead to the production of diverse patterns of interaction and outcome, and possibly to normative equilibria. In general, any established social relationship among players interacts with the particular situational conditions. Thus, an interaction situation with prisoners dilemma (PD) features would not entail a "dilemma" for players with a solidary relationship since they tend to strongly prefer mutual cooperation over other patterns of interaction. Nor would such a situation pose a dilemma for enemies, who would expect and enact noncooperative patterns. However, for players who are rationally indifferent to one another, there is a dilemma, as discussed below.

Given a modified PD situation (see Table 2), the game processes and outcomes will be a function not only of the immediate or given payoff matrix but also of the particular social relationship among the players involved. Here we focus on the following relationships: solidary relationships as among friends or relatives; anomie, or relations of indifference, among the players; competitive relationships as among rivals (that is, not simply "situationally competitive "); relationships of hate and enmity; (5) status or authority relationships. Each of these relationships implies certain core values and norms, which the players in performing their roles, try to realize or satisfy in the given situation S. In making assessments of actions and outcomes, they are assumed to do so largely from the perspective of their particular positions in the relationship, that is, the value complex associated with or central to the position takes precedence over other values or value complexes exogenous to the relationship (there are limits to this commitment, as analyzed elsewhere [6,8]). These are the primary values or norms in the situation for one or several of the players.

Let the PD group $I = \{1,2\}$. Their particular social relationship is played out in interaction situations with certain action opportunity structures (configuration of actions) as well as outcome structures. Consider the two-person PD type game in Table 2. Player 1 chooses the rows and receives the first payoff listed in the box, and player 2 chooses the column and receives the second one. The payoffs may be interpreted as money or real resources.

---

involvement of others) in the course of her interactions. She tends to do this in accordance with the norms and values relevant to her role at t. The applications of GGT to open as well as closed games illustrates the concrete effects of social context on game structures and processes, in particular, the impact of social relationships on the interaction patterns and outcomes.

		Player 2	
		b_1	b_2
Player 1	a_1	+16, +16	−10, +10
	a_2	+20, −20	−4, −4
	a_3	+8, −40	−6, −10

Table 2. A two-person PD type game

In the following, we apply our analyses to determining fuzzy Nash and normative equilibria, given different social relationships in which players' value orientations and motives vary.

Situation 1: Rational Egoists in an Anomic Social Relationship.

The value complexes of rational egoists are characterized by an orientation to outcomes affecting self and to maximizing individual payoffs [5]. They make judgments in these terms generating, for the game above, the following membership functions: $M_{J(i,t)} : A \times A \to [0,1]$, where $i \in \{1,2\}$, which are represented by matrices $[J(i,t)_{jk}]$ and where $J(1,t)_{jk} = M_{J(1,t)}(a_j,a_k), J(2,t)_{jk} = M_{J(2,t)}(b_j,b_k)$ shows the fuzzy evaluative judgements or preference relationships among the options for actors 1 and 2, respectively [see (4)]. Thus, we obtain

$$[J(1,t)_{jk}]_{3\times3} = \begin{bmatrix} 1 & 0 & 0.5 \\ 1 & 1 & 1 \\ 0.5 & 0 & 1 \end{bmatrix} \text{ and } [J(2,t)_{jk}]_{2\times2} = \begin{bmatrix} 1 & 1-c \\ c & 1 \end{bmatrix},$$

where $c \in (0.5,1)$.

Observe that

$$M_{J(1,t)}(a_2,a_k) \geq 0.5 \text{ for all } a_k \in ACT(1,t), \text{ and}$$
$$M_{J(2,t)}(b_2,b_k) \geq 0.5 \text{ for all } b_k \in \mathrm{ACT}(2,t).$$

Thus, the two-tuple strategy (a_2,b_2) is a Nash but not a normative equilibrium (based on each player's judgments in the situation). From the individual perspective, the outcome is nonnormative in that it is a nonoptimal result, whereas the players are oriented to achieving optimal results.

Another variant of this entails players who use a minmax strategy (which generates other patterns of fuzzy preference relations in the given PD type game). The rules for calculating or determining action (modalities) make a difference [5,8].

Determining their actions according to the rule, "use the minmax strategy," the players make judgments and generate the following membership functions: $M_{J(i,t)} : A \times A \to [0,1]$, where $i \in \{1,2\}$, which may be represented by matrices $[J(i,t)_{jk}]$, where

$$J(1,t)_{jk} = M_{J(1,t)}(a_j,a_k), J(2,t)_{jk} = M_{J(2,t)}(b_j,b_k).$$

Thus,

$$[J(1,t)_{jk}]_{3\times 3} = \begin{bmatrix} 1 & 0 & 1-c \\ 1 & 1 & 1 \\ c & 0 & 1 \end{bmatrix}, \text{ where } c \in (0.5,1)$$

and

$$[J(2,t)_{jk}]_{2\times 2} = \begin{bmatrix} 1 & 0 \\ 1 & 1 \end{bmatrix}.$$

Observe that

$$M_{J(1,t)}(a_2,a_k) \geq 0.5 \text{ for all } a_k \in ACT(1,t), \text{ and}$$
$$M_{J(2,t)}(b_2,b_k) \geq 0.5 \text{ for all } b_k \in ACT(2,t).$$

Thus, the two-tuple strategy (a_2,b_2) is a Nash equilibrium state (with respect to the players' judgments). Moreover, it is also a classical Nash equilibrium in pure strategies.

Remark 5. Rational egoists experience a genuine dilemma in the PD game. Their calculus predisposes them to the second option which is noncooperative, but the resulting pattern $(-4,-4)$ is suboptimal, which is obviously a problematic result for instrumentally oriented players. For this reason, the noncooperative outcome in the PD game is an unstable state. The players are disposed to try, to the extent possible, to transform the game into a "more rational one" enabling them to attain the optimal (that is, the mutual cooperative) outcome (a_1,b_1), other things being equal. For instance, through direct negotiation and norm formation — or other transformations of the game or social relationship configurations — they may guarantee the mutual beneficial interaction (a_1,b_1) (and success at this would contribute to building further trust and eventually a solidary relationship with a normative basis).

Remark 6. Consider that the payoffs in Table 2 are in "cents" (parts of a US dollar). Such small payoffs might be interpreted or understood as no differences in "gains" or "losses" at all: all are judged similarly, and the actors are indifferent to the various outcomes. So the players would not generate any predictable patterns whatsoever.

Along similar lines, Schelling [26] pointed out that a game where the differences in payoffs between cooperation and defection were roughly the same or similar would result in the cooperative pattern and would not create a dilemma for rational egoists, and much less for players in a solidary relationship. Nevertheless, defection might imply improper behavior and risk and generate negative reactions from the other. Mutual cooperation would be, under such circumstances, the preferred choice.

Situation 2: Solidary Relationship with Cooperative Value Orientations and Motives.

The value complexes of players in a solidary relationship typically contain norms of cooperation as well as a principle of symmetrical distributive justice (for instance, the sharing of gains and losses). The appropriate decision modality would be collective or joint decision making — unless the players are forced by circumstances to act separately, in which case, each would expect to take the other and to be taken into account in turn.

In a strong solidary relationship, the "payoffs" of the game will be transformed into evaluations or preferences based on the relationship's prevailing norm(s) or value(s) r_I: in particular, payoffs that are equal and collectively and/or individually gainful are better than unequal payoffs or negative payoffs for either or both of the actors. The fuzzy judgment function would be as follows for $i \in \{1,2\}$:

$M_{J(i,t)}$ (actions in the context of the solidary relationship) =

- 1 for actions whose payoffs are judged totally consistent with the norm of solidarity or reciprocity, r_I, for instance, aiming for substantial more or less equal gains for both.
- $c \in (0.5,1)$ for actions whose payoffs moderately satisfy the norm of solidarity but are not outstanding, that is, they are judged to be relatively close to r_I, but not perfectly fitting. This could be payoffs that are relatively good for both but are better for one than the other.
- 0.5 for payoffs that satisfy minimum standards of solidarity or reciprocity. They are borderline.
- 0 for actions or expressions that fail to satisfy minimum standards or criteria of the norm r_I.

The payoffs $(-10,10),(8,-40),(-6,-10)$, and $(20,-20)$ would not satisfy the equality criterion, and the payoffs for mutual noncooperation $(-4,-4)$ would not satisfy the criterion of maximum gains for both. Thus, judgment based on outcome analysis — driven by a desire to maximize the fit or degree of membership in a player's value complex — would result in the players's choice of mutual cooperation (a_1,b_1), which gives an optimum payoff to both.

Acting according to the application of r_I, as indicated above, the players would generate the following fuzzy preference relationships: $M_{J(i,t)} : A \times A \to [0,1]$, where $i \in \{1,2\}$ which may be represented by matrices

$$[J(i,t)_{jk}],$$

where $J(1,t)_{jk} = M_{J(1,t)}(a_j,a_k), J(2,t)_{jk} = M_{J(2,t)}(b_j,b_k)$. Thus,

$$[J(1,t)_{jk}]_{3\times 3} = \begin{bmatrix} 1 & 1 & 1 \\ 0 & 1 & c \\ 0 & 1-c & 1 \end{bmatrix}, \text{ where } c \in (0.5,1),$$

and

$$[J(2,t)_{jk}]_{2\times 2} = \begin{bmatrix} 1 & 1 \\ 0 & 1 \end{bmatrix}.$$

Observe that $M_{J(1,t)}(a_1, a_k) \geq 0.5$ for all $a_k \in \mathrm{ACT}(1,t)$ and $M_{J(2,t)}(b_1, b_k) \geq 0.5$ for all $b_k \in \mathrm{ACT}(2,t)$. The two-tuple strategy (a_1, b_1) is a normative Nash equilibrium state (with respect to each player's judgment).

Such fuzzification of games may also be developed in relation to fuzzy judgments of the intrinsic qualities of the options, for instance, their "degree of cooperation" [19]. Consider that the fuzzy membership function instead of being applied to outcomes is applied to the action alternatives in terms of their characteristics or intrinsic qualities of "cooperativeness." Fuzzy distinctions are made between actions judged fully fitting or a complete member ($M_{J(i,t)} = 1$) and actions judged as failing to satisfy the minimum standards or criteria of a norm of cooperation ($M_{J(i,t)} = 0$). Between these limits, a player may have a number of intermediate values. For each player i, the fuzzy evaluative judgment function, $J(i,t)$, at time t expresses judgments of actions such as "moderately satisfying" or "minimally satisfying" with respect to the standards of cooperation.

Each of the players would judge her own behavior as well as that of the other player in these terms. Applying this norm in the PD game to the actions themselves (assuming that, for instance, the outcomes are unknown or excessively complex) would result in judgments such as the following: if a_1 and b_1 have satisfactory qualities of "cooperation" and the other options do not, then, (a_1, b_1) pattern would be the normative Nash equilibrium for the game played by players with such a solidary relationship.

This is precisely the result obtainable in GGT for crisp games. Then, how would fuzzy judgment processes make a difference? Consider again the game in Table 2 where player 1 has three options, characterized by "cooperation," "sort of cooperation," and "noncooperation." The actual outcomes of the options are not known; judgments are then based on characteristics or qualities of the actions themselves. Consider that player 2 classifies the options differently from player 1, for instance, that 2 applies a stricter standard of cooperation than 1 so that 2 considers or would consider what 1 defines as "sort of cooperation" as "noncooperation" or "disloyalty," whereas player 2 considers that it satisfies minimum standards of "cooperation." Thus, because of ignorance, haste, tiredness, or other pressures, 1 enacts what she considers moderate cooperation. As they play the game, they observe one another's behavior. Though 1 would judge the interaction pattern as a cooperative pattern, player 2 would see it as asymmetrical, noncooperation/cooperation, and feel cheated or betrayed. According to 2, the primary norm of the relationship would be violated or betrayed. A serious enough breach would result in termination of the relationship — as the experience of betrayal in social relationships often witnesses. However, if the result is only some expressions of disappointment or anger and is seen as part of "learning to know one another," then, repetition of the game is likely to entail player 1 applying 2's standard on future occasions to avoid hurting the other (that is, as an expression of the sentiment of empathy) or to avoid negative reactions from the other in subsequent interactions.

In sum, players with a solidary egalitarian relationship would in the context of a PD type game experience no dilemma in a wide range of situations, other things being equal. Mutual cooperation is a right and proper interaction, that is a normative equilibrium for their particular relationship (see below). Although the normative equilibrium (a_1, b_1) realizes the norm of solidarity r_I, it may fail with respect to another key or salient norm or value r_I*, for instance, with respect to an ambition level of high mutual gain. In this case, (a_1, b_1) would not satisfy r_I*. (However, this pattern is right and proper only up to some limit ([8]).[13] Also, there are fuzzy limits to the application of the solidary norm r_I, as analyzed elsewhere [8] (in many instances, solidarity is not an absolute principle). Other values or interests come into play so that one or both players choose the option "sort of cooperation." Of course, the players may try to fabricate "full cooperation" when actually doing less. These interactions typically risk negative judgment from the other and the development of mutual distrust. Fuzzy game analysis allows one to analyze and explain such patterns. Players do not play games precisely, but roughly. The predicted patterns of interaction as well as of stability and instability would differ in some cases from the classical theory, as already suggested above.

Remark 7. Consider players with multiple values (that is, with a probability of experiencing games as mixed motive situations) cooperating as well as being successful in economic terms. Each of the players in the game is oriented to acting in a cooperative way and would be disposed to choose (a_1, b_1). At the same time, each may be oriented to economic or other gains for herself. A matrix can be generated representing the players' *dual* subjective judgments of the degree of fit or membership of game consequences in terms of their relevant values in the situation. Depending on their anchoring points in their semantic judgment space, they would cooperate, given a wide range of payoff matrices. However, certain matrices would entail payoffs that for one or both players, exceed a threshold (anchoring point that triggers a shift in fuzzy judgment), and, therefore, would have a certain probability of defection or noncooperative action.

[13] GGT stresses the motivation of players to realize their goals or values and their tendency to experience frustration when they fail to do so. There is nothing *inherent* in the social relationship between rational egoists [analyzed in situation (1)] proscribing or inhibiting them from effectively exploiting opportunities to communicate and reach agreements in the pursuit of mutual gain, provided, of course, that institutional rules or ecological conditions permit. Hostile players, on the other hand, would have difficulty inherent in the relationship in using opportunities for communication, making and adhering to agreements, and pursuing opportunities for mutual gain.

Situation 3: Relationship of Rivals or Competitors.

The value complexes of players in competitive relationships entail contradictory values — each player is oriented to surpassing the other, e.g., maximizing the difference in gains (or losses) between self and other. According to the players' individual value orientations, the only acceptable outcome for each would be an asymmetrical one where self gains more (or loses less) than the other. In "civilized rivalry," there are typically norms applying, e.g., a norm of civility or prudence, constraining their actions vis-à-vis one another. Given a social relationship where the players compete

Table 3. First player's perceptions of payoff

	$\mathbf{b_1}$	$\mathbf{b_2}$
$\mathbf{a_1}$	0	-20
$\mathbf{a_2}$	40	0
$\mathbf{a_3}$	48	4

to receive the greatest difference between their payoffs, the original payoff matrix would take the following form (Table 3), as reconceptualized and perceived by the players: $c_{ij} = a_{ij} - b_{ij}$ for the first player and $d_{ij} = b_{ij} - a_{ij}$ for the second one and where $c_{ij} + d_{ij} = 0$. On the basis of the value orientation to "act to obtain the greatest difference between their payoffs," they make their judgments and generate the following memberships functions: $M_{J(i,t)} : A \times A \to [0,1]$, which may be represented for $i \in \{1,2\}$ by matrices $[J(i,t)_{jk}]$, and where

$$J(1,t)_{jk} = M_{J(1,t)}(a_j, a_k), J(2,t)_{jk} = M_{J(2,t)}(b_j, b_k).$$

For instance, we could have

$$[J(1,t)_{jk}]_{3\times 3} = \begin{bmatrix} 1\,0\,0 \\ 1\,1\,0 \\ 1\,1\,1 \end{bmatrix} \text{ and } [J(2,t)_{jk}]_{2\times 2} = \begin{bmatrix} 1\,0 \\ 1\,1 \end{bmatrix}.$$

Observe that $M_{J(1,t)}(a_3, a_k) \geq 0.5$ for all $a_k \in \mathrm{ACT}(1,t)$ and $M_{J(2,t)}(b_2, b_k) \geq 0.5$ for all $b_k \in \mathrm{ACT}(2,t)$. Thus, the two-tuple strategy (a_3, b_2) is a Nash equilibrium on the basis of the players' judgments. However, it is not a normative equilibrium, either on their respective individual levels or on the collective level (where in any case there is no shared norm or value applying).

Though there is a Nash equilibrium, there is no normative equilibrium, either on their respective individual levels or on the collective level. On the collective level, there is no norm or value applying and, therefore, no normative force or pressure toward convergence. On the individual level, each has the aim to outdo the other, but they cannot realize this goal at the same time. Hence, they would be dissatisfied and disposed to try to transform the game. This type of game played by rivals tends to be unstable.[14]

Situation 4: Players in a Status or Authority Relationship.

The value complexes of players in hierarchical relationships — for instance, an administrative or bureaucratic hierarchy — contain a primary or salient norm r_I specifying asymmetrical interaction and payoffs The superior dominates, and her subordinate shows deference and a readiness to comply with the wishes or demands of the superior in a certain sense (noncooperation on the part of the superior, in this case player 1, and cooperation on the part of the subordinate, player 2). Moreover, the payoffs should favor the high status player.

On the basis of the value orientation to "act to obtain a difference between players' payoffs favoring the high status player," the players make their judgments and generate the following memberships functions: $M_{J(i,t)} : A \times A \to [0,1]$, where $i \in \{1,2\}$, which may be represented by matrices $[J(i,t)_{jk}]$, where $J(1,t)_{jk} = M_{J(1,t)}(a_j, a_k)$ and $J(2,t)_{jk} = M_{J(2,t)}(b_j, b_k)$. For instance, the following might obtain:

$$[J(1,t)_{jk}]_{3\times 3} = \begin{bmatrix} 1\,0\,0 \\ 1\,1\,0 \\ 1\,1\,1 \end{bmatrix}, \text{ and } [J(2,t)_{jk}]_{2\times 2} = \begin{bmatrix} 1\,1 \\ 0\,1 \end{bmatrix}.$$

Observe that $M_{J(1,t)}(a_3, a_k) \geq 0.5$ for all $a_k \in \text{ACT}(1,t)$ and $M_{J(2,t)}(b_1, b_k) \geq 0.5$ for all $b_k \in \text{ACT}(2,t)$. Thus, the two-tuple strategy (a_3, b_1) is a Nash equilibrium as well as a normative equilibrium. The principle of distributive justice for a hierarchical relationship implies asymmetry. Right and proper outcomes entail greater gains for the dominant player 1 than for the subordinate 2. For players in such a relationship, the PD does *not* pose a dilemma, other things being equal. (a_3, b_1) is a normative equilibrium for this particular relationship. From a purely individual perspective, player 2 would not find it a normative equilibrium. She would personally prefer, for instance, the mutually cooperative outcome, although it would violate the norm r_I of her asymmetrical relationship with player 1. Under conditions where this individual orientation takes priority, the lower status player would be disposed to

[14] There are some similarities between the competitive relationship and a relationship of enmity (particularly if the players combine trying to cause harm to one another with trying to minimize loss). Consider the simple case where the players in an enmity relationship are oriented to, and make judgments on the basis of, the value to cause the greatest harm to the other. In general, then, in the case of agents who are enemies, the PD-type game poses no dilemma. Each can effectively punish the other through noncooperation. This pattern is a Nash equilibrium and is consistent with their relationship but lacks, of course, normative force. The Nash equilibrium would in this case also be (a_3, b_2), as in the previous case of rivalry. However, if the negative payoffs *to the other* (but not to self) are changed, for instance, a_2 were to have the consequences $(+1, -41)$ and $(-11, -11)$, then player 1 would prefer a_2 to a_3, giving (a_2, b_2) as the Nash equilibrium.

violate r_I. Player 1 is also typically subject to constraints on exploiting her situation in the form of laws as well as norms such as civility and *noblesse oblige*.

In our consideration of the prisoners' dilemma game from the GGT perspective, we have shown several of the ways that the social relational context of a game implies the application of distinct norms and values, and the generation of distinct value judgments and preferences as well as particular interaction patterns and outcomes. This approach also has implications for conceptualizing and determining game equilibria.

3.3 Fuzzy Bargaining in an Open Game Process

In open games, players construct and elaborate strategies and outcomes in the course of interaction, for instance, in a bargaining game in market exchange [8]. In bargaining games, there is a socially constructed "bargaining space" (settlement possibilities) varying as a function of the particular social relationship in whose context the bargaining interplay takes place.

Consider a buyer B and a seller S bargaining about the price p of a good or service X. The following is based on [8]. S has a reserve or minimum price $p_S(min)$ and B has a maximum price or value $p_B(max)$ where presumably $p_S(min) \leq p_B(max)$. We obtain bargaining space for the seller $P_S = [p_S(min), +\infty)$, buyer $P_B = [0, p_B(max)]$, and bargaining space for both $X = P_S \cap P_B = [p_S(min), p_B(max)]$, respectively. Each also has an idea or ambition level of a "good deal" to aim for in the particular situation: $p_S(ideal)$ and $p_B(ideal)$, where $p_S(ideal) \in P_S$ and $p_B(ideal) \in P_B$. Determination of these values is based on what they believe or guess about one another's limits (namely, the reserve price of the seller and the value of the buyer), or the determination can be based on other past experience (or on some theory, which may or may not be very accurate). Typically, these are adjusted as the bargaining process goes on. These "anchoring points" make up a *fuzzy semantic space*, which is basic to the judgment processes that go on in bargaining. *The players propose prices and may accept or reject one another's proposals based on their fuzzy judgments.* When one accepts the proposal of another, a deal is made. The proposal is the selling or final price.

Their fuzzy judgment functions concerning price levels can be generated as follows. For seller S, the fuzzy evaluative judgment function, $J(S,t)$, at time or context t is given by the membership function $M_{J(S,t)}$ which may be specified as follows:

$$M_{J(S,t)}(x) = \begin{cases} 1 & \text{for} \quad x \geq p_S(ideal) \\ c & \text{for} \quad p_s(\min) < x < p_s(ideal) \\ 0.5 & \text{for} \quad x = p_s(min) \\ 0 & \text{for} \quad x < p_s(min), \end{cases}$$

where x denotes an offer (or option) and $c \in (0.5, 1)$.

For buyer B, the fuzzy evaluative judgment function, $J(B,t)$, at time or context t is given by its membership function $M_{J(B,t)}$ which may be represented as follows:

$$M_{J(B,t)}(y) = \begin{cases} 1 & \text{for} \quad y \leq p_B(ideal) \\ d & \text{for} \quad p_B(ideal) < y < p_B(max) \\ 0.5 & \text{for} \quad y = p_B(max) \\ 0 & \text{for} \quad y > p_B(max), \end{cases}$$

where y denotes an offer (or option) and $d \in (0.5, 1)$.

Each player i is motivated or driven to maximize $M_{J(i,t)}$ (where $i \in \{S, B\}$) the fit or degree of membership in her judgment function which incorporates her underlying values and goals. Elsewhere [8], we have shown that any number of settlement results may obtain within the bargaining space, defined by the bargainer's limit. Also, their beliefs (or guesses) about one another's limits are key factors. In other words, if the buyer believes that the seller has a higher reserve price than she actually has, she is prepared to settle at a price between this estimated level and her ideal level. It is similar for the seller. The bargaining process entails not only communication of proposals and counterproposals but also adjustments of their estimates of one another's limits. These adjustments depend on their belief revision processes, the persuasiveness and bluffs of the players, and the time and resource constraints under which each is operating [7]:

1. Settlements are unambiguously reached (provided, of course, such an outcome exists) if $p_S(ideal) \leq p* \leq p_B(ideal)$.
2. No settlement or deal is reached because the offer is unambiguously unacceptable (given the particular anchoring points), that is, $M_{J(i,t)} = 0$ for either actor: $p^* < p_S(min)$ or $p^* > p_B(max)$.
3. A price agreement is attainable if the proposal p* satisfies the following conditions for buyer as well as seller: $p_S(min) \leq p* < p_S(ideal)$ and $p_B(ideal) < p* \leq p_B(max)$. (This may not be possible, but if so, in any case, the agreement would entail a high degree of ambiguity. A variety of prices satisfy these conditions and are a function of various contingencies and conditions).
4. Maximum ambiguity (maximum discontent agreement) obtains at the limits:

$$\text{for the seller, } p^* = p_S(min),$$
$$\text{for the buyer, } p^* = p_B(max).$$

Remark 8. In such bargaining processes, established social relationships among the players involved guide the construction of options and the patterns of interaction and outcomes [7].

Remark 9. Elsewhere [7], we have also shown that the particular social relationship — the particular social rules and expectations associated with the relationship — make for greater or lesser deception and communicative distortion, greater or lesser transaction costs, and the likelihood of successful bargaining. The difficulties — and transaction costs — of reaching a settlement are greatest for pure rivals. They would be more likely to risk missing a settlement than pragmatic "egoists" because rivals tend to suppress the potential cooperative features of the game situation in favor of pursuing their rivalry. Pure "egoists" are more likely to effectively resolve some of the collective action dilemmas in the bargaining setting. Friends may exclude bargaining altogether as a precaution against undermining their friendship. Or, if they do choose to conduct business together, their tendencies to self-sacrifice may make for bargaining difficulties (but different, of course, from those of rivals) and increased transaction costs in reaching a settlement [8].

Remark 10. Elsewhere, we have shown that the bargainers try to manipulate what the other believes about their limits. For instance, the seller convinces the buyer that $p_S(min)$ is much higher than it is, approaching or equaling $p_S(ideal)$. Similarly for the buyer. These processes of persuasion, deception, etc. draw out a bargaining process. They may also result in an aborted process.

Here, we have assumed that the anchoring points making up the fuzzy judgment function are stable. But, of course, players may change their judgment function, that is, by raising or lowering their "ideals." Working out the effects of this are straightforward.

In general, our analysis suggests a spectrum of settlement or equilibrium possibilities in negotiation games; exactly how wide or narrow and the particular space generated depends on players' ambition levels and limits. Our analysis also suggests that settlements depend in part on the players' beliefs or estimates of one another's anchoring points in their judgment systems. Finally, processes of persuasion and deception can be modeled and analyzed because the players operate with models of the situation that are constructions with incomplete and imperfect information (possibly with false information) [7,9].

4 Implications and Extensions

1. Because appropriate actions in a game are roughly categorized, there is variation in what is considered appropriate or right action. Fuzziness may make agreements and cooperation easier, in that a range of actions may fit or have a satisfactory degree of membership in goal and strategy sets. But we have also shown how differences in anchoring points in the judgment semantic space may lead to misunderstandings and conflicts, as in the case of persons with a solidary relation who have different standards (anchoring points) for the meaning of "cooperativeness," thus classifying differently an action as cooperative or not.

2. People in solidary relationships are often taking into account one another's standards or anchoring points, making adjustments so as to facilitate and stabilize interactions as smooth and peaceful. This process of mutual adjustment in the situation creates a stronger sense of consensus than obtains in fact.
3. Players may or may not share a common threshold function of similarity. If there are differences between actor i and j, then we would distinguish high overlap (or degree of consensus) situations from nonoverlap situations. With high overlap, there would be likely agreements across a range of behaviors. On the other hand, if the overlap is not large, then misunderstandings and conflicts are likely to emerge. This would be even more the case if the limited overlap were not known or recognized, but only discovered in the course of interaction.
4. However, if players know one another's anchoring points or standards and are prepared to adjust to one another, then the players maintain their behavior (in public situations) in the common or overlapping zone. This would entail a situational constraint on their behavior, as they maintain interaction patterns within their common overlap zone. Thus, they achieve normative equilibria, although there are risks that they may act and interpret action in nonoverlapping terms. Given a refusal to take one another into account — having no common sentiments or norms — they are then unlikely to achieve normative equilibria.
5. Thresholds tend to vary with context. i interacts with j, i knows j is stricter than others including herself and adjusts her situational thresholds accordingly, since, for instance, the matter is not as important to i as she believes that it is to j. Thus, i may operate in the presence of j with standards equally as strict as those she attributes to j.

Acknowledgments

We are grateful to Anna Gomolińska, Hannu Nurmi, and Andrzej Skowron for their comments and suggestions on an earlier draft of this paper.

References

1. J.P. Aubin. Cooperative fuzzy game. *Mathematics of Operations Research,* 6: 1–13, 1981.
2. A. Billot. *Economic Theory of Fuzzy Equilibria.* Springer, Berlin, 1992.
3. T.R. Burns, H. Flam. *The Shaping of Social Organization: Social Rule System Theory with Applications.* Sage, London, 1987, 2nd. ed., 1990.
4. T.R. Burns, A. Gomolińska. Modeling social game systems by rule complexes. In *Proceedings of the 1st International Conference on Rough Sets and Current Trends in Computing (RSCTC'1998),* LNAI 1424, 581–584, Springer, Berin, 1998.
5. T.R. Burns, A. Gomolińska. The theory of socially embedded games: The mathematics of social relationships, rule complexes, and action modalities. *Quality and Quantity: International Journal of Methodology,* 34(4): 379–406, 2000.

6. T.R. Burns, A. Gomolińska. The theory of social embedded games: Norms, human judgment, and social equilibria. Paper presented at *the Joint Conference of the American Sociological Association–International Sociological Association on Rational Choice Theory*, Washington, D.C., 2000. available at http://www.soc.uu.se/publications/fulltext/tb_generalized-game-theory.doc.
7. T.R. Burns, A. Gomolińska. Socio-cognitive mechanisms of belief change: Application of generalized game theory to belief revision, social fabrication, and self-fulfilling prophesy. *Cognitive Systems Research*, 2(1): 39–54, 2001.
8. T.R. Burns, A. Gomolińska, L.D. Meeker. The theory of socially embedded games: Applications and extensions to open and closed games. *Quality and Quantity: International Journal of Methodology*, 35(1):1–32, 2001.
9. T.R. Burns, E. Roszkowska. Rethinking the Nash equilibrium: The perspective on normative equilibria in the general theory of games. Paper presented at *the Group Processes/Rational Choice Mini-Conference*, American Sociological Association Annual Meeting, Anaheim, CA, 2001. available at http://www.soc.uu.se/publications/fulltext/tb_rethinking-nash-equil.doc.
10. D. Butnariu. Fuzzy games; a description of the concept. *Fuzzy Sets and Systems*, 4: 63–72, 1980.
11. M. Fedrizzi, J. Kacprzyk, H. Nurmi. Consensus degrees under fuzzy majorities and fuzzy preferences using OWA (ordered weighted average) operators. *Control and Cybernetics*, 22: 77–86, 1993.
12. A. Gomolińska. Rule complexes for representing social actors and interactions. *Studies in Logic, Grammar, and Rhetoric*, 3(16): 95–108, 1999.
13. A. Gomolińska. Derivability of rules from rule complexes. *Logic and Logical Philosophy*, 10: 21–44, 2002.
14. J. Kacprzyk, M. Fedrizzi, and H. Nurmi. Group decision making and consensus under fuzzy preferences and fuzzy majority. *Fuzzy Sets and Systems,* 49: 21–31, 1992.
15. R.D. Luce, H.Raiffa. *Games and Decisions*. Wiley, New York, 1957.
16. I. Nishizaki, M. Sakawa. *Fuzzy and Multiobjective Games for Conflict Resolution*. Physica, Heidelberg, 2001.
17. J.D. Montgomery. The self as a fuzzy set of roles, role theory as a fuzzy system. In *Sociological Methodology 2000*, Blackwell, Malden, MA, 2001.
18. J.F. Nash. Equilibrium points in n-person games. *Proceedings of the National Academy of Sciences (USA)*, 36: 48–49, 1950.
19. H. Nurmi. On fuzzy games. In *Progress in Cybernetics and Systems Research*, Vol. IV. Hemisphere, Washington, DC, 1978.
20. H. Nurmi. A fuzzy solution to a majority voting game. *Fuzzy Sets and Systems*, 5: 187–198, 1981.
21. H. Nurmi. Approaches to collective decision making with fuzzy preference relations. *Fuzzy Sets and Systems*, 6: 249–259, 1981.
22. H. Nurmi. Imprecise notions in individual and group decision theory: Resolution of Allais' paradox and related problems. *Stochastica*, 6: 283–303, 1982.
23. H. Nurmi. Resolving group choice paradoxes using probabilistic and fuzzy concepts. *Group Decision and Negotiation*, 7: 1–21, 2000.
24. C. Ponsard. Nash fuzzy equilibrium: Theory and application to a spatial duopoly. *European Journal of Operational Research*, 31: 376–384, 1987.
25. E. Roszkowska, T.R. Burns. Market transaction games and price determination. The perspective of the general theory of games. Paper presented at *the First World Congress of Game Theory*, Bilbao, Spain, 2000. available at http:www.soc.uu.se/publications/fulltext/tb_market-pricing-game.doc.

26. T.C. Schelling. *The Strategy of Conflict.* Harvard University Press, Cambridge, MA, 1963.
27. R. Sun. Robust reasoning: Integrated rule-based and similarity-based reasoning. *Artificial Intelligence,* 75(2): 241–295, 1995.
28. J. Von Neumann, O. Morgenstern. *Theory of Games and Economic Behaviour.* Princeton University Press, Princeton, NJ, 1972.
29. P. Winch. *The Idea of a Social Science and its Relation to Philosophy.* Routledge & Kegan, London, 1958.
30. L. Wittgenstein. *Remarks on the Foundations of Mathematics.* Blackwell, Oxford, 1956.

Chapter 18
Rough Neurons: Petri Net Models and Applications

James F. Peters,[1] Sheela Ramanna,[1] Zbigniew Suraj,[2] Maciej Borkowski[1]

[1] University of Manitoba, Winnipeg, Manitoba R3T 5V6, Canada
{jfpeters, sramanna, maciey}@ee.umanitoba.ca

[2] University of Information Technology and Management, Rzeszów, and University of Rzeszów, Poland
zsuraj@wenus.wsiz.rzeszow.pl

Summary. This chapter presents Petri net models for two forms of rough neural computing: training set production and approximate reasoning schemes (AR schemes) defined in the context of parameterized approximation spaces. The focus of the first form of rough-neural computing is inductive learning and the production of training (optimal feature set selection), using knowledge reduction algorithms. This first form of neural computing can be important in designing neural networks defined in the context of parameterized approximation spaces. A high-level Petri net model of a neural network classifier with an optimal feature selection procedure in its front end is given. This model is followed by the development of a number of Petri net models of what are known as elementary approximation neurons (EA neurons). The design of an EA neuron includes an uncertainty function that constructs a granule approximation and a rough inclusion (threshold activation) function that measures the degree to which granule approximation is part of a target granule. The output of an EA neuron is an elementary granule. There are many forms of elementary granules (e.g., conjunction of descriptors, rough inclusion function value). Each of the EA neurons considered in this chapter output a rough inclusion function value. An EA neuron can be designed so that it is trainable, that is, a feedback loop can be included in the design of the EA neuron so that one or more approximation space parameters can be tuned to improve the performance of the neuron. The design of three sample EA neurons is given. One of these neurons behaves like a high-pass filter.

1 Introduction

This chapter introduces a model of rough neurons in the context of rough set theory [16,17] and approximate reasoning schemes (AR schemes) [29]. Studies of neural networks in the context of rough sets [3,6,10,11,14,15,19,22,23,28,29,32,34,35,40], [44] and granular computing [26,29,32,33,35] are extensive. An intuitive formulation of information granulation was introduced by Zadeh [41,42]. Practical applications of rough-neural computing have recently been found in predicting urban highway traffic volume [11], speech analysis [14,15], classifying the waveforms of power system faults [6], signal analysis [20], and in assessing software quality [19]. In its most general form, rough-neural computing provides a basis for granular computing. A rough mereological approach to a rough-neural network springs from an interest in knowledge synthesized (induced) from successive granule approximations performed by neurons (cooperating agents) [27,28,43]. Such neurons resemble communicating agents described in [12,13]. The distributed agent model for a neural network leads naturally to nonlayered, neural network architectures, and the collective activities of interconnected neurons start to resemble a form of swarm intelligence [5], that is, it is possible for an agent (neuron) to communicate granules of knowledge to other agents (neurons) in its neighborhood rather than by means of the usual restricted model of movement of granules of information "upward" from neurons in one layer to neurons in a higher layer (e.g., in [1,6,8,19]).

Petri net models for two fundamental forms of rough-neural computing are presented: training set production and AR schemes defined in the context of parameterized approximation spaces. The focus of the first form of rough-neural computing is inductive learning and the production of training (optimal feature set selection) using knowledge reduction algorithms. This first form of neural computing can be important in designing neural networks defined in the context of parameterized approximation spaces. A high-level Petri net model of a neural network classifier with an optimal feature selection procedure in its front end is given. This model is followed by the development of a number of Petri net models of what are known as elementary approximation neurons (EA neurons). The design of a particular form of EA neuron is considered, namely, a neuron that includes an uncertainty (activation) function that constructs a granule approximation and a rough inclusion (threshold activation) function that measures the degree to which granule approximation is part of a target granule.

This chapter is organized as follows. An overview of set approximation, attribute reduction (derivation of minimal sets of attributes), decision rules, discretization, and rough membership set functions is presented in Sect. 2. The basic features of rough Petri nets are described in Sect. 3. A training set production model is briefly given in Sect. 4. A framework (scheme) for approximate reasoning in the context of parameterized approximation spaces is given in Sect. 5. A sample approximation space and an indistinguishability relation are also given in Sect. 5. Three elementary approximation neuron Petri net models are presented in Sect. 6. Also included in Sect. 6 is an elementary Petri net model for approximation neuron training.

2 Preliminaries

This section gives a brief overview of some fundamental concepts and features of rough set theory that are important to an understanding of the Petri net models given in this chapter.

2.1 Set Approximation

Rough set theory offers a systematic approach to set approximation [16]. To begin, let $S=(U, A)$ be an information system where U is a nonempty, finite set of objects and A is a nonempty, finite set of attributes, where $a : U \to V_a$ for every $a \in A$. For each $B \subseteq A$, there is associated an equivalence relation $\mathrm{Ind}_A(B)$ such that

$$\mathrm{Ind}_A(B) = \{(x,x') \in U^2 \mid \forall a \in B . a(x) = a(x')\}.$$

If $(x, x') \in \mathrm{Ind}_A(B)$, we say that objects x and x' are indiscernible from each other relative to attributes from B. The notation $[x]_B$ denotes equivalence classes of $\mathrm{Ind}_A(B)$. Further, partition $U/\mathrm{Ind}_A(B)$ denotes the family of all equivalence classes of relation $\mathrm{Ind}_A(B)$ on U. For $X \subseteq U$, the set X can be approximated only from information contained in B by constructing a B-lower and a B-upper approximation denoted by $\underline{B}X$ and $\bar{B}X$, respectively, where $\underline{B}X = \{x \mid [x]_B \subseteq X\}$ and $\bar{B}X = \{x \mid [x]_B \cap X \neq \emptyset\}$.

2.2 Attribute Reduction and Decision Rules

An approach to finding a subset of attributes (reduct) with the same classificatory power as the entire set of attributes in an information system is briefly described in this section. This leads to a brief discussion about the derivation of decision rules with minimal descriptions in their left-hand sides. In deriving decision system rules, the discernibility matrix and discernibility function are essential. Given an information system S=(U, A), the $n \times n$ matrix (c_{ij}) called the discernibility matrix of S (denoted M_S) is defined in (1).

$$c_{ij} = \{a \in A \mid a(x_i) \neq a(x_j)\}, \quad \text{for } i, j = 1, \ldots, n. \tag{1}$$

A discernibility function f_S relative to discernibility matrix M_S for an information system S is a Boolean function of m Boolean variables $a_1^*, \ldots, a_m^*$ corresponding to attributes $a_1, \ldots, a_m$, respectively, and defined in (2).

$$f_S(a_1^*, \ldots, a_m^*) =_{df} \wedge \left\{ \vee c_{ij}^* \middle| 1 \leqslant j \leqslant i \leqslant n,\, c_{ij} \neq \emptyset \right\}, \text{ where } c_{ij}^* = \left\{ a^* \middle| a \in c_{ij} \right\}. \tag{2}$$

Precise conditions for decision rules can be extracted from a discernibility matrix as in [30,31]. For the information system S, let $B \subseteq A$, and let $\wp(V_a)$ denote the power set of V_a, where V_a is the value set of a. For every $d \in A - B$, a decision function $d_d^B : U \to \wp(V_a)$ is defined in (3).

$$d_d^B(u) = \left\{ v \in V_d \mid \exists u' \in U,\, (u', u) \in \mathrm{Ind}_A(B),\, d(u') = v \right\}. \tag{3}$$

In other words, $d_d^B(u)$ is the set of all elements of the decision column of S such that the corresponding object is a member of the same equivalence class as argument u. The next step is to determine a decision rule with a minimal number of descriptors on the left-hand side. Pairs (a, v), where $a \in A$, $v \in V$, are called *descriptors*. A decision rule over the set of attributes A and values V is an expression of the form given in (4).

$$a_{i_1}(u_i) = v_{i_1} \wedge \ldots \wedge a_{i_j}(u_i) = v_{i_j} \wedge \ldots \wedge a_{i_r}(u_i) = v_{i_r} \underset{S}{\Rightarrow} d(u_i) = v, \tag{4}$$

where $u_i \in U$, $v_{i_j} \in V_{a_{i_j}}$, $v \in V_d$, $j = 1, \ldots, r$ and $r \leq \text{card}(A)$. The fact that a rule is true is indicated by writing it in the form given in (5).

$$(a_{i_1} = v_{i_1}) \wedge \ldots \wedge (a_{i_r} = v_{i_r}) \underset{S}{\Rightarrow} (a_p = v_p). \tag{5}$$

The set of all prime implicants of f_S determines the set of all reducts of S [9]. A reduct is a minimal set of attributes $B \subseteq A$ that can be used to discern all objects obtainable by all of the attributes of an information system [37]. The set of all reducts of S is denoted by RED(S). A method used to find a proper subset of attributes of A with the same classificatory power as the entire set A has been termed *attribute reduction* [36].

Let $R \in$ RED(S) be a reduct in the set of all reducts in an information system S. For information system S, the set of decision rules constructed with respect to a reduct R is denoted OPT(S, R). Then the set of all decision rules derivable from reducts in RED(S) is the set in (6).

$$\text{OPT}(S) = \cup\{\text{OPT}(S,R) \,|\, R \in \text{RED}(S)\}. \tag{6}$$

2.3 Discretization

Suppose that we need to obtain approximate knowledge of a continuum (e.g., behavior of a sensor signal over an interval of time) by considering parts of the continuum. Discretization of a continuum entails partition a particular interval into subintervals of reals. For example, consider the interval of reals $V_a = [v_a, w_a]$ for values of an attribute $a \in A$ in a consistent decision system S=(U, A). Discretization of V_a entails searching for a partition P_a of V_a (i.e., discovering a partition of the value sets of attributes into intervals). In rough set theory, discretization leads to partitions of value sets so that, if the name of the interval containing an arbitrary object is substituted for any object in place of its original value in S, a consistent information system is also obtained.

2.4 Rough Membership Set Functions

In this section, we consider a set function form of the traditional rough membership function, which was introduced in [17].

Definition 1. Let S=(U, A) be an information system, $B \subseteq A$, and let $[u]_B$ be an equivalence class of an object $u \in U$ of $\text{Ind}_A(B)$. A set function $\mu_u^B : \wp(U) \to [0, 1]$ is defined in (7).

$$\mu_u^B(X) = \frac{\text{card}(X \cap [u]_B)}{\text{card}([u]_B)}. \tag{7}$$

for any $X \in \wp(Y)$, $Y \subseteq U$, where $\wp(Y)$ denotes the power set of Y, called a *rough membership function* (rmf). A rough membership function provides a classification measure inasmuch as it tests the degree of overlap between the set X and an equivalence class $[u]_B$. The form of rough membership function in Definition 1 is slightly different from the classical definition [18] where the argument of the rough membership function is an object u and the set X is fixed. For example, let $X_{B_{\text{approx}}} \in \left\{\bar{B}X, \underline{B}X\right\}$ denote a set approximation. Then we compute the degree of overlap between $X_{B_{\text{approx}}}$ and $[u]_B$ in (8).

$$\mu_u^B(X_{B_{\text{approx}}}) = \frac{\text{card}([u]_B \cap X_{B_{\text{approx}}})}{\text{card}([u]_B)}. \tag{8}$$

3 Rough Petri Nets

Rough Petri nets were introduced in [21] and elaborated in [22]. A rough Petri net models a process that implements one or more features of rough set theory. Rough Petri nets are derived from colored and hierarchical Petri nets as well as from rough set theory. Colored Petri nets provide a well-understood framework for introducing computational mechanisms (data types, variables, constants, and functions) that are useful in describing processes that carry out set approximation, information granulation, and engage in approximate reasoning. The new form of Petri net uses rough set theory, multivalued logic, and receptor process theory to extend colored Petri nets. Three extensions of colored Petri nets are described in this chapter: (1) multivalued guards, (2) receptor processes, and (3) rough computing. Briefly, this is explained as follows. Boolean valued guards in traditional colored Petri nets are replaced by multivalued guards in rough Petri nets. Let T be a set of transitions. In a colored Petri net, a guard is a mapping $G: T \rightarrow \{0, 1\}$. A transition is enabled if G returns 1. In keeping with an interest in modeling approximate reasoning, we augment colored Petri nets with guards of the form $G: T \rightarrow [0, 1]$. With this form of guard, it is possible to model a level-of-enabling of a transition that is more general than the usual "on/off" enabling model. Places in a Petri net represent the states of a system. An input place is a source of input for a method associated with a transition. An output place is a repository for results computed by a transition method. In a rough Petri net, an input place can be a receptor process. This form of input place responds to each stimulus from the environment by measuring a stimulus and by making each measurement available to a transition method. This extension of the input place convention in colored Petri nets is needed to model dynamically changing systems that perform actions in response to sensor signals. There is an iteration "inside" a receptor process that idealizes a typical sensor-action system found in agents, that is, the intent behind a provision of input places that are receptor processes is to model a sensor that enables a response mechanism represented by a transition method each time the sensor is stimulated. Rough computation is the third extension of colored Petri nets considered in this chapter. This feature is the hallmark of a rough Petri net.

It is characterized by the design of transition methods that compute rough set structures (e.g., attribute reduction) as well as values of measurable functions and rough integrals. This feature is important to us because we are interested in modeling parts of intelligent systems such as neurons in a neural processing module and information granulation. A rough Petri net also includes a strength-of-connection mapping from arcs to weights. This feature of rough Petri nets is useful in modeling classical neural computation. A rough Petri net provides a basis for modeling, simulating, and analyzing approximate reasoning (especially in the context of rough-neural computing), decision, and control systems. In what follows, it is assumed that the reader is familiar with classical Petri nets in Petri [25] and colored Petri nets in Jensen [7].

Definition 2. *Rough Petri Net.* A rough Petri net (rPN) is a structure (Σ, P, T, A, N, C, G, E, I, W, $\mathfrak{R}$, ξ), where

- Σ is a finite set of nonempty data types called color sets.
- P is a finite set of places.
- T is a finite set of transitions.
- A is a finite set of arcs such that $P \cap T = P \cap A = T \cap A = \emptyset$.
- N is a 1–1 node function where $N : A \to (P \times T) \cup (T \times P)$.
- C is a color function where $C : P \to \Sigma$.
- G is a guard function where $G : T \to [0, 1]$.
- E is an arc expression function where $E : A \to Set\_of\_Expressions$ where $E(a)$ is an expression of type $C[p(a)]$ and $p(a)$ is the place component of $N(a)$.
- I is an initialization function where $I : P \to Set\_of\_Closed\_Expressions$ where $I(p)$ is an expression of type $C(p)$.
- W is a set of strengths-of-connections where $\xi : A \to W$.
- $\mathfrak{R} = \{\rho_\sigma \mid \rho_\sigma$ is a method that constructs a rough set structure or that computes a value$\}$.

A sample ρ_σ is a method that constructs a rough set structure (e.g., an upper approximation of a set X relative to a set of attributes B or the set OPT(S) of all rules derived from reducts of a decision system table for an information system S). Again, for example, ρ_σ can denote a rough membership function or a rough integral [17,22]. The availability of guards on transitions makes it possible to model sensor filters and various forms of fuzzy Petri nets. Higher order places representing receptor processes are part of rough Petri nets.

3.1 Receptor Processes

The notion of a receptor process comes from Dill [4]. In a rough Petri net, a receptor process is a higher order place that models a sensor. The input place labeled $?p1$ in Fig. 1(a), for example, represents a form of receptor process that accumulates a signal.

Definition 3. *Receptor Process.* A receptor process is a process that provides an interface between a system and its environment by recording its response to each stimulus in a finite set of sample sensor values (a signal) whenever stimuli are detected.

When an environment is a source of continuous stimulation, a receptor process continuously enqueues its responses and periodically enables any transition connected to it. The advantage in constructing such a Petri net model of sensor-dependent systems is that it facilitates reasoning about a system design and simulation of the responses of a system to stimuli. For example, let $?p1$ be a receptor process; X, a set of inputs (signal) produced by $?p1$, and let μ_u^e denote a rough membership function. Further, let $(Y, U - Y)$ be a partition defined by an expert e, and let $[u]_e$ denote a set in this partition containing u for a fixed $u \in U$ (see Fig. 1a).

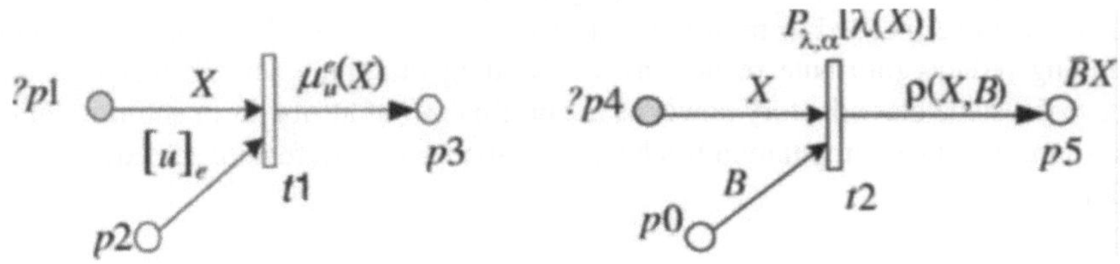

Fig. 1. **(a)** rPN without guard **(b)** guarded rPN

When transition $t1$ fires, $\mu_u^e(X)$ computes an rmf value. The label on transition $t2$ in Fig. 1b is an example of a Łukasiewicz guard (we explain this in this next section). Briefly, transition $t2$ is enabled, if $\alpha[\lambda(X)]$ holds for input X. The notation X denotes a set of values "produced" by the receptor process named $?p4$. The set X represents a signal or set of sample receptor process values that are accumulated over time. It is possible that $X = \{x\}$ (a restriction of X to a single sample receptor process value). In that case, the label x (rather than $\{x\}$) will be used to denote the input to a transition. The notation B on the arc from $p0$ to $t2$ represents a set of attributes that will be used by method ρ to construct a rough set structure. More details about receptor processes are given in [22].

3.2 Guarded Transitions

A guard $G(t)$ is an enabling condition associated with a transition t. In a rough Petri net, various families of guards can be defined that induce a level-of-enabling of transitions. Łukasiewicz guards were introduced in [24]. There are many forms of Łukasiewicz guards. Let $(L, \leq)$ be a lattice with a smallest element $\perp$ and with all other elements incomparable. In this chapter, $L - \{\perp\}$ is assumed to be equal to an interval of reals.

Definition 4. *Łukasiewicz Guard.* For a given function $\lambda(X)$ from the domain of variable X into $L - \{\perp\}$ and a condition α, i.e., a function from $L - \{\perp\}$ into $\{0, 1\}$, a Łukasiewicz guard $\mathrm{P}_{\lambda,\alpha}(X)$ is a function from the domain of the variable X into L defined by

$$\mathrm{P}_{\lambda,\alpha}(X) = \lambda(X), \text{ if } \alpha[\lambda(X)] = 1, \text{ and } \perp \text{ otherwise.}$$

We assume that a transition t labeled by $\mathrm{P}_{\lambda,\alpha}(X)$ is enabled if and only if $\alpha[\lambda(X)]$, i.e., $\alpha[\lambda(X)] = 1$. The value of $\lambda(X)$ of $\mathrm{P}_{\lambda,\alpha}(X)$ is a part of the output labeling the outgoing edges from t, if t is fired. Examples of conditions α are $0 < \lambda(X) \leq 1$ or $\lambda(X) \geq b > 0$ where b is a selected b in (0, 1].

4 Training Set Production

Training set production has been used by a number of researchers to reduce the size of the data set needed to train a neural network [32,38,39]. This approach entails finding reducts (attribute reduction), i.e., finding minimum sets of attributes that have the same classificatory power as the original set of attributes in an information system. A reduct computation has been described as a pure feature selection procedure [38].

Algorithm:
Basic signal classification based on selected reduct

Given:
Set of n-case stimuli (signals) containing an n-dimensional pattern with k real-valued attributes (sensors) A of the form $(a_1(x_i) \ldots a_k(x_i))$ for $a_i \in A,\ x_i \in X$.

1. For each $(a_1(x_i) \ldots a_k(x_i))$, determine decision value $v_d \in V_d$ (set of decision values).
2. Construct decision table $S_d = (X, A \cup \{d\})$, where each row of S_d is of the form $(a(x_i) \ldots a(x_i)\, v_d)$.
3. Compute RED(S_d) from S_d.
4. Select reduct $R \in$ RED(S_d). Assume that R has m attributes.
5. Compose reduced decision system $S_d = (X, A \backslash R \cup \{d\})$, where R is treated as a set of features describing all concepts in S_d.
6. Construct error back-propagation neural network (bpNN) over the reduced data set.
7. Train and test neural network bpNN.
8. Repeat steps 4 to 7 until the best classification result is obtained.

The notation *con*( ) in Fig. 2 denotes a procedure that performs signal conditioning. A sample form of *con*( ) based on a sensor fusion model using a discrete rough integral is given in [22].

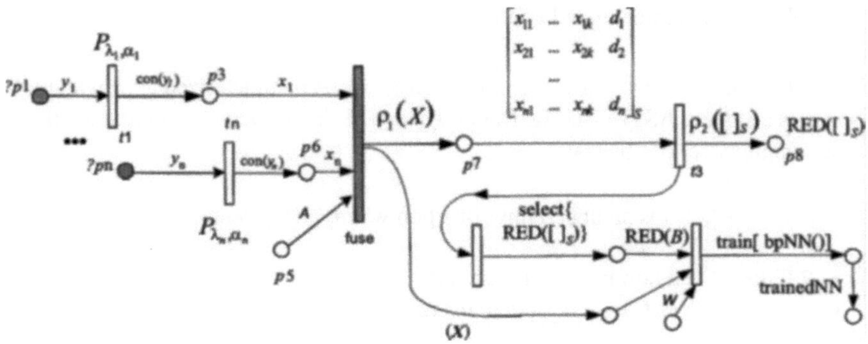

Fig. 2. Reduct-based neural net classifier model

5 Granular Computing Model

A brief introduction to a rough-neural computing Petri net model based on an adaptive calculus of granules is given in this section. Information granule construction and parameterized approximation spaces provide the foundation for the model of rough-neural computing [29]. A fundamental feature of this model is the design of neurons that engage in knowledge discovery. Mechanically, such a neuron returns granules (synthesized knowledge) derived from input granules. It has been pointed out that there is an analogy between calculi of granules in distributed systems and rough-neural computing [29]:

1. An agent with input and output ports providing communication links with other agents provides a model for a neuron η (analogously agent ag) with inputs supplied by neurons $\eta_1, \ldots, \eta_k$ (analogously agents $ag_1, \ldots, ag_k$), responds with output by η, and η is designed with a parameterized family of activation functions represented as rough connectives. In effect, a neuron resembles the model of an agent proposed by Milner [13].
2. Values of rough inclusions are analogous to weights in traditional neural networks.
3. Learning in a system governed by an adaptive calculus of granules is in the form of back propagation where incoming signals are assigned a proper scheme (granule construction) and a proper set of weights in negotiation and cooperation with other neurons.

5.1 Parameterized Approximation Spaces

A step toward the realization of an adaptive granule calculus in a rough-neural computing scheme is described in this section and is based on [31]. In a scheme for information granule construction in a distributed system of cooperating agents, weights are defined by approximation spaces. In effect, each agent (neuron) in such a scheme controls a local parameterized approximation space.

Definition 5. *Parameterized Approximation Space.* A parameterized approximation space is a system $AS_{\#,\$} = (U, I_{\#}, R, \nu_{\$})$ where #, $ denote vectors of parameters; U is a nonempty set of objects;

- $I_{\#} : U \rightarrow \wp(U)$ is an uncertainty function where $\wp(U)$ denotes the power set of U;
- $R \subseteq \wp(U)$ is a family of patterns; and
- $\nu_{\$} : \wp(U) \times \wp(U) \rightarrow [0, 1]$ denotes rough inclusion.

The uncertainty function defines a set of similarly described objects for every object x in U. A constructive definition of an uncertainty function can be based on the assumption that some metrics (distances) are given on attribute values. The family R describes a set of patterns (e.g., representing the sets described by the left-hand sides of decision rules). A set $X \subseteq U$ is definable on $AS_{\#,\$}$ if it is a union of some values of the uncertainty function. The rough inclusion function $\nu_{\$}$ defines the value of inclusion between two subsets of U. Using rough inclusion, the neighborhood $I_{\#}(x)$ can usually be defined as a collection of close objects. Also note that for some problems it is convenient to define an uncertainty set function of the form $I_{\#} : \wp(U) \rightarrow \wp(U)$. This form of uncertainty function works well in signal analysis, where we want to consider a domain over sets of sample signal values.

5.2 Sample Approximation Space

A very simple sample parameterized approximation space is given in this section.

Example 1. Threshold-Based Approximation Space This example is derived from [35]. Consider the information system $S = (U, A)$. Let $A = \{a\}$, where the attribute is real-valued, and let U be a nonempty set of reals with $x \in U$. Consider two elementary granules $a(x) \in [v_1, v_2]$ and $a(x) \in [v'_1, v'_2]$ for two ordinate sets of points defined with real numbers v_1, v_2, v'_1, v'_2 where $v_1 < v_2$ and $v'_1 < v'_2$. For simplicity, assume that the subinterval $[v_1, v_2]$ is contained in the interval $[v_1, v'_2]$. We want to measure the degree of inclusion of granule $a(x) \in [v_1, v_2]$ in granule $a(x) \in [v'_1, v'_2]$ (i.e., we assume that elements of R and neighborhoods of objects are such intervals). First, we introduce an overlapping range function r_a to measure the overlap between a pair of subintervals:

$$r_a\left([v_1, v_2], [v'_1, v'_2]\right) = \max\left(\left\{\min\left(\left\{v_2, v'_2\right\}\right) - \max\left(\left\{v_1, v'_1\right\}\right), 0\right\}\right).$$

The uncertainty function $I_{B,\delta} : U \rightarrow \wp(U)$ is defined as follows:

$$I_{B,\delta}(x) = \begin{cases} [v_1, v_2], \text{ if } \lfloor a(x)/\delta \rfloor \in [v_1, v_2] \\ [v'_1, v'_2], \text{ if } \lfloor a(x)/\delta \rfloor \in [v'_1, v'_2] \\ [0,0], \text{ otherwise,} \end{cases}$$

where $\lfloor a(x)/\delta \rfloor$ denotes the greatest integer less than or equal to $a(x)/\delta$. A rough inclusion function $\nu_{t_a} : \wp(U) \times \wp(U) \rightarrow [1, 0]$ is then defined as follows:

$$\nu_{t_a}\left[I_{B,\delta}(x), I_{B,\delta}(x')\right] = \frac{r_a\left[I_{B,\delta}(x), I_{B,\delta}(x')\right]}{v_2 - v_1} \in [0, t_a) \cup [t_a, 1].$$

Elementary granule $a(x) \in [v_1, v_2]$ is included in elementary granule $a(x) \in [v'_1, v'_2]$ in degree t_a if and only if

$$\frac{r_a\left[I_{B,\delta}(x), I_{B,\delta}(x')\right]}{v_2 - v_1} \geq t_a \qquad \text{(threshold criterion).}$$

This version of a parameterized approximation space depends on a threshold parameter t_a used as a criterion for the degree of inclusion of one interval in another interval at or above the threshold t_a whenever $\nu_{t_a}\left[I_{B,\delta}(x), I_{B,\delta}(x')\right] \geq t_a$. Changes in parameter δ in the uncertainty function $I_{B,\delta}$ and threshold parameter t_a in the inclusion model ν_{t_a} in this sample approximation space will change the result. With a threshold t_a, ν_{t_a} acts like a high-pass filter (it "passes" high values of ν above t_a and rejects low values of ν). The values of these parameters can be learned during training (i.e., during training, adjustments in the parameters are made to improve the classification of elementary input granules).

5.3 Indistinguishability Relation

To begin, let $S = (U, A)$ be an infinite information system where U is a nonempty subset of the reals $\Re$ and A is a nonempty, finite set of attributes, where $a : U \rightarrow V_a$ for every $a \in A$. Let $a(x) \geq 0$, $\delta > 0$, $x \in \Re$ (set of reals), and let $\lfloor a(x)/\delta \rfloor$ denote the greatest integer less than or equal to $a(x)/\delta$ ["floor" of $a(x)/\delta$] for attribute a. If $a(x) < 0$, then $\lfloor a(x)/\delta \rfloor = -\lfloor |a(x)|/\delta \rfloor$. The parameter δ serves as a "neighborhood" size on real-valued intervals. Reals within the same subinterval bounded by $k\delta$ and $(k+1)\delta$ are considered indistinguishable. For each $B \subseteq A$, there is associated an equivalence relation $\mathrm{Ing}_{A,\delta}(B)$ defined in (9).

$$\mathrm{Ing}_{A,\delta}(B) = \left\{ (x, x') \in \Re^2 \,|\, \forall a \in B.\ \lfloor a(x)/\delta \rfloor = \lfloor a(x')/\delta \rfloor \right\}. \tag{9}$$

If $(x, x') \in \mathrm{Ing}_{A,\delta}(B)$, we say that objects x and x' are indistinguishable from each other relative to attributes from B. A subscript Id denotes a set of identity sensors

$Id(x) = x$. They are introduced to avoid a situation, where there is more then one stimulus for which a sensor takes the same value. From (9), we can write

$$\text{Ing}_{A,\delta}(B \cup Id) = \left\{ \begin{array}{l} (x,x') \in \Re^2 | \lfloor x/\delta \rfloor = \lfloor x'/\delta \rfloor \wedge \\ \forall a \in B.\ \lfloor a(x)/\delta \rfloor = \lfloor a(x')/\delta \rfloor \end{array} \right\}.$$

The notation $[x]_B^\delta$ denotes equivalence classes of $\text{Ing}_{A,\delta}(B)$. Further, partition $U/\text{Ing}_{A,\delta}(B)$ denotes the family of all equivalence classes of relation $\text{Ing}_{A,\delta}(B)$ on U. For $X \subseteq U$, the set X can be approximated only from information contained in B by constructing a B-lower and a B-upper approximation denoted by $\underline{B}X$ and $\bar{B}X$, respectively, where $\underline{B}X = \left\{x \mid [x]_{B \cup Id}^\delta \subseteq X\right\}$ and $\bar{B}X = \left\{x \mid [x]_{B \cup Id}^\delta \cap X \neq \emptyset\right\}$. In some cases, we find it necessary to use a sensor reading y (an ordinate or "vertical" value) instead of stimulus x. In such cases, we create an equivalence class consisting of all points (ordinate values) for which sensor readings are "close" to y and define $[y]_B^\delta = \left\{ x \in \Re | \forall a \in B.\ \left\lfloor \frac{a(x)}{2\delta} \right\rfloor = \left\lfloor \frac{y}{2\delta} \right\rfloor \right\}$, where $[y]_B^\delta$ consists of equivalence classes $[x]_B^\delta$ such that $a(x) = y$. This is quite important when we want to extract information granules relative to sensor signals (sensor measurements rather than sensor stimuli then hold our attention).

Proposition 1. The relation $\text{Ing}_{A,\delta}(B)$ is an equivalence relation.

Note that it is also possible to define upper and lower approximations of a set in the context of the uncertainty function and rough inclusion in an approximation space, that is, we define an uncertainty function $I_\# : U \to \wp(U)$ and rough inclusion $\mu : X \times Y \to [0, 1]$. Then we obtain

$$\text{LOW}\left(AS_{\#,\$}, X\right) = \{x \in U \mid \nu_\$ [I_\#(x), X] = 1\},$$

$$\text{UPP}\left(AS_{\#,\$}, X\right) = \{x \in U \mid \nu_\$ [I_\#(x), X] > 0\}.$$

This is the approach taken in [29]. There is some flexibility in the way we define the uncertainty function for an approximation space. For some problems, it is more convenient to define a set function of the form $I_\# : \wp(U) \to \wp(U)$. This form of uncertainty function works well in signal analysis, where we want to consider a domain over sets of sample signal values. This change in $I_\#$ leads to slightly different definitions of upper and lower approximations in an approximation space:

$$\text{LOW}\left(AS_{\#,\$}, Y\right) = \{X \in \wp(U) \mid \nu_\$ [I_\#(X), Y] = 1\},$$

$$\text{UPP}\left(AS_{\#,\$}, Y\right) = \{X \in \wp(U) \mid \nu_\$ [I_\#(X), Y] > 0\}.$$

5.4 More Sample Approximation Spaces

In this section, we consider a parameterized approximation space defined relative to the indistinguishability relation and uncertainty set functions where the domain of such a function is the power set (set of subsets) of U.

Example 2. Indistinguishability-Based Approximation Space. Consider a parameterized approximation space with an uncertainty set function $I_{B,\delta}$ that constructs a granule (namely, an upper approximation) based on knowledge of B and indistinguishability relation with parameter δ and a rough inclusion function ν with threshold parameter t_0. We write $\nu_{t_0}(X,Y)$ to denote $\nu(X,Y) > t_0$. For simplicity, the traditional $\bar{B}X$ is constructed by a method named constructApprox as part of the definition of $I_{B,\delta}$, where $I_{B,\delta} : \wp(U) \to \wp(U)$ for $I_{B,\delta}(X) = \text{constructUPP}(X) = \bar{B}X$, and let $Y = [y]_B^\delta$. In this example, rough inclusion is defined as follows:

$$\nu : \wp(U) \times \wp(U) \to [0,1] \text{ for } \nu[I_{B,\delta}(X), Y] = \frac{\rho\left[I_{B,\delta}(X) \cap Y\right]}{\rho(Y)},$$

where

$$\nu(X,Y) = \frac{\rho\left(\bar{B}X \cap [y]_B^\delta\right)}{\rho\left([y]_B^\delta\right)} = \frac{\int_{\bar{B}X \cap [y]_B^\delta} 1 \, dx}{\int_{[y]_B^\delta} 1 \, dx}.$$

The rough inclusion of granule $I_{B,\delta}(X)$ is acceptable in granule Y, provided that the following constraint is satisfied:

$$\nu_{t_0}(X,Y), \text{ i.e., } \nu(X,Y) > t_0.$$

The essential thing to notice about this variant of $I_{B,\delta}$ in this example is that it constructs the granule $\bar{B}X$ from its domain X. The rough inclusion function then measures the degree of overlap between $\bar{B}X$ and a set represented by Y. The composition of the set Y is not treated in this example. The parameters for $I_{B,\delta}$ are δ (tolerance) and set of attributes (features) B. The parameter for ν is the threshold t_a.

6 Approximation Neuron Models

The parameters in a parameterized approximation space can be treated as an analogy to neural network weights, and each instance of such a granule-producing agent with a parameterized approximation space design parallels the architecture of neurons in a conventional neural network (see Fig. 1 in Chap. 2). To carry this analogy a step further, the parameters of an approximation space should be learned to induce the relevant information granules.

A neuron with a parameterized approximation space in its design is called an approximation neuron. In its simplest form, such a neuron constructs an elementary granule as a result of approximating the information received from its inputs. In more elaborate forms of an approximation neuron, for example, the output of the neuron may take the form of a rule derived from a condition vector of inputs, a reduct derived from a received decision table, or a set of rules derived from a received reduct and received decision table. The particular configuration of such a neuron depends on instantiation of the approximation space and particular activation functions used in designing the neuron. The design of an approximation neuron changes each time we modify the definition of the uncertainty function $I_{\#}$ and rough inclusion function $\nu_{\$}$ as well the parameters in # and $ chosen for these functions. In this section, we consider two fairly basic models of elementary approximation neurons (EA neurons). An EA neuron is a neuron that constructs an elementary granule as its output. An elementary information granule is an information granule that contains a single "piece" of information (e.g., an attribute, or sensor value, a condition for a rule, a measurement of rough inclusion). The output of such an approximation neuron is a rough inclusion value.

6.1 Threshold-Based Approximation Neuron

The approximation space in Example 1 suggests the possibility of designing a simple prototype neural network where changes in the parameters rather than changes in weights provide a basis for training. It is possible to design a simple prototype neural network where changes in the parameters rather than changes in weights provide a basis for training in the context of a parameterized approximation space. We want to consider a threshold-based approximation neuron with an elementary granule as its output, namely, $\nu\left[I_{B,\delta}(x), I_{B,\delta}(x')\right]$ (see Fig. 3). In Fig. 3, a Petri net is given to model an approximation neuron. This is an example of a rough Petri net. The label $?p_0$ in Fig. 3 denotes a receptor process (always input ready) "connected" to the environment of an agent.

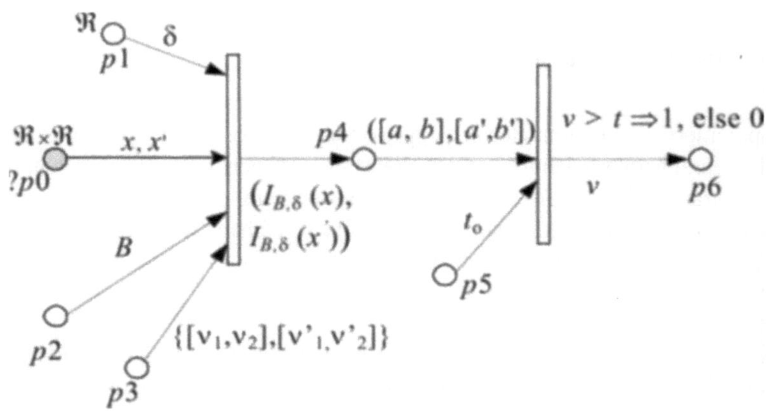

Fig. 3. Threshold-based neuron design

6.2 Indistinguishability-Based Approximation Neuron

It is also possible to design a prototype neural network based on the indistinguishability relation. In the form of neural computation described in this section, training entails changing δ (the interval width) until the rough inclusion function value exceeds the threshold t_0. An indistinguishability-based approximation neuron can be designed with an elementary granule as its output, namely, $v\left[I_{B,\delta}(X),Y\right]$, where $I_{B,\delta}(X)$ computes $\bar{B}X$, and on the output we have 1 if card($\bar{B}X \cap Y$)/card(Y) $>t_0$ and 0 otherwise (see Fig. 4).

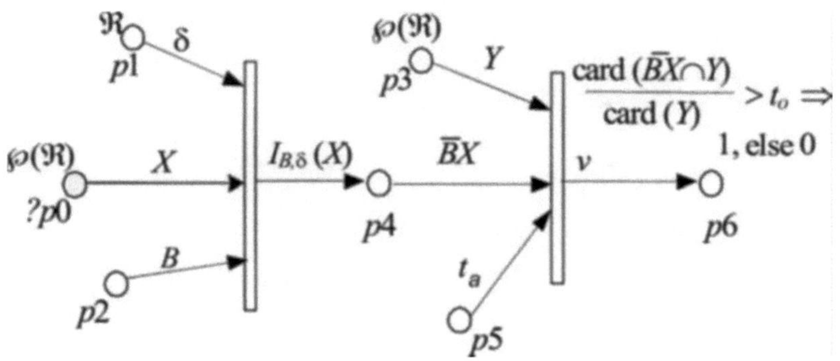

Fig. 4. Distinguishability-based neuron design

6.3 Approximation Neuron Training Model

In this section, a model for training is limited to a single approximation neuron. Up to this point, no guarded transitions have been used in Petri net models of approximation neurons. In this section, we consider a Petri net approximation neuron training model with three transitions, *neuron*, *train*, and a third transition named *approx* (see Fig. 5). Except for the guards and one extra communication transition, the Petri net in Fig. 3 is represented in Fig. 5 by the *neuron* transition together with its inputs and output. The firing of a *neuron* results in the computation of the rough inclusion of input X relative to relative to some set Y and either an initial value of δ or a changed value of δ "propagated back" from the transition train to place $p1$. Changes in δ occur during training and are the result of executing a procedure called *BP* (see Fig. 5). The term *back propagation* is typically used to describe training a multi-layer perceptron using a gradient descent applied to a sum-of-squares error function. Training in the basic neuron in Fig. 5 is much simpler since we are dealing only with the need to modify one parameter δ. If the transition train in Fig. 5 had more than one rough inclusion computation as input and more than one δ to adjust, then it would be appropriate to consider some form of traditional back-propagation method in adjusting the δ values. The transition *train* is enabled if $\nu[I_{B,\delta}(X),Y] - t_0 < 0$. In other words, this transition is enabled whenever $\nu[I_{B,\delta}(X),Y] < t_0$ (i.e., rough inclusion falls below threshold t_0). Each time the transition *train* fires, a new δ value is computed by the error function $\mathrm{BP}_\nu(\delta)$. The output place labeled $p1$ for the transition *train* in Fig. 5 is an alias for the input place $p1$ for the transition *neuron*. In the simple neural training model in Fig. 5, what happens to the neuron output when the rough inclusion value falls in the interval $[t_0,1]$ has been modeled with transition approx, which is enabled when the neuron output is at or above the required threshold. Transition approx serves as a high-pass filter, and only transmits ν-granules in $[t_0,1]$ to other agents.

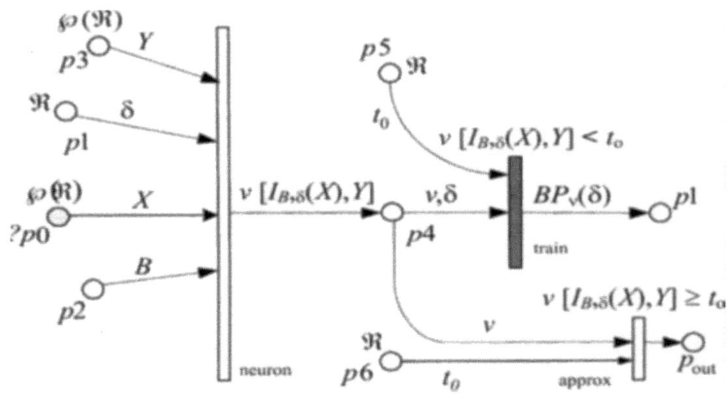

Fig. 5. Basic neuron training model

7 Conclusion

An approach to designing neurons using parameterized approximation spaces has been given in this chapter. In rough-neural computing, an AR scheme is defined in the context of one or more parameterized approximation spaces (denoted by $AS_{\#,\$}$). The methodology of an AR scheme is revealed by defining the universe U and the particular uncertainty and rough inclusion functions in the space $AS_{\#,\$}$. The scope of the AR scheme for a rough-neural network is determined by identifying the particular universe U, the set of attributes reflecting the local knowledge of an agent (neuron), and the specific parameters in the vectors #, \$. A parameterized approximation space for an AR scheme is defined in the context of traditional rough set theory (granule approximation) as well as the notion of being part of in a degree from approximate rough mereology. Training a rough neural network entails fine-tuning the parameters of its approximations spaces to make it possible for the network to achieve optimal classificatory power. A trainable AR scheme that consists of a single parameterized approximation space that constructs an information granule that approximates its inputs, outputs some form of information granule, and adjusts its parameters to reduce the error in its output is called an approximation neuron. In this chapter, Petri net models for what are known as elementary approximation neurons have been given. An important consideration not included in the approximation neuron models but treated independently is given in this chapter, namely, attribute reduction. It is well known that performance improvement results from the inclusion of a feature subset selection procedure (finding a minimum-length reduct) in the front end of a traditional back-propagation neural network classification system. A Petri net model of this form of neural network has also been given.

Acknowledgments

The research of James Peters and Sheela Ramanna has been supported by the Natural Sciences and Engineering Research Council of Canada (NSERC) research grant 185986 and research grant 194376, respectively. The research of Zbigniew Suraj was partially supported by grant #8T11C02519 from the State Committee for Scientific Research (KBN) in Poland. The research of Maciej Borkowski was supported by Manitoba Hydro.

References

1. M.A. Arbib. The artificial neuron. In E. Fiesler, R. Beale, editors, *Handbook of Neural Computation*, B1.7, Institute of Physics, Bristol, 1997.
2. I.M. Bochenski. *A History of Formal Logic*. Chelsea, New York, 1956.
3. B. Chakraborty. Feature subset selection by neuro-rough hybridization. In W. Ziarko, Y. Yao, editors, *Proceedings of the Second International Conference on Rough Sets and Current Trends in Computing (RSCTC 2000)*, LNAI 2005, 519–526, Springer, Berlin, 2001.
4. D.L. Dill. *Trace Theory for Automatic Hierarchical Verification of Speed-Independent Circuits*. MIT Press, Cambridge MA, 1989.
5. Bonabeau et al. *Swarm Intelligence: From Natural to Artificial Systems*. Oxford University Press, Oxford, 2000.
6. L. Han, J.F. Peters, S. Ramanna, R. Zhai. Classifying faults in high voltage power systems: A rough-fuzzy neural computational approach. In N. Zhong, A. Skowron, S. Ohsuga, editors, *New Directions in Rough Sets, Data Mining, and Granular-Soft Computing*, LNAI 1711, 47–54, Springer, Berlin, 1999.
7. K. Jensen. *Coloured Petri Nets—Basic Concepts, Analysis Methods and Practical Use*, Vol. 1. Springer, Berlin, 1992.
8. T. Kohonen. The self-organizing map. *Proceedings of the IEEE*, 78: 1464–1480, 1999.
9. J. Komorowski, Z. Pawlak, L. Polkowski, A. Skowron. Rough sets: A tutorial. In S.K. Pal, A. Skowron, editors, *Rough Fuzzy Hybridization: A New Trend in Decision-Making*, 3–98, Springer, Singapore, 1999.
10. P.J. Lingras. Comparison of neofuzzy and rough neural networks. *Information Sciences*, 110: 207–215, 1998.

11. P.J. Lingras. Fuzzy-rough and rough-fuzzy serial combinations in neurocomputing. *Neurocomputing: An International Journal*, 36: 29–44, 2001.
12. R. Milner. *Calculus of Communicating Systems*. Report number ECS-LFCS-86-7 of Computer Science Department, University of Edinburgh, Edinburgh, 1986.
13. R. Milner. *Communication and Concurrency*. Prentice-Hall, New York, 1989.
14. S. Mitra, P. Mitra, S.K. Pal. Evolutionary modular design of rough knowledge-based network with fuzzy attributes. *Neurocomputing: An International Journal*, 36: 45–66, 2001.
15. S. K. Pal, S. Mitra. Multi-layer perceptron, fuzzy sets and classification. *IEEE Transactions on Neural Networks*, 3: 683–697, 1992.
16. Z. Pawlak. *Rough Sets: Theoretical Aspects of Reasoning About Data*. Kluwer, Dordrecht, 1991.
17. Z. Pawlak, J.F. Peters, A. Skowron, Z. Suraj, S. Ramanna, M. Borkowski. Rough measures: Theory and applications. In S. Hirano, M. Inuiguchi, S. Tsumoto, editors, *Rough Set Theory and Granular Computing*, Vol. 5(1/2) of Bulletin of the International Rough Set Society, 177–184, 2001.
18. Z. Pawlak, A. Skowron. Rough membership functions. In R. Yager, M. Fedrizzi, J. Kacprzyk, editors, *Advances in the Dempster–Shafer Theory of Evidence*, 251–271, Wiley, New York, 1994.
19. W. Pedrycz, L. Han, J.F. Peters, S. Ramanna, R. Zhai. Calibration of software quality: Fuzzy neural and rough neural computing approaches. *Neurocomputing: An International Journal*, 36: 149–170, 2001.
20. J.F. Peters, L. Han, S. Ramanna. Rough neural computing in signal analysis. *Computational Intelligence*, 17(3): 493–513, 2001.
21. J.F. Peters, S. Ramanna. A rough set approach to assessing software quality: Concepts and rough Petri net model. In S.K. Pal, A. Skowron, editors, *Rough Fuzzy Hybridization: A New Trend in Decision-Making*, 349–380, Springer, Berlin, 1999.
22. J.F. Peters, S. Ramanna, M. Borkowski, A. Skowron, Z. Suraj. Sensor, filter and fusion models with rough Petri nets. *Fundamenta Informaticae*, 47: 307–323, 2001.
23. A. Skowron, J. Stepaniuk. Information granule decomposition. *Fundamenta Informaticae*, 47(3/4): 337–350, 2001.
24. J.F. Peters, A. Skowron, Z. Suraj, S. Ramanna. Guarded transitions in rough Petri nets. In *Proceedings of the 7th European Congress on Intelligent Systems and Soft Computing (EUFIT'99)*, 203–212, Verlag Mainz, Aachen, 1999.
25. C.A. Petri. *Kommunikation mit Automaten*. Institut für Instrumentelle Mathematik, Schriften des IIM Nr.3, Bonn, 1962.
26. L. Polkowski, A. Skowron. Rough mereological calculi of granules: A rough set approach to computation. *Computational Intelligence*, 17(3): 472–492, 2001.
27. L. Polkowski, A. Skowron. Rough mereology. In *Proceedings of the International Symposium on Methodology for Intelligent Systems (ISMIS'94)*, LNAI 869, 85–94, Springer, Berlin, 1994.
28. L. Polkowski, A. Skowron. Rough-neuro computing. In W. Ziarko and Y. Yao, editors, *Proceedings of the Second International Conference on Rough Sets and Current Trends in Computing (RSCTC 2000)*, LNAI 2005, 57–64, Springer, Berlin, 2001.
29. A. Skowron. Toward intelligent systems: Calculi of information granules. *Bulletin of the International Rough Set Society*, 5: 9–30, 2001.
30. A. Skowron, C. Rauszer. The discernibility matrices and functions in information systems. In R. Słowiński, editor, *Intelligent Decision Support: Handbook of Applications and Advances of the Rough Sets Theory*, 331–362, Kluwer, Dordrecht, 1992.

31. A. Skowron, J. Stepaniuk. Decision rules based on discernibility matrices and decision matrices. In *Proceedings of the 3rd International Workshop on Rough Sets and Soft Computing (RSSC'94)*, 602–609, San Jose, CA, 1994.
32. A. Skowron, J. Stepaniuk. Information granules in distributed environment. In N. Zhong, A. Skowron, S. Ohsuga, editors, *Proceedings of the 3rd International Workshop on Rough Sets, Data Mining, and Granular-Soft Computing (RSFDGrC'99)*, LNAI 1711, 357–365, Springer, Berlin, 1999.
33. A. Skowron, J. Stepaniuk. Information granules: Towards foundations of granular computing. *International Journal of Intelligent Systems*, 16: 57–1044, 2001.
34. A. Skowron, J. Stepaniuk, J.F. Peters. Extracting patterns using information granules. *Bulletin International Rough Set Society* 5: 135–142, 2001.
35. A. Skowron, J. Stepaniuk, J.F. Peters. Hierarchy of information granules. In L. Czaja, editor, *Proceedings of the Workshop on Concurrency, Specification and Programming (CSP 2001)*, 254–268, Warsaw, Poland, 2001.
36. R.W. Swiniarski. *RoughNeuralLab, software package*. Developed at San Diego State University, San Diego, CA, 1995.
37. R.W. Swiniarski, L. Hargis. Rough sets as a front end of neural networks texture classifiers. *Neurocomputing: An International Journal*, 36: 85–103, 2001.
38. R.W. Swiniarski, A. Skowron. Rough set methods in feature selection and recognition. *Pattern Recognition Letters*, 24(6): 833–849, 2003.
39. M.S. Szczuka. Rough sets and artificial neural networks. In L. Polkowski, A. Skowron, editors, *Rough Sets in Knowledge Discovery 2: Applications, Cases Studies and Software Systems*, 449–470, Physica, Heidelberg, 1998.
40. P. Wojdyłło. Wavelets, rough sets and artificial neural networks in EEG analysis. In *Proceedings of the 1st International Conference on Rough Sets and Current Trends in Computing (RSCTC'98)*, LNAI 1424, 444–449, Springer, Berlin, 1998.
41. L.A. Zadeh. Fuzzy logic = computing with words. *IEEE Transactions on Fuzzy Systems*, 4: 103–111, 1996.
42. L.A. Zadeh. The key roles of information granulation and fuzzy logic in human reasoning, concept formulation and computing with words. In *Proceedings of the 5th International Conference on Fuzzy Systems (FUZZ-IEEE'96)*, 8–11, New Orleans, LA, 1996.
43. N. Zhong, A. Skowron, S. Ohsuga, editors. *Proceedings of the 3rd International Workshop on New Directions in Rough Sets, Data Mining, and Granular Soft Computing (RSFDGrC'99)*, LNAI 1711, Springer, Berlin, 1999.
44. W. Ziarko, Y.Y. Yao , editors. *Proceedings of the International Conference on Rough Sets and Current Trends in Computing (RSCTC 2000)*, LNAI 2005, Springer, Berlin, 2001.

Chapter 19
Information Granulation and Approximation in a Decision-Theoretical Model of Rough Sets

Yiyu Yao

Department of Computer Science, University of Regina, Regina, Saskatchewan S4S 0A2, Canada
yyao@cs.uregina.ca; http://www.cs.uregina.ca/~yyao

Summary. Granulation of the universe and approximation of concepts in a granulated universe are two related fundamental issues in the theory of rough sets. Many proposals dealing with the two issues have been made and studied extensively. We present a critical review of results from existing studies that are relevant to a decision-theoretical modeling of rough sets. Two granulation structures are studied: one is a partition induced by an equivalence relation, and the other is a covering induced by a reflexive relation. With respect to the two granulated views of the universe, element oriented and granule oriented definitions and interpretations of lower and upper approximation operators are examined. The structures of the families of fixed points of approximation operators are investigated. We start with the notions of rough membership functions and graded set inclusion defined by conditional probability. This enables us to examine different granulation structures and approximations induced in a decision-theoretical setting. By reviewing and combining results from existing studies, we attempt to establish a solid foundation for rough sets and to provide a systematic way of determining the parameters required in defining approximation operators.

1 Introduction

The concept of information granulation was first introduced in 1979 by Zadeh in the context of fuzzy sets [44]. The basic ideas of information granulation have appeared in fields, such as interval analysis, quantization, rough set theory, the theory of belief functions, divide and conquer, cluster analysis, machine learning, databases, and many others [45]. There is fast growing and renewed interest in the study of information granulation and computations under the umbrella term of *granular computing* (GrC), covering theories, methodologies, techniques, and tools that use of granules in problem solving [4,7,8,15,17,18,21,22,28,34,43,46].

Granulation of a universe involves decomposing the universe into parts, or grouping individual elements or objects into classes, based on available information and knowledge. Elements in a granule are drawn together by indistinguishability, similarity, proximity, or functionality [45]. With the granulation of a universe, a subset of a universe may be considered a whole unit rather than individual elements. A set of indistinguishable objects is considered a granule of the universe. One can thus form a granulated view of the universe. A natural consequence of granulation is the problem of approximating concepts using granules. The theory of rough sets can be used for constructing a granulated view of the universe and for interpreting, representing, and processing concepts in the granulated universe. It offers a more concrete model of granular computing.

The basis of the theory of rough sets is the indiscernibility or indistinguishability of objects or elements in a universe of interest [11,12]. The standard approach for modeling indiscernibility of objects is through an equivalence relation defined on the basis of their attribute values with reference to an information table [12]. Two objects are equivalent if they have exactly the same description. The induced granulation is a partition of the universe, i.e., a family of pairwise disjoint subsets. It is a coarsening of the universe and is studied extensively in mathematics under the name of quotient set. The notion of indiscernibility can be generalized into similarity defined by a reflexive binary relation [9, 20, 23, 24, 30, 37] . The set of objects similar to an element can be viewed as an elementary granule or a neighborhood with the element as its center. Distinct elementary granules may have nonempty overlaps, and the family of elementary granules with respect to all elements of the universe forms a covering of the universe.

Each equivalence class can be viewed as a basic building block or an elementary granule. All other sets are to be represented in terms of equivalence classes (granules). Two formulations are particularly relevant to decision-theoretical modeling of rough sets [29,30,38–40]. The element oriented method is based on the notion of rough membership [13]. One can define a rough membership function with respect to a subset of the universe based on its overlaps with equivalence classes [13]. Rough membership functions can be viewed as a special type of fuzzy membership function. The core and support of the membership function are defined as the subsets of objects with full and nonzero memberships, respectively. They produce the lower and upper approximations [41]. The granule oriented method is based on the set inclusion relation. An equivalence class is in the lower approximation of a set if it is contained in the set, and the equivalence class is in the upper approximation of a set if it has a nonempty overlap with the set. These formulations can be extended to cases where nonequivalence relations are used [20,23,30,37].

The lower and upper approximation operators, as defined by the core and support of a rough membership function, represent only two extreme cases. One is characterized by full membership and the other by nonzero membership. They may be regarded as qualitative approximations of a set. The actual degree of membership is not taken into consideration. Likewise, the definition by set inclusion considers only full inclusion, without considering the degree of set inclusion. This makes the rough set approach very sensitive to the accuracy of input data and not suitable for processing noisy data.

Extended rough set approximations have been suggested to resolve the problem with qualitative approximations. For the element oriented method, Pawlak et al. [14] introduced the notion of probabilistic rough set approximations. The conditional probabilities used to define probabilistic rough sets are rough membership functions. The lower probabilistic approximation is defined as the set of elements with membership values greater than 0.5, and the upper probabilistic approximation is the set of elements with membership values greater than or equal to 0.5. Yao et al. [38,40] proposed and studied a more general type of probabilistic rough set approximation based on Bayesian decision theory. A pair of parameters (α, β) with $\alpha > \beta$ can be determined from a loss (cost) function. The lower probabilistic approximation is defined as the set of elements with membership values greater than or equal to α and the upper probabilistic approximation as the set of elements with membership values greater than β.

The main results of the decision-theoretical rough set model were later given and studied again by some authors based on graded set inclusion by extending the granule oriented definition. Ziarko [47] introduced a variable precision rough set model. A measure called the degree of misclassification is defined as the inverse of the conditional probabilities or the rough membership functions. A threshold value α is used. An equivalence class is in the lower approximation if its degree of misclassification is below or equal to α. Equivalently, this means that an element belongs to the lower approximation if its membership value is greater than or equal to $1-\alpha$. Similarly, an equivalence class is in the upper approximation if its degree of misclassification is less than $1-\alpha$, or equivalently an element belongs to the upper approximation if its membership value is greater than α. However, unlike the decision-theoretical model, there do not exist theoretical justification and a systematic way to determine the parameter α in the variable precision rough set model, except for the suggestion that the value of α must be in the range $[0, 0.5)$. The parameter of the variable precision rough set model can be easily interpreted in the decision-theoretical rough set model, as shown by Yao et al. [39].

Other extensions of the granule oriented definition are based on different measures of the degree of, or graded, inclusion of two sets. Skowron and Stepaniuk [20] introduced an abstract notion of vague inclusion. The measure of vague inclusion is a function that maps every two subsets to a value in the unit interval $[0, 1]$ and is characterized by monotonicity with respect to the second argument. An example of vague inclusion can be defined by the rough membership function [20]. By extending the notion of rough membership functions to the power set of the universe, Polkowski and Skowron [16,19] introduced the notion of rough inclusion which is characterized by additional properties. Bryniarski and Wybraniec-Skardowska [1] used a family of inclusion relations in defining the rough set approximation. By choosing a set of threshold values, one can easily obtain a set of inclusion relations from a measure of graded inclusion. All of those proposals for graded or degree of inclusion can be used to define rough set approximations, as is done in the variable precision rough set model. The introduction of parameterized rough set approximations offers more flexibility to the theory and extends its domain of applications. In general, the existence of parameterized measures is very useful. As there may not exist universally good measures for all data, one can fine-tune the parameters to search for relevant measures from a family of parameterized measures with respect to a given set of data [25,26]. When applying this general principle to parameterized rough set approximations, one can search for relevant approximations for a given set of data. To achieve this goal, we need to provide intuitive interpretations of the parameters and design a systematic way to fine-tune the parameters.

From the brief summary of studies related to decision-theoretical modeling of rough sets, we can now state the objective of this chapter. By reviewing and combining results from existing studies, we attempt to establish a solid foundation for rough sets and to provide a systematic way of determining the parameters required in defining rough set approximations. Intuitive arguments and experimental investigations are important, as reported and demonstrated by many studies on generalizing rough sets based on the degree of membership or the degree of set inclusion. A solid and sound decision-theoretical foundation may provide a convincing argument and guidelines for applying the theory.

Granular computing covers many more topics, such as fuzzy if-then rules and computing with words. Partitions and coverings represent very simple granulated views of the universe. For instance, objects of the universe can have complex structures. The indistinguishability and similarity of such objects should be defined by taking their structures into consideration. A related issue is the search for suitable similarity relations and granulations for a particular application. We will not deal with these advanced topics. Instead, our discussion is restricted to topics and simple granulation structures related to rough sets, and particularly related to a decision-theoretical model of rough sets [38,40]. More specifically, we deal only with two related fundamental issues, granulation and approximation. Nevertheless, the argument can be applied to granular computing in general.

The rest of the chapter is organized as follows. Section 2 gives a brief overview of two granulation structures on the universe. One is defined by an equivalence relation and the other by a reflexive relation. Section 3 focuses on two definitions of rough set approximations. One is based on rough membership functions and the other on the set inclusion relation between an equivalence class and the set to be approximated. Approximation structures are discussed. Section 4 discusses a decision-theoretical model of rough sets.

2 Granulation of a Universe

The notion of indiscernibility provides a formal way to describe the relationships among elements of the universe under consideration. In the theory of rough sets, indiscernibility is modeled by an equivalence relation. A granulated view of the universe can be obtained from the equivalence classes. By generalizing equivalence relations to similarity relations characterized only by reflexivity [23,37], one may obtain a different granulation of the universe.

2.1 Granulation by Equivalence Relations

Let $E \subseteq U \times U$ be an equivalence relation on a finite and nonempty universe U, that is, E is reflexive, symmetrical, and transitive. The equivalence relation can be defined on the basis of available knowledge. For example, in an information table, elements in the universe are described by a set of attributes. Two elements are said to be equivalent if they have the same values with respect to some attributes [12,42]. The equivalence class,

$$[x]_E = \{y \in U \mid yEx\}, \tag{1}$$

consists of all elements equivalent to x and is also the equivalence class containing x. The relation E induces a partition of the universe U:

$$U/E = \{[x]_E \mid x \in U\}, \tag{2}$$

that is, U/E is a family of pairwise disjoint subsets of the universe, and $\bigcup_{x \in U} [x]_E = U$. The partition, commonly known as the quotient set, provides a granulated view of the universe under the equivalence of elements. Intuitively speaking, the available knowledge allows us to talk only about an equivalence class as a single unit. In other words, under the granulated view, we consider an equivalence class as a whole, instead of individuals.

The pair $apr = (U,E)$ is referred to as an approximation space, indicating the intended application of the partition U/E for approximation [11]. Each equivalence class is called an elementary granule. The elementary granules, the empty set $\emptyset$, and unions of equivalence classes are called definable granules in the sense that they can be defined precisely in terms of equivalence classes of E. The meaning of definable sets will be clearer when we discuss approximations in the next section. Let $\mathrm{Def}(U/E)$ denote the set of all definable granules. It is closed under set complement, intersection, and union. In fact, $\mathrm{Def}(U/E)$ is a sub-Boolean algebra of the Boolean algebra formed by the power set 2^U of U and an σ-algebra of subsets of U generated by the family of equivalence classes U/E. In addition, U/E is the basis of the σ-algebra $\sigma(U/E)$.

2.2 Granulation by Similarity Relations

Indiscernibility, as defined by an equivalence relation, or equivalently a partition of the universe, represents a very restricted type of relationship among elements of the universe. In general, the notions of similarity and coverings of the universe may be used [9,10,20,23,30,37].

Suppose that R is a binary relation on universe U representing the similarity of elements in U. We assume that R is at least reflexive, i.e., an element must be similar to itself but not necessarily symmetrical and transitive [23]. For two elements $x, y \in U$, if xRy, then we say that x is similar to y. The relation R may be more conveniently represented by using the set of elements similar to x or the predecessor neighborhood [30], as follows:

$$(x)_R = \{y \in U \mid yRx\}. \tag{3}$$

The set $(x)_R$ consists of all elements similar to x. By the assumption of reflexivity, $x \in (x)_R$. When R is an equivalence relation, $(x)_R$ is the equivalence class containing x. The family of predecessor neighborhoods,

$$U/R = \{(x)_R \mid x \in U\}, \tag{4}$$

is a covering of the universe, $\bigcup_{x \in U}(x)_R = U$. For two elements $x, y \in U$, $(x)_R$ and $(y)_R$ may be different and have a nonempty overlap. This offers another granulated view of the universe.

Through the similarity relation, an element x is viewed by the set of elements similar to it, namely, $(x)_R$. We define an equivalence relation $\equiv_R$ on U, as follows [31,35]:

$$x \equiv_R y \Longleftrightarrow (x)_R = (y)_R. \tag{5}$$

Two elements are equivalent if they have exactly the same neighborhood. If R is an equivalence relation, then $\equiv_R$ is the same as R.

The pair $apr = (U, R)$ is referred to as a generalized approximation space. The neighborhood $(x)_R$ is called an elementary granule. Elementary granules, the empty set $\emptyset$, and unions of elementary granules are called definable granules in $apr = (U, R)$. Let $\mathrm{Def}(U/R)$ denote the set of all definable granules. It is closed under set union, and may not necessarily be closed under set complement and intersection. The set $\mathrm{Def}(U/R)$ contains both the empty set $\emptyset$ and the entire set U. From $\mathrm{Def}(U/R)$, we define

$$\mathrm{Def}^c(U/R) = \{A^c \mid A \in \mathrm{Def}(U/R)\}, \tag{6}$$

where A^c denotes the complement of A. The new system $\mathrm{Def}^c(U/R)$ contains $\emptyset$ and U and is closed under set intersection. It is commonly known as a closure system [33]. If the relation R is an equivalence relation, both systems become the same one. In general, the two systems are not the same.

3 Rough Set Approximations

In an approximation space or a generalized approximation space, a pair of rough set approximation operators, known as the lower and upper approximation operators, can be defined in many ways [11,12,29–31,41]. Two definitions are discussed in this section. The element oriented definition focuses on the belongingness of a particular element to the lower and upper approximations of a set. The granule oriented definition focuses on the belongingness of an entire granule to the lower and upper approximations [35]. The two definitions produce the same results in an approximation space $apr = (U, E)$, but they produce different results in a generalized approximation space $apr = (U, R)$.

We pay special attention to two families of subsets of a universe. One consists of those subsets whose lower approximations are the same as themselves, i.e., the fixed points of a lower approximation operator. The other consists of those subsets whose upper approximations are the same as themselves, i.e., the fixed points of an upper approximation operator. The structures of the two families show the structures and consequences of different granulation methods and may provide more insights into our understanding of approximation operators.

3.1 Rough Membership Functions

In an approximation space $apr = (U, E)$, an element $x \in U$ belongs to one and only one equivalence class $[x]_E$. For a subset $A \subseteq U$, a rough membership function is defined by [13]

$$\mu_A(x) = \frac{|[x]_E \cap A|}{|[x]_E|}, \tag{7}$$

where $|\cdot|$ denotes the cardinality of a set. The rough membership value $\mu_A(x)$ may be interpreted as the conditional probability that an arbitrary element belongs to A, given that the element belongs to $[x]_E$. Conditional probabilities were used earlier in developing a probabilistic rough set model [14,27,38,40].

Rough membership functions may be interpreted as fuzzy membership functions interpretable in terms of probabilities defined simply by the cardinalities of sets [32, 35, 41]. With this interpretation, one can define at most $2^{|U|}$ fuzzy sets. Two distinct subsets of U may derive the same rough membership function. By definition, the membership values are all rational numbers.

The theory of fuzzy sets is typically developed as an uninterpreted mathematical theory of abstract membership functions without the above limitations [6]. In contrast, the theory of rough sets provides a more specific and more concrete interpretation of fuzzy membership functions. The source of the fuzziness in describing a concept is the indiscernibility of elements. The limitations and constraints of such an interpreted subtheory should not be viewed as disadvantages of the theory. Such constraints suggest conditions that may be verified when applying the theory to real-world problems. It might be more instructive and informative if one knows that a certain theory cannot be applied. Explicit statements of conditions under which a particular model is applicable may prevent misuse of the theory.

When interpreting fuzzy membership functions in the theory of rough sets, we have these constraints:

$$\begin{array}{ll}
\text{(m1)} & \mu_U(x) = 1. \\
\text{(m2)} & \mu_\emptyset(x) = 0. \\
\text{(m3)} & y \in [x]_E \Longrightarrow \mu_A(x) = \mu_A(y). \\
\text{(m4)} & x \in A \Longrightarrow \mu_A(x) \neq 0. \\
\text{(m5)} & x \notin A \Longrightarrow \mu_A(x) \neq 1. \\
\text{(m6)} & \mu_A(x) = 1 \Longleftrightarrow [x]_E \subseteq A. \\
\text{(m7)} & \mu_A(x) > 0 \Longleftrightarrow [x]_E \cap A \neq \emptyset. \\
\text{(m8)} & A \subseteq B \Longrightarrow \mu_A(x) \leq \mu_B(x).
\end{array}$$

Property (m3) is particularly important. It shows that elements in the same equivalence class must have the same degree of membership, that is, indiscernible elements should have the same membership value. Such a constraint, which ties the membership values of individual elements according to their connections, is intuitively appealing. Although this topic has been investigated by some authors [2], there is still a lack of systematic study. Properties (m4) and (m5) state that an element in A cannot have a zero membership value and an element not in A cannot have full membership. They can be equivalently expressed as

$$\begin{array}{ll}
\text{(m4)} & \mu_A(x) = 0 \Longrightarrow x \notin A. \\
\text{(m5)} & \mu_A(x) = 1 \Longrightarrow x \in A.
\end{array}$$

According to properties (m6) and (m7), $\mu_A(x) = 1$ if and only if for all $y \in U, x \in [y]_E$ implies $y \in A$, and $\mu_A(x) > 0$ if and only if there exists a $y \in U$ such that $y \in A$ and $x \in [y]_E$. Since $x \in [x]_E$, property (m5) is a special case of (m6). Property (m8) suggests that a rough membership function is monotonic with respect to set inclusion.

In a generalized approximation space $apr = (U, R)$ defined by a reflexive relation, a rough membership function can be defined for a subset A of a universe, by substituting $[x]_E$ with $(x)_R$ in (7), as follows [31,32]:

$$\mu_A(x) = \frac{|(x)_R \cap A|}{|(x)_R|}. \tag{8}$$

By the reflexivity of R, one can verify that properties (m1), (m2), and (m4)–(m8) also hold, provided that $[x]_E$ is replaced by $(x)_R$. For (m3), we can have the weak version:

$$\text{(m3a)} \quad [y \in (x)_R, \mu_A(x) = 1] \Longrightarrow \mu_A(y) \neq 0.$$
$$\text{(m3b)} \quad [y \in (x)_R, \mu_A(x) = 0] \Longrightarrow \mu_A(y) \neq 1.$$

These two properties are qualitative. They state that membership values of related elements are related. If an element y is similar to another element x with full membership, then y cannot have null membership. Likewise, if y is similar to an element x with null membership, then y cannot have full membership. Properties (m3a) and (m3b) can also be expressed as

$$\text{(m3a)} \quad [y \in (x)_R, \mu_A(y) = 0] \Longrightarrow \mu_A(x) \neq 1.$$
$$\text{(m3b)} \quad [y \in (x)_R, \mu_A(y) = 1] \Longrightarrow \mu_A(x) \neq 0.$$

They can be similarly interpreted. With respect to the equivalence relation $\equiv_R$, we have the property

$$\text{(m3c)} \quad x \equiv_R y \Longrightarrow \mu_A(x) = \mu_A(y).$$

It is closer to the original (m3) of the standard rough membership function. All of these properties appear intuitively sound and meaningful.

A binary relation defines only a dichotomous relationship. Two elements are either related or not related. It is not surprising that we can draw conclusions only with respect to elements with null or full membership, as indicated by the previously stated properties.

The constraints on rough membership functions have significant implications for rough set operators. Rough membership functions corresponding to A^c, $A \cap B$, and $A \cup B$ must be defined using set operators and (7) or (8).

By the laws of probability,

$$\text{(o1)} \quad \mu_{A^c}(x) = 1 - \mu_A(x).$$
$$\text{(o2)} \quad \mu_{A \cup B}(x) = \mu_A(x) + \mu_B(x) - \mu_{A \cap B}(x).$$
$$\text{(o3)} \quad \max[0, \mu_A(x) + \mu_B(x) - 1] \leq \mu_{A \cap B}(x) \leq \min[\mu_A(x), \mu_B(x)].$$
$$\text{(o4)} \quad \max[\mu_A(x), \mu_B(x)] \leq \mu_{A \cup B}(x) \leq \min[1, \mu_A(x) + \mu_B(x)].$$

Unlike the commonly used fuzzy set operators typically defined by t-norms and t-conorms [6], the new intersection and union operators are not truth-functional, that is, it is impossible to obtain rough membership functions of $A \cap B$ and $A \cup B$ based solely on the rough membership functions of A and B. One must also consider their overlaps and their relationships to the equivalence class $[x]_E$ or the predecessor neighborhood $(x)_R$.

One can verify the following additional properties corresponding to the properties of t-norms and t-conorms:

(t1) Boundary conditions

$$[\mu_A(x)=0,\mu_B(x)=0] \Longrightarrow \mu_{A\cap B}(x)=0,$$
$$[\mu_A(x)=1,\mu_B(x)=a] \Longrightarrow \mu_{A\cap B}(x)=a,$$
$$[\mu_A(x)=a,\mu_B(x)=1] \Longrightarrow \mu_{A\cap B}(x)=a.$$

(t2) Monotonicity :

$$(A\subseteq C, B\subseteq D) \Longrightarrow \mu_{A\cap B}(x) \leq \mu_{C\cap D}(x).$$

(t3) Symmetry :

$$\mu_{A\cap B}(x)=\mu_{B\cap A}(x).$$

(t4) Associativity :

$$\mu_{A\cap(B\cap C)}(x)=\mu_{(A\cap B)\cap C}(x).$$

(s1) Boundary conditions :

$$[\mu_A(x)=1,\mu_B(x)=1] \Longrightarrow \mu_{A\cup B}(x)=1,$$
$$[\mu_A(x)=0,\mu_B(x)=a] \Longrightarrow \mu_{A\cup B}(x)=a,$$
$$[\mu_A(x)=a,\mu_B(x)=0] \Longrightarrow \mu_{A\cup B}(x)=a.$$

(s2) Monotonicity :

$$[A\subseteq C, B\subseteq D] \Longrightarrow \mu_{A\cup B}(x) \leq \mu_{C\cup D}(x).$$

(s3) Symmetry :

$$\mu_{A\cup B}(x)=\mu_{B\cup A}(x).$$

(s4) Associativity :

$$\mu_{A\cup(B\cup C)}(x)=\mu_{(A\cup B)\cup C}(x).$$

Boundary conditions follow from (o3) and (o4), and monotonicity follows from (m8). Although other properties are very close to the properties of t-norms and t-conorms, the monotonicity property is much weaker than the monotonicity of a t-norm t and a t-conorm s, i.e., $(a\leq c, b\leq d) \Longrightarrow t(a,b)\leq t(c,d)$ and $(a\leq c, b\leq d) \Longrightarrow s(a,b)\leq s(c,d)$. For four arbitrary sets A,B,C,D with $\mu_A(x)=a$, $\mu_B(x)=b$, $\mu_C(x)=c$, $\mu_D(x)=d$, $a\leq c$ and $b\leq d$, $\mu_{A\cap B}(x)\leq\mu_{C\cap D}(x)$ and $\mu_{A\cup B}(x)\leq\mu_{C\cup D}(x)$ may not necessarily hold.

3.2 Element Oriented Approximations

In an approximation space $apr=(U,E)$, we define a rough membership function μ_A for a subset $A\subseteq U$. By collecting elements with full and nonzero memberships, respectively, we obtain a pair of lower and upper approximations of A as follows:

$$\begin{aligned}\underline{apr}(A) &= \{x\in U \mid \mu_A(x)=1\} = core(\mu_A),\\ \overline{apr}(A) &= \{x\in U \mid \mu_A(x)>0\} = support(\mu_A).\end{aligned} \tag{9}$$

They are the core and support of the fuzzy set μ_A. An equivalent and more convenient definition without using membership functions is given by

$$\begin{aligned} \underline{apr}(A) &= \{x \in U \mid [x]_E \subseteq A\}, \\ \overline{apr}(A) &= \{x \in U \mid [x]_E \cap A \neq \emptyset\}. \end{aligned} \tag{10}$$

The lower and upper approximations can be interpreted as a pair of unary set-theoretical operators, $\underline{apr}, \overline{apr} : 2^U \longrightarrow 2^U$. They are dual operators in the sense that $\underline{apr}(A) = (\overline{apr}(A^c))^c$ and $\overline{apr}(A) = (\underline{apr}(A^c))^c$. Other properties of approximation operators can be found in many articles [5,11,12,31,37,39].

In this definition, we focus on whether a particular element is in the lower and upper approximations. It is thus referred to as the element oriented definition of rough set approximations. More specifically, an element $x \in U$ belongs to the lower approximation of A, if *all* of its equivalent elements belong to A. It belongs to the upper approximation of A, if at least *one* of its equivalent elements belongs to A. The element oriented interpretation of approximation operators is related to the interpretation of the necessity and possibility operators in modal logic [29,37].

So far, we have shown that, as a consequence of granulation, a set A is viewed differently. The fuzzification of A leads to a rough membership function, and the approximation of A leads to a pair of sets. Moreover, approximations of a set can be viewed as a qualitative characterization of a rough membership function using the core and support. A study of families of sets that are invariant under fuzzification and approximation may bring more insights into the understanding of granulation structures.

A set A is said to be a lower exact set if $A = \underline{apr}(A)$, an upper exact set if $A = \overline{apr}(A)$, and a lower and an upper exact set if $\underline{apr}(A) = A = \overline{apr}(A)$. Lower exact sets are fixed points of the lower approximation operator $\underline{apr}$, and upper exact sets are fixed points of the upper approximation operator $\overline{apr}$. Let

$$\begin{aligned} \mathrm{E}(\underline{apr}) &= \{A \subseteq U \mid A = \underline{apr}(A)\}, \\ \mathrm{E}(\overline{apr}) &= \{A \subseteq U \mid A = \overline{apr}(A)\} \end{aligned} \tag{11}$$

be the set of lower exact sets and the set of upper exact sets, respectively. By definition, we immediately have the following results:

Theorem 1. *In an approximation space $apr = (U,E)$,*

$$\mathrm{E}(\underline{apr}) = \mathrm{E}(\overline{apr}) = \mathrm{Def}(U/E). \tag{12}$$

Theorem 2. *In an approximation space $apr = (U,E)$,*

$$\mu_A(x) = \chi_A(x), \text{ for all } x \in U, \tag{13}$$

if and only if $A \in \mathrm{Def}(U/E)$, where χ_A is the characteristic function of A defined by $\chi_A(x) = 1$ if $x \in A$ and $\chi_A(x) = 0$ if $x \notin A$.

Theorem 1 shows that a set in $\mathrm{Def}(U/E)$ is both lower and upper exact, and only a set in $\mathrm{Def}(U/E)$ has such a property. For this reason, a set in $\mathrm{Def}(U/E)$ is called a definable set. Theorem 2 states that μ_A is a crisp set if and only if $A \in \mathrm{Def}(U/E)$. All other subsets of U will induce noncrisp fuzzy sets. The fuzziness is a natural consequence of the indiscernibility of elements.

In a generalized approximation space $apr = (U,R)$ defined by a reflexive relation R, rough set approximations can be defined by replacing $[x]_E$ with $(x)_R$ in (9) and (10), as follows:

$$\begin{aligned} \underline{apr}(A) &= \{x \in U \mid (x)_R \subseteq A\} \\ &= \{x \in U \mid \mu_A(x) = 1\} = core(\mu_A), \\ \overline{apr}(A) &= \{x \in U \mid (x)_R \cap A \neq \emptyset\} \\ &= \{x \in U \mid \mu_A(x) > 0\} = support(\mu_A). \end{aligned} \tag{14}$$

The results regarding fuzzification, as well as lower and upper exact sets, are summarized in the following theorems:

Theorem 3. *In a generalized approximation space $apr = (U,R)$ defined by a reflexive relation R,*

1. $A = \underline{apr}(A)$ *if and only if* $A = \bigcup_{x \in A} (x)_R$,
   $A = \overline{apr}(A)$ *if and only if* $A = \bigcap_{x \notin A} ((x)_R)^c$.
2. $\mathrm{E}(\underline{apr}) \subseteq \mathrm{Def}(U/R)$, $\mathrm{E}(\overline{apr}) \subseteq \mathrm{Def}^c(U/R)$.
3. $\mathrm{E}(\underline{apr})$ *and* $\mathrm{E}(\overline{apr})$ *are closed under* $\cap$ *and* $\cup$.
4. $A \in \mathrm{E}(\underline{apr})$ *if and only if* $A^c \in \mathrm{E}(\overline{apr})$.
5. $\mathrm{E}(\underline{apr}) \cap \mathrm{E}(\overline{apr})$ *is a sub – Boolean algebra of* 2^U.

Theorem 4. *In a generalized approximation space $apr = (U,R)$ defined by a reflexive relation R,*

$$\mu_A(x) = \chi_A(x), \text{ for all } x \in U, \tag{15}$$

if and only if $A \in \mathrm{E}(\underline{apr}) \cap \mathrm{E}(\overline{apr})$.

The sets $\mathrm{E}(\underline{apr})$ and $\mathrm{E}(\overline{apr})$ are not necessarily the same and may not be closed under set complement. Although $\mathrm{E}(\underline{apr})$ is a subfamily of $\mathrm{Def}(U/R)$ closed under both $\cap$ and $\cup$, $\mathrm{E}(\overline{apr})$ is a subfamily of $\mathrm{Def}^c(U/R)$ closed under both $\cap$ and $\cup$. A lower exact set must be expressed as a union of some elementary granules. However, not every union of elementary granules is a lower exact set. A set A is lower and upper exact if and only if μ_A is a crisp set.

In defining rough set approximations, only the two extreme points of the unit interval $[0,1]$ are used: zero is used for upper approximations and one for lower approximations. In general, we can use a pair of values (α,β) with $\alpha > \beta$ to define a pair of graded lower and upper approximations,

$$\begin{aligned}\underline{apr}_{\alpha}(A) &= \{x \in U \mid \mu_A(x) \geq \alpha\} = (\mu_A)_{\alpha},\\ \overline{apr}_{\beta}(A) &= \{x \in U \mid \mu_A(x) > \beta\} = (\mu_A)_{\beta^+},\end{aligned} \tag{16}$$

where $(\mu_A)_{\alpha}$ denotes the α-cut of the fuzzy set μ_A and $(\mu_A)_{\beta^+}$ the strong β-cut of μ_A. The condition $\alpha > \beta$ implies that $\underline{apr}_{\alpha} \subseteq \overline{apr}_{\beta}(A)$. When $\alpha = \beta$, to keep this property, we define

$$\begin{aligned}\underline{apr}_{\alpha}(A) &= \{x \in U \mid \mu_A(x) > \alpha\} = (\mu_A)_{\alpha^+},\\ \overline{apr}_{\alpha}(A) &= \{x \in U \mid \mu_A(x) \geq \alpha\} = (\mu_A)_{\alpha}.\end{aligned} \tag{17}$$

By imposing an additional condition $\alpha + \beta = 1$, we can obtain a pair of dual operators [47,39]. For standard approximation operators,

$$\begin{aligned}\underline{apr}(A) &= \underline{apr}_1(A),\\ \overline{apr}(A) &= \overline{apr}_0(A).\end{aligned} \tag{18}$$

The probabilistic rough set approximation operator proposed by Pawlak et al. [14] is given by $(\underline{apr}_{0.5}, \overline{apr}_{0.5})$. Properties of rough set approximations under the pair of parameters (α,β) can be found in [37–40,47].

We can define the notions of lower exact sets and upper exact sets of graded approximation operators. In an approximation space $apr = (U,E)$, from property (m3) we can conclude that an entire equivalence class is either in or not in a lower or an upper approximation. This implies that $\underline{apr}_{\alpha}(A)$ and $\overline{apr}_{\beta}(A)$ must be in $\mathrm{Def}(U/E)$. Conversely, if A is in $\mathrm{Def}(U/E)$, $\underline{apr}_{\alpha}(A) = A$ and $\overline{apr}_{\beta}(A) = A$ for $\alpha \in (0,1]$ and $\beta \in [0,1)$. Therefore it follows that graded approximations do not change families of lower and upper exact sets.

Theorem 5. *In an approximation space $apr = (U,E)$, for $\alpha \in (0,1]$ and $\beta \in [0,1)$,*

$$\begin{aligned}\mathrm{E}(\underline{apr}_{\alpha}) &= \{A \subseteq U \mid A = \underline{apr}_{\alpha}(A)\} = \mathrm{E}(\underline{apr}) = \mathrm{Def}(U/E),\\ \mathrm{E}(\overline{apr}_{\beta}) &= \{A \subseteq U \mid A = \overline{apr}_{\beta}(A)\} = \mathrm{E}(\overline{apr}) = \mathrm{Def}(U/E).\end{aligned} \tag{19}$$

The result of Theorem 5 cannot be easily extended to a generalized approximation space $apr = (U,R)$ defined by a reflexive relation R. The characterization of families of graded lower exact sets and graded upper exact sets in a generalized approximation space is an interesting problem.

Theorem 6. *In a generalized approximation space $apr = (U,R)$ defined by a reflexive relation R, for $\alpha \in (0,1]$ and $\beta \in [0,1)$,*

$$\begin{aligned}\mathrm{E}(\underline{apr}_{\alpha}) &= \{A \subseteq U \mid A = \underline{apr}_{\alpha}(A)\} \subseteq \mathrm{Def}(U/\equiv_E),\\ \mathrm{E}(\overline{apr}_{\beta}) &= \{A \subseteq U \mid A = \overline{apr}_{\beta}(A)\} \subseteq \mathrm{Def}(U/\equiv_R),\end{aligned} \tag{20}$$

where $\mathrm{Def}(U/\equiv_R)$ *is the family of definable sets defined by the equivalence relation* $\equiv_R$.

The theorem easily follows from the property (m3c). The families $\mathrm{E}(\underline{apr}_\alpha)$ and $\mathrm{E}(\overline{apr}_\beta)$ may not necessarily be closed under $\cap$ and $\cup$.

3.3 Graded Inclusion of Sets

A rough membership function is defined on the basis of the relationship between two sets; one is the equivalence class $[x]_R$ (or the neighborhood $(x)_R$) of an element x and the other is a set A. For the maximum membership value one, $[x]_E \subseteq A$, i.e., $[x]_E$ is a subset of A. For the minimum membership value zero, $[x]_E \cap A = \emptyset$, or equivalently $[x]_E \subseteq A^c$, i.e., $[x]_E$ is totally not a subset of A. For a value between zero and one, it may be interpreted as the degree to which $[x]_E$ is a subset of A. By extending the notion of rough membership functions to the power set of a universe, one obtains a measure of the graded inclusion of two sets [19,20]:

$$v(A,B) = \frac{|A \cap B|}{|A|}. \tag{21}$$

When $A = \emptyset$, we define $v(\emptyset, B) = 1$ and $v(\emptyset,\emptyset) = 1$, i.e., the empty set is a subset of any set.

The value $v(A,B)$ can be interpreted as the conditional probability that a randomly selected element from A belongs to B. It may be used to measure the degree to which A is a subset of B. There is a close connection between graded inclusion and fuzzy set inclusion [20].

Measures related to v have been proposed and used by many authors. Ziarko [47] used the measure,

$$c(A,B) = 1 - v(A,B) = 1 - \frac{|A \cap B|}{|A|}, \tag{22}$$

in a variable precision rough set model. One can easily obtain the same results by using v. Skowron and Stepaniuk [20] suggested that graded (vague) inclusion of sets may be measured by a function,

$$v : 2^U \times 2^U \longrightarrow [0,1] \tag{23}$$

with monotonicity regarding the second argument, i.e., for $A,B,C \subseteq U$, $v(A,B) \leq (A,C)$ for any $B \subseteq C$. The function defined by equation (21) is an example of such a measure. In fact, (21) considers only the overlap with the first argument, but not the size of the second argument.

Starting from rough membership functions, Skowron and Polkowski [19] introduced the concept of rough inclusion defined by a function $v: 2^U \times 2^U \longrightarrow [0,1]$ satisfying more properties, in addition to monotonicity with respect to the second argument. The unit interval $[0,1]$ can also be generalized to a complete lattice in the definition of rough inclusion [16]. Rough inclusion is only an example for measuring degrees of inclusion in rough mereology. A more detailed discussion on rough mereology and related concepts can be found in [16, 17]. Instead of using a measure of graded inclusion, Bryniarski and Wybraniec-Skardowska [1] proposed using a family of inclusion relations called context relations, indexed by a bounded and partially ordered set called a rank set. The unit interval $[0,1]$ can be treated as a rank set. From a measure of graded inclusion, a context relation with respect to a value $\alpha \in [0,1]$ can be defined by

$$\subseteq_\alpha = \{(A,B) \mid v(A,B) \geq \alpha\}. \tag{24}$$

In other words, v may be interpreted as a fuzzy relation on 2^U, and $\subseteq_\alpha$ may be interpreted as an α-cut of the fuzzy relation. The use of a complete lattice, or a rank set, corresponds to lattice based fuzzy relations in the theory of fuzzy sets.

3.4 Granule Oriented Approximations

In an approximation space $apr = (U,E)$, an equivalence class $[x]_E$ is treated as a unit. A granule oriented definition of approximation operators can be used. Approximations of a set are expressible in terms of unions of equivalence granules:

$$\begin{aligned} \underline{apr}(A) &= \bigcup\{[x]_E \mid [x]_E \subseteq A\}, \\ \overline{apr}(A) &= \bigcup\{[x]_E \mid [x]_E \cap A \neq \emptyset\}. \end{aligned} \tag{25}$$

The lower approximation $\underline{apr}(A)$ is the union of those equivalence granules that are subsets of A. The upper approximation $\overline{apr}(A)$ is the union of those equivalence granules that have nonempty intersections with A. This definition is equivalent to the element oriented definition.

In a generalized approximation space $apr = (U,R)$, granule oriented rough set approximations can be defined by generalizing (25). The equivalence class $[x]_E$ is replaced by the neighborhood $(x)_R$. One such generalization is [30]

$$\begin{aligned} \underline{apr}'(A) &= \bigcup\{(x)_R \mid x \in U, (x)_R \subseteq A\}, \\ \overline{apr}'(A) &= [\underline{apr}'(A^c)]^c. \end{aligned} \tag{26}$$

We generalize the lower approximation as a union of elementary granules and define the upper approximation through duality. The lower approximation is the union of some granules in U/R, but the upper approximation cannot be expressed in this way [30]. The approximation operators $\underline{apr}'$ and $\overline{apr}'$ are different from the element oriented definition. The lower exact sets and upper exact sets are related to $\mathrm{Def}(U/R)$ and $\mathrm{Def}^c(U/R)$.

Theorem 7. *In a generalized approximation space* $apr = (U,R)$ *defined by a reflexive relation* R,

$$\begin{aligned} \mathrm{E}(\underline{apr}') &= \{A \subseteq U \mid A = \underline{apr}'(A)\} = \mathrm{Def}(U/R), \\ \mathrm{E}(\overline{apr}') &= \{A \subseteq U \mid A = \overline{apr}'(A)\} = \mathrm{Def}^c(U/R). \end{aligned} \quad (27)$$

This theorem generalizes the result of Theorem 1 in the sense that Theorem 1 considers a subclass of reflexive relations.

Granule oriented approximations can be generalized into graded approximations through graded inclusion measures. We can replace the relation $\subseteq$ by a relation $\subseteq_\alpha$ for $\alpha \in (0,1]$. For an approximation space $apr = (U,E)$, we define a lower approximation operator as

$$\underline{apr}_\alpha(A) = \bigcup\{[x]_E \mid [x]_E \subseteq_\alpha A\} = \bigcup\{[x]_E \mid v([x]_E, A) \geq \alpha\}. \quad (28)$$

The graded upper approximation operator can be defined by another parameter $\beta \in [0,1)$. If a pair of dual operators is needed, the corresponding graded upper approximation operator can be defined by the dual of $\underline{apr}_\alpha$. Granule oriented graded approximations are the same as those obtained from the element oriented definition.

For a generalized approximation space $apr = (U,R)$, with respect to a value $\alpha \in (0,1]$, we define

$$\underline{apr}'_\alpha(A) = \bigcup\{(x)_R \mid (x)_R \subseteq_\alpha A\} = \bigcup\{(x)_R \mid v((x)_R, A) \geq \alpha\}. \quad (29)$$

The graded upper approximation can be defined by duality. The granule oriented definition produces different approximations from the element oriented definition. By definition, the graded lower approximation of a set can be expressed as the union of some granules. However, not every union of granules can be the lower approximation of a certain set.

Theorem 8. *In a generalized approximation space* $apr = (U,R)$ *defined by a reflexive relation* R, *for* $\alpha \in (0,1]$,

$$\begin{aligned} \mathrm{E}(\underline{apr}'_\alpha) &= \{A \subseteq U \mid A = \underline{apr}'_\alpha(A)\} \subseteq \mathrm{E}(\underline{apr}') = \mathrm{Def}(U/R), \\ \mathrm{E}(\overline{apr}'_{1-\alpha}) &= \{A \subseteq U \mid A = \overline{apr}'_{1-\alpha}(A)\} = \mathrm{E}(\overline{apr}') = \mathrm{Def}^c(U/R). \end{aligned} \quad (30)$$

Both families $\mathrm{E}(\underline{apr}'_\alpha)$ and $\mathrm{E}(\overline{apr}'_{1-\alpha})$ may not necessarily be closed under $\cap$ and $\cup$. In general, $\mathrm{E}(\underline{apr}') \neq \mathrm{E}(\underline{apr}'_\alpha)$.

This section not only summarizes the main results from existing studies on rough set approximations in a unified framework but also presents many new results. From the discussion, we can conclude that parameterized rough set approximations are useful and need further investigation.

4 A Decision-Theoretical Model of Rough Sets

In this section, the basic notions of the Bayesian decision procedure for classification is briefly reviewed [3]. Rough set approximation operators are formulated as classifying objects into three disjoint classes, the positive, negative, and boundary regions.

For clarity, we consider only the element oriented definition with respect to the granulated view of a universe induced by an equivalence relation. The same argument can easily be applied to other cases.

4.1 An Overview of the Bayesian Decision Procedure

Let $\Omega = \{w_1, \ldots, w_s\}$ be a finite set of s states, and let $\mathcal{A} = \{a_1, \ldots, a_m\}$ be a finite set of m possible actions. Let $P(w_j|\mathbf{x})$ be the conditional probability that an object x is in state w_j, given that the object is described by $\mathbf{x}$. In the following discussions, we assume that these conditional probabilities $P(w_j|\mathbf{x})$ are known.

Let $\lambda(a_i|w_j)$ denote the loss, or cost, of taking action a_i when the state is w_j. For an object with description $\mathbf{x}$, suppose action a_i is taken. Since $P(w_j|\mathbf{x})$ is the probability that the true state is w_j, given $\mathbf{x}$, the expected loss associated with taking action a_i is given by

$$R(a_i|\mathbf{x}) = \sum_{j=1}^{s} \lambda(a_i|w_j) P(w_j|\mathbf{x}). \tag{31}$$

The quantity $R(a_i|\mathbf{x})$ is also called the conditional risk. Given description $\mathbf{x}$, a decision rule is a function $\tau(\mathbf{x})$ that specifies which action to take, that is, for every $\mathbf{x}$, $\tau(\mathbf{x})$ assumes one of the actions, $a_1, \ldots, a_m$. The overall risk $\mathbf{R}$ is the expected loss associated with a given decision rule. Since $R[\tau(\mathbf{x})|\mathbf{x}]$ is the conditional risk associated with action $\tau(\mathbf{x})$, the overall risk is defined by

$$\mathbf{R} = \sum_{\mathbf{x}} R[\tau(\mathbf{x})|\mathbf{x}] P(\mathbf{x}), \tag{32}$$

where the summation is across the set of all possible descriptions of objects, i.e., the knowledge representation space. If $\tau(\mathbf{x})$ is chosen so that $R[\tau(\mathbf{x})|\mathbf{x}]$ is as small as possible for every $\mathbf{x}$, the overall risk $\mathbf{R}$ is minimized.

The Bayesian decision procedure can be formally stated as follows: For every $\mathbf{x}$, compute the conditional risk $R(a_i|\mathbf{x})$ for $i = 1, \ldots, m$ defined by (31), and then select the action for which the conditional risk is minimum. If more than one action minimizes $R(a_i|\mathbf{x})$, any tiebreaking rule can be used.

4.2 Rough Set Approximation Operators

In an approximation space $apr = (U, E)$, with respect to a subset $A \subseteq U$, one can divide the universe U into three disjoint regions, the positive region $\mathrm{POS}(A)$, the negative region $\mathrm{NEG}(A)$, and the boundary region $\mathrm{BND}(A)$:

$$\begin{aligned} \mathrm{POS}(A) &= \underline{apr}(A), \\ \mathrm{NEG}(A) &= U - \overline{apr}(A), \\ \mathrm{BND}(A) &= \overline{apr}(A) - \underline{apr}(A). \end{aligned} \tag{33}$$

The lower approximation of a set is the same as the positive region. The upper approximation is the union of the positive and boundary regions, $\overline{apr}(A) = \mathrm{POS}(A) \cup \mathrm{BND}(A)$. One can say with certainty that any element $x \in \mathrm{POS}(A)$ belongs to A and that any element $x \in \mathrm{NEG}(A)$ does not belong to A. One cannot decide with certainty whether or not an element $x \in \mathrm{BND}(A)$ belongs to A.

In an approximation space $apr = (U, E)$, an element x is viewed as $[x]_E$, that is, the equivalence class containing x is considered to be a description of x. The classification of objects according to approximation operators can easily be fitted into the Bayesian decision-theoretical framework. The set of states is given by $\Omega = \{A, \neg A\}$ indicating that an element is in A and not in A, respectively. We use the same symbol to denote both a subset A and the corresponding state. With respect to three regions, the set of actions is given by $\mathcal{A} = \{a_1, a_2, a_3\}$, where a_1, a_2, and a_3 represent the three actions in classifying an object, deciding $\mathrm{POS}(A)$, deciding $\mathrm{NEG}(A)$, and deciding $\mathrm{BND}(A)$, respectively.

Let $\lambda(a_i|A)$ denote the loss incurred for taking action a_i when an object belongs to A, and let $\lambda(a_i|\neg A)$ denote the loss incurred for taking the same action when the object does not belong to A. The rough membership values $\mu_A(x) = P(A|[x]_E)$ and $\mu_{A^c}(x) = P(\neg A|[x]_E) = 1 - P(A|[x]_E)$ are the probabilities that an object in the equivalence class $[x]_E$ belongs to A and $\neg A$, respectively. The expected loss $R(a_i|[x]_E)$ associated with taking the individual actions can be expressed as

$$\begin{aligned} R(a_1|[x]_E) &= \lambda_{11} P(A|[x]_E) + \lambda_{12} P(\neg A|[x]_E), \\ R(a_2|[x]_E) &= \lambda_{21} P(A|[x]_E) + \lambda_{22} P(\neg A|[x]_E), \\ R(a_3|[x]_E) &= \lambda_{31} P(A|[x]_E) + \lambda_{32} P(\neg A|[x]_E), \end{aligned} \tag{34}$$

where $\lambda_{i1} = \lambda(a_i|A)$, $\lambda_{i2} = \lambda(a_i|\neg A)$, and $i = 1, 2, 3$. The Bayesian decision procedure leads to the following minimum-risk decision rules:

(P) If $R(a_1|[x]_E) \leq R(a_2|[x]_E)$ and $R(a_1|[x]_E) \leq R(a_3|[x]_E)$, decide $\mathrm{POS}(A)$.

(N) If $R(a_2|[x]_E) \leq R(a_1|[x]_E)$ and $R(a_2|[x]_E) \leq R(a_3|[x]_E)$, decide $\mathrm{NEG}(A)$.

(B) If $R(a_3|[x]_E) \leq R(a_1|[x]_E)$ and $R(a_3|[x]_E) \leq R(a_2|[x]_E)$, decide $\mathrm{BND}(A)$.

Tiebreaking rules should be added so that each element is classified into only one region. Since $P(A|[x]_E) + P(\neg A|[x]_E) = 1$, the above decision rules can be simplified such that only the probabilities $P(A|[x]_E)$ are involved. We can classify any object in the equivalence class $[x]_E$ based only on the probabilities $P(A|[x]_E)$, i.e., the rough membership values and the given loss function λ_{ij} $(i = 1,2,3;\ j = 1,2)$.

Consider a special kind of loss function with $\lambda_{11} \leq \lambda_{31} < \lambda_{21}$ and $\lambda_{22} \leq \lambda_{32} < \lambda_{12}$, that is, the loss of classifying an object x belonging to A into the positive region $\mathrm{POS}(A)$ is less than or equal to the loss of classifying x into the boundary region $\mathrm{BND}(A)$, and both of these losses are strictly less than the loss of classifying x into the negative region $\mathrm{NEG}(A)$. The reverse order of losses is used for classifying an object that does not belong to A. For this type of loss function, the minimum-risk decision rules (P)–(B) can be written as follows:

(P) If $P(A|[x]_E) \geq \gamma$ and $P(A|[x]_E) \geq \alpha$, decide $\mathrm{POS}(A)$,

(N) If $P(A|[x]_E) \leq \beta$ and $P(A|[x]_E) \leq \gamma$, decide $\mathrm{NEG}(A)$,

(B) If $\beta \leq P(A|[x]_E) \leq \alpha$, decide $\mathrm{BND}(A)$,

where

$$\alpha = \frac{\lambda_{12} - \lambda_{32}}{(\lambda_{31} - \lambda_{32}) - (\lambda_{11} - \lambda_{12})},$$
$$\gamma = \frac{\lambda_{12} - \lambda_{22}}{(\lambda_{21} - \lambda_{22}) - (\lambda_{11} - \lambda_{12})},$$
$$\beta = \frac{\lambda_{32} - \lambda_{22}}{(\lambda_{21} - \lambda_{22}) - (\lambda_{31} - \lambda_{32})}. \tag{35}$$

By the assumptions $\lambda_{11} \leq \lambda_{31} < \lambda_{21}$ and $\lambda_{22} \leq \lambda_{32} < \lambda_{12}$, it follows that $\alpha \in (0,1]$, $\gamma \in (0,1)$, and $\beta \in [0,1)$.

A loss function should be chosen to satisfy the condition $\alpha \geq \beta$. This ensures that the results are consistent with rough set approximations. More specifically, the lower approximation is a subset of the upper approximation, and the boundary region may be nonempty.

Theorem 9. *If a loss function with $\lambda_{11} \leq \lambda_{31} < \lambda_{21}$ and $\lambda_{22} \leq \lambda_{32} < \lambda_{12}$ satisfies the condition*

$$(\lambda_{12} - \lambda_{32})(\lambda_{21} - \lambda_{31}) \geq (\lambda_{31} - \lambda_{11})(\lambda_{32} - \lambda_{22}) \tag{36}$$

then, $\alpha \geq \gamma \geq \beta$.

Let $l = (\lambda_{12} - \lambda_{32})(\lambda_{21} - \lambda_{31})$ and $r = (\lambda_{31} - \lambda_{11})(\lambda_{32} - \lambda_{22})$. l is the product of the differences between the cost of making an incorrect classification and the cost of classifying an element into the boundary region, and r is the product of the differences between the cost of classifying an element into the boundary region and the cost of a correct classification. A larger value of l, or equivalently a smaller value of r, can be obtained if we move λ_{32} away from λ_{12} or move λ_{31} away from λ_{21}. The condition can be intuitively interpreted as saying that the cost of classifying an element in the boundary region is closer to the cost of a correct classification than to the cost of an incorrect classification. Such a condition seems reasonable.

When $\alpha > \beta$, $\alpha > \gamma > \beta$. After tiebreaking, we obtain the decision rules:

(P1) If $P(A|[x]_E) \geq \alpha$, decide $\text{POS}(A)$.
(N1) If $P(A|[x]_E) \leq \beta$, decide $\text{NEG}(A)$.
(B1) If $\beta < P(A|[x]_E) < \alpha$, decide $\text{BND}(A)$.

When $\alpha = \beta$, $\alpha = \gamma = \beta$. In this case, we use the decision rules:

(P2) If $P(A|[x]_E) > \alpha$, decide $\text{POS}(A)$.
(N2) If $P(A|[x]_E) < \alpha$, decide $\text{NEG}(A)$.
(B2) If $P(A|[x]_E) = \alpha$, decide $\text{BND}(A)$.

For the second set of decision rules, we use a tiebreaking criterion so that the boundary region may be nonempty.

The value of α should be in the range $[0.5, 1]$, in addition to constraint $\alpha \geq \beta$, as suggested by many authors [20,38,40,47]. The following theorem gives the condition for $\alpha \geq 0.5$.

Theorem 10. *If a loss function with $\lambda_{11} \leq \lambda_{31} < \lambda_{21}$ and $\lambda_{22} \leq \lambda_{32} < \lambda_{12}$ satisfies the condition*

$$\lambda_{12} - \lambda_{32} \geq \lambda_{31} - \lambda_{11}, \tag{37}$$

then, $\alpha \geq 0.5$.

Condition (37) says that the difference between the cost of classifying an element not in A in the positive region and the cost of classifying the element into the boundary region is more than the difference between the cost of classifying an element in A into the boundary region and a correct classification. It forms part of condition (36). However, they do not imply each other. By combining results from Theorems 9 and 10, we have the condition for $\alpha \geq 0.5$ and $\alpha \geq \beta$.

Corollary 1. *If a loss function with* $\lambda_{11} \leq \lambda_{31} < \lambda_{21}$ *and* $\lambda_{22} \leq \lambda_{32} < \lambda_{12}$ *satisfies the conditions,*

$$\begin{aligned} &\lambda_{12} - \lambda_{32} \geq \lambda_{31} - \lambda_{11}, \\ &(\lambda_{12} - \lambda_{32})(\lambda_{21} - \lambda_{31}) \geq (\lambda_{31} - \lambda_{11})(\lambda_{32} - \lambda_{22}), \end{aligned} \tag{38}$$

then, $\alpha \geq 0.5$ *and* $\alpha \geq \beta$.

If dual approximation operators are required, one needs to impose additional conditions on a loss function [39].

Theorem 11. *If a loss function with* $\lambda_{11} \leq \lambda_{31} < \lambda_{21}$ *and* $\lambda_{22} \leq \lambda_{32} < \lambda_{12}$ *satisfies the condition,*

$$(\lambda_{12} - \lambda_{32})(\lambda_{32} - \lambda_{22}) = (\lambda_{31} - \lambda_{11})(\lambda_{21} - \lambda_{31}), \tag{39}$$

then, $\beta = 1 - \alpha$.

Condition (39) does not guarantee that $\alpha \geq \beta = 1 - \alpha$, or equivalently, $\alpha \geq 0.5$. The condition for $\alpha = 1 - \beta \geq 0.5$ can be obtained by combining conditions (36) and (39) or by combining conditions (37) and (39).

Corollary 2. *If a loss function with* $\lambda_{11} \leq \lambda_{31} < \lambda_{21}$ *and* $\lambda_{22} \leq \lambda_{32} < \lambda_{12}$ *satisfies the two sets of equivalent conditions,*

$$\begin{aligned} \text{(i)} \quad &(\lambda_{12} - \lambda_{32})(\lambda_{21} - \lambda_{31}) \geq (\lambda_{31} - \lambda_{11})(\lambda_{32} - \lambda_{22}), \\ &(\lambda_{12} - \lambda_{32})(\lambda_{32} - \lambda_{22}) = (\lambda_{31} - \lambda_{11})(\lambda_{21} - \lambda_{31}), \end{aligned} \tag{40}$$

$$\begin{aligned} \text{(ii)} \quad &\lambda_{12} - \lambda_{32} \geq \lambda_{31} - \lambda_{11}, \\ &(\lambda_{12} - \lambda_{32})(\lambda_{32} - \lambda_{22}) = (\lambda_{31} - \lambda_{11})(\lambda_{21} - \lambda_{31}), \end{aligned} \tag{41}$$

then, $\alpha = 1 - \beta \geq 0.5$.

Based on the results obtained so far, we can now investigate loss functions producing existing rough set approximation operators.

Consider the loss function

$$\lambda_{12} = \lambda_{21} = 1, \quad \lambda_{11} = \lambda_{22} = \lambda_{31} = \lambda_{32} = 0. \tag{42}$$

There is a unit cost if an object belonging to A is classified in the negative region or if an object not belonging to A is classified in the positive region; otherwise, there is no cost. This loss function satisfies the conditions given in Corollary 2. A pair of dual approximation operators can be obtained. From (35), $\alpha = 1 > \beta = 0$, $\alpha = 1 - \beta$, and $\gamma = 0.5$. According to decision rules (P1)–(B1), we obtain the standard rough set approximations [11,12].

Consider another loss function,

$$\lambda_{12} = \lambda_{21} = 1, \quad \lambda_{31} = \lambda_{32} = 0.5, \quad \lambda_{11} = \lambda_{22} = 0, \tag{43}$$

that is, a unit cost is incurred if the system classifies an object belonging to A in the negative region or if an object not belonging to A is classified in the positive region. Half of a unit cost is incurred if any object is classified in the boundary region. For other cases, there is no cost. The loss function satisfies the conditions given in Corollary 2. In fact, the loss function makes all $\geq$ relations in these conditions become $=$. By substituting these λ_{ij}'s in (35), we obtain $\alpha = \beta = \gamma = 0.5$. By using decision rules (P2)–(B2), we obtained the probabilistic rough set approximation proposed by Pawlak et al. [14].

The loss function,

$$\lambda_{12} = \lambda_{21} = 4, \quad \lambda_{31} = \lambda_{32} = 1, \quad \lambda_{11} = \lambda_{22} = 0, \tag{44}$$

states that there is no cost for a correct classification, four units of cost for an incorrect classification, and one unit cost for classifying an object in a boundary region. It also satisfies the conditions in Corollary 2. From (35), $\alpha = 0.75$, $\beta = 0.25$, and $\gamma = 0.5$. By decision rules (P1)–(B1), we have a pair of dual approximation operators, $\underline{apr}_{0.75}$ and $\overline{apr}_{0.25}$.

In general, the relationships between the loss function λ and the pair of parameters (α, β) are summarized as follows:

Theorem 12. *For a loss function with $\lambda_{11} \leq \lambda_{31} < \lambda_{21}$ and $\lambda_{22} \leq \lambda_{32} < \lambda_{12}$,*

1. *α is monotonically nondecreasing with respect to λ_{12} and monotonically nonincreasing with respect to λ_{32}.*
2. *If $\lambda_{11} < \lambda_{31}$, α is strictly monotonically increasing with respect to λ_{12} and strictly monotonically decreasing with respect to λ_{32}.*
3. *α is strictly monotonically decreasing with respect to λ_{31} and strictly monotonically increasing with respect to λ_{11}.*
4. *β is monotonically nonincreasing with respect to λ_{21} and monotonically nondecreasing with respect to λ_{31}.*
5. *If $\lambda_{22} < \lambda_{32}$, β is strictly monotonically decreasing with respect to λ_{21} and strictly monotonically increasing with respect to λ_{31}.*
6. *β is strictly monotonically increasing with respect to λ_{32} and strictly monotonically decreasing with respect to λ_{22}.*

The connection between threshold values of parameterized rough set approximations and the loss function has significant implications in applying the decision-theoretical model of rough sets. For example, if we increase the cost of an incorrect classification λ_{12} and keep other costs unchanged, the value α would not be decreased. Unlike the variable precision rough set model, the decision-theoretical model requires a loss function. Parameters α and β are determined from the loss function. One may argue that the loss function may be considered a set of parameters. However, in contrast to standard threshold values, they have an intuitive interpretation. The connections given in the theorem show the consequences of a loss function and provide an interpretation of the parameters required in terms of a more realistic concept of loss or cost. One can easily interpret and measure loss or cost in a real application.

5 Conclusion

Successful applications of the theory of rough sets depend on a clear understanding of the various concepts involved. For this purpose, a decision-theoretical model of rough sets is studied in this chapter by focusing on two related fundamental issues, granulation of a universe and approximation in a granulated universe. The decision-theoretical model not only provides a sound basis for rough set theory but also provides a unified framework in which many existing models of rough sets can be derived. The decision model can also be interpreted in terms of a more familiar and interpretable concept known as loss, or cost.

Two granulation structures are examined. An equivalence relation induces a partition of a universe, and a reflexive relation induces a covering of a universe. Under a granulated view of a universe, a subset of the universe can be fuzzified and approximated. Fuzzification of a set leads to a rough membership function, which is a special type of fuzzy membership function. Approximations of a set can be defined in two ways. The element oriented formulation is based on a rough membership function and is related to the notion of α-cut in fuzzy sets. The granule oriented formulation is based on the set inclusion relation and, in general, based on a graded set inclusion relation related to a rough membership function. The two formulations produce the same results when a universe is granulated by an equivalence relation, and produce different results when a universe is granulated by a reflexive relation.

The families of fixed points of lower and upper approximation operators are studied and provide insights into our understanding of granulation structures and induced approximation structures. With a partition defined by an equivalence relation, the families of fixed points are related to a sub-Boolean algebra of the power set of the universe. With a covering defined by a reflexive relation, the families of fixed points are related to closure systems.

The conditions on a loss function are investigated. In particular, we explicitly state the connections between the parameters required for defining graded approximation operators and losses for various classification decisions. This provide an interpretation for parameters used in other models of rough sets. We also identify conditions on a loss function so that other rough set approximation operators, such as the standard approximation operators, probabilistic approximation operators, and variable precision approximation operators, can be obtained. The decision-theoretical model, is therefore, more general than other models.

Acknowledgments

The author wishes to thank Sai Ying and an anonymous referee for critical comments and suggestions on the chapter.

References

1. E. Bryniarski, U. Wybraniec-Skardowska. Generalized rough sets in contextual space. In T. Y. Lin, N. Cercone, editors, *Rough Sets and Data Mining*, 339–354, Kluwer, Boston, 1997.
2. D. Dubois, H. Prade. Similarity-based approximate reasoning. In J.M. Zurada, R.J. Marks II, C.J. Robinson, editors, *Computational Intelligence: Imitating Life*, 69–80, IEEE Press, New York, 1994.
3. R.O. Duda, P. E. Hart. *Pattern Classification and Scene Analysis*. Wiley, New York, 1973.
4. S. Hirano, M. Inuiguchi, S. Tsumoto. *Proceedings of the International Workshop on Rough Set Theory and Granular Computing (RSTGC 2001)*, Vol. 5(1/2) of *Bulletin of International Rough Set Society*, 2001.
5. J. Komorowski, Z. Pawlak, L. Polkowski, A. Skowron. Rough sets: A tutorial. In S. K. Pal, A. Skowron, editors, *Rough Fuzzy Hybridization: A New Trend in Decision Making*, 3–98, Springer, Singapore, 1998.
6. G.J. Klir, B. Yuan. *Fuzzy Sets and Fuzzy Logic: Theory and Applications*. Prentice–Hall, Englewood Cliffs, NJ, 1995.
7. T.Y. Lin. Granular computing on binary relations I: Data mining and neighborhood systems, II: Rough set representations and belief functions. In L.Polkowski, A. Skowron, editors, *Rough Sets in Knowledge Discovery* 1, 107–140, Physica, Heidelberg, 1998.
8. T.Y. Lin, Y.Y. Yao, L.A. Zadeh, editors. *Rough Sets, Granular Computing and Data Mining*. Physica, Heidelberg, 2001.
9. E. Orlowska. Logic of indiscernibility relations. *Bulletin of the Polish Academy of Sciences. Mathematics*, 33: 475–485, 1985.
10. E. Orlowska, Z. Pawlak. Measurement and indiscernibility. *Bulletin of the Polish Academy of Sciences. Mathematics*, 32: 617–624, 1984.
11. Z. Pawlak. Rough sets. *International Journal of Computer and Information Sciences*, 11: 341–356, 1982.
12. Z. Pawlak. *Rough Sets: Theoretical Aspects of Reasoning about Data*. Kluwer, Dordrecht, 1991.
13. Z. Pawlak, A. Skowron. Rough membership functions. In R.R. Yager, M. Fedrizzi, and J. Kacprzyk, editors, *Advances in the Dempster–Shafer Theory of Evidence*, 251–271, Wiley, New York, 1994.

14. Z. Pawlak, S.K.M. Wong, W. Ziarko. Rough sets: Probabilistic versus deterministic approach. *International Journal Man-Machine Studies*, 29: 81–95, 1988.
15. W. Pedrycz. *Granular Computing: An Emerging Paradigm*. Springer, Berlin, 2001.
16. L. Polkowski, A. Skowron. Rough mereology: A new paradigm for approximate reasoning. *International Journal Approximate Reasoning*, 15: 333–365, 1996.
17. L. Polkowski, A. Skowron. Towards adaptive calculus of granules. *Proceedings of 1998 IEEE International Conference on Fuzzy Systems (FUZZ–IEEE'98)*, 111–116, Anchorage, AK, 1998.
18. A. Skowron. Toward intelligent systems: calculi of information granules. *Bulletin International Rough Set Society*, 5: 9–30, 2001.
19. A. Skowron, L. Polkowski. Rough mereology and analytical morphology. In E. Orlowska, editor, *Incomplete Information: Rough Set Analysis*, 399–437, Physica, Heidelberg, 1998.
20. A. Skowron, J. Stepaniuk. Tolerance approximation spaces. *Fundamenta Informaticae*, 27: 245–253, 1996.
21. A. Skowron, J. Stepaniuk. Information granules and approximation spaces. *Proceedings of the 7th International Conference on Information Processing and Management of Uncertainty in Knowledge-Based Systems (IPMU'98)*, 354–361, Paris, 1998.
22. A. Skowron, J. Stepaniuk. Information granules: Towards foundations of granular computing. *International Journal for Intelligent Systems*, 16: 57–85, 2001.
23. R. Slowinski, D. Vanderpooten. Similarity relation as a basis for rough approximations. In P.P. Wang, editor, *Advances in Machine Intelligence & Soft-Computing IV*, 17–33, Durham, NC, 1997.
24. R. Slowinski, D. Vanderpooten. A generalized definition of rough approximations based on similarity. *IEEE Transactions on Data and Knowledge Engineering*, 12: 331–336, 2000.
25. D.H. Wolpert. Off-training set error and a priori distinctions between algorithms. Technical Report number SFI-TR-95-01-003 of the Santa Fe Institute, Santa Fe, NM, 1995. Available at http://acoma.santafe.edu/sfi/publications/Working-Papers/95-01-003.ps.
26. D.H. Wolpert, W.G. Macready. No free lunch theorems for search. Technical Report number SFI-TR-95-02-010 of the Santa Fe Institute, Santa Fe, NM, 1996. Available at http://acoma.santafe.edu/sfi/publications/Working-Papers/95-02-010.ps.
27. S.K.M. Wong, W. Ziarko. Comparison of the probabilistic approximate classification and the fuzzy set model. *Fuzzy Sets and Systems*, 21: 357–362, 1987.
28. R.R. Yager, D. Filev. Operations for granular computing: mixing words with numbers. In *Proceedings of 1998 IEEE International Conference on Fuzzy Systems (FUZZ-IEEE'98)*, 123–128, Anchorage, AK,1998.
29. Y.Y. Yao. Two views of the theory of rough sets in finite universes. *Int. J. Approx. Reason.*, 15: 291–317, 1996.
30. Y.Y. Yao. Relational interpretations of neighborhood operators and rough set approximation operators. *Inf. Sci.*, 111: 239–259, 1998.
31. Y.Y. Yao. Generalized rough set models. In L. Polkowski, A. Skowron, editors, *Rough Sets in Knowledge Discovery* 1, 286–318, Physica, Heidelberg, 1998.
32. Yao, Y.Y. A comparative study of fuzzy sets and rough sets. *Information Sciences*, 109: 227–242, 1998.
33. Y. Y. Yao. On generalizing Pawlak approximation operators. In *Proceedings of the 1st International Conference on Rough Sets and Current Trends in Computing (RSCTC'98)*, LNAI 1424, 298–307, Springer, Berlin, 1998.

34. Y.Y. Yao. Granular computing: Basic issues and possible solutions. In *Proceedings of the 5th Joint Conference on Information Sciences (JCIS 2000)*, 186–189, Atlantic City, NJ, 2000.
35. Y.Y. Yao. Rough sets and interval fuzzy sets. In *Proceedings of the 20th International Conference of the North American Fuzzy Information Processing Society (NAFIPS 2001)*, 2347–2352, Vancouver, Canada, 2001.
36. Y.Y. Yao. Information granulation and rough set approximation. *International Journal for Intelligent Systems*, 16: 87–104, 2001.
37. Y.Y. Yao, T.Y. Lin. Generalization of rough sets using modal logic. *International Journal for Intelligent Automation and Soft Computing*, 2: 103–120, 1996.
38. Y. Y. Yao, S.K.M. Wong. A decision theoretic framework for approximating concepts. *International Journal Man-Machine Studies*, 37: 793–809, 1992.
39. Y.Y. Yao, S.K.M. Wong, T.Y. Lin. A review of rough set models. In T.Y. Lin, N. Cercone, editors, *Rough Sets and Data Mining: Analysis for Imprecise Data*, 47–75, Kluwer, Boston, 1997.
40. Y.Y. Yao, S.K.M. Wong, P. Lingras. A decision–theoretic rough set model. In Z.W. Ras, M. Zemankova, M.L. Emrich, editors, *Methodologies for Intelligent Systems*, 17–24, North-Holland, New York, 1990.
41. Y.Y. Yao, J.P. Zhang. Interpreting fuzzy membership functions in the theory of rough sets. In *Proceedings of the 2nd International Conference on Rough Sets and Current Trends in Computing (RSCTC 2000)*, LNAI 2005, 82–89, Springer, Berlin, 2001.
42. Y.Y. Yao, N. Zhong. Granular computing using information tables. In T.Y. Lin, Y.Y. Yao, L.A. Zadeh, editors, *Data Mining, Rough Sets and Granular Computing*, Physica, Heidelberg, 2001.
43. Y.Y. Yao, N. Zhong. Potential applications of granular computing in knowledge discovery and data mining. In *Proceedings of World Multiconference on Systemics, Cybernetics and Informatics (SCI'99)*, 573-580, Orlando, FL, 1999.
44. L.A. Zadeh. Fuzzy sets and information granularity. In N. Gupta, R. Ragade, R.R. Yager, editors, *Advances in Fuzzy Set Theory and Applications*, 3–18, North-Holland, Amsterdam, 1979.
45. L.A. Zadeh. Towards a theory of fuzzy information granulation and its centrality in human reasoning and fuzzy logic. *Fuzzy Sets and Systems*, 19: 111–127, 1997.
46. N. Zhong, A. Skowron, and S. Ohsuga, editors. *New Directions in Rough Sets, Data Mining, and Granular-Soft Computing*, LNAI 1711, Springer, Berlin, 1999.
47. W. Ziarko. Variable precision rough set model. *Journal of Computer and Systems Science*, 46: 39–59, 1993.

Part IV

Case Studies

Introduction to Part IV

Part IV, *Case Studies*, has chapters concerned with applications of rough set based granular computing and rough-neural computing to specific case studies, including audio signal acquisition, imbalanced data mining, surround sound perception, handwritten digit recognition, rule discovery in data, image processing, pattern recognition, and signal classification.

The chapter, *Intelligent Acquisition of Audio Signals Employing Neural Networks and Rough Set Algorithms*, by Czyżewski provides a study on intelligent techniques for signal acquisition. Among the proposed algorithms are (i) a purely neural net algorithm (ii) a rough set based algorithm, and (iii) a rough-neural algorithm. The results of the algorithms are reported. Engineering aspects of the respective systems are also described.

The case of mining an imbalanced data set is analyzed by Grzymala-Busse, Goodwin, and Xinqun Zheng in *An Approach to Imbalanced Data Sets Based on Changing Rule Strength.* Preterm birth data and breast cancer data are considered for analysis. It is shown that introducing strength multipliers increases the sensitivity factor of data bearing out the validity of the technique.

Kostek's chapter, *Rough-Neural Approach to Testing the Influence of Visual Cues on Surround Sound Perception*, uses rough-neural computing as a technique for evaluating the influence of visual cues on the perception of surrounding sound. To this end, genetic algorithm-based systems and a system combining neural networks and rough sets are applied. Experimental results are reported.

The problem of optical character recognition is addressed by Trung Nguyen in *Handwritten Digit Recognition Using Adaptive Classifier Construction Techniques,* in which an adaptive recognition system is examined on real data. The system is based on similarity and dissimilarity measures and the results of experiments are presented.

Skarbek's article, *From Rough Through Fuzzy to Crisp Concepts: Case Study on Image Color Temperature Description,* is concerned with an aspect of image processing related to color temperature detection using a process of granulation. Starting from an information system built from votes on the category to which a given object belongs, roughness degrees are used to construct an approximation to a fuzzy membership function. From this, an optimal family of partitions of the range of features is found, and these are used to induce crisp decision rules. The results of experiments are reported and discussed.

The chapter, *Information Granulation and Pattern Recognition*, by Skowron and Swiniarski provides a detailed analysis of applications of information granulation in pattern recognition. They discuss feature selection involving a rough set granulation paradigm. In particular, they introduce a hybrid system combining principal component analysis and rough set methods for reduct determination, aimed at selecting discriminatory features among principal components. Numerical experiments on face recognition are reported. In the second half of the chapter, a knowledge base of soft rules expressed in natural language across a database of concepts and relations is introduced. This is discussed from the point of view of rough-neural computing, leading to the concept of a rough neural network.

The chapter, *Computational Analysis of Acquired Dyslexia of Kanji Characters Based on Conventional and Rough Neural Networks*, by Tsumoto introduces two computational models to explain Japanese dyslexia syndrome. The experimental results show that Iwata's model can explain dylexia syndromes better than the classical model, although the computational model is a little weaker in explaining the all of aspects of acquired Japanese dyslexia. Furthermore, the results suggest that positive neurons in the Iwata model may play an important role in achieving the cognitive functions of Kanji characters. This chapter provides a preliminary study on simulation of neurological diseases based on rough neural networks.

WaRS: A Method of Signal Classification by Wojdyłło describes a hybrid system in which signal analysis by wavelet techniques is coupled with postprocessing by rough set methods to induce classifying rules. The results of experiments are described and discussed.

Zhong, Liu, Dong, and Ohsuga present in their chapter, *A Hybrid Model for Rule Discovery in Data*, a general hybrid model for rule discovery in data. In this approach, a generalized distribution table (GDT), a kind of transition matrix, representing probabilistic relations among concepts and instances, is used as a probabilistic search space. By using, a GDT along with a transition matrix in a stochastic process, various methods of symbolic reasoning such as abduction, deduction, induction as well as soft computing methods based on rough sets, fuzzy sets, inductive logic programming, and granular computing are involved in the search process. In particular, the variant GDT-RS coupling, integrating GDT with rough set methodology, is discussed. Experimental results with real data are provided.

Chapter 20
Intelligent Acquisition of Audio Signals, Employing Neural Networks and Rough Set Algorithms

Andrzej Czyżewski

Gdańsk University of Technology, Sound and Vision Engineering Department,
Narutowicza 11/12, 80-952 Gdańsk, Poland
andrzej@akustyka.com

Summary. The algorithms stemming from the rough-neural computing approach were applied to digital acquisition of audio signals with regard to automatic localization of sound sources in the presence of noise and a parasite echo. The application of neural networks to the automatic detection of the sound arrival direction was tested first; then, it was followed by some experiments employing rough sets, and finally the rough-neural approach to this problem solving was examined. The output of each algorithm tested was supposed to provide information about the direction of arriving sound. In the rough-neural algorithm, the result of its action can also be available in the form of words defining the direction of arriving sound. Some details of the engineered systems and results of their experimental verification are compared and discussed.

1 Introduction

Sound source localization plays an important role in spatial-filtering techniques applied to many telecommunication systems. Speech signals arriving from various directions may interfere with the target signal but also can mask it. Consequently, the main purpose of the spatial filtering technique applied to telecommunication systems or to hearing aids is to attenuate the unwanted signal coming from directions other that the desired. Another issue addressed by this kind of sound processing algorithm is automatic sound source tracking that is applicable to advanced video teleconferencing systems employing automatically turned video cameras.

Numerous source localization methods were investigated by various researchers [1,3,4,5]. These include frequency-domain processing introduced at the Sound and Vision Engineering Department [6,7, 17–20, 28]. Most of systems are based on digital signal processing technology and are computationally intensive.

This chapter presents alternative method applications employing neural networks and rough set based decision modules. The computer simulators (Matlab and Rosetta) were used in the experiments. The new methods were investigated using some prerecorded sound excerpts that provided an experimental data stream.

The description of the proposed algorithms and some results of experiments are included. Some conclusions were drawn on the basis of experiments concerning learning algorithm applications to spatial filtering of sound.

2 Experimental Setup

Human hearing is the most effective detector of the direction of arriving sounds. Therefore, to make signal processing algorithms more efficient, their operating principle should follow the properties of human hearing. Despite the great development of science in the field of human perception, issues related to sound localization are not finally recognized hence the underlying phenomena are still the subject of intense research [2,7,13]. According to the present state of knowledge, the perception of sound direction by the human binaural system is based on the following two principal entities [13]:

- Interaural level difference (ILD): the difference in the intensities of waveforms in the left and right ears.
- Interaural time difference (ITD): the difference of arrival times of relevant waveforms in both ears, which is equivalent to the phase difference in the waveforms.

In digital signal processing, sound source localization can be performed by a microphone array which can be either linear or non-linear [16,19]. The layout of a sound acquisition system is presented in Fig. 1. Eight microphones placed symmetrically on a 30 cm diameter rim were used.

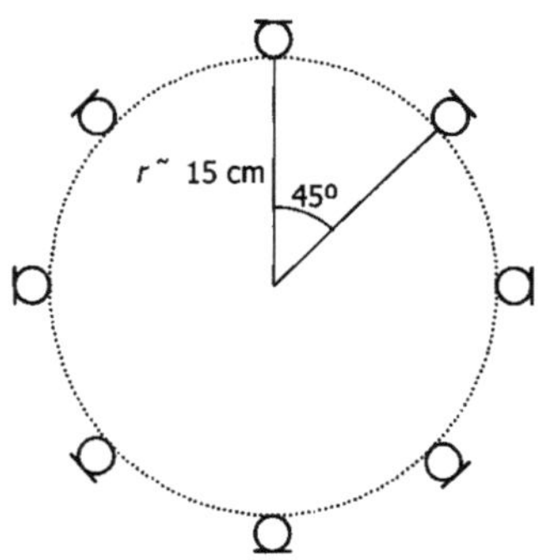

Fig. 1. Diagram of the microphone array shown in the horizontal plane

Under ideal conditions, the signal received from the ith microphone of the circular array of microphones at the t-th moment of time can be described as follows:

$$x_i(t) = \alpha \cdot s[t - (i-1) \cdot \tau],$$

where α_i is the attenuation coefficient for the ith microphone, $s(t)$ is the source signal, and τ is the time delay of the acoustic wave between adjoining microphones.

Thus the estimation of the source location based on the processing of acoustic signals with a microphone array provides a deterministic problem.

However, under real conditions, various distortions occur along with interference signals, such as background noise and reverberated sounds. Hence, the signals received by a circular microphone array are expressed by the following relationships:

$$\left\{\begin{array}{c} x_1(t) = \alpha_1 \cdot h_1(t) * s(t) + n_1(t) \\ x_2(t) = \alpha_2 \cdot h_2(t) * s(t-\tau) + n_2(t) \\ \vdots \\ x_i(t) = \alpha_i \cdot h_i(t) * s(t-(i-1)\cdot\tau) + n_i(t) \\ \vdots \end{array}\right\},$$

where $h_i(t)$ is the impulse response of the reverberant channel associated with ith microphone, $n_i(t)$ is the ambient noise received by ith microphone, and * is the convolution operator.

The impulse response $h_i(t)$ is acquired in a room so that a loud impulse is produced (e.g., a pistol shot), and then the reverberant sound is recorded until the acoustic energy decreases to an undetectable level.

These conditions make the task of sound source localization more complex, and therefore, a number of methods have been proposed to solve the problem. Most of them are based on an estimate of the sound source position on the basis of signals received by microphones in the matrix, including cross-correlation techniques [3], adaptive filtration [5], and computation of relevant eigenvalue vectors and matrices [1]. In turn, tracking or localizing a number of sources, maximum likelihood-based methods are exploited [31]. More details can be found in the abundant literature on localization of acoustic sources for multimedia applications [15,16,23,29,31].

2.1 Sound Acquisition Setup

As was said, the array consisting of 8 electret microphones was set on the circumference of a 15 cm radius (30 cm in diameter) rim. The set of microphones shown in Fig. 2 was fixed 1.58 m from the floor. The recording parameters were as follows: 16 bit/sample, sampling frequency equal to 48 kHz. There was one male speaker, distanced 1.5 m from the array. The speaker read a logatom list from consecutive spots differing in 5°. As a result, 72 eight-track recordings were made, and every recording lasted approximately 55 s. For the purposes of the experiments, eight additional excerpts were also prepared representing the sound direction from –270° to +270°, recorded every 15°. Another set of signals recorded for experimental purposes was derived from angles 0°–90° with a resolution of 50°.

Fig. 2. The microphone matrix used for sound acquisition

2.2 Extracting Feature Vectors

A direct processing of a sound sample stream by learning algorithms might be impractical because the volume of data can be very high, and there are too many indirect dependencies between consecutive sample packets which may not be interpreted easily. Therefore, during the feature extraction process, the signal was divided into frames of length N equal to 512, 1024, and 2048 samples and then processed by some feature extraction algorithms. As verified earlier, in the practical application of spatial filtration (beamforming) in hearing aids, the following parameters can be efficiently exploited [18]:

$$M_i = \frac{\min(|L_i|, |R_i|)}{\max(|L_i|, |R_i|)},$$

$$D_i = \frac{|L_i - R_i|}{|L_i| + |R_i|},$$

$$A_i = |\angle L_i - \angle R_i|,$$

where L_i and R_i are magnitudes of the ith spectral bin for the left and right channels, respectively.

Considering that the above parameters concern pairs of channels i and j, the parameters Ch_i^k and Ch_j^k for the kth spectral bin can be rewritten as:

$$M_{ij}^k = \frac{\min\left(\left|Ch_i^k\right|, \left|Ch_j^k\right|\right)}{\max\left(\left|Ch_i^k\right|, \left|Ch_j^k\right|\right)},$$

$$D_i = \frac{Ch_i^k - Ch_j^k}{\left|Ch_i^k\right| + \left|Ch_j^k\right|},$$

$$A_i = \left|\angle Ch_i^k - \angle Ch_j^k\right|.$$

It can be shown that the parameters are in a simple functional relationship, and therefore one of them is superfluous and can be dropped. In such a case, parameters representing a single spectral bin are as follows:

$$M_{ij}^k = \frac{\min\left(\left|Ch_i^k\right|, \left|Ch_j^k\right|\right)}{\max\left(\left|Ch_i^k\right|, \left|Ch_j^k\right|\right)},$$

$$A_i = \left|\angle Ch_i^k - \angle Ch_j^k\right|.$$

As was said, in the experiments, eight-channel signals were examined thus the following sets of parameters can be considered:

Type A is all mutual combinations of channels, yielding 56 parameters per spectral bin;

type B is a combination of opposite channels, yielding eight parameters per spectral bin.

Due to the fact that the above parameters are to be fed to a learning algorithm, they are grouped into input vectors. The following three types of such vectors can be considered:

Type $V1$ is if all spectral bins are included in a vector;

type $V2$ is an input vector that consists of parameters for a single bin and the additional information on the bin's frequency;

type $V3$ is an input vector that consists only of parameters for a single spectral bin.

In this case, the learning algorithm assumes a modular structure where a separate subsystem is dedicated for each spectral bin. The final decision is made on the basis of the interpreting of output values of all subalgorithms.

Of particular interest is the modular concept related to the feature vectors of the type $V3$. The description of data to be fed to the modular decision algorithm in this case is gathered in Table 1.

Table 1. Analysis of training conditions for the input vector $V3$

N=512; N/2=256	N=1024; N/2=512	N=2048; N/2=1024
A:vector Size=56	*A*:vector Size=56	*A*:vector Size=56
B:vector Size=8	*B*:vector Size=8	*B*:vector Size=8
256 decision module	512 decision module	1024 decision module
material for training: 186 vectors/s	material for training: 92 vectors/s	material for training: 45 vectors/s

3 Application of Neural Networks

3.1 Training Algorithms

Some heuristic algorithms were chosen for neural network training , namely the general and simplified Fahlman's algorithm (QuickPROP) [24] and the resilient PROPagation (RPROP) [25]. In the general Fahlman algorithm (denoted further as Fahlman I), the weight update rule for a single weight in the *k*th cycle is for the Fahlman algorithm computed as:

$$\Delta w_{ij}^{k} = -\eta^{k} \cdot S_{ij}^{k} + \alpha_{ij}^{k} \cdot \Delta w_{ij}^{k-1},$$

where the error gradient term S_{ij}^{k} assumes

$$S_{ij}^{k} = \nabla E\left(\Delta w_{ij}^{k}\right) + \gamma \cdot \Delta w_{ij}^{k}, \qquad \gamma = 10^{-4}$$

and the learning rate η^{k} and the momentum ratio α_{ij}^{k} vary according to formulas that can be found in the literature [24].

In the simplified Fahlman algorithm denoted further as Fahlman II, the weight update rule is modified, so that it is expressed by the following relationship:

$$\Delta w_{ij}^{k} = \left\{ \begin{array}{c} \alpha_{ij}^{k} \cdot \Delta w_{ij}^{k-1} \text{ for } \Delta w_{ij}^{k-1} \neq 0 \\ -\eta_0 \cdot \nabla E\left(\Delta w_{ij}^{k}\right) \text{ otherwise, i.e.: } \Delta w_{ij}^{k-1} = 0 \end{array} \right\},$$

where the momentum ratio α_{ij}^{k} changes according to the expression:

$$\alpha_{ij}^{k} = \min\left\{ \frac{\nabla E\left(\Delta w_{ij}^{k}\right)}{\nabla E\left(\Delta w_{ij}^{k-1}\right) - \nabla E\left(\Delta w_{ij}^{k}\right)}, \alpha_{\max} \right\},$$

and the constant values of the training parameters are the same as in the general QuickPROP, i.e., $0.01 \leq \eta_0 \leq 0.6$, $\alpha_{max} = 1.75$.

In the RPROP algorithm, the weight update rule is given by the following formula based on the signum function:

$$\Delta w_{ij}^k = -\eta_{ij}^k \cdot sgn\left[\nabla E\left(\Delta w_{ij}^k\right)\right],$$

where the learning rate η_{ij}^k assumes values according to the rules described in the literature [25].

3.2 Results of Modular Neural Network Application

Tables 2–4 present the results of the experiments with neural networks applied to detecting the direction of the incoming sound. These results are presented with regard to the training algorithm and sound arrival direction. Additionally, the length of the sample packet and the number of training and testing vectors are shown in these tables. It was assumed that the maximum value found on neural networks outputs identifies the particular network which detected the appropriate direction of the arriving sound. The percentage of accurate scores obtained with this method is shown in the tables.

Table 2. Results of direction detection for vector type $V3$, $N = 512$; parameter type C, training/testing vectors are 1042/446

Direction	Fahlman I		Fahlman II		RPROP	
	Epochs	Scores	Epochs	Scores	Epochs	Scores
−45°	16.335	82%	15.893	83%	23.119	85%
−30°	14.239	84%	18.991	83%	21.092	80%
−15°	15.268	78%	16.453	80%	19.672	82%
0°	17.218	79%	18.002	81%	20.999	81%
15°	16.001	81%	19.979	83%	21.017	82%
30°	19.965	80%	21.310	82%	20.775	82%
45°	18.342	79%	21.367	78%	25.901	80%

Table 3. Results of direction detection for vector type $V3$, $N = 1024$; parameter type A; training/testing vectors are 515/221

Direction	Fahlman I		Fahlman II		RPROP	
	Epochs	Scores	Epochs	Scores	Epochs	Scores
−45°	27.890	90%	37.199	92%	41.092	89%
−30°	32.893	89%	34.269	87%	39.501	88%
−15°	32.672	88 %	31.474	89 %	42.277	90 %
0°	29.994	90%	35.892	90%	37.512	88%
15°	30.173	86%	40.866	82%	50.899	85%
30°	29.980	85%	30.101	85%	35.924	84%
45°	27.559	87%	38.943	88%	39.994	88%

Table 4. Results of direction detection for vector type $V3$, $N = 2048$; parameter type A; training/testing vectors are 252/108

Direction		Fahlman I		Fahlman II		RPROP
	Epochs	Scores	Epochs	Scores	Epochs	Scores
−45°	21.218	86%	22.190	85%	30.168	86%
−30°	19.900	87%	26.886	87%	27.110	86%
−15°	20.457	87%	21.106	88%	28.249	88%
0°	28.189	88%	27.000	89%	39.271	89%
15°	25.190	85%	32.981	86%	31.992	87%
30°	23.267	86%	24.119	85%	28.428	86%
45°	24.219	87%	31.148	87%	28.550	87%

As results from the data presented, the direct application of the modular network structure allowed detecting of the direction of the arriving sound with quite good accuracy, however, not better than 92%.

4 Application of Rough Sets

The rough set based decision making system was proposed to allow better sound source localization performance when noisy and distorted signals introduce much uncertainty related to the decision-making process. A block diagram of the sound source localization method introduced is shown in Fig. 3.

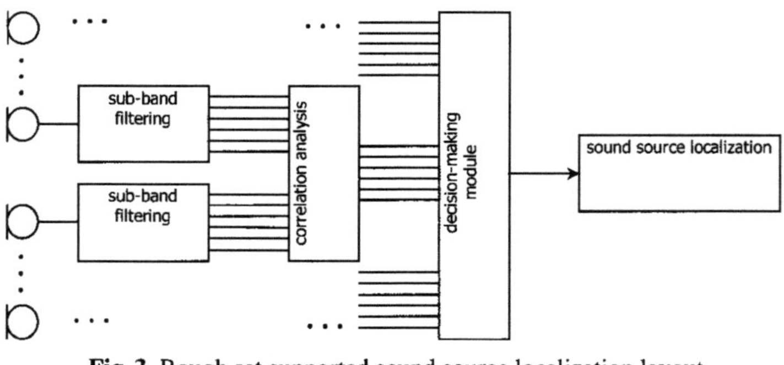

Fig. 3. Rough set supported sound source localization layout

The input signal of each microphone is first passed through a set of band-pass filters. Subsequently, correlation analysis is performed for each pair of microphone signals and each subband. As an output, a set of correlation parameters is calculated. The sound source is then localized using a decision making unit upon a set of correlation parameters for subsequent pairs of microphones and following frequency subbands. Since the input signal may contain noise and distortions, a rule-based rough set algorithm was employed for this task.

The processing might be performed for all combinations of microphones within the array. However, such an approach demands significant computing power to perform correlation analysis and tends to load a large amount of data into the decision making module. Therefore, in the experiments performed, only pairs of counter positioned microphones were considered. This resulted in a significant reduction in computing power requirements without losing performance capability.

4.1 Correlation Parameters

Correlation parameters are calculated for each pair of counter positioned microphones within subsequent frequency subbands. Correlation analysis is performed within the octave subbands. The boundary frequencies of subbands are presented in Table 5.

Table 5. Boundary frequencies of subbands

Band	Lower boundary	Higher boundary
1	20 Hz	100 Hz
2	100 Hz	200 Hz
3	200 Hz	400 Hz
4	400 Hz	800 Hz
5	800 Hz	1600 Hz
6	1600 Hz	3200 Hz
7	3200 Hz	6400 Hz
8	6400 Hz	20000 Hz

Subband filtering was performed using the spectral filtering method. The *Mathematica* notebook performing spectral filtration was developed. The following signal processing constraints were observed:

sampling frequency is 48 kHz;
window size is 2048 samples;
overlap is 1024 samples, and
windowing function is Hamming type.

Initially, a standard autocorrelation function was applied according to Pearson's formula:

$$\rho(n) = \sum_i \frac{[x(t)_i - \bar{x}(t)]\,[y(t-n)_i - \bar{y}(t+n)]}{\sqrt{\sum_i [x(t)_i - \bar{x}(t)]^2}\sqrt{\sum_i [y(t-n)_i - \bar{y}(t+n)]^2}}.$$

Its simplified form is as

$$\rho(n) = \frac{\sum_i x(t)_i y(t-n)_i}{\sum_i x(t)_i y(t)_i}.$$

The correlation maximum should correspond to time alteration of signals between microphones of concern at the given moment. However, since a speech signal may include significant energy alterations, correlation function maxima may correspond to energy peaks. Therefore, as an alternate solution, the alteration maxima discerning function (AMDF) was introduced to allow correlation analysis. The AMDF function is given as

$$\mathrm{AMDF}(n) = \sum_{i} |x(t)_i y(t-n)_i| .$$

Based on the AMDF function, a signal time lag between microphones can be estimated upon locating the global minimum. An example of the AMDF function plot for a speech signal within the fifth subband (see Table 5) is presented in Fig. 4. To allow a better illustration, the reverse signed AMDF denoted as (–AMDF) was shown. Note that the zero-lag location corresponds to 100 on the n axis.

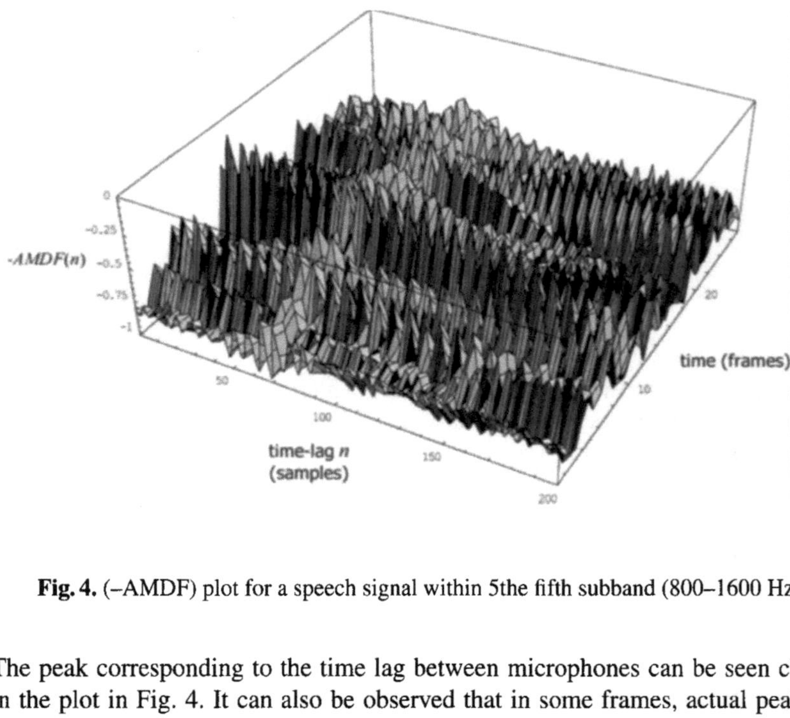

Fig. 4. (–AMDF) plot for a speech signal within 5the fifth subband (800–1600 Hz)

The peak corresponding to the time lag between microphones can be seen clearly in the plot in Fig. 4. It can also be observed that in some frames, actual peaks do not correspond to the time lag. Consequently, the AMDF function is accumulated through frames of processing according to the formula

$$(\mathrm{AMDF})_{\mathrm{ac}}(n) = \frac{1}{M} \sum_{m=1}^{M} (\mathrm{AMDF})_m(n),$$

where M is the number of frames of analysis. A time lag between microphones can be estimated upon location of the minimum within the accumulated AMDF function. However, according to speech signal characteristics as well as the presence of potential noise and distortions such an estimate may lead to erroneous source localization. Therefore, for each pair of microphones and subsequent subbands, the accumulated AMDF function is represented using two parameters: location of the minimum of the accumulated AMDF within a subband and an average signal energy within the subband defined as

$$E_b = \frac{1}{N}\sum_{n=1}^{N} \left|(\mathrm{AMDF})_b(n)\right|,$$

where N is the number of samples and b is the current subband.

These parameters are then provided for further processing using the decision system.

4.2 Rough Decision System

The method presented is based on the rule-based rough set decision system. For the purpose of the experiments presented, the rough set software toolbox Rosetta was used [34].

Learning data are processed in the following way. First, the data set (knowledge base) is acquired. The knowledge base consists of objects that are represented by using conditional attributes and decision parameters. As an input for the decision making system, a set of correlation parameters, specifically, the location of AMDF minima (denoted Dn) along with signal energy (denoted An) within subbands, are used. For each pair of microphones, each subband is represented by two parameters, giving a total amount of 64 parameters representing the input pattern. The input file for the rough-set processing consists of a header and a data set, as shown in Table 6.

Table 6. Data layout for rough set based sound source localization

Parameter	D1	A1	D2	A2	D3	A3	...	Angle
type	integer	float(4)	integer	float(4)	integer	float(4)	...	string(3)
#1	−30	32.0292	−52	112.24	4	180.875	...	000
#2	−37	32.1163	−87	96.1503	−37	181.063	...	090
...	...	...	...	...	...	...	...	...
#N	−60	35.5952	−48	151.269	−40	395.7	...	270

Consequently, acquired data are quantized to convert real attribute values into discretized form allowing further rule-based processing. Based on the discrete values, attributes are analyzed in terms of discernibility. Sets of attributes allowing partition of object classes are then revealed. These sets provide reducts. Consequently, rules are generated upon reducts.

The Rosetta system supports a variety of quantization as well as reduct and rule generation procedures; however, details of these procedures were beyond the scope of this chapter [34]. For the purpose of the experiments presented, the following processing parameters were used:

discretization: equal frequency binding using three intervals, and

reduct and rule generation: an object related genetic algorithm producing a set of rules via minimal attribute subsets that discern object classes; reducts and rules are generated upon analysis of all learning patterns.

These processing parameters were chosen during preliminary research aimed at optimizing system efficiency and generalization ability.

4.3 Experiments on Sound Source Localization

For the purpose of the experiments presented, speech recordings made within an anechoic chamber were used. The experiments were divided into three subsequent parts: "low-resolution source localization", "low-resolution source localization in the presence of wide band noise" and "high-resolution source localization".

Low-Resolution Source Localization Experiments on low-resolution source localization were performed. In this phase, a sound source was located at angles of 0°, 90°, 180° and 270°, respectively. Five sound examples for each angle were used in the experiments. Two series of experiments were performed. In the first phase, one instance representing each angle was used for training, whereas the trained system was tested using the other patterns. In the second phase, the patterns used previously for testing were used for training and vice versa. Source localization scores for both phases of the experiments are shown in Table 7. The numbers given in brackets represent the number of properly classified examples versus the number of all tested examples.

Table 7. Sound localization accuracy for low-resolution analysis

Accuracy	1st Phase	2nd Phase
Minimum	68.75%(11/16)	100%(4/4)
Maximum	100%(16/16)	100%(4/4)
Average	80%(64/80)	100%(20/20)

An example of sound localization rule-based decisions is shown in Fig. 5 based on Rosetta program window display.

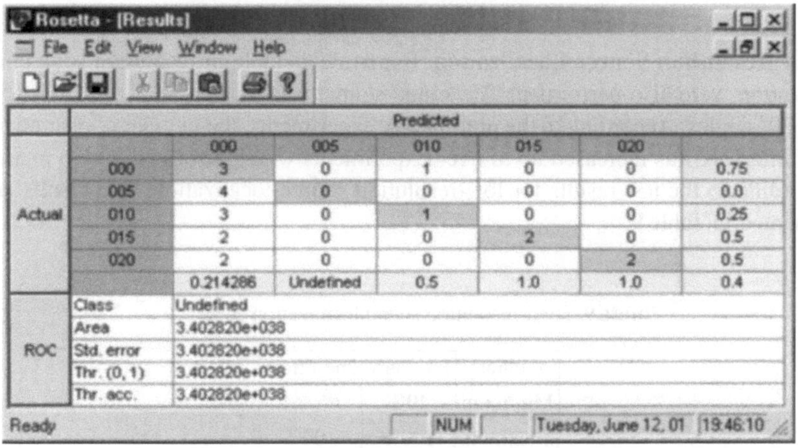

Fig. 5. Example of sound localization decisions for low-resolution analysis. In the region of the window denoted as "Actual", the predicted angle values versus existing ones are shown, and the number of relevant cases is displayed.

Low-Resolution Source Localization of Noisy Signal For the purpose of the experiments on source localization of a noisy signal, white noise was mixed with the sound samples. Experiments were performed for noise levels relative to the maximum level of the speech signal as follows: 0, –20 and –40 dB. Experiments were performed in two phases, according to the procedure illustrated above for low-resolution source localization without noise. The results of the experiments are presented in Table 8 (again, the numbers given in brackets represent properly classified examples versus all tested examples).

Table 8. Sound localization accuracy for low-resolution analysis

Accuracy	1st Phase	2nd Phase
Minimum	40%(8/20)	60%(3/5)
Maximum	100%(20/20)	100%(5/5)
Average	63%(63/100)	88%(22/25)

Comparing results gathered in Table 7 and Table 8 allows one to draw the conclusion that the presence of strong noise influences the accuracy of detecting the direction of sound arrival. However, the rough set algorithm handles noisy examples quite well, still ensuring 88% average accuracy and 100% maximum accuracy because this algorithm makes correct decisions while processing uncertain data patterns.

High-Resolution Source Localization Experiments on high-resolution source localization were also performed. The sound source was located at 0°, 5°, 10°, 15°, and 20° angles. According to the preliminary experiments, the number of discretization intervals was increased up to five. Experiments were performed in two phases according to the test results for low-resolution source localization. The results are presented in Table 9.

Table 9. High-resolution sound localization accuracy

Accuracy	1st Phase	2nd Phase
Minimum	40%	66.6%
Maximum	100%	100%
Average	63%	89.2%

An exemplary decision set for high-resolution source localization is presented in Fig. 6.

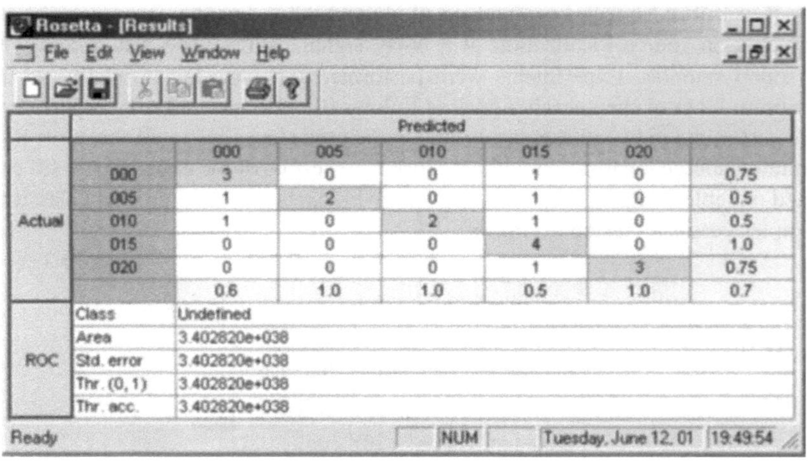

Fig. 6. Sound source localization decisions for high-resolution analysis. In the region of the window denoted as "Actual," the predicted angle values versus existing ones are shown, and the number of relevant cases is displayed, as previously.

All results presented were obtained using a randomly selected syllable, meanwhile the system learned by employing all other prerecorded syllables. As the results from the data presented show, the application of the rough set algorithm allowed high-resolution detection of the direction of arriving sound with good accuracy, in some cases even equal to 100%.

5 Application of the Neuro-Rough Algorithm

The last group of experiments employed a hybridized neuro-rough system, as presented in the block diagram in Fig. 7. The system consists of some consecutive blocks described below. Application of the modular multilayer neural network, in which every module is responsible for processing a different spectral component, allowed estimating sound arrival directions independently for each frequency. The overall decision was then provided by the rough set algorithm on the basis of the analysis of all quantized outputs of individual neural networks.

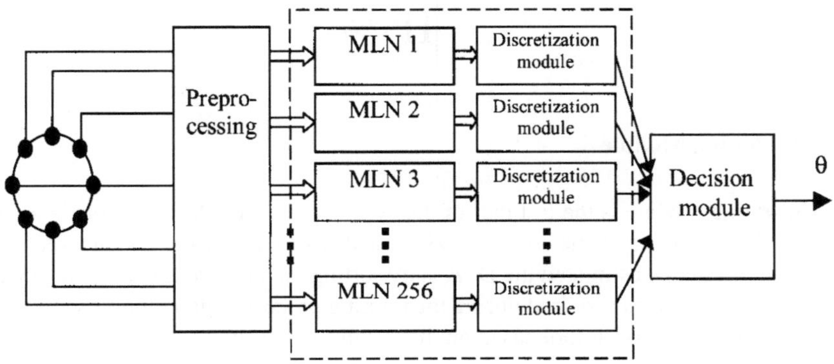

Fig. 7. Block diagram of speaker localization system based on neuro-rough approach.

Multilayer Neural Network The system uses 256 separated multilayer neural networks (MLNs) because the resolution of spectral analysis is limited to 256 spectral lines (512 sample buffer lengths). Each MLN is related to a separate spectral component. Each MLN has an input layer, a hidden layer, and an output layer. The number of input-layer neurons was set at 56 because the number of different pairs selected from eight microphones equals 56. The number of neurons in the hidden layer changed from 21 to 45. The output layer included five neurons [providing encoded binary representation of the direction of sound arrival (DSA)]. Each neuron adopted a continuous nonlinear function varying in the range from 0 to 1. The MLNs were trained employing the resilient back-propagation method.

5.1 Classification Module

The DSA was encoded in 5-bit vectors, and this representation was used to train MLNs. As a result of DSA estimation by 256 MLNs, 256 vectors were obtained consisting of five elements having values between zero and one. The discretization module was connected to each neuron in the output layer of each of the 256 MLNs and realized a division of real numbers into some subintervals. The division $\prod_A$ on $[a,b]$ is defined as the set of k subintervals:

$$\prod_A = \{[a_0,a_1),[a_1,a_2),\ldots,[a_{k-1},a_k]\},$$

where $a_0 = a, a_{i-1} < a_i, i = 1,\ldots,k, a_k = b$.

This approach to quantization is based on calculating division points a_i. After quantization, the parameter value is transformed into the number of the subinterval to which this value belongs. In most experiments, the simplest division called binary quantization was used, where

$$\left|\prod_A\right| = 2.$$

5.2 Voting Module

The voting module is the last part of the system presented. Its main role is to interpret binary vectors as the concrete DSAs, and store each of the 256 answers in the memory. Finally, estimated the DSA is by sorting the table contents of the DSA according to the rules acquired during the training process. Again, the Rosetta system was used as a decision tool based on the rough set theory.

5.3 Tests and Results

The set of 18 directions (DSAs) defined as rear left, left,... front, front right, right,... rear right was defined. The results of the first tests are presented in Table 10. The 15 learning vectors for each of the DSAs were used (15 vectors times 18 DSA = 270 learning vectors). A continuous linear unipolar, neuron activation function was used. The three different numbers of neurons in the hidden layer were tested: 21, 35, and 45. The percentage value in this table indicates the range of wrong answers of 256 classification modules estimating a single DSA.
The next experiments were made with an increased number of learning vectors. The results derived from the system with 36 learning vectors for each DSA is presented in Table 11.

Subsequent experiments were organized employing 54 learning vectors for each DSA. The MLNs in the decision module included 45 neurons in the hidden layer. In this test, three boundary values in the classification module were used. This means that rough rules with strength lower than these values were rejected. The results obtained are shown in Table 12.

Table 10. Percentage ranges of wrong decisions among individual networks contained in the modular structure; employment of 15 learning vectors for each DSA; various numbers of neurons number in hidden layer

Number of neurons in hidden layer		21	35	45
Crisp	Loud sounds	63.5-67.66%	56.58-65.84%	61.78-66.28%
	Quiet sounds	71.72-77.19%	70.29-74.98%	69.49-75.48%
	Noise	89.08-91.78%	88.72-91.45%	88.98-91.75%
Uncertain	Loud sounds	43.45-48.78%	41.82-46.07%	40.65-46.27%
	Quiet sounds	53.08-56.21%	49.85-54.06%	48.94-54.64%
	Noise	75.5-79.9%	76.22-78.58%	75.35-79.82%

Table 11. Percentage ranges of wrong decisions among individual networks contained in the modular structure; employment of 36 learning vectors for each DSA

Number of neurons in hidden layer		21
Crisp	Loud sounds	52.76-54.56%
	Quiet sounds	61.81-68.29%
	Noise	87.35-89.54%
Uncertain	Loud sounds	34.64-37.70%
	Quiet sounds	43.10-47.48%
	Noise	74.33-76.56%

Table 12. Percentage ranges of wrong decisions among individual networks contained in the modular structure; employment of 54 learning vectors for each DSA

Boundary value		0.4	0.5	0.6
Crisp	Loud Sounds	50.24-53.26%	49.63-52.97%	49.44-53.54%
	Quiet sounds	58.29-65.65%	58.12-65.71%	58.01-65.80%
	Noise	87.52-89.78%	87.50-89.50%	87.65-89.56%
Uncertain	Loud sounds	32.86-37.13%	32.51-37.28%	32.51-37.74%
	Quiet sounds	40.71-46.05%	40.47-45.70%	40.71-45.96%
	Noise	73.35-77.34%	73.44-77.24%	73.46-77.06%

The results obtained indicate a high correlation between the number of learning vectors and the decision quality. It was necessary to increase the number of neurons in the hidden layer simultaneously with an increasing number of learning vectors employed during the learning phase. In the test with 54 learning vectors, about 50% of the 256 outputs of the decision module provided correct answers. With this value, it was possible to create a properly working DSA recognition system. As observed, the number of correct answers was higher than the number of answers indicating other (wrong) DSAs. Taking advantage of this dependence, a voting module was created. The rough rules were employed for interpreting the current state of the modular neural network outputs. Although a considerable portion of the individual neural network outputs might provide inaccurate decisions, by using this simple method, it was possible to get 100% efficiency of the correct DSA recognition in each test run. The best solution was to use 54 learning vectors to train MLNs and 45 neurons in a hidden layer in each MLN. Such a system is characterized by the best results in DSA estimation. On the other hand, the whole decision system is quite computationally intensive. The work presented shows that localization of DSA with a modular neural network structure hybridized with rough sets is possible and provides perfect results.

6 Conclusions

The experiments performed proved that sound source position can be localized successfully using neural networks or employing a combination of correlation analysis and rough set rule-based processing or modular neural networks plus rough set decision modules (the most effective solution). The rough set approach provides generalization ability to the system allowing proper source localization even with noisy and reverberated sound patterns.

The results obtained demonstrate also that nonlinear filters based on learning decision algorithms may provide quite an effective tool for detecting sound source position in the presence of noncorrelated noise. The problem analyzed is highly nondeterministic in this case. Consequently, intelligent filters used in a sound acquisition system can result in a significant improvement in speech intelligibility and an increase in the signal-to-noise ratio. The results also open the possibility of employing intelligent sound localization algorithms in experimental teleconference systems. Additionally, they provide a practical example how, by using neuro-rough hybridization, real numbers representing angles of direction of sound arrival could be associated with words describing directions in natural language (for example, rear left, left, front left, front, front right, ..., rear right). Therefore, the problem discussed and its solution demonstrate another way of computing with words.

Acknowledgments

The research was sponsored by the Committee for Scientific Research, Warsaw, Poland, Grant No. 8 T11D 00218 and No. 4 T11D 014 22.

References

1. B. Berdugo, M.A. Doron, J. Rosenhouse, and H. Azhari. On direction finding of an emitting source from time delays. *Journal of the Acoustical Society of America*, 106: 3355–3363, 1999.
2. M. Bodden. Modeling human sound-source localization and the cocktail-party-effect. *Acta Acustica*, 1: 43–55, 1993.
3. M.S. Brandstein. A pitch-based approach to time-delay estimation of reverberant speech. In *Proceedings of the IEEE Workshop on Applications of Signal Processing to Audio and Acoustics*, Mohonk, New Paltz, NY, 1997.
4. W.-F. Chang, M.W. Mak. A conjugate gradient learning algorithm for recurrent neural networks. *Neurocomputing: An International Journal*, 24: 173–189, 1999.
5. S.-J. Chern, S.-H. Lin. An adaptive time delay estimation with direct computation formula. *Journal of the Acoustical Society of America*, 96: 811–820, 1994.
6. J. Czerniawski. *The Verification of New Algorithms for Identifying Sound Source Position and Sound Acquisition Methods Based on Spatial Filtering*. Masters Thesis, Technical University of Gdansk, 2001 (in Polish).
7. A. Czyzewski, R. Krolikowski. Neuro-rough control of masking thresholds for audio signal enhancement. *Neurocomputing: An International Journal*, 36: 5–27, 2001.
8. M.S. Datum, F.Palmieri, A. Moiseff. An artificial neural network for sound localization using binaural cues. *Journal of the Acoustical Society of America*, 100: 3372–3383, 1996.
9. S.P. Day, M.R. Davenport. Continuous-time temporal back-propagation with adaptable time delays. In *IEEE Transactions on Neural Networks*, 4: 348–354, 1993.

10. J.L. Elman. Finding structure in time. *Cognitive Science*, 14: 179–211, 1990.
11. M.W. Goudreau, C.L. Giles, S. T. Chakradhar, D. Chen. First-order vs. second-order single layer recurrent neural networks. *IEEE Transactions on Neural Networks*, 5: 511–518, 1994.
12. M.W. Goudreau, C.L. Giles. Using recurrent neural networks to learn structure of interconnection networks. *IEEE Transactions on Neural Networks*, 8: 793–820, 1995.
13. M.W. Hartmann. How we localize sound. *Physics Today*, 11: 24–29, 1999.
14. B.G. Horne, C.L. Giles. An experimental comparison of recurrent neural networks. In G. Tesauro,D. Touretzky, T. Leen, editors, *Advances in Neural Information Processing Systems*, Vol. 7, 697–705, MIT Press, Cambridge, MA, 1995.
15. G. Jacovitti, G. Scarano. Discrete time techniques for time delay estimation. In *IEEE Transactions on Signal Processing*, 41: 525–533, 1993.
16. F. Khalil, J.P. Lullien, A. Gilloire. Microphone array for sound pickup in teleconference systems. *Journal of the Audio Engineering Society*, 42: 691–700, 1994.
17. A. Czyzewski, B. Kostek, J. Lasecki. Microphone array for improving speech intelligibility. In *Proceedings of the 20th Tonmeistertagung, International Convention on Sound Design*, 428–434, Karlsruhe, 1998.
18. B. Kostek, A. Czyzewski, J. Lasecki. Spatial filtration of sound for multimedia systems. In *Proceedings of the Copenhagen Workshop on Multimedia Signal Processing (MMSP'99)* (CD-ROM), 209–213, IEEE Signal Processing Society, Piscataway, NJ, 1999.
19. B. Kostek, A. Czyzewski, J. Lasecki. Computational approach to spatial filtering. In *Proceedings of the 7th European Congress on Intelligent Techniques and Soft Computing (EUFIT'99)* (CD-ROM), 242, Aachen, 1999.
20. J. Lasecki, B. Kostek, A. Czyzewski. Neural network-based spatial filtration of sound. In *Proceedings of the 106th Audio Engineering Society Convention*, Preprint No. 4918, Munich, 1999.
21. T. Lin, C.L. Giles, B.G. Horne, S.Y. Kung. A delay damage model selection algorithm for NARX neural networks. In *IEEE Transactions on Signal Processing* (special issue on neural networks), 11: 2719–2730, 1997.
22. T. Lin, B.G. Horne, P. Tino, C.L. Giles. Learning long-term dependencies in NARX recurrent neural networks. In *IEEE Transactions on Neural Networks*, 7: 1329–1351, 1996.
23. Y. Mahieux, G. le Tourneur, A. Saliou. A microphone array for multimedia workstations. *Journal of the Audio Engineering Society*, 44: 365–372, 1996.
24. S. Fahlman. An empirical study of learning speed in back-propagation networks. Technical Report CMU-CS-88-162 of Carnegie Mellon University, Pittsburgh, PA, 1988.
25. M. Riedmiller, H. Braun. A direct adaptive method for faster backpropagation learning: The RPROP Algorithm. In *Proceedings of the IEEE International Conference on Neural Networks*, 586-591, San Francisco, 1993.
26. H.T. Siegelmann, B.G. Horne, C.L. Giles. Computational capabilities of recurrent NARX neural networks. In *IEEE Transactions on Systems, Man and Cybernetics* Part B: Cybernetics, 2: 208–228, 1997.
27. J.P.F. Sum, W.-K. Kan, G.H. Young. A note on the equivalence of NARX and RNN. *Neural Computing and Applications*, 8: 33–39, 1999.
28. M. Szczerba. *Sound Source Localization Based on Rough-Set Approach*. Technical Report, Sound & Vision Engineering Department, Technical University Gdansk, 2001.
29. H. Wang, P. Chu. Voice source localization for automatic camera pointing system in videoconferencing. In *Proceedings of the IEEE Workshop on Applications of Signal Processing to Audio and Acoustics*, Mohonk, New Paltz, NY, 1997.

30. R.J. Williams, D. Zipser. A learning algorithm for continually running fully recurrent neural networks. *Neural Computation*, 1: 270–280, 1989.
31. M. Zhang, M.H. Er. An alternative algorithm for estimating and tracking talker location by microphone arrays. *Journal of the Audio Engineering Society*, 44: 729–736, 1996.
32. I. Ziskind, M. Wax. Maximum likelihood localization of multiple sources by alternating projection. In *IEEE Transactions on Acoustics, Speech and Signal Processing*, 36: 1553–1560, 1988.
33. J.M. Zurada. *Introduction to Artificial Neural Networks*. West, St. Paul, MN, 1992.
34. A. Øhrn. *Discernibility and Rough Sets in Medicine: Tools and Applications*. Ph.D. Dissertation, NTNU Report 1999: 133, IDI Report (1999), Department of Computer and Information Science, Norwegian University of Science and Technology, Trondheim, 1999.

Chapter 21
An Approach to Imbalanced Data Sets Based on Changing Rule Strength

Jerzy W. Grzymala-Busse,[1] Linda K. Goodwin,[2] Witold J. Grzymala-Busse,[3] Xinqun Zheng[1]

[1] Department of Electrical Engineering and Computer Science, University of Kansas, Lawrence, KS 66045, USA
jerzy@lightning.eecs.ukans.edu

[2] Department of Information Services and the School of Nursing, Duke University, Durham, NC 27710, USA
goodw010@mc.duke.edu

[3] RS Systems, Inc., Lawrence, KS 66047, USA
jerzy@lightning.eecs.ukans.edu

Summary. This chapter describes experiments with a challenging data set describing preterm births. The data set, collected at the Duke University Medical Center, was large but at the same time many attribute values were missing. However, the main problem was that only 20.7% of the total number of cases represented the important preterm birth class. Thus, the data set was imbalanced. For comparison, we include results of experiments on another imbalanced data set, the well-known breast cancer data set.

Our approach to dealing with this imbalanced data set was to induce a rule set using our standard procedure: the LEM2 algorithm of the LERS rule induction system and then increase the rule strength for all rules describing preterm births by multiplying all such rule strengths by the same number called a strength multiplier. The rule strengths for any rule describing the majority class, full-term birth, remained unchanged. The optimal strength multiplier was determined experimentally using our optimality criterion: the maximum of the sum of sensitivity and specificity.

1 Introduction

Approximately one of every 10 infants is born preterm (premature). Preterm birth is the leading cause of death in infants, and those who survive frequently suffer from lifelong handicaps and require health care that costs about $1 million in the first year of life. Creasy and Herron [3] developed a manual preterm risk scoring tool that was widely used for nearly a decade but later evaluated as ineffective for accurate identification of most preterm births [4]. A decade of manual preterm risk-scoring tools yielded only 17–38% positive predictive values [10]; thus, data-driven decision-support tools are needed to improve diagnosis in this complex domain.

This chapter describes a series of experiments with preterm birth data provided by the Duke University Medical Center. Duke's, data set includes, a sample of, 19970 women that is ethnically diverse and includes 1229 variables. Duke's data subset was partitioned into two parts: training (14977 cases) and testing (4993 cases). The prenatal data set collected at the Duke University Medical Center is associated with many technical challenges. First of all, it is large. Second, the data set is missing many attribute values. For example, an average attribute of the training data set is not specified for 32.7% of cases. Even worse, for the two mutually disjoint subsets of the main set (1229 attributes) identified by experts from the Duke University Medical Center as important, the first set containing 52 attributes and the second set containing 54 attributes, named Duke-1 and Duke-2, respectively, are missing even more attribute values.

The Duke-1 data set contains laboratory test results. The Duke-2 data set represents the most essential remaining attributes that, according to experts, should be used in diagnosing is preterm birth. Duke-1 is missing 64.8% attribute values, Duke-2 is missing 36.1% attribute values.

There are many approaches to handling missing attribute values in data mining [5,8,11,15]. So far, we experimented with the closest fit algorithm for missing attribute values, based on replacing a missing attribute value with an existing value of the same attribute in another case that resembles, as much as possible, the case that is missing attribute values [8].

Furthermore, both data sets, Duke-1 and Duke-2, are imbalanced because only 3103 training cases are preterm; all remaining 11874 cases are full term. Similarly, in the testing data set, there are only 1023 preterm cases, whereas the number of full term cases is 3970. Since both data sets, Duke-1 and Duke-2, yield similar results, for brevity we will present results only for Duke-1.

The data sets are further complicated by numerical attributes. Usually data with numerical attributes are consistent, i.e., for any two cases with the same vectors of attribute values, the outcome is the same. This is not so with Duke's data set. Even with all 1229 attributes, the training data set is inconsistent. Thus, discretization, the process of converting numerical attributes into symbolic attributes, is a difficult problem for this preterm-birth data. Our solution is based on preserving the existing rate of conflicting cases (i.e., keeping the same inconsistency level). Following this approach, the numerical attribute values of the training data set were sorted for every attribute. Every value v was replaced by the interval $[v, w)$, where w was the next value bigger than v in the sorted list. This approach to discretization is very cautious since, in the training data set, we put only one attribute value in each interval. For testing data sets, values were replaced by the corresponding intervals taken from the training data set. It is possible that a few values come into the same interval. This method was selected to keep the same inconsistency level of the data.

2 Rule Induction

In our research, the main data mining tool was LERS (learning from examples based on rough sets), developed at the University of Kansas [6]. LERS has proven its applicability having been used for years by NASA Johnson Space Center (Automation and Robotics Division) as a tool to develop expert systems of the type most likely to be used in medical decision making on board the International Space Station. LERS was also used to enhance facility compliance under Sections 311, 312, and 313 of Title III, the Emergency Planning and Community Right to Know. The project was funded by the U. S. Environmental Protection Agency. System LERS was used in other areas as well, e.g., in the medical field to compare the effects of warming devices for postoperative patients and to assess preterm birth [16].

LERS handles inconsistencies using rough set theory. The main advantage of rough set theory, introduced by Pawlak in 1982 [12–14], is that it does not need any preliminary or additional information about data (like probability in probability theory, grade of membership in fuzzy set theory, etc.). In the rough set theory approach, inconsistencies are not removed from consideration. Instead, lower and upper approximations of the concept are computed. On the basis of these approximations, LERS computes two corresponding sets of rules: certain and possible, using algorithm LEM2 [6].

3 Classification

For classification of unseen cases, the LERS system uses a modified "bucket brigade algorithm" [2,9]. In this approach, the decision to which concept an example belongs is made by using two factors, *strength* and *support*. They are defined as follows: The strength factor measures how well the rule has performed during training. The second factor, support, is related to a concept and is defined as the sum of scores of all matching rules from the concept. The concept getting the largest support wins the contest.

In LERS, the strength factor is adjusted to be the *strength* of a rule, i.e., the total number of examples correctly classified by the rule during training. The concept C for which support, i.e., the sum of all strengths (R) for all matching rules R describing concept C is the largest, is a winner, and the example is classified as a member of C.

If an example is not completely matched by any rule, some classification systems use *partial matching*. The quasi-linear algorithm system (AQ15), during partial matching, uses a probabilistic sum of all measures of fit for rules [11]. In the original bucket brigade algorithm, partial matching is not considered a viable alternative for complete matching. The bucket brigade algorithm depends on a default hierarchy instead [9]. In LERS, partial matching does not rely on the user's input. If complete matching is impossible, all partially matching rules are identified. These are rules with at least one attribute-value pair matching the corresponding attribute-value pair of an example.

For any partially matching rule R, the additional factor, called *matching factor* (R), is computed. Matching factor (R) is defined as the ratio of the number of matched attribute-value pairs of a rule R with the case to the total number of attribute-value pairs of the rule R. In partial matching, concept C for which the sum of all products of matching factor (R) and strength (R) for all partially matching rules R describing concept C is the greatest, is the winner, and the example is classified as a member of C.

4 Sensitivity and Specificity

In many applications, e.g., the medical area, we distinguish between two classes, basic and complementary. The basic class is more important, e.g., in the medical area, it is defined as the class of all cases that should be diagnosed as affected by a disease or other medical condition, e.g., preterm birth.

The set of all correctly classified (preterm) cases from the basic class is called true-positives, incorrectly classified basic cases (i.e., classified as full term) are called false-negatives, correctly classified complementary (full term) cases are called true-negatives, and incorrectly classified complementary (full term) cases are called false-positives.

Sensitivity is the conditional probability of true-positives, given the basic class, i.e., the ratio of the number of true-positives to the sum of the number of true-positives and false-negatives.

Specificity is the conditional probability of true-negatives, given the complementary class, i.e., the ratio of the number of true-negatives to the sum of the number of true-negatives and false-positives.

5 Data Sets

In this chapter, we present the results of experiments on three data sets. The first two data sets were collected at the Duke University Medical Center. The only difference between these two data sets is the approach used to guess missing attribute values. The first data set, Duke-TR, was obtained from the original data set Duke-1 by splitting Duke-1 into training (75%) and testing (25%) data sets, then the training data set was preprocessed using a closest fit approach to missing attribute values. In the closest fit algorithm for missing attribute values, a missing attribute value is replaced by an existing value of the same attribute in another case that resembles, as much as possible, the case with the missing attribute values. To search for the closest fit case, we need to compare two vectors of attribute values of the given case with missing attribute values of a searched case. There are many possible variations of the closest fit idea. In this chapter we will restrict our attention to the closest fitting cases within the same class. This algorithm is a part of the system OOMIS. During the search, the entire training set within the same class is scanned; for each case, a proximity measure is computed; the case for which the proximity measure is the largest is the closest fitting case that is used to determine the missing attribute values.

The testing data set of Duke-TR is missing 64.5% attribute values. During matching of testing cases against rules in the classification process, missing attribute values are ignored for matching.

On the other hand, another data set, Duke-ALL, was obtained from the original data set Duke-1 – first, by preprocessing using a closest fit approach to missing attribute values and then, by splitting for training (75%) and testing (25%) data sets.

The quality of the first data set, Duke-TR, is poor, whereas the quality of the second set, Duke-ALL, is high; therefore, we experimented on a third data set of medium quality, also imbalanced (with 29.7% of cases in the basic class), called Breast-Cancer. This is a well-known data set accessible from the Data Repository of the University of California at Irvine. It is missing only 0.35% attribute values.

Table 1. Results of experiments

		Duke-TR	Duke-ALL	Breast-cancer
	Error rate in %	21.19	3.18	24.0
Initial	Sensitivity in %	0.59	85.54	33.33
	Specificity in %	98.97	99.67	93.01
	Average rule strength:			
	Basic class	29.42	23.85	2.46
	Complementary class	104.67	77.53	5.20
	Strength multiplier	5.552	44.0	4.0
	Error rate in %	44.98	2.90	35.0
Optimal	Sensitivity in %	53.40	96.73	61.40
	Specificity in %	61.29	97.19	66.43

The results of our experiment are cited in the Table 1 and presented in Figs. 1–6. In the charts in Figs. 4–6, series 1 represents sensitivity, series 2 represents specificity, and series 3 represents the error rate (the ratio of the total number of incorrectly classified cases from both classes to the total number of cases).

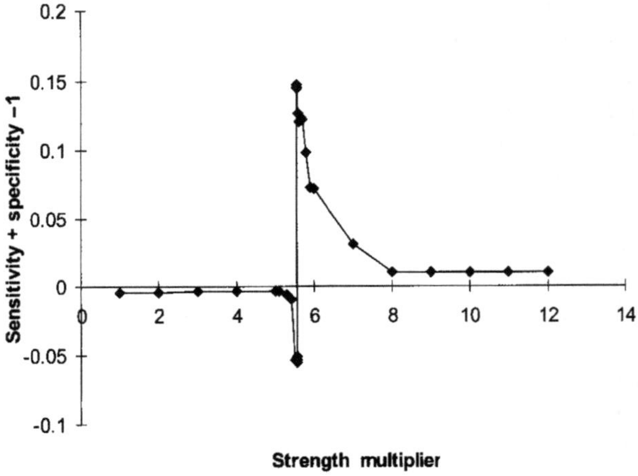

Fig. 1. Duke-TR data

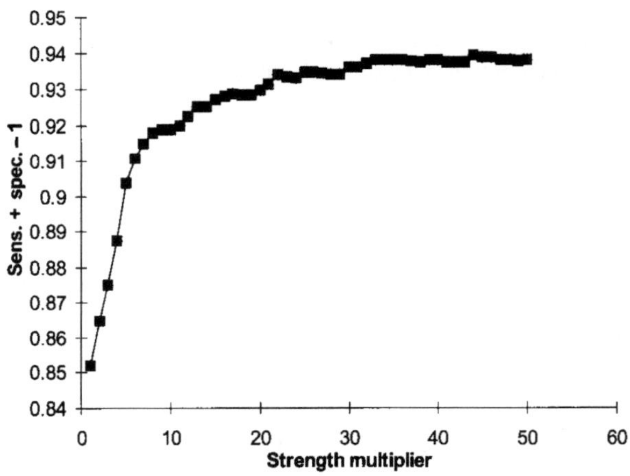

Fig. 2. Duke-ALL data

6 Strength Multipliers

In imbalanced data sets with two classes (concepts), one class is represented by the majority of cases, whereas the other class is a minority. Unfortunately, in medical data, the smaller class, as a rule, is more important. In Duke's perinatal training data, only 20.7% of the cases represent the basic class, preterm birth. During rule induc-

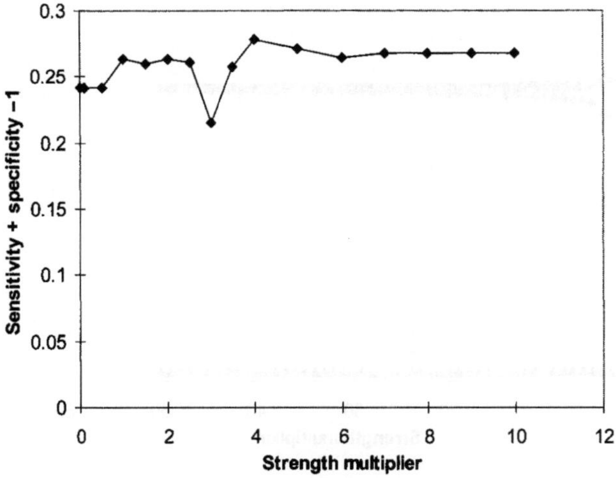

Fig. 3. Breast cancer data

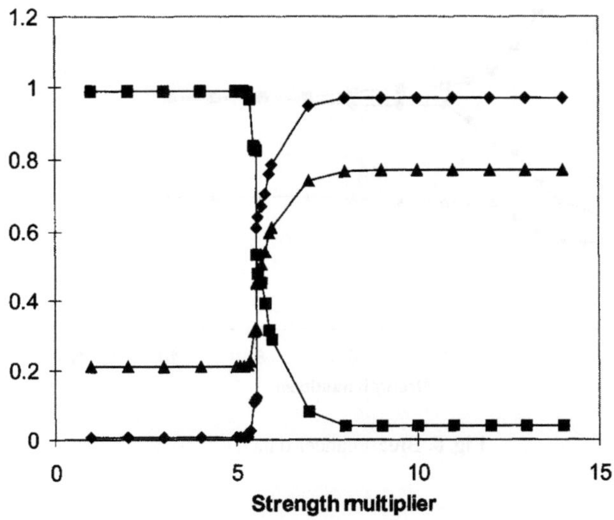

Fig. 4. Duke-TR data

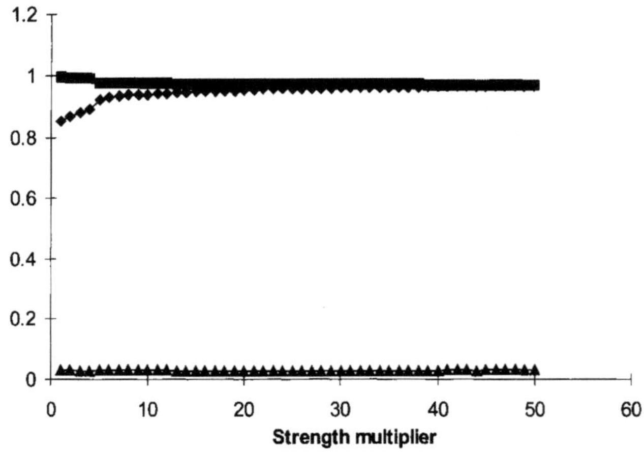

Fig. 5. Duke-ALL data

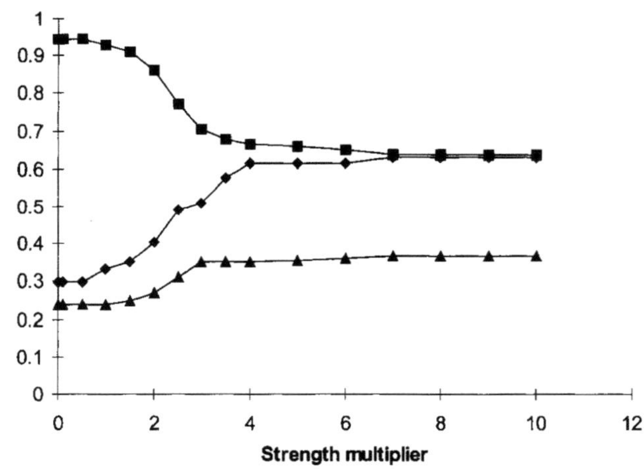

Fig. 6. Breast cancer data

tion, the average of all rule strengths for the bigger class is also greater than the average of all rule strengths for the more important but smaller basic class. During classification of unseen cases, rules matching a case and voting for the basic class are outvoted by rules voting for the bigger, complementary class. Thus, the sensitivity is poor, and the resulting classification system would be rejected by diagnosticians.

Therefore, it is necessary to increase the sensitivity. The simplest idea is to add cases to the basic class in the data set, e.g., by adding duplicates of the available cases. The total number of training cases will increase, hence the total running time of the rule induction system will also increase. However, adding duplicates will not change the knowledge hidden in the original data set, but it may create a balanced data set so that the average rule set strength for both classes will be approximately equal. The same effect may be accomplished by increasing the average rule strength for the basic class. In our research, we selected the optimal rule set by multiplying the rule strength for all rules describing the basic class by the same real number, called a *strength multiplier* [7].

In general, the sensitivity increases with the increase in the strength multiplier. At the same time, the specificity decreases. It is difficult to estimate the optimal value of the strength multiplier. In our experiments, the choice of the optimal value of the strength multiplier was based on an analysis presented by Bairagi and Suchindran [1]. Let p be a probability of the correct prediction, i.e., the ratio of all true-positives and all false-positives to the total number of all cases. Let P be the probability of an actual basic class, i.e., the ratio of all true-positives and all false-negatives to the total number of all cases. Then,

$$p = Sensitivity \times P + (1 - Specificity) \times (1 - P).$$

As Bairagi and Suchindran observed [1], we would like to see the change in p as large as possible with a change in P, i.e., we would like to maximize

$$\frac{dp}{dP} = Sensitivity + Specificity - 1.$$

Thus the optimal value of the strength multiplier is the value that corresponds to the maximal value of $Sensitivity + Specificity - 1$.

7 Conclusions

The results of our experiments show that specificity may be increased by changing strength multipliers for rules describing the basic class and by using the LERS classification system.

For poor quality data (Duke-TR), by increasing the strength multiplier (until it reaches an optimal value), a large increase in sensitivity (by 52.81%) is achieved, but specificity decreases significantly (by 37.68%); thus the total error rate increases (by 23.79%).

For high-quality data (Duke-ALL), under the same circumstances, a significant increase in sensitivity (by 11.19%) and a small decrease in specificity (2.48%) resulted in a decrease in the total error rate (by 0.28%).

And, finally, medium-quality data (breast cancer) are characterized by a large increase in sensitivity (by 28.07%) with a decrease in specificity (26.58%), and, at the same time, an increase in the total error rate (by 11%).

For many important applications, e.g., the medical area, an increase in sensitivity is crucial, even if it is achieved at the cost of specificity. Thus, the suggested method of increasing the strength multiplier may be successfully applied to machine learning from imbalanced data.

References

1. R. Bairagi, C.M. Suchindran. An estimator of the cutoff point maximizing sum of sensitivity and specificity. *Sankhya, Series B, Indian Journal of Statistics*, 51: 263–269, 1989.
2. L.B. Booker, D.E. Goldberg, J.F. Holland. Classifier systems and genetic algorithms. In J. G. Carbonell, editor, *Machine Learning: Paradigms and Methods*, 235–282, MIT Press, Cambridge, MA, 1990.
3. R.K. Creasy, M.A. Herron. Prevention of preterm birth. *Seminars in Perinatology*, 5: 295–302, 1981.
4. R.K. Creasy. Preterm birth prevention: Where are we? *American Journal of Obstetrics & Gynecology*, 168: 1223–1230, 1993.
5. J.W. Grzymala-Busse. On the unknown attribute values in learning from examples. In *Proceedings of the 6th International Symposium on Methodologies for Intelligent Systems (ISMIS'91)*, LNAI 542, 368–377, Springer, Berlin, 1991.
6. J.W. Grzymala-Busse. LERS — A system for learning from examples based on rough sets. In R. Słowiński, editor, *Intelligent Decision Support: Handbook of Applications and Advances of the Rough Sets Theory*, 3–18, Kluwer, Dordrecht, 1992.
7. J.W. Grzymala-Busse, L.K. Goodwin, X. Zhang. Increasing sensitivity of preterm birth by changing rule strengths. In *Proceedings of the 8th Workshop on Intelligent Information Systems (IIS'99)*, 127–136, Institute of Fundamentals of Computer Science of the Polish Academy of Sciences, Warsaw, 1999.
8. J.W. Grzymala-Busse, W.J. Grzymala-Busse, L.K. Goodwin. A closest fit approach to missing attribute values in preterm birth data. In *Proceedings of the 7th International Workshop on Rough Sets, Fuzzy Sets, Data Mining and Granular-Soft Computing (RSFDGrC'99)*, LNAI 1711, 405–413, Springer, Berlin, 1999.
9. J.H. Holland, K.J. Holyoak, R.E. Nisbett. *Induction: Processes of Inference, Learning, and Discovery*. MIT Press, Cambridge, MA, 1986.
10. M. McLean, W.A. Walters, R. Smith. Prediction and early diagnosis of preterm labor: A critical review. *Obstetrical & Gynecological Survey*, 48: 209–225, 1993.
11. R.S. Michalski, I. Mozetic, J. Hong, N. Lavrac. The AQ15 inductive learning system: An overview and experiments. Report number UIUCDCD-R-86-1260 of the Department of Computer Science, University of Illinois, 1986.
12. Z. Pawlak, J.W. Grzymala-Busse, R. Słowiński, W. Ziarko. Rough sets. *Communications of the ACM*, 38: 89–95, 1995.
13. Z. Pawlak. Rough sets. *International Journal of Computer and Information Sciences*, 11: 341–356, 1982.

14. Z. Pawlak. *Rough Sets: Theoretical Aspects of Reasoning about Data.* Kluwer, Dordrecht, 1991.
15. J.R. Quinlan. *C4.5: Programs for Machine Learning.* Morgan Kaufmann, San Mateo CA, 1993.
16. L.K. Woolery, J. Grzymala-Busse. Machine learning for an expert system to predict preterm birth risk. *J. Amer. Med. Inf. Assoc.*, 1: 439–446, 1994.

Chapter 22
Rough-Neural Approach to Testing the Influence of Visual Cues on Surround Sound Perception

Bożena Kostek

Gdańsk University of Technology, Sound and Vision Engineering Department,
Narutowicza 11/12, 80-952 Gdańsk, Poland
bozenka@sound.eti.pg.gda.pl

Summary. This chapter aims at revealing in which way and how surround sound interferes or is associated with visual context. Such parameters as distance, angle, or level of sound source were tested with and without a video image on the screen. For that purpose, a subjective testing was applied. Processing of the results obtained was done by employing genetic algorithms and combined neural network and rough set systems. The main task of the experiments was the application of modular neural networks to quantize surround sound parameter values. A rough set algorithm was used to make decisions showing the influence of visual cues on the perception of surround sound.

1 Introduction

The study presented shows the methodology of testing the influence of a video image on surround sound perception. Discovering such a relationship may result in formulating some rules for surround sound production accompanied by a video image. One can find references to the literature concerning audio-visual perception, but they are in most cases related to classical studies on this subject, including stereo sound systems for HDTV (high definition television) [4,10,15,34,35]. At present, digital video, film, or multimedia presentations are often accompanied by surround sound. Home theater systems win popularity. Meanwhile, only few researchers have made an effort to explore the influence of video on surround source perception and vice versa [2,6,19,28,36]. However, there is still no clear answer to the question how video influences the localization of virtual sound sources in multichannel surround systems, e.g., DTS (digital theater system), and in most references, one can find a list of problems to be solved while testing relevant intermodal relations [7,8,28,29]. Therefore, several problems should be addressed: What is the optimum width of the surround panorama for individual kinds of music? What kind of audio material should rear loudspeakers transmit? What changes in sound mix (if any) should be made when video zoom is modified? Such experiments should be based on subjective testing procedures [7,17], in which experts should listen to the sound without and with a video image and provide assessments.

Experiments related to examination of human perception of surround audio and accompanying visual cues may be divided into several categories: testing without or with a video image, and the low-level or high-level multimodal interrelation between an audio and video signal. The first category of tests may be used for calibration tasks. In the second category of experiments, one can use abstract audio-visual objects, e.g., synthesized graphical primitive objects (numbers, circles, lines, ping–pong ball, etc.) and synthesized artificial sounds. More complex objects may belong to the third category, for example, an audio–video recording of a speaker, a soloist, etc. The reason for using simple objects instead of complex ones is the need to discover and describe basic mechanisms underlying the audio-visual perception of human beings. In a simple way, one may examine the influence of the shape, color, and movement of a visual object on the localization of the sound in a surround system. On the other hand, high-level multimodal interrelation experiments employing more complex objects may be used as a basis for adjusting audio panorama settings during audio-visual postproduction. The results of such experiments can show in which cases and in what way video will affect the localization of virtual sound sources. In most cases, video "attracts" the attention of listeners, and as a consequence, they localize sound closer to the screen center. Therefore, this is called the image proximity effect or the sound localization shifting effect [7].

The main goal of this research study was to discover dependencies between reactions of sight and hearing due to the perception of visual stimuli accompanied by surrounding sound. To achieve this aim, a number of psychoacoustic experiments was conducted on a group of properly trained experts, and as a result of those experiments, a collection of data was created. Those data were then analyzed by modern techniques of intelligent data exploration and knowledge discovery. The problem concerns mainly finding hidden relations between semantic descriptors of subjective impressions (in the form of words). Thus, for the purpose of this research study, various soft computing techniques employing genetic algorithms, modular neural networks, and rough sets were used. A detailed description of the results of those experiments is given in this chapter.

2 Experimental Layout

2.1 Test Principles

Testing the influence of a video image on surround sound localization perception is a subjective process. It can be assigned to the category of object evaluation processes. This means that during the evaluation process a number of properly trained experts experienced in critical listening should take part by filling in a given questionnaire [17].

There are several issues to be checked during subjective test sessions. First of all, the software package along with the sound system should be calibrated because some phase shifting or delays may be present in electroacoustic channels. Also, tests should be carried out without a video image on the screen. Then, in the next step, a video image should appear together with sound. Other factors that should be checked are image size and image stability. Also, differences in sound level presentations may affect sound perception. Obviously, to maintain test reliability, subjects' preferences should also be taken into account. To that end, factors such as a subject's gender, experience, etc. are of importance. Since there is a large number of interrelated factors underlying tests, they will result in a huge number of experts' answers, in some cases contradictory to each other. To discover dependencies among data obtained, some techniques belonging to the soft computing domain will be necessary.

Subjective testing sessions took place in two rooms, which were acoustically separated (see Fig. 1). The expert's seat was positioned in the so-called "sweet spot", the best place for listening.

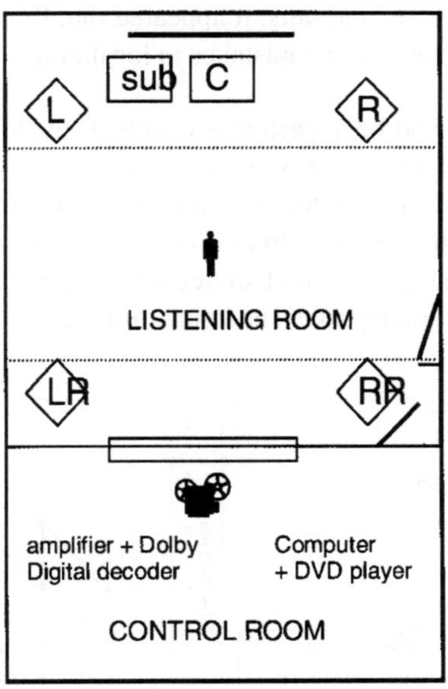

Fig. 1. Setup used during listening tests

The whole experimental setup consisted of a software package allowing for audio and video encoding, an AC-3 encoder, a computer with a built–in DVD player, an amplifier with a Dolby Digital decoder, a video projector, a screen, and loudspeakers working in a 5.1 system. During the tests, files were used with audio encoded in the AC-3 (Dolby Digital) format and video encoded in MPEG2 standard. All audio files contained five channels (without an ultra-low frequency channel) and a bit rate equal to 448 kbit/s. Sound files were prepared with Samplitude 2496 software and then exported to the AC-3 encoder (A.Pack). The video files were prepared with the Adobe Premiere software. All video files had resolution of 720 times 576 and relevant quality. It prevented the possible influence of video quality on experts' judgment [3,12,11]. After encoding, audio and video files were multiplexed into vob format files.

2.2 Test Procedure Setup

The first stage of experiments consisted of a series of five audio presentations without accompanying video. These tests, called mapping tests, aimed at checking both the correctness of the directional hearing of an expert and the expert's reliability. As was said, sounds were presented without an image — the screen was blank. If listeners did not localize sounds properly, their results not qualified for further processing. The number of experts participating in experiments was 34. This group consisted of staff members and students of the Sound and Vision Engineering Department. After statistically checking experts' answers, it appeared that four of them should be excluded from this group due to some mistakes in localizing sound arrival direction.

After the mapping tests had been performed another test took place. It was aimed at checking whether and how experts react to video presented synchronously with a sound sample. In addition, this test was performed to see whether loudspeakers can be replaced by phantom sources created in the software. In Fig. 2, two diagrams show the angle and distance of sound source localization with and without video employing both existing loudspeakers and phantom sound sources.

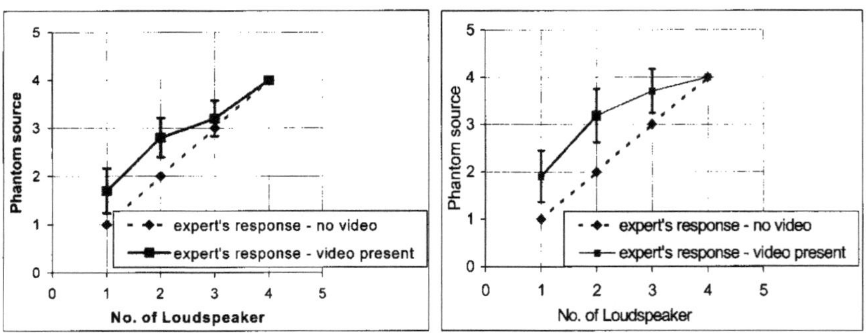

Fig. 2. Comparison of answers for two types of experiments: sound source angle localization shift caused by image appearance (left hand) and sound source distance localization shift caused by the image appearance (right hand); loudspeaker No. 4 was closest to the screen

The arrangement of loudspeakers in the experiment, the results of which are shown in Fig. 2, was as follows: four loudspeakers were aligned along the left-hand side of the screen (angle localization) and in the second case positioned between the listener and the screen (distance localization). In the first case, loudspeaker No. 1 was placed at the edge of the room, whereas the fourth one was positioned directly under the screen. In the second case, loudspeaker No. 1 was closest to the listener. While using phantom sources, the arrangement of loudspeakers was as shown before in Fig. 1. First, experts listened to sound samples (no video), and their task was to determine from which loudspeaker a particular sound sample was heard. Then, in the next phase of experiments, an object was displayed on the screen with a synchronously generated sound sample.

As seen from Fig. 2, results lying on the diagonal of the diagrams refer to the situation in which there was no video while listening to sound samples. On the other hand, there is a shift caused when the image appears. This means that the sound sample transmitted, for example, from loudspeaker No. 1 (the most distant from the screen) was perceived as the one transmitted closer to the screen. As indicated by the standard deviation measures in the diagrams, experts differently localized perceived sound samples, thus it may be concluded that this effect is both expert- and sound-type-dependent. On the other hand, experts have no difficulties while perceiving a sound sample either from a loudspeaker or listening to the phantom source.

In the experiments, apart from mapping tests, 10 abstraction tests and 15 or 20 high-level abstraction tests were presented to the experts. The audio-visual signals used in the experiments are presented in Table 1. They are both low and high level. The abstraction tests used simple objects instead of complex ones. That was due to the need to discover and describe basic mechanisms underlying audio-visual perception.

Table 1. Description of audio-visual signals used in tests

Low-level intermodal relations –abstraction tests	High-level intermodal relations –thematic tests
Video: blinking circle Audio: amplitude-modulated tone	Video: talking speaker Audio: speaker's voice
Video: circle with modulated colors Audio: amplitude-modulated tone	Video: musician playing solo Audio: musical excerpt
Video: bouncing ping–pong ball Audio: ping–pong ball sound	Video: musical band playing Audio: musical excerpt
Video: metronome Audio: metronome sound	Video: musical video clip Audio: music from a video clip
Video: vertical bar moving from right to left Audio: filtered noise	Video: film excerpt Audio: sound track from a movie

The questionnaire for assigning the arrival direction of the sound in the mapping tests is shown in Fig. 3. While considering a spherical space around the listener's head, one can sample it at different elevations (from below the horizontal plane to directly overhead). In addition, at each elevation, the full 360° of azimuth can be sampled in equally sized increments. A total of some hundreds of locations can be obtained in this way. However, in the experiments, only the horizontal plane was considered, of which the division of angles is seen in Fig. 3.

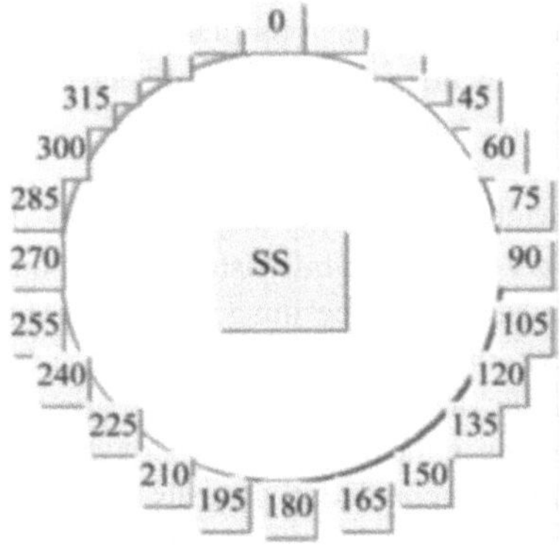

Fig. 3. Questionnaire for assigning direction of sound arrival

Even though surround sound systems allow creating phantom sound sources in a 360° range, using too many sound sources may introduce some errors, due to inaccuracy positioning in phantom sound sources . Thus, the number of sources was limited to the following angles: 0° (central loudspeaker), 22.5°, 45° (front right loudspeaker), 90°, 135° (rear right loudspeaker), 180°, 225° (rear left loudspeaker), 270°, 315° (front left loudspeaker), 338°. To increase the number of possible answers, experts could also choose other angles: 7.5°, 15°, 30°, 37.5°, 60°, 75°, 105°, 120°, 150°, 165°, 195°, 210°, 240°, 255°, 285°, 300°, 322.5°, 330°, 345°, and 352.5°.

Furthermore, to allow an expert to express a more spatial-like impression, not only those angle-oriented but also some angle-group oriented entities were added, such as L+C+R, wide central base (315°+0°+45°); WF, wide front base (315°+45°); WR, wide right base (45°+135°); WB, wide rear base (135°+225°); WL, wide left base (225°+315°); SS, Sweet Spot; ALL, all five channels playing simultaneously. In this way, attributes defining the sound domain space were assigned.

The visual domain space was described with only one attribute assigned to thematic tests indicating whether video was present or not. In the abstraction tests, several attributes were added describing, for example, how the line was moving on the screen: L2R, from left to right; R2L, from right to left; D2U, up; U2D, down. All of those given attribute sets served as a basis for determining the structure of decision rules discovered by the data mining system.

It is important to point out that the assumption was made that all parameters in both visual and sound domains could contain only binary data. This means that a given angle could be either completely included in the perception of a surrounding sound or completely excluded from it. Similarly, images could be used for a given test or not.

3 Genetic Algorithm-Based Processing of Listening Test Results

3.1 Knowledge Base

Methodology based on searching for repetitive patterns existing in data and association rule generation from those patterns was used for data mining in this research study. Data were represented as a simple information system. An example of a data record from the information system is shown in Fig. 4. A record in the abstraction test database (Fig. 4) contains values of one at the 4th, 11th, 13th, and 22nd positions (zeros elsewhere). This means that a sound stimulus presented at 90° (4th attribute) accompanied by an image (11th attribute) of a vertical line moving from the right to the left side of the screen was actually localized by an expert at 45° (22nd attribute).

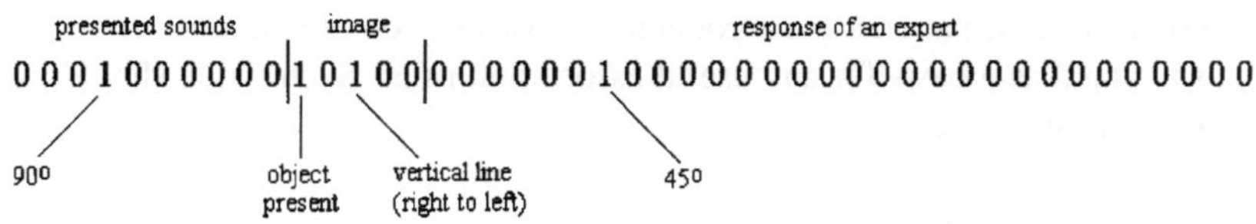

Fig. 4. Example of a record in the database (abstract test case)

After creating the appropriate data sets, it was possible to explore and analyze the data. The aim was to discover the influence of visual stimuli on the perception of a sound in surround space, thus, searching for association rules was performed. The genetic algorithm was employed for this task. Since genetic algorithms belong to the most often used soft computing methods, their principles will be not reviewed here.

In this research study, the chromosomes that are being produced and modified during the evolutionary process represent patterns covering records in the data set. Each of them has the length of the number of attributes describing the data (specific for the type of the tests, abstraction vs. thematic), and the alleles of the chromosome are constrained by the domains of those attributes. An allele of such a chromosome can either contain a value that is valid for a corresponding attribute in the data set (in this case ones, all zeros can be omitted since such testing is aimed at interrelation of angle and image) or a "don't care" asterisk which means that this attribute is not important and will not be used to generate a rule [8]. An example of a chromosome is presented in Fig. 5. Each of such patterns has possible coverage in the

$$\{***1******1*1********1***********************\}$$

Fig. 5. Example of a chromosome (set positions: 4th, 11th, 13th, 22nd)

data (support) which is given by the number of records matching the pattern (i.e., having given values at the set position). For the above example, it will be all records containing '1' at the 4th, 11th, 13th, and 22nd positions, regardless of other values. Obviously, one should look for patterns that have relatively high support, and this can form the basis of the fitness function used for this algorithm. The desired level of support in data can be adjusted by setting the epsilon value, which stands for the percentage of maximum allowed error in terms of pattern coverage (the higher the epsilon, the lower the minimum support required) [8,32].

Although the support of a pattern is a basic feature of the fitness function implemented in the algorithm, it cannot be its ultimate characteristic. The number of "set" positions (not the "don't care" asterisks) is also very important. For example, a pattern consisting only of asterisks will gain the support of 100% of the data records, but it has no meaning in terms of knowledge discovery. The structure of the IF-THEN rules generated afterward is also very important, and from the practical point of view, patterns must contain at least two (or even three) set attribute values to stand as a basis for any useful association rules. Such a rule should have the following structure:

$$\{\text{presented sound}\} \cup \{\text{image}\} \Rightarrow \{\text{response of an expert}\}.$$

Obviously, not all chromosomes will have physical coverage in the available data set. Some of them (especially those with a relatively large number of set positions) might not have support at all; however, some parts of them (subsets of values) still can be very useful and after an application of some genetic operators (i.e., crossover and mutation), may produce the desired result. It is crucial then to appropriately treat all those chromosomes and assign them some "credit" in terms of the fitness function, even though they do not have support in the data as a whole.

Based on the above discussion, all chromosomes (potential solutions) should be awarded or punished according to the criteria during the evolutionary process. Thus, the fitness function can be completely described as a multilayer estimate of the fitness of the chromosomes in terms of their partial support in the data at first, and then total coverage of the data weighted by the number of set positions.

Another very important feature of the genetic algorithm used here is a multipoint crossover option. In many experiments mining patterns in different types of data, this approach was found much more effective with regard to both the number of patterns discovered and the time of convergence. On the basis of empirical premises, the maximum number of cuts (crossover points) was set to 1 for every 10 attributes. In the example given in Fig. 6, there are three crossover points, and the arrows point to the genetic material that will be exchanged and thus will create two new chromosomes.

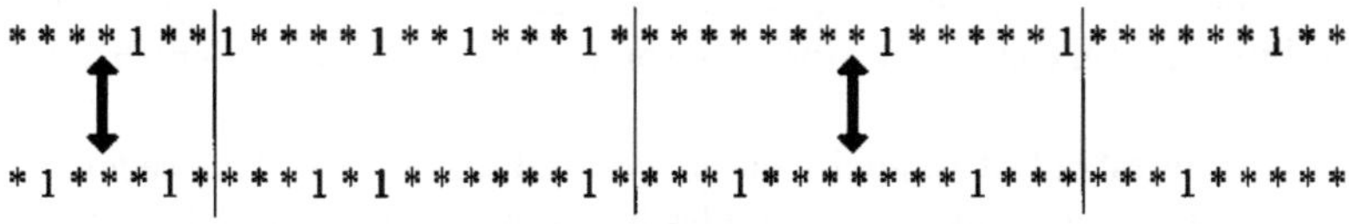

Fig. 6. Example of a multipoint crossover (three-point)

As an outcome of several evolutions modeled by this genetic algorithm, a set of data patterns was created. Those patterns along with the information about the level of their support were then used as input to the application generating association rules. Association rules determine the existence of some relations between attributes in data or values of those attributes. Basically, they are simple IF-THEN type rules that, for binary domain of values, can be considered statements [8]:

IF attributes from the premise part of the rule have values of one, THEN the attributes included in the consequent part also tend to have values of one.

In the case discussed, rules should be of the following type:

IF a given set of angles was used to reproduce a sound and an image was/was not present, THEN the experts tended to localize the sound source at a particular angle/set of angles.

Association rules are characterized both by their support in data (number of cases to which a given rule applies, or how "popular" the rule is) and the confidence (the ratio of the support of the rule to the number of cases that contains its premise part, revealing how sure one can be that the rule is correct by judging on the basis of the values from the premise part of the rule).

An algorithm of searching for association rules consists of two parts: searching for patterns hidden in data (in this project, this was achieved initially by applying the genetic algorithm) and generating rules based on those patterns. The idea of the algorithm for rule generation in this research study is relatively simple. Basically, it takes "not asterisk" values of each of the patterns, divides them into subsets, and by moving those subsets from the premise to the consequent part (according to the specified constraints), creates all possible rules based on the given pattern. The algorithm is quite resource consuming; thus it removes all records that are covered by any others (i.e., those that are subsets of another set). This decreases the computational complexity of the algorithm and, together with the support and confidence parameter, limits the number of rules generated.

3.2 Pattern Searching

To increase the variety of patterns, the algorithm was launched on several computers simultaneously. Because of the randomness aspects of genetic algorithms, the results differed from each other. However, some of the results were duplicated.

The support threshold of desired patterns was reduced to 5%. This seems to be extremely low, but it is valid because rules based on patterns with relatively small support in data still may have quite a large level of confidence. As a result of several evolutions of the genetic algorithm, 806 distinct patterns for the abstraction test and 890 for the subject test were found. Some of those patterns were characterized by including set values only in the range of generated locations (angles), not those that were responses from experts. This was quite obvious, taking into consideration the fact that a big part of the tests generated consisted of different angles at the same time, e.g., WF, L+C+R, etc. (see the description of the sound space), and an appropriately engineered algorithm should definitely find them. However, some patterns that were satisfactory in terms of the rule definition were also discovered (i.e., they consisted of generated locations that were perceived by an expert, as well as the information about the image presented). Some examples of those patterns are given below.

(Abstraction tests; ABSTRACT space, 45 attributes), support [18/300]:

$$\{1*1***1*1*******1**********1***1*****1*******\}.$$

(Thematic tests; THEMATIC space, 41 attributes), support [16/465]:

$$\{11***11*1**************1*****************\}.$$

3.3 Rule Generation

Patterns discovered and prepared in the previous step were then used as a basis for associative rule generation. At this level, sets of attributes were divided into premise (angles generated along with the information about an image) and consequent (response from an expert) parts. After removing duplicated patterns, 49 effective patterns for abstraction and 23 for thematic tests were preserved. On the basis of this final set of patterns, a number of rules of given support and confidence was generated. Some rules indicated a lack of any influence of the image on the perceived localization of sound, and this was usually connected to sounds perceived behind the listener. Nevertheless, most rules proved an existing interrelation between the auditory and visual senses. A sample of such a rule is presented below for the abstraction tests. This is a rule with a clear indication of audio-visual dependencies:

IF i045=1 AND i135=1 AND P=1, THEN 45=1 [s=6%] [c=66%].

(IF sound is presented at angles of 45° AND 135° AND the image is present, THEN the perceived angle is 45° WITH support of 6% AND confidence of 66%.)

IF i225=1 AND i315=1 AND P=1 AND D2U=1, THEN 315=1 [s=4%] [c=75%].

(IF sound is presented at angles of 225° and 315° AND there is an image of a horizontal line moving from down to up, THEN the perceived angle is 315° WITH support of 4% AND confidence of 75%.)

Based on the experiments performed it may be concluded that rules generated by the genetic algorithm proved the existence of the proximity effect while perceiving sound in the presence of a video image. However, the support for these rules is so low that it is difficult to conclude whether these rules are valid, even if the confidence related to such rules is quite high. That is why, in the next section, another approach to processing data obtained in subjective tests will be presented. It concerns a hybrid system consisting of a modular neural network and rough set based inference system.

4 Rough-Neural Hybridization

4.1 Hybrid Neural Networks

The enumerated applications of artificial neural networks to various fields drove development in theory. Due to this fact, some new trends in this domain appeared. One of these trends involves the compound structure of neural networks, so-called hierarchical neural networks. The basic network structure is composed of numbers of subnetworks. These subnetworks have a common input layer. Their middle layers are independent of one another. Every subnetwork has an assigned output node [18].

Another trend that differs much from the all-class-one-network is the modular neural network concept. In this case, information supplied by the outputs of subnetworks can be fused by applying either the fuzzy or rough set approach. Hybrid methods have been developed by integrating the merits of various paradigms to solve problems more efficiently. It is often pointed out that hierarchical or modular neural networks are especially useful while discussing complex classification tasks involving a large number of similar classes. In such a case, one can refer to some sources that appeared recently in the literature [5,20,21,30,33].

Feature subset selection by neuro-hybridization was presented as one of the most important aspects in machine learning and data mining applications by Chakraborty [5]. He engineered the neuro-rough hybrid algorithm that uses rough set theory in the first stage to eliminate redundant features. Then, a neural network used in the second stage operates on a reduced feature set. On the other hand, Auda and Kamel proposed a modular neural network that consists of an unsupervised network to decompose the classification task across a number of neural subnetworks. Then, information from the outputs of such modules are integrated via a multimodule decision making strategy that can classify a tested sample as "vague class" or the boundary between two or more classes [1].

The paper by Peters et al. reviews the design and application of neural networks with two types of rough neurons: approximation neurons and decider neurons [26]. The paper particularly considers the design of rough-neural networks based on rough membership functions, the notion introduced recently by Pawlak and Skowron [25]. A so-called rough membership neural network consists of a layer of approximation neurons that construct rough sets. The output of each approximation neuron is computed with a rough membership function. Values produced by the layer of approximation neurons provide condition vectors. The output layer is built of a decider neuron that is stimulated by each new condition vector. A decider neuron compares the new condition vector with existing ones extracted from decision tables and returns the best fit. The decider neuron enforces rules extracted from decision tables. Information granules in the form of rules are extracted from decision tables using the rough set method [26].

Other approaches based on modular and complex integral neural networks are also widely used in various problems as robust search methods, especially for uncertainty and redundancy in data [24,27,31].

4.2 Rough-Neural System Principles

As mentioned before, at least three factors should be taken into account while testing surround sound perception accompanied by video: sound arrival angle, distance, and sound level. It is obvious that these three might be interrelated; however, as was shown in a previous study employing subjective tests based on fuzzy logic technique [17], it was sufficient to work with a single function separately and then to interrelate these factors in some rule premises [7]. Rule premises contained the above mentioned factors and assessed descriptors assigned to them during subjective testing sessions, and the consequence (decision) resulted from these test data. However, rules that were formulated were experts hard to verify. Therefore, in this study, a new concept of rule discovery was conceived. For this purpose, a modular rough-neural system was engineered that is described further in (Fig. 7).

As seen in Fig. 7, the two main blocks of the rough-neural system are related to data processing. These are neural network modules that quantize numerical data and the

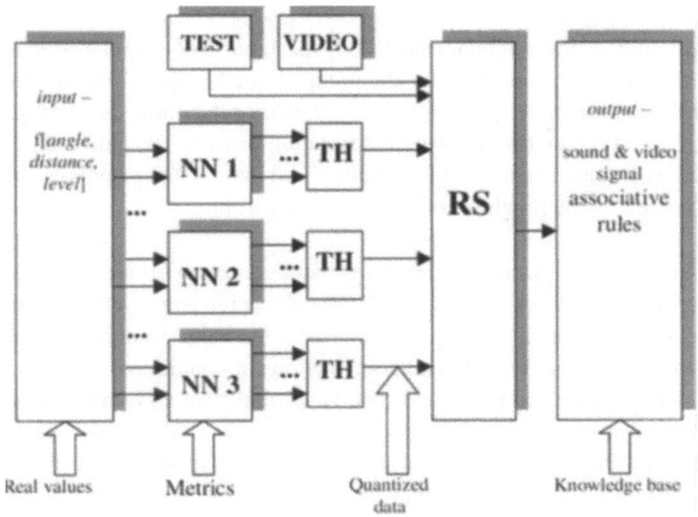

Fig. 7. Rough-neural system layout

rough set based engine that extracts rules from data. The elements of the input vector shown in Fig. 7 are numbers representing the realm of angles, distances, and sound loudness values, whereas a rough set-based decision system requires quantized data. Consequently, for quantization purposes, a self-organizing neural net was proposed, similar to other experiments [9]. For this purpose, the self-organizing map (SOM) introduced by Kohonen was chosen [13]. This is one of the best known neural network clustering algorithms, which assigns data to one of specified subsets according to clusters detected in a competitive learning process. During this learning, only the weight vector that is most similar to a given input vector is accepted for weight building. Since data can be interpreted as points in a metric space, each pattern is considered an N-dimensional feature vector, and patterns belonging to the same cluster are expected to possess strong interval similarity, according to the measure of similarity chosen. Typically, the Euclidean metric is used in SOM implementations [13].

Using the SOM as a data quantizer, a scalar and a vector quantization can be taken into account. In the first case, the SOM is supplied to a single element of the key vector. In the second case, a few attributes can constitute input vectors, which reduces the number of attributes and helps to avoid a large number of attribute combinations in the rough set inference. The SOM of the Kohonen type defines mapping of N-dimensional input data onto a two-dimensional regular array of units, and the SOM operation is based on competition between the output neurons due to any stimulation by the input vector [13]. As a result of the competition, this cth output unit wins, provided that the following relations are fulfilled.

In the structure of the implemented SOM, the input and output nodes are fully connected, whereas the output units are arranged in a hexagonal lattice. The initial value for the learning rate $\eta^{(0)}$ was equal to 0.95. For the purposes of rough-neural hybridization, at the end of the weight adaptation process, the output units should be labeled with some symbols to assign quantized input data to symbols, which are to be processed by the rough set rule induction algorithm.

The engineered rule induction algorithm is based on rough set methodology that is well described in the literature [16,22,23]. The algorithm employed aimed to reduce the computational complexity [9]. This concerns the values reduction of attributes and searching for reducts, so that all combinations of conditional attributes are analyzed at a reasonable computational cost. Particularly, for a given sorted table, the optimum number of sets of attributes A ($A \subseteq C = \{a_1, \ldots, a_i, \ldots, a_{|C|}\}$), subsets of conditional attributes C, can be analyzed using a special way of attribute sorting [9]. The algorithm splits the decision table into two tables consisting only of certain rules and uncertain ones. There is additional information associated with every object in them. The information concerns the minimal set of indispensable attributes and the rough measure μ_{RS}. The latter case is applied only for uncertain rules. More details corresponding to the rough set algorithm can be found in the literature [9].

4.3 Experiments

The results from test sessions gathered in a database (see Table 2) were then further processed. The type of test provides, therefore, one of the attributes that are contained in the decision table. Other attributes included in the decision table are *angle*, *distance*, *level*, *video* and a decision attribute called the *proximity effect*. To differentiate between attributes resulting from experts' answers and actual values of angle, distance, and level known to the experimenter, two adjectives, namely, *subjective* and *objective* were added to attribute names and from this six attributes resulted that are contained in Table 2.

As mentioned before, during the test session, experts were asked to fill in questionnaires, an example of which was shown in Fig. 3. As there are numerical values gathered using questionnaires, values indicated by experts then form a feature vector that is fed to the neural network (NN) modules. The neural network module assigns each numerical value indicated by an expert to one of the clusters corresponding to semantic descriptors. The selection of the strongest output of the neural network is done by adding a threshold function operating in the range of (–1,1) to the system shown in Fig. 7. These threshold filters connected to outputs of the NN can be realized in practice by the output neuron transfer function. Their role is to choose only the strongest values obtained in the clustering process. Therefore, Table 2 contains descriptors resulting from neural network based quantization related to *angle subjective*, *distance subjective*, and *level subjective* attributes. Semantic descriptors related to the *angle subjective* attribute are as follows: *none*, *front*, *left front*, *left*, *left rear*, *rear*, *right rear*, *right*, *right front*. All but one is an obvious descriptor. The *none* descriptor is related to the case when distance equals zero, thus, a phantom source is positioned in the so-called *sweet spot* (expert's head position). This means that sound is subjectively perceived as directly transmitted to the head of an expert so there is no angle of its arrival defined. In addition, *distance subjective* is quantized by the NN module as *none* (*sweet spot* location), *close* (large distance from the screen), *medium*, *far* (smaller distances from the screen), and correspondingly *level subjective* is denoted as *low*, *medium*, *high*.

Table 2. Decision table (fragment)

Experts' answers	*Angle subj.*	*Distance subj.*	*Level subj.*	*Angle object.*	...	*Test*	*Video*	*Decision -proximity effect*
e_1	*Front*	*Close*	*Medium*	0°	...	*Abstract.*	*Static*	*No shift*
e_2	*Left front*	*Close*	*High*	60°	...	...	...	*Medium shift*
...	...	...	...	...	...	...	...	...
e_n	*Left rear*	*Far*	*Medium*	315°	...	*Thematic*	*Dynamic*	*Strong shift*

Values of angle, distance, and level of the phantom sound source are given numerically by the experimenter; however, they are quantized values (angles in degrees, distance in centimeters, and level in dB). The range of angle attribute was already shown in Sect. 2.2. The quantization resolution of angles and distance was directly related to the practically available resolution of phantom sound sources in the applied Samplitude 2496 software. Level values were quantized in the range of 50 to 100 dB with 10 dB steps. The problem of quantization of level, distance, and level attributes is further complicated because of some acoustical principles, which will not be reviewed here. The values of these attributes were left in numerical form, because in this case, rules will be easily understandable. On the other hand, descriptors related to *test*, *video* attributes, and the proximity effect attribute were set as semantic descriptors. Therefore *sound* and *video* attributes can have values such as *abstraction*, *thematic*, and correspondingly *no video*, *static*, *dynamic*. The decision attribute can be read as *no shift*, *slight shift*, *medium shift* and *strong shift*, and these descriptors will appear in the consequence part of a rule. Some exemplary rules that can be derived from the decision table are presented here:

IF *angle subjective* = *front* AND *distance subjective* = *close* AND *level subjective* = *medium* AND *angle objective* = 0° AND *distance objective* = *20* AND *level objective* = *70* AND *test* = *abstraction* AND *video* = *static*, THEN *proximity effect* = *no shift*.

IF *angle subjective* = *left front* AND *distance subjective* = *close* AND *level sub-*

jective = *medium* AND *angle objective* = 60° AND *distance objective* = 20 AND *level objective* = 70 AND *test* = *thematic* AND *video* = *static*, THEN *proximity effect* = *slight shift.*

IF *angle subjective* = *front* AND *distance subjective* = *far* AND *level subjective* = *high* AND *angle objective* = 45° AND *distance objective* = 20 AND *level objective* = 90 AND *test* = *abstraction* AND *video* = *static*, THEN *proximity effect* = *strong shift.*

Rules that will have a high value of the rough set measure can be included in the knowledge base to be used for investigating psychological principles of sound and vision interaction.

5 Conclusions

The subjective listening tests proved that visual objects could influence the subjective localization of sound sources. Measurement data showed that visual objects may "attract" listeners' attention; thus, in some cases, sound sources are then localized closer to the screen. It was found that the image proximity effect is listener dependent, which is probably related to some psychological processes occurring in individual human brains. As seen from the concepts and experiments presented, numerical values and subjective descriptors gathered in the decision table can be processed by the hybridized rough-neural algorithm. In this way, a concept of simultaneous computing with numerical data and with words was applied allowing for processing data obtained from both objective values and their subjective counterparts. On the basis of the experiments described in this chapter and the opinions of experts taking part in them, it can be stated that subjective listening tests and soft computing processing of their results seem appropriate for analyzing of hearing and sight hidden relations. They create an environment for automatic exploration of data derived from psychoacoustic experiments with surround sound and accompanying vision, employing knowledge discovery based on soft computing oriented methodologies. The results of such experiments and their analysis could yield recommendations to sound engineers producing surround movie sound tracks, digital video, and multimedia content.

Acknowledgments

The research is sponsored by the Committee for Scientific Research, Warsaw, Poland, Grant No. 8 T11D 00218 and No. 4 T11D 014 22.

References

1. G. Auda, M. Kamel. A modular neural network for vague classification. In *Proceedings 2nd International Conference on Rough Sets and Current Trends in Computing (RSCTC 2000)*, LNAI 2005, 584–589, Springer, 2001.

2. S. Bech, V. Hansen, W. Woszczyk. Interactions between audio-visual factors in a home theater system: Experimental results. In *Proceedings of the 99th Audio Engineering Society Convention*, Preprint No. 4096, New York, 1995.
3. J.G. Beerends, F.E. de Caluwe. The influence of video quality on perceived audio quality and vice versa. *Journal of the Audio Engineering Society*, 47(5): 355–362, 1999.
4. M. Brook, L. Danilenko, W. Strasser. Wie bewertet der Zuschauer das stereofone Fernsehes? In *Proceedings of the 13 Tonemeistertagung; Internationaler Kongres*, 367–377, 1984.
5. B. Chakraborty. Feature subset selection by neuro-rough hybridization. In *Proceedings of the 2nd International Conference on Rough Sets and Current Trends in Computing (RSCTC 2000)*, LNAI 2005, 519–526, Springer, Berlin, 2001.
6. A. Czyzewski, B. Kostek, P. Odya, S. Zielinski. Influence of visual cues on the perception of surround sound. In *Proceedings of the 139th Meeting of the Acoustical Society of America*, Vol. 5 of *Journal of the Acoustical Society of America*, 107, 3aPP14, 2000.
7. A. Czyzewski, B. Kostek, P. Odya, S. Zielinski. Determining influence of visual cues on the perception of surround sound using soft computing. In *Proceedings of the 2nd International Conference on Rough Sets and Current Trends in Computing (RSCTC 2000)*, LNAI 2005, 545–552, Springer, Berlin, 2001.
8. A. Czyzewski, B. Kostek, P. Odya, T. Smolinski. Discovering the influence of visual stimuli on the perception of surround sound using genetic algorithms. In *Proceedings of the 19th International Audio Engineering Society Conference*, 287–294, Germany, 2001.
9. A. Czyzewski, R. Krolikowski. Neuro-rough control of masking thresholds for audio signal enhancement. *Neurocomputing: An International Journal*, 36: 5–27, 2001.
10. M.B. Gardner. Proximity image effect in sound localization. *Journal of the Acoustical Society of America*, 43: 163, 1968.
11. M.P. Hollier, R. Voelcker. Objective performance assessment: Video quality as an influence on audio perception. In *Proceedings of the 103rd Audio Engineers Society Convention*, Preprint No. 4590, New York, 1997.
12. J. Kaminski, and M. Malasiewicz. *Investigation of Influence of Visual Cues on Perceived Sound in the Surround System.* Master's thesis, Sound and Vision Engineering Department, Technical University of Gdansk, 2001 (in Polish).
13. T. Kohonen. *The Self-Organizing Map. Proceedings IEEE*, 78: 1464–1477, 1990.
14. T. Kohonen, E. Oja, O. Simula, A. Visa, J. Kangas. Engineering applications of the self-organizing map. *Proceedings IEEE*, 84: 1358–1384, 1996.
15. S. Komiyama. Subjective evaluation of angular displacement between picture and sound directions for HDTV sound systems. *Journal of the Audio Engineering Society*, 37: 210, 1989.
16. J. Komorowski, Z.Pawlak, L. Polkowski, A. Skowron. Rough sets: A tutorial. In S. K. Pal, A. Skowron, editors, *Rough Fuzzy Hybridization: A New Trend in Decision-Making*, 3–98, Springer, Singapore, 1999.
17. B. Kostek. *Soft Computing in Acoustics, Applications of Neural Networks, Fuzzy Logic and Rough Sets to Musical Acoustics.* Physica, Heidelberg, 1999.
18. Z. Liqing. A new compound structure of hierarchical neural networks. In *Proceedings of the IEEE World Congress on Computational Intelligence (ICEC'98)*, 437–440, Anchorage, AK, 1998.
19. D.J. Meares. Perceptual attributes of multichannel sound. In *Proceedings of the Audio Engineering Society 12th International Conference*, 171–179, Copenhagen, 1993.
20. S. Mitra, S.K. Pal, M. Banerjee. Rough–fuzzy knowledge-based network – A soft computing approach. In S.K. Pal, A. Skowron, editors, *Rough–Fuzzy Hybridization: A New Trend in Decision-Making*, 428–454, Springer, Heidelberg, 1999.

21. Y. Pan, H. Shi, L. Li. The behavior of the complex integral neural network. In *Proceedings of the 2nd International Conference on Rough Sets and Current Trends in Computing (RSCTC 2000)*, LNAI 2005, 624–631, Springer, Berlin, 2001.
22. Z. Pawlak. Rough sets. *International Journal of Computer and Information Sciences*, 11(5): 341–356, 1982.
23. Z. Pawlak. *Rough Sets: Theoretical Aspects of Reasoning about Data.* Kluwer, Dordrecht, 1991.
24. Z. Pawlak. Rough sets and decision algorithms. In *Proceedings of the 2nd International Conference on Rough Sets and Current Trends in Computing (RSCTC 2000)*, LNAI 2005, 30–45, Springer, Berlin, 2001.
25. Z. Pawlak, A. Skowron. Rough membership functions. In R.R. Yager, M. Fedrizzi, J. Kacprzyk, editors, *Advances in the Dempster–Shafer Theory of Evidence*, 251–271, Wiley, New York, 1994.
26. J.F Peters, A. Skowron, L. Han, S. Ramanna. Towards rough neural computing based on rough membership functions: Theory and application. In *Proceedings of the 2nd International Conference on Rough Sets and Current Trends in Computing (RSCTC 2000)*, volume 2005 of *Lecture Notes in Artificial Intelligence*, 611–618, Springer, Berlin, 2001.
27. L. Polkowski, A. Skowron. Rough-neuro computing. In *Proceedings of the 2nd International Conference on Rough Sets and Current Trends in Computing (RSCTC 2000)*, LNAI 2005, 57–64, Springer, Berlin, 2001.
28. N. Sakamoto, T. Gotoh, T. Kogure, M. Shimbo. Controlling sound-image localization in stereophonic reproduction. *Journal of the Audio Engineering Society*, 29(11): 794–798, 1981.
29. N. Sakamoto, T. Gotoh, T. Kogure, and M. Shimbo. Controlling sound-image localization in stereophonic reproduction: Part II. *Journal of the Audio Engineering Society*, 30(10): 719–721, 1982.
30. M. Sarkar, B. Yegnanarayana. Application of fuzzy-rough sets in modular neural networks. In S. K. Pal, A. Skowron, editors, *Rough–Fuzzy Hybridization: New Trend in Decision-Making*, 410–427, Springer, Heidelberg, 1999.
31. A. Skowron, J. Stepaniuk, J.F. Peters. Approximation of information granule sets. In *Proceedings of the 2nd International Conference on Rough Sets and Current Trends in Computing (RSCTC 2000)*, LNAI 2005, 65–72, Springer, Berlin, 2001.
32. T. Smolinski, T. Tchorzewski. *A System of Investigation of Visual and Auditory Sensory Correlation in Image Perception in the Presence of Surround Sound.* Master's thesis, Polish-Japanese Institute of Information Technology, Warsaw, 2001 (in Polish).
33. M.S. Szczuka. Rough sets and artificial neural networks. In L. Polkowski, A. Skowron, editors, *Rough Sets in Knowledge Discovery 2. Applications, Case Studies and Software Systems*, 449–470, Physica, Heidelberg, 1998.
34. G.J. Thomas. Experimental study of the influence of vision on sound localization. *Journal of Experimental Psychology*, 28: 163–177, 1941.
35. M. Wladyka. *Examination of Subjective Localization of Two Sound Sources in Stereo Television Picture.* Master's thesis, Sound Engineering Department, Technical University of Gdansk, 1987 (in Polish).
36. W. Woszczyk, S. Bech, V. Hansen. Interactions between audio-visual factors in a home theater system: Definition of subjective attributes. In *Proceedings of the 99th Audio Engineers Society Convention*, Preprint No. 4133, New York, 1995.

Chapter 23
Handwritten Digit Recognition Using Adaptive Classifier Construction Techniques

Tuan Trung Nguyen

Institute of Mathematics, Warsaw University, ul. Banacha 2, 02-097 Warsaw, Poland
nttrung@mimuw.edu.pl

Summary. Optical character recognition (OCR) is a classic example of a decision making problem where class identities of image objects are to be determined. This concerns essentially finding a decision function that returns the correct classification of input objects. This chapter proposes a method of constructing such functions by using an adaptive learning framework based on a multilevel classifier synthesis schema. The schema's structure and the way classifiers on a higher level are synthesized from those on lower levels are subject to an adaptive iterative process that allows learning from input training data. Detailed algorithms and classifiers based on similarity and dissimilarity measures are presented. Also, results of computer experiments using the techniques described on a large handwritten digit database are included as an illustration of the application of the proposed methods.

1 Introduction

Optical character recognition (OCR) is an important field in pattern recognition (PR), where a class assignment function for input character images is to be determined. Historically, pattern recognition algorithms can be grouped within two major approaches: statistical (or decision-theoretical), which assumes an underlying and quantifiable statistical basis for generating a set of characteristic measurements from the input data that can be used to assign objects to one of *n* classes, and syntactic (or structural), which favors the interrelationships or interconnections of features that yield important structural descriptions of the objects concerned. Both approaches are widely used in pattern recognition in general, but in the particular field of optical character recognition, the structural approach, especially methods based on trees and attributed graphs, is gaining popularity [7].

Typically, a structural OCR system attempts to develop a descriptive language that can be used to reflect the structural characteristics of the input image objects. Once such a language has been established, it is used to describe the characteristic features of the target recognition classes so that new images can be assigned to one of them when checked against those features. Most existing systems employ some kind of hierarchical descriptions of complex patterns built from *primitives*, elemental blocks that can be extracted directly from input data. However, the vast majority of such systems assume a priori knowledge or a model of *primitives* and/or description hierarchy, leaving little room for adaptability to the input environment and, therefore, hindering the system's efficiency.

Based on the assumption that the construction of a recognition system itself needs to reflect the underlying nature of the input data, we propose a new framework in which the extraction of *primitives*, the development of the descriptive language, and the hierarchy of description patterns are all subject to an iterative adaptive process that learns from the input data. The framework is essentially based on the granular computing model, in which representational primitives equipped with similar measures play the part of information granules, whereas the pattern hierarchy implements the idea of the granular infrastructure comprising interdependencies among information blocks. (For a more comprehensive description of granular computing, see [6].) This allows for great flexibility of the system in response to the input data and as a consequence, a gradual improvement of the system's suitability to the underlying object domain. Moreover, it is noteworthy that the whole process is conducted automatically.

Later, we show that the same framework can also be used effectively to generate class dissimilarity functions that can be combined with similarity measures in the final recognition phase of the system; this makes our approach distinct from the majority of existing systems, which usually employ only class similarity when classifying new, unseen images.

Finally, we present the results of experiments on the large NIST 3 handwritten digit database, which confirm the effectiveness of the proposed methods.

2 Structural OCR Basics

Both statistical and structural approaches proved equally effective in PR in general, but the graphical nature of the input data in OCR intuitively favors employing structural methods. A major structural approach is the *relational graph* method, where the image objects from the training data set $U = \{u_1, u_2, \ldots, u_n\}$ are first converted to graphs $\{g_1, g_2, \ldots, g_n\}$, where specific image features are encoded in *nodes* and relations between them are represented by *edges*. Then, for each target class $\{CL_1, CL_2, \ldots, CL_c\}$, a set (library) of protopypical graphs $\{G_1, G_2, \ldots, G_c\}$ is developed, most often by means of some similarity measures.

These prototypes, also called *skeleton* graphs, are considered to contain characteristic traits for each target class and, in a way, represent images of that class. Now, given a new image object u_N, a representation graph g_N is extracted and *compared* with the skeleton graphs from each set G_i. The final class assignment can be denoted as

$$CL(u_N) = \Gamma[D_1(u_N, G_1), D_2(u_N, G_2), \ldots, D_c(u_N, G_c)],$$

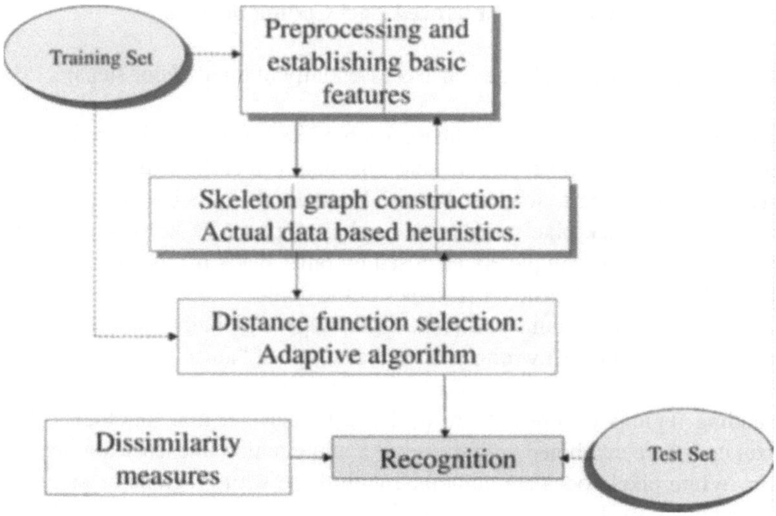

Fig. 1. Adaptive recognition system overview

where D_i are class similarity measures or distance functions and Γ reflects the chosen classification strategy [7].

It is obvious that the successful recognition depends on the choice of

- The graph model for the image data.
- Similarity measures used to build skeleton graphs.
- The distance functions and classification strategy in the final recognition phase.

Most existing OCR systems assume a priori domain knowledge about the input data and as a consequence, employ some or all these components as established in advance, with little or no feedback from the actual input training data. In this chapter, we shall show that all three components can be dynamically constructed by an adaptive process based extensively on the actual input image domain.

3 Relational Graph Model for Handwritten Digits

The development of OCR in general and handwritten digit recognition in particular over the years yielded many highly effective description models for analyzing digit images. For the research in this chapter we have chosen the enhanced loci coding scheme because of its simplicity, sophistication, low computational cost and its suitability to our higher level system components. The enhanced loci algorithm, though simple, has proved very successful in digit recognition [4].

3.1 Feature Extraction: The Enhanced Loci Coding Scheme

The enhanced loci coding scheme consists essentially of three major steps:

- *Feature extraction*: During this phase, each white pixel in the image is characterized by the black pixel regions that surround it in each of the four scanning directions: north, east, south and west. There are 16 possible code values for white pixels. Black pixels are labeled either as part of a horizontal, vertical, slanting edge or as completely enclosed by other black pixels. As a result, we have a description of basic local shapes of the digit.
- *Feature unification*: In this phase, the digit image is globally scanned to correct the pixels that had been wrongly labeled as result of "noise" in the picture.
- *Feature concentration*: In this phase, codes are combined with codes from surrounding regions to reflect both local and nonlocal region topologies. Black pixel codes are combined with the nearest white region in each scanning direction. White pixel codes are combined with those white regions outside of the nearest black strokes in the scanning direction.

Once the loci coding is done, the digit image is segmented into regions consisting of pixels with the same code value. These regions then serve as *primitives* to build the graph representation of the image. Suppose that an image I has been segmented into

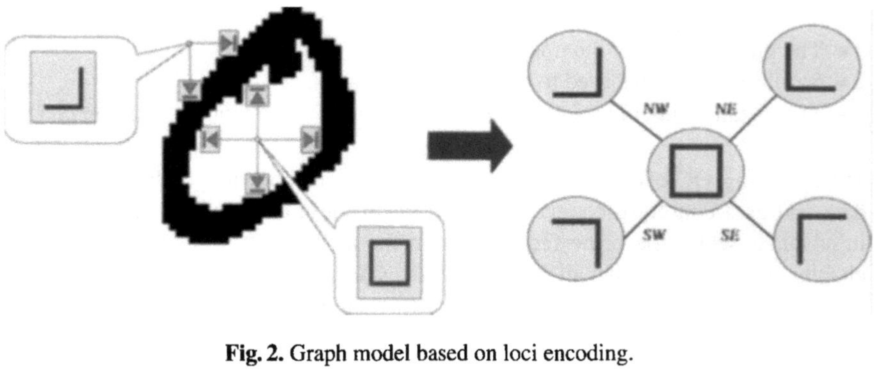

Fig. 2. Graph model based on loci encoding.

regions $R_1, R_2, \ldots, R_k$, each characterized by coordinates of the pixels contained and their shared loci code. The graph representation of the image is an attributed labeled graph denoted as

$$G_I = \{N, P, E\},$$

where

- $N = \{n_1, n_2, \ldots, n_k\}$ is a set of *nodes* representing $R_1, R_2, \ldots, R_k$.

- P is a set of *properties* of the nodes, containing, among other things, the loci code, the number of pixels, and the gravity center of the corresponding regions. We assume that the set of possible properties will remain the same for all digits from the input domain.
- E is a set of directed labeled edges between pairs of nodes, describing the relative direction between corresponding regions. These directions (labels) may be N, E, S, W, NE, SE, NW and SW. We also assume that the set of possible labels remains the same for all digits from the input domain.

It is easy to observe that such a graph G_I will contain local information about black strokes in nodes and nonlocal features about the shapes of the digit as edges (see also [5]).

3.2 Base Skeleton Graph Construction

Definition 1. A *base segment* $S_b = \{N_b, P, E_b\}$ is any two-node segment of any digit representation graph, i.e., $|N_b| = 2$ and $|E_b| = 1$. We shall say that a base segment S_b *matches* a graph G_I, or $match(S_b, G_I)$, if S_b is isomorphic to a subgraph of G_I.

Similar base segments can then be merged:

Definition 2. Given a set of base segments with common node and edge sets $S_1 = \{N, P_1, E\}, S_2 = \{N, P_2, E\}, \ldots, S_k = \{N, P_k, E\}$, a *base skeleton segment* is defined as

$$S_{bs} = \{N, P_{bs}, E\},$$

where P_{bs} is a combined set of properties

$$P_{bs} = \{(L_1, f_1), (L_2, f_2), \ldots, (L_k, f_k)\},$$

with loci code $L_i \in P_i$, and *the frequency of occurrence* of L_i,

$$f_i = \frac{|\{G_m : match(S_i, G_m)\}|}{|\{G_m : \exists 1 \leq j \leq k : match(S_j, G_m)\}|}.$$

We shall say that a base skeleton segment,

$$S_{bs} = \{N, \{(L_1, f_1), (L_2, f_2), \ldots, (L_k, f_k)\}, E\}$$

, *matches* a graph G_I, or $match(S_{bs}, G_I)$, if $\exists_{1 \leq j \leq k} : match(\{N, L_k, E\}, G_I)$.

Having constructed base skeleton segments, we can build *base skeleton graphs.*

Definition 3. A *base skeleton graph (BSG)* is any set of base skeleton segments.

Though this concept allows constructing of a broad range of skeleton graphs, a particular type, linked directly to the loci coding scheme, can be distinguished as highly useful:

Definition 4. A *loci base skeleton graph (LBSG)* is a BSG in which there exists one node, called a *center node*, which is connected to every other node. Furthermore, an *extended loci base skeleton graph (ELBSG)* is a graph in which several LBSGs are connected.

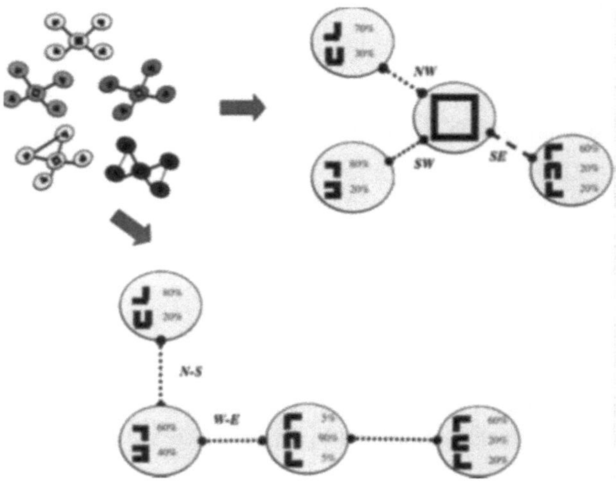

Fig. 3. Skeleton graph consolidation

LBSGs correspond to concentrated loci codes, where a region (represented by the center node) is characterized not only by its own topology, but also by topologies of all regions immediately surrounding it. ELBSGs, in turn, describe features that are more global, rather than direct neighborhoods.

One can look at BSGs as "soft" or "blurred" prototypical graphs that may be used to represent a class of digits. By fine-tuning various parameters of the model, e.g., the set of properties or connection labels, or by imposing various cutout thresholds, we can dynamically control the primitives extraction process.

3.3 Graph Similarity Measures

To build prototype libraries for each digit class from the input domain, we need to be able to measure the similarity between representation and skeleton graphs.

Rather than concentrate on a specific, predefined measure, we aim to develop an entire framework for the flexible construction of families of similarity measures.

Given a BSG S and a digit representation graph G, similarity $\tau(S,G)$ is established as follows:

Definition 5. Suppose that for each node $n \in S$, $P(n) = \{(L_1, f_1), (L_2, f_2), \ldots, (L_{k_n}, f_{k_n})\}$ is the set of $(code, frequency)$ pairs at n. Then

if $match(S,G)$ **then for each** $n \in S$,

$$\tau_n(S,G) = \sum_{i=1}^{k_n} f_i \tau_C[L_i, L^n(G)],$$

where $L^n(G)$ is the loci code found at the node matching n in G, and τ_C is a code-defined similarity function that returns the similarity between two given loci codes:

$$\tau(S,G) = \sum_{n \in S} w_n \tau_n(S,G),$$

where w_n are connection-defined weight coefficients,

else

$$\tau(S,G) = 0.$$

This definition provides a tolerant matching scheme between representation graphs and skeleton graphs, which allows us to concentrate on the specific aspects of the graph description concepts at each given stage of the learning process. By fine-tuning code-defined and connection-defined weight coefficients, we can achieve significant flexibility in information granules construction.

Now, let $S = \{S_1, S_2, \ldots, S_k\}$ be an ELBSG where S_i are loci BSGs. The similarity $\tau(S,G)$ can be defined as

$$\tau(S,G) = \mathbb{F}[\tau_1(S_1,G), \tau_2(S_2,G), \ldots, \tau_k(S_k,G)],$$

where τ_i are single LBSG-defined similarity measures and $\mathbb{F}$ is a synthesis operator. The choice of $\mathbb{F}$ is greatly influenced by the actual structure of the granules' hierarchy, which in our case is the interconnections between local patterns in skeleton graphs. Both explicit a priori and data-driven learned synthesis types of operators can be employed.

Definition 6. A distance function between two graphs G_1, G_2 *with regard to* a skeleton graph S is defined as

$$d_S(G_1,G_2) = |\tau_S(G_1) - \tau_S(G_1)|.$$

Suppose that for a digit class k, a set (library) of prototypical graphs PG_k has been established. Then, we can consider different distance functions *with regard to* that class using various synthesis operators, e.g.,

$$d_k(G_1,G_2) = \max_{S\in PG_k} d_S(G_1,G_2),$$

$$d_k(G_1,G_2) = \sum_{S\in PG_k} w_S d_S(G_1,G_2),$$

where w_S are weight coefficients.

4 Adaptive Construction of Distance Functions

Based on the relational graph, similarity measure, and distance function models defined in previous sections, we can construct an iterative process that searches for an optimal classification model as follows:

Algorithm

Step 1 Initial skeleton graph set for each digit class k:

1. **select** a start node: $S = n_0$.
2. Among its neighboring node–candidates, **pick** a node n_i that
   - has class k as dominant among the supporting population, **or**
   - best splits class k from the others, namely, has the maximal number of pairs (G_1,G_2) of which only one belongs to class k and only one matching $S \cup \{n_i\}$.
3. $S := S \cup \{n_i\}$.
4. **repeat** 2–3 until *quality criteria* are met.
5. **output** S.

Also, for each digit class, Loci BSGs are constructed based on

- Frequency and histogram analysis.
- Adjustment of connection-defined weights using a **greedy clustering scheme** with the recognition rate as quality criteria.
- Manual selection of a number of core initial EBSGs using some domain knowledge.

The purpose of this step is to establish an initial BSG set that bears as much characteristic information as possible specific to each class. Manual selection allows for flexible integration of an expert's domain knowledge in the feature construction process.

Step 2 Distance function evaluation

With the established sets of skeleton graphs for each digit class, develop graph similarity measures and distance functions, as described in Sect. 3.3.

Based on developed distance functions, perform k-NN clustering on the input training data collection to obtain class separation.
Evaluate the recognition rate based on developed clusters.

Step 3 Adjustment of parameters

Using a **greedy strategy** with regard to the recognition rate, make adjustments to:

- Single code similarity function.
- Code-based and connection-based similarity weights.
- BSG-based distance function weight.

Reconstruct the skeleton graph set as needed.

Repeat steps 2–3 until quality criteria are met.

These two steps allow us to establish the hierarchy of information granules through distance functions. The clustering-based learning on both infrastructure building and granule construction levels ensures that the final structure reflects the nature of the input domain.

End algorithm

It can be observed that this is an adaptive iterative process with a two-layered k-NN clustering scheme, aimed at optimizing three components crucial to the recognition process:

- Primitives extraction process, implemented by loci coding scheme, code-defined and connection-defined similarity measures.
- Similarity measures model, represented by base skeleton graphs.
- Class distance functions (discriminants), synthesized from extended skeleton graphs.

5 Dissimilarity Measures

So far, similarity measures are used to construct libraries of prototypes so that future input data may be checked against them. Thus, the recognition process relies on how a new object resembles those that had been learned. However, sometimes it could really help if we knew whether an object u *does not* belong to a class c. At times in the recognition process, it might be crucial to know that, for example, a digit is neither a '8' nor a '9,' though we still do not know for sure if it is a '1' or a '7.' Furthermore, there may be situations when, based on similarity analysis, we know that a digit may be a '0' or a '1,' and at the same time, by using dissimilarity measures, we can state that it cannot be a '0.' Hence, dissimilarity measures are very useful for constructing classification discriminants.

In our relational graph model, dissimilarity measures can be defined as follows:

Antigraphs

Definition 7. Let $G = \{V, p, E, l\}$ where V is a set of nodes, E is a set of connections, p is an node attribute assignment function, and l and is a connection label assignment function. Further, let P be a set of all possible node properties and L be a set of all possible connection labels. Then, graphs

$$G_V = \{V, p_V, E, l\} \text{ where } p_V(n) = P \setminus \{p(n)\} \, \forall n \in V,$$

and

$$G_E = \{V, p, E, l_E\} \text{ where } l_E(e) = L \setminus \{l(e)\} \, \forall e \in E,$$

are called *antigraphs* of G.

Then, for instance, the dissimilarity to a digit representation graph may be defined as similarity to its antigraph.

Based on the same framework described in Sect. 4, we can construct antigraph skeleton sets for each digit target class or several target classes and use them as discriminants to improve the classification quality. In this way, similarity and dissimilarity measures can be effectively combined, using a common framework, to augment the final recognition rate.

6 Results of Experiments

Extensive testing has been conducted to verify the developed methods. We chose the U.S. National Institute of Standards and Technology (NIST) Handwritten Segmented Character Special Database 3, a major reference base within the handwritten character recognition community, as the main data collection. The base contains 223,125 128×128 normalized binary images with isolated handwritten digits from 2100 different people. (For details, see [3])

As a reference experiment's data collection, we chose a random portion of the whole base that contained

Fig. 4. Samples from the NIST SD 3 collection

- 44,000 digits as a training table, of which 4000 have been separated for tests during the learning process.
- 4000 digits for the final test table

Most skeleton graphs are simple in size (5 to 19 nodes). Complexity varies from class to class; the maximum connection depth does not exceed 5. The LBSG-type graphs are the main engine for the main recognition phase, whereas general BSG-type and antigraphs were particularly useful in fine-tuning postprocessing.

The system has been implemented as a Win32 application under Windows NT 4.0 using Microsoft Visual C++ 6.0. The learning phase took, on average, 4 hours 15 minutes on a Intel Pentium III 733MHz computer with 128MB RAM. The final recognition speed was 0.045 s per digit.

It is interesting to observe that there were no rejections, which showed that the chosen graph model and loci coding were highly appropriate for the digits concerned.

The results obtained qualify our system close to the leading recognition packages tested at NIST, of which the average zero-rejection error rates were 1.70% (see [3]).

Table 1. Adaptive distance function recognition results.

Class	No. of skeleton graphs	No. of digits	Misclassified	Reject
0	8	439	1.6%	0%
1	7	328	1.5%	0%
2	12	417	6.7%	0%
3	11	375	6.1%	0%
4	14	421	5.5%	0%
5	9	389	9.8%	0%
6	11	397	5.3%	0%
7	8	366	3.3%	0%
8	13	432	4.9%	0%
9	9	436	6.9%	0%
	Total	4000	**5.2** %	0 %

Table 2. Recognition results with dissimilarity improvement.

Class	No. of skeleton graphs	No. of digits	Misclassified	Reject
0	8	439	0.91%	0%
1	7	328	1.52%	0%
2	12	417	2.16%	0%
3	11	375	4.53%	0%
4	14	421	3.80%	0%
5	9	389	2.57%	0%
6	11	397	2.02%	0%
7	8	366	1.37%	0%
8	13	432	2.78%	0%
9	9	436	1.83%	0%
	Total	4000	**2.35**%	0%

7 Conclusion

We presented a uniform framework for automatically constructing classifiers based on an adaptive scheme. A model for synthesizing similarity measures from input data primitives through higher level features has been proposed. The method allows flexible learning from the input training data during the construction phase and proved effective. The same framework can be used to develop dissimilarity measures that are highly useful in improving the classification quality. Experiments conducted on a large handwritten digit database showed that the method can be applied to practical problems with encouraging results. The framework can easily be adapted to recognizing other structured objects such as handprinted characters, fingerprints, iris images, and human faces.

References

1. M. R. Anderberg. *Cluster Analysis for Applications*. Academic Press, New York, 1973.
2. J. Bazan, H. S. Nguyen, T. T. Nguyen, J. Stepaniuk, A. Skowron. Application of modal logics and rough sets for classifying objects. In M. De Glas, Z. Pawlak, editors, *Proceedings of the Second World Conference on the Fundamentals of Artificial Intelligence (WOCFAI'95)*, 15–26, Ankor, Paris, 1995.
3. J. Geist, R. A. Wilkinson, S. Janet, P. J. Grother, B. Hammond, N. W. Larsen, R. M. Klear, C. J. C. Burges, R. Creecy, J. J. Hull, T. P. Vogl, C. L. Wilson. The second census optical character recognition systems conference. NIST Technical Report NISTIR 5452, 1–261, 1994.
4. K. Komori, T. Kawatani, K. Ishii, Y. Iida. A feature concentrated method for character recognition. In B. Gilchrist, editor, *IFIP Proceedings*, North Holland, Amsterdam, 29–34, 1977.
5. H. S. Nguyen, T. T. Nguyen. An approach to the handwriting digit recognition problem based on modal logic. Master's Thesis, Institute of Mathematics, Warsaw University, 1993.
6. L. Polkowski, A. Skowron. Towards adaptive calculus of granules. In L.A. Zadeh, J. Kacprzyk, editors, *Computing with Words in Information/Intelligent Systems*, 201–227, Physica, Heidelberg, 1999.
7. R. J. Schalkoff. *Pattern Recognition: Statistical, Structural and Neural Approaches*. Wiley, New York, 1992.
8. Y. Kodratoff, R. Michalski. *Machine Learning: An Artificial Intelligence Approach*, Vol. 3. Morgan Kaufmann, San Francisco, 1990.

Chapter 24
From Rough through Fuzzy to Crisp Concepts: Case Study on Image Color Temperature Description

Władysław Skarbek

Department of Electronics and Information Technology, Warsaw University of Technology, Nowowiejska 15/19, 00-665 Warsaw, Poland
and
Multimedia Group, Altkom Akademia S.A., Stawki 2, 00-193 Warsaw, Poland
w.skarbek@ire.pw.edu.pl

Summary. This chapter proposes a framework for designing interval-based classifiers for fuzzy categories based on rough information systems. The information system is given by joining objective measurements of a quantized scalar feature t for a class of objects with subjective decisions (votes) regarding the category c, to which objects belong. Using the roughness degree, we estimate at sparse points unknown fuzzy membership functions for n categories. Having such sparse membership functions and the original information system, we find a family of optimal partitions of a feature's range for two important cost functions. In practice, only interval partitions are useful for fast decisions. The algorithm generating optimal intervals is given and applied to extracting of image color temperature descriptions.

1 Introduction and Background

Fuzzy concepts are also used to model subjective judgments of people related to the human perception of an external stimulus [8]. HOT, WARM, NEUTRAL, and COLD are categories that are usually considered in the context of the feeling temperature for air or touched objects. The same categories are used in physics in the context of the color temperature of objects.

Similar to air temperature, color temperature is perceived by humans individually, and it strongly depends on accompanying circumstances such as viewing and lighting conditions.

For both concepts of temperature, there exist precise physical procedures for objective evaluation of the temperature for a given object temperature. However, in color, the procedure is not applicable to all possible real objects, for instance to a perfect black body [7].

In the latter case, we can also compute an average color temperature (CT) for the given image of a complex real or virtual scene [1]. Though such a computational notion of the temperature correlates only with its physical counterpart, we consider CT an objective measurement procedure applied to color images. As a result of this measurement process, for any image, we get its CT as a real number $t \in [t_{\min}, t_{\max}]$. Color temperature categories, such as hot and warm images, are important for painters and people of art. Computer graphics people, in the near future, while designing a new virtual scene will likely issue a query to a web search engine: *please give me a few images similar to the given one with respect to the image texture and the color temperature.* This kind of query for a search by content must consider the subjective feeling of color temperature rather than its exact value.

Therefore, we need a fast procedure that converts the objective value into the subjective category, HOT, WARM, NEUTRAL, and COLD. This chapter describes such a procedure which is based on the partition of the CT range into four disjoint intervals. The subdivision into intervals is optimal according to intuitive cost functions.

The situation where the subjective categories are based on objective object attributes, appears to be common while comparing the interpretation of human sensors with physical sensors, and therefore it is worth to generalize the case. In Sect. 2, a generalized information model is described. In Sect. 3 the indiscernibility relation of rough set theory is used to measure the roughness degree for each quantized CT value [5], [3]. Additionally it is shown that the fuzzy membership functions for fuzzy concepts are equal to related roughness degree functions. The crisp set model is considered in Sect. 4 in the context of the conformance measure and the maximum conformance error measure. The maximum conformance partition of the range is given, and the algorithm producing the optimal interval subdivision is developed in Sect. 5 for both cost functions: the maximum conformance measure and the minimum maximum conformance error (equivalent to a maximum minimum conformance measure).

2 Information Model

Our information model includes attributes specific to subjective and objective experiments conducted with physical objects of interest.

The experiment participant with `personId` is given an object with `objectId` to assign `category` from n fuzzy categories, On the other hand, the objective procedure (the single one) assigns `measurement` to the same object.

As a result of the join operation using the `objectId` attribute, we get the information model with attributes: `personId, objectId, measurement, category.`

We can combine `personId` and `objectId` into one attribute called `voteId.` This leads to the information model with only three attributes : `voteId, measurement, category.` We assume that

- `measurement` is a scalar attribute with real values in $t \in T \doteq [t_{\min}, t_{\max}]$.
- `category` is an enumeration attribute with integral values in $c \in C_n \doteq [1, n]$.
- `voteId` is also an enumeration attribute with positive integral values $v \in V \subset \mathcal{N}$.

In our case study, `measurement` is `colorTemperature` with values quantized to integers in the range $[1667, 25,000]$ (the unit is the Kelvin degree), and we have $n = 4$ fuzzy categories (HOT, WARM, NEUTRAL, COLD). In the experiments, 38 viewers observed about 2000 images, which gives about 68,000 votes.

It appeared that further quantization (requantization) of measurement values accelerate optimal partition design with small loss of optimality. The uniform quantization with step Q transforms t into $t_Q = \lfloor (t - t_{\min})/Q \rfloor$. The dequantization process reconstructs $\tilde{t} = t_{\min} + Q(t_Q + 0.5)$.

Hence, using the abbreviated notation, the information model has three attributes (va, t_Qa, ca) where for any vote object o, the vote identifier $va(o) \in \mathcal{N}$, the measurement level $t_Qa(o) \in T_Q$, and the category identifier $ca(o) \in [1, n]$. The corresponding variable values from the above ranges are denoted by v, t, and c.

In conclusion, the information model discussed here is the relation (data table) $R \subset V \times T \times C_n$ with column names (va, ta, ca) and row values (v, t, c).

3 Rough and Fuzzy Set Models

Let I_1 denotes the indiscernibility relation relative to t, i.e., two objects with identifiers v_1 and v_2 are in relation I_1 $(v_1 I_1 v_2)$ if and only for the certain measurement $t \in T$ and categories c_1 and c_2 : $(v_1, t, c_1) \in R$ and $(v_2, t, c_2) \in R$.

The class of abstraction for v denoted $[v]_{I_1}$ depends only on fixed value t and by this observation, we introduce a symbol $V_t \doteq [v]_{I_1}$.

Similarly, let I_2 denote the indiscernibility relation relative to c. Then $V^c \doteq [v]_{I_2}$ denotes the class of abstraction for I_2 in which all objects are assigned to the category c. We can say that the information model is represented by information granules $V_c^t \doteq V_t \cap V^c$.

The object v is *rough* if and only if there exists t and $c_1 \neq c_2$ such that $v \in V_t$ and $V^{c_1} \cap V_t \neq \emptyset$ and $V^{c_2} \cap V_t \neq \emptyset$. The object v is *crisp* if and only it is not *rough.* The set of objects $V' \subset V$ is *rough* if and only if there exists a rough object $v \in V'$.

For instance, in our case study, votes on a neutral image create a rough set. Each vote on such an image is rough, too.

We can measure the *roughness degree* of the object $v \in V_t$ with respect to category c by using the following formula:

$$\rho_c(v) \doteq \frac{|V_t^c|}{|V_t|}. \quad (1)$$

The object is crisp if and only if there exists a category c such that $\rho_c(v) = 1$. On the other hand, the object is rough if and only if there exists a category c such that $0 < \rho_c(v) < 1$.

Let us observe that $\rho_c(v)$ depends on t, but not on $v \in V_t$. Now we are ready to link with rough objects, the *fuzziness* $\mu_c(t)$ of their measurements with respect to category c:

$$\mu_c(t) \doteq \rho_c(v) \text{ where } v \in V_t. \quad (2)$$

If $V_t = \emptyset$, then $\mu_c(t)$ is not defined. In practice, measurements are sparse in their domain. Therefore, μ is the partial function in T_Q, and only at high quantization thresholds Q, the domain of μ equals to T_Q. Let us denote the domain of μ_c by $\text{dom}(\mu)$ since it is independent of c.

We see that the roughness of a particular object with respect to a given category is equal to the fuzziness of the measurement value that is obtained for this object. Hence, if the rough set is the primary concept, then the partial fuzzy set is the secondary one. On the other hand, if the fuzzy set is the primary concept, then the roughness is the secondary one as the formula below shows for $v \in V_t$:

$$\rho_c(v) = \frac{\mu_c(t)}{\sum_{d=1}^{n} \mu_d(t)}. \quad (3)$$

We can easily check the consistency of the above formulas, by substituting fuzziness (2) into (3), we obtain the definition (1) for roughness.

Table 1 shows a small part of the results of voting for color temperature category. Votes are given for objects (images), but we move them to the temperature domain by summation of all votes for images with the same color temperature value.

Let us emphasize that partial fuzzy sets with a sparse domain cannot be used for classifying new objects as new measurements may hit points outside of the domain. Interpolation is possible to restore the missing data, but due to the sparseness of the fuzzy features in the discrete domain included in the continous range of objective attributes, it can lead to a high percentage of *wrong crisp decisions.* In the next section, the notion of crisp decision and the meaning of wrong and right decisions will be given.

4 Crisp Set Model

The crisp set model for our rough set model and its fuzzy counterpart model are related to crisp decisions that are finally undertaken in real-life systems. For instance,

Table 1. Results of voting for the CT category.

t	$\|V_t\|$	$\|V_t^1\|$	$\mu_1(t)$	$\|V_t^2\|$	$\mu_2(t)$	$\|V_t^3\|$	$\mu_3(t)$	$\|V_t^4\|$	$\mu_4(t)$
3041	38	2	0.053	34	0.895	2	0.053	0	0.000
3841	38	1	0.026	18	0.474	15	0.395	4	0.105
4938	38	0	0.000	24	0.632	13	0.342	1	0.026
6385	38	0	0.000	2	0.053	24	0.632	12	0.316
6991	38	0	0.000	0	0.000	21	0.553	17	0.447
7308	38	0	0.000	4	0.105	27	0.711	7	0.184
9768	38	0	0.000	2	0.053	9	0.237	27	0.711
16936	38	1	0.026	2	0.053	11	0.290	24	0.632
25000	38	0	0.000	0	0.000	6	0.158	32	0.842
25000	38	0	0.000	0	0.000	2	0.053	36	0.947
25000	38	0	0.000	1	0.026	4	0.105	33	0.868

the MPEG-7 search for an engine to answer the content-based query, "for the given input image, give me k images from the same subjective color temperature category and with the nearest values of the color temperature," must rapidly establish for the input image, its color temperature and the input image category. Objective temperature measurement is a known procedure, but the approximation of subjective temperature category needs the elaboration of a new tool. This task will be implemented in this and in the next section.

The crisp decision δ assigns to the object measured value t a label for one of the fuzzy categories:

$$\delta : T \rightarrow C_n = \{1,\ldots,n\}. \tag{4}$$

This implies a set partition $\pi(\delta) = (T_1,\ldots,T_n)$ of the range T, where

$$T_i \doteq \delta^{-1}\{i\}. \tag{5}$$

What is the right decision and what is the wrong one for fuzzy categories? The answer is not as obvious as for crisp categories. In the fuzzy case, we have to consider the quality of any crisp decision function in the context of an objective cost (quality) function γ. Then instead of assessing individual decisions for individual objects, we will be able to assess the whole decision function, and possibly we will find the one that gives an optimum value for γ.

At least in theory, there are many possible cost functions for the given class of crisp decision functions. In the context of rough sets, it seems that quality functions based on the notion of decision conformance with subjective votes should be relevant. We measure the *conformance* γ for δ as the conformance for the set partition $\pi(\delta)$. The conformance for the set partition is the sum of conformances for its set elements. The conformance for the set T_c is the sum of the conformances for all measurements assigned to the category c. Finally, the conformance for measurement t with respect to category c is the fraction of all votes confirming the decision function, i.e., $|V_t^c|/|V|$. Hence,

$$\gamma(\delta) \doteq \frac{\sum_{c=1}^{n} \sum_{t \in T_c} |V_t^c|}{|V|}. \tag{6}$$

It can be easily verified that

$$0 \leq \gamma(\delta) \leq 1. \tag{7}$$

Let us observe that $\gamma(\delta) = 1$ implies that the information model is crisp, i.e., all objects are crisp; $\gamma(\delta) = 0$ means that none of the votes agrees with the crisp decision.

We are interested in optimal decision functions δ for which the maximum conformance is obtained, i.e., $\gamma(\delta)$ achieves its maximum value. Usually, there are many optimal decision functions. Let us denote by $B(\gamma)$ the set of all best decision functions for the measure γ. In more formal notation,

$$B(\gamma) \doteq \text{all} \arg\max_{\delta} \gamma(\delta). \tag{8}$$

As a dual measure to conformance measure, we have the *conformance error* measure,

$$\bar{\gamma}(\delta) \doteq \frac{\sum_{c=1}^{n} \sum_{t \in T_c} \sum_{d \neq c} |V_t^d|}{|V|}. \tag{9}$$

By proper summations, we can easily show the following duality equation:

$$\gamma(\delta) + \bar{\gamma}(\delta) = 1. \tag{10}$$

Therefore, whenever we find a decision function δ which is optimal for the conformance, i.e., it gives the maximum possible conformance, then it is optimal for the conformance error, i.e., it gives the minimum possible conformance error. In our notation, this statement is equivalent to the equality $B(\gamma) = B(\bar{\gamma})$, where

$$B(\bar{\gamma}) \doteq \text{all} \arg\min_{\delta} \bar{\gamma}(\delta).$$

It appears that using set partitions for the range T, we can describe exactly the form of all optimal decision functions for the measure γ. If $\delta \in B(\gamma)$, then its partition $(T_1, \ldots, T_n)$ is of the form

$$T_1 = T_1^0 \cup U_1, \ \ldots \ , T_n = T_n^0 \cup U_n, \tag{11}$$

where $(U_1, \ldots, U_n)$ is the partition (with possible empty components) for the complementary set $U \doteq T - (T_1^0 \cup \ldots T_n^0) = U_1 \cup \cdots \cup U_n$. The canonical sets $T_1^0, \ldots, T_n^0$ are defined recursively for $j = 1, \ldots, n$:

$$T_i^0 \doteq \{t \in \text{dom}(\mu) - \bigcup_{j<i} T_j^0 : \ \mu_i(t) = \max_{j \geq i} \mu_j(t)\}. \tag{12}$$

A more conservative approach is measuring the *minimum of conformance* computed separately in classes:

$$\gamma_{\min}(\delta) \doteq \frac{\min_{c=1}^{n} \sum_{t\in T_c} |V_t^c|}{|V|}. \tag{13}$$

A dual measure to the minimum of conformance is the *maximum of conformance error* computed separately in classes:

$$\overline{\gamma}_{\max}(\delta) \doteq \frac{\max_{c=1}^{n} \sum_{t\in T_c} \sum_{d\neq c} |V_t^d|}{|V|}. \tag{14}$$

Then, the duality equation has the form:

$$\gamma_{\min}(\delta) + \overline{\gamma}_{\max}(\delta) = 1. \tag{15}$$

It also implies that whenever we find a decision function δ that is optimal for the minimum of conformance in categories, i.e., it gives the maximum for the minimum of conformance in classes, then it is optimal for the maximum of conformance error in categories, i.e., it gives the minimum for the maximum of conformance error in classes. Again in our notation, this statement is equivalent to the equality $B(\overline{\gamma}_{\max}) = B(\gamma_{\min})$, where

$$B(\overline{\gamma}_{\max}) \doteq \text{all} \arg\min_{\delta} \overline{\gamma}_{\max}(\delta),\; B(\overline{\gamma}_{\min}) \doteq \text{all} \arg\max_{\delta} \overline{\gamma}_{\min}(\delta).$$

Contrary to $B(\gamma)$, no characterizations of $B(\gamma_{\min})$ that are analogous to (12) are known to the author.

Though in the discrete case of bounded range measurements, we can effectively represent one of the optimal partitions given by (11) and (12), the computation of partition membership for any measurement is costly. Moreover, such representation depends strongly on the initial data table. Interval partitions are most feasible in practice. The next section describes algorithms searching for interval-based decision functions.

5 Interval Data Model

We obtain interval data model by considering interval-based decision functions. Suppose that the discrete range T is partitioned into closed discrete intervals using cut points $t_0 = t_{\min} < t_1 < \cdots < t_{n-1} < t_n = t_{\max}$:

$$T = [t_{\min}, t_1] \cup \bigcup_{i=2}^{n} [t_{i-1} + 1, t_i].$$

Then, the interval-based decision function is defined as follows for $t > t_{\min}$:

$$\delta(t) = i \text{ if and only } t_{i-1} < t \leq t_i. \tag{16}$$

Obviously, $\delta(t_{\min}) = 1$. For both cost functions (6) and (13), we are looking now for cut points that define an optimal interval-based decision function.

Let us denote by $\bar{t}_c$ the centroid of the fuzzy set μ_c :

$$\bar{t}_c \doteq \frac{\sum_{t \in T} t |V_t^c|}{\sum_{t \in T} |V_t^c|}.$$

We assume that

$$\bar{t}_1 < \bar{t}_2 < \cdots < \bar{t}_n.$$

Then, the algorithm looks for the cut point t_i ($i = 1, \ldots, n-1$) in the discrete interval $[\bar{t}_i, \bar{t}_{i+1})$.

Initially, we take the left end points of these intervals as cut points. It is possible to generate the next configuration of cut points from the previous one by changing only one cut point by ± 1. This allows incremental calculation of the cost functions with only few arithmetic operations per interval partition. While comparing cost function values, the common denominator $|V|$ can be ignored. Therefore, we use integral arithmetic in computing optimal intervals.

The above idea for an algorithm was implemented in a Java application that finds optimal intervals for fuzzy categories defined on the color temperature range.

The following Java method controls generation of cut points (h, w, n) by moving them along a 3-D serpentine that scans the discrete 3-D volume:

$$[\bar{\mu}_1, \bar{\mu}_2) \times [\bar{\mu}_2, \bar{\mu}_3) \times [\bar{\mu}_3, \bar{\mu}_4),$$

where $\bar{\mu}$ denotes the centroid of the corresponding membership function rounded to the nearest integral number.

The scan is implemented by three nested loops. The external loop is controlled by the variable h. The inner loops are controlled by w and n. Whenever one of these variables reaches the end point of its range, the direction of its change is reversed. In this way, a continuous discrete scan is achieved.

It is assumed that cut points belong to intervals to which their mnemonics belong, i.e., h is in the hot interval, etc.

```
public CToptimalIntervals getCToptimalIntervals(String optiTech)
                                        throws DataException {
```

```
    OptimizationTechnique ot=null;
    // ...
    int havg=getHotCentroidBin(), wavg=getWarmCentroidBin(),
        navg=getNeutralCentroidBin(), cavg=getColdCentroidBin();
    int hbest, wbest, nbest;
    int wmin=wavg, wmax=navg-1, nmin=navg, nmax=cavg-1;
    int dw=1, dn=1, w=wmin, n=nmin;
    for (int h=havg; h<wavg; h++) {
        while (true) { // w loop
            while (true) { // n loop
                if (ot.betterMeasure(h,w,n)) {
                    hbest=h; wbest=w; nbest=n;
                }
                n+=dn;
                if (n<nmin || n>nmax) { dn=-dn; n+=dn; break; }
            }
            w+=dw;
            if (w<wmin || w>wmax) { dw=-dw; w+=dw; break; }
        }
    }
    //...
}
```

The second Java program fragment illustrates the incremental calculation of the conformance measure that accelerates the original algorithm by two orders of magnitude.

The quality measure for the given interval subdivision, i.e., for the given cut points, is computed using the value of the measure found for the previous cut points on the serpentine. Therefore, the previous cut points are stored in the temporary variables.

Because we scan the 3-D volume by steps always along one axis, it is enough to check which of the cut points is changed in the positive or negative direction and then apply incremental formulas for computing measures. For instance, moving with n to the right, we expand a neutral interval by the point n, and at the same time, we remove it from the cold interval. Therefore, the conformance for the neutral category is increased by votes for this category of images with the CT value quantized to the nth interval. However, at the same time, the conformance for the cold category is decreased by votes for this category of images with the CT value quantized to the nth interval.

It is interesting that our optimization problem has more than one optimal solution. This follows from the fact that certain CT *bins* are empty, i.e., no votes were registered for the given quantized interval in the CT range. The algorithm finds all optimal solutions and returns their average rounded to the nearest integral number.

```
private long total;
private boolean firstTime=true;


private int hOld,wOld,nOld;
public boolean betterMeasure(int h, int w, int n) {
    if (firstTime) {
    //...
    }
    if (n!=nOld) {
        if (n>nOld)
            total+=neutralV[n]-coldV[n];
        else
            total+=coldV[n+1]-neutralV[n+1];
        nOld=n;
        if (total>bestTotal) {
            bestTotal=total; return true;
        }
        else return false;
    }   else
    if (w!=wOld) {
    //...
    }   else
    //...
}
```

Selected results of the experiments are collected in Table 2. The left part shows the cut points obtained for the maximum conformance measure at different quantization thresholds. The right part includes results for the maximum of minimum conformance across category intervals. Despite the sparseness of data points (less than 10% of the range points), we observe the high stability of cut points in functions of quantization step Q. Note that cut points are converted from the quantized domain to the CT domain.

Table 2. Selected results of experiments

Q	t_{min}	t_1	t_2	t_3	t_1	t_2	t_3	t_{max}
30	1667	2327	4187	8027	2207	3887	7667	25,000
20	1667	2247	4167	8027	2207	3887	7667	25,000
10	1667	2247	4167	8027	2197	3887	7657	25,000
5	1667	2252	4167	8062	2192	3882	7657	25,000
2	1667	2251	4165	8063	2191	3881	7657	25,000
1	1667	2251	4165	8063	2191	3881	7656	25,000

For the MPEG-7 standard, the following intervals were proposed in the author's contribution [6]: HOT, [1667, 2250]; WARM, [2251, 4170]; NEUTRAL, [4171, 8060];

COLD, [8061,25,000].

The above intervals were used by the Samsung Company for image browsing. The MPEG-7 contribution [2] of the SAIT Institute confirms that the above intervals improved browsing accuracy by more than 10% relative to the intervals proposed earlier by SAIT in [1].

Conclusion

The proposed framework for the design of interval-based classifiers for fuzzy categories was based on rough information systems. It appeared effective in the case study on image color temperature descriptions. The study shows the interaction between rough and fuzzy concepts when we consider subjective assessments for objectively measured features. The rough set model has shown an advantage over the fuzzy approach, making the search for optimal intervals simple and efficient.

Acknowledgments

The research has been supported in part by the KBN grant 7 T08A 05016.

References

1. S.K. Kim, D.S. Park. Report of vce-6 on MPEG-7 color temperature browsing descriptors. *ISO/IEC JTC1/SC29/WG11*, MPEG-57, Sydney, 2001.
2. S.K. Kim, D.S. Park, Y. Choi. Report of vce-6 on MPEG-7 color temperature browsing descriptors. *ISO/IEC JTC1/SC29/WG11*, MPEG-58, Pattaya, 2001.
3. J. Komorowski, Z. Pawlak, L. Polkowski, A. Skowron. Rough sets: A tutorial. In *[4]*, 3–98, 1999.
4. S.K. Pal, A. Skowron, editors. *Rough Fuzzy Hybridization: A New Trend in Decision–Making*. Springer, Singapore, 1999.
5. Z. Pawlak. *Rough Sets: Theoretical Aspects of Reasoning about Data*. Kluwer, Dordrecht, 1991.
6. W. Skarbek. Optimal intervals for fuzzy categories of color temperature. *ISO/IEC JTC1/SC29/WG11*, MPEG-58, Pattaya, 2001.
7. G. Wyszecki, W.S. Stiles. *Color Science*. Wiley, New York, 1982.
8. L.A. Zadeh. Fuzzy logic = computing with words. *IEEE Transactions on Fuzzy Systems*, 4: 103–111, 1996.

Chapter 25
Information Granulation and Pattern Recognition

Andrzej Skowron,[1] Roman W. Swiniarski[2]

[1] Institute of Mathematics, Warsaw University, Banacha 2, 02-097 Warsaw, Poland
skowron@mimuw.edu.pl
[2] San Diego State University, Department of Mathematical and Computer Sciences, 5500 Campanile Drive, San Diego, CA 92182, USA
rswiniar@sciences.sdsu.edu

Summary. We discuss information granulation applications in pattern recognition. The chapter consists of two parts. In the first part, we present applications of rough set methods for feature selection in pattern recognition. We emphasize the role of different forms of reducts that are the basic constructs of the rough set approach in feature selection. In the overview of methods for feature selection, we discuss feature selection criteria based on the rough set approach and the relationships between them and other existing criteria. Our algorithm for feature selection used in the application reported is based on an application of the rough set method to the result of principal component analysis used for feature projection and reduction. Finally, the first part presents numerical results of face recognition experiments using a neural network, with feature selection based on proposed principal component analysis and rough set methods. The second part consists of an outline of an approach to pattern recognition with the application of background knowledge specified in natural language. The approach is based on constructing approximations of reasoning schemes. Such approximations are called approximate reasoning schemes and rough neural networks.

1 Introduction

Reduction of pattern dimensionality via feature extraction and feature selection [9,17,21,22] is among the most fundamentals steps in data preprocessing. We present rough sets methods and principal components analysis (PCA) in the context of feature selection in pattern recognition.

The chapter begins with a short introduction to rough set theory [28]. We emphasize the special role of reducts in feature selection, including dynamic reducts [2,4,5]. Then, we present a short overview of a feature selection problem including open-loop and closed-loop feature selection methods [9]. This section focuses the discussion on feature selection criteria, including rough set based methods. The next section presents a short description of principal component analysis [9] as a method of feature projection and reduction. It also contains a description of rough set-based methods, proposed jointly with principal component analysis, for feature projection and reduction. The following section describes the results of numerical experiments of face recognition using rough set based methods for feature selection and neural networks. This section also contains a short description of feature extraction from facial images using singular value decomposition (SVD).

The second part of the chapter consists of an outline of an approach, called the rough-neural computing approach (see Chaps. 2 and 3), for pattern recognition with an application of background knowledge specified in natural language. The approach is based on a rough mereological approach (see, e.g., [35]) for information granule calculi. The goal of information granule calculi is to make it possible to imitate reasoning in natural language by means of information granules. Such granules have a complex information structure representing approximations of reasoning schemes in natural language over vague concepts. Reasoning schemes in natural language are built over vague concepts and relations between them creating ontologies. In natural language we call them approximation reasoning schemes (AR schemes) or rough-neural networks. Our approach to pattern recognition is based on searching for clusters of objects close to a given standard (prototype) to a given degree. Using the rough set approach, one can interpret such standards as the lower approximations of concepts. Moreover, methods for extracting special relationships, called productions, between such clusters are emphasized. They correspond to local relationships between concepts from background knowledge. They make it possible to conclude that a target concept is satisfied to a satisfactory degree for a given object if the input concepts are satisfied to some satisfactory degree by input patterns related to the object. A special method for composing such productions leads to derivations of robust AR schemes. Any AR scheme guarantees that the target concept of such a scheme is satisfied to a satisfactory degree for a given object if the input concepts for this scheme are satisfied to some satisfactory degree by the object. AR schemes are then used to induce approximations of more complex concepts from a knowledge base, assuming that classifiers representing some primitive concepts have been constructed. In the second part of the chapter, we outline the approach, and we present an illustrative example.

2 Preliminaries of Rough Sets

Rough set theory was introduced by Zdzisław Pawlak (see, e.g., [18,28]) to deal with imprecise or vague concepts. In recent years, we have witnessed a rapid growth of interest in rough set theory and its applications worldwide (see, e.g., [18, 30, 33, 34, 45, 52].

In this section, we present the basic concepts of rough set theory and some of its extensions. A variety of methods for generating decision rules, reduct computation, and continuous variable discretization are very important issues not discussed here. We emphasize only the developed methodology based on discernibility and Boolean reasoning for efficient computation of different constructs, including reducts and decision rules.

Many other important issues are not covered here. Let us mention some of them. The relationship of rough set theory to many other theories has been extensively investigated. In particular, its relationships to fuzzy set theory, the theory of evidence, Boolean reasoning methods, statistical methods, and decision theory have been clarified and seem to be thoroughly understood. There are reports on many hybrid methods obtained by combining the rough set approach with others, such as fuzzy sets, neural networks, genetic algorithms, principal component analysis, and singular value decomposition [27]. Recently, it has been shown that the rough set approach can be used for synthesizing concept approximations in a distributed environment of intelligent agents. These issues related to various logics related to rough sets and many advanced algebraic properties of rough sets are also not covered here. Readers interested in these issues are advised to consult [18,32,33,44] and the bibliography included in these books and articles.

2.1 Basic Approach

The rough set approach is founded on the assumption that with every object of a universe of discourse, we associate some information (data, knowledge). For example, if objects are patients suffering from a certain disease, then the symptoms of the disease form information about patients. Objects characterized by the same information are indiscernible (similar) in view of the available information about them. The indiscernibility relation generated in this way is the mathematical basis of rough set theory.

Any set of all indiscernible (similar) objects is called an elementary set and forms a basic granule (atom) of knowledge about a universe. Any union of some elementary sets is referred to as crisp (precise) set — otherwise, the set is rough (imprecise, vague).

Consequently, each rough set has boundary-line cases, i.e., objects that cannot be classified with certainty either as members of the set or of its complement. Obviously, crisp sets have no boundary-line elements at all. That means that boundary-line cases cannot be properly classified by employing the available knowledge.

Thus, the assumption that objects can be "seen" only through the information available about them leads to the view that knowledge has a granular structure. Due to the granularity of knowledge, some objects of interest cannot be discerned and appear the same (or similar). As a consequence, vague concepts (in contrast to precise or crisp concepts) cannot be characterized in terms of information about their elements. Therefore, in the proposed approach, we assume that any vague concept is replaced by a pair of precise concepts — called the lower and the upper approximations of the vague concept. The lower approximation consists of all objects that surely belong to the concept, and the upper approximation contains all objects that possibly belong to the concept. Obviously, the difference between the upper and the lower approximations constitutes the boundary region of the vague concept. Approximations are two basic operations in rough set theory.

2.2 Approximations and Rough Sets

We have mentioned in Sect. 2.1 that the starting point of rough set theory is the indiscernibility relation, generated by information about objects of interest. The indiscernibility relation is intended to express the fact that due to the lack of knowledge, we are unable to discern some objects by employing the available information. It means that, in general, we are unable to deal with each particular object, but we have to consider clusters of indiscernible objects as fundamental concepts of our theory.

Suppose that we are given two finite, nonempty sets U and A, where U is the *universe* of *objects, cases*, and A is a set of *attributes, features*. The pair $IS = (U,A)$ is called an *information table*. With every attribute $a \in A$, we associate a set V_a, of its *values*, called the *domain* of a. By $\mathbf{a}(x)$ we denote a data pattern $(a_1(x),\dots,a_n(x))$ defined by the object x and attributes from $A = \{a_1,\dots,a_n\}$. A data pattern of IS is any feature value vector $\mathbf{v} = (v_1,\dots,v_n)$ where $v_i \in V_{a_i}$ for $i = 1,\dots,n$ such that $\mathbf{v} = \mathbf{a}(x)$ for some $x \in U$.

Any subset B of A determines a binary relation $I(B)$ on U, called the *indiscernibility relation*, defined by

$$xI(B)y \text{ if and only if } a(x) = a(y) \text{ for every } a \in B, \tag{1}$$

where $a(x)$ denotes the value of attribute a for object x.

Obviously $I(B)$ is an equivalence relation. The family of all equivalence classes of $I(B)$, i.e., the partition determined by B, will be denoted by $U/I(B)$, or simply U/B; an equivalence class of $I(B)$, i.e., the block of the partition U/B containing x, will be denoted by $B(x)$.

If $(x,y) \in I(B)$, we will say that x and y are *B-indiscernible*. Equivalence classes of the relation $I(B)$ (or blocks of the partition U/B) are referred to as *B-elementary sets*. In the rough set approach, elementary sets are the basic building blocks (concepts) of our knowledge about reality. The unions of *B-elementary sets* are called *B-definable sets*.

The indiscernibility relation will be further used to define basic concepts of rough set theory. Let us define now the following two operations on sets:

$$B_*(X) = \{x \in U : B(x) \subseteq X\}, \tag{2}$$

$$B^*(X) = \{x \in U : B(x) \cap X \neq \emptyset\}, \tag{3}$$

assigning to every subset X of the universe U two sets $B_*(X)$ and $B^*(X)$ called the *B-lower* and the *B-upper approximation* of X, respectively. The set,

$$BN_B(X) = B^*(X) - B_*(X), \tag{4}$$

will be referred to as the *B-boundary region* of X.

If the boundary region of X is the empty set, i.e., $BN_B(X) = \emptyset$, then the set X is *crisp* (*exact*) with respect to B; in the opposite case, i.e., if $BN_B(X) \neq \emptyset$, the set X is referred to as *rough* (*inexact*) with respect to B.

A rough set can be also characterized numerically, e.g., by the following coefficient:

$$\alpha_B(X) = \frac{|B_*(X)|}{|B^*(X)|}, \tag{5}$$

called the *accuracy of approximation*, where $|X|$ denotes the cardinality of $X \neq \emptyset$. Obviously, $0 \leq \alpha_B(X) \leq 1$. If $\alpha_B(X) = 1$, then X is *crisp* with respect to B (X is *precise* with respect to B), and otherwise, if $\alpha_B(X) < 1$, then X is *rough* with respect to B (X is *vague* with respect to B).

Several generalizations of the rough set approach based on approximation spaces defined by (U,R), where R is an equivalence relation (called the indiscernibility relation) in U, have been reported in the literature (for references, see the papers and bibliography in [18,32,33,44]. Let us mention two of them.

A generalized approximation space can be defined as $AS = (U,I,\nu)$ where I is the *uncertainty function* defined on U with values in the power set $P(U)$ of U [$I(x)$ is the *neighborhood* of x] and ν is the *inclusion function* defined on the Cartesian product $P(U) \times P(U)$ with values in the interval $[0,1]$ measuring the degree of inclusion of sets. The lower AS_* and upper AS^* approximation operations can be defined in AS by

$$AS_*(X) = \{x \in U : \nu(I(x),X) = 1\}, \tag{6}$$
$$AS^*(X) = \{x \in U : \nu(I(x),X) > 0\}. \tag{7}$$

In the case discussed above, $I(x)$ is equal to the equivalence class $B(x)$ of the indiscernibility relation $I(B)$; when a tolerance (similarity) relation $\tau \subseteq U \times U$ is given, we let $I(x) = \{y \in U : x\tau y\}$, i.e., $I(x)$ is equal to the tolerance class of τ defined by x. The standard inclusion relation is defined by $\nu(X,Y) = \frac{|X \cap Y|}{|X|}$ if X is nonempty, and otherwise, $\nu(X,Y) = 1$. For applications, it is important to have some constructive definitions of I and ν.

One can consider another way to define $I(x)$. Usually, together with AS, we consider some set F of formulas describing sets of objects in the universe U of AS

defined by semantics $\|\cdot\|_{AS}$, i.e., $\|\alpha\|_{AS} \subseteq U$ for any $\alpha \in F$. Now, one can take the set,

$$N_F(x) = \{\alpha \in F : x \in \|\alpha\|_{AS}\} \text{ and } I(x) = \|\alpha\|_{AS}, \tag{8}$$

where α is selected or constructed from $N_F(x)$. Hence, more general uncertainty functions having values in $P[P(U)]$ can be defined (see also Chap. 3). The parametric approximation spaces are examples of such approximation spaces. These spaces have interesting applications. For example, by tuning their parameters, one can search for the optimal, under chosen criteria (e.g., the minimal description length), approximation space for a concept description.

The approach based on inclusion functions has been generalized to the *rough mereological approach*. The *inclusion relation* $x\mu_r y$ with intended meaning *x is part of y to a degree r* has been taken as the basic notion of *rough mereology* that is a generalization of Leśniewski mereology. Rough mereology offers a methodology for synthesizing and analyzing objects in a distributed environment of intelligent agents, in particular, for synthesizing of objects satisfying a given specification in satisfactory degree or for control in such complex environment. Moreover, rough mereology has been recently used for developing foundations of *information granule calculus*, an attempt toward formalization of the computing with words paradigm recently formulated by Lotfi Zadeh [58]. Research on rough mereology has shown the importance of another notion, namely, the *closeness* of complex objects (e.g., concepts). This can be defined by $xcl_{r,r'}y$ if and only if $x\mu_r y$ and $y\mu_{r'}x$. The inclusion and closeness definitions of complex information granules are dependent on applications. However, it is possible to define the granule syntax and semantics as a basis for the inclusion and closeness definitions.

Finally, let us mention that approximation spaces are usually defined as parameterized approximation spaces. In the simplest case, the parameter set is defined by the power set of a given feature set. By parameter tuning, the relevant approximation space is selected for a given data set and target task.

2.3 Rough Sets and Membership Function

Rough sets can also be introduced by using a *rough membership function*, defined by

$$\mu_X^B(x) = \frac{|X \cap B(x)|}{|B(x)|}. \tag{9}$$

Obviously, $0 \leq \mu_X^B(x) \leq 1$. Hence, the value of the membership function for a given object x can be interpreted as the degree of overlap between the indiscernibility class of x and the set X. One can also interpret this value as the conditional probability that an object from the indiscernibility class defined by x belongs to X.

The rough membership function can be used to define approximations and the boundary region of a set, as shown here:

$$B_*(X) = \{x \in U : \mu_X^B(x) = 1\}, \tag{10}$$

$$B^*(X) = \{x \in U : \mu_X^B(x) > 0\}, \tag{11}$$

$$BN_B(X) = \{x \in U : 0 < \mu_X^B(x) < 1\}. \tag{12}$$

2.4 Decision Tables and Decision Rules

Sometimes, in an information table (U,A), it is useful to distinguish a partition of A into two classes $C,D \subseteq A$ of attributes, called *condition* and *decision* (*action*) attributes, respectively. The tuple $DT = (U,C,D)$ is called a *decision table (system)*. Any such decision table where $U = \{u_1,\ldots,u_N\}$, $C = \{a_1,\ldots,a_n\}$ and $D = \{d_1,\ldots,d_k\}$ can be represented by a data sequence (also called data set) of data patterns $((\mathbf{v}_1, \mathbf{target}_1),\ldots,(\mathbf{v}_N, \mathbf{target}_N))$, where $\mathbf{v}_i = \mathbf{C}(x_i)$, $\mathbf{target}_i = \mathbf{D}(x_i)$, and $\mathbf{C}_i = (a_1(x_i),\ldots,a_n(x_i))$, $\mathbf{D}_i = (d_1(x_i),\ldots,d_k(x_i))$, for $i = 1,\ldots,N$. It is obvious that any data sequence also defines a decision table. The equivalence classes of $I(D)$ are called decision classes.

Let $V = \bigcup\{V_a \mid a \in C\} \cup V_d$. Atomic formulas over $B \subseteq C \cup D$ and V are expressions $a = v$ called *descriptors* (*selectors*) over B and V, where $a \in B$ and $v \in V_a$. The set $\mathcal{F}(B,V)$ of formulas over B and V is the least set containing all atomic formulas over B and V and closed with respect to the propositional connectives $\wedge$ (conjunction), $\vee$ (disjunction) and $\neg$ (negation).

By $\|\varphi\|_{DT}$, we denote the meaning of $\varphi \in \mathcal{F}(B,V)$ in the decision table DT which is the set of all objects in U with the property φ. These sets are defined as follows: $\|a = v\|_{DT} = \{x \in U \mid a(x) = v\}$, $\|\varphi \wedge \varphi'\|_{DT} = \|\varphi\|_{DT} \cap \|\varphi'\|_{DT}$; $\|\varphi \vee \varphi'\|_{DT} = \|\varphi\|_{DT} \cup \|\varphi'\|_{DT}$; $\|\neg\varphi\|_{DT} = U - \|\varphi\|_{DT}$.

The formulas from $\mathcal{F}(C,V)$, $\mathcal{F}(D,V)$ are called *condition formulas of DT* and *decision formulas of DT*, respectively.

Any object $x \in U$ belongs to a *decision class* $\| \bigwedge_{a \in D} a = a(x)\|_{DT}$ of DT. All decision classes of DT create a partition of the universe U.

A *decision rule* for DT is any expression of the form $\varphi \Rightarrow \psi$, where $\varphi \in \mathcal{F}(C,V)$, $\psi \in \mathcal{F}(D,V)$, and $\|\varphi\|_{DT} \neq \emptyset$. Formulas φ and ψ are referred to as the *predecessor* and the *successor* of decision rule $\varphi \Rightarrow \psi$. Decision rules are often called "*IF*...*THEN*..." rules.

Decision rule $\varphi \Rightarrow \psi$ is *true* in, DT if and only if $\|\varphi\|_{DT} \subseteq \|\psi\|_{DT}$. Otherwise, one can measure its *truth degree* by introducing some inclusion measure of $\|\varphi\|_{DT}$ in $\|\psi\|_{DT}$ (see Chap. 3).

Each object x of a decision table determines a *decision rule*,

$$\bigwedge_{a \in C} a = a(x) \Rightarrow \bigwedge_{a \in D} a = a(x). \tag{13}$$

Decision rules corresponding to some objects can have the same condition parts but different decision parts. Such rules are called *inconsistent* (*nondeterministic, conflicting, possible*); otherwise, the rules are referred to as *consistent* (*certain, sure, deterministic, nonconflicting*) rules. Decision tables containing inconsistent decision rules are called *inconsistent* (*nondeterministic, conflicting*); otherwise, the table is *consistent* (*deterministic, nonconflicting*).

When a set of rules has been induced from a decision table containing a set of training examples, they can be inspected to see if they reveal any novel relationships between attributes that are worth pursuing for further research. Furthermore, the rules can be applied to a set of unseen cases to estimate their classificatory power. For a systematic overview of rule application methods, the reader is referred to bibliographies included in [18,32,33,44].

2.5 Dependency of Attributes

Another important issue in data analysis is discovering dependencies between attributes. Intuitively, a set of attributes D depends totally on a set of attributes C, denoted $C \Rightarrow D$, if the values of attributes from C uniquely determine the values of attributes from D. In other words, D depends totally on C, if there exists a functional dependency between values of C and D.

Formally, dependency can be defined in the following way. Let D and C be subsets of A.

We will say that D *depends on* C to a *degree* k $(0 \leq k \leq 1)$, denoted $C \Rightarrow_k D$, if

$$k = \gamma(C,D) = \frac{|POS_C(D)|}{|U|}, \tag{14}$$

where

$$POS_C(D) = \bigcup_{X \in U/D} C_*(X), \tag{15}$$

called a *positive region* of the partition U/D with respect to C, is the set of all elements of U that can be uniquely classified in blocks of the partition U/D by means of C. If $k = 1$, we say that D *depends totally* on C, and if $k < 1$, we say that D *depends partially* (to a *degree* k) on C. The coefficient k expresses the ratio of all elements of the universe, which can be properly classified in blocks of the partition U/D, employing attributes C and will be called the *degree of the dependency.* It can be easily seen that if D depends totally on C, then $I(C) \subseteq I(D)$. This means that the partition generated by C is finer than the partition generated by D. Notice that the concept of dependency discussed above corresponds to that considered in relational databases. Summing up D, is *totally* (*partially*) dependent on C, if *all* (*some*) elements of the universe U can be uniquely classified in blocks of the partition U/D, employing C. The coefficient $1-\gamma(C,D)$ can be called the inconsistency degree of the DT [24].

2.6 Discernibility and Boolean Reasoning

The ability to discern between perceived objects is important in constructing many entities such as reducts, decision rules, and decision algorithms. In the classical rough set approach, the *discernibility relation* $DIS(B) \subseteq U \times U$ is defined by

$$xDIS(B)y \text{ if and only if } non\,[xI(B)y]\,. \tag{16}$$

However, this is generally not the case for generalized approximation spaces [one can define indiscernibility by $x \in I(y)$ and discernibility by $I(x) \cap I(y) = \emptyset$ for any objects x, y].

Boolean reasoning [7,8,42] is based on constructing for a given problem P, a corresponding Boolean function f_P with the following property: the solutions of problem P can be decoded from prime implicants of the Boolean function f_P. Let us mention that to solve real-life problems, it is necessary to deal with Boolean functions that have a huge size and a large number of variables.

A successful methodology based on the discernibility of objects and Boolean reasoning has been developed for computing many important, for applications entities such as reducts and their approximations (see the following section), decision rules, association rules, discretization of real value attributes, symbolic value grouping, searching for new features defined by oblique hyperplanes or higher order surfaces, pattern extraction from data as well as conflict resolution or negotiation (for references, see the papers and bibliographies in [18,32,33,44]).

Most of the problems related to generating of the above mentioned entities are NP-complete or NP-hard [46]. However, it was possible to develop efficient heuristics returning suboptimal solutions of the problems. The results of experiments on many data sets are very promising. They show very good quality of solutions generated by the heuristics in comparison with other methods reported in the literature (e.g., with respect to the classification quality of unseen objects). Moreover, they are very efficient from the point of view of time necessary for computing the solution.

It is important to note that the methodology allows us to construct heuristics having a very important *approximation property* which can be formulated as follows: expressions generated by heuristics (i.e., implicants) *close* to prime implicants define approximate solutions for the problem.

2.7 Reduction of Attributes

We often face the question whether we can remove some data from a data table and preserve its basic properties, that is, whether a table contains some superfluous data. Let us express this idea more precisely.

Given an information system IS, a *reduct* is a minimal set of attributes $B \subseteq A$ such that $I(A) = I(B)$. In other words, a reduct is a minimal set of attributes from A that preserves the original classification defined by the set A of attributes. Finding a minimal reduct is NP-hard; one can also show that for any m (sufficiently large), there exists an information system with m attributes having a number of reducts exponential in m. There exist, fortunately, good heuristics that compute sufficiently many reducts with the required properties (e.g., related to their length) in an acceptable time.

Let IS be an information system with n objects. The *discernibility matrix* of IS is a symmetrical $n \times n$ matrix with entries c_{ij} as given below. Each entry consists of the set of attributes upon which objects x_i and x_j differ.

$$c_{ij} = \{a \in A \mid a(x_i) \neq a(x_j)\} \quad \text{for} \quad i,j = 1,\ldots,n. \tag{17}$$

A *discernibility function* f_{IS} for an information system IS is a Boolean function of m Boolean variables $a_1^*,\ldots,a_m^*$ (corresponding to the attributes $a_1,\ldots,a_m$) defined by

$$f_{IS}(a_1^*,\ldots,a_m^*) = \bigwedge\left\{\bigvee c_{ij}^* \mid 1 \leq j \leq i \leq n,\, c_{ij} \neq \emptyset\right\}, \tag{18}$$

where $c_{ij}^* = \{a^* \mid a \in c_{ij}\}$. In the sequel, we will write a_i instead of a_i^*.

The discernibility function f_{IS} describes constraints which should be preserved if one would like to preserve discernibility between all pairs of discernible objects from IS. It requires us to keep at least one attribute from each nonempty entry of the discernibility matrix, i.e., corresponding to any pair of discernible objects. One can show [46] that the sets of all minimal sets of attributes preserving discernibility between objects, i.e., reducts correspond to prime implicants of the discernibility function f_{IS}.

The intersection of all reducts is the so-called *core*. It is well known that choosing a random reduct as a relevant set of features in an information system will give rather poor results. Hence, several techniques have been developed to select relevant reducts or their approximations. Among them is one based on so-called *dynamic reducts* [2,4]. The attributes are considered relevant if they belong to dynamic reducts with a sufficiently high stability coefficient, i.e., they appear with sufficiently high frequency in random samples extracted from a given information system.

There are several kinds of reducts considered for decision tables. We will discuss one of them. Let $\mathcal{A} = (U, A, d)$ be a decision system (i.e., we assume, for simplicity of notation that the set D of decision attributes consists of only one element d, $D = \{d\}$ and $C = A$). The *generalized decision in* $\mathcal{A}$ is the function $\partial_A : U \longrightarrow \mathcal{P}(V_d)$ defined by

$$\partial_A(x) = \{i \mid \exists x' \in U \; x' \, IND(A) \, x \text{ and } d(x') = i\}. \tag{19}$$

A decision system $\mathcal{A}$ is called *consistent (deterministic)*, if $|\partial_A(x)| = 1$ for any $x \in U$, otherwise, $\mathcal{A}$ is *inconsistent (nondeterministic)*. Any set consisting of all objects with the same generalized decision value is called a *generalized decision class*. Decision classes are denoted by C_i, where the subscript denotes the decision value.

It is easy to see that a decision system $\mathcal{A}$ is consistent if and only if $POS_A(d) = U$. Moreover, if $\partial_B = \partial_{B'}$, then $POS_B(d) = POS_{B'}(d)$ for any pair of nonempty sets $B, B' \subseteq A$. Hence, the definition of a decision-relative reduct: a subset $B \subseteq A$ is a *relative reduct* if it is a minimal set such that $POS_A(d) = POS_B(d)$. Decision-relative reducts may be found from a discernibility matrix $M^d(\mathcal{A}) = (c^d_{ij})$ assuming

$$c^d_{ij} = \begin{cases} c_{ij} - \{d\} \text{ if } (|\partial_A(x_i)| = 1 \text{ or } |\partial_A(x_j)| = 1) \\ \emptyset \qquad\qquad\qquad \text{otherwise.} \end{cases} \tag{20}$$

Matrix $M^d(\mathcal{A})$ is called *the decision-relative discernibility matrix of* $\mathcal{A}$. Construction of *the decision-relative discernibility function* from this matrix follows the construction of the discernibility function from the discernibility matrix. One can show that the set of *prime implicants* of $f^d_M(\mathcal{A})$ defines the set of all *decision-relative reducts* of $\mathcal{A}$.

Since the core is the intersection of all reducts, it is included in every reduct, i.e., each element of the core belongs to some reduct. Thus, the core is the most important subset of attributes since none of its elements can be removed without affecting the classification power of attributes.

Yet another kind of reduct, called reduct relative to objects, can be used for generating minimal decision rules from decision tables ([18,44]).

In some applications, instead of reducts, we prefer to use their approximations called α-reducts, where $\alpha \in [0,1]$ is a real parameter. For a given information system $\mathcal{A} = (U, A)$, the set of attributes $B \subseteq A$ is called α-reduct if B has a nonempty intersection with at least $\alpha \cdot 100\%$ of nonempty sets $c_{i,j}$ of the discernibility matrix of $\mathcal{A}$.

Different kinds of reducts and their approximations are discussed in the literature as basic constructs for reasoning about data represented in information systems or decision tables (see, e.g., [3,48,49]). It turns out that they can be efficiently computed using heuristics based on the Boolean reasoning approach.

3 Feature Selection

Feature selection is a process of finding a subset of features from the original set of features forming patterns in a given data set, optimal according to the given goal of processing and the criterion. An optimal feature selection is a process of finding a subset,

$$A_{opt} = \{a_{1,opt}, a_{2,opt}, \ldots, a_{m,opt}\}, \tag{21}$$

of A, which guarantees accomplishing a processing goal by minimizing a defined feature selection criterion $J_{\text{feature}}(A_{\text{feature\_subset}})$. A solution of an optimal feature selection does not need to be unique.

One can distinguish two paradigms in data model building and potentially, in an optimal feature selection (*minimum construction paradigms*): *the Occam's razor* and *minimum description length principle* [40].

By virtue of the minimum construction idea, one of the techniques for best feature selection could be based on choosing a minimal feature subset that fully describes all concepts (for example, classes in prediction-classification) in a given data set [1,28]. Let us call this paradigm *a minimum concept description*. However, this approach, good for a given (possibly limited) data set, may not be appropriate for processing unseen patterns. A robust processing algorithm with an associated set of features (reflecting complexity) is a trade-off between the ability to process a given data set versus generalization ability.

The second general paradigm of optimal feature selection, mainly used in classifier design, relates to selecting a feature subset that guarantees the maximal between-class separability for reduced data sets. This relates to the discriminatory power of features.

Feature selection methods consists of two main streams [6,11,13,16]: *open-loop methods* and *closed-loop methods*.

Open loop methods (*filter method*) are based mostly on selecting features using a between-class separability criterion [9,11]. They do not use feedback from predictor quality for the feature selection process.

Closed-loop methods [16] also called *wrapper methods*, are based on feature selection using *predictor (classifier) performance* (and thus forming feedback in processing) as a criterion of feature subset selection. A selected feature subset is evaluated using as a criterion, $J_{\text{feature}} = J_{\text{predictor}}$ a performance evaluation $J_{\text{predictor}}$ of a whole prediction algorithm for the reduced data set containing patterns with the selected features as patterns elements.

Let us consider the problem of defining a feature selection criterion for a prediction task based on an original data set T containing N cases $(\mathbf{a}, \mathbf{target})$ constituted of *n-dimensional* input patterns **a** and a **target** pattern of output. Assume that the m-feature subset $A_{\text{feature}} \subseteq A$ ought to be evaluated on the basis of the closed-loop type criterion. A reduced data set T_{feature}, with patterns containing only m-features from the subset A_{feature}, should be constructed. Then, a type of predictor PR_{feature} (for example, k-nearest neighbors, or neural network), used for feature quality evaluation, should be decided. This predictor ideally should be the same as a final predictor PR for a whole design; however, in a simplified suboptimal solution, a computationally less expensive predictor can be used only for feature selection. Let us assume that, for the feature set A considered, a reduced feature data set A_{feature} has been selected and a predictor algorithm PR_{feature} based on A_{feature}, used for feature evaluation, decided. Then, evaluation of feature quality can be provided by using one of the methods used for the final predictor evaluation. This will require defining a performance criterion, $J_{PR_{\text{feature}}}$, of a predictor PR_{feature}, and an error counting method that will show how to estimate performance by averaging results. Consider as an example, a holdout error counting method for predictor performance evaluation. To evaluate the performance of a predictor PR_{feature}, an extracted feature data set T_{feature} is split into a N_{tra} case training set $T_{\text{feature,tra}}$ and a N_{test} case test set $T_{\text{feature,test}}$ (holdout for testing). Each case $(\mathbf{a}_f^i, \mathbf{target}^i)$ of both sets contains a feature pattern $\mathbf{a}_f^i$ labeled by $\mathbf{target}^i$. The evaluation criteria can be defined separately for prediction classification and prediction regression.

We will consider a defining feature selection criterion for a prediction classification task, when a feature subset T_{feature} case contains pairs $(\mathbf{a}_f, c_{\text{target}})$ of a feature input pattern $\mathbf{a}_f$ and a categorical type target c_{target} taking a value corresponding to one of the possible r decision classes C_i. The quality of classifier PR_{feature}, computed on the basis of the limited size test set $T_{\text{feature,test}}$ with N_{test} patterns, can be measured by using the following performance criterion $J_{PR_{\text{feature}}}$ (here equal to a feature selection criterion J_{feature}):

$$J_{PR_{\text{feature}}} = \hat{J}_{\text{all miscl}} = \frac{n_{\text{all miscl}}}{N_{\text{test}}} \cdot 100\%, \tag{22}$$

where $n_{\text{all miscl}}$ is the number of all misclassified patterns and N_{test} is the number of all tested patterns. This criterion estimates the probability of error from the relative frequency of error. Usually, cross-validation techniques are used to obtain better estimation of predictor quality.

An overview of feature selection methods can be found in [22,23]. Let us only mention that several methods of feature selection are inherently built into a predictor design procedure [39] and some methods of feature selection merge feature extraction with feature selection. A feature reduction (pruning) method for a self-organizing neural network map, based on concept description, is suggested in [25].

We will concentrate in this chapter on the rough set approach to feature selection and on some relationships of rough set methods with existing ones.

3.1 Feature Selection Based on Rough Sets

The rough set approach to feature selection can be based on the minimal description length principle [40] and methods for tuning parameters of approximation spaces to obtain high-quality classifiers based on selected features. We have mentioned before an example of such parameter with possible values in the power set of the feature set, i.e., related to feature selection. Other parameters can be used, e.g., to measure the closeness of concepts [44].

One can distinguish two main steps in this approach.

In the first step, by using Boolean reasoning, relevant kinds of reducts from given data tables are extracted. These reducts preserve exactly the discernibility (and some other) constraints (e.g., reducts relative to objects for minimal decision rule generation).

In the second step, reduct approximations are extracted by parameter tuning. These reduct approximations allow shorter concept description than the exact reducts, and they still preserve the constraints to a sufficient degree to guarantee, e.g., sufficient approximation quality of the described (induced) concept [44].

In using rough sets for feature selection, two cases can be distinguished, global and local feature selection schemes. In the former case, the relevant attributes for the whole data table are selected, whereas in the latter case the descriptors of the form, (a,v) where $a \in A$ and $v \in V_a$, are selected for a given object. In both cases, we are searching for relevant features for object classification. In the global case, we are searching for features defining a partition (or covering) of the object universe. This partition should be relevant for describing the approximation of a partition (or part of it) defined by decision attribute. In the local case, we are extracting descriptors defining a relevant neighborhood for a given object with respect to a decision class.

Using rough sets [2,4,28,52] for feature selection was proposed in several contributions (see, e.g., [53,54]). The simplest approach is based on calculation of a core for a discrete attribute data set containing strongly relevant features and reducts containing a core plus additional weakly relevant features, such that each reduct is satisfactory for description of concepts in the data set. Based on a set of reducts for a data set, some criteria for feature selection can be formed, for example, selecting features from a minimal reduct, i.e., a reduct containing a minimal set of attributes. Dynamic reducts were proposed to find a robust (well-generalizing) feature subset [2,4]. The selection of a dynamic reduct is based on the cross-validation method. Methods of dynamic reduct generation have been applied to relevant feature extraction, e.g., for dynamic selection of features represented in discretization as well as in the process of inducing relevant decision rules. Some other methods based on noninvasive data analysis and rough sets are reported in [12]. Let us now summarize the applications of rough set methods for feature selection in a closed loop. The method is based on searching first for short (dynamic) reducts or reduct approximations. This step can be realized using, for example, software systems such as ROSETTA (see http://www.idi.ntnu.no/~ aleks/rosetta/rosetta.html) or RSES (see alfa.mimuw.edu.pl). It can be based on genetic algorithms with the fitness function measuring the quality of the selected reduct approximation B-dependent, among others, on (1) the quality of the reduct approximation by the set B; (2) the cardinality of the feature set B; (3) the discernibility power of the feature set B with respect to the discernibility between decision classes measured, e.g., by means of the approximation quality of a D-reduct by B; (4) the number of equivalence classes created by a feature set on a given data set and/or the number of rules generated by this set [57]; (5) the closeness of concepts [44]; and (6) the conflict resolution strategy [55]. The parameters used to specify and compose the above components into a fitness function are tuned in an evolutionary process to obtain the classifier of the highest quality using the feature set B. The classifier quality is measured by means of the quality of new object classification. Let us finally mention recently reported results based on ensembles of classifiers constructed on the basis of different reducts (see, e.g., [57]). For more details on the application of rough sets to feature selection in a closed loop, refer to Chap. 3.

In the following sections, we point out some relationships of the rough set approach with existing methods for feature selection. The conclusion is that these methods are strongly related to extracting different kinds of reducts.

3.2 Relevance of Features

There have been both deterministic and probabilistic attempts to define *feature relevancy* [1,16,28].

Let us denote by $\mathbf{a}_i$ a vector of features (attributes),

$$(a_1, a_2, \ldots, a_{i-1}, a_{i+1}, \ldots, a_n),$$

obtained from the original feature vector $\mathbf{a}$ by removing a_i. By $\mathbf{v}_i$ is denoted a value of $\mathbf{a}_i$ ([16]).

A feature a_i is *relevant* if there exists some value v_i of that feature, a decision value (predictor output) v, and value $\mathbf{v}_i$ (generally a vector) for which $P(a_i = v_i) > 0$ such that

$$P(d = v, \mathbf{a}_i = \mathbf{v}_i | a_i = v_i) \neq P(d = v, \mathbf{a}_i = \mathbf{v}_i). \tag{23}$$

In the light of this definition, a feature a_i is relevant if the probability of a **target** (given all features) can change if we remove knowledge about a value of that feature.

In [16], other definitions of *strong* and *weak relevance* were introduced.

A feature a_i is *strongly relevant* if there exists some value of that feature v_i, a value v (predictor output) of decision d and a value $\mathbf{v}_i$ of a $\mathbf{a}_i$ for which $P(a_i = v_i, \mathbf{a}_i = \mathbf{v}_i) > 0$ such that

$$P(d = v | \mathbf{a}_i = \mathbf{v}_i, a_i = v_i) \neq P(d = v | \mathbf{a}_i = \mathbf{v}_i). \tag{24}$$

Strong relevance implies that a feature is indispensable, i.e., its removal from a feature vector will change prediction accuracy.

Let us assume that $DT = (U, A, d)$ is a decision table where $V_d = \{1, \ldots, r\}$. The decision d defines the (target) decision classes $DC_s = \{x \in U | d(x) = s\}$ for $s = 1, \ldots, r$. We define a new decision table $DT_d = (U, A, d_A)$ assuming

$$d_A(x) = \left(\mu^A_{C_1}(x), \ldots, \mu^A_{C_s}(x)\right) \text{ for } x \in U. \tag{25}$$

It means that the new decision is equal to the probability distribution defined by the case (object) x in decision table DT. Now, one can show that the reducts relative to such a decision, called frequency related reducts [50], are reducts of the type discussed above.

One can also define reducts corresponding to the relevant features specified by means of the following definition of a relevant feature.

A feature a_i is *weakly relevant* if it is not strongly relevant, and there exists a subsequence $\mathbf{b}_i$ of $\mathbf{a}_i$, for which there exist some value of that feature v_i, a decision value (predictor output) v of d, and a value $\mathbf{v}_i$ of vector $\mathbf{b}_i$, for which $P(a_i = v_i, \mathbf{b}_i = \mathbf{v}_i) > 0$ such that

$$P(d = v | \mathbf{b}_i = \mathbf{v}_i, a_i = v_i) \neq P(d = v | \mathbf{b}_i = \mathbf{v}_i). \tag{26}$$

We can observe that weak relevance indicates that a feature might be dispensable (i.e., not relevant); however, sometimes (combined with some other features), it may improve prediction accuracy.

A feature is *relevant* if it is either *strongly relevant* or *weakly relevant*, otherwise, it is *irrelevant*. We can see that irrelevant features will never contribute to prediction accuracy and thus can be removed.

It has been shown in [16] that for some predictor designs, feature relevancy (even strong relevancy) does not imply that a feature must be in an optimal feature subset.

3.3 Criteria Based on Mutual Information

Entropy can be used as a *mutual information measure* of a data set for feature selection. Let us consider a decision table (data set) $DT = (U,A,d)$. Assume that $A = \{a_1,\ldots,a_n\}$. Then any *n-dimensional* pattern vector $\mathbf{a}(x) = (a_1(x),\ldots,a_n(x))$, where $x \in U$ is labeled by a decision class from $DC = (DC_1,\ldots,DC_r)$. The value of a mutual information measure for a given feature set $B \subseteq A$ can be understood as the suitability of feature subset B for classification. If initially only probabilistic knowledge about classes is given, then the uncertainty associated with the data can be measured by the entropy,

$$E(DC) = -\sum_{i=1}^{r} P(DC_i)\, log_2\, P(DC_i), \tag{27}$$

where $P(DC_i)$ is the a priori probability of a class DC_i occurrence. It is known that entropy $E(DC)$ is an expected amount of information needed for class prediction.

As a measure of uncertainty, the conditional entropy $E(C|B)$ upon the subset of features B can be defined for discrete features as

$$E(DC|B) = -\sum_{all\ \mathbf{v}} P(\mathbf{v}) \left[\sum_{i=1}^{r} P(DC_i|\mathbf{v})\, log_2\, P(DC_i|\mathbf{v})\right]. \tag{28}$$

More generally, for continuous features,

$$E(DC|B) = -\int_{all\ \mathbf{v}} p(\mathbf{v}) \left[\sum_{i=1}^{r} P(DC_i|\mathbf{v})\, log_2\, P(DC_i|\mathbf{v})\right], \tag{29}$$

where $p(\mathbf{v})$ is a probability density function. The mutual information $MI(C,B)$ between the classification and feature subset B is measured by a decrease in uncertainty about the prediction of classes, given knowledge about patterns $\mathbf{v}$ formed from features B

$$J_{\text{feature}}(B) = MI(DC,B) = E(DC) - E(DC|B). \tag{30}$$

One can consider entropy related reducts [50] and Boolean reasoning to extract relevant feature sets with respect to the entropy measure. Moreover, using Boolean reasoning, one can search for frequency related reducts that preserve probability distributions to a satisfactory degree.

3.4 Criteria Based on an Inconsistency Count

An example of criteria for feature subset evaluation can be the *inconsistency measure* [24,28].

The idea of attribute reduction can be generalized by introducing a concept of *significance of attributes* that enables us to evaluate attributes not only in the two-valued scale *dispensable–relevant (indispensable)* but also in the multivalue case by assigning to an attribute a real number from the interval $[0,1]$ that expresses the importance of an attribute in the information table.

The significance of an attribute can be evaluated by measuring the effect of removing the attribute from an information table. It was shown previously that the number $\gamma(C,D)$ expresses the degree of dependency between attributes C and D or the accuracy of the approximation of U/D by C. It may now be checked how coefficient $\gamma(C,D)$ changes when attribute a is removed. In other words, what the difference is between $\gamma(C,D)$ and $\gamma(C-\{a\},D)$. The difference is normalized, and the significance of attribute a is defined by

$$\sigma_{(C,D)}(a) = \frac{\gamma(C,D)-\gamma(C-\{a\},D)}{\gamma(C,D)} = 1 - \frac{\gamma(C-\{a\},D)}{\gamma(C,D)}. \tag{31}$$

Coefficient $\sigma_{C,D}(a)$ can be understood as a classification error which occurs when attribute a is dropped. The significance coefficient can be extended to sets of attributes as follows:

$$\sigma_{(C,D)}(B) = \frac{\gamma(C,D)-\gamma(C-B,D)}{\gamma(C,D)} = 1 - \frac{\gamma(C-B,D)}{\gamma(C,D)}. \tag{32}$$

The *inconsistency rate* used in ([24]) for a reduced data set can be expressed by $J_{\text{inc}}(B) = \sigma_{(C,D)}(B)$.

Another possibility is to consider as relevant the features that come from approximate reducts of sufficiently high quality.

Any subset B of C is called an *approximate reduct* of C and the number,

$$\varepsilon_{(C,D)}(B) = \frac{\gamma(C,D)-\gamma(B,D)}{\gamma(C,D)} = 1 - \frac{\gamma(B,D)}{\gamma(C,D)}, \tag{33}$$

is called an *error of reduct approximation*. It expresses how exactly the set of attributes B approximates the set of condition attributes C with respect to determining D.

Several other methods of reduct approximation based on measures different from the positive region have been developed. All experiments confirm the hypothesis that by tuning the level of approximation the quality of the classification of new objects may be increased in most cases. It is important to note that it is once again possible to use Boolean reasoning to compute the different types of reducts and to extract relevant approximations from them.

3.5 Criteria Based on Interclass Separability

Some of the criteria for feature selection that are based on *interclass separability* are based on the idea of Fisher's linear transformation: a good feature (with high discernibility power) should cause a small within-class scatter and a large between-class scatter [9,11,13].

The rough set approach also offers methods for dealing with interclass separability. In [43], so-called *D*-reducts have been investigated. These reducts preserve not only discernibility between required pairs of cases (objects), but they also allow us to keep the distance between objects from different decision classes above a given threshold (if this is possible).

3.6 Criteria Based on a Minimum Concept Description

Open-loop type criteria of feature selection based on a minimum construction paradigm were studied [1] in machine learning and in statistics for discrete features of noise-free data sets. The straightforward technique of best feature selection could choose a minimal feature subset that fully describes all concepts (for example, classes in classification) in a given data set (see, e.g., [1,28]). Here a criterion of feature selection could be defined as Boolean function $J_{\text{feature}}(B)$ with value one if a feature subset B is satisfactory for describing all concepts in a data set; otherwise, it has a value of zero. The final selection would based on choosing a minimal subset for which a criterion gives a value of one.

The idea of feature selection, with the minimum concept description criterion, can be extended by using the concept of reduct defined in the theory of rough sets [28,44]. A reduct is a minimal set of attributes that describes all concepts. However, a data set may have many reducts. If we use the definition of the above open-loop feature selection criterion, we can see that for each reduct B, we have the maximum value of the criterion $J_{\text{feature}}(B)$. Based on a paradigm of the minimum concept description, we can select a minimum length reduct as the best feature subset. However, the minimal reduct is good for ideal situations, where a given data set fully represents a domain of interest. For real-life situations and limited-size data sets, other reducts (generally other feature subsets) might be better for generalizing prediction. A selection of a robust (generalizing) reduct, as a best open-loop feature subset, can be supported by introducing the idea of a dynamic reduct [2,4] or by an ensemble of classifiers defined by reducts [57].

3.7 Feature Selection with Individual Feature Ranking

One straightforward feature selection procedure is based on an evaluation of the predictive power of individual features, then ranking such evaluated features, and eventually choosing the first best m features [20]. A criterion applied to an individual feature could be of either the open-loop or closed-loop type. This algorithm has limitations and assumes independence of features. It also relies on an assumption that the final selection criterion can be expressed as the sum or products of the criteria evaluated for each feature independently. It can be expected that a single feature alone may have very low predictive power, whereas this feature, when put together with others, may demonstrate significant predictive power.

One can attempt to select a minimal number $\hat{m}$ of the best ranked features that guarantees performance better or equal to a defined level according to a certain criterion $J_{\text{feature,ranked}}$. One criterion for evaluating the predictive power of a feature could be defined by the rough set *measure of significance* of the feature (attribute), discussed before.

4 Principal Component Analysis and Rough Sets for Feature Projection, Reduction, and Selection

Orthonormal projection and reduction of pattern dimensionality may improve the recognition process by considering only the most important data representation, possibly with uncorrelated elements retaining maximum information about the original data and with possible better generalization abilities.

We will discuss PCA for feature projection and reduction, followed by the joint method of feature selection using PCA and the rough set method.

4.1 Principal Component Analysis for Feature Projection and Reduction

We generally assume that our knowledge of a domain is represented as a limited-size sample of N random *n-dimensional patterns* $\mathbf{x} \in \mathbf{R}^n$ representing extracted object features. We assume that an unlabeled training data set $T = \{\mathbf{x}^1, \mathbf{x}^2, \ldots, \mathbf{x}^N\}$ can be represented as an $N \times n$ data pattern matrix $\mathbf{X} = \left[\mathbf{x}^1, \mathbf{x}^2, \ldots, \mathbf{x}^N\right]^T$. The training data set can be characterized by the square $n \times n$ dimensional *covariance* matrix $\mathbf{R}_x$. Assume that the eigenvalues of the covariance matrix $\mathbf{R}_x$ are arranged in the decreasing order $\lambda_1 \geq \lambda_2 \geq \ldots \lambda_n \geq 0$ (with $\lambda_1 = \lambda_{max}$), with the corresponding orthonormal eigenvectors $\mathbf{e}^1, \mathbf{e}^2, \ldots, \mathbf{e}^n$. Then the optimal linear transformation

$$\mathbf{y} = \hat{\mathbf{W}}\mathbf{x}, \tag{34}$$

is provided using the $m \times n$ optimal Karhunen-Loéve transformation matrix $\hat{\mathbf{W}}$ (denoted also by $\mathbf{W}_{\text{KLT}}$),

$$\hat{\mathbf{W}} = \left[\mathbf{e}^1, \mathbf{e}^2, \ldots, \mathbf{e}^m\right]^T, \tag{35}$$

composed of m rows that are the first m orthonormal eigenvectors of the original data covariance matrix $\mathbf{R}_x$. The optimal matrix $\hat{\mathbf{W}}$ transforms the original *n-dimensional* patterns $\mathbf{x}$ into *m-dimensional* ($m \leq n$) feature patterns $\mathbf{y}$,

$$\mathbf{Y} = (\hat{\mathbf{W}}\mathbf{X}^T)^T = \mathbf{X}\hat{\mathbf{W}}^T, \tag{36}$$

minimizing the mean least square reconstruction error. The PCA method can be effectively used for feature extraction and dimensionality reduction by forming the *m-dimensional* ($m \leq n$) feature vector $\mathbf{y}$ containing only the first m most dominant principal components of $\mathbf{x}$. The open question remains, which principal components to select as the best for a given processing goal. One of the possible methods (criteria) for selecting a dimension of a reduced feature vector $\mathbf{y}$ is to choose a minimal number of the first m most dominant principal components $y_1, y_2, \ldots, y_m$ of $\mathbf{x}$ for which the mean square reconstruction error is less than the heuristically set error threshold ε. Another method may assume selecting the minimal number of the first m most dominant principal components for which a percentage V of a sum of unused eigenvalues of a sum of all eigenvalues,

$$V = \frac{\sum_{i=m+1}^{n} \lambda_i}{\sum_{i=1}^{n} \lambda_i} 100\%, \tag{37}$$

and is less than a defined threshold ζ.

We have applied PCA, with the resulting Karhunen–Loéve transformation (KLT) [9,11,6], for orthonormal projection (and reduction) of reduced singular value decomposition (SVD) patterns $\mathbf{x}_{svd,r}$ representing recognized face images.

The selection of the best principal components for classification is yet another feature selection problem. In the next section, we will discuss an application of rough sets to feature selection/reduction.

4.2 Application of Rough Set Based Reducts for Selecting of Discriminatory Features from Principal Components

The PCA, with the resulting linear Karhunen–Loéve projection, provides feature extraction and reduction optimal from the point of view of minimizing the reconstruction error. However, PCA does not guarantee that selected first principal components, as a feature vector, will be adequate for classification. Nevertheless, the projection of high-dimensional patterns into lower dimensional orthogonal principal component feature vectors might help to provide better classification for some data types.

In many applications of PCA, an arbitrary number of the first dominant principal components is selected as a feature vector. However, these methods do not cope with the selection of the most discriminative features well suitable for classification. Even assuming that the Karhunen–Loéve projection can help in classification and can be used as a first step in the feature extraction/selection procedure, still an open question remains, "which principal components to choose for classification?"

One of the possibilities for selecting features from principal components is to apply rough set theory [28,44]. Specifically, defined in rough sets, the computation of a reduct can be used for selecting some principal components. Thus, these principal components will describe all concepts in a data set. For a suboptimal solution, one can choose the minimal length reduct or dynamic reduct as a selected set of principal components forming a selected, final feature vector. The following steps can be proposed for the PCA and rough sets based procedure for feature selection. Rough sets assume that a processed data set contains patterns labeled by associated classes with the discrete values of its elements (attributes, features). We know that PCA is predisposed to transform patterns with real-valued features (elements) optimally. Thus, after realizing the Karhunen–Loéve transformation, the resulting projected pattern features must be discretized by some adequate procedure. The resulting discrete attribute valued data set (an information system) can be processed using rough set methods.

Let us assume that we are given a limited-size data set T, containing N cases labeled by associated classes,

$$T = \{(\mathbf{x}^1, c^1_{\text{target}})(\mathbf{x}^2, c^2_{\text{target}}), \ldots, (\mathbf{x}^N, c^N_{\text{target}})\}. \tag{38}$$

Each case $(\mathbf{x}^i, c^i_{\text{target}})$ $(i = 1, 2, \ldots, N)$ is constituted of an *n-dimensional* real-valued pattern $\mathbf{x}^i \in \mathbf{R}^n$ with corresponding categorical target class c^i_{target}. We assume that a data set T contains N_i $(\sum_i^l N_i = N)$ cases from each categorical class c_i, with the total number of classes denoted by l.

Since PCA is an unsupervised method, first, from the original, class labeled data set T, a pattern part is isolated as an $N \times n$ data pattern matrix,

$$\mathbf{X} = \begin{bmatrix} \mathbf{x}^1 \\ \mathbf{x}^2 \\ \cdots \\ \mathbf{x}^N \end{bmatrix}, \tag{39}$$

which each row contains one pattern. The PCA procedure is applied to the extracted pattern matrix $\mathbf{X}$, with a resulting full size an $n \times n$ optimal Karhunen–Loéve matrix $\mathbf{W}_{\text{KL}}$ (where n is the length of the original pattern $\mathbf{x}$). Now, according to the designer decision, the number $m \leq n$ of first dominant principal components has to be selected. Then, the reduced $m \times n$ Karhunen–Loéve matrix $\hat{\mathbf{W}}_{\text{KL}}$, containing only the first m rows of the full size matrix $\mathbf{W}$, is constructed. Applying the matrix $\mathbf{W}_{\text{KL}}$ the original *n-dimensional* pattern $\mathbf{x}$ can be projected, using transformation

$\mathbf{y} = \hat{\mathbf{W}}_{\text{KL}}\mathbf{x}$, into the reduced *m-dimensional* pattern $\mathbf{y}$ in the principal component space. The entire projected $N \times m$ matrix $\mathbf{Y}$ of patterns can be obtained by the formula $\mathbf{Y} = \mathbf{X}\hat{\mathbf{W}}_{\text{KL}}^T$.

At this stage, the reduced, projected data set, represented by $\mathbf{Y}$ (with real-valued attributes), has to be discretized. As a result, the discrete attribute data set represented by the $N \times m$ matrix Y_d is computed. Then, the patterns from $\mathbf{Y}_d$ are labeled by corresponding target classes from the original data set T. They form a decision table DT_m with *m-dimensional* principal component related patterns. From the decision table DT_m, one can compute the selected reduct $A_{\text{feature,reduct}}$ of size l (for example, minimal length or dynamic reduct) as a final selected attribute set. Here, a reduct computation is a pure feature selection procedure.

Once the selected attribute set has been found (as a selected reduct), the final discrete attribute decision table $DT_{f,d}$ is composed. It consists of those columns from the discrete matrix $\mathbf{Y}_d$ that are included in the selected feature set $A_{\text{feature,reduct}}$. Each pattern in $DT_{f,d}$ is labeled by the corresponding target class. Similarly, one can obtain a real-valued resulting reduced decision table $DT_{f,l}$ extracting (and adequately labeling by classes) those columns from the real-valued projected matrix $\mathbf{Y}$ that are included in the selected feature set $A_{\text{feature,reduct}}$. Both resulting reduced decision tables can be used for classifier design.

Algorithm: Feature extraction/selection using PCA and rough sets.
Given: An *N-case* data set T containing *n-dimensional* patterns, with real-valued attributes, labeled by l associated classes $\{(\mathbf{x}^1, c^1_{target}), (\mathbf{x}^2, c^2_{target}), \ldots, (\mathbf{x}^N, c^N_{target})\}$.

1. Isolate from the original class labeled data set T a pattern part as an $N \times n$ data pattern matrix $\mathbf{X}$.
2. Compute covariance matrix $\mathbf{R}_x$ for matrix $\mathbf{X}$.
3. Compute the eigenvalues and corresponding eigenvectors for matrix $\mathbf{R}_x$, and arrange them in descending order.
4. Select the reduced dimension $m \leq n$ of a feature vector in principal component space using a defined selection method, which may be based on the judgment of the ordered values of the computed eigenvalues.
5. Compute the optimal $m \times n$ Karhunen–Loéve transform matrix $\hat{\mathbf{W}}_{\text{KL}}$ based on eigenvectors of $\mathbf{R}_x$.
6. Transform original patterns from $\mathbf{X}$ into *m-dimensional* feature vectors in the principal component space by formula $\mathbf{y} = \hat{\mathbf{W}}_{\text{KL}}\mathbf{x}$ for a single pattern, or formula $\mathbf{Y} = \mathbf{X}\hat{\mathbf{W}}_{\text{KL}}$ for a whole set of patterns (where $\mathbf{Y}$ is an $N \times m$ matrix).
7. Discretize the patterns in $\mathbf{Y}$ with the resulting matrix $\mathbf{Y}_d$.
8. Compose the decision table DT_m constituted of the patterns from matrix $\mathbf{Y}_d$ with the corresponding classes from the original data set T.
9. Compute a selected reduct from the decision table DT_m treated as a selected set of features $A_{\text{feature,reduct}}$ describing all concepts in DT_m.

10. Compose the final (reduced) discrete attribute decision table $DT_{f,d}$ containing those columns from the projected discrete matrix $\mathbf{Y}_d$ that correspond to the selected feature set $A_{\text{feature,reduct}}$. Label patterns by corresponding classes from the original data set T.
11. Compose the final (reduced) real-valued attribute decision table $DT_{f,r}$ containing those columns from the projected discrete matrix $\mathbf{Y}_d$ that correspond to the selected feature set $A_{\text{feature,reduct}}$. Label patterns by corresponding classes from the original data set T.

The results of the method of feature extraction/selection discussed depend on the data set type and three designer decisions:

1. Selection of dimension $m \leq n$ of the projected pattern in the principal component space.
2. Discretization method (and resulting quantization) of the projected data.
3. Selection of a reduct.

First, for the selected dimension m, the applied quantization method may lead to an inconsistent decision table DT_m for which no reduct exists (preserving discernibility between all pairs of objects from different decision classes). Then, a designer should return to the discretization step and select another discretization. Even if a reduct cannot be found for all possible discretization attempts, a return is realized to the stage of selecting a dimension m of the reduced feature vector $\mathbf{y}$ in the principal component space. It means that possibly the projected vector does not contain a satisfactory set of features. In this situation, a design procedure should provide the next iteration with a selected larger value of m. If a reduct cannot be found for $m = n$, a data set is not classifiable in a precise deterministic sense. Last, selection of a reduct will impact the ability of a classifier designed to generalize predictions for unseen objects.

5 Numerical Experiments — Face Recognition

As a demonstration of the role of rough set methods in feature selection/reduction, we have carried out numerical experiments of face recognition. We considered the ORL (see www.cam-orl.co.uk/facedatabase.html) face database [41] gray-scale face image data sets. We provided, separately, recognition experiments for 10 category data sets and 40 category data sets of face images. Each category was represented by 10 instances of face images. Each gray-scale face image had the dimensions of 112×92 pixels. Feature extraction from face images was provided by SVD.

Face images were classified with a single, hidden-layer error back-propagation neural network, learning vector quantization neural network (LVQ) and rule-based rough set classifier.

5.1 Singular Value Decomposition for Feature Extraction from Face Images

Singular value decomposition can be used to extract features from images [14,53]. A rectangular $n \times m$ real image represented by an $n \times m$ matrix $\mathbf{A}$, where $m \leq n$, can be transformed into a diagonal matrix by SVD. Assume that the rank of matrix $\mathbf{A}$ is $r \leq m$. The matrices $\mathbf{A}\mathbf{A}^T$ and $\mathbf{A}^T\mathbf{A}$ are nonnegative and symmetrical and have the identical eigenvalues λ_i. For $m \leq n$, there are at most $r \leq m$ nonzero eigenvalues. The SVD transform decomposes matrix $\mathbf{A}$ into the product of two orthogonal matrices, Ψ of dimension $n \times r$, and Φ of dimension $m \times r$, and a diagonal matrix $\Lambda^{1/2}$ of dimension $r \times r$. The *singular value decomposition* (SVD) of a matrix (image) $\mathbf{A}$ is given by

$$\mathbf{A} = \Psi \Lambda^{1/2} \Phi^T = \sum_{i=1}^{r} \sqrt{\lambda_i} \psi_i \varphi_i^T, \tag{40}$$

where the matrix Ψ and Φ have r orthogonal columns $\psi_i \in \mathbf{R}^n$, $\varphi_i \in \mathbf{R}^m$ $(i = 1, \ldots, r)$, respectively (representing orthogonal eigenvectors of $\mathbf{A}\mathbf{A}^T$ and $\mathbf{A}^T\mathbf{A}$). The square matrix $\Lambda^{1/2}$ has diagonal entries defined by

$$\Lambda^{1/2} = diag(\sqrt{\lambda_1}, \sqrt{\lambda_2}, \ldots, \sqrt{\lambda_r}), \tag{41}$$

where $\sigma_i = \sqrt{\lambda_i}$ $(i = 1, 2, \ldots, r)$ are the *singular values* of matrix $\mathbf{A}$. Each λ_i, $(i = 1, 2, \ldots, r)$ is the nonzero eigenvalue of $\mathbf{A}\mathbf{A}^T$ (as well as $\mathbf{A}^T\mathbf{A}$). Given a matrix $\mathbf{A}$ (an image) decomposed $\mathbf{A} = \Psi \Lambda^{1/2} = \Phi^T$ and since Ψ and Φ have orthogonal columns, thus the *singular value decomposition transform* (SVD transform) of the image $\mathbf{A}$ is defined as

$$\Lambda^{1/2} = \Psi^T \mathbf{A} \Phi. \tag{42}$$

If matrix $\mathbf{A}$ represents an $n \times m$ image, then r singular values $\sqrt{\lambda_i}$ $(i = 1, 2, \ldots, r)$ from the main diagonal of the matrix $\Lambda^{1/2}$ can be considered extracted features of the image. These r singular values can be arranged as an image feature vector (SVD pattern) $\mathbf{x}_{svd} = [\sqrt{\lambda_1}, \sqrt{\lambda_2}, \ldots, \sqrt{\lambda_r}]^T$ of an image.

Contrary to principal component analysis, SVD is a purely matrix processing technique, not a direct statistical technique. SVD decomposition is applied to each face image separately as a face feature extraction, whereas eigenfaces [56] are obtained by projecting face vectors into principal component space derived statistically from the covariance matrix of the set of images.

Despite the expressive power of the SVD transformation [14], it is difficult to say arbitrarily how powerful the SVD features could be for classification of face images.

The *r-element* SVD patterns can be heuristically reduced by removing their r_r trailing elements whose values are below the heuristically selected threshold $\varepsilon_{\mathrm{svd}}$. This can result in $n_{\mathrm{svd},r} = r - r_r$ element reduced SVD patterns $\mathbf{x}_{\mathrm{svd},r}$. In the next sections, we discuss techniques of finding a reduced set of face image features.

5.2 ORL Data Sets

The entire image data set was divided into training and test sets: 70% of these images were used for the training set. Given the original face image set, we applied feature extraction using SVD of matrices representing image pixels. As a result, we obtained for each image a 92-element $\mathbf{x}_{\mathrm{svd}}$ SVD pattern where the features were the singular values of an object matrix ordered in the descending order. In the next step, we carried out several simple classification experiments using SVD patterns of different lengths to estimate the suboptimal reduction of these patterns. These patterns are obtained by cutting trailing elements from the original 92-element SVD pattern.

These experiments helped to select 60-element reduced SVD patterns $\mathbf{x}_{\mathrm{svd},r}$. Then, according to the proposed method, we applied PCA for feature projection/reduction based on the reduced SVD patterns from the training set. Similarly to the reduction for the SVD pattern, we provided several classification experiments for different lengths of reduced PCA patterns. These patterns are obtained by considering only a selected number of the first principal components. Finally, the projected 60-element PCA patterns were in this way heuristically reduced to 20-element reduced PCA patterns $\mathbf{x}_{\mathrm{svd},r,\mathrm{pca},r}$. In the last preprocessing step, the rough set method was used for the final feature selection/reduction of the reduced PCA continuously valued patterns. To discretize the continuously reduced PCA features we applied the method of dividing each attribute value range into 10 evenly spaced bins. The discretized training set was used to find relevant reducts, e.g., the minimal reduct [18]. This reduct was used to form the final pattern. The training and the test sets (decision tables) with real-value pattern attributes were reduced according to the selected reduct.

In this chapter, we describe the simplest approach to relevant reduct selection. Existing rough set methods can be used to search for other forms of relevant reducts. Among them are those based on ensembles of classifiers [10]. In our approach, first a set of reducts of high quality is induced. This set is used to construct a set of predictors, and next from such predictors, the global predictor is constructed using an evolutionary approach (for details, see [57]). Predictors based on these more advanced methods make possible to achieve predictors of better quality. Certainly, the whole process of inducing such classifiers needs more time.

In all of these cases, statistical methods, e.g., cross-validation techniques, are used to estimate the robustness of the predictors constructed.

5.3 Neural Network Classifier

The error back-propagation neural network classifier designed was composed of an input layer, one hidden layer, and an output layer followed by a class-choosing module. The network learning algorithm had momentum and adaptive learning techniques built into it. First, we studied a 10-category data set with 90% of the cases in the training set and 10% cases in the test set. We selected the five element reduct based on the reduced 20-element PCA pattern of the training set. A neural network with 50 neurons in the hidden layer was designed. The number of hidden neurons was chosen on the basis of the experiments performed. The neural network provided 99% correct classification of the test set. The rough set rule based classifier for the discretized data set restricted to the attributes from the five element reduct has exhibited 100% accuracy.

We also studied a 40-category data set with a total number of 400 cases. For this data set, we selected seven element reduct of the 320-case training set as a base for the final feature selection of reduced PCA patterns. An error back-propagation neural network with 300 neurons was designed. The number of neurons in the hidden layer was chosen experimentally. The neural network provided 96.25% correct classification of the 320-case training set and 75.5% accuracy for the 80-case test set. We applied the resilient back-propagation algorithm as a network training function that updates weight and bias values, with a performance criterion goal of 0.000299. The rough set rule based classifier for the discretized data set restricted to the attributes from the seven element reduct exhibited 94.5% accuracy for the 80-case test set.

The learning vector quantization (LVQ) neural network, trained for the training set with reduced final patterns, provided 95.8% accuracy for the test set with 28 cases. The network was trained for 200 code-book vectors and $k = 4$ neighbors.

The SVD has demonstrated a potential as a feature extraction method for face images. The processing sequence SVD, PCA with Karhunen–Loéve transformation, and the rough set approach created possibilities for a significant reduction of pattern dimensionality with an increase in classification accuracy and generalization. The classifiers considered have demonstrated the ability to recognize face images after such substantial reduction of pattern length.

6 AR Schemes and Rough Neural Networks

In the previous sections, we discussed hybrid methods for classifier construction with the application of the rough set approach, soft computing methods (e.g., neural networks), and classical statistical methods (e.g., PCA). Two basic steps can be distinguished in the methods presented: (i) reduction in preprocessing of data dimensions (number of features) and (ii) inducing classifier descriptions from reduced data. The methods are not supported by background knowledge, which could help to construct classifiers.

Now, we would like to outline an approach based on soft background knowledge represented in natural language which can be used in searching for complex classifiers. We assume that background knowledge consists of some vague concept representations and relations between them, i.e., we assume that an ontology of concepts relevant to a given problem is specified. Using ontology one can derive some reasoning schemes over vague concepts. We will consider complex information granules representing approximations of such reasoning schemes. We call them AR schemes. In a distributed environment, i.e., when information about concepts is exchanged between different agents (sources of information), its is necessary to add one more component to AR schemes that is responsible for approximate translation of information granules received by agents from other agents. AR schemes extended to a distributed environment of agents are called rough-neural networks.

The AR schemes are discussed in Chap. 3. We consider a special case where standards are represented by vague concepts expressed in natural language, and we outline an approach based on AR schemes. We consider applications of AR schemes for complex networks of classifiers constructed by means of experimental data and soft background knowledge.

6.1 Classifiers as Information Granules

An important class of information granules creates classifiers. One can observe that sets of decision rules generated from a given decision table $DT = (U,A,d)$ (see, e.g., [18]) can be interpreted as information granules. Classifier construction from a DT can be described as follows:

1. First, one can construct granules G_j corresponding to each particular decision $j = 1,\ldots,r$ by taking a collection $\{g_{ij} : i = 1,\ldots,k_j\}$ of left-hand sides of decision rules for a given decision.
2. Let E be a set of elementary granules (e.g., defined by conjunction of descriptors [18]) over $IS = (U,A)$. We can now consider a granule denoted by $Match(e,G_1,\ldots,G_r)$ for any $e \in E$ that is a collection of coefficients ε_{ij} where $\varepsilon_{ij} = 1$ if the set of objects defined by e in IS is included in the meaning of g_{ij} in IS, i.e., $Sem_{IS}(e) \subseteq Sem_{IS}(g_{ij})$; and zero, otherwise. Hence, the coefficient ε_{ij} is equal to one if and only if granule e matches granule g_{ij} in IS.
3. Let us now denote by $Conflict\_res$ an operation (resolving conflict between decision rules recognizing elementary granules) defined on granules of the form $Match(e,\ G_1,\ \ldots,\ G_r)$ with values in the set of possible decisions $1,\ldots,r$. Hence,

   $$Conflict\_res\,[Match(e,G_1,\ldots,G_r)]\,,$$

   is equal to the decision predicted by the classifier,

   $$Conflict\_res\,[Match(\bullet,G_1,\ldots,G_r)]\,,$$

   on the input granule e.

Hence, classifiers are special cases of information granules. Parameters to be tuned are voting strategies, matching strategies of objects against rules as well as other parameters like closeness of granules in the target granule.

Classifier construction is illustrated in Fig. 1, where three sets of decision rules are presented for the decision values 1,2, and 3, respectively. Hence, $r = 3$. In the figure, to omit too many indexes, we write α_i instead of g_{i1}, β_i instead of g_{i2}, and γ_i instead of g_{i3}, respectively. Moreover, $\varepsilon_1, \varepsilon_2, \varepsilon_3$, denote $\varepsilon_{1,1}, \varepsilon_{2,1}, \varepsilon_{3,1}$; $\varepsilon_4, \varepsilon_5, \varepsilon_6, \varepsilon_7$ denote $\varepsilon_{1,2}, \varepsilon_{2,2}, \varepsilon_{3,2}, \varepsilon_{4,2}$; and $\varepsilon_8, \varepsilon_9$ denote $\varepsilon_{1,3}, \varepsilon_{2,3}$, respectively. The reader can

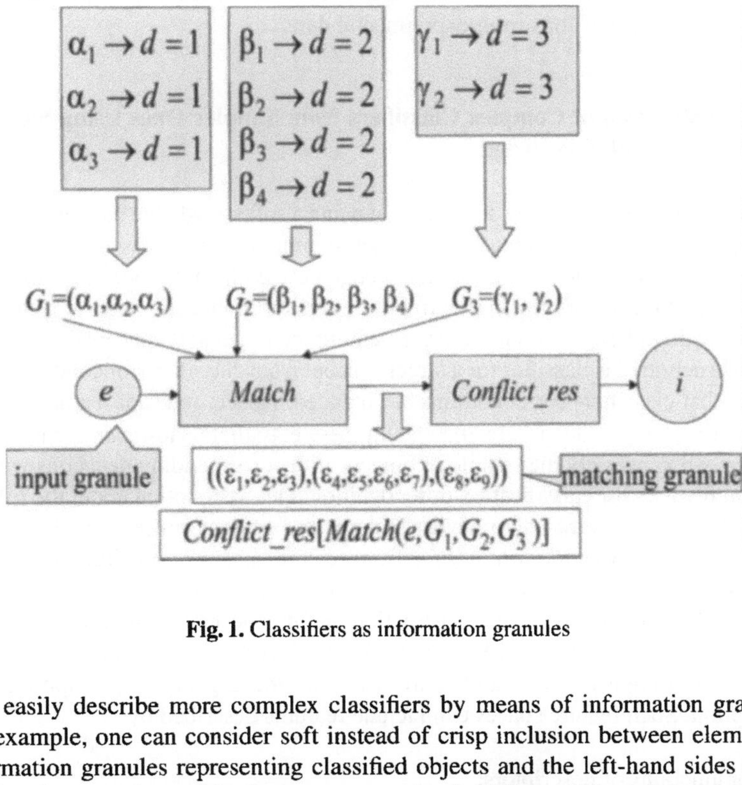

Fig. 1. Classifiers as information granules

now easily describe more complex classifiers by means of information granules. For example, one can consider soft instead of crisp inclusion between elementary information granules representing classified objects and the left-hand sides of decision rules or soft matching between recognized objects and the left-hand sides of decision rules.

6.2 Soft Background Knowledge

We are assuming a knowledge base formulated by means of soft concepts, and relations between them are given. Such background knowledge is called soft ontology and is represented in natural language. One can construct derivations over such ontologies. They represent reasoning schemes in natural language. We are interested in derivations with conclusions representing decision classes. Such derivations can be treated as soft descriptions of cases. A set of such derivations is called a knowledge base. It consists of soft information about conclusions related to complex concepts, assuming that some simpler or elementary ones are satisfied. We are going to show how such knowledge bases can be used in searching for complex classifiers. Moreover, we present an outline for carrying out reasoning from measurements to conclusions about complex soft concepts using AR schemes and rough-neural networks. One can treat our approach as a search method for relevant features supported by background knowledge represented in natural language. Our methodology can be treated as one for building interfaces between experimental knowledge and expert knowledge represented in natural language. The aim is to use the background knowledge to derive conclusions from experimental data.

6.3 Construction of Complex Classifiers from Simpler Ones Using Soft Background Knowledge

In this section, we discuss the possibility of using a soft knowledge base as a guide in searching for relevant features for constructing more complex classifiers from simpler ones. Any soft rule from a soft knowledge base with a left-hand side consisting of a conjunction of soft conditions (representing soft concepts) and the right-hand side consisting of target condition (representing the target soft concept) can be used for construction of a classifier for a target concept from classifiers for conditions. We assume that classifiers for conditions are induced. Hence, relevant features for approximating these concepts are encoded in these classifiers. However, for the target concept, we know only the sample of objects with corresponding decisions without relevant features for classification or recognition. Our assumption about the rules in a knowledge base is that the relevant features for a target concept can be discovered in feature spaces *that are not far* from feature spaces of condition classifiers. Using Boolean reasoning one can measure the distance between feature spaces by means of the complexity of the construction that it is necessary to perform to reach one such space from another. We would like to illustrate this intuition by presenting several examples of relevant feature spaces for the target concept's approximate description. Such feature spaces can include features described by

1. conjunctions of descriptors;
2. disjunctions of conjunctions of descriptors;
3. disjunctions of conjunctions of descriptor disjunctions; and
4. characteristic functions of clusters.

In all cases, the descriptors are selected from the feature spaces of input classifiers from which a new target classifier is constructed. One can observe that features described by descriptor disjunctions correspond to a symbolic value grouping of nominal features or a discretization of real value features [18]. Disjunctions of conjunctions of such features describe higher level patterns generalized next to clusters.

The clusters are constructed by means of such patterns and by an appropriately chosen similarity measure between patterns. The similarity measure should make it possible to generalize the previously defined patterns to clusters. The clusters are next used for defining features relevant to the new classifier construction. A mechanism for measuring the degree of closeness of input patterns to such clusters should be developed for computing of degrees to which analyzed objects are included in the cluster. Such degrees are treated as values of features defined by clusters.

One can interpret the process of searching for relevant features as a searching process for descriptors corresponding to such clusters. Such descriptors should satisfy the following constraint: sets of objects defined by descriptor conjunctions should be included to a satisfactory degree in a given concept (e.g., decision class) and should be supported by sufficiently many objects. In this way, such descriptors are making it possible to obtain short descriptions of concept approximations.

Certainly, one can use some more sophisticated operations transforming the feature spaces of condition classfiers into feature spaces of target classifiers. Evolutionary computing [19] can search for relevant features in such feature spaces.

The basic assumption is that using the soft knowledge base can help us to discover relevant features for more complex classifiers.

The approach discussed does not yet guarantee the robustness of classifiers, i.e., preserving the high quality of new object classification (or recognition) under acceptable deviations of information about objects. We propose an approach making it possible to eliminate this drawback. The approach is based on methods for constructing reasoning clusters constructed *along* derivations in natural language. These reasoning clusters link pattern clusters consisting of patterns sufficiently included in so-called standards (prototypes) or close to each other. The inclusion (closeness) degree of patterns in clusters is controlled to guarantee that under deviations of input patterns, the deviation of output patterns still returns acceptable solutions. This idea is formalized by using AR schemes and rough-neural networks. In the following section, we outline a solution based on AR schemes and rough-neural networks, and we emphasize their possible applications in pattern recognition.

6.4 AR Schemes and Rough-Neural Networks

In this section, we briefly recall an approach for approximate reasoning based on AR schemes (see Chap. 3). We use terminology from the multiagent area [15].

We assume each agent $ag \in Ag$ is equipped with a system of information granules $S(ag)$. Using such a system, the agent ag creates a representation for all of its components.

Agents are able to extract local approximate reasoning schemes, called productions, from such representations. Algorithmic methods for extracting such productions from data are discussed, e.g., in [30,45,47]. They are based on decomposition strategies.

The right-hand side of any *production* for decomposition of condition α at *ag* is of the form

$$\alpha, \varepsilon^{(i)}, \tag{43}$$

and the left-hand side is of the form

$$\alpha_1, \varepsilon_1^{(i)}; \ldots; \alpha_n, \varepsilon_n^{(i)}, \tag{44}$$

where $i = 1, \ldots, k$ for some k.

Such a production represents information about an operation o that can be performed by an agent *ag*. In the production, n denotes the arity of operation. The operation o represented by the production transforms standard (prototype) input information granules represented by $\alpha_1, \ldots, \alpha_n$ into standard (prototype) information granule α. Moreover, if input information granules $g_1, \ldots, g_n$ are included (close) to $\alpha_1, \ldots, \alpha_n$ to degrees $\varepsilon_1^{(i)}, \ldots, \varepsilon_n^{(i)}$, then the result of operation o on information granules $g_1, \ldots, g_n$ is included (close) in the standard α to a degree at least $\varepsilon^{(i)}$, where $1 \leq i \leq k$. Standard (prototype) granules can be interpreted in different ways. In particular, in the applications discussed for pattern recognition, they describe the centers of discovered clusters. In more general cases, standards correspond to concept names expressed in natural language.

Sample productions are basic components of a reasoning system related to the agent set Ag. An important property of such productions is that they are expected to be discovered from available experimental data and background knowledge. Let us observe also that the degree structure is not necessarily restricted to positive reals from the interval $[0,1]$. The inclusion degrees can be complex information granules used to represent the degree of inclusion.

It is worthwhile mentioning that productions can also be interpreted as constructive descriptions of some operations on fuzzy sets (see Chap. 3). The methods for such constructive descriptions are based on rough sets and Boolean reasoning (see, e.g., [18,28]).

Reasoning in multiagent system can be represented as a process of constructing information granules. This process is not restricted to internal operations performed by agents. The agents can communicate. In this process, they exchange some information granules. It is important to note that any agent possesses her/his own information granule system. Hence, a granule received by one agent from another agent cannot be, in general, understood precisely by the receiving agent. We assume that associated with the jth argument of any operation o performed by an agent ag, there is an approximation space $AS(ag)^j$ (see, e.g., [38,47]) making it possible to construct relevant approximations of the received information granules used next as operation arguments. The result of approximation is an information granule in the information granule system of agent ag. In some cases, approximation can be induced using rough set methods (see, e.g., [47]). In general, constructing information granule approximations is a complex process because, for instance, a high-quality approximation of concepts often can be obtained only through dialog (including negotiations, conflict resolution, and cooperation) among agents. In this process, the approximation can be constructed gradually when dialog is progressing.

Approximation spaces are usually parameterized. This means that it is necessary to tune their parameters to find suboptimal approximations of information granules. This observation was the starting point for the rough-neural computing paradigm (see [26,38,45,47] and Chap. 3).

In general, the inputs of rough neurons are derived from information granules instead of real numbers, and parameterized approximation spaces correspond to real weights in the classical neuron. The result of operation o depends on the parameters chosen for approximation spaces. The process of tuning the parameters of such approximation spaces corresponds to the process of weight tuning in classical neurons (see Fig. 1 in Chap. 2).

Now, we are able to discuss one of the main concepts of our approach, approximate reasoning schemes (AR schemes). They can be treated as derivations obtained by using the productions of different agents. Assume, for simplicity of consideration, that agents are working using the same system of information granules, i.e., they do not use approximation spaces to approximate granules received from other agents. The approach can be extended to the more general case. The relevant derivations defining AR schemes satisfy a so-called robustness (or stability) condition, that is, at any node of a derivation, the inclusion (or closeness) degree of a constructed granule (to a given standard) is higher than required by the production to which the result should be sent. This makes it possible to obtain a sufficient robustness condition for the whole derivation. For details refer to [31,34–37] and to chapters in this book discussing the foundations of rough-neural computing approach. In the general case, i.e., when it is necessary to use approximation spaces, the AR schemes can be interpreted as rough neural networks. When standards are interpreted as concept names in natural language and there is given a reasoning scheme in natural language over such standards, the corresponding rough-neural network represents a cluster of reasoning constructions approximately following (in other information granule systems) the reasoning given in natural language.

Let us observe that AR schemes are not classical proofs defined by means of deductive systems. They are approximate reasoning schemes discovered from data and background knowledge. The notion of classical proof is substituted by derivations defining AR schemes, i.e., derivations satisfying some constraints. Deductive systems are substituted by productions systems of agents linked by approximation spaces, communication strategies, and mechanisms deriving AR schemes. This revision of classical logical notions seems to be important for solving complex pattern recognition problems.

6.5 Illustrative Example

Let us consider a very simple illustrative example of the face-recognition problem. Assume that among the concepts of a given knowledge base are the following concepts:

1. $exactly\_one\_ear\_visible = yes$.
2. $nose\_shape\_visible = yes$.
3. $nose\_shape\_in\_front\_view = sharp$.
4. $nose\_shape = sharp$.
5. $face\_in\_side\_view = yes$.

and rules

1. **If** $exactly\_one\_ear\_visible = yes$ **and** $nose\_shape\_visible = yes$,
   then $face\_in\_side\_view = yes$.
2. **If** $face\_in\_side\_view = yes$
   and $nose\_shape = sharp$,
   then $nose\_shape\_in\_front\_view = sharp$.

First, a classifier for the concept

$$face\_in\_side\_view = yes,$$

is constructed in the context of classifiers for

$$exactly\_one\_ear\_visible = yes, \quad nose\_shape\_visible = yes,$$

and next a classifier for $nose\_shape\_in\_front\_view = sharp$ is constructed in the context of its sensory classifiers,

$$face\_in\_side\_view = yes, \quad nose\_shape = sharp.$$

Then, productions for such rules are induced. Finally, they are used to derive robust AR schemes. From such schemes, one can predict on the basis of estimates from sensory classifiers that the *nose_shape_in_front_view* should be to a high degree *sharp* for a given object x if sensory properties for this object x are satisfied to sufficient degrees:

exactly_one_ear_visible = *yes*, *nose_shape_visible* = *yes*, and *nose_shape* = *sharp*.

This can be confronted with another conclusion of approximate reasoning on objects from the database, e.g., the face is to a sufficient degree *in_the_front_view* and the *nose_shape* is to a sufficient degree *non_sharp*. Such objects can be eliminated from candidates identifying x in the database.

7 Conclusions

We have presented a rough set method and its role in feature selection for pattern recognition.

In the first part, we proposed a sequence of data mining steps, including application of SVD, PCA, and rough sets, for feature selection. This processing sequence has shown a potential for feasible feature extraction and feature selection in designing neural network classifiers for face images. The method discussed provides a substantial reduction of pattern dimensionality. Rough set methods have shown the ability to reduce significantly pattern dimensionality and have proven to be viable data mining techniques as the front end of neural network classifiers.

In the second part, we discussed an approach to pattern recognition based on rough-neural computing with the application of soft knowledge bases. This research direction seems to be promising for complex pattern recognition problems, such as identification of objects and path planning by autonomous systems.

Acknowledgments
The research has been partially supported by the COBASE project from NSF National Research Council, USA, National Academy of Sciences, USA and Poland 2000–2001. Moreover, the research of Andrzej Skowron has been partially supported by the State Committee for Scientific Research of the Republic of Poland (KBN), research grant 8 T11C 025 19, and by a Wallenberg Foundation grant.

References

1. H. Almuallim, T.G. Dietterich. Learning with many irrelevant features. In *Proceedings of the Ninth National Conference on Artificial Intelligence*, 574–552, AAAI Press, Menlo Park, CA, 1991.
2. J. Bazan. A comparison of dynamic and non-dynamic rough set methods for extracting laws from decision system. In *[32]*, 321–365, 1998.
3. J. Bazan, S.H. Nguyen, H.S. Nguyen, P. Synak, J. Wróblewski. Rough set algorithms in classification problems. In *[29]*, 49–88, 2000.

4. J. Bazan, A. Skowron, P. Synak. Dynamic reducts as a tool for extracting laws from decision tables. In *Proceedings of the Symposium on Methodologies for Intelligent Systems (ISMIS'94)*, LNAI 869, 346–355, Springer, Berlin, 1994.
5. J. Bazan, A. Skowron, P. Synak. *Market data analysis: A rough set approach.* Report number 6 of the Institute of Computer Science, Warsaw University of Technology, 1994.
6. C.M. Bishop. *Neural Networks for Pattern Recognition.* Oxford University Press, Oxford, 1995.
7. G. Boole. *An Investigation of the Laws of Thought on which are Founded the Mathematical Theories of Logic and Probabilities.* Walton and Maberley, London, 1854.
8. F.M. Brown. *Boolean Reasoning.* Kluwer, Dordrecht, 1990.
9. K. Cios, W. Pedrycz, R. Swiniarski. *Data Mining Methods for Knowledge Discovery.* Kluwer, Boston, 1998.
10. T.G. Dietterich. Machine learning research: Four current directions. *AI Magazine*, 18(4): 97–136, 1997.
11. R.O. Duda, P.E. Hart. *Pattern Recognition and Scene Analysis.* Wiley, New York, 1973.
12. I. Duentsch, G. Gediga. Statistical evaluation of rough set dependency analysis. *International Journal of Human–Computer Studies*, 46: 589–604, 1997.
13. K. Fukunaga. *Introduction to Statistical Pattern Recognition.* Academic Press, New York, 1990.
14. Z.Q. Hong. Algebraic feature extraction of image for recognition. *Pattern Recognition*, 24(3): 211–219, 1991.
15. M.N. Huhns, M.P. Singh, editors. *Readings in Agents.* Morgan Kaufmann, San Mateo, CA, 1998.
16. G. John, R. Kohavi, K. Pfleger. Irrelevant features and the subset selection problem. In *Machine Learning: Proceedings of the 11th International Conference (ICML'94)*, 121–129, Morgan Kaufmann, San Mateo, CA, 1994.
17. J. Kittler. Feature selection and extraction. In T.Y. Young, K.S. Fu, editors, *Handbook of Pattern Recognition and Image Processing*, 59–83, Academic Press, New York, 1986.
18. J. Komorowski, Z. Pawlak, L. Polkowski, A. Skowron. Rough sets: A tutorial. In *[27]*, 3–98, 1999.
19. J. Koza, editor. *Genetic Programming: On the Programming of Computers by Means of Natural Selection.* MIT Press, Cambridge, MA, 1992.
20. M. Kudo, J. Sklansky. Comparison of algorithms that select features for pattern classifiers. *Pattern Recognition*, 33: 25–41, 2000.
21. P. Langley, S. Sage. Selection of relevant features in machine learning. In *Proceedings of the AAAI Fall Symposium on Relevance*, 140–144, AAAI Press, Menlo Park, CA, 1994.
22. H. Liu, H. Motoda, editors. *Feature Extraction, Construction and Selection: A Data Mining Approach.* Kluwer, Boston, 1998.
23. H. Liu, H. Motoda. *Feature Selection for Knowledge Discovery and Data Mining.* Kluwer, Boston, 1998.
24. H. Liu, R. Setiono. A probabilistic approach to feature selection - a filter solution. In *Proceedings of the 13th International Conference on Machine Learning (ICML'96)*, 319–327, Springer, Heidelberg, 1996.
25. V. Lobo, F. Moura-Pires, R. Swiniarski. *Minimizing the number of neurons for a SOM-based classification, using Boolean function formalization.* Report number 08/4/97 of Department of Mathematical and Computer Sciences, San Diego State University, San Diego, CA, 1997.
26. S.K. Pal, W. Pedrycz, A. Skowron, R. Swiniarski, editors. Rough-neuro computing (special issue). Vol. 36 of *Neurocomputing: An International Journal*, 2001.

27. S.K. Pal, A. Skowron, editors. *Rough Fuzzy Hybridization: A New Trend in Decision–Making*. Springer, Singapore, 1999.
28. Z. Pawlak. *Rough Sets: Theoretical Aspects of Reasoning about Data*. Kluwer, Dordrecht, 1991.
29. L. Polkowski, Y.Y. Lin, S. Tsumoto, editors. *Rough Set Methods and Applications: New Developments in Knowledge Discovery in Information Systems*. Physica, Heidelberg, 2000.
30. L. Polkowski, A. Skowron. Rough mereological approach to knowledge–based distributed AI. In J.K. Lee, J. Liebowitz, J.M. Chae, editors, *Proceedings of the 3rd World Congress on Expert Systems*, 774–781, Cognizant Communication Corporation, New York, 1996.
31. L. Polkowski, A. Skowron. Rough mereological foundations for design, analysis, synthesis, and control in distributed systems. *Information Sciences An International Journal*, 104(1-2): 129–156, 1998.
32. L. Polkowski, A. Skowron, editors. *Rough Sets in Knowledge Discovery 1: Methodology and Applications*. Physica, Heidelberg, 1998.
33. L. Polkowski, A. Skowron, editors. *Rough Sets in Knowledge Discovery 2: Applications, Case Studies and Software Systems*. Physica, Heidelberg, 1998.
34. L. Polkowski, A. Skowron. Grammar systems for distributed synthesis of approximate solutions extracted from experience. In G. Paun, A. Salomaa, editors, *Grammar Models for Multiagent Systems*, 316–333, Gordon and Breach, Amsterdam, 1999.
35. L. Polkowski, A. Skowron. Towards adaptive calculus of granules. In L.A. Zadeh, J. Kacprzyk, editors, *Computing with Words in Information/Intelligent Systems 1*, 201–227, Physica, Heidelberg, 1999.
36. L. Polkowski, A. Skowron. Rough mereology in information systems. A case study: Qualitative spatial reasoning. In *[29]*, 89–135, 2000.
37. L. Polkowski, A. Skowron. Rough mereological calculi of granules: A rough set approach to computation. *Computational Intelligence*, 17(3): 472–492, 2001.
38. L. Polkowski, A. Skowron. Rough-neuro computing. In W. Ziarko, Y.Y. Yao, editors, *Proceedings of the 2nd International Conference on Rough Sets and Current Trends in Computing (RSCTC 2000)*, LNAI 2005, 57–64, Springer, Berlin, 2001.
39. J.R. Quinlan, editor. *C4.5: Programs for Machine Learning*. Morgan Kaufmann, San Mateo, CA, 1993.
40. J. Rissanen. Modeling by shortest data description. *Automatica*, 14: 465–471, 1978.
41. F. Samaria, A. Harter. Parameterization of stochastic model for human face identification. In *Proceedings of IEEE Workshop on Application of Computer Vision*, 1994. Available at www.cam-orl.co.uk/facedatabase.html.
42. B. Selman, H. Kautz, A. McAllester. Ten challenges in propositional reasoning and search. In *Proceedings of IJCAI'97*, 50–54, Morgan Kaufmann, San Francisco, 1997.
43. A. Skowron. Extracting laws from decision tables. *Computational Intelligence*, 11(2): 371–388, 1995.
44. A. Skowron. Rough sets in KDD. In Z. Shi, B. Faltings, M. Musem, editors, *16th World Computer Congress (IFIP 2000):Proceedings of Conference on Intelligent Information Processing (IIP 2000)*, 1–17, Publishing House of Electronic Industry, Beijing, 2000 (plenary talk).
45. A. Skowron. Toward intelligent systems: Calculi of information granules. In S. Hirano, M. Inuiguchi, S. Tsumoto, editors, *Proceedings of International Workshop on Rough Set Theory and Granular Computing (RSTGC-2001)*, Vol. 5(1/2) of *Bulletin of International Rough Set Society*, 9–30, 2001 (keynote speech).

46. A. Skowron, C. Rauszer. The discernibility matrices and functions in information systems. In *[51]*, 331–362, 1992.
47. A. Skowron, J. Stepaniuk. Information granules: Towards foundations of granular computing. *International Journal of Intelligent Systems*, 16(1): 57–86, 2001.
48. D. Ślęzak. Approximate reducts in decision tables. In *Proceedings of the 6th International Conference on Information Processing and Management of Uncertainty in Knowledge-Based Systems (IPMU'96)*, Vol. 3, 1159–1164, Universidad da Granada, Granada, 1996.
49. D Ślęzak. Various approaches to reasoning with frequency based decision reducts: A survey. In *[29]*, 235–285, 2000.
50. D. Ślęzak. *Approximate Decision Reducts*. Ph.D. Dissertation, Faculty of Mathematics, Informatics and Mechanics, Warsaw University, 2002 (in Polish).
51. R. Słowiński, editor. *Intelligent Decision Support: Handbook of Applications and Advances of the Rough Sets Theory*. Kluwer, Dordrecht, 1992.
52. R. Swiniarski. Introduction to rough sets. In *Materials of The International Short Course on Neural Networks, Fuzzy and Rough Systems. Theory and Applications*, 1–24, San Diego State University Press, San Diego, CA, 1993.
53. R. Swiniarski, J. Nguyen. Rough sets expert system for texture classification based on 2D spectral features. In *Proceedings of the 3rd Biennial European Joint Conference on Engineering Systems Design and Analysis (ESDA'96)*, 3–8, Montpellier, France, 1996.
54. R. Swiniarski, F. Hunt, D. Chalvet, D. Pearson. Feature selection using rough sets and hidden layer expansion for rupture prediction in a highly automated production system. In *Proceedings of the 12th International Conference on Systems Science*, 12–15, Wrocław, Poland, 1995.
55. M. Szczuka. *Neural Networks and Symbolic Methods for Classifier Construction*. Ph.D. Dissertation, Faculty of Mathematics, Informatics and Mechanics, Warsaw University, 2000 (in Polish).
56. M.A. Turk, A.P. Pentland. Face recognition using eigenspaces. In *Proceedings of the 1991 IEEE Conference on Vision and Pattern Recognition (CVPR'91)*, 586–591, Maui, Hawaii,1991.
57. J. Wróblewski. *Adaptive Methods for Object Classification*. Ph.D. Dissertation, Faculty of Mathematics, Informatics and Mechanics, Warsaw University, 2002 (in Polish).
58. L.A. Zadeh. Fuzzy logic = computing with words. *IEEE Transactions on Fuzzy Systems*, 4: 103–111, 1996.

Chapter 26
Computational Analysis of Acquired Dyslexia of Kanji Characters Based on Conventional and Rough Neural Networks

Shusaku Tsumoto

Department of Medicine Informatics, Shimane Medical University, School of Medicine,
89-1 Enya-cho Izumo City, Shimane 693-8501, Japan
tsumoto@computer.org

Summary. Acquired dyslexia of Kanji characters is one of the most interesting research areas in neuropsychology. Although dyslexia of alphabets is known as the malfunction of the gyrus angularis (classical hypothesis), reading and writing of Chinese characters (Kanji characters) are intact in Japanese patients who suffered from cerebral infarction of the gyrus angularis. Thus, it has been pointed out that another group of neurons that integrates shape information should play an important role in recognition of Kanji characters (Iwata's model). In this chapter, two neural network models, based on the classical hypothesis and Iwata's model, are introduced to examine the validity of these two hypotheses. The computational experiments give the following three results: (i) Iwata's model learns Kanji characters much faster than the classical model; (ii) Iwata's model is more robust with respect to the malfunction of neurons; (iii) Iwata's model simulates the characteristics of the two types of Japanese dyslexia.

1 Introduction

The language system is one of the most important characteristics of human intelligence. This system includes not only auditory processing, but also symbol processing, which is discussed by Barsalou [1]. This neural mechanism for symbolic processing inspires an interesting topic in computing with words (CW) [9].

In neurology, the language system is studied by neurological symptoms, such as aphasia, dysarthria, and dyslexia, which are mainly observed in cerebral stroke, motor neuron diseases, and extrapyramidal disorders.

However, these studies are based on case studies, and the general explanation of the mechanism of these disorders are still difficult even for neurologists and neuropsychologists because of the complexities of neurological systems.

Though in other medical domains, such as physiology, numerical simulation based on mathematical modeling was applied, simulating neurological disorders by conventional mathematical modeling has been very difficult because the brain consists of large neural networks. However, due to the growth of computational power, neural networks have become a good tool for simulating neurological disorders, and some successful simulation results have been reported in [8].

One of the interesting language disorders is dyslexia where a patient cannot read or recognize symbols. Although most human beings in the world usually use phenological characters, dyslexia is observed as the disability to read symbols. However, since Japanese languages use both phenological and semantic symbols, two kinds of dyslexia have been observed in the literature [6].

According to neuropsychological studies, two hypotheses have been proposed: one hypothesis (classic hypothesis) was introduced by Geschwind [4], in which the dyslexia of alphabets is known as the malfunction of the gyrus angularis. The other (Iwata's model) was introduced by Iwata [5], in which the gyrus angularis controls the recognition of phenological characters and the left posterior inferior temporal gyrus play an important role in recognizing semantic characters.

In this chapter, two neural network models, which correspond to the classical hypothesis and Iwata's hypothesis, are introduced to examine the validity of these hypotheses. Computational experiments give the following three results: (i) Iwata's model learns Kanji characters much faster than classical model; (ii) Iwata's model is more robust with respect to the malfunction of neurons; (iii) Iwata's model simulates the characteristics of the two types of Japanese dyslexia syndromes.

This chapter is organized as follows: Sections 2 and 3 describe Japanese dyslexia syndromes and their foci. Section 4 introduces two computational models for dyslexia. Section 5 presents experimental results and their discussion. Finally, Sect. 6 concludes this chapter.

2 Japanese Characters and Dyslexia

The Japanese language uses three kinds of characters, Hiragana, Katakana, and Kanji. The former two sets of characters are used as syllabograms, each of which corresponds to the Japanese syllable. For example, "い" and "イ" correspond to the syllable, "i." The main difference between Hiragana and Katakana is that the former set is used to represent ordinary sentences, whereas the latter is used to describe loan words, such as "ラジオ", which is borrowed from the English word, "radio." Since they are phenological characters, Hiragana and Katakana are sometimes grouped into one category, "Kana."

On the other hand, Kanji is used as a combination of morphograms and syllabograms. For example, "犬" means a dog, which corresponds to "いぬ," represented by Hiragana. The reason we use Kanji characters is that Hiragana or Katakana characters are too weak to represent the meanings of Japanese words.

Since Japanese syllables are restricted to only 50 kinds, the Japanese language has many homonyms. For example, "去ぬ" has the following two meanings: one is a dog, and the other one is a verb "leave." Japanese uses Kanji characters to differentiate these two meanings: "犬" and "去ぬ." In this way, Hiragana and Kanji characters have different functions, which causes acquired dyslexia of the Japanese language to be more difficult than in other languages.

Although dyslexia of alphabets is known as malfunction of the gyrus angularis (the classical hypothesis) [4], reading and writing of Chinese characters (Kanji characters) are intact in Japanese patients who suffered from cerebral infarction of the gyrus angularis. Thus, it has been pointed out that another group of neurons, which functions to integrate shape information, should play an important role in the recognition of Kanji characters (Iwata's model) [5].

3 Acquired Dyslexia

3.1 Classical Hypothesis

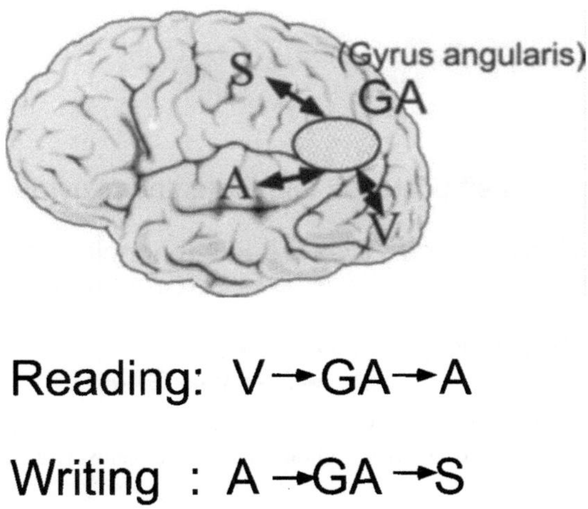

Fig. 1. Classic hypothesis for pure dyslexia. S, GA, A, and V denote the somatosensory area, the gyrus angularis, the auditory language area, and the visual cortex area, respectively. GA is a central area that integrates all the information about phenomic characters.

The classical hypothesis for reading and writing characters is illustrated in Fig. 1, where S, A, V and GA denote the somatosensory area, the auditory language area, the visual cortex, and the gyrus angularis, respectively. In the reading process, first, information on characters is captured by the visual cortex. Then, the information obtained is transmitted to the gyrus angularis, which integrates all of the information about characters. Finally, integrated information is sent to the auditory language area, which is closely connected with speaking languages.

This hypothesis proposed by Geschwind [4], explains the clinical cases of pure dyslexia reported by Dejerine [2] very well.

However, it is notable that this hypothesis is introduced to explain pure dyslexia with respect to alphabets, a kind of syllabogram. Recently, it has been reported that this hypothesis does not explain dyslexia syndromes in Japanese whose patients cannot read Hiragana but can read Kanji characters, or cannot read Kanji characters but can read Hiragana [5].

3.2 Hypothesis for Japanese Language

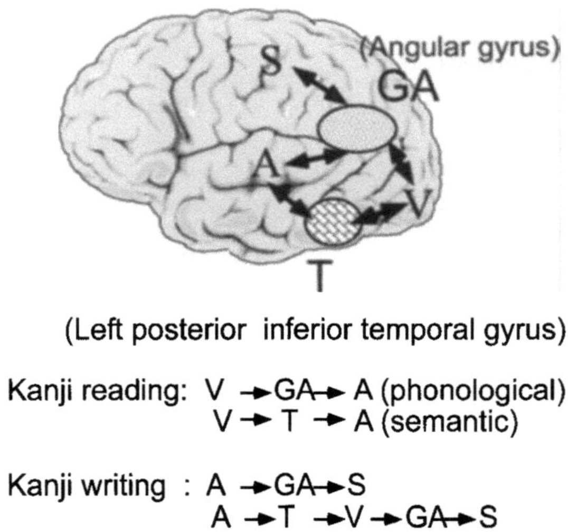

Fig. 2. Hypothesis for Japanese dyslexia. Note that Japanese characters have two pathways. Although Kana characters (phenomics) follows the classic pathway, Kanji characters may have their own pathway.

Iwata proposes a new hypothesis for the Japanese reading and writing process based on clinical cases of Japanese dyslexia syndrome [5]. Close clinicopathological examinations show that left posterior inferior temporal gyrus (T) is closely involved with integrating semantic information on Kanji characters, that is, a patient whose T region is damaged can read Kanji characters and Hiragana but cannot understand these meanings, but a patient whose GA region is damaged can read Kanji characters and understand these meanings but cannot read Hiragana. Thus, for reading Kanji characters, two pathways should be considered, illustrated in Fig. 2. As to the phonological aspects, the reading process is the same as that in Hiragana. On the other hand, for semantic aspects, visual information is transmitted from the visual cortex to the T region. Then, integrated information is sent from the T region to the auditory language area (V → T → A).

3.3 Dyslexia of Kanji Characters

In dyslexia of Chinese characters, the following four types of syndromes are reported.

The first is *disorder of capturing a global structure*, which means that a patient can read only some parts of Chinese characters. For example, when "青"(blue) is shown, a patient misread it as "月"(moon), which is only the lower part of "青." This syndrome can be viewed as damage to integrating partial information on Kanji characters.

The second is *confusion of morphology*, where the patient confuses a Kanji character with a character of a similar shape. For example, when "鍵" is shown, a patient misread it as "銀"(silver), whose left part is the same as that of the character "鍵"(key). This syndrome can be viewed as damage to integrating shape information on Kanji characters.

The third is *semantic misleading*, where a patient misread a character as one whose meaning is very similar, although morphological characteristics are not similar. For example, when "馬"(horse) is shown, a patient read it as "犬"(dog). The shared characteristic of both characters is that they describe animals. This syndrome can be viewed as damage to integrating semantic information.

Finally, the fourth is *disorder of the choice of auditory representation*. In this syndrome, a patient makes an error in choosing the phonological representation of Kanji characters. For example, in "言葉", "言" should be read as "こと" (koto), but a patient read it as "げん" (gen). This syndrome can be viewed as damage to integrating phonological information.

Thus, these types suggest that dyslexia of Kanji characters is observed when neurons integrating information on Kanji characters are damaged.

4 Computational Model

Computational models are based on three-layer neural networks, whose learning algorithms are back-propagation [7]. For simplicity, we use the following ten Kanji characters for inputs: 木(tree), 林(grove), 森(forest), 桜(cherry tree), 桃(peach tree), 松(pine tree), 梅(plum tree), 橋(bridge), 柱(pillar), and 竹(bamboo), all of which are related to "wood." Each character is represented by a 16×16 square dot matrix. Furthermore, to make a phonological representation of each Kanji character, we also use 16 Kana characters (see Fig. 3).

き (ki)	は (ha)	や (ya)	し (shi)
も (mo)	り (ri)	さ (sa)	く (ku)
ら (ra)	と (to)	う (u)	ま (ma)
つ (tsu)	め (me)	た (ta)	け (ke)

Fig. 3. Kana characters

For outputs, each category is set to each Kanji character. Thus, four bit strings are used to represent each category, which corresponds to the meaning of each character. Furthermore, for simplicity, the auditory representation is set to the same representation as the above category, that is, four bit strings are also used to represent the pronunciation of each Kanji character, and constrain that this representation and semantic representation is the same. This is a kind of simplification of the interaction between auditory representation and semantic representation.

Using these representations, we introduce two computational models to examine the effect of damage to neurons in dyslexia syndrome, as follows.

4.1 Classical Model

A computational model based on the classical hypothesis is illustrated by Fig. 4.

This model consists of one large component of neural networks. The visual representation corresponds to the input layer. Then, the intermediate layer, corresponding to the GA region, is composed of 100 neurons. Finally, auditory representation corresponds to the output layer.

In each learning step, we give one Kanji character and its corresponding Kana representation. For example, when "木" is input, its corresponding Kana representation "き"(ki) is given as an answer. For each learning step, the output of neural network is compared with its answer, and the total calculation is terminated after the total error rate is saturated.

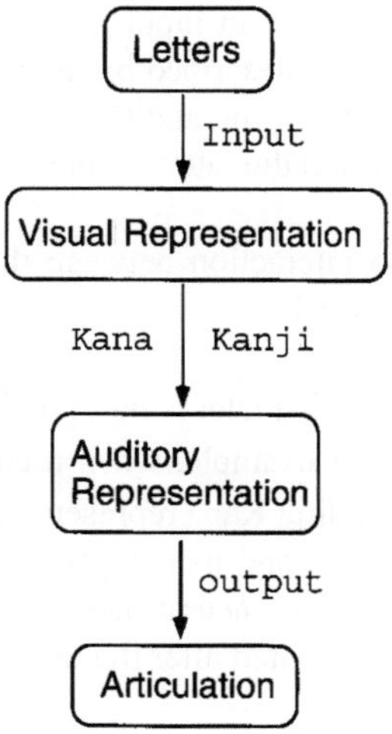

Fig. 4. Classical model

4.2 Iwata's Model

A computational model based on Iwata's model is given in Fig. 5.

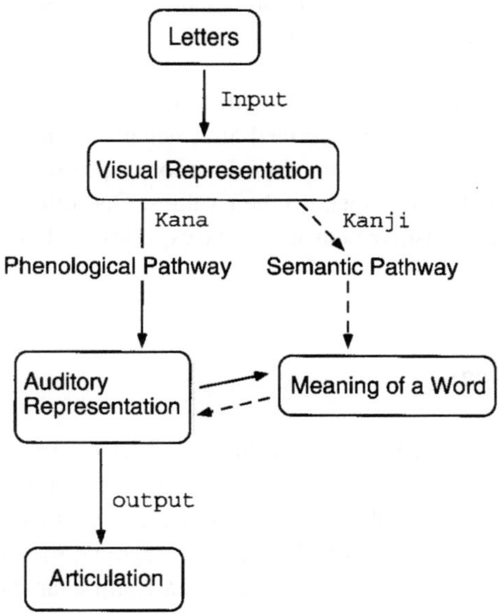

Fig. 5. Iwata's model

This model consists of the following two neural networks: one describes the phonological pathway, and the other represents the semantic pathway. For both models, visual representation corresponds to an input layer. Then, the phonological pathway and the semantic pathway are described by intermediate layers, each of which is composed of 50 neurons. Finally, the auditory representation and meaning of a word correspond to the former and the latter output layers, respectively.

It is assumed that there is no interaction between the phonological pathway and the semantic pathway in this model.

In each learning step, one Kanji character is input with its corresponding Kana representation and its category. For example, when a character "犬" is input for the semantic pathway, its corresponding Kana representation "き"(ki) is given as an input for the phonological pathway, and its category "1" is given as an answer. For each learning step, the output of the neural network is compared with its answer, and the whole calculation is terminated after the total error rate is saturated.

4.3 Computational Models with Rough Neurons

After learning steps, neurons in intermediate layers can be classified into three types: the first are neurons that do not contribute much to learning a given character (called negative neurons). The second are neurons that contribute somewhat to learning (called boundary neurons). Finally, the third are indispensable to learning (called positive neurons). This classification corresponds to a rough classification of neurons with respect to learning characters.

In this chapter, computational models with rough neurons are defined as models with positive neurons, which are derived after negative and boundary neurons are removed from neurons after learning. For selection of positive neurons, the weights for neurons are sorted with respect to their values. One-third of the best weighted neurons are selected as positive neurons. In the experiments below, about 20 neurons were selected. [1]

4.4 Focal Lesioning

After learning steps, we make the following "focal lesioning" steps for each model to simulate the effect of damaged neurons on misreading. First, one neuron is randomly selected from the neural networks. And then, its neighbors (within *k* neurons) are set to the neurons malfunctioned. Then, we examine the misclassification rate of this damaged network by using the above ten Kanji characters for each model.

[1] Readers may argue that this selection is ad-hoc. However, to rank the weight of neurons is one of the hardest tasks. Even if the selection of neurons is based on statistics of weights, the validity of selection is very difficult to argue. It will be our future work to look for good classification methods for neurons in neural networks.

These two steps are iterated for 100 steps, and then the statistical significance is tested by the t-test with respect to the error rate obtained.

In these procedures, first, k is set to 1, and then increased with step one. Then the above test procedures are repeated until the misclassification rate is equal to 1.0.

5 Experimental Results

We made the following two experiments using the aforementioned computational model. First, we compared learning steps needed for termination for each computational model. For this experiment, we changed the order of Kanji characters and started learning steps. These steps were repeated 1000 times, and the steps obtained were averaged and tested using the t-test. Second, we compared the behavior of each network using focal lesioning with respect to a parameter k. From these experiments, the following three interesting results were obtained.

5.1 Learning Steps for Termination

The statistics of this experiments are shown in Table 1. The convergence of Iwata's model is higher than that of the classical model, which is statistically significant with $p < .01$. This result suggests that processing of Kana and Kanji should be separated. Concerning error rates in the second column, both models are not significantly different from other models. The third column shows the error rates obtained by positive neurons (rough neurons). Surprisingly, Iwata's model was much more robust than the classical model, which suggests that although positive neurons play important roles in Iwata's model, both boundary and positive neurons play important roles in the classical model.

Table 1. Learning steps needed for termination

Model	Learning steps	Error rate	Positive neuron's error rate
Classical model	66700 ± 1319	87.5 ± 8.5 (%)	79.7 ± 8.2 (%)
Iwata's model	56400 ± 795	89.2 ± 9.2 (%)	84.5 ± 7.2 (%)
p value	0.00947	0.341	0.0129

5.2 Behavior of Neural Networks

In conventional neural networks, phase-transition behavior was observed with respect to k for each model. Figure 6 shows this behavior: accuracy shows how many learned characters can be recognized correctly after damage. The curves with circle points and with plus points depict the change of accuracy of Iwata's model and the classical model, respecively, when the value of k is increased.

When k was larger than the critical value, the error rate of a model suddenly became nearly 0.0 (Fig. 6). The main difference between the two models was that Iwata's model was more robust than the classical model.

On the other hand, in rough neural networks, the Iwata model was less robust than the classic model (Fig. 6).

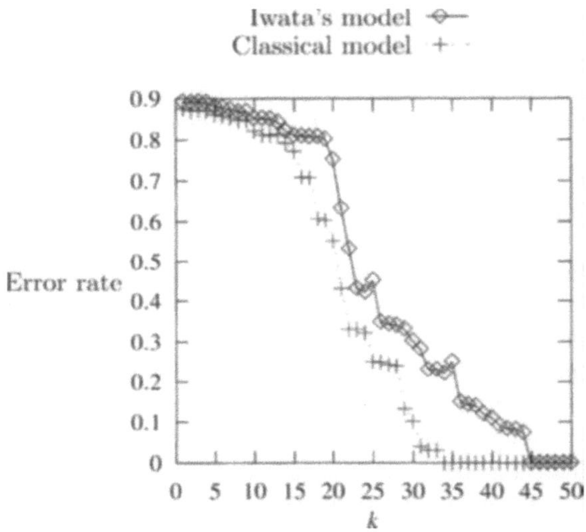

Fig. 6. Effect of parameter k on error rate (conventional neural networks)

These two different behaviors in conventional and rough neural networks suggest that positive neurons play more important roles in Iwata's model than in the classic model.

5.3 Simulated Dyslexia Syndromes

The computational model based on Iwata's model was able to simulate two types of dyslexia: disorder of capturing global structure and confusion of morphology. For example, "森" was almost always misread as "木" when k was larger than 30, and "桜" was misread as "桃" when k was larger than 20. Thus, these results suggest that the former type of dyslexia is usually observed when the focus of infarction is very large.

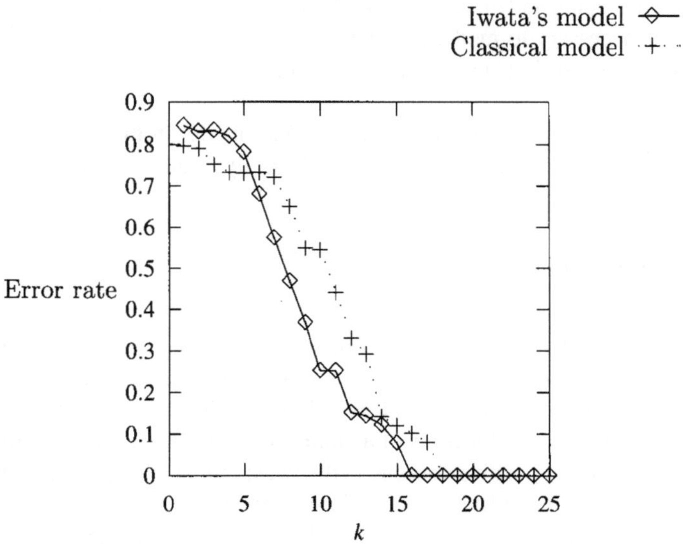

Fig. 7. Effect of parameter *k* on error rate (rough–neural networks)

6 Discussion

These three results suggest that Iwata's model explains dylexia syndromes better than the classical one for the Japanese language. However, this computational model can explain only two types of Japanese dyslexia syndromes, disorder of capturing global structure and confusion of morphology. It cannot explain the other two types, semantic misleading and disorder of the choice of auditory representation, which shows that our model is still weak.

One reason is that this neural network includes only the model of pattern recognition (recognition letters) and does not include the semantic model. Recently, Gardenfors introduced a new concept, "conceptual spaces", in [3], which is inspired by his research on how children learn language. He argues that concept understanding consists of three levels of representation: symbolic representations, subconceptual representations, and conceptual representations. Then, he discusses that although subconceptual representations are closely related to neural networks, not only the aspects of these levels, but also associations between two levels are important for language understanding.

According to his discussion, the computational studies in this chapter show how subconceptual representations are important for recognition of symbols, which shows the relations between the symbol level and the subconceptual level. However, since the proposed model does not consider the relations between the subconceptual level and the conceptual level, it cannot capture the semantic aspects of dyslexia (the semantic misleading and disorder of the choice of the auditory representation). Therefore, it is necessary to model the relations between subconceptual and conceptual representations to simulate semantic misleading.

It is our future work to refine a computational model from the viewpoint of conceptual spaces to explain the other two types of dyslexia syndrome.

7 Conclusion

In this chapter, two computational models are introduced to explain Japanese dyslexia syndromes. The experimental results show that Iwata's model explains dyslexia syndrome better than the classical model, although the computational model is a little weak to explain all aspects of acquired Japanese dyslexia. Furthermore, those results suggest that positive neurons in the Iwata model may play important roles in achieving the cognitive functions for Kanji characters. This chapter is a preliminary study of the simulation of neurological diseases based on rough neural networks. Although the discussions are very intuitive and not mathematically rigorous, this simple model captures the essential nature of acquired dyslexia syndrome. More formal analysis will appear in the future work.

References

1. L.W. Barsalou. Perceptual symbol systems. *Behavioral and Brain Sciences*, 22: 577–660, 1999.
2. J. Dejerine. Sur un cas de cécité verbale avec agraphie, suivi d'autopsie. *Comptes rendus de la Société de biologie*, 43: 197–201, 1891.
3. P. Gardenfors. *Conceptual Spaces*. MIT Press, Cambridge, MA, 2000.
4. N. Geschwind. Disconnexion syndromes in animals and man. *Brain*, 88: 237–294, 585–644, 1965.
5. M. Iwata. Neural mechanism of reading and writing in the Japanese language. *Functional Neurology*, 1: 43–52, 1986.
6. M. Kawamura, K. Hirayama, K. Hasegawa, N. Takahashi, A. Yamaura. Alexia with agraphia of Kanji (Japanese morphograms). *Journal of Neurology, Neurosurgery, and Psychiatry*, 50: 1125–1129, 1987.
7. J. McClelland, D. Rumelhart, et al. , editors. *Parallel Distributed Processing*. MIT Press, Cambridge, MA, 1986.
8. J. Reggia, R.S. Berndt, C.L. D'Autrechy. Connectionist models in neuropsychology. In F. Boller, J. Grafman, editors, *Handbook of Neuropsychology*, Vol. 9, 297–333, Elsevier, Amsterdam, 1994.
9. L.A. Zadeh. A new direction in AI: Toward a computational theory of perceptions. *AI Magazine*, 22(1): 73–84, 2001.

Chapter 27
WaRS: A Method for Signal Classification

Piotr Wojdyłło

Department of Mechanical and Aeronautical Engineering, University of California, One Shields Avenue, Davis, CA 95616, USA
pwoj@mimuw.edu.pl

Summary. Rough hybrid methods are used worldwide for pattern recognition and classification problems. In this chapter, WaRS, a combination of wavelets with rough set tools is presented, and examples of its application to problems of artificial and real-life (biomedical) origin are supplied.

1 Introduction

The WaRS method was successfully applied in signal classification, see [32,34]. The standard problem is the following. We have a decision system with a special series of data provided for each object, call it a " signal." Its characteristics are that it is large and, despite the important information, much of it is not useful information, called noise. (One should be aware that we are speaking about noise from the viewpoint of a decision system, which may be different from that generally accepted in signal analysis.) Wavelets were already applied in biomedical sciences, e.g. [28,30]. Wavelet-fed neural networks are discussed in [26].

The main objective of signal processing in WaRS is removal of informational noise and selection of important classification features. This cannot be done, in the author's opinion, without the effectiveness of wavelet tools. Algorithms of frequential analysis and other helpful heuristics were developed in previous works [32,34]. The applied methods lead to high classification ratios and a fairly simple structure of decision systems obtained with few predicates in the description. The use of dynamic reducts and conflict solving weights [3] appeared to be of importance here.

After coefficient selection, which may be considered scaling by sophisticated hyperplanes in high-dimensional space, the decision system is built on the basis of classical rough set methods in the Rough Set Exploration System (RSES).

The four problems provided in this chapter serve to illustrate the method. The first two that appear in Sect. 6 and 7 were discussed earlier in [32] and [34], respectively. They deal with the problems of classifying EEG recordings of discerning two groups of epilepsy (Sect. 6) and improving classification in the presence of additional noise

(Sect. 7). The problem discussed in Sect. 8 has its origin in experimental biology and deals with emotions in animal fear conditioning. The data were available courtesy of Prof. Stefan Kasicki, Institute of Experimental Biology, Polish Academy of Sciences. In this case, the main purpose of the analysis is to create a decision system with a simple structure that covers most of the "emotional" cases. The paradigm of the experiment is that after preparation, the rat is trained to react with fear to a light signal. This is done in the first day of the experiment. Then the outer setting moves its attention to the tone signal that relieves it from the shock.

The last one is the classification problem formulated in Breiman [6]. It was recently dealt with in Saito's thesis [29] with a result very close to the theoretical minimum. Saito chooses the most accurate orthonormal basis for the system analysis in the sense of entropy-type measure minimization. Comparison of the results presented here leads to the conclusion that connection with wavelet preprocessing enhances the efficiency of rough set methods in this problem. But achieving a result that is close to the theoretical minimum is a problem of further optimization and development of the method and will be the subject of research in the future.

2 Rough Sets

2.1 The Main Notions

The structure of the data in this chapter is a *decision table* [25]. A decision table is a pair of the form $\mathbf{A} = (U, A \cup \{d\})$, where U is a *universe* of *objects* and $A = (a_1, ..., a_m)$ is a set of *conditional attributes*, i.e., mappings of the form $a_i : U \to V_a$, where V_a is called the *value set* of the attribute a_i. d is a distinguished attribute related to as a *decision*. The ith *decision class* is a set of objects $C_i = \{o \in U : d(o) = d_i\}$, where d_i is the ith decision value taken from decision value set V_d. In our particular case, the set of conditional attribute values will be either finite (in scaled or rule-based tables) or contained in some interval of a real domain (in original and wavelet-processed data).

For any subset of attributes $B \subset A$, *indiscernibility relation* IND(B) is defined as

$$xIND(B)y \Leftrightarrow \forall_{a \in B} a(x) = a(y), \tag{1}$$

where $x, y \in U$.

Having the indiscernibility relation, we may define the reduct. $B \subset A$ is a *reduct* of an information system if $IND(B) = IND(A)$ and no proper subset of B has this property. Intuitively, a reduct is the minimal set of attributes that allows us to preserve the ability to distinguish objects in the same way as in the original decision table.

A *decision rule* is a formula φ of the form,

$$(a_{i_1} = v_1) \wedge \ldots \wedge (a_{i_k} = v_k) \Rightarrow d = v_d, \tag{2}$$

where $1 \leq i_1 < \ldots < i_k \leq m$, $v_i \in V_{a_i}$. The set of all rules for a particular decision table $\mathbf{B} \subset \mathbf{A}$ is denoted by $RUL(\mathbf{B})$. Atomic subformulas $(a_{i_1} = v_1)$ are called *conditions.* We say that rule r is *applicable* to an object, or alternatively, the object *matches* the rule, if its attribute values satisfy the premise of the rule. We can connect some characteristics with the rule . *Support* denoted as $Supp_{\mathbf{A}}(r)$ is equal to the number of objects from $\mathbf{A}$ for which rule r applies correctly, i.e., the premise of the rule is satisfied, and the decision given by the rule is similar to that preset in the decision table. $Match_{\mathbf{A}}(r)$ is the number of objects in $\mathbf{A}$ for which rule r applies in general. Analogously, the notion of a matching set for a collection of rules may be introduced. By $Match_{\mathbf{A}}(R, o)$, we denote the subset M of rule set R such that rules in M are applicable to the object $o \in U$. The rule is said to be *optimal* if removal of any of its conditions causes decrease in its support. Support and matching are also used to define coefficient of consistency $\mu_{\mathbf{A}}(r)$ for a rule as being equal to $\mu_{\mathbf{A}}(r) = \frac{Supp_{\mathbf{A}}(r)}{Match_{\mathbf{A}}(r)}$.

In our study, we refer only to the rules that are derived using knowledge contained in the data. In that spirit, we may introduce the *meaning* of the premise of rule r in the decision table $\mathbf{A}$. The meaning of $Pred(r)$ will be denoted by $|Pred(r)|_{\mathbf{A}}$ and defined inductively in the following way:

1. if $Pred(r)$ is of the form $a = v$, then, $|Pred(r)|_{\mathbf{A}} = \{o \in U : a(o) = v\}$.
2. $|Pred(r) \wedge Pred(r')|_{\mathbf{A}} = |Pred(r)|_{\mathbf{A}} \cap |Pred(r')|_{\mathbf{A}}$.
3. $|Pred(r) \vee Pred(r')|_{\mathbf{A}} = |Pred(r)|_{\mathbf{A}} \cup |Pred(r')|_{\mathbf{A}}$.
4. $|\neg Pred(r)|_{\mathbf{A}} = U - |Pred(r)|_{\mathbf{A}}$.

For the classification systems built in this chapter, the dynamic rule [2] is a key notion. If we consider a family $\mathbf{F}$ of the subsets (subtables) of $\mathbf{A}$ [$\mathbf{F} \subset \mathbf{P}(\mathbf{A})$], then the rule $r \in \bigcup_{\mathbf{B} \in \mathbf{F}} RUL(\mathbf{B})$ is $\mathbf{F}$-*dynamic* (usually simply *dynamic*) if and only if

$$|Pred(r)|_{\mathbf{B}} \neq \emptyset \Rightarrow r \in RUL(\mathbf{B}), \text{ for any } \mathbf{B} \in \mathbf{F}. \tag{3}$$

In our further study, we will rely on certain numerical characteristics of dynamic rules. One important characteristic is the *stability coefficient* of the dynamic rule r relative to $\mathbf{F}$ denoted by $SC_{\mathbf{A}}^{\mathbf{F}}$ and defined as

$$SC_{\mathbf{A}}^{\mathbf{F}}(r) = \frac{card(\{\mathbf{B} \in \mathbf{F} : r \in RUL(\mathbf{B})\})}{card(\{\mathbf{B} \in \mathbf{F} : |Pred(r)|_{\mathbf{B}} \neq \emptyset\})}. \tag{4}$$

This coefficient reflects the frequency of occurrence of a particular rule in the set of rules generated by subsequent steps of the rule-generation algorithm. The more frequent the rule [the higher $SC_{\mathbf{A}}^{\mathbf{F}}(r)$], the better its reliability. For further details, consult [2] and [3].

2.2 Conflict Among Rules

$SC^{\mathbf{F}}_{\mathbf{A}}(r)$ in our study is used for resolving conflicts among rules. If for some object $o \in U$ there exist two (or more) rules $r_i, r_j \in R$ such that $r_i, r_j \in Match_{\mathbf{A}}(R,o)$, where R is a set of rules, and r_i points at a decision class different from r_j, then *conflict* occurs. To make a decision, we have to choose which rules should be trusted more in a particular situation. To grade rules for conflict resolving, we connect weights with groups of them. The weight introduced in [3] was especially successful in conflict solving:

$$W(B_i,o) = \frac{\sum\limits_{r \in Match(B_i,o)} Supp_{\mathbf{A}}(r) \cdot SC^{\mathbf{P}(\mathbf{A})}_{\mathbf{A}}(r)}{\sum\limits_{r \in B_i} Supp_{\mathbf{A}}(r) \cdot SC^{\mathbf{P}(\mathbf{A})}_{\mathbf{A}}(r)}, \tag{5}$$

where B_i is the set of rules connected with the ith decision class ($i = 1,2$) and o is the object to be classified. When the denominator in the above equals zero the weight is set to zero, too. Instead of family $\mathbf{F}$, we use the entire $\mathbf{P}(\mathbf{A})$. Clearly, in the above formula, the value of the weight depends on the number of rules that match the object to be classified and on their stability as described by the stability coefficient. The final decision is taken by comparing summarized weights for both decision classes $W(B_1,o)$ and $W(B_2,o)$ and choosing the one that has a higher value.

2.3 Attribute Scaling

When our attributes are mapping objects into a potentially infinite (in our case real) domain, we have to use some methods in order to avoid time-consuming computation. For that purpose, several methods known as *discretization, quantization*, and *scaling* were proposed by many researchers [1,24]. In this work, we rely on methods proposed in [2,24]. To give the reader some idea we introduce basic keywords with some explanation.

Let us consider an attribute $a_i : U \to V_a$, where V_a is an interval on the real axis. By *cut* c_i for the attribute a_i we mean any real number belonging to V_a. By using c_i, we can exchange a_i with a new binary attribute $\overline{a_i}$ defined by

$$\overline{a_i}(o) = \begin{cases} 0 \text{ iff } a_i(o) < c_i \\ 1 \text{ otherwise.} \end{cases} \tag{6}$$

The potential problem is how to search for cuts that will really help us in the classification task. The heuristics we use are based on discernibility. The cut is selected if the attribute obtained by applying this cut allows us to discern the highest number possible of objects not yet discerned in the decision table. One cut is chosen at each step . By repeating this procedure until full discernibility is achieved, we select a set of new binary attributes for further computations. In this way, we have a binary decision table with a number of attributes usually smaller than in the original.

This method, although elegant, is usually too sensitive in the presence of noise in data. Therefore, in our approach, we usually choose several best cuts according to Johnson's heuristic [15], as it proved useful in [24]. In our particular application, it turned out from experiments that this number should be around 100.

Yet another method we use is dynamic scaling, as proposed in [3]. The idea of this method has a lot in common with dynamic reduct and rule calculation. Let us consider a gigantic decision table that is constructed from the initial one by taking as attributes all possible intervals in which a cut may be located. Of course, in the nontrivial case, such a table is practically unmanageable since it has a number of elements proportional to the square of that of the initial table. But interesting for us is the fact that the set of cuts constructed as presented in the previous paragraphs is a reduct in this huge table. This reduct, however, because of its limited applicability to new cases, is not always the best for our needs. Therefore, we extend the set of cuts by adding several, possibly redundant cuts (unnecessary with regard to discernibility) for higher flexibility. The optimal criteria for adding a particular cut have not yet been identified. For details of the heuristics for this task, review [3,24]. Having the widened set of cuts, we perform dynamic reduct calculation using the attributes generated by them. The best dynamic reduct is chosen, and final scaling of the decision table is determined by cuts belonging to this reduct. All of the process in terms of time complexity and quality of result relies strongly on the number of redundant cuts we use. Therefore, it is sometimes necessary to make several trials to establish the cut set that satisfies the demands.

3 Wavelets

3.1 Choice of Wavelet

For the ongoing examples, we have chosen a Daubechies wavelet of the sixth order. There was the following reason to do so: Daubechies wavelets are smooth with a coefficient $\gamma \approx 0.2$ - the wavelet of the nth order is in the class of smoothness $C^{\gamma n}$; the wavelet of the sixth order is in class C^1 (is continuous together with its derivative). The graph of the wavelet is shown in Fig. 1 along with an appropriate scaling function. The role of this latter object shall be explained in the following subsection.

Most often, one uses smooth wavelets to deal with smooth signals. In this case, however, we also stuck to smooth wavelets following the research intuition that no essential information is contained in the high-oscillation component of the signal.

Certainly, the higher the order of the wavelet, the higher the level of smoothness. One should be careful, however, so as not to overdo, higher order wavelets require longer filters or longer support of the wavelet, depending on the method chosen. Therefore, wavelets of a higher order always relate to a higher computation cost.

3.2 Construction of a Wavelet

To construct a wavelet, one needs a *scaling function*. (This is the so-called multiresolution analysis approach, very popular and of highest practical importance, since there are few wavelets not originating from it and these do not have good properties for applications.)

Definition 1. Let $\varphi \in L^2(\mathbb{R})$ and $(a_k) \in l^2(\mathbb{Z})$. If

$$\sum_{k\in\mathbb{Z}} a_k \, \overline{a_{k+2n}} = \frac{1}{2}\delta_{no}$$

and

$$\varphi(x) = 2\sum_{k\in\mathbb{Z}} a_k \, \varphi(2x-k), \tag{7}$$

then φ is called a scaling function. It is also known that in this case, the system $(\varphi(x-n))_{n\in\mathbb{Z}}$ is orthonormal [33].

The values of nonzero a_k for the Daubechies wavelet of the sixth order are given in Table 1.

Equation (7) is a *scaling equation*. Our candidate for the wavelet will be a function,

$$\psi(x) = 2\sum_{k\in\mathbb{Z}} (-1)^k a_{1-k}\varphi(2x-k). \tag{8}$$

In this setting [with the assumption that $[\varphi(x-n)]_{n\in}$ is an orthonormal basis], it is quite easy to prove that the system $(\psi_{jk})_{j,k\in\mathbb{Z}}$ defined as

$$\psi_{jk}(x) := 2^{j/2}\psi(2^j x - k),$$

is an orthonormal basis in $L^2(\mathbb{R})$ [20,33]. There are two nontrivial points here: First, do (a_k) like that exist? Second, even *if* they exist, does φ (and what follows ψ) have any useful properties, such as being continuous, once/twice differentiable and/or decaying fast? There is also a third point (purely practical but of importance): Is the algorithm for finding ψ fast enough for signal/image processing and other time-critical applications?

3.3 Intermezzo

For the first two points, the answer was believed to be *"no"* except in trivial cases (a Haar function as a wavelet, satisfying all properties but noncontinuous). The amazing solution with the positive answer started a revolution in wavelets, when in 1986, Meyer proved the existence of arbitrarily smooth wavelets decaying reasonably fast [21]. A further breakthrough was made by Daubechies showing that there

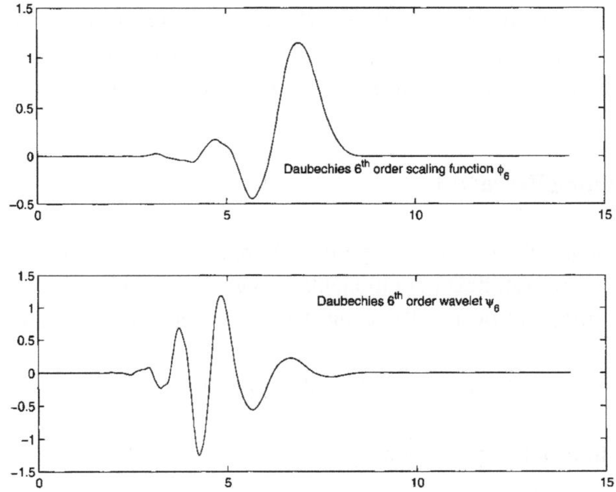

Fig. 1. Daubechies scaling function (top) and wavelet (bottom) of 6th order.

are smooth wavelets with compact support. The longer the support of the wavelet, the smoother it is [8].

This was also an answer to the third point: compact support means only finitely many coefficients a_k, and the computational cost of the processing is low. Afterwards, Coifman showed that one can choose coefficients a_k to obtain almost symmetrical wavelets, Beylkin and Rokhlin developed the theory for the computational tool called fast wavelet transform [5]. It will be a great omission not to mention others' work and the well-developed field of Gabor analysis, but space is limited here, and we ask the interested reader to consult the recent books in the Academic Press series *Wavelets and Their Applications* and the comprehensive book by Chui with contemporary references [7]. For Gabor analysis, see the books by Feichtinger and Strohmer [14] and Groechenig [16] with references therein.

3.4 Convergence of Approximation

After this intermezzo, let us return to our discussion of the wavelet and the scaling function. Translations of the function $[\varphi(x-n)]_{n\in\mathbb{Z}}$ span the space, whose norm-closure is denoted by V_0. The orthogonal projection on this space corresponds to the *trend component* of the signal. The *fluctuation part* is contained in its orthogonal complement. Then let us dilate the space V_o by means of the operator $\mathcal{D}f(x) = 2^{1/2}f(2x)$, call the other's closed span V_1, and continue the process. Since V_o is contained in V_1 [and that it is so follows from scaling equation (7)], the projection of the signal $P_{V_1}f$ is the finer approximation of f as $P_{V_o}f$. If the technical condition

$\overline{\bigcup_j V_j} = L^2(\mathbb{R})$ holds, we get the converging sequence of finer and finer approximations $\left(P_{V_j} f\right)_{j \in \mathbb{Z}_+}$. The differences $P_{V_j} f - P_{V_{j-1}} f$ are "fluctuations" with respect to the trend parts contained in the subsequent V_js. The wavelet is constructed so that $\psi(x-n)$ spans the space whose closure is the orthogonal complement of V_o in V_1.

3.5 Thresholding Procedure

Real-life experience shows that fluctuations of "real-world" vectors (signals or images) have very few significant components in their basis. This is the way how one achieves high compression results using wavelets. The reconstruction formula is thus

$$f \approx \sum_{j,k \in \mathbb{Z}} h(\langle f, \psi_{jk} \rangle_{L^2(\mathbb{R})}) \psi_{jk}, \tag{9}$$

where h is the thresholding function.

For wavelet coefficients, one applies the thresholding procedure and this is the moment when compression gain is achieved. There are methods for thresholding. Their effectiveness depends on the type of signal/images processed. We have decided to use the hard thresholding procedure,

$$h(a) = \begin{cases} a & |a| > \theta \\ 0 & |a| \leq \theta, \end{cases}$$

cutting off all coefficients a below the value θ. For the simple argument that dyadic dilations and integral translations of this function $\psi_{jk}(x) = 2^{j/2} \psi\left(2^j x - k\right)$ with $j,k \in \mathbb{Z}$ are an orthonormal basis for $L^2(\mathbb{R})$, see [33]. This is equivalent to the following formula:

$$f = \sum_{j,k \in \mathbb{Z}} \langle f, \psi_{jk} \rangle \psi_{jk}.$$

The main tool in processing wavelet coefficients is the hard thresholding procedure investigated thoroughly by Donoho, Johnstone et al. [11–13]. Let us consider the set,

$$A_\vartheta = \left\{ (j,k) \in \mathbb{Z}^2 : \left| \langle f, \psi_{jk} \rangle \right| \geq \vartheta \right\},$$

for given and nonnegative ϑ. Then we sum up all coefficients with indexes in A_ϑ:

$$\widetilde{f} = \sum_{(j,k) \in A_\vartheta} \langle f, \psi_{jk} \rangle \psi_{jk}.$$

The effectiveness of hard thresholding is closely related to unconditionality of wavelet bases in many functional spaces: Lebesgue spaces $L^p(\mathbb{R})$, Sobolev spaces, and Besov–Triebel–Lizorkin spaces. Unconditionality is shown in [9,21,35]. It was also proved that among all orthonormal bases, unconditional bases are optimal for the hard thresholding procedure [10].

We apply the following procedure of wavelet analysis to electric brain wave data. We find the form of the wavelet ψ related to the sequence (c_n) by formulas (1–8). Then, we compute the scalar products $\langle f, \psi_{jk} \rangle$.

The sole decomposition into trend and fluctuation parts has a perfect reconstruction property. Certainly, in the thresholding intermittent step, the property is "violated" in the strict sense. Nevertheless, both the application to real signal/images as well as some theoretical work show that the similarity of the signal reconstructed after compression to the original is very high.

4 Quadrature Mirror Filtering

The ideal scheme for subband coding consists of first filtering the incoming signal into two frequency channels associated with the intervals $[0,\pi]$, $[\pi,2\pi]$ and then subsampling the corresponding outputs, retaining only one point in every two. This operation, which consists of restricting a sequence defined on $\mathbb{Z}$ to $2\mathbb{Z}$, is called decimation and is denoted by $2 \downarrow 1$ or D. The graphical representation of the decomposition scheme is given in Fig. 2, and the synthesis scheme in Fig. 3.

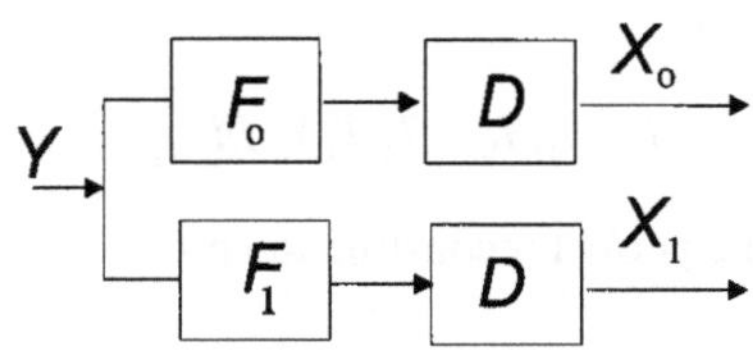

Fig. 2. The decomposition of the signal Y into a low-frequency part X_o (approximation) and a high-frequency part X_1 (detail). F_o and F_1 are the filtering operators for low- and high-pass filters, respectively. D is the decimation operator

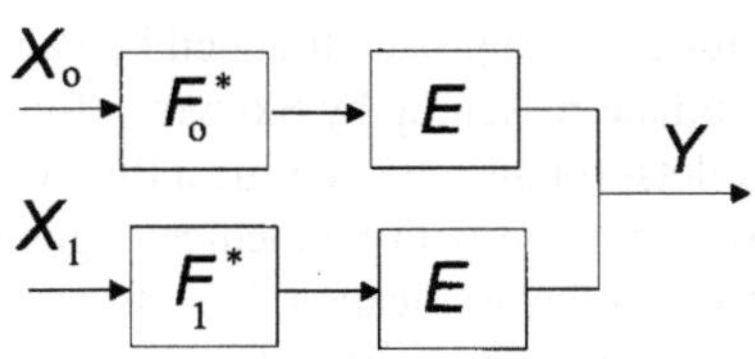

Fig. 3. Reconstruction of the signal from the approximation (X_o) and detail (X_1) parts. F_i^* are the adjoints of the corresponding F_i operators, and E is the extension operator adjoint to D

The scheme for reconstructing the original signal is the dual analysis scheme. We began by extending the sequences (y^1_{2n}), (y^2_{2n}) by inserting zeros at all odd integers. Next, we filter this "absurd decision," again using filters F_o, F_1. This output returns

(x_n).

We are going to describe in detail subband coding using two frequency channels. Assume that the input signals (x_n) have already been sampled and are defined on the integers $n \in \mathbb{Z}$; they are arbitrary sequences with finite energy:

$$\sum_{n \in \mathbb{Z}} |x_n|^2 < \infty.$$

We denote by $D : l^2(\mathbb{Z}) \to l^2(2\mathbb{Z})$ the decimation operator, which retains only the terms with even indexes in the sequence (x_n). The adjoint operator

$$E = D^* : l^2(2\mathbb{Z}) \to l^2(\mathbb{Z})$$

is the crudest possible extension operator. It inserts zeros at odd indexes.

Definition 2. The two filters F_o and F_1 are called quadrature mirror filters if for all signals X of finite energy,

$$\|DF_o(X)\|^2 + \|DF_1(X)\|^2 = \|X\|^2.$$

Denote by $T_i = DF_i : l^2(\mathbb{Z}) \to l^2(2\mathbb{Z})$. Then the assertion of the definition may be formulated equivalently as

$$I = T_0^* T_0 + T_1^* T_1 = Id_{l^2(\mathbb{Z})}.$$

This property is called the perfect reconstruction property.

5 The WaRS Method in Detail

The objective of this section is to present in detail a new and reliable method of preprocessing real-world signal data based on the rapidly growing wavelet method in combination with the rough set approach founded by Pawlak [25] and developed in the direction of approximate reasoning by Skowron (for a survey, see e.g. [19]). The WaRS method (wavelets+rough sets) significantly reduces the dimensionality of the problem, which is very large with biomedical signals, and keeps all essential information that allows good classification. Rough set based classifiers built on the data yield usually only a few (about 10) rules that are easy to interpret and, in particularly interesting cases, they have often the very simple form of a one-term expression, e.g.,

$$(a_7 = 0) \Rightarrow (d = 0),$$

where a_i is the scaled wavelet coefficient (see further text)and $d = 0$ (decision) refers to the emotional state of the rat, for instance. Thus the real-world data of high dimension is significantly reduced to a small number of binary attributes that are necessary

Table 1. Coefficients of the Daubechies wavelet of the sixth order and of the appropriate low-pass filter

# of coefficient(n)	the value of coefficient (c_n)
1	0.111541
2	0.494624
3	0.751133
4	0.315250
5	-0.226264
6	-0.129766
7	0.097501
8	0.027522
9	-0.031582
10	0.000553
11	0.004777
12	-0.001077

to apply the rules computed.

Then, there is a question of appropriate conflict solving when rules give contradictory results. In this case, we got the best result by using a weight based on the stability coefficient (4); see [3]. In Sect. 9 we point out that the WaRS method enhances the behavior of a straightforwardly applied rough set classifier. Let us briefly describe details of the WaRS method and refer the interested reader to [34] for variants and extensions.

5.1 Algorithm

The input of the algorithm is a decision table whose objects are real or artificial time-series generators with appropriately assigned decisions. The attributes are values of the time series in subsequent equidistant moments of time. It may happen that these values are vectors in multidimensional spaces (if a few measurement are taken at the same time, e.g., when recording electric brain activity at 21 points on the scalp).

Let the members of the universe be different generators (patients, animals etc.) or a few of them but underlying the changing conditions that evoke a change in decision. It is certainly assumed that the recognized feature (decision) has its representation in the time series generated. Thus $a_i : U \to \mathbb{R}^m$, where $a_i(o) = f_o(i)$. For each coordinate f_o^s of f_o, the wavelet coefficients are computed. Thus we arrive at the modified set of attributes,

$$\widetilde{a}_{jk} : U \longrightarrow \mathbb{R}^m,$$

where

$$\widetilde{a}_{jk}(o) = \left[\langle f_o^s(\cdot), \psi_{jk}(\cdot) \rangle \right]_{s=1\ldots m}.$$

Then we use the thresholding procedure described in Sect. 3.5, and the new set of attributes is given by

$$a_{jk}^* : U \longrightarrow \mathbb{R}^m,$$

$$\left[a_{jk}^*\right]^s(o) = h_\theta \left[\langle f_o^s(\cdot), \psi_{jk}(\cdot) \rangle \right],$$

where

$$h_\theta(a) = \begin{cases} a & |a| \geq \theta \\ 0 & |a| < \theta. \end{cases}$$

To apply discretization procedures and other data mining techniques, we need to choose as an attribute,

$$(j,k) \longmapsto \langle f, \psi_{jk} \rangle,$$

where f is the electric brain wave signal from the nth recording.

The straightforward procedure of taking all pairs (j,k) giving at least one coefficient with an absolute value above $\vartheta = 1.0$ would result in a large number of attributes. Therefore, we apply *frequential analysis of occurrences* [34].

5.2 Frequential Analysis of Occurrences

For frequential analysis of occurrences, we choose wavelet coefficients $\langle f, \psi_{jk} \rangle$ related to pairs (j,k) such that the number of their representatives greater than ϑ for all recordings is greater than or equal to the certain threshold M, usually chosen proportionally to the number of all recordings (objects); the coefficient of this proportion α is taken to be around 0.7 for two decision classes and around 0.6 for three decision classes. In other words,

$$F_M = \left\{ (j,k) : \# \left\{ (e,p) : \left| \langle f_{e,p}, \psi_{jk} \rangle \right| \geq \theta \right\} \geq M \right\},$$

or

$$F_\alpha = \left\{ (j,k) : \# \left\{ ob : \left| \langle f_{ob}, \psi_{jk} \rangle \right| \geq \theta \right\} \approx \alpha Ob \right\},$$

where f_{ob} is the wave signal from the obth measurement and Ob is the number of all measurements.

5.3 Completion of the Algorithm

Finally, we consider the universe U with attributes $\left(a^*_{jk}\right)_{(j,k)\in F_\alpha}$ and the decision from the original setting. This decision table is used thereafter to develop a rough set classifier. Artificial neural networks are used for slight modification of weights in the conflict-solving part of the classification process.

The RSES methods that cooperate with WaRS best are attribute scaling with the number of cuts much higher than required for discernibility, conflict-solving weight (see 4), dynamic reducts, and rule shortening with a low shortening coefficient (around 0.5).

With this method, we have worked on epilepsy type data (Sect. 6), rat local field potential data (cooperative research with the group of Prof. Andrzej Wróbel, Institute of Experimental Biology, Polish Academy of Sciences, is in progress). In these cases, we use the WaRS method with inner products of the signal and appropriate wavelets.

5.4 Variant

Instead of computing wavelet coefficients of the signal, one may filter it with quadrature mirror filters described in Sect. 4. The thresholding procedure and frequential analysis of occurrences applies as usual. This variant of the original method is implemented in the Breiman example in Sect. 9 and for the rat reaction to the light and tone problem in Section 8.

In the rat reaction model, we modify the approach a little further, introducing a special feature related to the saw-like complexes present in the filtered signal in the crucial moments of the experiment (see Sect. 8).

6 Hybrid WaRS Epilepsy Classifier

The problem of epilepsy diagnosis is important in EEG analysis. The research is carried out in two directions. The first is related to the detection of characteristic patterns in an epileptic EEG. This automated analysis of signals helps physicians to look quickly through a great amount of data. Some results of the research in this direction are reported in [28] and [30]. They are devoted to differentiating single-focal and multifocal epilepsy. Another approach is based on a search for relevant features that could support a diagnosis. Our chapter is relevant to the second approach. In this chapter, we propose to use wavelet analysis and to hybridize it with some rough set methods or neural network based methods. Wavelets have already proved their efficiency in this field [30]. The objective of this section is to solve the following problem for EEG signals. There are two groups of children suffering from epilepsy. The first, according to clinical treatment, suffers posttraumatic epilepsy (denoted B) and consists of recordings from 11 subjects. The epilepsy of children in the second group (denoted A) is due to other facts, and there are 25 children in this group. There are also two additional groups *probably* belonging to the basic ones. For the purpose of this research, they were added to appropriate basic groups. For every patient, we have at hand an EEG score of 21 electrodes of 2.5 seconds sampled at a frequency of 102.4 Hz. The problem is

Question 1. Using only these EEG scores and no other clinical information, how to distinguish between posttraumatic epilepsy and the other group?

The experiments show that hybrid methods using wavelets in combination with rough set methods and neural networks can give satisfactory results. The initial data has 5736 real-valued attributes for 44 objects. Direct application of methods for decision rule generation cannot give a satisfactory solution because of this large number of attributes with many values. Hence, in preprocessing, compression of data by wavelet methods is used. In the next phase, data received from wavelet analysis are used as input for rough set or neural network methods. In many cases, hybrid methods can give better results (see, for example, [31]).

After computing the coefficients, the next step is thresholding the signals, which results in a 20-fold decrease of the number of wavelet coefficients above the threshold $\vartheta = 1.0$ in comparison with all coefficients:

$$\frac{\#A_\vartheta}{\#A_0} \approx 0.05.$$

Because of the unconditionality of the wavelet basis,

$$\tilde{f} = \sum_{(j,k)\in A_\vartheta} \langle f, \psi_{jk} \rangle \psi_{jk}$$

is very close to the original f that keeps much of the visual information contained in the brain wave signal.

For quality assessment, we have used a cross-validation scheme. For rough set methods, the classification was done in a five cross-validation scheme, whereas for an artificial neural network (ANN), a two cross-validation scheme was applied. It was confirmed experimentally that dividing data into two subgroups for an ANN is a better setup than five cross-validation. Both methods analyze one subgroup and then classify the remainder. We present here a survey of the most important results. Table 1 summarizes the efficiency of the classifiers obtained in this experimental setting. If the reader is inclined to a more thorough study, we advise consulting the original paper [34]. In the first experiment (No. 1 in Table 2), we took the four most frequent

Table 2. Results of hybrid classification method

No.	Experiment	Efficiency (A)	Efficiency (B)
1.	M = 660 (RSES dynamic scaling)	75%	56%
2.	means M≈ 600 (ANN)	63%	54%
3.	translation of global cuts	93%	73%
4.	50 best cuts (RSES)	84%	64%
5.	pattern identification	68.5% (A)	67% (B)

coefficients with M = 660, and the dynamical scaling led to the best result in this section.

In the second experiment (No. 2 in Table 2), we tried finding features of the signal translated in time. We did it considering 13 coefficients from five levels (five different values of j) and consecutively located in time (subsequent values of k). These coefficients have M≈600. As input data, we took averages of these 13 coefficients in levels. Here, the artificial neural networks showed its efficiency yielding the result of **63%** (A) and **54%** (B) correctly classified objects in the two cross-validation setting.

6.1 Global Cuts

The basic data D (with $M = 200$) was analyzed by RSES procedures for discretization [24]. The interesting fact is that only four cuts are enough to split all data D into subsets of decision classes. Therefore, the efficiency of this classification is 100% (A) and 100%(B). We claim that they may be relevant features of an EEG. The following experiment confirms it. We translated the cuts from their attributes to the appropriate attributes from the next electrode. Explicitly, we changed each of the rules yielded by the cuts:

$$\langle f_p^e, \psi_{jk} \rangle > \theta \Rightarrow (p \in A),$$

to the analogous rule involving the signal from the next electrode,

$$\langle f_p^{e+1}, \psi_{jk} \rangle > \theta \Rightarrow (p \in A).$$

For the second class, the change was analogous. This attempt resulted in good efficiency of classification, namely, 93% (A) and 73% (B).

6.2 Best Cuts

Our next approach (experiment No.4 in Table 2) was to scale data before analysis to ensure its greater stability. Using discretization procedures from the RSES library, a fixed number of cuts discerning the largest possible number of pairs for the population were generated. Data scaled using these cuts and then analyzed by the RSES system and an ANN. The best results were obtained when scaled with 50 cuts and then analyzed by the RSES system, namely, **84**% and **64**%.

6.3 Pattern Identification

This experiment (No. 5 in Table 2) comes from medical intuition about the problem. It consists of seeking a pattern in an EEG score or in its elements. We have chosen four triples (j, k, ϑ_{jk}), and data were scaled putting one if $\langle f, \psi_{jk} \rangle > \vartheta_{jk}$ and zero otherwise. For each electrode, we count ones and their number is one of 21 attributes. Thus, we compare wavelet components of the signal with the pattern and the number of ones is the quality of their match. These four triples correspond to the pattern $\sum_{j,k} \vartheta_{jk} \psi_{jk}$ given in Fig. 5. The classification quality in this case was **68,5**% (A) and **67**% (B) by means of an ANN.

6.4 Discussion

1. The best results are **75**% (A) and **56**% (B) for frequential analysis, **84**% (A) and **64**% (B) for the best cuts approach, **85**% (A) and **69**% (B) for the best cuts approach with additional scaling and **68.5**% (A) and **67**% (B) for pattern identification. The results obtained allow us to suggest that the hybrid method described can be treated as a promising tool for classifying objects. However, further work in this direction is needed.
2. The comparison of results shows that improvement of classification in group B yields worse results on group A. One of the possible causes may be the fact that part of the data from group B does not include information necessary for efficient classification.
3. The whole population can be divided into subsets of decision classes by using only four cuts. Hence, the decision classes can be defined by these four cuts. Our experiments with translated cuts are showing that they carry a real piece of information from an EEG.
4. Further work on rough set methods should allow us to discover features of signal discriminating groups. Then, wavelet analysis may be concentrated on preprocessing, efficient in denoising to make these features detectable by rough set methods.
5. The analysis of attribute dissimilarities for patients from different decision classes is a promising field of investigation.
6. The next direction of further research is optimization of wavelets with respect to essential signal features finding a compression ratio or reconstruction accuracy. It requires interpolating wavelets. This enables us to use genetic algorithms for wavelet optimization.

7 Noise-Resistant Epilepsy Classifier

Although valid in some cases, the straightforward constructed classifier (WaRS) has some features that, in our opinion, may show as inconvenient or troublesome. In particular, we have to be aware of the facts that

1. The amount of data (number of cases) compared to the size of the data (number of measurements) is relatively small, and therefore some features of even a single object may significantly influence the overall performance.
2. The presence of a significant amount of noise in EEG data makes rules obtained by the initial algorithm doubtful. According to the experiments, those rules do not show necessary resistance to noised data, and the preprocessing stage is not designed to eliminate every possible kind of distortion.
3. The classifier uses rules and weights for them that are statically set during the construction process. No universal method is known for making modifications to those weights without complete reconstruction of the classifier (i.e. rule and weight calculation).
4. Weights that are given by stability coefficients usually span across a wide range. In some cases, the value of stability coefficient for a rule may be as small as $const \cdot 10^{-16}$ or as big as $const \cdot 10^{7}$. This forces us to use high numerical precision in our algorithms and makes those algorithms more complicated and costly.
5. The process of decision making is not very straightforward. It comprises checking an object against the rules and then establishing and comparing summarized weights. It requires three basic steps, and to do so, we have to store the set of rules and weights.

To achieve a more significant classifier that is better justified by the data, at the first step, we decided to extend our set of data. Instead of the original EEG score, we used its noised copies. We called this part of process *noisification.*

Let us consider an EEG signal for the pth patient measured at the eth electrode $f_p^e(t)$. We add to it a white noise that is

$$f_{p,i}^e(t) = f_p^e(t) + \alpha\varepsilon_t, \tag{10}$$

where ε_t is a random variable with a normal distribution $N(0,1)$ independent for different t, and α describes the level of noise added. In the next stage, we process these signal by the WaRS method. Replacement of the original signals with their "noisified" copies gives the advantage that classifier based on them does not rely on the peculiarities of the EEG signal, which might have (and often has) a casual character. For the group A, we used 11 copies of each EEG score and 25 copies for group B. The additional point of this procedure is that the number of 'noisified' copies from each group is equal. We applied two main levels of noise: $\alpha = 0.1$ and $\alpha = 0.2$. Their influence can be seen mostly on high-frequency wavelet coefficients (see Fig. 4 and 5).

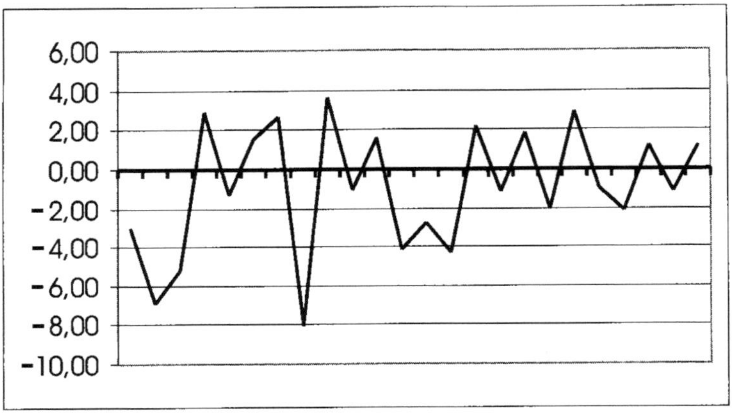

Fig. 4. Wavelet coefficients of noised EEG signal from Fig. 3 after thresholding. Noise level $\alpha = 0.1$

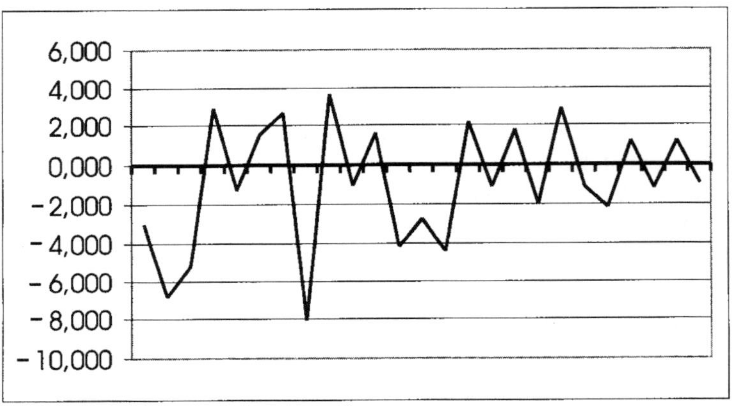

Fig. 5. Wavelet coefficients of noised EEG signal from Fig. 3 after thresholding. Noise level $\alpha = 0.2$

The procedure specified above gives us a larger set of objects, and therefore, the regularities in the data that we search and exploit for classification purposes may be better funded and verified.

To achieve a higher level of confidence in classifier validity and flexibility, we decided to employ methods from the area of adaptive learning. The approach that seemed most promising was that of an ANN. Neural network based systems usually show good behavior with regard to the noise and the possibility of adding new data. In the first step, we want to replace this part of the initial classifier that is connected with weights. Instead of examining the output of all rules and summarizing them using weights, we treat those outputs as an input vector x to a simple neural network.

Let us explain now how we use the neural network to construct a noise-resistant classifier.

First, with every object, we connect a vector of values obtained by applying all of the rules to that object. More precisely, for a given object o_i we introduce vector $\overrightarrow{ro(i)} = (ro(i)_1, ..., ro(i)_k)$, where k is the number of rules we have, according to the formula,

$$ro(i)_j = \begin{cases} -1 \text{ if } r_j \text{ applies to } o_i \text{ and decision is } 0 \\ 0 \text{ if } r_j \text{ do not apply to } o_i \\ 1 \text{ if } r_j \text{ applies to } o_i \text{ and decision is } 1, \end{cases} \tag{11}$$

where $r_1, ..., r_k$ are the rules that we computed.

Second, by applying the rules to all available objects, we get a new training set for use with a neural network. It has a number of attributes equal to the cardinality of the set of rules. The network itself is very simple and contains only one neuron equipped with $k+1$ inputs corresponding to rule output vector $\overrightarrow{ro(i)}$ and bias. Our neuron uses either a sigmoid or hyperbolic tangent as its activation function. This network is further trained to recognize objects from a set of examples (constructed in step 3). Gradient based methods, such as gradient descent with regularization and momentum or Lavenberg–Marquardt are used to perform neural network learning (refer to [4]).

Let us observe that with a set of rules and a single neuron, we can always achieve at least the level of accuracy that is given by a rule-based system using weights based on stability coefficients (5). We can just simulate the process of rule weighting. If we take a single, nonbiased neuron that has threshold activation function φ such that

$$\varphi(I) = \begin{cases} 0 \text{ iff } I \leq 0 \\ 1 \text{ otherwise,} \end{cases} \tag{12}$$

and we connect its inputs with the attributes produced in step 3 of the procedure specified above, then, by setting every jth ($j = 1, ..., k$) weight equal [according to (5) and previously introduced notation],

$$w_j = \frac{Supp_{\mathbf{A}}(r) \cdot SC_{\mathbf{A}}^{\mathbf{P}(\mathbf{A})}(r)}{\sum\limits_{r \in B_i} Supp_{\mathbf{A}}(r) \cdot SC_{\mathbf{A}}^{\mathbf{P}(\mathbf{A})}(r)}, \tag{13}$$

where $i \in \{1,2\}$, we got the effect required. Thanks to the way we introduced the attributes $ro(.)_1, ..., ro(.)_k$ in (11), the actual output of the neuron can be represented as

$$y = \varphi \left[\frac{\sum\limits_{r \in B_2} Supp_{\mathbf{A}}(r) \cdot SC_{\mathbf{A}}^{\mathbf{P}(\mathbf{A})}(r)}{\sum\limits_{r \in B_2} Supp_{\mathbf{A}}(r) \cdot SC_{\mathbf{A}}^{\mathbf{P}(\mathbf{A})}(r)} - \frac{\sum\limits_{r \in B_1} Supp_{\mathbf{A}}(r) \cdot SC_{\mathbf{A}}^{\mathbf{P}(\mathbf{A})}(r)}{\sum\limits_{r \in B_1} Supp_{\mathbf{A}}(r) \cdot SC_{\mathbf{A}}^{\mathbf{P}(\mathbf{A})}(r)} \right], \tag{14}$$

which is exactly the formula for decision making in a rough set rule-based system, just expressed in slightly different terms. In that way the behavior of the rule and the stability coefficient based classifier may be preserved.

The advantages of putting a neural network into our classification scheme are

1. Such a solution is more data driven and reveals more information that is contained in the data. It gives a potential advantage over methodologies that rely on preset, statical rules of precedence not always adequate in the actual situation.
2. There is a possibility of classifier modification without the necessity of total reconstruction. If a new object was misclassified, it is added as a new learning example. As long as this new example is not completely different from the previous ones and does not contradict all the knowledge we have learned so far, it is possible to add the knowledge it contains to our system by simply adjusting (learning) weights in the neural network. Only when a new object is obviously contradictory to our prior findings do we totally reconstruct the classifier.
3. The proposed solution allows more convenient and versatile representation of our classification system. It also opens the way for performing modifications to our system that go deeper and touch not only neuron learning but underlying rules and cuts, too.
4. The relatively small size of our classifier's top layer allows us to interpret it in ways that have real meaning. We may speak of the influence that particular features have on our final decision based on the neuron weights. Since there are no more than 10 of them, how they behave is quite transparent. Moreover, the features we look at may be tracked back through the steps of frequential and wavelet analysis so they can be connected with elements of the actual EEG signal.

The modified version of the classifier underwent several tests to examine its accuracy and flexibility. Additional tests with noisified data were also performed to have a comparison with other methods as well as to find out which step in our modified procedure is most important. We wanted to check whether or not scaling, dynamic scaling, rule selection, or rule weights may be omitted in classification. We were also interested in verifying how manipulating some coefficients may improve or spoil the clarity of our classifier. To do so, we constructed classifiers for the unscaled data and without the use of rules. Below we discuss, one by one, the results on consecutive data sets.

The results we present underwent thorough verification. They were averaged across several repetitions of the cross-validation test. Typically, we used a fivefold cross-validation technique, so we had 80% of the examples for learning and 20% for testing. The distribution of classes A and B between the training and testing samples was random and, therefore, we distinguished the performance on those classes (see tables below). In classifiers based on a neural network, we allowed the output of our classifier to differ from the exact value. If the difference between the classifier output and the required value of decision was less than some preset value, we regarded it as a correct one. This tolerance margin was decided during experimental evaluation, and finally, we chose the value 0.15. So, outputs greater than 0.85 are treated as one and those less than 0.15, are considered equal to zero.

The first experiment was performed with the use of data received from wavelet and frequential analysis with the data unscaled. For this experiment, we used 378 real-valued attributes that were chosen from 1806 wavelet coefficients at the stage of frequential analysis. To get an idea of how complicated it is to learn these data, we attempted to construct a neural network that classifies them well. It turned out that a simple, one-neuron classifier can achieve only limited accuracy. This neuron used biased, sigmoidal activation and was trained using gradient descent methods with an adaptive learning rate and momentum. Since this result was quite accurate, we checked it against generality. However, the one-neuron solution showed major overfitting. It was basically impossible to add even a few new cases to a trained network without loss of overall quality. An attempt to learn five new examples using a previously trained network resulted in a drop in accuracy by 25% on class A and 17% on class B. Therefore, we also constructed a more complicated network for the task. The smallest, empirically constructed network that achieved good results had one hidden layer containing 55 biased, sigmoidal neurons and one neuron of the same kind in the output layer. The network was fully connected, biased, and trained with the Lavenberg–Marquardt (LM) version of a back-propagation algorithm. We also made an experiment using the gradient descent learning method (GD) for comparison. In both cases, the average accuracy was almost the same, but the GD method was slightly faster and less memory-consuming. Unfortunately, in both simple and more complicated neural networks we got no clue about the main mechanisms that drive decision making. The total number of weights and their rather insignificant distribution does not allow identifying of most important features in our data efficiently. Table 3 shows the accuracy achieved by these networks.

Table 3. Experimental data without scaling (real-valued attributes)

Method	Training (A)	Training (B)	Testing (A)	Testing (B)
Single neuron	89%	90.2%	87.7%	90.1%
55-1 Network (LM)	93.1%	90.7%	92.8%	94%
55-1 Network (GD)	92.2%	91.4%	92.9%	92.4%

The next two experiments involved data that was scaled using 100 additional, redundant cuts. The data used for learning contained 105 binary attributes. The difference between those two runs was the amount of noise used in the process of data preprocessing with wavelets. Rough set rules and weights were computed. Application of rule shortening with a factor between 0.3 and 0.6 gave us a set of rules with a single conditional term. There were between 7 and 16 rules in consecutive runs of the cross-validation test.

In the first of the experiments, the accuracy on the training set for the rough set (RSES) method (rules with weights) was 100%, but not for the testing. For the rules computed at this stage, we constructed a neuron that learned weights for those rules ("Rules+neuron" in Tables 4 and 5). The biased, sigmoidal neuron learned using gradient descent methods with the addition of momentum and adaptive learning rate selection. For comparison, we also performed learning for a single neuron having an architecture as above in the table before rule application (105 binary attributes). Accuracy on the training set was 100%. Average accuracy on the testing set was also very good. Unfortunately, the process of learning in the latter was significantly slower, and the network obtained that way had no intuitive explanation. Table 4 summarizes the results of this experiment. The next experiment was basically the same as the previous one. The difference was that the data had been prepared with more noise added than in the previous case. The noisification coefficient [as in formula (10)] was equal to $\alpha = 0.2$ ($\alpha = 0.1$ in the previous experiment). Distortion in the data affected the whole process of classifier establishment. Although the number of binary attributes after scaling was the same (105), their setting was significantly different. This fact affected the extraction of a larger number of shortened, rough set rules. Although in the previous experiment (less noisified data) this number was between 7 and 13 (9 on average), now this limit shifted up to between 9 and 16 (with the average around 11). The results from the rough set classifier and a single neuron trained on rule-based attributes are presented in Table 5. As in the previous case, the neural network was tested on the data before rule application for comparison purposes.

Table 4. Experimental results with scaled data and noise level $\alpha = 0.1$.

Method	Training (A)	Training (B)	Testing (A)	Testing (B)
RSES	100%	100%	57%	91,5%
Rules+neuron	92%	84%	91,4%	83%
Neuron (105 attr.)	100%	100%	99,8%	99,9%

Table 5. Experimental results with scaled data and noise level $\alpha = 0.2$

Method	Training (A)	Training (B)	Testing (A)	Testing (B)
RSES	100%	100%	71.4%	79.5%
Rules+neuron	90%	91%	89.8%	90.7%
Neuron (105 attr.)	100%	100%	99.5%	99.5%

The last experiment followed the pattern of the previous ones. The data preparation was performed with the setting $\alpha = 0.1$. The important difference was the way in which the training and testing sets were constructed. Let us remember that every object in the data table has, due to the noisification procedure, several slightly modified versions. In this experiment, once the object is randomly selected for the testing set, all of its modified (noisified) versions are automatically added to the same sample. Therefore, no information relative to this object is present in the training set. As expected, the results were significantly below the previous.

The experimental results have positively verified some of our expectations. One conclusion that comes from the results presented above is that we can produce quite a simple classifier for a basically complicated task. The classifier construction is a trade-off between complexity and accuracy. If we compare the accomplishments of a single neuron with (7 to 16 inputs) and without (105 binary inputs) implementing the rules to the classifier, we can see that in this latter case it behaves better in classification accuracy (a difference up to 10%). This result comes mostly from higher flexibility and expressive power connected with the larger number of inputs. The correlations that are present in the data (105 attributes) also contribute to this result. On the other hand, this more accurate classification system has rather limited meaning to us. The principles of classification are formulated using a language composed of 105 attribute-weight pairs not to mention bias. This is basically not what we are looking for while constructing our classifier. There are no weights in the trained neuron that are prevailing or negligible, so we get no clear suggestion of the importance of particular features.

The comparison with a rough set based classifier, using weights given by stability coefficients, also shows some interesting properties of our proposed system. In the experiments, the rough set classifier (built on the RSES library) achieved similar quality on the training set. But the other approaches performed relatively better on the testing data. There were no longer cases of major disproportion between accuracy on class A and B, which is very valuable. Generally, the proposed methods showed greater stability of results which suggests that they can describe more general and relevant properties that exist inside our information. It can be seen that adaptive selection of weights for rules results in better flexibility and generality of the derived solution. Since EEG is one of the most important sources of information in the therapy of epilepsy, several researchers tried to address the issue of decision support for such data. In our work, we try to establish a tool for noise-resistant classification of EEG signals. The data we deal with are connected to the treatment of different kinds of epilepsy. By identifying features in the signal, we want to provide an automatic system that will support a physician in the diagnostic process. By applying wavelets, frequential analysis, rough sets, and dynamic scaling in connection with a simple neural network we obtained a novel and reliable classifier architecture. Experiments prove that the proposed methods provide extended robustness and generalization abilities as well as the possibility of directly interpreting the results obtained.

8 Rat Emotion Experiment

The sample of data from the laboratory of Prof. Kasicki was used in this chapter as an example of the successful application of the methods presented herein. Preliminary results of the experiment were already presented as an abstract [17]. Detais of the full experiment together with a detailed analysis and discussion of the biological consequences are presented in a separate paper [18].

The experiment deals with the emotions in animal fear conditioning. In this case, the main purpose of the analysis is to create the decision system of the simple structure that covers most of the "emotional" cases. The paradigm of the experiment is that after preparation, the rat is trained to react with fear to a light signal. This is done in the first day of the experiment. Then, the outer setting moves its attention to the tone signal that relieves it from a shock.

This second training takes a 12-day period, and our results show that its reactions are really as expected by the researchers. We point to the part of its electrical brain wave which is responsible for this reaction, and we follow with a detailed study of the moment that the rat is "in fear" or "relaxed." Summarizing, on the first day, rat is treated with the sequence of a light and a shock. During each of the next days, five trials from group A are performed with the sequence of a light followed by a shock and the other 5 from group B with the sequence of a light followed by a tone and no shock. The type of a trial is chosen on random basis. Thus, we observe the following states of the rat: relaxes before all signals, fear after the light, and relief after the tone or an increase in fear when the tone is not sounded (see Fig. 6 and 7).

The time before $t = 500$ is considered a leisure time for the rat. At $t = 500$, the light is on. In half of the experiments (group A), after the next 5 seconds ($t = 1000$), the electric shock is issued. In the others (group B), after 3 seconds, the *tone* is released, and the shock is *not* issued. Thus, in the first part ($t < 500$), a rat has a leisure time. In the second part ($500 < t < 800$), it is in fear. Then, after the tone, it should calm down. In the 'light only' case, it should keep being aggressive. In the final part ($t > 1000$), measurable only for 'tone' cases, the animal should calm almost completely.

One should observe that the aggressive reaction to the light appears after the preconditioning phase in the first day of the experiment.

Note. The signal is sampled at the rate of 100 Hz. Whenever we refer to the time in absolute units, we mean the number of samples, i.e., t=600 means 6 s after the start of the recording.

8.1 Data Processing

Figure 9 shows rat reactions to the light only, and Fig. 10 its reactions to the sequence of the light in 5 seconds, followed by the tone in 8.

Figures 9 and 10 contain the result of quadrature mirror filtering of rat brain waves

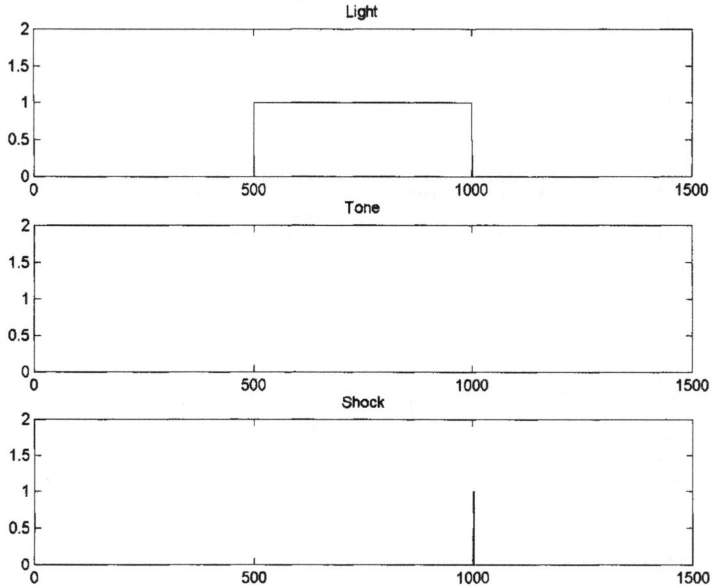

Fig. 6. The stimuli received by the rat during the experiment on group A (light only).

for the first four scales. The light lines show the low-frequency part; the high frequency is drawn with the darker line. Subsequent subgraphs refer to subsequent levels of the analysis – i increases as one moves down the figure. There are some saw-like complexes visible in the detailed part in the third row (darker line).

In the third box from the top after the number 100, corresponding to the 8th second, in the darker line, one distinguishes in Fig. 9 the saw-like phenomenon which is not present in Fig. 10 where the tone is sounded. Subsequent subgraphs refer to subsequent levels of the analysis – i increases as one moves down the figure. A magnified version of this phenomenon compared to a less varying wave for the light-and-tone group is shown in Fig. 11 for the convenience of the reader.

The first box shows the result of quadrature mirror filtering of the brain wave recording: the approximation part is graphed with the light line; the detailed part is drawn in the darker one. The input for the next stage of quadrature mirror filtering is the approximation line from the previous one,

$$a_{i+1} = T_o a_i.$$

The detailed part (the darker light) is the result of high-pass filter action and downsampling of the approximation part from the previous box:

$$d_{i+1} = T_1 a_i.$$

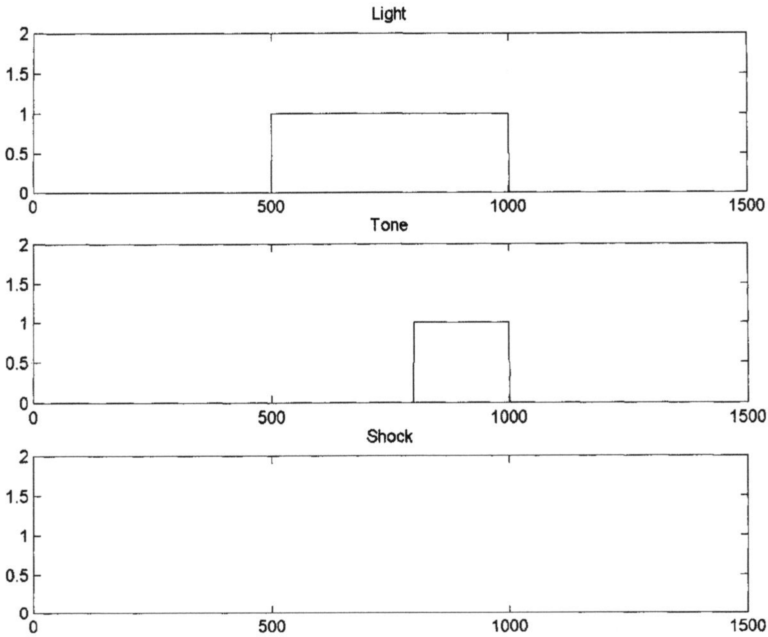

Fig. 7. The stimuli received by the rat during the experiment on group B (light + tone)

The time 120 in the third box corresponds to the start of the shock.

The magnitude of the saw-like phenomena was taken as a measure of the rat's fear (speaking imprecisely). In the strict sense, we reduce the phenomenon to analysis of the sequential values

$$m_t = \sum_{j=1}^{9} |d_3(t+j+1) - d_3(t+j)|,$$

for t=1, 10, 19, ... We consider the couple of (m_t, m_{t+9}) of two consecutive non-overlapping values of m_t as state of the rat. This couple corresponds to a period of approximately 1.4 s. Thus, we classify brain wave intervals of this length. There are six such intervals in each recording.

The first three are the leisure time for the rat (it even does not know that the brain wave measurement started); the next two are a time of uncertainty, behavior under a threat of shock, signaled by the light turned on; the next one is that the rat can find out if the shock will come (the tone was sounded or not), but in the first experiments it still does not know that it is the case. The difference in the rat's electric brain activity can be, as we show later, an example of a learning process.

Our objective is to discover two reactions confronting two states. First, let us con-

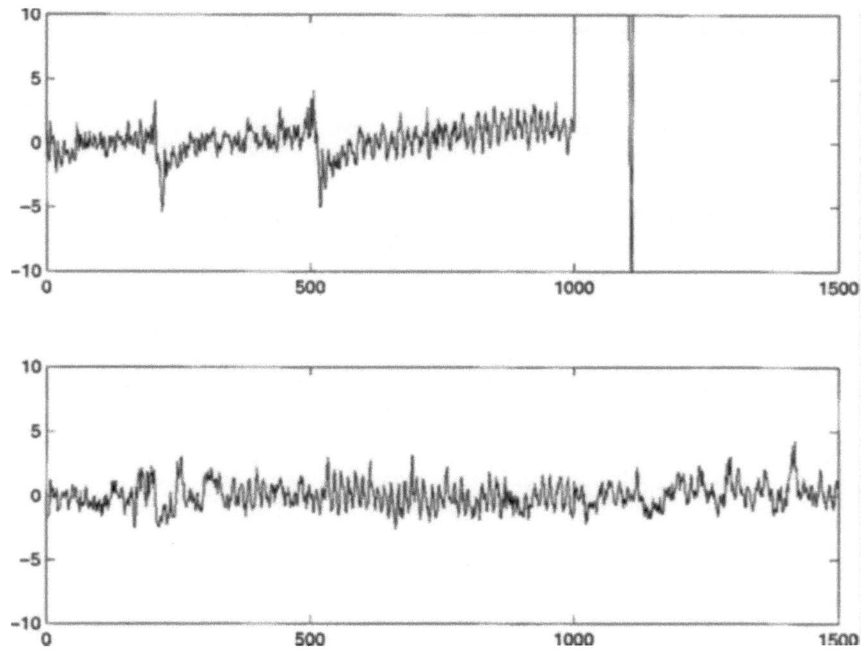

Fig. 8. Examples of rat brain wave sampled at a frequency of 100 Hz. The top one is the 'light-only' example and the square part after $t = 1000$ is the result of the electric impulse coming across the skin of the animal, which is a consequence of the shock issued

sider the reaction to light, and we observe it in the intervals of leisure time and those of uncertainty that follow the light going on. As the second, let us consider the reaction to the tone in the intervals when it should calm down or should not (after the 8th second). The time interval $I = [140k, 140(k+1)]$ is represented as a pair

$$S_I = (m_t, m_{t+9}) = \left(S_I^1, S_I^2\right).$$

Thus the rat learns *two things*; first, sensitivity to the light linked to the shock; second, that the tone supersedes the light. It is the tone that 'rules and decides' whether or not the shock will appear. Examples of the behavior of the m_t-couples are given in Fig. 12 – 15.

Figures 12 and 13 depict the reaction to the light in terms of S_I in the 13 days of the experiment. The points correspond to the relaxed state 'o' ($k = 1,2,3$ before the 5th second) or to the fear state 'x' ($k = 4,5$ after the 5th second and before the 8th). The first graph (day 1) refers to the preconditioning phase, where there were no tone recordings. The graphs are labeled with the day number accordingly.

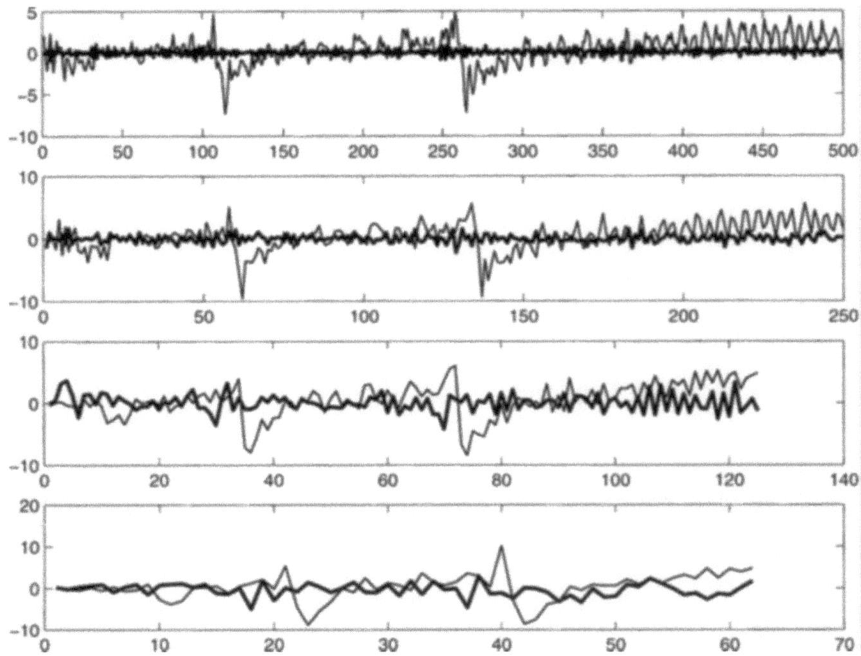

Fig. 9. The result of quadrature mirror filtering of rat brain waves for the first four scales. The recording from group A (light only)

The reaction to the tone is shown in Fig. 14–15. The points representing pairs S_I are plotted in the subgraphs. The points correspond to the "should calm down" state "+" ($k = 6$ after the 8th second, tone sounded) or to the "should not calm down" state 'o' ($k = 6$ after the 8th second, no tone). The first graph (day 1) refers to the preconditioning phase, where there were no 'tone' recordings. The graphs are labeled with the day number as on the previous graph.

Usually, when both S_I^1, S_I^2 are small, the rat is relaxed. If both are big, then it is aggressive. This rule can be observed as the first day of the experiment is analyzed (there were only the light and the electric shock paradigm then; see Fig. 12).

As the experiment continues, the regions of the reaction to the light transcend (being visibly separated at the beginning), and in turn the regions of the reaction to the tone slide away more and more. As a measure of their discernibility, we introduce classification efficiency by the RSES-based classification system with the parameter settings: dynamic reduct generation, shortening rule level 0.5, all rules generated, conflict-solving measure (5). Since there are only 10 recordings, we use the 'drop-one' method to validate the classification system.

The efficiency of classification of S_I was given in Fig. 16 and 17. The efficiency

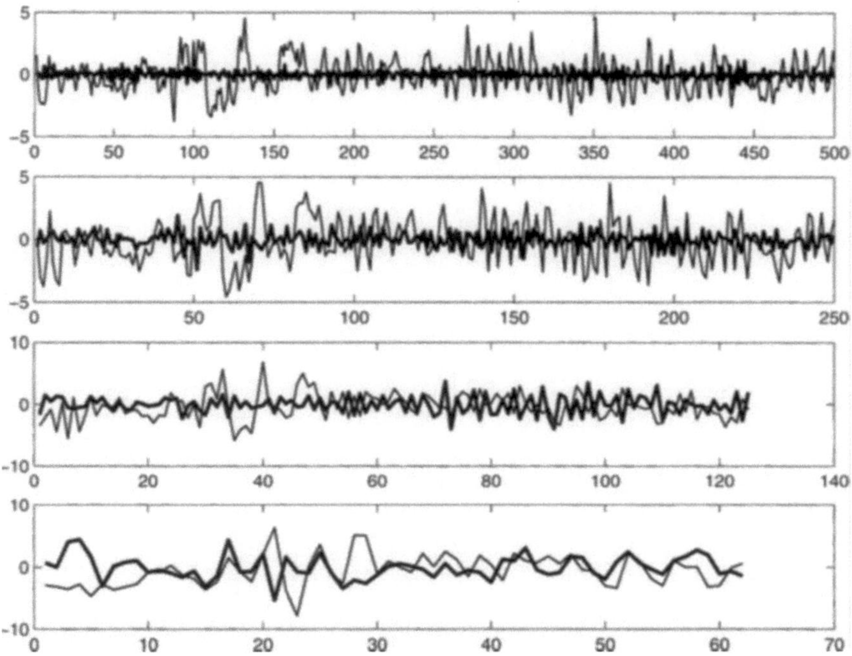

Fig. 10. The result of quadrature mirror filtering of rat brain waves for the first four scales. The recording from group B (light+tone)

related to the reaction to the light is printed at the top and the reaction to the tone at the bottom of Fig. 16. Their sum is plotted in Fig. 17 as a joint measure of the "rat's concentration."

The interesting point is that its slope coincides with that of the typical learning curve with the minimum in the tenth or eleventh day around the reasonable number of 10%. The last point looks like an overfitting phenomenon, where the rat reacts only to the tone and neglects the light signal. In the 12-day period of the learning process, a full reversal occurred in the accuracy of reaction. At the beginning the rat *precisely* responds to the light (10% of errors) and very vaguely to the tone (30% of errors); at the end of the experiment, precisely to the tone (10%) and vaguely to the light (30%).

Special attention should be paid to the localization of the points corresponding to the time before and after the light signal in the second stage of the experiment (days 10 to 14, Fig. 13). They move aside to the left bottom corner of the graph, which is the relaxed state, thus leading to the conclusion that in this phase, the rat almost neglects the light.

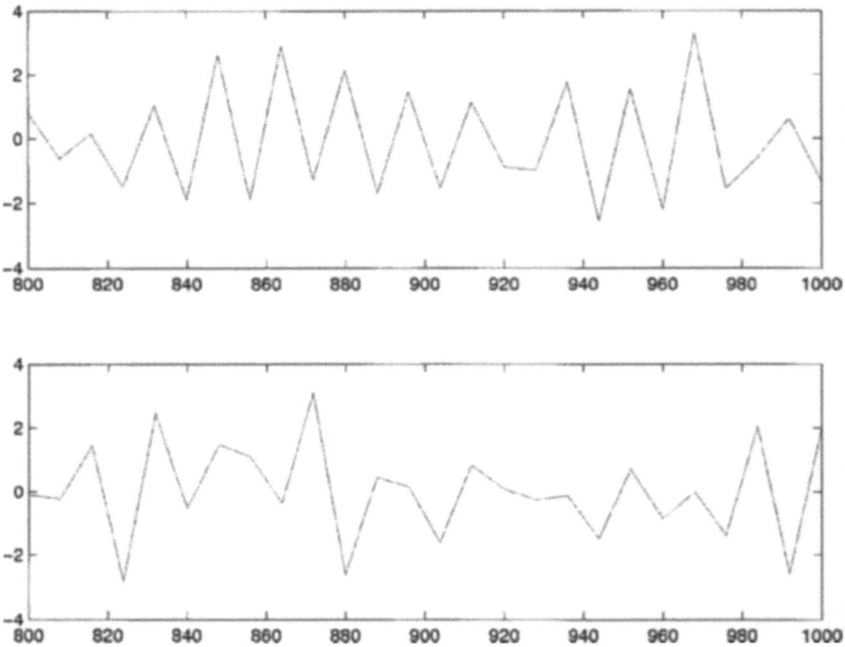

Fig. 11. The magnified version of the coefficients for the light-only (top) and tone (bottom) experiment near $t = 800$. The saw-like components in the 'light-only' example are more accountable

The other point that requires attention is that of the reaction to the tone in days 12 to 13 (Fig. 15). If the tone is sounded, corresponding S_l (pluses) are concentrated under the line $S_l^2 < 20$, whereas S_l^1 varies in a wide range. This bottom half of the graph is not the relaxed state nor the fear state. One could describe it as a vigilance state, that is to say, a quick switching between these states to adapt the reaction promptly.

8.2 Comments

The discernibility of the brain reaction to the light is at the top and to the tone at the bottom. The reaction to the light is less and less specific, whereas to the tone it is

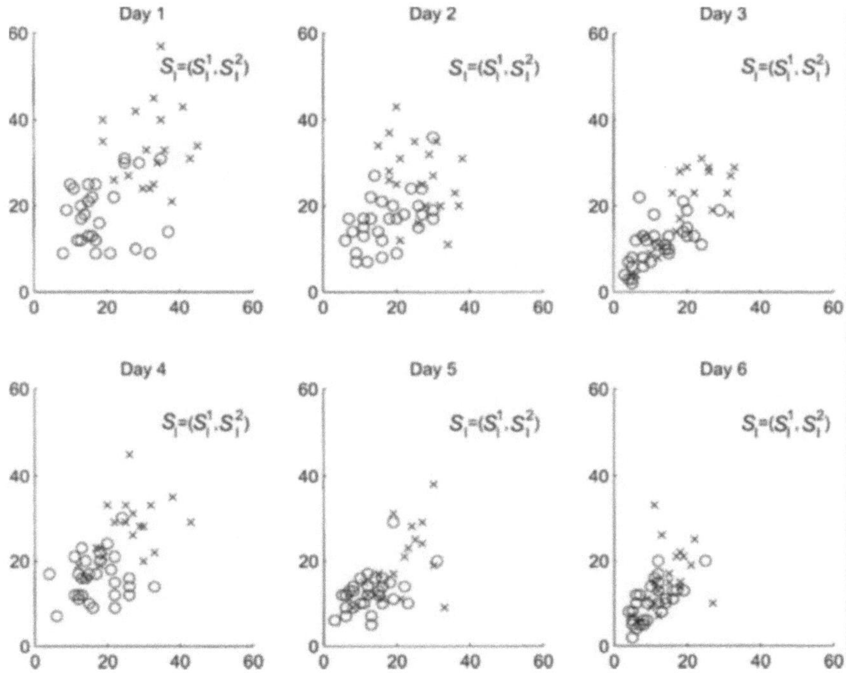

Fig. 12. The reaction to the light in terms of S_I during the experiment (days 1 to 6)

more and more precise. The interesting point is that the curve is the sum of two error ratios and looks like a classical learning curve that goes down oscillating and even reflects the phenomenon of overfitting at the last point which reveals a high joint error ratio. Comparison with Fig. 16 shows that the error rate in the recognition of the tone is almost as low as the reaction to the light at the beginning.

9 Artificially Generated Problem

Let us consider the three classes of problems based on waveforms, i.e., $h_1(t)$, $h_2(t)$, and $h_3(t)$ as linear splines. They have a triangle form with the peaks of height 6 at points 7, 11, and 15, respectively. The slope of the triangle is 45°. Each class consists of a random convex combination of two of these waveforms sampled at the integers with the white noise added. The convexity coefficient u is uniformly distributed on the interval (0,1). The sampling consists of 21 points distributed at equal distances. See Fig. 18 for the graphs of h_i functions and Figs. 19–21 for examples of the signals belonging to the appropriate decision classes.

Breiman et al. proposed this problem in their classic book, *Classification and Regression Trees* [6]. They tested the methods of classification tree generation on this

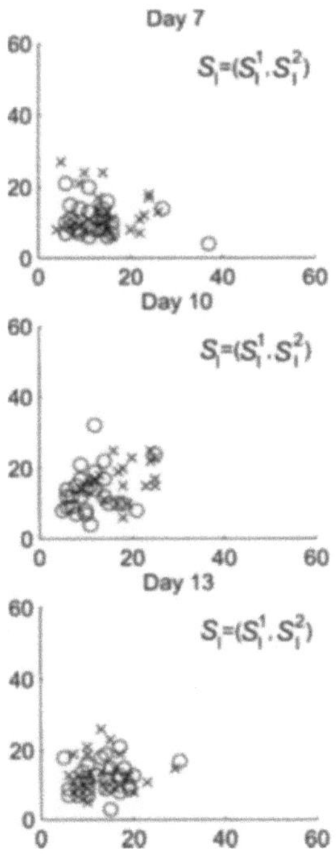

Fig. 13. The reaction to the light in terms of S_I during the experiment (days 7 to 13)

example. For this problem they applied the Gini splitting criterion, and the final tree was selected by pruning and cross-validation techniques. The cross-validation estimate of efficiency obtained has error rate 28%. The research on linear combinations of attributes and their optimization resulted in a classification ratio of 20%. An analytic expression for the Bayes rule can be derived in this problem. Using the rule obtained on a test sample of size 5000 gave an error rate of 14%.

The same example was used by Saito, who in his thesis written under the supervision of R.R. Coifman, developed the local discriminant basis method. The algorithm screens through the large number of known wavelet bases assessing which gives the largest value of an entropy-based measure for the problem. Then a certain number of the largest wavelet coefficients is picked up serving as an input to the CART algorithms. The different schemes resulted in different error rates; the best one was 15.9%, the others ranged from 20.9 to 32.9% [29].

9.1 The results

This example was tackled with the WaRS method in its quadrature mirror filtering version. For the sake of comparison, the input was initially fed into the RSES classi-

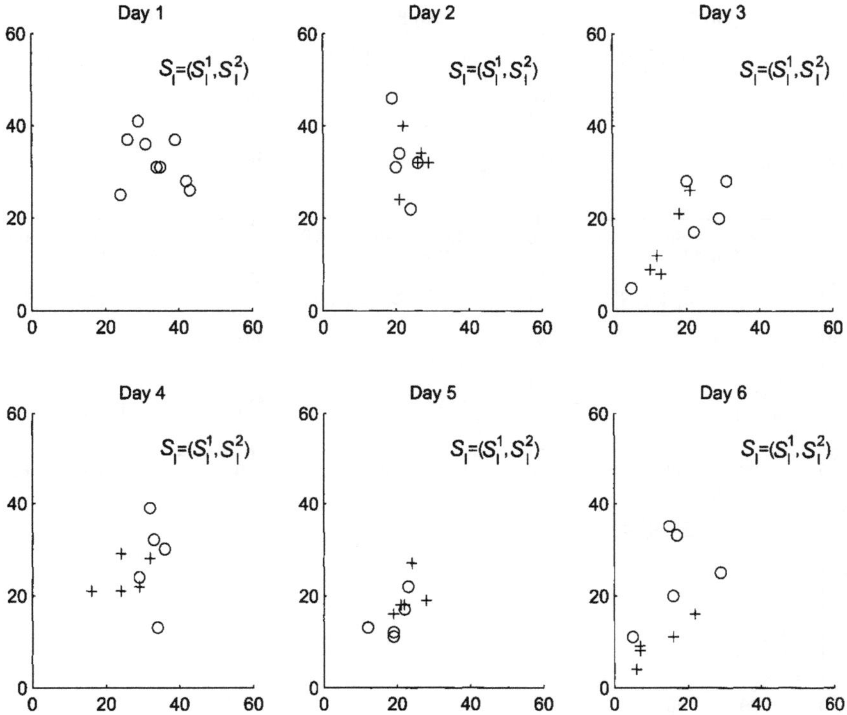

Fig. 14. The reaction to the tone in terms of S_I during the experiment (days 1-6)

fication system. The experiment yielded a 31.5% error rate. The possible reason for this confusion is related, in the author's opinion, to the essential amount of noise in the data, which seems to be one of the important points in the development of rough set classifiers.

The WaRS method was run with parameters as follow. The quadrature mirror filtering was done with both filters applied on each stage of processing. The low and high-pass filters are of the Daubechies wavelet of the sixth order. The coefficients obtained were compressed with the threshold $\vartheta = 1.0$. Frequential analysis was performed with a range of mean occurrence number of [0.55, 0.72]. The minimal error rate was obtained for the second approximation part $a_2(\cdot)$ of the signal. Its value was 24.4%. Thus, we obtained an improvement in the pure rough set classifier, even for low-dimensional data.

9.2 Discussion

The results presented show a significant improvement in the rough set classifiers by introducing wavelet preprocessing supplied with additional tools. However, they are far from optimal. To this aim, effort should be made in both the domain of

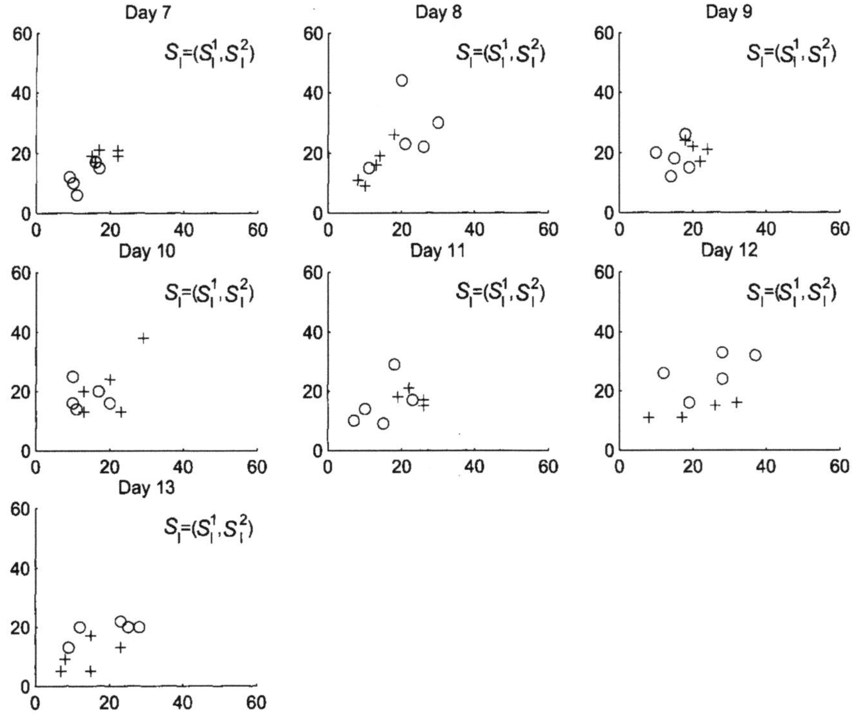

Fig. 15. The reaction to the tone in terms of S_I during the experiment (days 7-13)

noise-resistant rough set classifiers and the more sophisticated methods of wavelet preprocessing. On the one hand, the 'optimization-by-screening' methods have a heavy load of computational cost, and from this point of view, the implemented version of the WaRS method with its simplicity and speed possesses significant potential advantages. Also optical classification based on attributes from frequential analysis suggests a system of a few clear rules, which is not the case when one tries to implement them. On the other hand, rough set based methods give slightly poorer classification on artificial data, but achieve very good results on real-world data and supply the researcher with a set of transparent rules. A sample of it was given in this chapter. An extensive study and comparison of the rough set approach with CART methods may be found in [23].

Acknowledgements

The rough set algorithms used in the experimental work are the research results of my colleagues and friends working in the group of Prof. Andrzej Skowron, Jan Bazan, Son Hung Nguyen and Sinh Hoa Nguyen. The wavelet algorithms were built by the author using known approaches from the literature, as indicated in appropriate

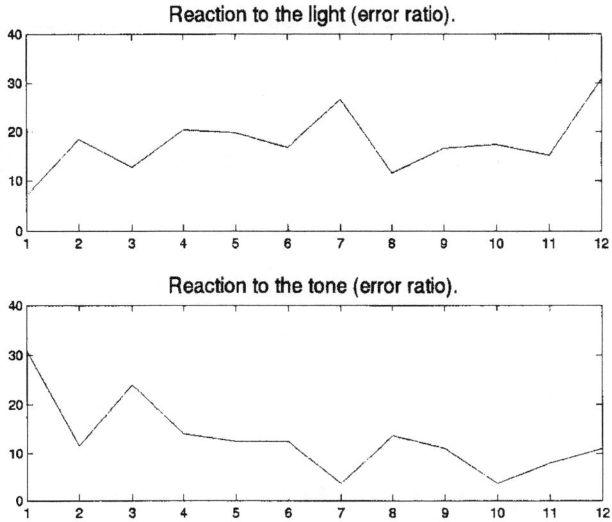

Fig. 16. Comparison of the classification ratio in the drop-one setting for the reaction to the light (top) and to the tone (bottom)

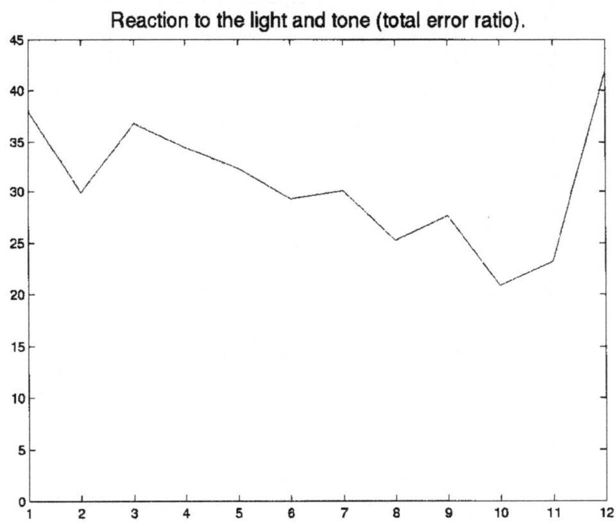

Fig. 17. The joint classification ratio (sum of error ratio) for the light and tone reactions. The interesting point is the resemblance to the classical learning curve (see text)

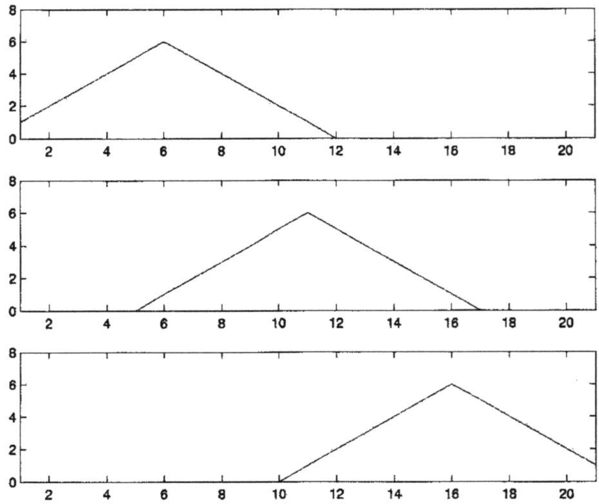

Fig. 18. Functions h_1, h_2, and h_3 (ordered from the top to the bottom)

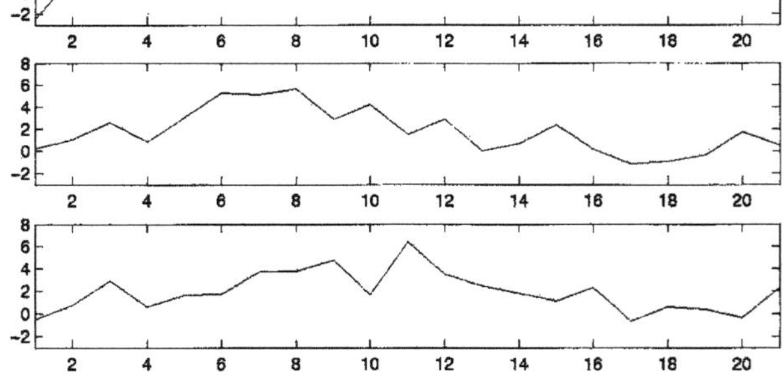

Fig. 19. Examples of signals belonging to class 1 related to functions h_1 and h_3

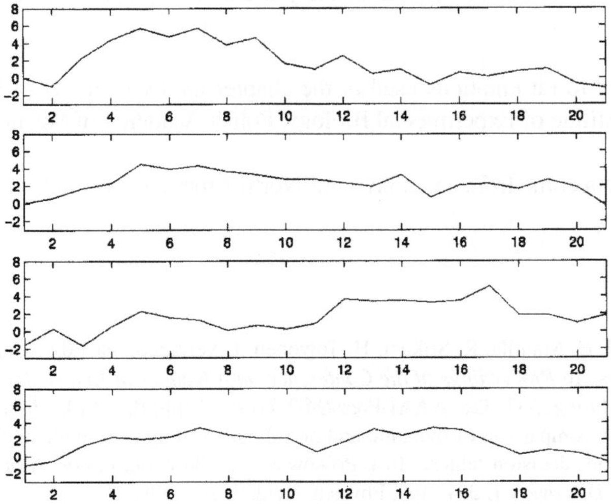

Fig. 20. Examples of signals belonging to class 2 related to functions h_1 and h_2

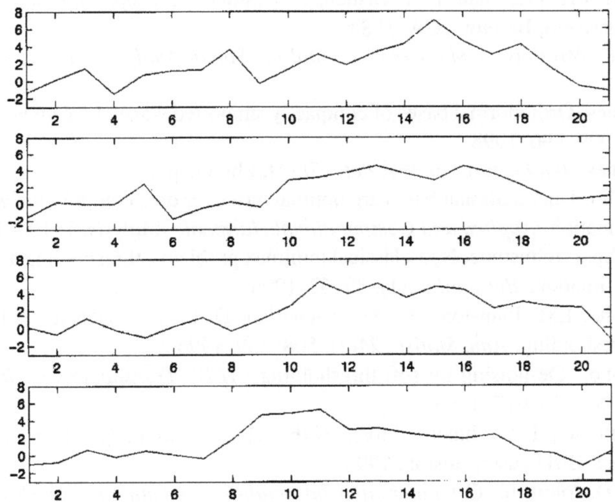

Fig. 21. Examples of signals belonging to class 3 related to functions h_2 and h_3

sections. The heuristics proposed in the frequential analysis or thresholding procedure as well as the method of frequential analysis are an original contribution by the author.

The data related to rat emotions used in the chapter are by courtesy of Prof. Stefan Kasicki, Institute of Experimental Biology, Polish Academy of Sciences.

The epilepsy data come from an anonymous North American hospital.

References

1. R. Agrawal, H. Manilla, R. Srikant, H. Toivonen, I. Verkamo. Fast discovery of association rules. In *Proceedings of the Conference on Advances in Knowledge Discovery and Data Mining*, 307–328, AAAI-Press/MIT Press, Cambridge, MA, 1996.
2. J. Bazan. A comparison of dynamic and non-dynamic rough set methods for extracting laws from decision tables. In L.Polkowski, A. Skowron, editors, *Rough Sets in Knowledge Discovery 1*, 321–365, Physica, Heidelberg, 1998.
3. J. Bazan. *Approximate Reasoning Methods for Synthesis of Decision Algorithms*. Ph. D. dissertation, Institute of Mathematics, Warsaw University, Warsaw 1998.
4. M. Beale, H.B. Demuth. *Neural Network Toolbox*. The MathWorks Inc., Natick, MA, 1997.
5. G. Beylkin, R. Coifman, V. Rokhlin. Fast wavelet transforms and numerical algorithms I. *Comm. Pure Appl. Math.*, 43: 141–183, 1991.
6. L. Breiman, J.H. Friedman, R.A. Olshen, C. J. Stone. *Classification and Regression Trees*. Wadsworth, Belmont, CA, 1984.
7. C. K. Chui. *Wavelets: A Mathematical Tool for Signal Analysis*. *SIAM*, Philadelhia, 1997.
8. I. Daubechies. Orthonormal bases of compactly supported wavelets. *Comm. Pure Appl. Math.*, 41: 909–996, 1998.
9. I. Daubechies. *Ten Lectures on Wavelets*. *SIAM*, Philadelphia, 1992.
10. D.L. Donoho. Unconditional bases are optimal bases for data compression and for statistical estimation. *Applied and Computational Harmonic Analysis*, 1: 100–115, 1993.
11. D.L. Donoho, I. Johnstone. Neo-classical minimax problems, thresholding and adaptive function estimation. *Bernoulli*, 2(1): 39–62, 1996.
12. D.L. Donoho, I.M. Johnstone, G. Kerkyacharian, D. Picard. Density estimation by wavelet thresholding. *Ann. Statist.*, 24(2): 508–539, 1996.
13. D.L. Donoho. De-noising by soft-thresholding. *IEEE Transactions on Information Theory*, 41(3): 613–627, 1995.
14. H.G. Feichtinger, T. Strohmer, editors. *Gabor Analysis and Algorithms. Theory and Applications*. Birkhauser, Boston, 1998.
15. M. Garey, D. Johnson. *Computers and Intractability: A Guide to the Theory of NP-Completeness* ,(12th ed.). W.H. Freeman, San Francisco, 1998.
16. K. Gröchenig. *Foundations of Time-Frequency Analysis*. Birkhäuser, Boston, 2001.
17. P. Jelen, U. Slawinska, S. Kasicki. Danger and safety signals differentiate hippocampal theta activity in classical conditioning in rats. *Acta Neurobiol. Exp.*, 61(3): 232, 2001.
18. P. Jelen, U. Slawinska, S. Kasicki. Danger and safety signals differentiate hippocampal theta activity in classical conditioning in rats. Preprint, 2001.

19. J. Komorowski, Z. Pawlak, L. Polkowski, A.Skowron. Rough sets: A tutorial. In S.K. Pal, A. Skowron, editors, *Rough–Fuzzy Hybridization: A New Trend in Decision Making*, 3–98, Springer, Singapore, 1999.
20. S. Mallat. A theory for multiresolution signal decomposition: The wavelet representation. *IEEE Transactions on Pattern, Analysis and Machine Intelligence*, 11: 674–693, 1989.
21. Y. Meyer. *Wavelets and Operators*. Cambridge University Press, Cambridge, 1992.
22. Y. Meyer. *Wavelets: Applications and Algorithms. SIAM*, Philadelphia, 1993.
23. H.S. Nguyen. Discretization of real value attributes. Boolean reasoning approach. Ph. D. Dissertation, Faculty of Mathematics, Informatics and Mechanics, Warsaw University, 2002.
24. S.H. Nguyen, H.S. Nguyen. Discretization methods in data mining. In L. Polkowski, A. Skowron, editors, *Rough Sets in Knowledge Discovery 1*, 451–482, Physica, Heidelberg, 1998.
25. Z. Pawlak. *Rough Sets: Theoretical Aspects of Reasoning about Data*. Kluwer, Dordrecht, 1991.
26. A. Petrosian, D. Prokhorov, R. Homan, D. Wunsch. Recurrent neural network based prediction of epileptic seizures in intra- and extracranial EEG. *Neurocomputing: An International Journal*, 30: 201–218, 2000.
27. C. Rauszer, A. Skowron. The discernibility matrices and functions in information systems. In R. Słowiński, editor, *Intelligent Decision Support: Handbook of Advances in Rough Sets Theory*, 331–362, Kluwer, Dordrecht, 1992.
28. E. Rodin, M. Litzinger, J. Thompson. Complexity of focal spikes suggests relative epileptogenicity. *Epilepsia*, 36(11): 1078-1083, 1995.
29. N. Saito. *Local Feature Extraction and Its Applications Using a Library of Bases*. Ph.D. Dissertation, Department of Mathematics, Yale University, 1994.
30. L. Senhadji, J. Dillenseger, F. Wendling, C. Rocha, A. Kinie. Wavelet analysis of EEG for three-dimensional mapping of epileptic events. *Annals of Biomedical Engineering*, 23(5): 543–552, 1995..
31. R.W. Świniarski. Rough sets and principal component analysis and their applications in data model building and classification. In S.K. Pal, A. Skowron, editors, *Rough–Fuzzy Hybridization: A New Trend in Decision Making*, 275–300, Springer, Singapore 1999.
32. M. Szczuka, P. Wojdyłło. Neuro-wavelet classifiers for EEG signals based on rough set methods. *Neurocomputing: An International Journal*, 36: 103–122, 2001.
33. P. Wojdyłło. Wavelets and Mallat's multiresolution analysis. *Fundamenta Informaticae*, 34: 469–474, 1998.
34. P. Wojdyłło. Wavelets, rough sets and artificial neural networks in EEG analysis. In *Proceedings of the 1st International Conference on Rough Sets and Current Trends in Computing (RSCTC'98)*, LNAI 1424, 444–449, Springer, Berlin, 1998.
35. P. Wojtaszczyk. *A Mathematical Introduction into Wavelets*. Cambridge University Press, Cambridge, 1997.

Chapter 28
A Hybrid Model for Rule Discovery in Data

Ning Zhong,[1] Chunnian Liu,[2] Ju-Zhen Dong,[1] Setsuo Ohsuga[3]

[1] Department of Information Engineering, Maebashi Institute of Technology, 460-1, Kamisadori-Cho, Maebashi-City, 371, Japan
zhong@maebashi-it.ac.jp

[2] School of Computer Science, Beijing Polytechnic University, Beijing 100022, P.R. China
bpvliu@public.bta.net.cn

[3] Department of Information and Computer Science, School of Science and Engineering, Waseda University, 3-4-1 Okubo Shinjuku-Ku, Tokyo 169, Japan
ohsuga@fd.catv.ne.jp

Summary. This chapter presents a hybrid model for rule discovery in real-world data with uncertainty and incompleteness. The hybrid model is created by introducing an appropriate relationship between deductive reasoning and a stochastic process and extending the relationship to include abduction. Furthermore, a generalization distribution table (GDT), which is a variant of the transition matrix of a stochastic process, is defined. Thus, the typical methods of symbolic reasoning such as deduction, induction, and abduction, as well as the methods based on soft computing techniques such as rough sets, fuzzy sets, and granular computing can be cooperatively used by taking the GDT and/or the transition matrix of a stochastic process as media. Ways of implementing the hybrid model are also discussed.

1 Introduction

Massive data sets in the real world have driven research, applications, and tool development in business, science, government, and academia. The continued growth in data collection in all of these areas ensures that the fundamental problem that KDD (knowledge discovery in databases) addresses, namely, how does one understand and use one's data, will continue to be of critical importance across a large swath of organizations [5].

Although simple statistical techniques and machine learning for data analysis were developed long ago, advanced techniques for intelligent data analysis are not yet mature. KDD can be regarded as a new generation of information processing technology for real- world applications rather than the toy examples. KDD is a typical interdisciplinary field. To solve real-world problems, not only the component techniques developed in the AI community, and also from other related communities in computer science, statistics, and cognitive science are required.

To deal with the complexity of the real world, we argue that an ideal knowledge (rule) discovery system should have these features:

- Imperfect data can be handled effectively, and the accuracy affected by imperfect data can be explicitly represented in the strength of the rule.
- The use of background knowledge can be selected according to whether or not background knowledge exists. On the one hand, background knowledge can be used flexibly in the discovery process; on the other hand, if no background knowledge is available, it can also work.
- Biases can be flexibly selected and adjusted for constraint and search control.
- Data change can be processed easily. Since the data in most databases are ever changed (e.g., data are often added, deleted, or updated), a good method for real applications has to handle data change conveniently.
- The discovery process can be performed in a distributed cooperative mode.

It is clear that no method can contain all of the above characteristics. A unique method is to combine several techniques to construct a hybrid approach. We argue that the hybrid approach is an important way to deal with real-world problems. Here, "hybrid" means combining many advantages of existing methods and avoiding their disadvantages or weaknesses when the existing methods are used separately. There are ongoing efforts to integrate logic (including nonclassical logic), artificial neural networks, probabilistic and statistical reasoning, fuzzy set theory, rough set theory, genetic algorithms and other methodologies in a soft computing paradigm [2,28,16].

In this chapter, a hybrid model is proposed for discovering *if-then* rules in data in an environment with uncertainty and incompleteness. The central feature of the hybrid model is the generalization distribution table (GDT) that is a variant of the transition matrix of a stochastic process. We will also discuss ways for implementing the hybrid model and some experimental results.

2 A Hybrid Intelligent Model

2.1 An Overview

In general, hybrid models involve a variety of different types of processes and representations in both learning and performance. Hence, multiple mechanisms interact in a complex way in most models.

Figure 1 shows a hybrid intelligent model for discovering *if-then* rules in data, which is created by introducing an appropriate relationship between deductive reasoning and a stochastic process [16] and extending the relationship so as to include abduction [28]. Then, a generalization distribution table (GDT), which is a variant of the transition matrix (TM) of a stochastic process, is defined [28,29]. Thus, the typical methods of symbolic reasoning such as deduction, induction, and abduction, as well as methods based on soft computing techniques such as rough sets, fuzzy sets, and

(A, A -> B => B)

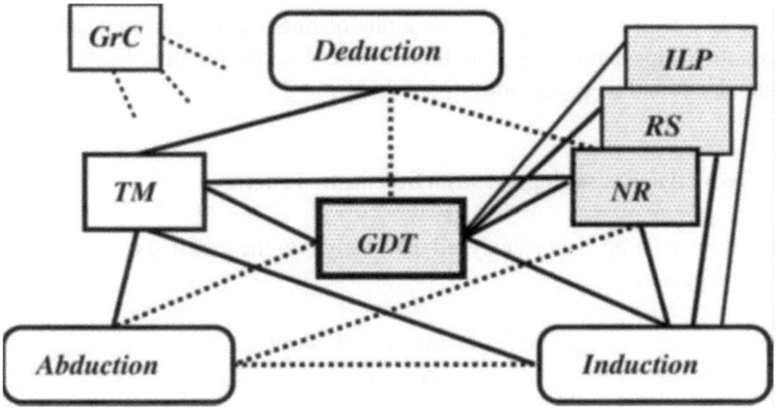

(B, A -> B => A) ***(A, B => A -> B)***

TM : Transition Matrix *GDT: Generalization Distribution Table*
RS: Rough Sets *NR: Networks Representation*
GrC: Granular Computing *ILP: Inductive Logic Programming*

Fig. 1. A hybrid model for knowledge discovery

granular computing can be cooperatively used by taking GDT and/or the transition matrix of a stochastic process as media.

The shadowed parts in Fig. 1 are the major parts of our current study. The central idea of our methodology is to use GDT as a hypothesis search space for generalization, in which the probabilistic relationships between concepts and instances across discrete domains are represented. The main features of GDT are the following:

- Biases for search control can be selected in a flexible way. Background knowledge can be used as a bias to control the initiation of a GDT and in the rule discovery process.
- The rule discovery process is oriented toward inducing rules with a high quality of classification of unseen instances. The rule uncertainty, including the ability to predict unseen instances, can be explicitly represented by the rule strength.

Based on the GDT, we have developed or are developing three hybrid systems. The first is called GDT-NR, which is based on the network representation of a GDT; the second is called GDT-RS, which is based on a combination of a GDT and rough set theory; the third is called GDT-RS-ILP, which is based on a combination of a GDT, rough set theory, and inductive logic programming (ILP), for extending GDT-RS for relation learning. Furthermore, granular computing (GrC) can be used as a preprocessing step to change granules of individual objects to deal with continuous values and imperfect data [24,25].

In Fig. 1, we further distinguish two kinds of lines: solid lines and dotted lines. The relationships denoted by solid lines will be described in this chapter, but those denoted by dotted lines will not.

2.2 Deductive Reasoning Versus a Stochastic Process

Deductive reasoning can be analyzed ultimately into the repeated application of the strong syllogism:

$$\frac{\begin{array}{c}\text{If A is true, then B is true.}\\ \text{A is true.}\end{array}}{\text{Hence, B is true,}}$$

that is, $[A \wedge (A \rightarrow B) \Rightarrow B]$ in short, where A and B are logic formulas. Let us consider two predicates F and G, and let d be a finite set. For simplicity, we assume that F and G are single-place predicates. They give descriptions of an object in d. Or, in other words, F (the definition for G is the same as that for F from now on) classifies all elements in the set d into two classes:

$$\{x \in d \mid F(x) \text{ is true (or x satisfies F)}\}, \text{ and}$$
$$\{x \in d \mid F(x) \text{ is false (or x does not satisfy F)}\}.$$

In the following, $F(x)$ and $\overline{F}(x)$ mean "$F(x)$ is true (or x satisfies F)" and "$F(x)$ is false (or x does not satisfy F)," respectively, for $x \in d$. Thus, one of the most useful forms of $A \rightarrow B$ can be denoted by multilayer logic [15,16] into

$$[\forall X/d][F(X) \rightarrow G(X)],$$

that is, the multilayer logic formula reads "for any X belonging to d, if $F(X)$ is true, then $G(X)$ is also true." Notice that since the set d can be looked upon as an ordered set, F is represented by a sequence of n binary digits. Thus, the ith binary digit is one or zero for $F(a_i)$ or $\overline{F}(a_i)$ corresponding to the ith element a_i of the set d. Based on the preparation, several basic concepts are described first for creating an appropriate relationship between deductive reasoning and a stochastic process.

1. Expansion function
   If the domain set d of formula F is finite, $d = \{a_1, a_2, \ldots, a_n\}$, then the multilayer logic formulas $[\forall X/d]F(X)$ and $[\exists X/d]F(X)$ can be expanded as

   $F(a_1) \wedge F(a_2) \wedge \ldots \wedge F(a_n)$ [or $\wedge_{a_i \in d} F(a_i)$ in short], and
   $F(a_1) \vee F(a_2) \vee \ldots \vee F(a_n)$ [or $\vee_{a_i \in d} F(a_i)$ in short], respectively.

These are called *expansion functions* of multilayer logic formulas. An expansion function is used to extract from a set the elements that possess specified properties. Furthermore, the multilayer logic formulas

$$[\forall X\#/d][F(X) \to G(X)], \text{ and } [\exists X\#/d][F(X) \to G(X)],$$

can be expanded as

$$\wedge_{a_i \in d}[F(a_i) \to G(a_i)], \text{ and } \vee_{a_i \in d}[F(a_i) \to G(a_i)], \text{ respectively.}$$

2. States of d with respect to F
   Let $d = \{a_1, a_2, \ldots, a_n\}$ be a finite set. The *states* of d with respect to F are defined as the conjunctions of either $F(a_i)$ or $\overline{F}(a_i)$ for every element a_i in d, that is, the states of d corresponding to F are

$$S_1(d,F) : \overline{F}(a_1) \wedge \overline{F}(a_2) \wedge \ldots \wedge \overline{F}(a_{n-1}) \wedge \overline{F}(a_n),$$
$$S_2(d,F) : \overline{F}(a_1) \wedge \overline{F}(a_2) \wedge \ldots \wedge \overline{F}(a_{n-1}) \wedge F(a_n),$$
$$\ldots$$
$$S_{2^n}(d,F) : F(a_1) \wedge F(a_2) \wedge \ldots \wedge F(a_{n-1}) \wedge F(a_n).$$

   Let $Prior\_S(d,F) = \{S_1(d,F), \ldots, S_{2^n}(d,F)\}$. Each $S_i(d,F)$ in $Prior\_S(d,F)$ is called a prior state of d with respect to F. For example, if $d = \{a_1, a_2, a_3\}$, its possible prior and posterior states are

$$Prior\_S(d,F) = \overline{F}(a_1) \wedge \overline{F}(a_2) \wedge \overline{F}(a_3), \overline{F}(a_1) \wedge \overline{F}(a_2) \wedge$$
$$\wedge F(a_3), \ldots, F(a_1) \wedge F(a_2) \wedge F(a_3),$$
$$Posterior\_S\{[\forall X/d]F(X)\} = F(a_1) \wedge F(a_2) \wedge F(a_3),$$
$$Posterior\_S\{[\exists X/d]F(X)\} = \overline{F}(a_1)\overline{F} \wedge (a_2) \wedge F(a_3), \ldots, F(a_1) \wedge$$
$$\wedge F(a_2) \wedge F(a_3).$$

   Using binary digits one and zero instead of $F(a_i)$ and $\overline{F}(a_i)$, the above states can be expressed as follows:

$$Prior\_S(d,F) = \{000, 001, 010, \ldots, 111\},$$
$$Posterior\_S\{[\forall X/d]F(X)\} = \{111\},$$
$$Posterior\_S\{[\exists X/d]F(X)\} = \{001, 010, \ldots, 111\}.$$

3. Probability vector of state occurring
   A probability vector with respect to $[\forall X/d]F(X)$ is
   $P^{[\forall X/d]F(X)} = (0, 0, \ldots, 1)$, and
   a probability vector with respect to $[\exists X/d]F(X)$ is
   $P^{[\exists X/d]F(X)} = (0, \alpha, \ldots, \alpha)$
   where, $\sum \alpha = 1$, for any i.

Based on the basic concepts stated above, our purpose is to create an equivalent relationship between deductive reasoning and a stochastic process [16]. In other words, $[F \wedge (F \to G) \Rightarrow G]$ is explained by an equivalent stochastic process $P^{[\forall X/d]F(X)}T = P^{[\forall X/d]G(X)}$, where T is a transition matrix that is equivalent to $(F \to G)$. To create the equivalent relationship, the following three conditions must be satisfied in creating T :

- The elements t_{ij} of T are probabilities $p[S_j(d,G)|S_i(d,F)]$.
- $p[S_j(d,G)|S_i(d,F)]$ must satisfy the truth table of implicative relations.
- $\sum_{j=1}^{2^n} t_{ij} = 1$.

Here $S_i(d,F)$ denotes the ith prior state in $Prior\_S(d,F)$, and $S_j(d,G)$ denotes the jth prior state in $Prior\_S(d,G)$, that is, since

$$[\forall X/d][F(X) \rightarrow G(X)] \text{ (or } [\exists X/d][F(X) \rightarrow G(X)])$$

is equivalent to

$$\wedge_{a_i \in d}[F(a_i) \rightarrow G(a_i)] \text{ (or } \vee_{a_i \in d}[F(a_i) \rightarrow G(a_i)])$$

(i.e., by using the expansion function stated above). According to the truth table of implicative relations, if the value of $F(a_i)$ is known, to satisfy $F(a_i) \rightarrow G(a_i)$, it must follow that $\overline{F}(a_i) \vee G(a_i) = 1$. In other words, the creation of T must satisfy the condition,

if $F(a_i)$ is true, $G(a_i)$ is true; otherwise any value of $G(a_i)$ is correct.

Table 1 shows an example of the transition matrix corresponding to $[\forall X/d][F(X) \rightarrow G(X)]$ and $d = \{a_1, a_2, a_3\}$. In Table 1, the states in the left column denote, respectively,

000: $\overline{F}(a_1) \wedge \overline{F}(a_2) \wedge \overline{F}(a_3)$,
001: $\overline{F}(a_1) \wedge \overline{F}(a_2) \wedge F(a_3)$,
......
111: $F(a_1) \wedge F(a_2) \wedge F(a_3)$,

and the states in the top row denote, respectively,

000: $\overline{G}(a_1) \wedge \overline{G}(a_2) \wedge \overline{G}(a_3)$,
001: $\overline{G}(a_1) \wedge \overline{G}(a_2) \wedge G(a_3)$,
......
111: $G(a_1) \wedge G(a_2) \wedge G(a_3)$;

the elements t_{ij} of T denoted in the transition matrix are the probability distribution corresponding to $[\forall X/d][F(X) \rightarrow G(X)]$, and the elements of T not displayed are all zero. Furthermore, since any background knowledge is not used to create the probability distribution shown in Table 1, states occurrences are equiprobable. For example, if the state of F is $\{010\}$, to satisfy $F(X) \rightarrow G(X)$, the possible states of G are $\{010, 011, 110, 111\}$, and the probability of each of them is 1/4.

Table 1. A transition matrix equivalent to $[\forall X/d][F(X) \rightarrow G(X)]$

	000	001	010	011	100	101	110	111
000	1/8	1/8	1/8	1/8	1/8	1/8	1/8	1/8
001		1/4		1/4		1/4		1/4
010			1/4	1/4			1/4	1/4
011				1/2				1/2
100					1/4	1/4	1/4	1/4
101						1/2		1/2
110							1/2	1/2
111								1

2.3 Hypothesis Generation Based on a Transition Matrix

Our purpose is to create a hybrid model, as shown in Fig. 1, for rule discovery in data with uncertainty and incompleteness. For the purpose, we would like to discuss here a kind of weaker reasoning [18] (weaker syllogism):

If A is true, then B is true.
B is true.
Hence, A becomes more plausible,

that is, $[B \wedge (A \rightarrow B) \mapsto A]$ in short. The evidence does not prove that A is true, but verification of one of its consequences does give us more confidence in A. This is a kind of plausible reasoning for hypothesis generation, which is called "abduction." In other words, from the observed fact B and known rule $A \rightarrow B$, A can be guessed, that is, according to the transition matrix shown in Table 1, from each element $x \in d$ such that $G(x)$ is *true* and the rule $[\forall X/d][F(X) \rightarrow G(X)]$, $F(X)$ can be guessed. Thus, an appropriate relationship between deductive reasoning and abductive reasoning is created by using the transition matrix as a medium, as shown in Fig. 1.

2.4 Generalization Distribution Table (GDT)

The central idea of our methodology is to use a variant of a transition matrix, called a *generalization distribution table* (GDT), as a hypothesis search space for generalization, in which the probabilistic relationships between concepts and instances over discrete domains are represented [28]. Thus, the representation of the original transition matrix introduced in Sect. 2.2 must be modified appropriately, and some concepts must be described for our purpose.

A GDT is defined as consisting of three components. The first is *possible instances* that are all possible combinations of attribute values in a database. They are denoted in the top row of a GDT. The second is *possible generalizations* for instances that are all possible cases of generalization for all possible instances. They are denoted in the left column of a GDT. "$*$," which specifies a wild card, denotes the generalization for instances. For example, the generalization $*b_0c_0$ means the attribute a is unimportant for describing a concept.

The third component of the GDT is *probabilistic relationships* between the possible instances and the possible generalizations, which are represented in the elements G_{ij} of a GDT. They are the probabilistic distribution for describing the strength of the relationship between every possible instance and every possible generalization. The prior distribution is equiprobable, if any prior background knowledge is not used. Thus, it is defined by (1), and $\sum\limits_j G_{ij} = 1$:

$$G_{ij} = p(PI_j|PG_i) = \begin{cases} \dfrac{1}{N_{PG_i}} & \text{if } PI_j \in PG_i \\ 0 & \text{otherwise,} \end{cases} \tag{1}$$

where PI_j is the jth possible instance, PG_i is the ith possible generalization, and N_{PG_i} is the number of the possible instances satisfying the ith possible generalization, that is,

$$N_{PG_i} = \prod_{k \in \{l|\ PG[l]=*\}} n_k, \tag{2}$$

where $PG_i[l]$ is the value of the kth attribute in the possible generalization PG_i, $PG[l] = *$ means that PG_i doesn't contain attribute l.

Furthermore, for convenience, letting $E = \prod_{k=1}^{m} n_k$, (1) can be changed into the following form:

$$G_{ij} = p(PI_j|PG_i) = \begin{cases} \dfrac{\prod\limits_{k \in \{l|\ PG[l]\neq *\}} n_k}{E} & \text{if } PI_j \in PG_i \\ 0 & \text{otherwise,} \end{cases} \tag{3}$$

because of

$$\frac{1}{N_{PG_i}} = \frac{1}{\prod\limits_{k \in \{l|\ PG[l]=*\}} n_k} = \frac{\prod\limits_{k \in \{l|\ PG[l]\neq *\}} n_k}{\prod\limits_{k=1}^{m} n_k} = \frac{\prod\limits_{k \in \{l|\ PG[l]\neq *\}} n_k}{E}.$$

Since E is a constant for a given database, the prior distribution $p(PI_j|PG_i)$ is directly proportional to the product of the numbers of values of all attributes contained in PG_i.

Thus, in our approach, the basic process of hypothesis generation is to generalize the instances observed in a database by searching and revising the GDT. Here, two kinds of attributes need to be distinguished, *condition* attributes and *decision* attributes (sometimes called class attributes) in a database. Condition attributes as possible instances are used to create the GDT, but the decision attributes are not. The decision attributes are normally used to decide which concept (class) should be described in a rule. Usually, a single decision attribute is all that is required.

Table 2 is an example of the GDT, which is generated by using three condition attributes, a, b, and c, in a sample database shown in Table 3, and $a = \{a_0, a_1\}$, $b = \{b_0, b_1, b_2\}$, $c = \{c_0, c_1\}$. For example, the real meaning of these condition attributes can be respectively assigned as *Weather, Temperature, Humidity* in a weather forecast database, or *Temperature, Cough, Headache* in a medical diagnosis database. Attribute d in Table 3 is used as a *decision* attribute. For example, the real meaning of the decision attribute can be assigned as *Wind* or *Flu* corresponding to the assigned condition attributes.

2.5 Biases

Since our approach is based on the GDT, rule discovery can be constrained by three types of biases corresponding to the three components of the GDT defined in Sect. 2.4.

The first type of bias is related to the possible generalizations in a GDT. It is used to decide which concept description should be considered at first. To obtain the best concept descriptions, all possible generalizations should be considered, but not all of them need to be considered at the same time. Possible generalizations (concept descriptions) are divided into several levels of generalization according to the number of wild cards in a generalization: the greater the number of wild cards, the higher the level. Thus, it is clear that any generalization in a lower level is properly contained by one or more generalizations in an upper level. As the default, our approach prefers more general concept descriptions in an upper level to more specific ones in a lower level. However, if necessary, a metacontrol can be used to alter the bias, so that more specific descriptions are preferred to more general ones.

The second type of bias is related to the probability values denoted in G_{ij} in a GDT. It is used to adjust the strength of the relationship between an instance and a generalization. If no prior background knowledge as a bias is available, as a default, the probabilities of occurrence of all possible instances are equiprobable. However, a

Table 2. A GDT generated from the sample database shown in Table 3

PG \ *PI*	a0b0c0	a0b0c1	a0b1c0	a0b1c1	a0b2c0	a0b2c1	a1b0c0	...	a1b2c1
*b0c0	1/2						1/2		
*b0c1		1/2						...	
*b1c0			1/2					...	
*b1c1				1/2				...	
*b2c0					1/2			...	
*b2c1						1/2		...	1/2
a0*c0	1/3		1/3		1/3			...	
a0*c1		1/3		1/3		1/3		...	
a1*c0							1/3	...	
a1*c1								...	1/3
a0b0*	1/2	1/2						...	
a0b1*			1/2	1/2				...	
a0b2*					1/2	1/2		...	
a1b0*							1/2	...	
a1b1*								...	
a1b2*								...	1/2
**c0	1/6		1/6		1/6		1/6	...	
**c1		1/6		1/6		1/6		...	1/6
b0	1/4	1/4					1/4	...	
b1			1/4	1/4				...	
b2					1/4	1/4		...	1/4
a0**	1/6	1/6	1/6	1/6	1/6	1/6		...	
a1**							1/6	...	1/6

Table 3. A sample database

U \ *A*	*a*	*b*	*c*	*d*
u1	a_0	b_0	c_1	y
u2	a_0	b_1	c_1	y
u3	a_0	b_0	c_1	y
u4	a_1	b_1	c_0	n
u5	a_0	b_0	c_1	n
u6	a_0	b_2	c_1	n
u7	a_1	b_1	c_1	y

bias such as background knowledge can be used during the creation of a GDT. The distributions will be dynamically updated according to the real data in a database, and they will be unequiprobable. This issue will be further discussed in Sect. 2.6.

The third type of bias is related to the possible instances in a GDT. In our approach, the strength of the relationship between every possible instance and every possible generalization depends to a certain extent on the way the possible instances are defined and selected.

Furthermore, background knowledge can be used as a bias to constrain the possible instances and the prior distributions. For example, if the background knowledge,

> *"when the air temperature is very high, it is not possible there exists some frost at ground level,"*

is used to learn rules from an earthquake database in which there are attributes such as air temperature, frost at ground level, two centimeters below ground level, atmospheric pressure, etc., then we do not consider the possible instances that contradict this background knowledge. Thus, more refined results may be obtained by using background knowledge in the discovery process.

2.6 Adjusting the Prior Distribution by Background Knowledge

In general, informed knowledge discovery uses background knowledge about a domain to guide a discovery process toward finding interesting and novel rules hidden in data. Background knowledge may be of several forms, including rules already found, taxonomic relationships, causal preconditions, and semantic categories. This section describes a way of adjusting the prior distribution by background knowledge to discover much better knowledge.

As stated in Sect. 2.4, when no prior background knowledge as a bias is available, as a default, the occurrence of all possible instances is equiprobable, and the prior distribution of a GDT is shown in (1). However, the prior distribution can be adjusted by background knowledge and will be unequiprobable after the adjustment. Generally speaking, background knowledge can be given in

$$a_{i_1 j_1} \Rightarrow a_{i_2 j_2}, \qquad Q,$$

where $a_{i_1 j_1}$ is the $j_1 th$ value of attribute i_1 and $a_{i_2 j_2}$ is the $j_2 th$ value of attribute i_2. $a_{i_1 j_1}$ is called the *premise* of the background knowledge, $a_{i_2 j_2}$ is called the *conclusion* of the background knowledge, and Q is called the *strength* of the background knowledge. This means that Q is the probability of occurrence of $a_{i_2 j_2}$ when $a_{i_1 j_1}$ occurs. $Q = 0$ means that "*$a_{i_1 j_1}$ and $a_{i_2 j_2}$ never occur together*"; $Q = 1$ means that

"$a_{i_1 j_1}$ *and* $a_{i_2 j_2}$ *always occur at the same time;*" and $Q = 1/n_{i_2}$ means that the occurrence of $a_{i_2 j_2}$ is the same as that without background knowledge, where n_{i_2} is the number of values of attribute i_2. For each instance *PI* (or each generalization *PG*), let $PI[i]$ (or $PG[i]$) denote the entry of *PI* (or *PG*) corresponding to attribute i. For each generalization *PG* such that $PG[i_1] = a_{i_1 j_1}$ and $PG[i_2] = *$, the prior distribution between the *PG* and related instances will be adjusted. The probability of occurrence of attribute value $a_{i_2 j_2}$ is changed from $1/n_{i_2}$ to Q by background knowledge, so that, for each of the other values in attribute i_2, the probability of its occurrence is changed from $1/n_{i_2}$ to $(1-Q)/(n_{i_2}-1)$. Let the adjusted prior distribution be denoted by p_{bk}. The prior distribution adjusted by the background knowledge "$a_{i_1 j_1} \Rightarrow a_{i_2 j_2},\ Q$" is

$$p_{bk}(PI|PG) = \begin{cases} p(PI|PG) \times Q \times n_{i_2} & \text{if } PG[i_1] = a_{i_1 j_1}, PG[i_2] = *, \\ & \quad PI[i_2] = a_{i_2 j_2} \\ p(PI|PG) \times \dfrac{1-Q}{n_{i_2}-1} \times n_{i_2} & \text{if } PG[i_1] = a_{i_1 j_1}, PG[i_2] = *, \\ & \quad \exists j (1 \le j \le n_{i_2}, j \ne j_2)\ PI[i_2] = a_{i_2 j}, \\ p(PI|PG) & \textit{otherwise}, \end{cases} \tag{4}$$

where coefficients of $p(PI|PG)$, $Q \times n_{i_2}$, $\frac{1-Q}{n_{i_2}-1} \times n_{i_2}$, and 1 are called *adjusting factors* (AF for short) with respect to the background knowledge "$a_{i_1 j_1} \Rightarrow a_{i_2 j_2},\ Q$." They explicitly represent the influence of a piece of background knowledge on the prior distribution. Hence, the adjusted prior distribution can be denoted by

$$p_{bk}(PI|PG) = p(PI|PG) \times AF(PI|PG), \tag{5}$$

and the AF is

$$AF(PI|PG) = \begin{cases} Q \times n_{i_2} & \text{if } PG[i_1] = a_{i_1 j_1}, PG[i_2] = *, \\ & \quad PI[i_2] = a_{i_2 j_2}, \\ \dfrac{1-Q}{n_{i_2}-1} \times n_{i_2} & \text{if } PG[i_1] = a_{i_1 j_1}, PG[i_2] = *, \\ & \quad \exists j (1 \le j \le n_{i_2}, j \ne j_2)\ PI[i_2] = a_{i_2 j}, \\ 1 & \textit{otherwise}. \end{cases} \tag{6}$$

So far, we have explained how the prior distribution is influenced by only one piece of background knowledge. We then consider the case that there are several pieces of background knowledge such that for each i $(1 \le i \le m)$ and each j $(1 \le j \le n_i)$, there is at most only one piece of background knowledge with a_{ij} as its conclusion.

Let S be the set of all pieces of background knowledge to be considered. For each generalization PG, let

$$
\begin{aligned}
B[S,PG] = \{ & i \in \{1,\ldots,m\} | \\
& \exists i_1 (1 \le i_1 \le m)\ \exists j_1 (1 \le j_1 \le n_{i_1})\ \exists j (1 \le j \le n_i) \\
& [(\text{there is a piece of background knowledge in } S \text{ with } a_{i_1 j_1} \\
& \ \text{as its premise and with } a_{ij} \text{ as its conclusion}) \\
& \ \& \ PG[i_1] = a_{i_1 j_1} \ \& \ PG[i] = *,] \},
\end{aligned}
$$

and for each $i \in B[S,PG]$, let

$$
\begin{aligned}
J[S,PG,i] = \{ & j \in \{1,\ldots n_i\} |\ \exists i_1 (1 \le i_1 \le m)\ \exists j_1 (1 \le j_1 \le n_{i_1}) \\
& [(\text{there is a piece of background knowledge in } S \text{ with } a_{i_1 j_1} \\
& \ \text{as its premise and with } a_{ij} \text{ as its conclusion}) \\
& \ \& \ PG[i_1] = a_{i_1 j_1} \ \& \ PG[i] = *] \}.
\end{aligned}
$$

Then, we must use the following *adjusting factors* AF_S with respect to all pieces of background knowledge:

$$
AF_S(PI|PG) = \prod_{i=1}^{m} AF_i(PI|PG), \tag{7}
$$

where

$$
AF_i(PI|PG) \tag{8}
$$

$$
= \begin{cases}
Q_{ij} \times n_i & \text{if } i \in B[S,PG],\ j \in J[S,PG,i], \text{ and } PI[i] = a_{ij}, \\[2ex]
\dfrac{1 - \sum\limits_{j \in J[S,PG,i]} Q_{ij}}{n_i - |J[S,PG,i]|} \times n_i & \begin{array}{l} \text{if } i \in B[S,PG], \\ \forall j (j \in J[S,PG,i])[PI[i] \ne a_{ij}], \end{array} \\[2ex]
1 & \text{otherwise,}
\end{cases}
$$

where for each i $(1 \le i \le m)$ and each j $(1 \le j \le n_i)$, Q_{ij} denotes the strength of the background knowledge (in S) with a_{ij} as its conclusion.

3 Ways of Implementation

We have tried several ways of implementing some aspects of the hybrid model stated in the above section. One possible way is to use the transition matrix of a stochastic process as a medium for implementing a hybrid system [16]. Let us assume that some causal relation seems to exist between observations. Let the observations be classified in finite classes and represented by a state set. The scheme of the transition process stated in Sect. 2.2 is used as a framework to represent the causal relation. Through learning in this framework, a tendency of the transition between input and output is learned. If the transition matrix reveals the complete or approximate equivalence with a logical inference, the logical expression can be discovered. Furthermore, the transition process can be represented by a connectionist network for solving the complexity of the space.

Another way is to use a GDT as a medium for generalizing and dealing with uncertainty and incompleteness. The following sections discuss several ways of implementing hybrid systems based on a GDT.

3.1 GDT-NR

Based on the hybrid model and the GDT methodology, two hybrid systems, GDT-NR and GDT-RS, have been developed [29,30]. This section introduces GDT-NR.

We know that the connectivity of a network represented by a network drawing (a network representation, for short) can be naturally represented in a matrix (a matrix representation, for short) [21]. In contrast, a matrix representation can obviously be changed into a network representation. Since the GDT is a variant of a transition matrix, it can be represented by networks [30]. The connectionist networks consist of three layers: the *input unit* layer, the *hidden unit* layer, and the *output unit* layer.

A unit that receives instances from a database is called an *input unit*, and the number of input units corresponds to the number of condition attributes. A unit that receives a result of learning in a hidden unit, which is used as one of the rule candidates discovered, is called an *output unit*. A unit that is neither an input nor an output unit is called a *hidden unit*. Let the hidden unit layer be further divided into *stimulus units* and *association units*. Since the *stimulus units* are used to represent the possible instances, such as the top row in GDT, and the *association units* are used to represent the possible generalizations of instances, such as the left column in GDT, they are also called *instance units* and *generalization units*, respectively. Furthermore, there is a link between a stimulus unit and an association unit if the association unit represents a possible generalization of a possible instance represented by the stimulus unit. Moreover, the probabilistic relationship between a possible instance and a possible generalization is represented in the weight of the link, and the initial weights are equiprobable, like the G_{ij} of an initial GDT.

The network representation of a GDT provides many advantages such as

- A large GDT can be conveniently decomposed into smaller ones and the computation of weights can be done in a parallel-distributed mode. In the representation, a network is used only to represent the possible generalizations for the possible instances in one of the levels of generalization that were defined in

Sect. 2.4. Hence, we need to create n networks if there are n levels of generalization. Thus, the number N_{cns} of networks that should be created, is

$$N_{\text{cns}} = N_{\text{attrs}} - 2, \tag{9}$$

where N_{attrs} is the number of attributes in a database. The merit of this approach is obviously that rule candidates can be generated in parallel. For example, we need to use two networks for the sample database shown in Table 3.

- Since the creation of networks is based on the GDT, the meaning of every unit in the networks can be explained clearly. Thus, not only the trained results in the networks are explicitly represented in the form of an *if-then* rules with *strength*, but background knowledge can also be used for dynamically revising and changing the network representation in the discovery process. Hence, the networks are called *knowledge-oriented* networks.
- The networks do not need to be explicitly created in advance. They can be embodied in the search algorithm, and we need only to record the weights related to stimulus units activated by instances. In other words, although the weights of inhibitory links also need to be calculated and revised in principle, it is not necessary for our rule discovery process currently. The instances collected in many real-world databases are generally a small subset of all possible instances. Hence, the real size of the networks is much smaller than the size of the corresponding GDT.

We have developed an effective method for incremental rule discovery based on a network representation of a GDT [30]. One advantageous feature of our approach is that every instance in a database is searched only once, and if the data in a database are changed (added to, deleted, or updated), we need only to modify the networks and the discovered rules related to the changed data; the database is not searched again. Unlike the back-propagation networks, our approach does not need multiple passes across the training data set. We argue that this is a very important feature for discovering rules in very large databases.

Here, we would like to stress that the network representation of a GDT does not need to be explicitly created in advance. It can be embodied in the learning algorithm, and we need only to record the weights and the units stimulated by instances in a database. However, the recording number still is quite large when dealing with large, complex databases. We need to find a much better way to solve the problem.

3.2 GDT-RS

To solve the problem stated above, we tried another way of implementing the GDT methodology by combining GDT with *rough set* methodology (GDT-RS). Using the *rough set* theory as a methodology of rule discovery is effective in practice [8,17]. The discovery process based on rough set methodology is that of knowledge reduction so that the decision specified can be made by using a minimal set of conditions.

The process of knowledge reduction is similar to the process of generalization in a hypothesis search space. By combining a GDT with rough sets, we can first find the rules with larger strengths from possible rules and then find minimal relative reducts from the set of rules with larger strengths. Thus, by using GDT-RS, a minimal set of rules with larger strengths can be acquired from databases with noisy, incomplete data.

In the rest of this section, rough set methodology for rule discovery is outlined, and representation of rules discovered by GDT-RS is described. Then how to combine GDT with rough set methodology is discussed.

Rough Set Methodology In *rough set* methodology for rule discovery, a database is regarded as a decision table, which is denoted

$$T = (U, A, \{V_a\}_{a \in A}, f, C, D),$$

where U is a finite set of instances (or objects), called the *universe*; A is a finite set of attributes; each V_a is the set of values of attribute a; f is a mapping from $U \times A$ to $V (= \bigcup_{a \in A} V_a)$; and C and D are two subsets of A, called the sets of *condition* attributes and *decision* attributes, respectively, such that $C \cup D = A$ and $C \cap D = \emptyset$. By $IND(B)$, we denote the indiscernibility relation defined by $B \subseteq A$, $[x]_{IND(B)}$ denotes the indiscernibility (equivalence) class defined by x, and U/B the set of all indiscernibility classes of $IND(B)$ [17,22,8].

The process of rule discovery is that of simplifying a decision table and generating a minimal decision algorithm. In general, an approach to decision table simplification consists of the following steps:

1. Elimination of duplicate condition attributes. This is equivalent to eliminating some column from the decision table.
2. Elimination of duplicate rows.
3. Elimination of superfluous values of attributes.

A representative approach for computing reducts of condition attributes is to represent knowledge in the form of a discernibility matrix [22,17]. The basic idea can be briefly presented as follows.

Let $T = (U, A, \{V_a\}_{a \in A}, f, C, D)$ be a decision table with $U = \{u_1, u_2, \ldots, u_n\}$. By a *discernibility matrix* of T, denoted $M(T)$, we will mean an $n \times n$ matrix defined as

$$m_{ij} = \begin{cases} \{c \in C : c(u_i) \neq c(u_j)\} & \text{if } \exists d \in D[d(u_i) \neq d(u_j)] \\ \lambda & \text{if } \forall d \in D[d(u_i) = d(u_j)], \end{cases}$$

for $i, j = 1, 2, \ldots, n$ such that u_i or u_j belongs to the C-positive region of D.

Thus, entry m_{ij} is the set of all condition attributes that classify objects u_i and u_j into different decision classes in U/D. Since $M(T)$ is symmetrical and $m_{ii} = \emptyset$, $M(T)$ is represented only by elements in the lower triangle, that is, m_{ij} with $1 \leq j < i \leq n$. Furthermore, $m_{ij} = \lambda$ denotes that this case does not need to be considered. Hence, it is interpreted as logic truth.

The *discernibility function* f_T for T is defined as follows.

For any $u_i \in U$,

$$f_T(u_i) = \bigwedge_j \{\bigvee m_{ij} : j \neq i, j \in \{1,2,\ldots,n\}\},$$

where

- $\bigvee m_{ij}$ is the disjunction of all variables a such that $c \in m_{ij}$, if $m_{ij} \neq \emptyset$;
- $\bigvee m_{ij} = \bot(false)$, if $m_{ij} = \emptyset$;
- $\bigvee m_{ij} = \top(true)$, if $m_{ij} = \lambda$.

Each logical product in the minimal disjunctive normal from (DNF) of $f_T(u_i)$ is called a *reduct* of instance u_i.

Generating minimal decision algorithm means eliminating the superfluous decision rules associated with the same decision class. It is obvious that some decision rules can be dropped without disturbing the decision making process, since some other rules can take over the job of the eliminated rules.

Rule Strength In GDT-RS, the rules are expressed in the following form:

$$P \rightarrow Q \quad \text{with} \quad S$$

that is, "*if P, then Q* with the strength S," where P denotes a conjunction of conditions, Q denotes a concept that the rule describes, and S is a "measure of strength" of the rule defined by

$$S(P \rightarrow Q) = s(P) \times [1 - r(P \rightarrow Q)], \tag{10}$$

where $s(P)$ is the strength of the generalization P (i.e., the condition of the rule) and r is the noise rate function. The strength of a given rule reflects the incompleteness and uncertainty in the process of rule inducing influenced by both unseen instances and noise.

The strength of the generalization $P = PG$ is given by (11) under the assumption that the prior distribution is uniform, or by Eq. (12) when the prior distribution is unequiprobable (i.e., background knowledge is used):

$$s(P) = \sum_l p(PI_l|P) = card([x]_{IND(P)}) \times \frac{1}{N_P}, \tag{11}$$

where $card([x]_{IND(P)})$ is the number of observed instances satisfying the generalization P.

$$s(PG) = \sum_l p_{bk}(PI_l|PG) = \left[\sum_l BKF(PI_l|PG)\right] \times \frac{1}{N_{PG}}. \quad (12)$$

The strength of generalization P represents explicitly the prediction for unseen instances. On the other hand, the noise rate is given by (13)

$$r(P \to Q) = 1 - \frac{card([x]_{IND(P)} \cap [x]_{IND(Q)})}{card([x]_{IND(P)})}, \quad (13)$$

where $card([x]_{IND(Q)})$ is the number of all instances from class Q within the instances satisfying generalization P. It shows the quality of classification measured by the number of instances satisfying generalization P which cannot be classified into class Q. The user can specify an allowed noise level as a threshold value. Thus, rule candidates with noise level larger than a given threshold value will be deleted.

Simplifying a Decision Table by GDT-RS By using a GDT, it is obvious that one instance can be expressed by several possible generalizations and several instances can be generalized into one possible generalization. Simplifying a decision table by GDT-RS leads to a minimal set of generalizations, which contains all of the instances in a decision table. The method of computing the reducts of condition attributes in GDT-RS, in principle, is equivalent to the discernibility matrix method [22], [17]. However, we won't find dispensable attributes because

- Finding dispensable attributes does not benefit the best solution acquisition. The larger the number of dispensable attributes, the more difficult it is to acquire the best solution.
- Some values of a dispensable attribute may be indispensable for some values of a decision attribute.

For a database with noises, the generalization that contains instances in different decision classes should be checked by (13). If a generalization X contains instances belonging to a decision class corresponding to Y more than those belonging to other decision classes and the noise rate (of $X \to Y$) is smaller than a threshold value, then generalization X is regarded as a consistent generalization of the decision class corresponding to Y, and "$X \to Y$ with $S(X \to Y)$" becomes a candidate rule. Otherwise, the generalization X is contradictory to all decision classes, and so no rule with X as its premise is generated.

Rule Selection There are several possible ways to select rules. For example,

- selecting rules that contain as many instances as possible;

- selecting rules in levels as high as possible, according to the first type of bias stated above;
- selecting rules with larger strengths.

Here, we would like to describe a method of rule selection for our purpose, as follows:

- Since our purpose is to simplify the decision table, the rules that contain fewer instances will be deleted if a rule that contains more instances exists.
- Since we prefer simpler results of generalization (i.e., more general rules), we first consider the rules corresponding to an upper level of generalization.
- If two generalizations in the same level have different strengths, the one with larger strength will be selected first.

We have developed two algorithms, called *Optimal Set of Rules* and *Suboptimal Solution,* respectively, for implementing GDT-RS [4]. We describe *Optimal Set of Rules*, below.

Algorithm (Optimal Set of Rules) Let T_{noise} be the expected threshold value.

Step 1. Create one or more GDTs.
If prior background knowledge is not available the prior distribution of a generalization is calculated using (1) and (2).

Step 2. Consider the indiscernibility classes with respect to the condition attribute set C (such as u_1, u_3, and u_5 in the sample database of Table 3) as one instance, called a *compound instance* (such as $u'_1 = [u_1]_{IND(a,b,c)}$ in the following table). Then, the probabilities of generalizations can be calculated correctly.

U \ A	a	b	c	d
$u'_1, \{u_1, u_3, u_5\}$	a_0	b_0	c_1	y,y,n
u_2	a_0	b_1	c_1	y
u_4	a_1	b_1	c_0	n
u_6	a_0	b_2	c_1	n
u_7	a_1	b_1	c_1	y

Step 3. For any compound instance u' (such as the instance u'_1 in the table above), let $d(u')$ be the set of decision classes to which the instances in u' belong. Furthermore, let $X_v = \{x \in U : d(x) = v\}$ be the decision class corresponding to the decision value v. The rate r_v can be calculated by (13). If there exist a $v \in d(u')$ such that $r_v(u') = min\{r_{v'}(u') | v' \in d(u')\} < T_{noise}$, then we let the compound instance u' point to the decision class corresponding to v. If any $v \in d(u')$ such that $r_v(u') < T_{noise}$ does not exist, we treat the compound instance u' as a contradictory one and set the decision class of u' to $\perp (uncertain)$. For example,

U \ A	a	b	c	d
$u'_1\{u_1,u_3,u_5\}$	a_0	b_0	c_1	$\perp$

Let U' be the set of all the instances except the contradictory ones.

Step 4. Select one instance u from U'. Using the idea of a discernibility matrix, create a discernibility vector (that is, a row or column corresponding to u in the discernibility matrix) for u. For example, the discernibility vector for instance $u_2 : a_0b_1c_1$ is as follows:

U \ U	$u'_1(\perp)$	$u_2(y)$	$u_4(n)$	$u_6(n)$	$u_7(y)$
$u_2(y)$	b	λ	a,c	b	λ

Step 5. Compute all of the so-called local relative reducts for instance u by using the discernibility function. For example, from instance u_2:$a_0b_1c_1$, we obtain two reducts $\{a,b\}$ and $\{b,c\}$:

$f_T(u2) = (b) \wedge \top \wedge (a \vee c) \wedge (b) \wedge \top = (a \wedge b) \vee (b \wedge c).$

Step 6. Construct rules from the local reducts for instance u, and revise the strength of each rule using (10). For example, the following rules are acquired:

$\{a_0b_1\} \rightarrow y$ with $S = 1 \times \frac{1}{2} = 0.5$, and

$\{b_1c_1\} \rightarrow y$ with $S = 2 \times \frac{1}{2} = 1$ for instance u_2:$a_0b_1c_1$.

Step 7. Select the best rules from the rules (for u) obtained in *Step* 6 according to its priority. For example, the rule "$\{b_1c_1\} \rightarrow y$" is selected for instance u_2:$a_0b_1c_1$ because it matches more instances than the rule "$\{a_0b_1\} \rightarrow y$."

Step 8. $U' = U' - u$. If $U' \neq \emptyset$, then go back to *Step* 4. Otherwise go to *Step* 9.

Step 9. If any rule selected in *Step* 7 covers exactly one instance, then STOP; otherwise, select a minimal set of rules covering all instances in the decision table.

The following table gives the result learned from the sample database shown in Table 3.

U	Rules	Strengths
u_2,u_7	$b_1 \wedge c_1 \rightarrow y$	1
u_4	$c_0 \rightarrow n$	0.167
u_6	$b_2 \rightarrow n$	0.25

The time complexity of algorithm 1 is $O(mn^2Nr_{\max})$, where n is the number of instances in a database, m is the number of attributes, and $Nr_{\max}$ is the maximal number of reducts for instances. We can see that the algorithm is not suitable for a database with a lot of attributes. A possible method of solving the issue is to find a suboptimal solution so that the time complexity of such an algorithm decreases to $O(m^2n^2)$ [4].

3.3 Comparisons

The main features of GDT-RS and GDT-NR, as well as a comparison, are summarized in Table 4. It shows that GDT-RS is better than GDT-NR for large, real-world applications, although both of them are very soft techniques for rule discovery in data.

Table 4. Comparison of GDT-RS and GDT-NR

GDT-RS	GDT-NR
The GDT is embodied in the search algorithm.	A large number of weights needs to be recorded.
Learning means finding a minimal set of generalizations.	Learning means revising weights incrementally.
???	Performing in a parallel distributed cooperative mode.
It can work well for a database with a large number of attributes.	It is not suitable for a database with a large number of attributes.
Speed is quite high.	Speed is not high.

Furthermore, we compare GDT-RS with C4.5 (ID3) that is one of the major systems for inductive learning nowadays. C4.5 [20] is a descendant of the ID3 algorithm [19], which builds decision trees top-down and prunes them. The tree is constructed by finding the best single attribute test to conduct at the root node of the tree. After the test is chosen, the instances are split according to the test, and the subproblems are solved recursively.

In C4.5, the order of selected attributes may affect the final result. Therefore, the best rules may not be discovered. Moreover, unseen instances affect the accuracy of the rules, but in C4.5, unseen instances are not considered. In C4.5, background knowledge cannot be used for search control in the learning process, and the learning process cannot be performed in a distributed cooperative mode. The main features of GDT-RS and C4.5 (ID3) are summarized in Table 5.

Finally, we compare GDT-RS with discriminant analysis that is a typical statistical method [6]. Table 6 summarizes the main features of GDT-RS and discriminant analysis and compares them. We can see that discriminant analysis algebraically describes the differential features of instances. The values of the variables must be numerical. The symbol data must be quantized. The discriminant equation is an algebraic expression, but its meaning cannot be explained clearly.

Table 5. Comparison of GDT-RS and C4.5 (ID3)

GDT-RS	C4.5 (ID3)
Background knowledge can be used easily.	Difficult to use background knowledge.
The stability and uncertainty of a rule can be expressed explicitly.	The stability and uncertainty of a rule cannot be explained clearly.
Unseen instances are considered.	Unseen instances are not considered.
A minimal set of rules containing all instances can be discovered.	Does not consider whether the rules discovered are the minimal set covering all instances.

Table 6. Comparison of GDT-RS and discriminant analysis

GDT-RS	Discriminant Analysis
If-then rules.	Algebraic expressions.
Multiclass, high-dimension, large-scale data can be processed.	Difficult to deal with the data with multiclass.
Background knowledge can be used easily.	Difficult to use background knowledge.
The stability and uncertainty of a rule can be expressed explicitly.	The stability and uncertainty of a rule cannot be explained clearly.
Continuous data must be discretized.	Symbolic data must be quantized.

3.4 GDT-RS-ILP

GDT-RS and GDT-NR stated above are two systems belonging to *attribute-value learning*, which is the main stream in inductive learning and data mining communities to date. Another type of inductive learning is *relation learning*, called *inductive logic programming* (ILP) [7,12].

ILP is a relatively new method in machine learning. ILP is concerned with learning from examples within the framework of predicate logic. ILP is relevant to data mining, and compared with attribute-value learning methods, it possesses the following advantages:

- ILP can learn knowledge which is more expressive than that by attribute-value learning methods because the former is in predicate logic, whereas the latter is usually in propositional logic.
- ILP can use background knowledge more naturally and effectively because in ILP, the examples, the background knowledge, the learned knowledge are all expressed within the same logic framework.

However, when applying ILP to large real-world applications, we can identify some weak points compared with the attribute-value learning methods, such as

- It is more difficult to handle numbers (especially continuous values) prevailing in real-world databases, because predicate logic lacks effective means for this.
- The theory, techniques, and experiences are much less mature for ILP to deal with imperfect data (uncertainty, incompleteness, vagueness, impreciseness, etc. in examples, background knowledge, as well as learned rules) than in the traditional attribute-value learning methods (see [7,23], for instance).

The discretization of continuous valued attributes, which is a kind of granular computing, as a preprocessing step, is a solution for the first problem mentioned above [13]. Another way is to use *constraint inductive logic programming* (CILP), an integration of ILP and CLP (*constraint logic programming*) [9].

For the second problem, a solution is to combine GDT (also GDT-RS) with ILP, that is, use GDT-ILP and GDT-RS-ILP to deal with some kinds of imperfect data that occur in large real-world applications [10].

Normal Problem Setting for ILP We follow the notations of [14]. Especially, supposing C is a set of clauses $\{c_1, c_2, ...\}$, we use $\overline{C}$ to denote the set $\{\neg c_1, \neg c_2, ...\}$. The normal problem setting for ILP can be stated as follows:

Given positive examples E^+ and negative examples E^- (both are sets of clauses), and the background knowledge B (a finite set of clauses), ILP is to find a theory H (a finite set of clauses) that is correct with respect to E^+ and E^-. That demands

1. $\forall_{e \in E^+} H \cup B | = e$ *(completeness w.r.t. E^+).*
2. $H \cup B \cup \overline{E^-}$ *is satisfiable (consistency w.r.t. E^-).*

The above ILP problem setting is somewhat too general. In most of the ILP literature, the following simplifications are assumed:

- Single predicate learning. The concept to be learned is represented by a single predicate p (called the *target predicate*). Examples are instances of target predicate p, and the induced theory is the defining clauses of p. Only background knowledge B may contain definitions of other predicates that can be used in the defining clauses of the target predicate.
- Restricted within definite clauses. All clauses contained in B and H are definite clauses, and the examples are ground atoms of the target predicate. We can prove:

supposing that Σ is a set of definite clauses and E^- is a set of ground atoms, then Σ is consistent with respect to E^- if and only if $\forall_{e \in E^-} \Sigma| \neq e$.

So in this case, the second condition of correctness (*consistency*: $H \cup B \cup \overline{E^-}$ is satisfiable) can be replaced by a simpler form $\forall_{e \in E^-} H \cup B| \neq e$. Clearly, this simpler form is more operational (easier to test) than the general one.

Here we restate the (simplified) normal problem setting for ILP in a more formal way:

Given:

- The target predicate p.
- Positive examples E^+ and negative examples E^- (two sets of ground atoms of p).
- Background knowledge B (a finite set of definite clauses).

To find:

- Hypothesis H (the defining clauses of p) that is *correct* with respect to E^+ and E^-, i.e.
  1. $H \cup B$ is *complete* with respect to E^+ (that is, for all $e \in E^+$, $H \cup B$ implies e). We also say that $H \cup B$ *covers* all positive examples.
  2. $H \cup B$ is *consistent* with respect to E^- (that is, for no $e \in E^-$, $H \cup B$ implies e). We also say that $H \cup B$ *rejects* any negative examples.

To make the ILP problem meaningful, we assume the following *prior* conditions:

1. B is not complete with respect to E^+. (Otherwise, there will be no learning task at all because the background knowledge itself is the solution).
2. $B \cup E^+$ is consistent with respect to E^-. (Otherwise, there will be no solution to the learning task).

Note that almost all practical ILP systems impose some restrictions on the hypothesis space and/or on the search strategies to solve real-world problems with reasonable computing resources (time and/or space). These restrictions are called *declarative bias*. Bias setting is a key factor for an ILP system to be successful in practice. We can cite the famous statement: "No bias, no prediction" [11]. Without bias or with too weak bias, an ILP system may show poor performance or produce too many solutions that are not interesting to the user. On the other hand, if the bias is too strong, we may miss some useful solutions or have no solution at all.

In the above normal problem setting of ILP, everything is assumed correct and perfect. But in large, real-world empirical learning, data are not always perfect. To the contrary, uncertainty, incompleteness, vagueness, impreciseness, etc. are frequently observed in the input to ILP, the training examples and/or background knowledge. Imperfect input, in addition to improper bias setting, will induce imperfect hypotheses. Thus, ILP has to deal with imperfect data. In this aspect, the theory, techniques, measurement, and experiences are much less mature in ILP than in the traditional attribute-value learning methods (compare with [23], for example).

At the current stage, we apply GDT and rough set theory to ILP to deal with some kinds of imperfect data occurring in large real-world applications. We concentrate on incomplete background knowledge (where essential predicates/clauses are missing), indiscernible data (where some examples belong to both sets of positive and negative training examples), missing classification (where some examples are unclassified), and too strong a declarative bias (hence the failure in searching for solutions). Although imperfect data handling is too vast a task , we observe that many problems concerning imperfect input or too strong a bias in ILP have a common feature. In these situations, though it is impossible to differentiate distinct objects, we may consider *granules* — sets of objects drawn together by similarity, indistinguishability, or functionality. The emerging theory of *granular computing* (GrC) (see [25–27]) grasps the essential concept — granules, and uses them in general problem solving. The main idea is that, when we use granules instead of individual objects, we are actually relaxing the strict requirements in the standard normal problem setting for ILP. Based on this idea, the GDT-RS-ILP system is under development.

4 Experiments

Some databases such as opticians, slope collapse, meningitis, postoperative patients, mushrooms, earthquakes, bacterial examination, and cancer have been tested for our approach. This section discusses the results of discovering rules from databases on opticians and meningitis by using GDT-NR and GDT-RS, as examples, and compares our approach with C4.5 (ID3).

4.1 Experiment 1

We first use an opticians decisions database for comparing our methods (GDT-RS and GDT-NR) with C4.5. This example concerns an opticians decisions as to whether or not a patient is suited to contact lens use [3,17]. Table 7 shows this database in which A, B, C, and D are condition attributes, whereas E is a decision attribute.

The attribute E represents the optician's decisions, which are the following:

1. The patient should be fitted with hard contact lenses.

2. The patient should be fitted with soft contact lenses.
3. The patient should not be fitted with contact lenses.

These decisions are based on some facts concerning the patient, which are expressed by the condition attributes given below together with corresponding attribute values:

A: age (1: young, 2: prepresbyopic, 3:presbyopic).
B: spectacle (1: myope, 2: hypermytrope).
C: astigmatic (1: no, 2: yes).
D: tear production rate (1: reduced, 2: normal).

Table 7. Optician's decisions

U	*A*	*B*	*C*	*D*	*E*	*U*	*A*	*B*	*C*	*D*	*E*
u1	1	1	1	1	3	u13	2	2	1	1	3
u2	1	1	1	2	2	u14	2	2	1	2	2
u3	1	1	2	1	3	u15	2	2	2	1	3
u4	1	1	2	2	1	u16	2	2	2	2	3
u5	1	2	1	1	3	u17	3	1	1	1	3
u6	1	2	1	2	2	u18	3	1	1	2	3
u7	1	2	2	1	3	u19	3	1	2	1	3
u8	1	2	2	2	1	u20	3	1	2	2	1
u9	2	1	1	1	3	u21	3	2	1	1	3
u10	2	1	1	2	2	u22	3	2	1	2	2
u11	2	1	2	1	3	u23	3	2	2	1	3
u12	2	1	2	2	1	u24	3	2	2	2	3

Tables 8 and 9 show the results of our method and C4.5, respectively. In the tables, *Instances used* is the number of instances covered by the rule; *Accuracy* denotes how many instances within all instances covered by the rule are classified correctly; *Strength* indicates the strengths of the generalizations.

We can see that the result of our method is not the same as that of C4.5. The rule set generated by GDT-RS and GDT-NR is the same. And all instances shown in Table 7 are covered by the rule set generated by GDT-RS and GDT-NR. However, instances u_8, u_{16}, and u_{24} are not covered by any rule generated by C4.5.

Furthermore, the strengths and accuracy of all rules discovered by our method are 1 and 100%, respectively, because the data set shown in Table 7 is complete (i.e., there are no incomplete or uncertain data in this data set). Unfortunately, the accuracy of $rule_2$ generated by C4.5 is not 100%, even if the data set is complete.

Here, we would like to stress that most data sets in the real world are incomplete. Suppose that instances u_{23} and u_{24} were not collected yet. Using our method, the

Table 8. Results obtained by GDT-RS and GDT-NR

No.	Rules	Strength	Instances used		Accuracy
			+	−	
1	$B(1) \wedge C(2) \wedge D(2) \rightarrow E(1)$	1	3	0	100%
2	$A(1) \wedge C(2) \wedge D(2) \rightarrow E(1)$	1	2	0	100%
3	$A(1) \wedge C(1) \wedge D(2) \rightarrow E(2)$	1	2	0	100%
4	$B(2) \wedge C(1) \wedge D(2) \rightarrow E(2)$	1	3	0	100%
5	$A(2) \wedge C(1) \wedge D(2) \rightarrow E(2)$	1	2	0	100%
6	$D(1) \rightarrow E(3)$	1	12	0	100%
7	$A(2) \wedge B(2) \wedge C(2) \rightarrow E(3)$	1	2	0	100%
8	$A(3) \wedge B(1) \wedge C(1) \rightarrow E(3)$	1	2	0	100%
9	$A(3) \wedge B(2) \wedge C(2) \rightarrow E(3)$	1	2	0	100%

Table 9. Results obtained by C4.5

No.	Rules	Instances used		Accuracy
		+	−	
1	$B(1) \wedge C(2) \wedge D(2) \rightarrow E(1)$	3	0	100%
2	$C(1) \wedge D(2) \rightarrow E(2)$	5	1	83.3%
3	$D(1) \rightarrow E(3)$	12	0	100%

rule set generated from this incomplete data set is the same as that shown in Table 8 except that the strength of $rule_6$ is changed from 1 to 22/24 because two unseen instances are not collected yet and $rule_9$ is not generated because no instance should be covered by this rule. In contrast, using C4.5, $rule_1$ in Table 9 is replaced by the following:

$$C(2) \wedge D(2) \rightarrow E(1),$$

and the accuracy dropped from 100% to 80%.

This example shows that the accuracy and stability of the rule set generated by our method (GDT-RS and GDT-NR) are better than that generated by C4.5.

4.2 Experiment 2

This section describes the results of discovering rules from a slope collapse database in GDT-RS. The slope collapse database collected data on dangerous natural steep slopes in Yamaguchi, Japan. There are 3436 instances in this database. Among them, 430 places were collapsed, and 3006 were not. There are thirty two condition attributes and one decision attribute. The task is to find out why the slope collapsed.

The attributes are listed in Table 10, *collapse* is a decision attribute, and the remaining thirty two attributes are condition attributes.

Table 10. Condition attributes in the slope-collapse database

Attribute name	Number of values
Extension of collapsed steep slope	Continuous
Gradient	Continuous
Altitude	Continuous
Slope azimuthal	9
Slope shape	9
Direction of high rank topography	10
Shape of transverse section	5
Transition line	3
Position of transition line	5
Condition of the surface of the Earth	5
Thickness of surface of soil	Continuous
Condition of ground	6
Condition of base rock	4
Relation between slope and unsuccessive face	7
Fault, broken region	4
Condition of weather	5
Kind of plant	6
Age of tree	7
Condition of lumbering	4
Collapse history of current slope	3
Condition of current slope	5
Collapse history of adjacent slope	3
Condition of adjacent slope	6
Spring water	4
Countermeasure work	3
State of upper part of countermeasure work	5
State of upper part of countermeasure work2	6
State of upper part of countermeasure work3	7
No. of active faults	Continuous
Active fault traveling	7
Distance between slope and active fault	Continuous
Direction of slope and active fault	9

Table 11 shows what kinds of conditions cause slope collapse. Since the rules are too many, we just list part of them with higher strengths. In the table, *Used* denotes the number of instances covered by the rule, *Strength* indicates the strengths of the generalization (conditions), which can be obtained by (11). $E = \prod_{i=1}^{m} n_i$, where n_i is the number of values of the ith condition attribute, $n = [2, 27, 9, 9, 10, 5, 5, 2, 6, 3]$.

Table 11. The results of slope collapse

Conditions	Used	Strength
s_azimuthal(2) ∧ s_shape(5) ∧ direction_high(8) ∧ plant_kind(3)	5	(4860/E)
altitude[21,25) ∧ s_azimuthal(3) ∧ soil_thick(≥ 45)	5	(486/E)
s_azimuthal(4) ∧ direction_high(4) ∧ t_shape(1) ∧ tl_position(2) ∧ s_f_distance(≥ 9)	4	(6750/E)
altitude[16,17) ∧ s_azimuthal(3) ∧ soil_thick(≥ 45) ∧ s_f_distance(≥ 9)	4	(1458/E)
altitude[20,21) ∧ t_shape(3) ∧ tl_position(2) ∧ plant_kind(6) ∧ s_f_distance(≥ 9)	4	(12150/E)
altitude[11,12) ∧ s_azimuthal(2) ∧ tl_position(1)	4	(1215/E)
altitude[12,13) ∧ direction_high(9) ∧ tl_position(4) ∧ s_f_distance[8,9)	4	(4050/E)
altitude[12,13) ∧ s_azimuthal(5) ∧ t_shape(5) ∧ s_f_distance[8,9)	4	(3645/E)
altitude[36,37) ∧ plant_kind(5)	3	(162/E)
altitude[13,14) ∧ s_shape(2) ∧ direction_high(4)	3	(2430/E)
altitude[8,9) ∧ s_azimuthal(3) ∧ s_shape(2)	3	(2187/E)
altitude[18,19) ∧ s_shape(4) ∧ plant_kind(2)	3	(1458/E)

For the continuous attributes, the number of values has been changed after preprocessing by discretization [13].

The results have been evaluated by an expert who also did the same work on similar data using discriminant analysis. He picked out the important factors (attributes) about "collapse" from the same data. According to the opinions of experts, the results are reasonable and interesting. The more important factors (attributes) are used as conditions that caused the slope to collapse.

4.3 Experiment 3

This section describes an experimental result in which background knowledge is used in the learning process to discover rules from a meningitis database [30]. The database was collected at the Medical Research Institute, Tokyo Medical and Dental University. It has 140 records each described by 38 attributes that can be categorized into present history, physical examination, laboratory examination, diagnosis, therapy, clinical course, final status, risk factor, etc. The task is to find important factors for diagnosis (bacteria and virus, or their more detailed classifications) and predicting prognosis. A more detailed explanation of this database can be found at http://www.kdel.info.eng.osaka-cu.ac.jp/SIGKBS.

For each of the decision attributes, DIAG2, DIAG, CULTURE, C_COURSE, and COURSE(Grouped), we ran our GDT-RS on it twice, using background knowledge and without using background knowledge, to acquire the rules, respectively. Automatic discretization is used to discretize continuous attributes [13].

Some background knowledge given by a medical doctor is described as follows:

- Never occurring together:
  EEG_WAVE(normal) $\Leftrightarrow$ *EEG_FOCUS(+)*
  CSF_CELL(low) $\Leftrightarrow$ *Cell_Poly(high)*
  CSF_CELL(low) $\Leftrightarrow$ *Cell_Mono(high)*
  ....
- Occurring with lower possibility:
  WBC(low) $\Rightarrow$ *CRP(high)*
  WBC(low) $\Rightarrow$ *ESR(high)*
  WBC(low) $\Rightarrow$ *CSF_CELL(high)*
  ......
- Occurring with higher possibility:
  WBC(high) $\Rightarrow$ *CRP(high)*
  WBC(high) $\Rightarrow$ *ESR(high)*
  WBC(high) $\Rightarrow$ *CSF_CELL(high)*
  WBC(high) $\Rightarrow$ *Cell_Poly(high)*
  WBC(high) $\Rightarrow$ *Cell_Mono(high)*
  BT(high) $\Rightarrow$ *STILL(high)*
  BT(high) $\Rightarrow$ *LASEGUE(high)*
  ....

"High" in brackets denoted in the background knowledge means that the value is greater than the maximal one in the normal values, and "low" means that the value is less than minimal in the normal values. For example, the first one of the background knowledge is read:

If the brain wave (EEG-WAVE) is normal,
the focus of brain wave (EEG_FOCUS) is never abnormal.

Although similar results can be obtained from the meningitis database by using GDT-RS and C4.5 if such background knowledge is not used [30], it is difficult to use such background knowledge in C4.5 [20]. By using the background knowledge in GDT-RS, we got some interesting results.

First, some candidates of rules, which are deleted due to lower strengths during rule discovery without background knowledge, are selected. For example, r_1 : is deleted when no background knowledge is used, but after using the background knowledge stated above, it is reserved because its strength increased four times.

r_1 : *ONSET(acute)* $\wedge$ *ESR*($\leq$ *5)* $\wedge$ *CSF_CELL*($>$ *10)* $\wedge$ CULTURE(-) $\rightarrow$ *VIRUS(E)*.

Without using background knowledge, the strength S of $rule_1$ is 30*(384/E). In the background knowledge given above, there are two clauses related to this rule:

- Never occurring together:
  CSF_CELL(low) $\Leftrightarrow$ *Cell_Poly(high)*
  CSF_CELL(low) $\Leftrightarrow$ *Cell_Mono(high)*.

By using automatic discretization of continuous attributes *Cell_Poly* and *Cell_Mono*, the attribute values of each of the attributes are divided into two groups, high and low. Since the high groups of *Cell_Poly* and *Cell_Mono* do not occur if *CSF_CELL* (low) occurs, the product of the numbers of attribute values is decreased to $E/4$, and the strength S is increased to $S = 30 * (384/E) * 4$.

Second, using background knowledge also causes some rules to be replaced by others. For example, the rule

r_2 : *DIAG(VIRUS(E))* $\wedge$ *LOC[4, 7)* $\rightarrow$ *EEG_abnormal* with $S = 30/E$,

can be discovered without background knowledge, but if some background knowledge stated above is used, it is replaced by

$r_{2'}$: *EEG_FOCUS(+)* $\wedge$ *LOC[4, 7)* $\rightarrow$ *EEG_abnormal* with $S = (10/E) * 4$.

The reason is that both of them cover the same instances, but the strength of $r_{2'}$ becomes larger than that of r_2.

The results have been evaluated by a medical doctor. In his opinion, both r_2 and $r_{2'}$ are reasonable, but the $r_{2'}$ is much better than the r_2.

This example shows that our approach is a *soft* one that can use background knowledge as a bias for controlling the creation of GDT and the discovery process.

5 Conclusion

In this paper, a hybrid model for rule discovery in real-world data with uncertainty and incompleteness was presented. The central idea of our methodology is using the GDT as a hypothesis search space for generalization, in which the probabilistic relationships between concepts and instances across discrete domains are represented. By using the GDT as a probabilistic search space, (1) unseen instances can be considered in the rule discovery process and the uncertainty of a rule, including its ability to predict unseen instances, can be explicitly represented in the strength of the rule; and (2) biases can be flexibly selected for search control, and background knowledge can be used as a bias to control the creation of a GDT and the rule discovery process. Several hybrid discovery systems, which are based on the hybrid model, have been and continue to be developed.

The ultimate aim of the research project is to create an *agent-oriented* and *knowledge-oriented* hybrid intelligent model and system for knowledge discovery and data mining in an evolutionary, parallel-distributed cooperative mode. In this model and system, the typical methods of symbolic reasoning, such as deduction, induction, and abduction, as well as methods based on soft computing techniques, such as rough sets, fuzzy sets, and granular computing, can be cooperatively used by taking GDT and the transition matrix of stochastic process as media. The work that we are doing takes but one step toward this model and system.

Acknowledgments

The authors would like to thank Prof. S. Tsumoto, Prof. H. Nakamura, Mr. K. Kuramoto, and Mr. H. Sakakibara for providing the medical databases, the slope collapse database, background knowledge, and evaluating the experimental results. The authors also would like to thank Prof. K. Inoue for valuable comments and help.

References

1. R. Andrews, J. Diederich, A. B. Tickle. Survey and critique of techniques for extracting rules from trained artificial neural networks. *Knowledge-Based Systems*, 8(6), 373–389, 1995.
2. M. Banerjee, S. Mitra, S. K. Pal. Rough fuzzy MLP: Knowledge encoding and classification. *IEEE Transactions on Neural Networks*, 9(6): 1203–1216, 1998.
3. J. Cendrowska. PRISM: An algorithm for inducing modular rules. *International Journal of Man-Machine Studies*, 27: 349–370, 1987.
4. J.Z. Dong, N. Zhong, S. Ohsuga. Probabilistic rough induction: the GDT-RS methodology and algorithms. In Z.W. Ras, A. Skowron, editors, *Foundations of Intelligent Systems*, LNAI 1609, 621–629, Springer, Berlin, 1999.
5. U.M. Fayyad, G. Piatetsky-Shapiro, P. Smyth, R. Uthurusamy, editors. *Advances in Knowledge Discovery and Data Mining*. AAAI Press, Menlo Park, CA, 1996.
6. R.A. Johnson, D.W. Wichern. *Applied Multivariate Statistical Analysis*. Prentice-Hall, Englewood Cliffs, NJ, 1992.
7. N. Lavrac, S. Dzeroski, I. Bratko. Handling imperfect data in inductive logic programming. In L. de Raedt, editor, *Advances in Inductive Logic Programming*, 48–64, IOS, Amsterdam, 1996.
8. T.Y. Lin, N. Cercone. *Rough Sets and Data Mining: Analysis of Imprecise Data*. Kluwer, Dordrecht, 1997.
9. C. Liu, N. Zhong, S. Ohsuga. Constraint ILP and its application to KDD. In *Proceedings of the International Joint Conference on Artificial Intelligence (IJCAI'97). The Workshop on Frontiers of ILP*, 103–104, Nagoya, Japan, 1997.
10. C. Liu, N. Zhong. Rough problem settings for inductive logic programming. In N. Zhong, A. Skowron, S. Ohsuga, editors, *New Directions in Rough Sets, Data Mining, and Granular-Soft Computing*, LNAI 1711, 168–177, Springer, Berlin, 1999.

11. T.M. Mitchell. *Machine Learning*. McGraw-Hill, New York, 1997.
12. S. Muggleton. Inductive logic programming. *New Generation Computing*, 8(4): 295–317, 1991.
13. S.H. Nguyen, H.S. Nguyen. Quantization of real value attributes for control problems. In *Proceedings of the 4th European Congress on Intelligent Techniques and Soft Computing (EUFIT'96)*, 188–191, Verlag Mainz, Aachen, 1996.
14. S.H. Nienhuys-Cheng, R.Wolf, editors. *Foundations of Inductive Logic Programming*. LNAI 1228, Springer, Berlin, 1997.
15. S. Ohsuga, H. Yamauchi. Multi-layer logic: A predicate logic including data structure as a knowledge representation language. *New Generation Computing*, 3(4): 403–439, 1985.
16. S. Ohsuga. Symbol processing by non-symbol processor. In *Proceedings of the 4th Pacific Rim International Conference on Artificial Intelligence (PRICAI'96)*, 193–205, 1996.
17. Z. Pawlak. *Rough Sets: Theoretical Aspects of Reasoning about Data*. Kluwer, Dordrecht, 1991.
18. G. Polya. *Mathematics and Plausible Reasoning: Patterns of Plausible Inference*. Princeton University Press, Princeton, NJ, 1968.
19. J. R. Quinlan. Induction of decision trees. *Machine Learning*, 1: 81–106, 1986.
20. J. R. Quinlan. *C4.5: Programs for Machine Learning*. Morgan Kaufmann, San Mateo, CA, 1993.
21. D.E. Rumelhart, G.E. Hinton, R.J. Williams. Learning internal representations by back-propagation errors. *Nature*, 323: 533–536, 1986.
22. A. Skowron, C. Rauszer. The discernibility matrices and functions in information systems. In R. Słowi'nski, editor, *Intelligent Decision Support: Handbook of Advances of Rough Sets Theory*, 331–362, Kluwer, Dordrecht, 1992.
23. Y.Y. Yao, N. Zhong. An analysis of quantitative measures associated with rules. In *Proceedings of the Pacific-Asia Conference on Knowledge Discovery and Data Mining (PAKDD'99)*, LNAI 1574, 479–488, Springer, Berlin, 1999.
24. Y.Y. Yao, N. Zhong. Potential applications of granular computing in knowledge discovery and data mining. In *Proceedings of the 5th International Conference on Information Systems Analysis and Synthesis (IASA'99)*, edited in the invited session on Intelligent Data Mining and Knowledge Discovery, 573–580, International Institute of Informatics and Systemics, Orlando, Fl, 1999.
25. Y.Y. Yao. Granular computing: Basic issues and possible solutions. In *Proceedings of the Joint Conference on Intelligence Systems (JCIS 2000)*, invited session on Granular Computing and Data Mining, 1: 186–189, Association for Intelligent Machinery, Atlantic City, NJ, 2000.
26. L.A. Zadeh. Fuzzy sets and information granularity. In N. Gupta, R. Ragade, R.R. Yager, editors, *Advances in Fuzzy Set Theory and Applications*, 3–18, North–Holland, Amsterdam, 1979.
27. L.A. Zadeh. Toward a theory of fuzzy information granulation and its centrality in human reasoning and fuzzy logic. *Fuzzy Sets and Systems*, 90: 111–127, 1997.
28. N. Zhong, S. Ohsuga. Using generalization distribution tables as a hypothesis search space for generalization. In *Proceedings of the 4th International Workshop on Rough Sets, Fuzzy Sets, and Machine Discovery (RSFD'96)*, 396–403, University of Tokyo, Tokyo, Japan, 1996.
29. N. Zhong, J.Z. Dong, S. Ohsuga. Data mining: A probabilistic rough set approach. In *Proceedings of the 1st International Conference on Rough Sets and Current Trends in Computing (RSCTC'98)*, LNAI 1424, 127–146, Springer, Berlin, 1998.

30. N. Zhong, J.Z. Dong, S. Fujitsu, S. Ohsuga. Soft techniques for rule discovery in data. *Transactions of Information Processing Society of Japan*, 39(9): 2581–2592, 1998.

Author Index

Index

Cognitive Technologies

Cognitive Technologies

T. Frühwirth, S. Abdennadher: Essentials of Constraint Programming.
IX, 144 pages. 2003

J. W. Lloyd: Logic for Learning.
X, 256 pages. 2003

S. K. Pal, L. Polkowski, A. Skowron (Eds.): Rough-Neural Computing.
XXV, 734 pages. 2004

H. Prendinger, M. Ishizuka (Eds.): Life-like Characters.
IX, 477 pages. 2004

MIX
Papier aus verantwortungsvollen Quellen
Paper from responsible sources
FSC® C105338

If you have any concerns about our products,
you can contact us on
ProductSafety@springernature.com

In case Publisher is established outside the EU,
the EU authorized representative is:
Springer Nature Customer Service Center GmbH
Europaplatz 3, 69115 Heidelberg, Germany

Printed by Libri Plureos GmbH
in Hamburg, Germany